Comprehensive Analytical Chemistry
Volume 67

Persistent Organic Pollutants (POPs): Analytical Techniques, Environmental Fate and Biological Effects

Advisory Board

Comprehensive Analytical Chemistry

Volume 67

Persistent Organic Pollutants (POPs): Analytical Techniques, Environmental Fate and Biological Effects

Edited by

Eddy Y. Zeng
School of Environment,
Jinan University,
Guangzhou, China

ELSEVIER

AMSTERDAM • BOSTON • HEIDELBERG • LONDON
NEW YORK • OXFORD • PARIS • SAN DIEGO
SAN FRANCISCO • SINGAPORE • SYDNEY • TOKYO

Elsevier
Radarweg 29, PO Box 211, 1000 AE Amsterdam, Netherlands
The Boulevard, Langford Lane, Kidlington, Oxford OX5 1GB, UK
225 Wyman Street, Waltham, MA 02451, USA

ISBN: 978-0-444-63299-9
ISSN: 0166-526X

Contents

3. Fast Analytical Techniques Based on Microextraction

Ruifen Jiang and Gangfeng Ouyang

4. Application of Passive Sampling Techniques in Measurement of HOCs in Aquatic Environments

Lian-Jun Bao and Eddy Y. Zeng

5. Assessment of Sediment Toxicity with SPME-Based Approaches

Jing You, Huizhen Li and Michael J. Lydy

Contributors to Volume 67

Bilqees M. Adedigba, Lancaster Environment Centre, Lancaster University, Lancaster, UK

Cristina Ávila, GEMMA–Group of Environmental Engineering and Microbiology, Department of Hydraulic, Maritime and Environmental Engineering, Universitat Politècnica de Catalunya-BarcelonaTech, Barcelona, Spain

Lian-Jun Bao, State Key Laboratory of Organic Geochemistry, Guangzhou Institute of Geochemistry, Chinese Academy of Sciences, Guangzhou, China

Christian Bogdal, Institute for Chemical and Bioengineering, ETH Zürich, Switzerland

Eva Brorström-Lundén, IVL-Swedish Environmental Research Institute, Aschebergsgatan, Göteborg, Sweden

Robert M. Burgess, Office of Research and Development, US Environmental Protection Agency, Narragansett, RI, USA

Mark G. Cantwell, Office of Research and Development, US Environmental Protection Agency, Narragansett, RI, USA

David O. Carpenter, Institute for Health and the Environment, University at Albany, Rensselaer, NY, USA

Jingwen Chen, Key Laboratory of Industrial Ecology and Environmental Engineering (Ministry of Education), School of Environmental Science and Technology, Dalian University of Technology, Dalian, Liaoning, China

Li Chen, Department of Environmental Health, School of Public Health, Fudan University, Shanghai, China

Adrian Covaci, Toxicological Centre, University of Antwerp, Universiteitsplein, Wilrijk, Belgium

Jordi Dachs, Department of Environmental Chemistry, Institute for Environmental Assessment and Water Research (IDAEA-CSIC), Barcelona, Spain

Alin C. Dirtu, Toxicological Centre, University of Antwerp, Universiteitsplein, Wilrijk, Belgium; Department of Chemistry, "Al. I. Cuza" University of Iasi, Iasi, Romania

Trine Eggen, Norwegian Institute of Agriculture & Environment Research, Ås, Norway

Steven J. Eisenreich, Department of Hydrology and Hydraulics, and IUPWARE Programme, Vrije Universiteit Brussel (VUB), Brussels, Belgium

Beate I. Escher, The University of Queensland, National Research Centre for Environmental Toxicology (Entox), Brisbane, QLD, Australia; UFZ - Helmholtz Centre for Environmental Research, Cell Toxicology, Leipzig, Germany; Eberhard Karls University Tübingen, Environmental Toxicology, Center for Applied Geosciences, Tübingen, Germany

Joan García, GEMMA–Group of Environmental Engineering and Microbiology, Department of Hydraulic, Maritime and Environmental Engineering, Universitat Politècnica de Catalunya-BarcelonaTech, Barcelona, Spain

Caroline Gaus, The University of Queensland, National Research Centre for Environmental Toxicology (Entox), Brisbane, QLD, Australia

Ying Guo, New York State Department of Health and Department of Environmental Health Sciences, Wadsworth Center, School of Public Health, State University of New York at Albany, Albany, NY, USA

Kimberly J. Hageman, Department of Chemistry, University of Otago, Dunedin, New Zealand

S. Michele Harmon, Department of Biology and Geology, University of South Carolina Aiken, Aiken, SC, USA

Hayley Hung, Air Quality Research Division, Science and Technology Branch, Environment Canada, Toronto, Ontario, Canada

Ruifen Jiang, School of Chemistry and Chemical Engineering, Sun Yat-sen University, Guangzhou, China

Javier Castro Jiménez, Department of Environmental Chemistry, Institute for Environmental Assessment and Water Research (IDAEA-CSIC), Barcelona, Spain

Ling Jin, The University of Queensland, National Research Centre for Environmental Toxicology (Entox), Brisbane, QLD, Australia

Roland Kallenborn, Institute for Chemistry Biotechnology and Food Sciences (IKBM), Faculty of Veterinary Medicine and Biosciences (VetBio), Norwegian University of Biosciences, Ås, Norway

Kurunthachalam Kannan, New York State Department of Health and Department of Environmental Health Sciences, Wadsworth Center, School of Public Health, State University of New York at Albany, Albany, NY, USA

James C.W. Lam, State Key Laboratory in Marine Pollution, City University of Hong Kong, Hong Kong SAR, China; Research Centre for the Oceans and Human Health, Shenzhen Key Laboratory for Sustainable Use of Marine Biodiversity, City University of Hong Kong, Shenzhen Research Institute Building, Shenzhen, PR China

Paul K.S. Lam, State Key Laboratory in Marine Pollution, City University of Hong Kong, Hong Kong SAR, China; Research Centre for the Oceans and Human Health, Shenzhen Key Laboratory for Sustainable Use of Marine Biodiversity, City University of Hong Kong, Shenzhen Research Institute Building, Shenzhen, PR China; Department of Biology and Chemistry, City University of Hong Kong, Hong Kong SAR, China

Fei Li, Key Laboratory of Industrial Ecology and Environmental Engineering (Ministry of Education), School of Environmental Science and Technology, Dalian University of Technology, Dalian, Liaoning, China; Key Laboratory of Coastal Environmental Processes and Ecological Remediation, Yantai Institute of Coastal Zone Research, Chinese Academy of Sciences, Yantai Shandong, China; Shandong Provincial Key Laboratory of Coastal Environmental Processes, Yantai Institute of Coastal Zone Research, Chinese Academy of Sciences, Yantai Shandong, China

Huizhen Li, State Key Laboratory of Organic Geochemistry, Guangzhou Institute of Geochemistry, Chinese Academy of Sciences, Guangzhou, China

Lucy Li, Department of Environmental Health, School of Public Health, Fudan University, Shanghai, China

Felichesmi Lyakurwa, Key Laboratory of Industrial Ecology and Environmental Engineering (Ministry of Education), School of Environmental Science and Technology, Dalian University of Technology, Dalian, Liaoning, China

Michael J. Lydy, Center for Fisheries, Aquaculture and Aquatic Sciences and Department of Zoology, Southern Illinois University, Carbondale, IL, United States

Govindan Malarvannan, Toxicological Centre, University of Antwerp, Universiteitsplein, Wilrijk, Belgium

Hong-Gang Ni, Peking University Shenzhen Graduate School, Shenzhen, China

Gangfeng Ouyang, School of Chemistry and Chemical Engineering, Sun Yat-sen University, Guangzhou, China

Martin Scheringer, Institute for Chemical and Bioengineering, ETH Zürich, Switzerland; Institute for Sustainable and Environmental Chemistry, Leuphana University, Lüneburg, Germany

Kirk T. Semple, Lancaster Environment Centre, Lancaster University, Lancaster, UK

Julia C. Sullivan, Oak Ridge Institute for Science and Education, Narragansett, RI, USA

Christian Vogelsang, Norwegian Institute for Water Research, Oslo, Norway

Liesbeth Weijs, Toxicological Centre, University of Antwerp, Universiteitsplein, Wilrijk; Systemic Physiological and Ecotoxicological Research (SPHERE), Department of Biology, University of Antwerp, Groenenborgerlaan, Antwerp, Belgium; National Research Centre for Environmental Toxicology (Entox), The University of Queensland, Coopers Plains, QLD, Australia

Xianhai Yang, Key Laboratory of Industrial Ecology and Environmental Engineering (Ministry of Education), School of Environmental Science and Technology, Dalian University of Technology, Dalian, Liaoning, China

Jing You, State Key Laboratory of Organic Geochemistry, Guangzhou Institute of Geochemistry, Chinese Academy of Sciences, Guangzhou, China

Eddy Y. Zeng, School of Environment, Jinan University, Guangzhou, China

Yunhui Zhang, Department of Environmental Health, School of Public Health, Fudan University, Shanghai, China

Series Editor's Preface

Our first volume on environmental analytical chemistry was Volume 32. After its publication in 1997, I remember well being appointed editor-in-chief of the Comprehensive Analytical Chemistry book series (CAC). After that, several volumes on environmental contaminants were published, focusing on specific group of pollutants, such as Volume 40 on surfactants and more recently Volumes 50 and 62 on pharmaceutical residues. In addition, other environmental applications were included in state-of-the-art books on analytical techniques, like the use of advanced gas chromatography–mass spectrometry (GC-MS) and time-of-flight, two-dimensional GC and passive sampling techniques, in CAC Volumes 61, 55, and 48, respectively.

This book, edited by Eddy Zeng, is an excellent cocktail of novel analytical techniques and applications to the trace determination of persistent organic pollutants (POP) as well as of pharmaceuticals and personal care products (PPCP). The general introduction covers a comprehensive variety of analytical techniques, including field application of passive sampling, microextraction, and bioanalytical approaches for the determination of POPs and PPCPs in the environment. Occurrence, fate, and removal of selected pollutants in a broad range of matrices are reported in different chapters, including e-waste, wastewaters, soil/sediments, the atmosphere, biota, and human health samples. Finally risk assessment chapters dealing with ecological effects in the aquatic environment, reproductive toxicity, and in-silico toxicity models, are also reported. The 19 chapters of this book make it a very comprehensive title in our series to better understand the analysis, fate, and toxicity of POPs and PPCPs in the environment.

I am convinced that this book will be today the reference book in the analytical and environmental chemistry community of POPs and PPCPs. The book is specially suited for newcomers who want to become familiar with this research field and it can also be used for advanced training courses. Finally I would like to thank not only the editor of the book but also the various authors for compiling such a world-class book on environmental organic contaminants.

D. Barceló
IDAEA-CSIC, Barcelona, and ICRA, Girona
October 10, 2014.

Volume Editor's Preface

The initial persistent organic pollutants (POPs), targeted by the Stockholm Convention in 2001, specifically include a group of pesticides, industrial chemicals, and unintended by-products, i.e., aldrin, chlordane, DDT, dieldrin, endrin, heptachlor, hexachlorobenzene, mirex, toxaphene, polychlorinated biphenyls (PCBs), polychlorinated dibenzo-*p*-dioxins (PCDDs), and polychlorinated dibenzofurans (PCDFs), often referred to as the "Dirty Dozen." In 2009, nine new POPs were added to the target list of the Stockholm Convention, including chlordecone, alpha hexachlorocyclohexane, beta hexachlorocyclohexane, hexabromobiphenyl, hexabromodiphenyl and heptabromodiphenyl ethers, lindane, pentachlorobenzene, perfluorooctane sulfonic acid, its salts and perfluorooctane sulfonyl fluoride, and tetrabromodiphenyl and pentabromodiphenyl ethers. In a broader sense, however, any organic chemicals that are persistent and widespread in the environment, able to bioaccumulate and biomagnify in living organisms including humans, and toxic to both humans and wildlife, may be regarded as POPs. Such a broad perception of POPs has been widely used in the scientific literature, and is beneficial in recognizing and consequently minimizing the negative environmental and health impacts of organic contaminants in general. With this in mind, the current book not only focuses on the POPs targeted by the Stockholm Convention, but also covers nontarget and/or emerging organic contaminants.

The subject matters of the 19 chapters constituting this comprehensive book are divided in three categories, i.e., sampling and analytical methods, environmental fate, and biological effects, as explicitly revealed in the book's title. These topics reflect the main components of past and present research activities on POPs worldwide, and are expected to continuously dictate the research trends for many years to come.

Chapters 1–5 fall into the first category. Chapter 1 summarizes the important features of sampling and chemical analytical procedures for measuring POPs in complex environmental matrices. Chapter 2 reviews the current biological approaches for assessing the chemical exposure and its toxicological implications and points to the need for establishing an integrated framework for linking the two. Chapter 3 previews the recent advances in microextraction techniques for sample preparation and analysis for a variety of POPs. Chapter 4 evaluates the applications of passive sampling techniques in field-measuring dissolved hydrophobic organic chemicals in situ, particularly the utility of passive samplers in determining the escaping/depositing fluxes of HOCs between the sediment–water interface. Chapter 5 synopsizes the latest developments in application of solid-phase microextraction methods for evaluating the toxicity of sediment-associated organic contaminants to benthic organisms.

The second category includes Chapters 6–14, comprising the book's main content, which may be consistent with the dominance of the global POPs research output in the same area. Chapter 6 deals with the removal of pharmaceuticals and personal care products (PPCPs) from wastewater by constructed wetlands. Chapter 7 discusses the fate of PPCPs during wastewater treatment processes and influencing factors. Chapter 8 recaps the progress in measurements of dry and wet depositional fluxes and diffusive air–water exchange of selected POPs. Chapter 9 discusses the likelihood for electronic waste (e-waste) as a new and important source of halogenated organic contaminants and metals, based on an assimilation of current literature. Chapter 10 also deals with e-waste, but focuses on human exposure to emerging contaminants released from e-waste recycling activities and related health effects. Chapter 11 assimilates available information about the long-range atmospheric transport of POPs and methods for assessing sources of POPs in remote areas. Chapter 12 describes the state of coastal contamination by emerging contaminants, particularly halogenated polybrominated diphenyl ether (PBDE) alternatives, as characterized by their occurrences in sediment and marine mammals. Chapter 13 analyzes the current monitoring data about the long-range atmospheric transport of POPs to polar regions and suggests the need for more long-term monitoring efforts. Chapter 14 presents evidence that brominated flame retardants are similar to POPs in terms of the potential to bioaccumulate and biotransform.

The themes of Chapters 15–19 are within the third and final category. Chapter 15 reviews the current approaches for measuring bioavailability of organic contaminants in soil and sediment and how bioavailability and bioaccessibility have been used in risk assessment and bioremediation of contaminated soil. Chapter 16 examines the environmental occurrence and ecological effects of benzotriazoles, a group of organic compounds with similar characteristics to POPs and of emerging concern. Chapter 17 demonstrates the utility of models based on quantitative structure–activity relationships in predicting the thyroid hormone effects of PBDE derivatives. Chapter 18 presents an integrated report on the fate, transport, and toxicity of selected POPs, such as PCBs, PCDDs, PCDFs, perfluorinated chemicals, and organobromide compounds, in aquatic environments. Chapter 19 reports an integrated health-based risk assessment of PBDEs, using China as a case study.

Aside from providing a comprehensive coverage of past research efforts on POPs, this book also demonstrates the hard work and fruitful outcome of the international team involved. The authorship of this book includes scientists from 13 countries around the world, i.e., Australia, Belgium, Canada, China, Germany, New Zealand, Norway, Romania, Spain, Sweden, Switzerland, the United Kingdom, and the United States. It is truly a great example of a successful multicontinental collaboration, which is also mirrored in the global battle against the potential adverse effects of POPs. I feel extremely honored to have the privilege to work with such a highly respectable group of scientists, without whom this book could not have possibly been completed. Thanks also go to Series Editor, Dr Damià Barceló, the Editorial Project Manager at Elsevier, Mr Derek Coleman, and the production team for their professionalism during the entire process. Finally, I would be very gratified if readers find this book helpful and/or useful in their pursuits of information about POPs.

Eddy Y. Zeng
School of Environment, Jinan University,
Guangzhou, China

Chapter 1

Analytical Methods for the Measurement of Legacy and Emerging Persistent Organic Pollutants in Complex Sample Matrices

Ying Guo and Kurunthachalam Kannan[1]
New York State Department of Health and Department of Environmental Health Sciences, Wadsworth Center, School of Public Health, State University of New York at Albany, Albany, NY, USA
[1]*Corresponding author: E-mail: kkannan@wadsworth.org*

Chapter Outline

1. INTRODUCTION

Persistent organic pollutants (POPs) are bioaccumulative and toxic chemicals that resist degradation in the environment, are prone to long-range transport, and have the potential to elicit adverse effects on environmental and human health. Efforts to eliminate or restrict the production and use of POPs have resulted in a global treaty led by the United Nations Environment Programme (UNEP), the Stockholm Convention on Persistent Organic Pollutants, which

Comprehensive Analytical Chemistry, Vol. 67. http://dx.doi.org/10.1016/B978-0-444-63299-9.00001-6

was ratified by over 150 countries since May 2004. Currently, over 20 POPs are listed in the Stockholm Convention (Figure 1). A few others are proposed as potential POPs and include short-chain chlorinated paraffins, chlorinated naphthalenes, hexachlorobutadiene, and pentachlorophenol. In addition, polycyclic aromatic hydrocarbons (PAHs) and brominated flame retardants (BFRs) as well as organometallic compounds, such as tributyltin, are also listed as POPs due to their high toxicity and widespread environmental distribution.

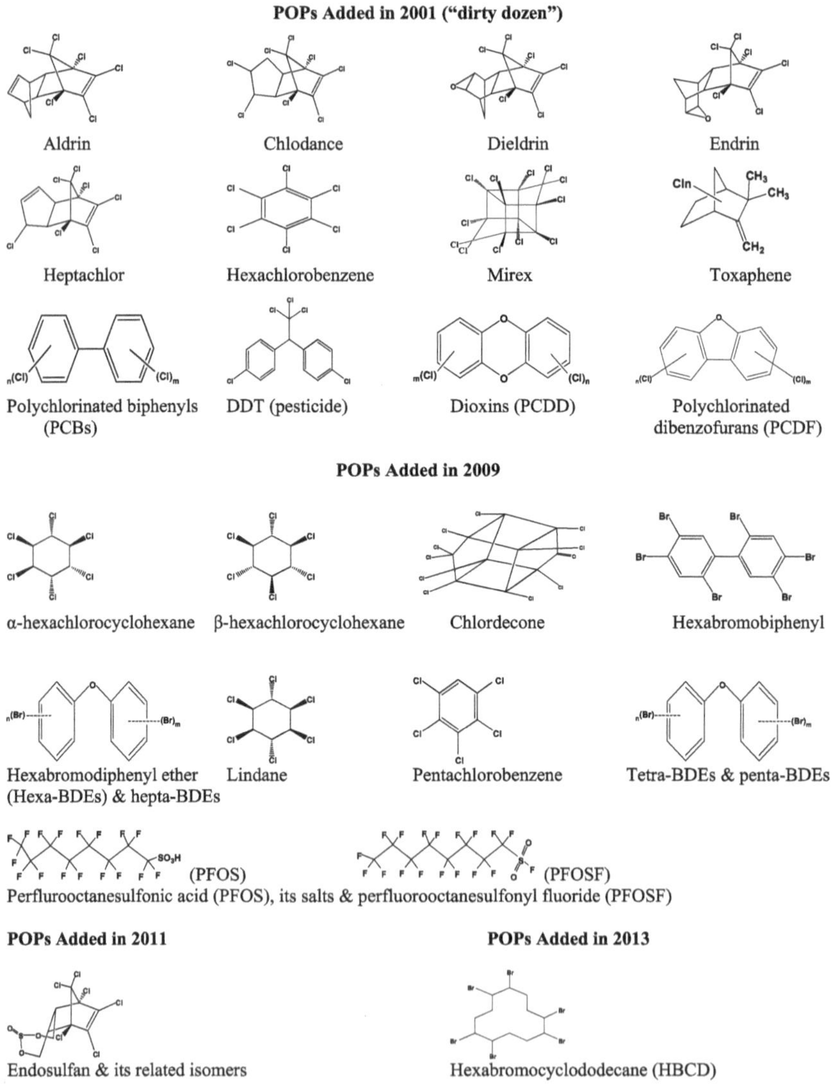

FIGURE 1 Persistent organic pollutants listed in the Stockholm Convention.

Most POPs are man-made compounds and include pesticides and industrial chemicals; a few other POPs are unintentional by-products of industrial processes. Among the "dirty dozen" Stockholm Convention POPs, aldrin, chlordane, DDT, dieldrin, endrin, heptachlor, hexachlorobenzene, mirex, and toxaphene are organochlorine pesticides (OCPs); PCBs are industrial chemicals; and PCDDs or PCDFs are by-products of incomplete combustion.

Halogenation of (F, Cl, or Br) aliphatic or aromatic rings is the main characteristic of POPs (Figure 1) and renders them low water solubility, high lipid solubility, and high resistance toward chemical, biological, and photolytic degradation. POPs are prone to long-range transportation and bioaccumulation in the food web and are ubiquitous in the environment, even in remote polar regions, including the Arctic Ocean [1–3]. For instance, the water solubility of penta-BDEs (bromo diphenyl ether) is low, at ~6.0 μg/L, but its octanol–water partition coefficient (K_{ow}) is high (log K_{ow} is ~6.5) [4,5]. The estimated characteristic travel distance of penta-BDE ranges from 608 to 1349 km from the sources of release [6]. Penta-BDE congeners (typically BDE-99 and BDE-100) have been found to biomagnify in both the marine [7–9] and terrestrial food chain [10,11].

Measurement of trace levels of POPs in various environmental matrices is the first step to monitor and elucidate distribution and fates in the environment. Because POPs are present at ultratrace levels (parts per trillion to parts per million) in environmental samples, precise analysis of these chemicals requires careful planning and execution of analysis. A reliable analytical method for the determination of POPs should be composed of three main steps (Figure 2), which include sampling, transport and storage, and analysis, in addition to

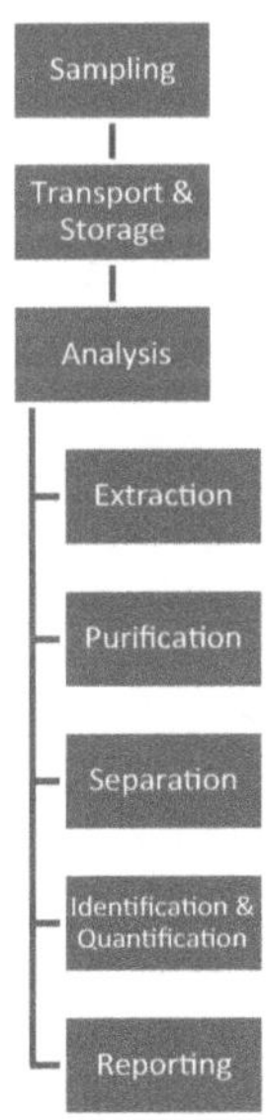

FIGURE 2 Main steps in the analysis of environmental pollutants.

a good quality assurance (QA) and quality control (QC) system. Hundreds of publications on the analysis of POPs have appeared during the past three decades, and several earlier reviews have described the methods for the analysis of POPs. Several interlaboratory comparison studies and proficiency testing (PT) programs are in place throughout the world. Nevertheless, there are imminent challenges in accurate and precise analysis of POPs in complex environmental matrices.

This chapter will provide a brief and systematic description of the methods for the measurement of legacy (such as OCPs) and emerging POPs (such as BFRs) in complex sample matrices, with the emphasis on analytical methods. We have also listed a large number of reviews that have been published on this topic.

2. ANALYTICAL METHODS

2.1 Sampling

2.1.1 Sampling Design

A well-designed research plan is crucial to execute a study in a systematic manner and to address problems in a logical manner to accomplish goals. The study design should raise the questions "why" (objective), "what" (target population and unit), and "how" (sampling method). Sampling is an integral part of any environmental research. The sampling design specifies the number, type, and location (spatial and temporal) of the sampling units [12]. There are several commonly used sampling methodologies, including judgmental, simple random, stratified, systematic and grid, and composite sampling.

Judgmental sampling is a nonprobability sampling technique and the researchers select the units to be sampled based on their knowledge, professional experience, and judgment. Simple random sampling involves a subset of a unit chosen from a population. Each individual is chosen randomly and has the same probability of being chosen at any stage during the sampling process. Simple random sampling is the most useful when the population of interest is relatively homogeneous.

Systematic and grid sampling is a type of probability sampling method in which sample units from a larger population are selected according to a random starting point and a fixed, periodic interval. Here, samples are taken at regularly spaced intervals over space or time. An initial location or time is chosen at random, and the remaining sampling locations are defined so that all the locations are at regular intervals over an area (grid) or time (systematic) [12].

Stratified sampling is a method to randomly or systematically collect units from different homogeneous subgroups of a target population that has been divided by prior information. Composite sampling is a method to physically combine and homogenize samples that are collected from the same location chosen by another sampling design, based on a fixed composition scheme.

TABLE 1 Various Sampling Methods in Persistent Organic Pollutant Research

Sampling Method	Research Purpose
Judgment sampling	A screening situation for small-scale problem with small budgets or limited schedule in order to assess whether further investigation is warranted
Random sampling	Estimating the prevalence of a rare trait
Systematic sampling	Understanding of when contamination is present; estimating a population mean
Grid sampling	Understanding of where contamination is present; estimating a population mean
Composite sampling	Estimating a population mean; identifying population unit that has a rare trait; identifying population unit that has the highest contamination levels
Stratified sampling	Estimating a population mean or proportion based on spatial or temporal information on contaminant patterns

No single method applies to all research assessment needs. An appropriate sampling design is chosen based on research purpose and/or budget (Table 1) [12]. Thus, it is important to establish goals and objectives of a study, on which the sampling design will be based.

2.1.2 Sampling of Environmental Matrices for Analysis of POPs

POPs are ubiquitous in the environment and undergo biogeochemical cycles, as illustrated in Figure 3. POPs have been detected in solid environmental matrices, including biota (vegetation and animal as well as human samples) and soil (including sediment and dust), water (freshwater, groundwater, wastewater, seawater, rain, snow), and air. For the examination of environmental transport, transformation, and fate of POPs, sample media selected for analysis must be faithfully representative. Depending on the understanding of the spatial and temporal distribution as well as variability of the distribution of POPs in the environmental matrix, subsequent analysis of samples and further refinement of objectives and significance of the project are needed.

2.1.2.1 Plant and Animal Sampling for Analysis of POPs

Small organisms, such as zooplankton [13,14] and phytoplankton [15], and large vertebrates, such as whales [16] and polar bears [17–21], have been analyzed for POPs. These studies used animals or plants as indicators to monitor POP contamination in their living environment, to estimate bioaccumulation or biomagnification of POPs in the food chain, to explore the fate of POPs

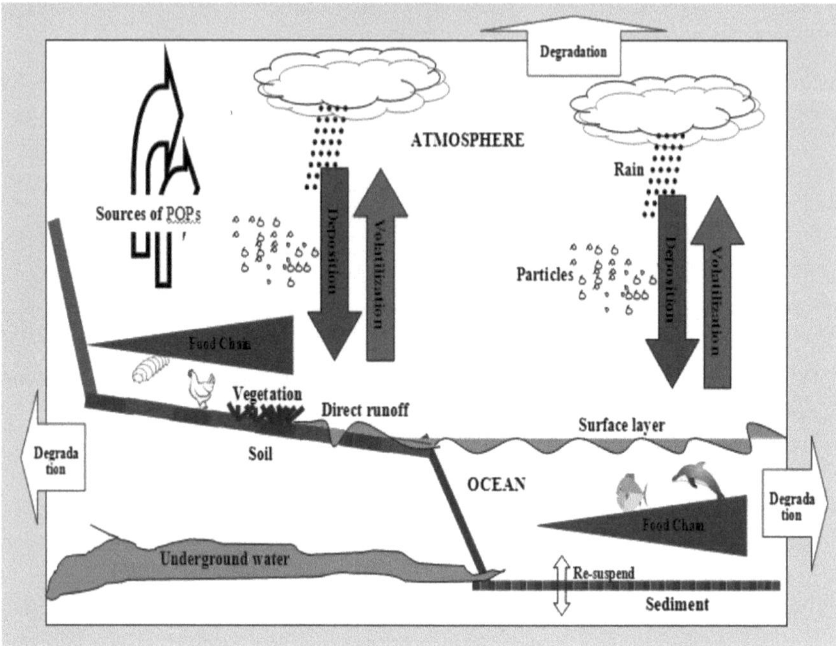

FIGURE 3 A schematic illustration of the fate and transport of persistent organic pollutants in the environment.

in the environment, and to estimate human exposure to POPs. Whereas whole body homogenates can be used in the analysis of POPs for small organisms, for large vertebrates, such as whales and dolphins, lipid-rich tissues often have been used. Unless adequate care is taken, this may pose problems in the calculation of biomagnification factors. This is because comparison of POP levels in the whole body of one organism with a specific tissue in other organisms may not provide an accurate indication of biomagnification.

Aquatic ecosystems are ultimate sinks for many POPs [22]. Shellfish, fish, and marine mammals have been used as monitors of POPs in aquatic ecosystems. The Mussel Watch Program used mussels and oysters to monitor for PBDEs and PCBs in coastal waters [23]. In terrestrial ecosystem, earthworms, birds, and bird eggs have been widely used to monitor for POPs. The eggs of the peregrine falcon (*Falco peregrinus*) were used to monitor PBDE contamination trends in the northern United States [24].

2.1.2.2 Sampling of Human Specimens for Analysis of POPs

Human exposure to POPs is well documented, and such exposures arise from dietary sources, inhalation, and/or dermal absorption. Occurrence of POPs in human tissues is an indication of exposure and accumulation. Because many POPs are lipophilic, breast milk and blood, which contain lipids, have been the most commonly used specimens in human biomonitoring studies of POPs. Adipose fat is an ideal matrix for the analysis of lipophilic chemicals, but the collection of such matrices is very invasive; however, samples from cadavers or individuals who undergo liposuction or other surgical procedures have been used in analysis of POPs [25]. Breast milk has been used to monitor the time trends of POPs in several countries, including Canada [26], Korea [27], Sweden [28,29], Japan [30], Serbia [31], China [32], Czech Republic [33], Rome [34], New Zealand [35], Italy [36], and Spain [37,38]. Breast milk was also used to compare spatial trends of POPs in several Asian countries [39] as well as perfluorinated compounds (PFCs), (PCDD/PCDFs), polybrominated diphenyl ethers (PBDEs), and OCPs from all over the world [40–43]. In addition to breast milk and blood, adipose tissue, placenta, urine, semen, and hair have been analyzed for POPs [44–50]. The choice of human specimens for analysis of POPs is dictated by ease of sampling and the study objectives.

Many human biomonitoring studies involve samples that come from donors who are seeking medical or therapeutic treatments in clinical settings or from volunteer donors. One of the major considerations currently faced with regard to human specimen collection is the need for approval from ethical committees.

2.1.2.3 Water Sampling for Analysis of POPs

POPs, in general, have low water solubility; thus, their concentrations in water, as well as in suspended particulate matter (SPM) in water bodies, are usually in the low picograms per liter to nanograms per liter range. For the analysis of dissolved phase concentrations of POPs in water, especially in oceanic water, a large sample volume (several tens to hundreds of liters) is required. Recently, passive samplers are gaining popularity in water sampling. One measure that can be used to determine global distribution of POPs is to take samples from remote marine locations and analyze them for POPs.

Surface waters include streams, rivers, lakes, ponds, creeks, lagoons, estuaries, surface impoundments, and coastal waters. Because water bodies can be dynamic, e.g., rivers in which concentrations can spike during certain times of the year, sampling should be planned well. A grab sample is an individual sample collected randomly at a specific time point. A composite sample is a representative sample prepared from a combination of multiple grab sample aliquots collected at periodic time intervals or locations [51]. Most POPs in water are sampled by the use of a conventional bottle sampling technique, whereby a precleaned sampling bottle is immersed into a target water body, usually with the help of a water pump. When sampling shallow water, contamination from disturbed sediment should be avoided. Sampling location, water quality parameters (depth, temperature, pH), and surrounding circumstances should be carefully recorded. Sampling near the edges of water bodies should be avoided.

For the collection of SPM in a water column, methods such as sediment traps (containers at a predetermined depth in the water column), time-integrative SPM samplers (sedimentation boxes/devices), large-volume sampling (pumping a large volume of water through a filtration system), and continuous flow centrifuge (centrifuging a large amount of water) are used.

2.1.2.4 Soil and Sediment Sampling for Analysis of POPs

Depending on the purpose of research, soils are collected usually at the top six inches, and the number of soil samples collected varies, depending on the variability of POP concentrations within a location and the data needs of the program. Prior to sampling, it is important to determine the depth at which the soil samples will be collected. Surface soil is usually collected by a spade or a scoop. Subsurface soil is commonly collected by tube/core samplers. Tube samplers are hollow tubes that are driven or screwed into soil. The soil fills the tube and is retained when the sampler is withdrawn. Accurate recording of the sampling location (with a global positioning system), surrounding environmental settings, soil physical characteristics, and types of the soils, e.g., color, mottles, soil texture, particle shape, structure, consistency, and horizon thickness are important for interpretation of the results.

Sediment samples, including surface and deep sediment core, have been used to study the distribution and historical profiles of POPs in rivers, lakes, ponds, lagoons, and marine/estuarine areas. Information on sampling location, historical sources of contamination, hydrodynamic conditions of water bodies as related to physical sediment fate and transport, and human or biota utilization of the sampling site should be recorded for further interpretation of the occurrence of POPs. Generally, a grab sampler is used if a large volume or a large surface area of surficial sediment is needed, and a core sampler is used if the aims are to compare concentrations of POPs in surficial and historically deposited sediment, to reduce sediment disturbance or to reduce oxygen exposure, or if sediment is soft and fine-grained [51].

Pore water samples in sediment have been useful for the assessment of POP levels and associated toxicity to benthic organisms due to the thermodynamic equilibrium of POPs between sediment and pore water. Pore water samples are collected from sediment by suction devices or piezometers; passive pore water samplers have also been used (based on the equilibrium partitioning of contaminant between pore water and organic sorbents). The most commonly used passive samplers in analysis of POPs include semipermeable membrane devices (SPMD), solid-phase microextraction (SPME) devices, and polyethylene samplers.

2.1.2.5 Air Sampling for Analysis of POPs

Most POPs are semivolatile and can be present both in the gaseous phase and in SPM or aerosol in air. Although air is a relatively homogeneous medium, air pollution can differ significantly, depending on the sampling site and time (spatial and temporal). The air sampling location and duration should represent the ambient conditions, and additional meteorological parameters, such as wind direction, can be helpful in the interpretation of results.

The solid sorbent sampling method (active and passive) is the most commonly used technique for the collection of POPs in the gas phase. Active air sampling draws air through a tube filled with solid polymeric adsorbent material, to which airborne POPs get absorbed. The high-volume (Hivol) sampler is used for active sampling. A calibrated personal sampling pump is used to draw a known volume of air, and different types of sorbent materials, such as polyurethane foam (PUF) and XAD-2® resin, are used to capture airborne POPs. With the use of a Hivol sampler, a high sampling rate can be achieved, and the sampling time can be shortened, which ultimately increases the accuracy of POP measurements and decreases the opportunity for degradation of POPs during sampling. Hivol samplers, however, are expensive and require electricity for operation, and transportation can be tedious. Thus, they are not amenable to simultaneous monitoring of POPs in a large area or in remote areas.

Passive air samplers (PAS) absorb POPs onto a sorbent media by physical processes, such as diffusion by Fick's first law, instead of actively pulling the air with a pump [52]. The sorbent medium captures POPs from the ambient air. The evolution of PAS in recent years is considered a milestone in the development of air sampling techniques. PAS is inexpensive, discreet, and easy to deploy and can be used for long-term sampling. PAS with different sampling devices, such as PUF disks and XAD resin, are now widely used in air sampling. However, to quantify concentrations of airborne POPs sampled by a PAS, the theoretical uptake rate for each POP by the sorbent has to be experimentally determined, and that rate may not be valid under the environmental conditions encountered during field sampling. In addition, the application of PAS is affected by environmental factors, such as wind speed, temperature, and humidity [53–55]. However, PAS can be used to measure relative concentrations of POPs on a spatial scale and to identify hot spots of contamination.

Some studies have compared the results of the two (active and passive) air sampling methods [56–58]. Hayward et al. [56] determined pesticides in the atmosphere of rural southern Ontario collected by four methods: two active samplers (Hivol and low-volume sampler) and two passive samplers (PUF disks and XAD resin). The average concentrations of OCPs obtained by different sampling techniques were comparable and improved with an increase in the length of the comparison period, especially for pesticides with high temporal variability. Gouin et al. [57] showed agreement within a factor of two or three for concentrations of OCPs, PCBs, and PBDEs, collected by Hivol and PAS, in the atmosphere of the Laurentian Great Lakes. These studies suggested that there was no significant difference in POP concentrations collected by active or passive air sampling techniques.

When selecting an appropriate air sampling method, both the benefits and challenges of active and passive sampling methods should be considered. A summary of recent studies of airborne POPs collected by different air samplers is presented in Table 2.

TABLE 2 Studies of Persistent Organic Pollutants (POPs) in Atmosphere Collected by Different Sampling Techniques

Technique/ Equipment	Sampling Location	Target POPs	References
AAS/Hivol	Shipboard air over the Yellow Sea, east	OCPs	
	China Sea and South China Sea, China		[59]
AAS/Hivol	Indoor and outdoor	PBDEs and PFAs	[60]
AAS/Hivol	Urban site in Beijing	Short-chain chlorinated paraffins	[61]
AAS/Hivol	Cities in Yangtze river Delta	PBDEs, PCBs, and PAHs	[62]
PAS/PUF disks and AAS/Hivol	Densely populated tropical Andean city of Manizales, Colombia	PCDD/Fs and DI-PCBs	[58]
PAS/PUF disks	Rural, urban, and industrial sites, Chile	OCPs, PCBs, PBDEs, and PAHs	[63]
PAS/PUF disk	Semiurban/industrial site, Algeria	OCPs, PCDD/Fs and DI-PCBs	[64]
PAS/PUF disk	Across the Tuscany region, Italy	PBDEs, PCBs, OCPs, and PAHs	[65]
PAS/PUF disk	Mountains of southeast and southern Brazil	PBDEs and PCBs	[66]
PAS/XAD resin and PUF disks	All seven continents (GAPS network)	OCPs	[67,68]

AAS, Active air sampling; GAPS, Global atmospheric passive sampling.

The majority of the POPs in the air are expected to be present in particulate matter due to their high sorption coefficients. Particulate matter is usually sampled by filtration using a glass fiber filter (GFF) or a filter paper by inserting the selected filter into a holder and drawing the air. The selected filter is cleaned, balanced, and weighed before sampling, and then balanced and weighed again to obtain the mass of particulate matter in the air.

2.1.3 QA/QC

The QC procedure is an integral part of a good field and laboratory practice. The QC procedure includes all the procedures applied for data collection and generation of activities to achieve and maintain the level of preestablished data quality [51]. The QC protocol should include all technical controls, e.g., sampling and analytical methods, use of field blanks, field duplicate samples, inclusion of performance testing or reference samples, and statistical analysis. The controls start with the regulatory requirements of the data acquisition and carry through the data reporting and completion of all documentation.

QA refers to the procedures used by the management to ensure the required QC and its adherence at any point in the project. QA constitutes the overview and monitoring processes designed to ensure that the quality of the data generated meets the desired levels as established by the management. Field duplicates and field blanks are the most commonly used QC samples during sampling. Field duplicates are two separate samples taken from the same source and are used to determine data repeatability based on field sampling and laboratory analysis. Field blanks are prepared and analyzed to check the cleanliness of sample containers, environmental contamination, and purity of reagents and solvents used in the field [51].

Several QA/QC protocols have been recommended by the UNEP and other standardization organizations. One important aspect of QA/QC is the demonstration of proficiency of the laboratory in the analysis of POPs. Several national and international organizations offer PT programs and provide certified standard reference materials. Participation in PT programs and analysis of certified reference materials will assure confidence in the analysis of POPs in environmental samples.

2.2 Transport and Storage of Samples

Samples should be securely and immediately transported to the laboratory after collection. For transportation and storage of samples, appropriate container, preservation condition, and holding time should be considered. This is to ensure that POPs do not undergo any chemical changes until laboratory analysis. Additionally, a detailed sampling and sample information and a clear, undetachable sample label are important during transport and storage.

Usually, glass containers stored at cool temperatures (<6 °C) and kept away from light are the basic sample preservation conditions for preserving POPs in environmental samples (Table 3). However, even under proper preservation and storage, concentrations or profiles of POPs in samples can change over time. Thus, sample preparation and instrumental analysis of POPs must be completed as soon as possible. Low temperature (e.g., −20 °C or −80 °C for biota and solids) may increase the holding time. One study showed that there was no significant difference in the concentrations of PCBs and DDTs in fish tissue stored at −20 °C and −70 °C for over 2 years [69]. Water/liquid samples should not be frozen for analysis of POPs, as POPs may precipitate upon thawing. The holding time may differ according to individual compound and the environmental matrix of interest (Table 3).

TABLE 3 Preservation Method and Holding Time for Persistent Organic Pollutants (POPs) in environmental Matrix [51]

POPs	Matrix	Preservation (Keep Away From Light)	Holding Time
OCPs	Drinking water	Residual chlorine in the sample should be reduced by adding 50 mg/L of sodium sulfite Adjust the sample pH to <2 by adding 6N HCl Cool to ≤6 °C	Extract within 14 days; analyze within 40 days of extraction
OCPs	Sediment, soils, sludge, and other nonaqueous samples	Cool to ≤6 °C	
OCPs	Wastewater, surface water, and storm water	NaOH or H_2SO_4 to pH 5–9 Cool to ≤6 °C	Extract within 7 days; analyze within 40 days of extraction
OCPs	Groundwater	Cool to ≤6 °C	
PCBs	Drinking water	If chlorine is present, first add 50 mg/L of sodium sulfite Adjust the sample to pH <2 by adding 6 N HCl Cool to ≤6 °C	Extract within 1 year; analyze within 1 year of extraction
PCBs	Groundwater, other aqueous waste	If chlorine is present, add 3 mL 10% $Na_2S_2O_3$ per gallon Cool to ≤6 °C	
PCBs	Sediment, soils, sludge, and other nonaqueous samples	Cool to ≤6 °C	
PCBs	Wastewater	Cool to ≤6 °C If residual chlorine is present, add 0.008% $Na_2S_2O_3$	

TABLE 3 Preservation Method and Holding Time for Persistent Organic Pollutants (POPs) in environmental Matrix [51]—cont'd

POPs	Matrix	Preservation (Keep Away From Light)	Holding Time
PAHs	Groundwater	Cool to ≤6 °C If chlorine is present, add 3 mL 10% $Na_2S_2O_3$ per gallon	Extract within 7 days; analyze within 40 days of extraction
PAHs	Wastewater	Cool to ≤6 °C If residual chlorine is present, add 0.008% $Na_2S_2O_3$	
PCDD/Fs	Groundwater	Cool to ≤6 °C If chlorine is present, add 0.008% $Na_2S_2O_3$ If pH is greater than 9, adjust to pH 7–9 with sulfuric acid	Extract within 30 days; analyze within 45 days of extraction
PCDD/Fs	Sediment, soils, and sludge	Cool to ≤6 °C	
PCDD/Fs	Wastewater	Cool to ≤6 °C If chlorine is present, add 0.008% $Na_2S_2O_3$ If pH is greater than 9, adjust to pH 7–9 with H_2SO_4	1 year

Availability of archived samples for environmental analysis can save a significant amount of time and cost in research programs. An increasing number of programs and institutions throughout the world are opting to set up environmental and biological specimen banks for long-term archival and systematic storage of specimens for unspecified future use. Environmental samples and human specimens are collected and stored at low temperatures so that they can be used in the future for retrospective studies. Environmental specimen banks are important resources of research materials that are used to document geographic and temporal trends in environmental pollutants. The National Aquatic Biological

Specimen Bank (NABSB), which collects fish tissue samples in the Great Lakes Basin in Canada, has been in place since 1977. The NABSB employs standardized banking protocols, computerized sample tracking, and maintenance of all data and metadata associated with each specimen. By 2010, the NABSB had held over 52,000 biological samples from 18,749 species of top predator fish, forage fish, plankton, and other invertebrates collected for over 32 years of environmental monitoring in the Canadian waters of the Great Lakes. Numerous publications on contamination, including legacy and emerging POPs, in the aquatic ecosystems have been published [70]. It is recommended that the samples collected for POP analysis be given consideration for archival for future use. As such, some thought should be given to sampling and storage requirements of specimens for future analysis of novel, yet unknown, contaminants.

2.3 Analysis of POPs in Environmental Matrices

In general, the laboratory analytical steps for POPs include sample preparation (extraction and purification), instrumental analysis (separation, identification, and quantification), and data reporting.

2.3.1 Sample Preparation in Analysis of POPs

Due to the ultratrace levels of POPs in environmental samples (e.g., parts per trillion to parts per million), and the complexity of environmental matrices, samples must be properly prepared before instrument analysis as a means to concentrate/enrich POPs in the final extract, remove interferences, reduce sample weight or volume for transportation and storage, and protect analytical instrumentation. The following instructions for sample preparation are based on the assumption that the samples were taken from an ambient environment and are relatively clean and homogenized properly.

2.3.1.1 Extraction

Because the majority of POPs are lipophilic, the extraction methods are based on the isolation of lipids from the sample matrix, controlled both kinetically and thermodynamically. A large number of publications [71–81] contain reviews of the extraction methods of POPs from environmental matrices. The most commonly used extraction methods are Soxhlet, pressurized liquid extraction (PLE, also known as accelerated solvent extraction (ASE)), supercritical fluid extraction (SFE), matrix solid-phase dispersion (MSPD), microwave-assisted extraction (MAE), ultrasonic-assisted extraction (UAE), liquid–liquid extraction (LLE), solid-phase extraction (SPE), and SPME. Each extraction method has its own merits and demerits; further details of these methods can be found elsewhere [72,82–84].

Soxhlet extraction, invented in 1879 by Franz von Soxhlet, is one of the most frequently used liquid–solid extraction methods. POPs are isolated by the Soxhlet procedure from a wide variety of environmental samples, including sediments, soils, dust, tissues of biota, air filters, or other solid samples [76,85]. The main advantages of Soxhlet extraction are that the extracts need not be filtered or centrifuged and that a large amount of samples can be extracted. The extraction method is also robust and is applicable to a wide range of environmental chemicals. The Soxhlet procedure, however, is time and solvent consuming. A conventional Soxhlet method usually requires large volumes of organic solvent (>100 mL) and a relatively long extraction time (12–48 h).

A relatively more efficient PLE and SFE have been developed in the past 20 years. The main advantages of PLE and SFE are short extraction time and low solvent consumption. PLE employs solvents at high pressures (500–3000 psi) and/or high temperatures (50–200 °C), which can enhance the solubilization and desorption of POPs from the sample matrix, thereby accelerating the speed of the extraction process [86,87]. SFE extraction of POPs from solid matrix uses a supercritical fluid as an extractant (usually carbon dioxide) [88–90]. The requirement for high pressures increases the operational cost of SFE compared to conventional Soxhlet extraction.

MSPD was introduced in 1989 by Barker [91] and is widely used for solid samples, including human biospecimens. MSPD is a simple and fast extraction procedure in which samples are homogenized with a bonded phase (such as Florisil or silica), transferred into a glass column (using more bonded phase, if necessary), and then eluted with solvents [73,91,92]. MAE is a process of heating solvents in contact with a sample with microwave energy to partition analytes from the sample matrix into the solvent [93]. UAE is used for extraction of analytes from solid samples with organic solvents by the application of ultrasonic radiation in water bath or with other devices, such as probes, sonoreactors, or microplate horns.

SPE and SPME are alternatives to the time- and solvent-consuming LLE. With SPE, analytes are captured in a bonded phase packed in a cartridge and eluted with a solvent. In recent years, SPE has been widely used as a method of choice for analysis of POPs in human fluids (serum or saliva). The SPME method involves capturing of analytes, in both liquid and gas phases, by a fiber coated with an extraction phase (solid or liquid) and then directly desorbing the analytes in the injection port of the analytical instruments. SPME is simple, fast, solvent free, reliable, and flexible for the analysis of POPs in air and water samples.

In general, an extraction process should consider the following aspects in sample preparation, which dictate the extraction methods that can be applied: interferences removing capacity, analyte recoveries, convenience of operation, cost, and effects on environment and human health. POP researchers have used different extraction methods for various environmental matrices in the past 5 years, as summarized in Table 4.

In general, water/moisture, lipids, proteins, and sulfur should be eliminated from sample extracts during sample preparation. Freeze-drying and anhydrous sodium sulfate are used to remove water; sulfuric acid and gel permeation

TABLE 4 Sample Preparation Method and Instrumental Analysis for Persistent Organic Pollutants in Environmental Matrixes in the Recent 5 years (2009–2014)

Environmental Matrix	Analytes	Extraction Method	Clean-up Method	Separation and Instrumental Analysis	References
Soil	OCPs and PCBs	Soxhlet with DCM	Silica–alumina column	HRGC–HRMS	[94]
Soil	OCPs, PAHs, and PCBs	UAE with HEX: DCM (1:1)	For PAHs, silica column For OCPs, activated Florisil column For PCBs, multilayer silica column (silver nitrate silica, activated silica, basic silica, and acid silica)	GC–MS	[95]
Soil	PBDEs	ASE with toluene	Acid treatment with 98% H_2SO_4, multilayer silica column (acid silica, basic silica, and potassium hydroxide silica)	GC–MS	[96]
Soil	PAHs, PCBs and PBDEs	PCB and PBDEs: PFE and PAHs: LLE with HEX	PCBs and PBDEs: multilayer silica column (basic silica and acid silica) PAHs: basic silica column	GC–MS	[97]
Soil	OCPs, PBDEs and PCBs	Soxhlet with DCM	Alumina/silica column, then GPC	GC–MS	[98]
Soil	PCDD/Fs, DI-PCBs and PCBs	SFE	Multilayer silica column (silver nitrate silica, acid silica, and basic silica), then another multilayer silica column (silica, alumina, and PX21-carbon/celite)	HRGC–HRMS	[99]

Soil	PBDEs	Soxhlet with DCM: HEX (1:1)	Silica and alumina column	GC–MS	[100]
Sediment	PBDEs and PCBs	ASE with DCM: acetone (1:1)	Neutral silica column	HRGC–HRMS	[101]
Sediment	Emerging BFRs	Soxhlet with DCM: HEX (1:1)	Multilayer silica column (activated silica and acid silica)	GC–MS	[102]
Sediment	OCPs	Soxhlet with HEX: acetone (4:1)	Florisil column	GC–MS	[103]
Sediment	PAHs	Soxhlet with DCM	Activated silica column	HPLC–PAD	[104]
Sediment	PCBs and OCPs	MAE with HEX: DCM (1:1)	Silica and alumina column	HRGC–ECD	[105]
Sediment	PCDD/Fs, OCPs and PCBs	ASE with toluene	Multilayer silica column (acid silica, potassium hydroxide silica, silver nitrate silica, and activated silica), then basic silica column and carbon column	HRGC–HRMS	[106]
Sediment	Short-chain chlorinated paraffins	ASE with DCM: HEX (1:1)	Acid silica to remove lipid, then multilayer silica column (basic silica, activated silica, and acid silica)	GC–MS	[107]
Sediment	PFCs	LLE (methyl-tert-butyl ether, ion-pairing procedure)	By centrifugation	HPLC–MSMS	[108]

Continued

TABLE 4 Sample Preparation Method and Instrumental Analysis for Persistent Organic Pollutants in Environmental Matrixes in the Recent 5 years (2009–2014)—cont'd

Environmental Matrix	Analytes	Extraction Method	Clean-up Method	Separation and Instrumental Analysis	References
Sediment	PFCs	Ultrasonic extraction	SPE	HPLC–MSMS	[109]
Surface water	PBDEs	SPE (LC-C18)	Multilayer silica and alumina column (neutral alumina, basic silica, and acid silica)	GC–MS	[110]
Surface water	PCBs	SPE (C18 disk)	Acid silica column	GC–MS	[111]
Drinks, juices, and drinking water	OCPs, DI-PCBs, and PAHs	LLE with DCM	NA	GC–TOFMS	[112]
Surface water	PCBs and PAHs	LLE with DCM	NA	GC–MS	[113]
Surface water	OCPs	SPE (XAD-2)	Silica column	GC–MS	[114]
Surface water	PCBs and PBDEs	LLE with DCM	Neutral alumina and silica column	GC–ECD	[115]
Surface water	PFCs	LLE with DCM/ isopropyl alcohol (90/10)	Filtered before injection	LC–MSMS	[116]
Surface water	PAHs, PCBs, and OCPs	SPMDs, then extraction with HEX	Silica and alumina column, then SPE cartridge (C18)	HRGC–HRMS	[117]

Indoor air	Chlorinated paraffins	PUF and GFF were extracted with DCM in an ultrasonic bath	Acid silica cartridge	GC–MSMS for short and median chain GC–ECNI–MS for group patterns	[118]
Air	PCDD/Fs	US EPA method 1613B	Acid silica column, a sandwich column (acid and alkali silica), and an alumina column	GC–HRMS	[119]
Air	BRFs	Filters and PUF were extracted by ASE with HEX: acetone	Deactivated silica column, then an activated silica column	GC–MSD	[120]
Air	PCDD/Fs and PCBs	Filters and PUF were extracted by Soxhlet	Multilayer solid-phase chromatography column, then a basic alumina column	GC–MS for PCBs GC–HRMS for PCDD/Fs	[121]
Air	PBDEs	PUF and GFF was extracted by Soxhlet with acetone: HEX (1:1)	Activated silica gel column	GC–MS and GC/ECD (BDE-209)	[122]
Serum	Dioxin, furan, and PCBs	SPE (nonendcapped octadecyl disposable cartridges)	Multilayer silica column (acid silica, basic silica, and neutral silica), then basic alumina column with carbon	GC–MS GC–HRMS	[123]
Serum	PBDEs	SPE (C18)	Multilayer column (acid silica and activated silica)	GC–HRMS	[124]

Continued

TABLE 4 Sample Preparation Method and Instrumental Analysis for Persistent Organic Pollutants in Environmental Matrixes in the Recent 5 years (2009–2014)—cont'd

Environmental Matrix	Analytes	Extraction Method	Clean-up Method	Separation and Instrumental Analysis	References
Serum	OCPs	LLE (HEX:toluene (9:1))	Silica SPE cartridge	GC–μECD	[125]
Serum	PFCs	LLE (acetonitrile)	Oasis WAX SPE cartridge	HPLC–MSMS	[126]
Milk	PBDEs, PCBs, and PCDD/Fs	Dried, then Soxhlet with HEX: DCM (1:1)	GPC. Multilayer silica column and acid alumina column	GC–HRMS	[127]
Milk	OPCs	LLE (HE)	GPC. Silica column	GC–MS	[128]
Milk	OCPs, PCBs, and PBDEs	LLE (HEX, acetonitrile, and methanol)	Activated Florisil column	GC–ECD	[129]
Milk	PCDD/Fs	Dried, then Soxhlet with ethanol/toluene (7:3)	GPC. Acidic silica column, a Florisil column, and then Carbopack B or Carbopack C column	HRGC–HRMS	[130]
Milk	OCPs and PBDEs	SPE (C18)	SPE (Sep-Pak plus NH_2)	HRGC–HRMS	[131]

Milk	PCDD/Fs	LLE	Multilayer silica column (acid and potassium hydroxide), then silica, alumina, and carbon column	HRGC–HRMS	[132]
Milk	PBDEs	ASE with HEX: DCM (1:9)	Washed with 98% sulfuric acid, then Florisil column	LC–MSMS	[133]
Fish	PBDEs, PCBs, and OCPs	Soxhlet with DCM: HEX for 8 h	GPC, silica gel column	GC–ECD GC–MS	[134]
Fish	PFCs	Sonicated with methanol	SPE (WAX cartridge)	LC–MSMS	[135]
Fish	PCBs	MAE with DCM	Acidic silica gel column	GC–ECD	[136]
Fish	PCDD/Fs	EPA 1613B Soxhlet with HEX: DCM (1:1)	Sulfuric acid treatment, then multilayer silica column, and basic alumina column	GC–HRMS	[137]

HRGC, high-resolution gas chromatography; HPLC, high-performance liquid chromatography; NA, Not available.

chromatography (GPC) are used to remove lipids; formic acid and some organic solvents, e.g., acetonitrile, are used to denature proteins; and sulfuric acid and activated copper are used to remove sulfur from sample extracts.

The traditional extraction methods, such as Soxhlet, SPE, and LLE, have been widely used in the past 5 years for analysis of POPs (Table 4). Novel methods, such as PLE, SFE, MAE, UAE, and SPME have been developed and are applied in sample analysis. Solid samples, such as sediments, soils, tissues, and PUF, are extracted preferably by Soxhlet, ASE, and MAE, whereas liquid samples, such as water, serum, and milk, are extracted preferably by LLE, SPE, or SPME. The most commonly used solvents for the extraction of POPs are hexane (HEX), dichloromethane (DCM), acetone, or toluene and their mixtures, depending on the matrix and the target compounds.

The following section introduces information on novel extraction or cleanup methods in analysis of POPs for solid (MSPD and enhanced solvent extraction techniques) as well as liquid samples (sorption-based and solvent-based extraction techniques).

The MSPD is a relatively new method and is based on the SPE strategy. This method has been used in the preparation, extraction, and fractionation of solid, semisolid, and highly viscous biological samples [73,91,92,138,139]. MSPD has been used for the isolation of POPs from various matrices, including foodstuffs, biological tissues, sludge, and soils [138]. This method was used to extract and purify OCPs and PCBs in human lung tissue through the use of 3.5 g of Florisil and 15.5 mL of an organic solvent [140], to extract and purify OCPs in agricultural samples (produced similar results to ASE) with 1.5 g of Florisil and 15 mL of an organic solvent [141], and to extract PBDEs from fish with 3 g of silica gel, 3 g of acidic silica gel, and 20 mL of an organic solvent [142]. In MSPD, a fine dispersion of the matrix is mixed with select sorbent materials (Florisil, silica gel, alumina, C18) with a mortar and pestle and then packed into a glass column. The analytes are then eluted by a relatively small volume of a suitable solvent. One of the advantages of MSPD is the capacity to simultaneously accomplish extraction and purification in a single step, if an adsorbent, e.g., Florisil or silica gel, is packed at the bottom of the absorbent–matrix mixture (Figure 4).

The conventional extraction of POPs from biological or solid matrices usually begins with a homogenization step, followed by SLE or LLE with one or more steps of purification to remove coextracted interferences before instrumental analysis. Compared to the conventional extraction, MSPD is simple, inexpensive, and fast [81,138,139]. Lipophilic reversed phase-bonded materials have been widely used for MSPD, with an optimal preference for C8 and C18 bonded silica. The sample-to-sorbent ratio varies from 1:1 to 1:4 [81,138]. However, studies have shown that the extraction efficiency of MSPD is not as good as that of traditional methods, such as Soxhlet extraction [138].

The extraction efficiency can be enhanced by heating or shaking a dispersed sample or by use of a fluid or solvent with a high diffusion rate [81]. In recent

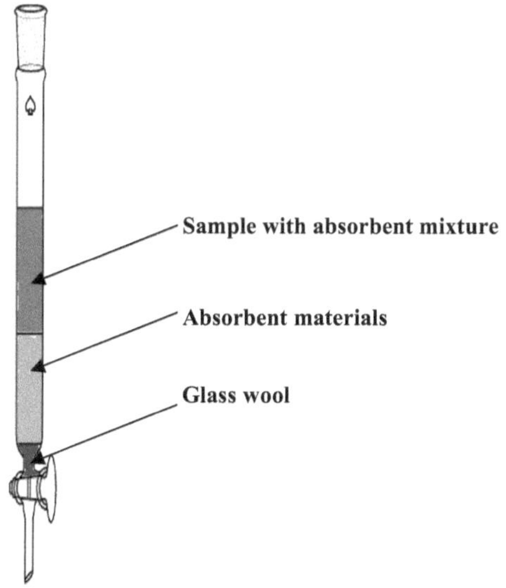

FIGURE 4 Glass column packing for extraction and purification of samples for analysis of persistent organic pollutants by matrix solid-phase dispersion.

years, enhanced solvent extraction techniques such as PLE (ASE), SFE, MAE, and UAE also have been applied in the extraction of POPs from environmental samples (Table 4).

PLE or ASE has been widely used in the extraction of POPs from solid or semisolid environmental samples, such as food and biological tissues [143], soil and sediment [86,144,145], and fish [146], since the mid-1990s [147]. The basic principles of PLE have been described in several earlier publications [143,148–150]. Briefly, PLE is a sample preparation technique that combines elevated temperature and pressure with liquid solvents, such as water, organic solvent, and dilute organic acid, to achieve fast and efficient extraction of the analytes from the solid matrix. PLE has been shown to have significant advantages over competing techniques with regard to time saving and solvent use, automation, and efficiency [143]. The high pressure implies a reduction in solvent viscosity, which thereby increases the solvent's ability to wet the matrix and to solubilize the target analytes. The high temperature also assists in breaking down analyte–matrix bonds and encourages analyte diffusion to the matrix surface. In addition, analytes' solubility in the solvent is improved at high temperatures [147].

Supercritical fluid extraction (SFE) is a relatively green extraction method that does not require the use of an organic solvent. Supercritical fluids have densities very similar to liquids but with lower viscosities, similar to gases, and high diffusion coefficients. This combination of properties makes the fluid more penetrating, affording it a high solvating power with the ability to extract solutes more quickly and efficiently than can liquid solvents [151–153]. Supercritical fluids have received interest due to their tunable solvent power and selectivity. The solvent power of supercritical fluids can be tuned by varying pressure and temperature. Thus, the same supercritical solvent under various operation conditions may yield different extraction or separation performances. SFE is suitable for the extraction of various analytes, especially nonpolar organic compounds, such as POPs, from different matrices. Over the past four decades, supercritical carbon dioxide has been used as an extractant due to its low cost, ease of reaching supercritical conditions under low pressure and temperature, and environment friendliness [153]. In SFE, for the extraction of analytes, the sample is placed in an extraction vessel and pressurized with carbon dioxide. The contents are transferred to a fraction collector and depressurized, thereby causing the carbon dioxide to lose its solvating power [80]. Thus, one advantage of SFE is that the extracts are very clean and require only moderate additional cleanup. However, the small volume of the extractor, which can hold only a few grams of the material, is a disadvantage when a larger sample mass is required to achieve a better limit of detection (LOD). Several reviews have summarized the principles and applications of SFE for solid matrices, food samples, aquaculture, and marine environmental samples for analysis of POPs [145,151,154,155].

The application of MAE in analytical laboratories dates back to the mid-1980s [156]. MAE is a process of using microwave energy to heat solvents in contact with a sample as a means to partition analytes from the sample matrix into the solvent [93]. Microwaves are electromagnetic waves at frequencies ranging from 300 MHz to 300 GHz. The principle of heating with microwave energy is based on the direct effect of microwaves on molecules by ionic conduction and dipole rotation. Type of solvents, matrix, power applied, temperature, and extraction time are the most important parameters that affect the extraction process. Moreover, the water content of the sample has to be carefully controlled to avoid excessive heating, allowing for reproducible results. A typical extraction procedure takes 15–30 min and uses a small solvent volume, in the range of 10–30 mL. These volumes are about 10 times smaller than the volumes used by conventional extraction techniques [93]. Recently, ASE has been used in the extraction of POPs from marine biota [157], sediments [158], soils [145,159], and vegetation [154]. Similar to MAE, UAE applies ultrasound radiation to increase the penetration of solvent into a sample and to improve the mass transfer.

Several extraction methods other than LLE have been used in the extraction of POPs from liquid or gaseous samples. These include sorption-based (SPME and stir bar sorptive extraction (SBSE)) and solvent-based extraction techniques.

SPME is a sample preparation technique used in both the laboratory and in situ sample analysis. This extraction is based on sorption theory and was developed in 1989 by Pawliszyn [81,160]. SPME is a simple, efficient, and solventless sample preparation method. All steps of conventional LLE, such as sampling, extraction, concentration, and transfer to the chromatography vial, are integrated into one step and one device. The SPME technique can be routinely used in combination with gas chromatography (GC), high-performance liquid chromatography (HPLC), and capillary electrophoresis and places no restriction on mass spectrometric (MS) [81,161] detection. There are three basic extraction models: direct extraction (fiber directly immersed into a liquid sample), headspace (fiber exposed in the vapor phase above a gaseous, liquid, or solid sample) and derivatization. Several papers reviewed the application of SPME in the analysis of POPs in plants [162], air [163,164], food samples [82], and the aquatic environment [165].

SBSE was developed by Baltussen and colleagues in 1999 [166,167]. Similar to SPME, SBSE is a solventless sample preparation technique. Instead of polymer-coated fibers, stir bars coated with 25–125 μL (0.3–1.0 mm layer) polydimethylsiloxane (PDMS) are used for the enrichment of organic compounds from aqueous matrices. SBSE of the headspace of a gas, liquid, or solid sample contained in a sealed vial also is possible [81]. The surface area and the amount of coating in SBSE are greater than those in SPME fiber coating. Therefore, SBSE offers higher recoveries and higher sensitivity than does SPME. Several review articles have summarized the application of SBSE in the analysis of POPs in environmental matrices [154,162,167–170].

Since its inception in 1995, based on the distribution principle, micro-LLE techniques have become increasingly popular. Similar to SPME, these methods are simple, inexpensive, and adaptable to a wide variety of samples matrices. There are two types of micro-LLE: exposed solvent and membrane-protected solvent techniques. The first one includes single-drop microextraction (SDME), liquid–liquid microextraction (LLME), liquid–liquid–liquid microextraction, and dispersive liquid–liquid microextraction (DLLME). The second one is hollow fiber-protected two- or three-phase microextraction (HF-LPME) [171]. Several recent articles have described the principles and applications of SDME [81,162,171,172], DLLME [81,162,171,173] and HF-LPME [81,162,171,174,175].

In the SDME procedure, a microdrop (1–3 μL) of the extraction phase is suspended from a syringe into the liquid or gaseous sample matrix. After extraction, the microdrop is retracted back into the microsyringe and then directly placed into the instrument for qualitative and quantitative analysis. In the DLLME, a mixture of the extraction phase (low solubility in water, density different from that of water, and the ability to extract analytes) and the dispersive solvent (miscible with both water and organic solvents, such as acetonitrile and methanol) is rapidly injected, using a syringe, into the aqueous sample, and tiny droplets are formed in the aqueous sample, which provide a vast interphase contact and accelerate the mass transfer of analytes [162,173,176]. In HF-LPME, the sample phase and the acceptor phases are separated by the porous hollow fiber, which also protects the extraction solvent without any loss and eliminates interference from large molecules and particles from the sample solution [162,175].

A few novel extraction techniques have been introduced in the analysis of POPs in environmental matrices in recent years. In 2011, Barletta and colleagues developed a new SBSE coating (PDMS/activated carbon) and used it for the extraction of six OCPs in sugarcane juice. The method achieved a limit of quantification (LOQ) of 0.5–40 μg/L, and the intraday precision ranged from 0.072% to 8.8% [177]. In 2013, PBDEs were analyzed in wastewater and sludge by SBSE coupled with GC–MS [178]. In 2013, PCBs and PBDEs were determined in sediments by UAE followed by SBSE coupled with thermal desorption (TD)–GC–MS. The accuracy and precision were in the ranges of 62.8–91.5% and 3.6–15.0%, respectively. The main advantage of this method is the shortening of the sample handling time and method miniaturization [179]. In 2013, OCPs were determined in water samples by SBSE coupled with LC–MS. The authors discussed the method validation and measurement uncertainty by comparing the results with those of conventional SPE-LC-MS/MS analysis, and the results from the two methods were similar [180]. In 2013, Assoumani and colleagues applied SBSE as a passive sampling technique for the determination of 19 pesticides in two rivers in France and showed that passive SBSE is useful for monitoring of moderately to highly hydrophobic organic pollutants in rivers [181].

Recently, a novel, simple, inexpensive, and fast LLME method was developed by Dasgupta and colleagues to detect OCPs, PCBs, and PAHs in water and other aqueous samples, e.g., soft drinks [112]. This method involved extraction of 10 mL sample with 250 μL chloroform by vortexing followed by centrifugation. The organic layer was evaporated and reconstituted in 20 μL of ethyl acetate and HEX mixture (1:9) for GC-TOFMS analysis. The recoveries of POPs were 76–120%, with a method detection limit (MDL) ranging from 1 to 250 ng/L. The MDLs were 400 times lower than the instrument quantification limits. A similar LLME method was used for the analysis of PCBs in water and wastewater samples, and the results were similar to those of the traditional LLE method [182].

Ionic liquids have potential as solvents in DLLME methods. Ionic liquid is an ionic media with some unique physicochemical properties, such as air and water stability, nonvolatility, thermal stability, tunable viscosity, and miscibility with water and organic solvents [183]. The potential of ionic liquid as a dispersive extraction solvent in the analysis of DDT and its metabolites from water samples has been investigated [183]. Limits of detection (LODs) as low as 0.33–0.63 μg/L were achieved. In another study, acetone was used as a disperser solvent in the analysis of PCBs and PBDEs in milk by the DLLME procedure [184]. Analytes in milk were extracted with HEX, purified by Florisil column chromatography, and then redissolved in acetone [184].

More than one extraction technique may be applied to evaluate the efficiency of extraction methods for POPs in environmental matrices. For example, DLLME followed by SPE was used in the extraction OCPs from natural water [185], MAE and DLLME were used in the isolation of PAHs in smoked fish [186], SFE and DLLME were used in the extraction OCPs from soil and marine sediment samples [187], LLE and DLLME were used in the extraction PBDEs in whole milk [188], and UAE, MSPD, and DLLME were used in the extraction OCPs in fish species [189].

2.3.1.2 Purification

Prior to instrumental analysis, appropriate purification of sample extracts is required. Several matrix constituents can be coextracted with target chemicals in the extraction procedure, e.g., sulfur in sediment, sewage sludge, dust or soil samples, and lipids in biota. These constituents, especially lipids, tend to reside in GC parts and affect sensitivity and reproducibility of results. In addition, some compounds that would interfere or coelute with the target analytes in the instrumental analysis can be coextracted from the matrix. Therefore, a thorough purification is required in ultratrace analysis of organic chemicals. Generally, a two-stage purification is followed, with the first step to remove lipids/sulfur and the second step to fractionate the target chemicals from other interferences or coeluting compounds.

Several nondestructive methods have been reported for the removal of lipids from crude sample extracts, including freezing followed by centrifugation, LLE, GPC, Florisil column chromatography, and SPE. The significant differences in melting points have been used in the separation of lipids and OCPs after the extraction of meat, fish, and milk. Lipids were precipitated by freezing at very low temperatures, while OCPs remained dissolved in cold organic solvent. LLE also has been used to remove lipids by altering the polarity of the extraction solvent. In this technique, the sample is allowed to equilibrate with a nonpolar solvent. The nonpolar hydrocarbon chain partitions into the solvent, leaving the polar analytes in the polar solvent. GPC is based on the size exclusion principle and is used to remove/fractionate high-molecular-weight compounds from low-molecular-weight chemicals in samples. Separation occurs via the use of porous beads packed in a glass column. The smaller molecular weight compounds can enter the pores more easily and, thus, have longer retention time in the beads than do the larger compounds. Therefore, compounds with different molecular weight can be separated by the GPC. For example, the most commonly used GPC for the analysis of POPs in fish extract utilizes a 700 mm × 25-mm glass column that contains 70 g of Bio-Rad® Envirobeads SX-3 resin with a mobile phase of 100 mL DCM or 60 g of Envirobeads SX-3 resin with a mobile phase of 100 mL 1:1 DCM:HEX with a flow rate of 5 mL/min. SPE with neutral adsorbents, such as silica gel, alumina, or Florisil, is widely used in the purification and fractionation steps. When GPC is not used in the removal of lipids, a combination of two or more adsorbents is recommended for efficient purification of extracts. Silica gel and Florisil have a better fat-retaining capacity as compared to alumina.

The most commonly applied destructive method of lipid removal is the treatment of sample extracts with concentrated sulfuric acid, either directly (for samples with high lipid content) or indirectly via sulfuric-acid-impregnated silica gel (for samples with low lipid content). Most halogenated flame retardants and

PCBs are stable under a strong acidic condition, but some OCPs, e.g., aldrin and heptachlor, will not survive sulfuric acid treatment.

None of the above methods can completely remove lipid from sample extracts that contain a high lipid content. In addition, other interfering substances are not completely removed by the lipid removal steps. Therefore, a second and third cleanup step is followed, with either commercially packed SPE cartridges or conventional glass columns packed with different sorbents, e.g., silica, Florisil, alumina, or C18 bonded silica.

For satisfactory purification of sample extracts, the choice of an appropriate sorbent bed and eluent solution is important and depends mainly on the chemical properties of the target analytes as well as the sample matrix. Several studies have used various combinations of sorbents and eluents to obtain reproducible results (Table 4). For example, in the analysis of dioxins (US Environmental Protection Agency (EPA) Method 1613), acidic (30% sulfuric acid with activated silica, w:w), neutral (silica baked at 180 °C for at least 1 h), and basic (23% sodium hydroxide with activated silica, w:w) silica gel; alumina (baked at 130 °C for at least 12 h); and Florisil (baked at 130–150 °C for at least 12 h) were used to remove nonpolar and polar interferences. Alumina and Florisil were used to remove chlorodiphenyl ether interferences in sample matrices. The selection of eluent has to achieve a balancing act between the solvent and the target analytes' polarities. For most separations, the solvent should be less polar than the target compounds. The compounds must also be soluble in the solvent so that they can be eluted from the column. The ability of the solvent to elute POP chemicals from columns packed with silica gel, alumina, or Florisil follows the sequence: HEX < carbon tetrachloride < trichloroethylene < toluene < benzene < DCM < chloroform < diethyl ether < ethyl acetate < acetone < methanol < water.

These principles of sorbent or solvent selection in the purification of extracts can be illustrated by two typical methods for simultaneous analysis of PCBs, PBDEs, and dioxins in environmental matrices (Figure 4). Silica gel has been found to retain PBDEs more strongly than the other halogenated compounds, such as PCBs and OCPs, thus allowing the fractionation of extracts based on the polarity of different classes of compounds. Neither Florisil nor alumina columns has the ability to separate PBDEs from PCBs or PCDD/Fs effectively. To determine dioxins, nearly all established methods involve combinations of cleanup methods developed for the analysis of PCBs and OCPs with an active carbon step to isolate them from chemical interferences (Figure 5(a)). Recently, a study from China reported separation of dioxins, PCBs, and PBDEs by a silica gel column packed with silica gel, including $AgNO_3$ silica [190] (Figure 5(b)).

In summary, the combination of sorbents and eluting solvents for the purification of sample extracts has to be chosen very carefully. Otherwise, POPs could either break down or adhere to the sorbent material, resulting in low or no recovery. Due to the wide range of polarities of POPs, it is difficult to find the best combination of sorbents and solvents for the determination of all POPs in a single procedure. Finding an optimal cleanup condition is, in itself, an art.

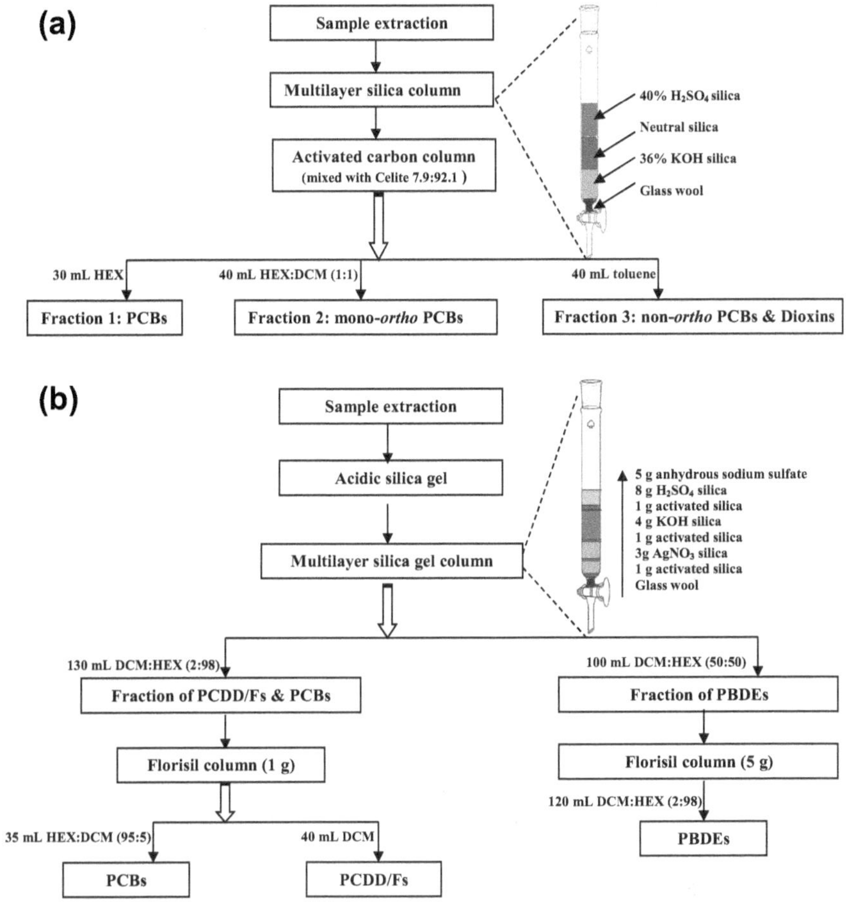

FIGURE 5 Typical cleanup procedures (based on EPA method 1613) in PCDD/Fs, PCBs and PBDEs analysis.

2.3.2 *Instrumental Analysis*

The separation and detection of most POPs are usually accomplished by GC coupled with several types of detectors, such as electron capture detector (ECD) and MS detector (Table 4). In addition, LC with MS detector and photodiode array detector have been used in the analysis of POPs [191–194]. Several reviews have summarized the instrumental methods of analysis of POPs in biological and environmental samples [72,75,79,82,195–201].

Fused silica open tubular capillary columns, coated with nonpolar or moderately polar chemical bonded liquid phases, are generally used to achieve separation of target chemicals through GC methods. Capillary columns with various lengths (15–60 m), internal diameters (0.15–0.25 mm), and film thicknesses (0.1–0.3 μm) have been used to achieve the required resolution.

For the separation of PCB congeners and OCPs, capillary columns packed with 100% dimethylpolysiloxane; methylpolysiloxane that contains 5, 35, or 50% phenylpolysiloxane; or 14% or 50% cyanopropylphenyl polysiloxane have been used as stationary phases [201,202]. PBDEs and other BFRs are usually separated by a column with a stationary phase composed of 5% phenyl polysiloxane/95% dimethyl polysiloxane, i.e., DB-5 MS from J&M or equivalent [196]. Short columns (10–15 m) with extremely thin film thickness (0.1 μm) permit short residence times for analytes in the column, which eliminates the chances of analyte degradation. However, longer columns (25–60 m) were used to achieve better analyte separation. For example, Vetter and Rosenfelde analyzed 122 brominated chemicals on a 30-m DB-5 GC column and separated 118 peaks [203]. For dioxins, the capillary column (fused silica with 0.25–0.32 mm internal diameter and 25–60 m length) is coated with moderate or high-polarity cyanopropyl stationary phase, e.g., DB-Dioxin: 44% methyl–28% phenyl–20% cyanopropylpolysiloxane. US EPA method 1613 recommends a DB-5 column (60 m, coated with 0.25 μm 5% phenyl, 94% methyl, 1% vinyl silicone) for the separation of 2,3,7,8-TCDD and a DB-225 column for the separation of 2,3,7,8-TCDF, as the latter cannot be separated from other coeluting isomers on a DB-5 column [204]. However, some capillary columns, such as DB-Dioxin and CP-Sil 88 (from Agilent J&M), designed specifically for dioxin analysis can resolve 2,3,7,8-TCDD and 2,3,7,8-TCDF from all other isomers in a single run [205].

Two-dimensional (2D) GC or comprehensive GC has been suggested as a powerful analytical technique to separate compounds in complex environmental and biological matrices and has been used in the analysis of POPs [206–210]. Two independent GC columns of different polarity are coupled through an interface. The elute from the first column is cut into small additional fractions and is subjected to further separation on the second column [206]. The second column has a much smaller diameter and shorter length, which results in a much sharper and smaller peak width and higher peak intensity. This technique has been used to simultaneously separate 59 compounds (including PBDEs, PCBs, PBBs, and OCPs) in human serum and milk [211]; 97 organic compounds (including OCPs, PCBs, PAHs, and others) in river water [212]; 17 2,3,7,8-substituted CDD/Fs and 12 dioxin-like PCBs in milk [213]; 160 pesticides and 25 POPs (including PCBs and PAHs) in grapes and wine [214]; 271 nonpolar, persistent, and bioaccumulative contaminants of 24 classes in Atlantic common dolphin blubber [215]; 34 pesticides and their metabolites in milk and cream [216]; and 200 of 209 PCBs in animal liver samples [217].

An ECD is still the most widely used detector for routine analysis of most PCBs and OCPs, due to its low cost and high sensitivity. An ECD is particularly useful for the detection of halogen-containing compounds. Other aromatic organic compounds also can be detected using an ECD. Coelution of several OCPs and PCBs is a significant problem in GC–ECD analysis, which can yield false-positive results. Therefore, confirmation of results by an MS detector is required. GC coupled with MS has revolutionized environmental organic analysis since the 1980s [218]. The MS analysis may resolve problems arising from coelution of analytes. The most commonly used MS instruments are either low-resolution MS (LRMS) or high-resolution MS (HRMS).

For the analysis of PCBs, PBDEs, and OCPs, a GC with LRMS or HRMS that operates under electron impact (EI) ionization or negative chemical ionization (NCI) can be used. The EI-MS offers great selectively and an increased ability to confirm a compound's identity using full-scan data. However, low-resolution EI-MS (with quadrupoles as a mass analyzer) is not routinely used for the analysis of BDE congeners with more than six bromine atoms [219]. The low sensitivity of EI can be overcome when an HRMS detector achieves a resolving power of over 10,000 [220,221]. In comparison with EI, NCI is a softer ionization method and is widely used in BFR analysis. Quantification with NCI is achieved using SIM mode by monitoring for the isotopes of bromide ions at m/z 79 and m/z 81. The sensitivity of this technique can be higher than that of EI, but the selectivity can be lower. Dioxins are analyzed mainly by the use of HRMS. HRMS at a resolving power of >10,000 under positive EI ionization allows for improved sensitivity in the lower femtogram range.

GC–MSMS is a technique that has been receiving considerable attention in recent years. MSMS can decrease matrix interferences and improve selectivity, thus improving the detection limit. Both tandem-in-time (ion trap) and tandem-in-space (triple quadrupoles) detectors have been applied to the analysis of POPs in various matrices. For example, GC–MSMS has been used in the detection of 16 PAHs in soil leachates [222]; 12 PBDEs in airborne particulate matter (the LOQ ranged from 0.063 $pg m^{-3}$ to 0.210 $pg m^{-3}$ when air volumes of 723 m^3 were collected) [223]; PAHs and OCPs in wastewater [224]; PCDD/Fs, PCBs, and PBDEs in food samples [225]; 18 PCBs, 16 OCPs, 14 PBDEs, and 25 PAHs in fish and fish feed (the method quantification limits were in the range 0.005–1 μg/kg and 0.05–10 μg/kg for fish muscle and fish feed, respectively) [226]; and PCBs and PBDEs in human serum and plasma [227–231].

Sample extracts are injected into the GC–MS or GC–MSMS systems by splitless and on-column injection methods. In the splitless mode, the liner (usually filled with a plug of glass wool) serves as the evaporation chamber in which solvent extract is rapidly volatilized. Degradation and memory effects can occur in the liner due to the presence of glass wool and an increase in active sites after multiple injections of dirty samples. These effects can be minimized by frequently replacing the liner. With on-column injection, the extract is injected directly into the column or into a glass insert. The extract is introduced directly into the front end of the GC column at room temperature, so no loss of analytes can occur [202,232]. However, the extract should be very clean, as the accumulation of dirt at the front end of the column can result in the deterioration of the column and can lead to active sites, which also leads to degradation of analytes. These phenomena can be suppressed by the application of an uncoated, deactivated retention gap. The accumulation of dirt from sample extract then occurs at the retention gap, which can be changed frequently [202,232].

Although GC with a capillary column and a mass selective detector is the instrument of choice for analysis of POPs, LC with various detectors, particularly with the introduction of the atmospheric pressure ionization (API), triple quadrupole mass analyzer has allowed the detection of several POPs that are relatively more polar or nonvolatile or have a high molecular weight [191–194,233–235]. In the API mode, electrospray (ESI) is particularly suited for the analysis of polar compounds [235,236], and atmospheric pressure chemical ionization (APCI) is highly effective in the analysis of moderate to less polar substances, such as BFRs. ESI or APCI produces protonated $[M+H]^+$ or deprotonated $[M–H]^-$ molecules. More recently, the possibility of coupling LC to an ion trap or the new generation of triple quadrupole and hybrid instruments, such as quadrupole/time-of-flight and linear ion traps [234,237], enabled robustness in the application of these novel technologies in the analysis of POPs. For example, 18 pesticides with different polarities (log K_{ow} between 0.6 and 4.7) were detected by LC–MSMS (ESI) in red wine [238]. Kittlaus et al. used a 2D LC (HILIC column and Agilent Poroshell 120EC-C18)–MSMS system to detect 300 pesticides in food samples [239]. In addition, atmospheric pressure photoionization (APPI) was used in the analysis of POPs in environmental samples [193]. For instance, Letcher and Chu analyzed high-molecular-weight BFRs in herring gull eggs using LC-APPI-TOFMS (using pure acetone as a dopant and LC mobile phase) [240].

In general, with the recent availability of labeled internal standards (ISs), quantification of POPs can be accomplished by isotope dilution, if MS is the choice of detection method. ISs are similar in analytical behavior to the target compounds. A known amount of IS is added into all samples or extracts, similar to that in calibration standards. The ratio of the peak area of the target compound to that of IS in the sample extract is compared to a similar ratio derived for calibration standard, i.e., isotope dilution. With the use of IS, variations in responses of an analytical instrument over time and the volume injected into the instrument can be compensated. In addition, the retention times of the target compound and the IS can be used to calculate the relative retention time of the target compound, which then can be used to compensate for small retention time shifts. It is ideal to use a labeled IS for every target analyte in the analysis. In some cases, more than one IS may be added to quantify target compounds at different retention times.

The isotope dilution method of quantification has several advantages over classical calibration strategies [32,55]: quantitative separation processes are not required, losses of substances are compensated, methodological calibration graphs are not needed, and excellent precision and accuracy can be achieved [218]. The isotope dilution method has been widely used in the analysis of POPs in environmental samples in recent years [211,218,241–245]. However, not all POPs have isotopically enriched standards. Further, the isotopically labeled standards are usually expensive.

2.3.3 Reporting

Reporting of data is the final step in sample analysis. As described by the UNEP Chemical Branch [246], reports should include the results of QA/QC and all the records of sample analysis, including the date, description of sample, method used, names of staff members who performed the analysis, and the signature of the QC officer.

QA/QC should include a number of actions from sample preparation to instrumental analysis, including recoveries (recoveries of target compounds in spiked blanks, in matrix spike sample and standard reference materials, and recoveries of IS or surrogate standards added into all samples) and consistency of reference materials or QC samples among different batches and blanks (method blank, solvent blank, and field blank).

LOD and LOQ values also need to be included. LOD is often defined as a signal that is three times the background, and the LOQ is normally three times higher than the LOD. There are several ways to calculate the LOQ and LOD; this has been described in detail elsewhere [247,248]. Compounds that are found at levels between LOD and LOQ can be reported as present or possibly as being present at an estimated concentration; in the latter case, however, the result has to be clearly marked as being below the LOQ. Results below the quantitation limit are often reported as <LOQ. In addition, approximated values, e.g., half the value of the detection limit, are used instead of <LOD in some reports or statistical analysis.

2.3.4 Review of Analysis of POPs

In Table 5, we have summarized several excellent reviews of the analysis of POPs in environmental matrices, from sampling design to sample preparation to instrumental analysis. It is clear that several new extraction and detection techniques have been developed for the analysis of POPs in the past several decades. As new POPs are added under the Stockholm Convention, and the analytical instrumentation and sensitivities continue to improve, this list is expected to grow. As can be seen in the table, the sample volume/mass used for extraction/analysis is decreasing over time, the volume of solvents used is decreasing, sensitivity and accuracy of the analytical instruments are increasing, and the methods are increasingly focused on high throughput and rapid analysis.

In summary, an overview of the methodology used in the analysis of POPs in environmental matrices, from sampling design to sample collection to sample analysis, is provided. An emphasis has been placed on the recently developed methods that are rapid, have high throughput, and minimize the generation of chemical wastes. We have summarized the large number of reviews on analysis of POPs (Table 5). Despite the evolution of methods for the analysis of POPs in complex matrices over four decades, accurate and precise determination of POPs continue to be a challenge, and the demand for the measurement of POPs in small sample volumes is increasing. We recommend that scientific communities pay utmost attention to the quality, reliability, and reproducibility of data

TABLE 5 Reviews of Analysis of Persistent Organic Pollutants in Environmental Matrices in the 1990 and 2000s

No.	Published Year	Description	References
Project Design or Sampling			
1	1997	Effects of storage conditions of biological materials on the contents of OCPs	[69]
2	2002	Guidance on choosing a sampling design for environmental data collection	[12]
3	2007	Guidance for analysis of POPs	[246]
4	2007	Sample handling strategies for the determination of POPs from biota samples	[73]
5	2009	Environmental and food sample handling challenges	[249]
6	2011	Sampling and analytical framework for routine analysis of organic pollutants	[74]
7	2013	Environmental field sampling handbook	[51]
Extraction and Cleanup			
1	1994	Chromatographic cleanup methods for the determination of POPs in aqueous environmental samples	[250]
2	1995	Critical review of analysis of non- and mono-*ortho*- chlorobiphenyls	[76]
3	2002	Extraction, cleanup, and detection methods of PCBs in food matrices	[156]
4	2002	Determination of BFRs in environmental samples	[251]
5	2003	Analysis of BFRs, with emphasis on PBDEs in environmental and human samples	[252]
6	2006	Analytical methods for PCBs and OCPs in environmental samples	[72]
7	2006	Chlorinated paraffins: a review of analysis and environmental occurrence	[253]

8	2007	Sample preparation techniques for contaminants in foods	[149]
9	2007	Recent development in the analysis of BFRs	[150]
10	2007	PAS in monitoring of organic constituents of air	[254]
11	2008	Analytical methods for the POPs in the atmosphere	[255]
12	2009	Analysis of PBDEs in environmental samples	[219]
13	2009	Sampling and analysis of pesticides in ambient air	[256]
14	2010	Evaluation of several sample extraction methods for the determination of PCBs and PBDEs in soil and fish samples	[77]
15	2010	PAHs in food and beverages: analytical methods and trends	[257]
16	2010	Advances in the analysis of short-chain chlorinated paraffins	[258]
17	2011	Analytical methods to determine pesticides in foods of animal origin	[82]
18	2011	Determination of OCPs in fatty foods	[201]
19	2011	Extraction and determination of PAHs in food and environmental samples	[259]
20	2012	Critical overview of selected contemporary sample preparation techniques	[81]
21	2012	Analytical methodology for the determination of OCPs in vegetation	[154]
22	2012	Sample preparation techniques for analysis of pesticide in soil	[145]
23	2012	Analytical methodology for the determination of OCPs in vegetation	[154]
24	2012	Methods for determination of PBDEs in environmental samples	[159]
25	2012	Sampling and analysis of PBDEs in environmental samples	[195]

Continued

TABLE 5 Reviews of Analysis of Persistent Organic Pollutants in Environmental Matrices in the 1990 and 2000s—cont'd

No.	Published Year	Description	References
26	2012	Analysis of novel BFRs	[196]
27	2012	Sample preparation for PBDEs and their metabolites in environmental matrices	[260]
28	2012	Analysis of emerging organic contaminants in environmental solid samples	[261]
29	2012	Sampling and analytical methods for POPs in air	[262]
30	2012	Overview of POP extraction and estimation	[263]
31	2012	Analytical methods in the determination of perfluoroalkyl acids in human milk	[264]
32	2013	Extraction and cleanup methods for OCP determination in milk	[71]
33	2013	Trends in analysis of POPs in human blood	[265]
34	2013	Methods for analysis of PFCs	[266]
35	2013	Analytical chemistry of the POPs identified in the Stockholm Convention	[79]
36	2013	Development in analysis of POPs under the Stockholm Convention	[80]
37	2014	New trends in sample preparation techniques for environmental analysis	[78]
38	1998	Review of Soxhlet extraction of solid materials	[85]
39	2000	Introduction of MAE	[93]
40	2007	Introduction of MSPD	[92]
41	2007	MSPD used for extracting contaminants from foodstuffs	[139]
42	2010	Recent developments in MSPD	[138]

43	1999	PLE used for extracting PCBs from solid environmental samples	[266]
44	2000	PLE used for extracting POPs from environmental samples	[267]
45	2005	PLE used for extracting from food and biological samples	[143]
46	2006	PLE used for extraction and cleanup of POPs from food and feed samples	[147]
47	2008	PLE used for extracting of PCDD/Fs and Dl-PCBs from environmental samples	[86]
48	2008	PLE used for extracting contaminants from environmental samples	[268]
49	2012	PLE used for extracting organic contaminants in food and feed	[269]
50	1996	Introduction of SFE	[152]
51	2006	SFE used for extracting contaminants from soil and sediment	[270]
52	2008	SFE used for extracting polyhalogenated pollutants from aquaculture and marine environmental samples	[151]
53	2010	Recent advances and applications of SFE	[155]
54	2014	SFE in analytical chemistry	[153]
55	2004	SPME as a powerful sample preparation tool	[161]
56	2008	New perspectives in the use of semipermeable membrane devices as PAS	[55]
57	2010	Green analytical methods for POPs in aquatic environment	[165]
58	2011	Developments in SPME for on-site sampling and sample preparation	[160]
59	2011	Solventless and solvent-minimized sample preparation techniques for pesticides in water samples	[170]

Continued

TABLE 5 Reviews of Analysis of Persistent Organic Pollutants in Environmental Matrices in the 1990 and 2000s—cont'd

No.	Published Year	Description	References
60	2011	Recent advances in applications of single-drop microextraction	[172]
61	2011	DLLME for determination of organic analytes	[173]
62	2013	Microextraction techniques for the determination of organic compounds from plants	[162]
63	2013	Advances in solvent microextraction techniques	[171]
64	1999	SBSE used for aqueous samples	[166]
65	2009	Development and applications of SBSE	[168]
66	2010	Novel applications, limitations, and potential solutions of SBSE	[169]
67	2012	Novel sorption-based methodologies: SBSE and related techniques	[167]
Instrumental Analysis			
1	2003	Modern developments in GC-MS-based environmental analysis	[199]
2	2006	Instrumental methods and challenges in quantifying PBDEs	[197]
3	2009	Application of MS in the analysis of PBDEs	[198]
4	2008	GC used for determination of environmentally relevant polybrominated compounds	[203]

5	2008	Recent developments in the application of GC × GC chromatography	[206]
6	2008	Advances in the GC determination of POPs in the aquatic environment	[202]
7	2008	GC × GC chromatography used for screening of POPs in environmental samples	[208]
8	2009	Review of GC × GC chromatography	[207]
9	2010	Isotope-labeled substances in analysis of POPs in environmental samples	[218]
10	2010	Recent trends in LC–MSMS of organic contaminants in environmental samples	[235]
11	2012	New development in LC–MS for the determination of micropollutants	[237]
12	2012	HRMS used in environmental sciences	[220]
13	2012	HRMS in food analysis	[221]
14	2012	MS and bioanalytical techniques in analysis of POPs in environmental samples	[271]
15	2013	GC–MSMS used for trace analysis of environmental pollutants in fish and fish feed	[226]
16	2013	Analysis comparison of dioxin by GC/MS–MS vs GC/HRMS	[272]

GC × GC, comprehensive GC.

as environmental analytical data are routinely used in many regulatory settings and in policy decision making. As the list of POPs continues to grow under the Stockholm Convention, more novel methods are expected to revolutionize our understanding of the fate and effects of POPs in the environment.

LIST OF ABBREVIATIONS

μECD	Micro electron capture detector
AAS	Active air sampling
BDE	Brominated diphenyl ether
CDD/F	Chlorinated dibenzo-*p*-dioxin/furan
DCM	Dichloromethane
DDT	Dichlorodiphenyltrichloroethane
Dl-PCB	Dioxin-like polychlorinated biphenyls
ESI	Electrospray ionization
GFF	Glass fiber filter
GPC	Gel permeation chromatography
HEX	Hexane
HILIC	Hydrophilic interaction liquid chromatography
HP-LPME	Hollow fiber-liquid phase microextraction
MSD	Mass spectrometry detector
PAD	Pulsed amperometric detection
PAHs	Polycyclic aromatic hydrocarbons
PBB	Polybrominated biphenyl
PCBs	Polychlorinated biphenyls
PCDD	Polychlorinated dibenzo-*p*-dioxins
PCDFs	Polychlorinated dibenzofurans
PFC	Perfluorochemicals
QA/QC	Quality assurance/quality control
SFE	Supercritical fluid extraction
SIM	Selected ion monitoring
SPM	Suspended particulate matter
SPMD	Semipermeable membrane device
TCDD	Tetrachloro dibenzo-*p*-dioxin
TCDF	Tetrachloro dibenzofuran
TD–GC–MS	Thermal desorption-gas chromatography-mass spectrometry
TOFMS	Time of flight mass spectrometer
U.S.EPA	United States Environmental Protection Agency
WAX	Weak anion exchange

REFERENCES

[1] C.A. de Wit, D. Herzke, K. Vorkamp, Brominated flame retardants in the Arctic environment – trends and new candidates, Sci. Total Environ. 408 (2010) 2885–2918.

[2] M. Gamberg, B. Braune, E. Davey, B. Elkin, P.F. Hoekstra, D. Kennedy, C. Macdonald, D. Muir, A. Nirwal, M. Wayland, B. Zeeb, Spatial and temporal trends of contaminants in terrestrial biota from the Canadian Arctic, Sci. Total Environ. 351–352 (2005) 148–164.

[3] M.S. Evans, D. Muir, W.L. Lockhart, G. Stern, M. Ryan, P. Roach, Persistent organic pollutants and metals in the freshwater biota of the Canadian Subarctic and Arctic: an overview, Sci. Total Environ. 351–352 (2005) 94–147.

[4] S.A. Tittlemier, T. Halldorson, G.A. Stern, G.T. Tomy, Vapor pressures, aqueous solubilities, and Henry's law constants of some brominated flame retardants, Environ. Toxicol. Chem. 21 (2002) 1804–1810.

[5] E. Braekevelt, S.A. Tittlemier, G.T. Tomy, Direct measurement of octanol-water partition coefficients of some environmentally relevant brominated diphenyl ether congeners, Chemosphere 51 (2003) 563–567.

[6] F. Wania, C.B. Dugani, Assessing the long-range transport potential of polybrominated diphenyl ethers: a comparison of four multimedia models, Environ. Toxicol. Chem. 22 (2003) 1252–1261.

[7] S. Burreau, Y. Zebuhr, D. Broman, R. Ishaq, Biomagnification of PBDEs and PCBs in food webs from the Baltic Sea and the northern Atlantic Ocean, Sci. Total Environ. 366 (2006) 659–672.

[8] B. Johnson-Restrepo, K. Kannan, R. Addink, D.H. Adams, Polybrominated diphenyl ethers and polychlorinated biphenyls in a marine foodweb of coastal Florida, Environ. Sci. Technol. 39 (2005) 8243–8250.

[9] S.D. Shaw, M.L. Berger, D. Brenner, K. Kannan, N. Lohmann, O. Papke, Bioaccumulation of polybrominated diphenyl ethers and hexabromocyclododecane in the northwest Atlantic marine food web, Sci. Total Environ. 407 (2009) 3323–3329.

[10] S. Voorspoels, A. Covaci, V.L.B. Jaspers, H. Neels, P. Schepens, Biomagnification of PBDEs in three small terrestrial food chains, Environ. Sci. Technol. 41 (2007) 411–416.

[11] L.-H. Yu, X.-J. Luo, J.-P. Wu, L.-Y. Liu, J. Song, Q.-H. Sun, X.-L. Zhang, D. Chen, B.-X. Mai, Biomagnification of higher brominated PBDE congeners in an urban terrestrial food web in North China based on field observation of prey deliveries, Environ. Sci. Technol. 45 (2011) 5125–5131.

[12] U.S. Environmental Protection Agency, Guidance on Choosing a Sampling Design for Environmental Data Collection. www.epa.gov/quality/qs-docs/g5s-final.pdf (accessed April, 2014).

[13] I.G. Hallanger, A. Ruus, N.A. Warner, D. Herzke, A. Evenset, M. Schoyen, G.W. Gabrielsen, K. Borga, Differences between Arctic and Atlantic fjord systems on bioaccumulation of persistent organic pollutants in zooplankton from Svalbard, Sci. Total Environ. 409 (2011) 2783–2795.

[14] N. Berrojalbiz, S. Lacorte, A. Calbet, E. Saiz, C. Barata, J. Dachs, Accumulation and cycling of polycyclic aromatic hydrocarbons in zooplankton, Environ. Sci. Technol. 43 (2009) 2295–2301.

[15] A. Gerofke, P. Komp, M.S. McLachlan, Bioconcentration of persistent organic pollutants in four species of marine phytoplankton, Environ. Toxicol. Chem. 24 (2005) 2908–2917.

[16] M. Lebeuf, B. Gouteux, L. Measures, S. Trottier, Levels and temporal trends (1988–1999) of polybrominated diphenyl ethers in Beluga whales (*Delphinapterus leucas*) from the St. Lawrence estuary, Canada, Environ. Sci. Technol. 38 (2004) 2971–2977.

[17] D.C.G. Muir, S. Backus, A.E. Derocher, R. Dietz, T.J. Evans, G.W. Gabrielsen, J. Nagy, R.J. Norstrom, C. Sonne, I. Stirling, M.K. Taylor, R.J. Letcher, Brominated flame retardants in polar bears (*Ursus maritimus*) from Alaska, the Canadian Arctic, East Greenland, and Svalbard, Environ. Sci. Technol. 40 (2006) 449–455.

[18] K. Kannan, S.H. Yun, T.J. Evans, Chlorinated, brominated, and perfluorinated contaminants in livers of polar bears from Alaska, Environ. Sci. Technol. 39 (2005) 9057–9063.

[19] T.W. Bentzen, D.C.G. Muir, S.C. Amstrup, T.M. O'Hara, Organohalogen concentrations in blood and adipose tissue of Southern Beaufort Sea polar bears, Sci. Total Environ. 406 (2008) 352–367.
[20] R. Dietz, F.F. Rigét, C. Sonne, R.J. Letcher, S. Backus, E.W. Born, M. Kirkegaard, D.C.G. Muir, Age and seasonal variability of polybrominated diphenyl ethers in free-ranging East Greenland polar bears (*Ursus maritimus*), Environ. Pollut. 146 (2007) 166–173.
[21] J. Verreault, G.V. Gabrielsen, S.G. Chu, D.C.G. Muir, M. Andersen, A. Hamaed, R.J. Letcher, Flame retardants and methoxylated and hydroxylated polybrominated diphenyl ethers in two Norwegian Arctic top predators: glaucous gulls and polar bears, Environ. Sci. Technol. 39 (2005) 6021–6028.
[22] B.G. Loganathan, K. Kannan, Global organochlorine contamination trends: an overview, Ambio 23 (1994) 187–191.
[23] K. Ramu, N. Kajiwara, A. Sudaryanto, T. Isobe, S. Takahashi, A. Subramanian, D. Ueno, G.J. Zheng, P.K.S. Lam, H. Takada, M.P. Zakaria, P.H. Viet, M. Prudente, T.S. Tana, S. Tanabe, Asian mussel watch program: contamination status of polybrominated diphenyl ethers and organochlorines in coastal waters of Asian countries, Environ. Sci. Technol. 41 (2007) 4580–4586.
[24] D. Chen, M.J. La Guardia, E. Harvey, M. Amaral, K. Wohlfort, R.C. Hale, Polybrominated diphenyl ethers in Peregrine Falcon (*Falco peregrinus*) eggs from the Northeastern US, Environ. Sci. Technol. 42 (2008) 7594–7600.
[25] B. Johnson-Restrepo, K. Kannan, D.P. Rapaport, B.D. Rodan, Polybrominated diphenyl ethers and polychlorinated biphenyls in human adipose tissue from New York, Environ. Sci. Technol. 39 (2005) 5177–5182.
[26] J.J. Ryan, D.F.K. Rawn, Polychlorinated dioxin, furans (PCDD/Fs), and polychlorinated biphenyls (PCBs) and their trends in Canadian human milk from 1992 to 2005, Chemosphere 102 (2014) 76–86.
[27] S.K. Kim, J. Yoon, Chronological trends of emission, environmental level and human exposure of POPs over the last 10years (1999–2010) in Korea: Implication to science and policy, Sci. Total Environ. 470–471 (2014) 1346–1361.
[28] J. Fang, E. Nyberg, A. Bignert, T. Bergman, Temporal trends of polychlorinated dibenzo-p-dioxins and dibenzofurans and dioxin-like polychlorinated diphenols in mothers' milk from Sweden, 1972–2011, Environ. Int. 60 (2013) 224–231.
[29] S. Lignell, M. Aune, P.O. Darnerud, S. Cnattingius, A. Glynn, Persistent organochlorine and organobromine compounds in mother's milk from Sweden 1996–2006: compound-specific temporal trends, Environ. Res. 109 (2009) 760–767.
[30] K. Akutsu, M. Kitagawa, H. Nakazawa, T. Makino, K. Iwazaki, H. Oda, S. Hori, Time-trend (1973–2000) of polybrominated diphenyl ethers in Japanese mother's milk, Chemosphere 53 (2003) 645–654.
[31] T. Vukavić, M. Vojinović, I. Mihajlović, A. Ristivojević, Human milk POPs and neonatal risk trend from 1982 to 2009 in the same geographic region in Serbia, Environ. Int. 54 (2013) 45–49.
[32] J. Zhou, X. Zeng, K. Zheng, X. Zhu, L. Ma, Q. Xu, X. Zhang, Y. Yu, G. Sheng, J. Fu, Musks and organochlorine pesticides in breast milk from Shanghai, China: levels, temporal trends and exposure assessment, Ecotoxicol. Environ. Safety 84 (2012) 325–333.
[33] O. Mikes, P. Cupr, L. Kohut, A. Krskova, M. Cerna, Fifteen years of monitoring of POPs in the breast milk, Czech Republic, 1994-2009: trends and factors, Environ. Sci. Pollut. Res. 19 (2012) 1936–1943.
[34] S. Alivernini, C.L. Battistelli, L. Turrio-Baldassarri, Human milk as a vector and an indicator of exposure to PCBs and PBDEs: temporal trend of samples collected in Rome, Bull. Environ. Contam. Toxicol. 87 (2011) 21–25.

[35] A. Mannetje, J. Coakley, P. Bridgen, C. Brooks, S. Harrad, A.H. Smith, N. Peace, J. Douwes, Current concentrations, temporal trends and determinants of persistent organic pollutants in breast milk of New Zealand women, Sci. Total Environ. 458–460 (2013) 399–407.

[36] A. Abballe, T.J. Ballard, E. Dellatte, A.d. Domenico, F. Ferri, A.R. Fulgenzi, G. Grisanti, N. Iacovella, A.M. Ingelido, R. Malisch, R. Miniero, M.G. Porpora, S. Risica, G. Ziemacki, E.D. Felip, Persistent environmental contaminants in human milk: concentrations and time trends in Italy, Chemosphere 73 (2008) S220–S227.

[37] A. Karrman, J.L. Domingo, X. Llebaria, M. Nadal, E. Bigas, B. van Bavel, G. Lindstrom, Biomonitoring perfluorinated compounds in Catalonia, Spain: concentrations and trends in human liver and milk samples, Environ. Sci. Pollut. Res. 17 (2010) 750–758.

[38] J.M. Llobet, R. Marti-Cid, V. Castell, J.L. Domingo, Significant decreasing trend in human dietary exposure to PCDD/PCDFs and PCBs in Catalonia, Spain, Toxicol. Lett. 178 (2008) 117–126.

[39] K. Haraguchi, A. Koizumi, K. Inoue, K.H. Harada, T. Hitomi, M. Minata, M. Tanabe, Y. Kato, E. Nishimura, Y. Yamamoto, T. Watanabe, K. Takenaka, S. Uehara, H.R. Yang, M.Y. Kim, C.S. Moon, H.S. Kim, P. Wang, A. Liu, N.N. Hung, Levels and regional trends of persistent organochlorines and polybrominated diphenyl ethers in Asian breast milk demonstrate POPs signatures unique to individual countries, Environ. Int. 35 (2009) 1072–1079.

[40] D. Smith, Worldwide trends in DDT levels in human breast milk, Int. J. Epidemiol. 28 (1999) 179–188.

[41] A. Kärrman, G. Lindström, Trends, analytical methods and precision in the determination of perfluoroalkyl acids in human milk, TrAC 46 (2013) 118–128.

[42] J.S. LaKind, Recent global trends and physiologic origins of dioxins and furans in human milk, J. Expo. Sci. Environ. Epidemiol. 17 (2007) 510–524.

[43] G.M. Solomon, P.M. Weiss, Chemical contaminants in breast milk: time trends and regional variability, Environ. Health Perspect. 110 (2002) A339–A347.

[44] S. Krol, B. Zabiegara, J. Namiesnik, Human hair as a biomarker of human exposure to persistent organic pollutants (POPs), TrAC 47 (2013) 84–98.

[45] O. Leino, H. Kiviranta, A.K. Karjalainen, C. Kronberg-Kippila, H. Sinkko, E.H. Larsen, S. Virtanen, J.T. Tuomisto, Pollutant concentrations in placenta, Food Chem. Toxicol. 54 (2013) 59–69.

[46] P. Dewan, V. Jain, P. Gupta, B.D. Banerjee, Organochlorine pesticide residues in maternal blood, cord blood, placenta, and breastmilk and their relation to birth size, Chemosphere 90 (2013) 1704–1710.

[47] D. Valvi, M.A. Mendez, D. Martinez, J.O. Grimalt, M. Torrent, J. Sunyer, M. Vrijheid, Prenatal concentrations of polychlorinated biphenyls, DDE, and DDT and overweight in children: a prospective birth cohort study, Environ. Health Perspect. 120 (2012) 451–457.

[48] J. Zheng, K.H. Chen, X.J. Luo, X. Yan, C.T. He, Y.J. Yu, G.C. Hu, X.W. Peng, M.Z. Ren, Z.Y. Yang, B.X. Mai, Polybrominated diphenyl ethers (PBDEs) in paired human hair and serum from e-waste recycling workers: source apportionment of hair PBDEs and relationship between hair and serum, Environ. Sci. Technol. 48 (2014) 791–796.

[49] P.Y. Liu, Y.X. Zhao, Y.Y. Zhu, Z.F. Qin, X.L. Ruan, Y.C. Zhang, B.J. Chen, Y. Li, S.S. Yan, X.F. Qin, S. Fu, X.B. Xu, Determination of polybrominated diphenyl ethers in human semen, Environ. Int. 42 (2012) 132–137.

[50] M. Wilhelm, U. Ewers, J. Wittsiepe, P. Furst, J. Hozer, G. Eberwein, J. Angerer, B. Marczynski, U. Ranft, Human biomonitoring studies in North Rhine-Westphalia, Germany, Int. J. Hyg. Environ. Health 210 (2007) 307–318.

[51] Department of Defense Environmental Data Quality Workgroup, DoD Environmental Field Sampling Handbook. www.denix.osd.mil (accessed April 2014).

[52] E.D. Palmes, A.F. Gunnison, Personal monitoring devices for gaseous contamninants, Am. Ind. Hyg. Assoc. J. 34 (1973) 78–81.

[53] J. Klanova, T. Harner, The challenge of producing reliable results under highly variable conditions and the role of passive air samplers in the Global Monitoring Plan, TrAC 46 (2013) 139–149.

[54] T. Harner, M. Bartkow, I. Holoubek, J. Klanova, F. Wania, R. Gioia, C. Moeckel, A.J. Sweetman, K.C. Jones, Passive air sampling for persistent organic pollutants: Introductory remarks to the special issue, Environ. Pollut. 144 (2006) 361–364.

[55] F.A. Esteve-Turrillas, V. Yusa, A. Pastor, M. de la Guardia, New perspectives in the use of semipermeable membrane devices as passive samplers, Talanta 74 (2008) 443–457.

[56] S.J. Hayward, T. Gouin, F. Wania, Comparison of four active and passive sampling techniques for pesticides in air, Environ. Sci. Policy 44 (2010) 3410–3416.

[57] T. Gouin, T. Harner, P. Blanchard, D. Mackay, Passive and active air samplers as complementary methods for investigating persistent organic pollutants in the Great Lakes Basin, Environ. Sci. Technol. 39 (2005) 9115–9122.

[58] J. Cortes, C.M. Gonzalez, L. Morales, M. Abalos, E. Abad, B.H. Aristizabal, PCDD/PCDF and dl-PCB in the ambient air of a tropical Andean city: passive and active sampling measurements near industrial and vehicular pollution sources, Sci. Total Environ. 491-492 (2014) 67–74.

[59] T. Lin, J. Li, Y. Xu, X. Liu, C. Luo, H. Cheng, Y. Chen, G. Zhang, Organochlorine pesticides in seawater and the surrounding atmosphere of the marginal seas of China: spatial distribution, sources and air-water exchange, Sci. Total Environ. 435–436 (2012) 244–252.

[60] M. Shoeib, T. Harner, M. Ikononou, K. Kannan, Indoor and outdoor air concentrations and phase partitioning of perfluoroalkyl sulfonamides and polybrominated diphenyl ethers, Environ. Sci. Technol. 38 (2004) 1313–1320.

[61] T. Wang, S. Han, B. Yuan, L. Zeng, Y. Li, Y. Wang, G. Jiang, Summer-winter concentrations and gas-particle partitioning of short chain chlorinated paraffins in the atmosphere of an urban setting, Environ. Pollut. 171 (2012) 38–45.

[62] L. Zhang, T. Zhang, L. Dong, S. Shi, L. Zhou, Y. Huang, Assessment of halogenated POPs and PAHs in three cities in the Yangtze River Delta using high-volume samplers, Sci. Total Environ. 454–455 (2013) 619–626.

[63] K. Pozo, T. Harner, A. Rudolph, G. Oyola, V.H. Estellano, R. Ahumada-Rudolph, M. Garrido, K. Pozo, R. Mabilia, S. Focardi, Survey of persistent organic pollutants (POPs) and polycyclic aromatic hydrocarbons (PAHs) in the atmosphere of rural, urban and industrial areas of Concepcin, Chile, using passive air samplers, Atmos. Pollut. Res. 3 (2012) 426–434.

[64] Y. Moussaoui, L. Tuduri, Y. Kerchich, B.Y. Meklati, G. Eppe, Atmospheric concentrations of PCDD/Fs, dl-PCBs and some pesticides in northern Algeria using passive air sampling, Chemosphere 88 (2012) 270–277.

[65] V.H. Estellano, K. Pozo, T. Harner, S. Corsolini, S. Focardi, Using PUF disk passive samplers to simultaneously measure air concentrations of persistent organic pollutants (POPs) across the Tuscany region, Italy, Atmos. Pollut. Res. 3 (2012) 88–94.

[66] R.O. Meire, S.C. Lee, A.C. Targino, J.P.M. Torres, T. Harner, Air concentrations and transport of persistent organic pollutants (POPs) in mountains of southeast and southern Brazil, Atmos. Pollut. Res. 3 (2012) 417–425.

[67] C. Shunthirasingham, C.E. Oyiliagu, X. Cao, T. Gouin, F. Wania, S.C. Lee, K. Pozo, T. Harner, D.C.G. Muir, Spatial and temporal pattern of pesticides in the global atmosphere, J. Environ. Monit. 12 (2010) 1650–1657.

[68] K. Pozo, T. Harner, S.C. Lee, F. Wania, D.C.G. Muir, K.C. Jones, Seasonally resolved concentrations of persistent organic pollutants in the global atmosphere from the first year of the GAPS Study, Environ. Sci. Technol. 43 (2009) 796–803.

[69] J. De Boer, F. Smedes, Effects of storage conditions of biological materials on the contents of organochlorine compounds and mercury, Mar. Pollut. Bull. 35 (1997) 93–108.

[70] D.J. GcGoldrick, M.G. Clark, M.J. Keir, S.M. Backus, M.M. Malecki, Canada's National Aquatic Biological Specimen Bank and Database, J. Great Lakes Res. 36 (2010) 393–398.

[71] J.G. Martins, A. Amaya Chavez, S.M. Waliszewski, A. Colin Cruz, M.M. Garcia Fabila, Extraction and clean-up methods for organochlorine pesticides determination in milk, Chemosphere 92 (2013) 233–246.

[72] D. Muir, E. Sverko, Analytical methods for PCBs and organochlorine pesticides in environmental monitoring and surveillance: a critical appraisal, Anal. Bioanal. Chem. 386 (2006) 769–789.

[73] N. Fidalgo-Used, E. Blanco-Gonalez, A. Sanz-Medel, Sample handling strategies for the determination of persistent trace organic contaminants from biota samples, Anal. Chim. Acta 590 (2007) 1–16.

[74] H.G. Ni, H. Zeng, E.Y. Zeng, Sampling and analytical framework for routine environmental monitoring of organic pollutants, TrAC 30 (2011) 1549–1559.

[75] G. Liu, M. Zheng, G. Jiang, Z. Cai, Y. Wu, Dioxin analysis in China, TrAC 46 (2013) 178–188.

[76] P. Hess, J.d. Boer, W.P. Cofino, P.E.G. Leonards, D.E. Wells, Critical review of analysis of non- and mono-ortho- chlorobiphenyls, J. Chromatogr. A 26 (1995) 417–465.

[77] P. Wang, Q. Zhang, Y. Wang, T. Wang, X. Li, L. Ding, G. Jiang, Evaluation of Soxhlet extraction, accelerated solvent extraction and microwave-assisted extraction for the determination of polychlorinated biphenyls and polybrominated diphenyl ethers in soil and fish samples, Anal. Chim. Acta 663 (2010) 43–48.

[78] C. Ribeiro, A.R. Riberio, A.S. Maia, V.M.F. Goncalves, M.E. Tiritan, New trends in sample preparation techniques for environmental analysis, Crit. Rev. Anal. Chem. 44 (2014) 142–185.

[79] W. Xu, X. Wang, Z. Cai, Analytical chemistry of the persistent organic pollutants identified in the Stockholm Convention: a review, Anal. Chim. Acta 790 (2013) 1–13.

[80] H.P.O. Tang, Recent development in analysis of persistent organic pollutants under the Stockholm Convention, TrAC 45 (2013) 48–66.

[81] L. Ramos, Critical overview of selected contemporary sample preparation techniques, J. Chromatogr. A 1221 (2012) 84–98.

[82] M. LeDoux, Analytical methods applied to the determination of pesticide residues in foods of animal origin. A review of the past two decades, J. Chromatogr. A 1218 (2011) 1021–1036.

[83] S. de Koning, H.-G. Janssen, U.A.T. Brinkman, Modern methods of sample preparation for GC analysis, Chromatographia 69 (2009) 33–78.

[84] M. Asensio-Ramos, A.V. Herrera-Herrera, M.A. Rodríguez-Delgado, J. Hernández-Borges, Sorbent-based techniques for the determination of pesticides in food, in: J. Pawliszyn, L. Mondello (Eds.), Comprehensive Sampling and Sample Preparation, Elsevier, Paola Dugo, 2012, pp. 263–312.

[85] M.D.L. de Castro, L.E. Garcia-Ayuso, Soxhlet extraction of solid materials: an outdated technique with promising innovative future, Anal. Chim. Acta 369 (1998) 1–10.

[86] P. Antunes, P. Viana, T. Vinhas, J.L. Capelo, J. Rivera, E.M.S.M. Gaspar, Optimization of pressurized liquid extraction (PLE) of dioxin-furans and dioxin-like PCBs from environmental samples, Talanta 75 (2008) 916–925.

[87] P. Suchan, J. Pulkrabova, J. Hajslova, V. Kocourek, Pressurized liquid extraction in determination of polychlorinated biphenyls and organochlorine pesticides in fish samples, Anal. Chim. Acta 520 (2004) 193–200.

[88] L. Bielska, K. Smidova, J. Hofman, Supercritical fluid extraction of persistent organic pollutants from natural and artificial soils and comparison with bioaccumulation in earthworms, Environ. Pollut. 176 (2013) 48–54.

[89] B. van Bavel, P. Dahl, L. Karlsson, L. Hardell, C. Rappe, G. Lindstöm, Supercritical fluid extraction of PCBs from human adipose tissue for HRGC/LRMS analysis, Chemosphere 30 (1995) 1229–1236.

[90] E. Bjorklund, T. Nilsson, S. Bowadt, K. Pilorz, L. Mathiasson, S.B. Hawthorne, Introducing selective supercritical fluid extraction as a new tool for determining sorption/desorption behavior and bioavailability of persistent organic pollutants in sediment, J. Biochem. Biophys. Methods 43 (2000) 295–311.

[91] S.A. Baker, A.R. Long, C.R. Short, Isolation of drug residues from tissues by solid phase dispersion, J. Chromatogr. 1989 (1989) 353–361.

[92] S.A. Baker, Matrix solid phase dispersion (MSPD), J. Biochem. Biophys. Methods 70 (2007) 151–162.

[93] C.S. Eskilsson, E. Bjorklund, Analytical-scale microwave-assisted extraction, J. Chromatogr. A 902 (2000) 227–250.

[94] N. Gai, J. Pan, H. Tang, S. Chen, D. Chen, X. Zhu, G. Lu, Y. Yang, Organochlorine pesticides and polychlorinated biphenyls in surface soils from Ruoergai high altitude prairie, east edge of Qinghai-Tibet Plateau, Sci. Total Environ. 478 (2014) 90–97.

[95] G.L. Yuan, H.Z. Wu, S. Fu, P. Han, X.X. Lang, Persistent organic pollutants (POPs) in the topsoil of typical urban renewal area in Beijing, China: status, sources and potential risk, J. Geochem. Explor. 138 (2014) 94–103.

[96] Y. Dong, L. Li, P. Bie, S. Jia, Q. Wang, Z. Huang, X. Qiu, J. Zhang, J. Hu, Polybrominated diphenyl ethers in farmland soils: source characterization, deposition contribution and apportionment, Sci. Total Environ. 466–467 (2014) 524–532.

[97] S.M. Rhind, C.E. Kyle, C. Kerr, M. Osprey, Z.L. Zhang, E.I. Duff, A. Lilly, A. Nolan, G. Hudson, W. Towers, J. Bell, M. Coull, C. McKenzie, Concentrations and geographic distribution of selected organic pollutants in Scottish surface soils, Environ. Pollut. 182 (2013) 15–27.

[98] X.P. Wang, J.J. Sheng, P. Gong, Y.G. Xue, T.D. Yao, K.C. Jones, Persistent organic pollutants in the Tibetan surface soil: spatial distribution, air-soil exchange and implications for global cycling, Environ. Pollut. 170 (2012) 145–151.

[99] M. Domotorova, Z. Stachova Sejakova, A. Kocan, K. Conka, J. Chovancova, A. Fabisikova, PCDDs, PCDFs, dioxin-like PCBs and indicator PCBs in soil from five selected areas in Slovakia, Chemosphere 89 (2012) 480–485.

[100] B. Gevao, A.N. Ghadban, S. Uddin, F.M. Jaward, M. Bahloul, J. Zafar, Polybrominated diphenyl ethers (PBDEs) in soils along a rural-urban-rural transect: sources, concentration gradients, and profiles, Environ. Pollut. 159 (2011) 3666–3672.

[101] S. Romano, R. Piazza, C. Mugnai, S. Giuliani, L.G. Bellucci, C.N. Huu, M. Vecchiato, S. Zambon, N.D. Hoai, M. Frignani, PBDEs and PCBs in sediments of the Thi Nai Lagoon (cnetrl Vietnam) and soil from its mainland, Chemosphere 90 (2013) 2396–2402.

[102] R.Q. Yang, H. Wei, J. Guo, A. Li, Emerging brominated flame retardants in the sediment of the Great Lakes, Environ. Sci. Technol. 46 (2012) 3119–3126.

[103] H. Kuranchie-Mensah, S.M. Atiemo, L.M.N. Palm, S. Blankson-Arthur, A.O. Tutu, P. Fosu, Determination of organochlorine pesticide residue in sediment and water from the Densu river basin, Ghana, Chemosphere 86 (2012) 286–292.

[104] K. Pozo, G. Perra, V. Menchi, R. Urrutia, O. Parra, A. Rudolph, S. Focardi, Levels and spatial distribution of polycyclic aromatic hydrocarbons (PAHs) in sediments from Lenga Estuary, central Chile, Mar. Pollut. Bull. 62 (2011) 1572–1576.

[105] I. Tolosa, M. Mesa-Albernas, C.M. Alonso-Hernamdez, Organochlorine contamination (PCBs, DDTs, HCBs, HCHs) in sediments from Cienfuegos Bay, Cuba, Mar. Pollut. Bull. 60 (2010) 1619–1624.

[106] K. Conka, J. ChovancovA, Z. Stachova Sejakova, M. Domotorova, A. Fabisikova, B. Drobna, A. Kocan, PCDDs, PCDFs, PCBs and OCPs in sediments from selected areas in the Slovak republic, Chemosphere 98 (2014) 37–43.

[107] X. Ma, C. Chen, H. Zhang, Y. Gao, Z. Wang, Z. Yao, J. Chen, J. Chen, Congener-specific distribution and bioaccumulation of short-chain chlorinated paraffins in sediments and bivalves of the Bohai Sea, China, Mar. Pollut. Bull. 79 (2014) 299–304.

[108] G. Perra, S.E. Focardi, C. Guerranti, Levels and spatial distribution of perfluorinated compounds (PFCs) in superficial sediments from the marine reserves of the Tuscan Archipelago National Park (Italy), Mar. Pollut. Bull. 76 (2013) 379–382.

[109] V.P. Beskoski, S. Takemine, T. Nakano, L. Slavkovic Beskoski, G. Gojgic-Cvijovic, M. Ilic, S. Miletic, M.M. Vrvic, Perfluorinated compounds in sediment samples from the wastewater canal of Pancevo (Serbia) industrial area, Chemosphere 91 (2013) 1408–1415.

[110] J. Wang, Z. Lin, K. Lin, C. Wang, W. Zhang, C. Cui, J. Lin, Q. Dong, C. Huang, Polybrominated diphenyl ethers in water, sediment, soil, and biological samples from different industrial areas in Zhejiang, China, J. Hazard. Mater. 197 (2011) 211–219.

[111] L. Zhang, S. Shi, L. Dong, T. Zhang, L. Zhou, Y. Huang, Concentrations and possible sources of polybrominated diphenyls in the surface water of the Yangtze River Delta, China, Chemosphere 85 (2011) 399–405.

[112] S. Dasgupta, K. Banerjee, S. Utture, P. Kusari, S. Wagh, K. Dhumal, S. Kolekar, P.G. Adsule, Extraction of pesticides, dioxin-like PCBs and PAHs in water based commodities using liquid-liquid microextraction and analysis by gas chromatography-mass spectrometry, J. Chromatogr. A 1218 (2011) 6780–6791.

[113] Z. Polkowska, K. Cichala-Kamrowska, M. Ruman, K. Koziol, W.E. Krawczyk, J. Namiesnik, Organic pollution in surface waters from the Fuglebekken Basin in Svalbard, Norwegian Arctic, Sensors 11 (2011) 8910–8929.

[114] G.-M. Momplaisir, C.G. Rosal, E.M. Heithmar, K.E. Varner, L.A. Riddick, D.F. Bradford, N.G. Tallent-Halsell, Development of a solid phase extraction method for agricultural pesticides in large-volume water samples, Talanta 81 (2010) 1380–1386.

[115] J. Ge, M. Liu, X. Yun, Y. Yang, M. Zhang, Q.X. Liu, J. Wang, Occurrence, distribution and seasonal variations of polychlorinated biphenyls and polybrominated diphenyl ethers in surface waters of the East Lake, China, Chemosphere 103 (2014) 256–262.

[116] C. Flores, F. Ventura, J. Martin-Alonso, J. Caixach, Occurrence of perflurooctane sulfonate (PFOS) and perfluorooctanoate (PFOA) in N.E. Spanish surface waters and their removal in drinking water treatment plant that combines conventional and advanced treatment in parallel lines, Sci. Total Environ. 461-462 (2013) 618–626.

[117] B. Karacik, O.S. Okay, B. Henkelmann, G. Pfister, K.W. Schramm, Water concentrations of PAH, PCB and OCP by using semipermeable membrane devices and sediments, Mar. Pollut. Bull. 70 (2013) 258–265.

[118] U.E. Friden, M.S. McLachlan, U. Berger, Chlorinated paraffins in indoor air and dust: concentrations, congener patterns, and human exposure, Environ. Int. 37 (2011) 1169–1174.

[119] Z. Ren, B. Zhang, P. Lu, C. Li, L. Gao, M. Zheng, Characteristics of air pollution by polychlorinated dibenzo-p-dioxins and dibenzofurans in the typical industrial areas of Tangshan City, China, J. Environ. Sci. 23 (2011) 228–235.

[120] K. Arinaitwe, D.C.G. Muir, B.T. Kiremire, P. Fellin, H. Li, C. Teixeira, Polybrominated diphenyl ethers and alternative flame retardants in air and precipitation samples from the northern Lake Victoria Region, East Africa, Environ. Sci. Technol. 48 (2014) 1458–1466.

[121] P. Bruckmann, E. Hiester, M. Klees, C. Zetzsch, Trends of PCDD/F and PCB concentrations and depositions in ambient air in Northwestern Germany, Chemosphere 93 (2013) 1471–1478.

[122] M. Yang, H. Qi, H.L. Jia, N.Q. Ren, Y.S. Ding, W.L. Ma, L.Y. Liu, H. Hung, E. Sverko, Y.F. Li, Polybrominated diphenyl ethers in air across China: levels, compositions, and gas-particle partitioning, Environ. Sci. Technol. 47 (2013) 8978–8984.

[123] R. Pieters, J. Focant, Dioxin, furan and PCB serum levels in a South African Tswana population: comparing the polluting effects of using different cooking and heating fuels, Environ. Int. 66 (2014) 71–78.

[124] C. Brasseur, C. Pirard, G. Scholl, E. De Pauw, J.F. Viel, L. Shen, E.J. Reiner, J.F. Focant, Levels of dechloranes and polybrominated diphenyl ethers (PBDEs) in human serum from France, Environ. Int. 65 (2014) 33–40.

[125] A. Saodi, N. Frery, A. Zeghnoun, M.L. Bidondo, V. Deschamps, T. Goen, R. Garnier, L. Guldner, Serum levels of organochlorine pesticides in the French adult population: the French National Nutrition and Health Study (ENNS), 2006–2007, Sci. Total Environ. 472 (2014) 1089–1099.

[126] H.Y. Kim, S.K. Kim, D.M. Kang, Y.S. Hwang, J.E. Oh, The relationships between sixteen perfluorinated compound concentrations in blood serum and food, and other parameters, in the general population of South Korea with proportionate stratified sampling method, Sci. Total Environ. 470–471 (2014) 1390–1400.

[127] H. Shen, G. Ding, Y. Wu, G. Pan, X. Zhou, J. Han, J. Li, S. Wen, Polychlorinated dibenzo-p-dioxins/furans (PCDD/Fs), polychlorinated biphenyls (PCBs), and polybrominated diphenyl ethers (PBDEs) in breast milk from Zhejiang, China, Environ. Int. 42 (2012) 84–90.

[128] Y. Fujii, Y. Ito, K.H. Harada, T. Hitomi, A. Koizumi, K. Haraguchi, Comparative survey of levels of chlorinated cyclodiene pesticides in breast milk from some cities of China, Korea and Japan, Chemosphere 89 (2012) 452–457.

[129] S.B. Hassine, W.B. Ameur, N. Gandoura, M.R. Driss, Determination of chlorinated pesticides, polychlorinated biphenyls, and polybrominated diphenyl ethers in human milk from Bizerte (Tunisia) in 2010, Chemosphere 89 (2012) 369–377.

[130] T.W. Wong, A.H.S. Wong, E.A.S. Nelson, H. Qiu, S.Y.K. Ku, Levels of PCDDs, PCDFs, and dioxin-like PCBs in human milk among Hong Kong mothers, Sci. Total Environ. 463–464 (2013) 1230–1238.

[131] S. Lee, S. Kim, H.K. Lee, I.S. Lee, J. Park, H.J. Kim, J.J. Lee, G. Choi, S. Choi, S. Kim, S.Y. Kim, K. Choi, S. Kim, H.B. Moon, Contamination of polychlorinated biphenyls and organochlorine pesticides in breast milk in Korea: time-course variation, influencing factors, and exposure assessment, Chemosphere 93 (2013) 1578–1585.

[132] A. Giovannini, G. Rivezzi, P. Carideo, R. Ceci, G. Diletti, C. Ippoliti, G. Migliorati, P. Piscitelli, A. Ripani, R. Salini, G. Scortichini, Dioxins levels in breast milk of women living in Caserta and Naples: assessment of environmental risk factors, Chemosphere 94 (2014) 76–84.

[133] M.A.E. Abdallah, S. Harrad, Polybrominated diphenyl ethers in UK human milk: Implications for infant exposure and relationship to external exposure, Environ. Int. 63 (2014) 130–136.

[134] P.M. Ondarza, M. Gonzalez, G. Fillmann, K.S.B. Miglioranza, PBDEs, PCBs and organochlorine pesticides distribution in edible fish from Negro River basin, Argentinean Patagonia, Chemosphere 94 (2014) 135–142.

[135] W.M. Young, P. South, T.H. Begley, G.O. Noonan, Determination of perfluorochemicals in fish and shellfish using liquid chromatography-tandem mass spectrometry, J. Agr. Food Chem. 61 (2013) 11166–11172.

[136] N. Bodin, N. Tapie, K. Le Menach, E. Chassot, P. Elie, E. Rochard, H. Budzinski, PCB contamination in fish community from the Gironde estuary (France): blast from the past, Chemosphere 98 (2014) 66–72.

[137] J.K.Y. Chan, Y.B. Man, G.H. Xing, S.C. Wu, M.B. Murphy, Y. Xu, M.H. Wong, Dietary exposure to polychlorinated dibenzo-p-dioxins and dibenzofurans via fish consumption and dioxin-like activity in fish determined by H4IIE-luc bioassay, Sci. Total Environ. 463–464 (2013) 1192–1200.

[138] A.L. Capriotti, C. Cavaliere, P. Giansanti, R. Gubbiotti, R. Semperi, A. Lagana, Recent developments in matrix solid-phase dispersion extraction, J. Chromatogr. A 1217 (2010) 2521–2532.

[139] S. Bogialli, A. Di Corcia, Matrix solid-phase dispersion as a valuable tool for extracting contaminants from foodstuffs, J. Biochem. Biophys. Methods 70 (2007) 163–179.

[140] G.N. Rallis, V.A. Sakkas, V.A. Boumba, T. Vougiouklakis, T.A. Albanis, Determination of organochlorine pesticides and polychlorinated biphenyls in post-mortem human lung by matrix solid-phase dispersion with the aid of response surface methodology and desirability function, J. Chromatogr. A 1227 (2012) 1–9.

[141] G. Fang, G. Min, J. He, C. Zhang, K. Qian, S. Wang, Multiwalled carbon nanotubes as matrix solid-phase dispersion extraction absorbents to determine 31 pesticides in agriculture samples by gas chromatography-mass spectrometry, J. Agr. Food Chem. 57 (2009) 3040–3045.

[142] P. Labadie, F. Alliot, C. Bourges, A. Desportes, M. Chevreuil, Determination of polybrominated diphenyl ethers in fish tissues by matrix solid phase dispersion and gas chromatography coupled to triple quadrupole mass spectrometry: case study on European ell (*Anguilla anguilla*) from Mediterranean coastal lagoons, Anal. Chim. Acta 675 (2010) 97–105.

[143] R. Carabias-Martinez, E. Rodriguez-Gonzalo, P. Revilla-Ruiz, J. Hernandez-Mendez, Pressurized liquid extraction in the analysis of food and biological samples, J. Chromatogr. A 1089 (2005) 1–17.

[144] L. Do, S. Lundstedt, P. Haglund, Optimization of selective pressurized liquid extraction for extraction and in-cell clean-up of PCDD/Fs in soils and sediments, Chemosphere 90 (2013) 2414–2419.

[145] J.L. Tadeo, R.A. Perez, B. Albero, A.I. Garcia-Valcarcel, C. Sanchez-Brunete, Review of sample preparation techniques for the analysis of pesticide residues in soil, J. AOAC Int. 95 (2012) 1258–1271.

[146] R. Ghosh, K.J. Hageman, E. Bjorklund, Selective pressurized liquid extraction of three classes of halogenated contaminants in fish, J. Chromatogr. A 1218 (2011) 7242–7247.

[147] E. Bjorklund, S. Sporring, K. Wiberg, P. Haglund, C.V. Holst, New strategies for extraction and clean-up of persistent organic pollutants from food and feed samples using selective pressurized liquid extraction, TrAC 25 (2006) 318–325.

[148] E. Bjorklund, T. Nilsson, Pressurised liquid extraction of persistent organic pollutants in environmental analysis, TrAC 19 (2000) 434–445.

[149] K. Ridgway, S.P.D. Lalljie, R.M. Smith, Sample preparation techniques for the determination of trace residues and contaminants in foods, J. Chromatogr. A 1153 (2007) 36–53.

[150] A. Covaci, S. Voorspoels, L. Ramos, H. Neels, R. Blust, Recent developments in the analysis of brominated flame retardants and brominated natural compounds, J. Chromatogr. A 1153 (2007) 145–171.

[151] D. Garcia-Rodriguez, A.M. Carro-Diaz, R.A. Lorenzo-Ferreira, Supercritical fluid extraction of polyhalogenated pollutants from aquaculture and marine environmental samples: a review, J. Sep. Sci. 31 (2008) 1333–1345.

[152] T.L. Chester, J.D. Pinkston, D.E. Raynie, Supercritical fluid chromatography and extraction, Anal. Chem. 68 (1996) 487R–514R.

[153] S.M. Pourmortazavi, M. Rahimi-Nasrabadi, S.S. Hajimirsadeghi, Supercritical fluid technology in analytical chemistry-review, Curr. Anal. Chem. 10 (2014) 3–28.

[154] E. Beceiro-Gonzalez, M.J. Gonzalez-Castro, S.M. Muniategcl-Lorenzo, P. Lopez-Mahia, D. Prada-Rodrigcez, Analytical methodology for the determination of organochlorine pesticides in vegetation, J. AOAC Int. 95 (2012) 1291–1307.

[155] M. Herrero, J.A. Mendiola, A. Cifuentes, E. Ibanez, Supercritical fluid extraction: recent advances and applications, J. Chromatogr. A 1217 (2010) 2495–2511.

[156] E. Bjorklund, C. von Holst, E. Anklam, Fast extraction, clean-up and detection methods for the rapid analysis and screening f seven indicator PCBs in food matrices, TrAC 21 (2002) 39–51.

[157] P.S. Dias, C.V.Z. Cipro, S. Taniguchi, R.C. Montone, Persistent organic pollutants in marine biota of Sao Pedro and Sao Paulo Archipelago, Brazil, Mar. Pollut. Bull. 74 (2013) 435–440.

[158] Y. Merdassa, J.F. Liu, N. Megersa, Development of a one-step microwave-assisted extraction method for simultaneous determination of organophosphorus pesticides and fungicides in soils by gas chromatography-mass spectrometry, Talanta 114 (2013) 227–234.

[159] I. Fulara, M. Czaplicka, Methods for determination of polybrominated diphenyl ethers in environmental samples – Review, J. Sep. Sci. 35 (2012) 2075–2087.

[160] C. Duan, Z. Shen, D. Wu, Y. Guan, Recent developments in solid-phase microextraction for on-site sampling and sample preparation, TrAC 30 (2011) 1568–1574.

[161] G. Vas, K. Vekey, Solid-phase microextraction: a powerful sample preparation tool prior to mass spectrometric analysis, J. Mass Spectrom. 39 (2004) 233–254.

[162] C. Yang, J. Wang, D. Li, Microextraction techniques for the determination of volatile and semivolatile organic compounds from plants: a review, Anal. Chim. Acta 799 (2013) 8–22.

[163] H.C. Menezes, L.C.A. Amorim, Z.L. Cardeal, Sampling and analytical methods for determining VOC in air by biomonitoring human exposure, Crit. Rev. Environ. Sci. Technol. 43 (2013) 1–39.

[164] M. Kosikowska, M. Biziuk, Review of the determination of pesticide residues in ambient air, TrAC 29 (2010) 1064–1072.

[165] M. Farre, S. Perez, M.F. Alpendurada, D. Barcelo, Green analytical chemistry in the determination of organic pollutants in the aquatic environment, TrAC 29 (2010) 1347–1362.

[166] E. Baltussen, P. Sandra, F. David, C. Cramers, Stir bar sorptive extraction (SBSE), a novel extraction technique for aqueous samples: theory and principles, J. Microcol. Sep. 11 (1999) 737–747.

[167] J.M.F. Nogueira, Novel sorption-based methodologies for static microextraction analysis: a review on SBSE and related techniques, Anal. Chim. Acta 757 (2012) 1–10.

[168] F.M. Lancas, M.E.C. Queiroz, P. Grossi, I.R.B. Olivares, Recent developments and applications of stir bar sorptive extraction, J. Sep. Sci. 32 (2009) 813–824.

[169] A. Prieto, O. Basauri, R. Rodil, A. Usobiaga, L.A. Fernandez, N. Etxebarria, O. Zuloaga, Stir-bar sorptive extraction: a view on method optimisation, novel applications, limitations and potential solutions, J. Chromatogr. A 1217 (2010) 2642–2666.

[170] M. Tankiewicz, J. Fenik, M. Biziuk, Solventless and solvent-minimized sample preparation techniques for determining currently used pesticides in water samples: a review, Talanta 86 (2011) 8–22.

[171] J.M. Kokosa, Advances in solvent-microextraction techniques, TrAC 43 (2013) 2–13.

[172] A. Jain, K.K. Verma, Recent advances in applications of single-drop microextraction: a review, Anal. Chim. Acta 706 (2011) 37–65.

[173] A.V. Herrera-Herrera, M. Asensio-Ramos, J. Hernandez-Borges, M.A. Rodrigues-Delgado, Dispersive liquid-liquid microextraction for determination of organic analytes, TrAC 29 (2010) 728–751.

[174] M. Ghambarian, Y. Yamini, A. Esrafili, Developments in hollow fiber based liquid-phase microextraction: principles and applications, Microchim. Acta 177 (2012) 271–294.

[175] S. Pedersen-Bjergaard, K.E. Rasmussen, Liquid-phase microextraction with porous hollow fibers, a miniaturized and highly flexible format for liquid-liquid extraction, J. Chromatogr. A 1184 (2008) 132–142.

[176] M. Rezaee, Y. Assadi, M.R.M. Hosseini, E. Aghaee, F. Ahmadi, S. Berijani, Determination of organic compounds in water using dispersive liquid-liquid microextraction, J. Chromatogr. A 1116 (2006) 1–9.

[177] J.Y. Barletta, P.C.F. De Lima Gomes, A.J. Dos Santo-Neto, F.M. Lancas, Development of a new stir bar sorptive extraction coating and its application for the determination of six pesticides in sugarcane juice, J. Sep. Sci. 34 (2011) 1317–1325.

[178] B. Rocha-Gutierrez, W.Y. Lee, Investigation of polybrominated diphenyl ethers in wastewater treatment plants along the U.S. and Mexico Border: a trans-boundary study, Water, Air, Soil Pollut. 224 (2013) 1398.

[179] P. Tolgyessy, B. Vrana, K. Silharova, An improved method for determination of polychlorinated biphenyls and polybrominated diphenyl ethers in sediment by ultrasonic solvent extraction followed by stir bar sorptive extraction coupled to TD-GC-MS, Chromatographia 76 (2013) 177–185.

[180] C. Margoum, C. Guillemain, X. Yang, M. Coquery, Stir bar sorptive extraction coupled to liquid chromatography-tandem mass spectrometry for the determination of pesticides in water samples: method validation and measurement uncertainty, Talanta 116 (2013) 1–7.

[181] A. Assoumani, S. Lissalde, C. Margoum, N. Mazzella, M. Coquery, In site application of stir bar sorptive extraction as a passive sampling technique for the monitoring of agricultural pesticides in surface waters, Sci. Total Environ. 463–464 (2013) 829–835.

[182] S. Ozcan, Analyses of polychlorinated biphenyls in waters and wastewaters using vortex-assisted liquid-liquid microextraction and gas chromatography-mass spectrometry, J. Sep. Sci. 34 (2011) 574–584.

[183] R.S. Zhao, L.L. Zhang, X. Wang, Dispersive liquid-phase microextraction using ionic liquid as extractant for the enrichment and determination of DDT and its metabolites in environmental water samples, Anal. Bioanal. Chem. 399 (2011) 1287–1293.

[184] X. Liu, A. Zhao, A. Zhang, H. Liu, W. Xiao, C. Wang, X. Wang, Dispersive liquid-liquid microextraction and gas chromatography-mass spectrometry determination of polychlorinated biphenyls and polybrominated diphenyl ethers in milk, J. Sep. Sci. 34 (2011) 1084–1090.

[185] V.G. Amelin, D.S. Bol'Shakov, A.V. Tret'Yakov, Dispersive liquid-liquid microextraction and solid-phase extraction of polar pesticides from natural water and their determination by micellar electrokinetic chromatography, J. Anal. Chem. 68 (2013) 386–397.

[186] V. Ghasemzadeh-Mohammadi, A. Mohammadi, M. Hashemi, R. Khaksar, P. Haratian, Microwave-assisted extraction and dispersive liquid-liquid microextraction followed by gas chromatography-mass spectrometry for isolation and determination of polycyclic aromatic hydrocarbons in smoked fish, J. Chromatogr. A 1237 (2012) 30–36.

[187] M.H. Naeeni, Y. Yamini, M. Rezaee, Combination of supercritical fluid extraction with dispersive liquid–liquid microextraction for extraction of organophosphorus pesticides from soil and marine sediment samples, J. Supercrit. Fluids 57 (2011) 219–226.
[188] Y. Han, X. Jia, T. Duan, H. Chen, Combination of saponification with in-tube liquid-liquid extraction and dispersive liquid-liquid microextraction for determination of polybrominated diphenyl ethers in whole milk by gas chromatography-mass spectrometry, Anal. Methods 3 (2011) 842–848.
[189] F. Rezaei, M.R.M. Hosseini, New method based on combining ultrasonic assisted miniaturized matrix solid-phase dispersion and homogeneous liquid–liquid extraction for determination of some organochlorinated pesticides in fish, Anal. Chim. Acta 702 (2011) 274–279.
[190] L. Zhou, H. Li, S. Ma, M. Ren, Z.-Q. Yu, G.-Y. Sheng, J.-M. Fu, Simultaneous separation and determination of PCDD/Fs, PBDD/Fs, PCBs and PBDEs in soil sample, J. Instrum. Anal. (in Chinese) (2013) 1160–1165.
[191] N.A. Al-Odaini, U.H. Yim, N.S. Kim, W.J. Shim, S.H. Hong, Isotopic dilution determination of emerging flame retardants in marine sediments by HPLC-APCI-MS/MS, Anal. Methods 5 (2013) 1771–1778.
[192] C. Garcia-Delgado, F. Yunta, E. Eymar, Methodology for polycyclic aromatic hydrocarbons extraction from either fresh or dry spent mushroom compost and quantification by high-performance liquid chromatography-photodiode array detection, Comm. Soil Sci. Plant Anal. 44 (2013) 817–825.
[193] S.N. Zhou, E.J. Reiner, C. Marvin, T. Kolic, N. Riddell, P. Helm, F. Dorman, M. Misselwitz, I.D. Brindle, Liquid chromatography-atmospheric pressure photoionization tandem mass spectrometry for analysis of 36 halogenated flame retardants in fish, J. Chromatogr. A 1217 (2010) 633–641.
[194] S. Chusaksri, S. Sutthuvaiyakit, P. Sutthivaiyakit, Confirmatory determination of organochlorine pesticides in surface waters using LC/APCI/tandem mass spectrometry, Anal. Bioanal. Chem. 384 (2006) 1236–1245.
[195] S. Krol, B. Zabiegala, J. Namiesnik, PBDEs in environmental samples: sampling and analysis, Talanta 93 (2012) 1–17.
[196] A. Papachlimitzou, J.L. Barber, S. Losada, P. Bersuder, R.J. Law, A review of the analysis of novel brominated flame retardants, J. Chromatogr. A 1219 (2012) 15–28.
[197] H.M. Stapleton, Instrumental methods and challenges in quantifying polybrominated diphenyl ethers in environmental extracts: a review, Anal. Bioanal. Chem. 386 (2006) 807–817.
[198] D. Wang, Q.X. Li, Application of mass spectrometry in the analysis of polybrominated diphenol ethers, Mass Spectrom. Rev. 29 (2009) 737–775.
[199] F.J. Santos, M.T. Galceran, Modern developments in gas chromatography-mass spectrometry-based environmental analysis, J. Chromatogr. A 1000 (2003) 125–151.
[200] J. de Boer, Persistent organic pollutants – are our methods sensitive and selective enough? Anal. Lett. 45 (2012) 485–494.
[201] S.W.C. Chung, B.L.S. Chen, Determination of organochlorine pesticide residues in fatty foods: a critical review on the analytical methods and their testing capabilities, J. Chromatogr. A 1218 (2011) 5555–5567.
[202] S.P.J. van Leeuwen, J. de Boer, Advances in the gas chromatographic determination of persistent organic pollutants in the aquatic environment, J. Chromatogr. A 1186 (2008) 161–182.
[203] W. Vetter, N. Rosenfelder, Gas chromatography retention data of environmentally relevant polybrominated compounds, Anal. Bioanal. Chem. 392 (2008) 489–504.
[204] N.C. Deziel, J.R. Nuckols, J.S. Colt, A.J. Roos, A. Pronk, C. Gourley, R.K. Severson, W. Cozen, J.R. Cerhan, P. Hartge, M.H. Ward, Determination of polychlorinated dibenzo-p-dioxins and polychlorinated dibenzofurans in house dust samples from four areas of the United States, Sci. Total Environ. 433 (2012) 516–522.

[205] E. Abad, J. Caixach, J. Rivera, Application of DB-5ms gas chromatography column for the complete assignment of 2,3,7,8,–substituted polychlorodibenzo-*p*-dioxins and polychlorodibenzofurans in samples from municipal waste incinerator emissions, J. Chromatogr. A 786 (1997) 125–134.

[206] M. Adahchour, J. Beens, U.A.T. Brinkman, Recent developments in the application of comprehensive two-dimensional gas chromatography, J. Chromatogr. A 1186 (2008) 67–108.

[207] H.J. CorteS, B. Winniford, J. Luong, M. Pursch, Comprehensive two dimensional gas chromatography review, J. Sep. Sci. 32 (2009) 883–904.

[208] L.R. Bordajandi, J.J. Ramos, J. Sanz, M.J. Gonzlez, L. Ramos, Comprehensive two-dimensional gas chromatography in the screening of persistent organohalogenated pollutants in environmental samples, J. Chromatogr. A 1186 (2008) 312–324.

[209] E. Leda, N. Ochiai, T. Miyawaki, T. Ohura, Y. Horri, Environmental analysis of chlorinated and brominated polycyclic aromatic hydrocarbons by comprehensive two-dimensional gas chromatography coupled to high-resolution time-of-flight mass spectrometry, J. Chromatogr. A 1218 (2011) 3224–3232.

[210] C. Meinert, U.J. Meierhenrich, A new dimension in separation science: comprehensive two-dimensional gas chromatography, Ange. Chem. – Int. Ed. 51 (2012) 10460–10470.

[211] J.F. Focant, A. Sjodin, W.I. Turner, D.G. Patterson Jr., Measurement of selected polybrominated diphenyl ethers, polybrominated and polychlorinated biphenyls, and organochlorine pesticides in human serum and milk using comprehensive two-dimensional gas chromatography isotope dilution time-of-flight mass spectrometry, Anal. Chem. 76 (2004) 6313–6320.

[212] V. Matamoros, E. Jover, J.M. Bayona, Part-per-trillion determination of pharmaceuticals, pesticides, and related organic contaminants in river water by solid-phase extraction followed by comprehensive two-dimensional gas chromatography time-of-flight mass spectrometry, Anal. Chem. 82 (2010) 699–706.

[213] P. Korytar, C. Danielsson, P.E.G. Leonards, P. Haglund, J. De Boer, U.A.T. Brinkman, Separation of seventeen 2,3,7,8-substituted polychlorinated dibenzo-p-dioxins and dibenzofurans and 12 dioxin-like polychlorinated biphenyls by comprehensive two-dimensional gas chromatography with electron-capture detection, J. Chromatogr. A 1038 (2004) 189–199.

[214] S. Dasgupta, K. Banerjee, S.H. Patil, M. Ghaste, K.N. Dhumal, P.G. Adsule, Optimization of two-dimensional gas chromatography time-of-flight mass spectrometry for separation and estimation of the residues of 160 pesticides and 25 persistent organic pollutants in grape and wine, J. Chromatogr. A 1217 (2010) 3881–3889.

[215] E. Hoh, N.G. Dodder, S.J. Lehotay, K.C. Pangallo, C.M. Reddy, K.A. Maruya, Nontargeted comprehensive two-dimensional gas chromatography/time-of-flight mass spectrometry method and software for inventorying persistent and bioaccumulative contaminants in marine environments, Environ. Sci. Technol. 46 (2012) 8001–8008.

[216] D.G. Hayward, T.S. Pisano, J.W. Wong, R.J. Scudder, Multiresidue method for pesticides and persistent organic pollutants (POPs) in milk and cream using comprehensive two-dimensional capillary gas chromatography-'time-of-flight mass spectrometry, J. Agr. Food Chem. 58 (2010) 5248–5256.

[217] D. Megson, R. Kalin, P.J. Worsfold, C. Gauchotte-Lindsay, D.G. Patterson, M.C. Lohan, S. Comber, T.A. Brown, G. O'Sulliva, Fingerprinting polychlorinated biphenyls in environmental samples using comprehensive two-dimensional gas chromatography with time-of-flight mass spectrometry, J. Chromatogr. A 1318 (2013) 276–283.

[218] A. Mechlinska, L. Wolska, J. Namiesnik, L. Wolska, Isotope-labeled substances in analysis of persistent organic pollutants in environmental samples, TrAC 29 (2010) 820–831.

[219] A.C. Dirtu, Analytical aspects for determination of polybrominated diphenyl ethers in environmental samples, Acta Chem. Iasi 17 (2009) 107–119.

[220] F. Hernández, J.V. Sancho, M. Ibáñez, E. Abad, T. Portolés, L. Mattioli, Current use of high-resolution mass spectrometry in the environmental sciences, Anal. Bioanal. Chem. 403 (2012) 1251–1264.

[221] A. Kaufmann, The current role of high-resolution mass spectrometry in food analysis, Anal. Bioanal. Chem. 403 (2012) 1233–1249.

[222] R.A. Pérez, B. Albero, J.L. Tadeo, M.V. Fraile, C. Sánchez-Brunete, Determination of PAHs in soil leachates by magnetic solid-phase extraction using nanoparticles and gas chromatography-tandem mass spectrometry, Anal. Methods 6 (2014) 1941–1950.

[223] M.I. Beser, J. Beltrán, V. Yusà, Design of experiment approach for the optimization of polybrominated diphenyl ethers determination in fine airborne particulate matter by microwave-assisted extraction and gas chromatography coupled to tandem mass spectrometry, J. Chromatogr. A 1323 (2014) 1–10.

[224] J. Robles-Molina, B. Gilbert-López, J.F. García-Reyes, A. Molina-Díaz, Comparative evaluation of liquid-liquid extraction, solid-phase extraction and solid-phase microextraction for the gas chromatography-mass spectrometry determination of multiclass priority organic contaminants in wastewater, Talanta 117 (2013) 382–391.

[225] M. Roszko, K. Szymczyk, R. JEdrzejczak, Separation of polychlorinated dibenzo-p-dioxins/furans, non-ortho/mono/di/tri/tetra-ortho-polychlorinated biphenyls, and polybrominated diphenyl ethers groups of compounds prior to their determination with large volume injection gas chromatography-Quadrupole ion storage tandem mass spectrometry, Anal. Chim. Acta 799 (2013) 88–98.

[226] K. Kalachova, J. Pulkrabova, T. Cajka, L. Drabova, M. Stupak, J. Hajslova, Gas chromatography-triple quadrupole tandem mass spectrometry: a powerful tool for the (ultra)trace analysis of multiclass environmental contaminants in fish and fish feed rapid detection in food and feed, Anal. Bioanal. Chem. 405 (2013) 7803–7815.

[227] J. Robles-Molina, B. Gilbert-López, J.F. García-Reyes, A. Molina-Díaz, Gas chromatography triple quadrupole mass spectrometry method for monitoring multiclass organic pollutants in Spanish sewage treatment plants effluents, Talanta 111 (2013) 196–205.

[228] K. Kalachova, T. Cajka, C. Sandy, J. Hajslova, J. Pulkrabova, High throughput sample preparation in combination with gas chromatography coupled to triple quadrupole tandem mass spectrometry (GC-MS/MS): a smart procedure for (ultra) trace analysis of brominated flame retardants in fish, Talanta 105 (2013) 109–116.

[229] Y. Sapozhnikova, S.J. Lehotay, Multi-class, multi-residue analysis of pesticides, polychlorinated biphenyls, polycyclic aromatic hydrocarbons, polybrominated diphenyl ethers and novel flame retardants in fish using fast, low-pressure gas chromatography-tandem mass spectrometry, Anal. Chim. Acta 758 (2013) 80–92.

[230] S. Walorczyk, D. Drodynski, B. Gnusowski, Multiresidue determination of 160 pesticides in wines employing mixed-mode dispersive-solid phase extraction and gas chromatography-tandem mass spectrometry, Talanta 85 (2011) 1856–1870.

[231] Y.P. Lin, I.N. Pessah, B. Puschner, Simultaneous determination of polybrominated diphenyl ethers and polychlorinated biphenyls by gas chromatography-tandem mass spectrometry in human serum and plasma, Talanta 113 (2013) 41–48.

[232] A. Binelli, C. Roscioli, L. Guzzella, Improvements in the analysis of decarbomodiphenyl ether using on-column injection and electron-capture detection, J. Chromatogr. A 1136 (2006) 243–247.

[233] L. Chen, L. Ding, H. Jin, D. Song, H. Zhang, J. Li, K. Zhang, Y. Wang, H. Zhang, The determination of organochlorine pesticides based on dynamic microwave-assisted extraction coupled with on-line solid-phase extraction of high-performance liquid chromatography, Anal. Chim. Acta 589 (2007) 239–246.

[234] M. Kuster, M.L. de Alda, D. Barcelo, Liquid chromatography-tandem mass spectrometric analysis and regulatory issues of polar pesticides in natural and treated waters, J. Chromatogr. A 1216 (2009) 520–529.

[235] M. Petrovic, M. Farre, M.L. de Alda, S. Perez, C. Postigo, M. Kock, J. Radjenovic, M. Gros, D. Barcelo, Recent trends in the liquid chromatography-mass spectrometry analysis of organic contaminants in environmental samples, J. Chromatogr. A 1217 (2010) 4004–4017.

[236] C. Coscolla, V. Yusa, M.I. Beser, A. Pastor, Multi-residue analysis of 30 currently used pesticides in fine airborne particulate matter (PM 2.5) by microwave-assisted extraction and liquid chromatography-tandem mass spectrometry, J. Chromatogr. A 1216 (2009) 8817–8827.

[237] Z. Sosa-Ferrera, C. Mahugo-Santana, J.J. Santana-Rodriguez, New development in liquid chromatography mass spectrometry for the determination of micropollutants, Chromatogr. Res. Int. 2012 (2012) 748989.

[238] M. Moeder, C. Bauer, P. Popp, M. van Pinxteren, T. Reemtsma, Determination of pesticide residues in wine by membrane-assisted solvent extraction and high-performance liquid chromatography-tandem mass spectrometry, Anal Bioanal. Chem. 403 (2012) 1731–1741.

[239] S. Kittlaus, J. Schimanke, G. Kempe, K. Speer, Development and validation of an efficient automated method for the analysis of 300 pesticides in food using two-dimensional liquid chromatography-tandem mass spectrometry, J. Chromatogr. A 1283 (2013) 98–109.

[240] R.J. Letcher, S. Chu, High-sensitivity method for determination of tetrabromobisphenol-S and tetrabromobisphenol-A derivative flame retardants in great lakes herring gull eggs by liquid chromatography-atmospheric pressure photoionization-tandem mass spectrometry, Environ. Sci. Technol. 44 (2010) 8615–8621.

[241] L.K. Ackerman, G.R. Wilson, S.L. Simonich, Quantitative analysis of 39 polybrominated diphenyl ethers by isotope dilution GC/low resolution MS, Anal. Chem. 77 (2005) 1979–1987.

[242] G. Diletti, G. Scortichini, R. Scarpone, G. Gatti, L. Torreti, G. Migliorati, Isotope dilution determination of polycyclic aromatic hydrocarbons in olive pomace oil by gas chromatography-mass spectrometry, J. Chromatogr. A 1062 (2005) 247–254.

[243] Q. Luo, Z.W. Cai, M.H. Wong, Polybrominated diphenyl ethers in fish and sediment from river polluted by electronic waste, Sci. Total Environ. 383 (2007) 115–127.

[244] Y. Tondeur, J. Hart, Ultratrace extraction of persistent organic pollutants, TrAC 28 (2009) 1137–1147.

[245] M. Alaee, Recommendations for monitoring of polybrominated diphenyl ethers in the Canadian environment, Environ. Monit. Assess. 88 (2003) 327–341.

[246] UNEP Chemicals Branch, Guidance for Analysis of Persistent Organic Pollutants (POPs). www.chem.unep.ch/POPs/laboratory/analytical_guidance_en.pdf (accessed April 2014).

[247] D.A. Armbruster, T. Pry, Limit of blank, limit of detection and limit of quantitation, Clin. Biochem. Rev. 29 (2008) S49–S52.

[248] L.A. Currie, Detection and quantification limits: origins and historical overview, Anal. Chim. Acta 391 (1999) 127–134.

[249] L.C. Mmualefe, N. Torto, Environmental and food sample handling challenges for developing countries, Toxicol. Environ. Chem. 91 (2009) 819–835.

[250] P. De Voogt, Chromatographic clean-up methods for the determination of persistent organic compounds in aqueous environmental samples, TrAC 13 (1994) 389–397.

[251] T. Hyötyläinen, K. Hartonen, Determination of brominated flame retardants in environmental samples, TrAC 21 (2002) 13–29.

[252] A. Covaci, S. Voorspoels, J. de Boer, Determination of brominated flame retardants, with emphasis on polybrominated diphenyl ethers (PBDEs) in environmental and human samples – a review, Environ. Int. 29 (2003) 735–756.

[253] S. Bayen, J.P. Obbard, G.O. Thomas, Chlorinated paraffins: a review of analysis and environmental occurrence, Environ. Int. 32 (2006) 915–929.
[254] M. Partyka, B. Zabiegala, J. Namiesnik, A. Przyjazny, Application of passive samplers in monitoring of organic constituents of air, Crit. Rev. Anal. Chem. 37 (2007) 51–78.
[255] Z. Xie, R. Ebinghaus, Analytical methods for the determination of emerging organic contaminants in the atmosphere, Anal. Chim. Acta 610 (2008) 156–178.
[256] V. Yusa, C. Coscolla, W. Mellouki, A. Pastor, M. de la Guardia, Sampling and analysis of pesticides in ambient air, J. Chromatogr. A 1216 (2009) 2972–2983.
[257] P. Plaza-Bolanos, A.G. Frenich, J.L.M. Vidal, Polycyclic aromatic hydrocarbons in food and beverages. Analytical methods and trends, J. Chromatogr. A 1217 (2010) 6303–6326.
[258] L. Ma, D. Xu, Y. Chen, Z. Chai, Advances in the analysis of short chain chlorinated paraffins, Prog. Chem. 22 (2010) 720–726.
[259] O.S. Fatoki, J.,X. Bhekumusa, B.O.O. Peolu, Polycyclic aromatic hydrocarbons (PAHs) in food and environmental samples: an overview, Fresenius Environ. Bull. 20 (2011) 2012–2020.
[260] J. Sun, J. Liu, Q. Liu, G. Qu, T. Ruan, G. Jiang, Sample preparation method for the speciation of polybrominated diphenyl ethers and their methoxylated and hydroxylated analogues in diverse environmental matrices, Talanta 88 (2012) 669–676.
[261] J.L. Tadeo, C. Sánchez-Brunete, B. Albero, A.I. García-Valcárcel, R.A. Pérez, Analysis of emerging organic contaminants in environmental solid samples, Cent. Eur. J. Chem. 10 (2012) 480–520.
[262] A. Cecinato, C. Balducci, D. Mastroianni, M. Perilli, Sampling and analytical methods for assessing the levels of organic pollutants in the atmosphere: PAH, phthalates and psychotropic substances: a short review, Environ. Sci. Pollut. Res. 19 (2012) 1915–1926.
[263] Z.A. Alothman, M.R. Khan, S.M. Wabaidur, M.R. Siddiqui, Persistent organic pollutants: overview of their extraction and estimation, Sensor Lett. 10 (2012) 698–704.
[264] A. Karrman, G. Lindstram, Trends, analytical methods and precision in the determination of perfluoroalkyl acids in human milk, TrAC 46 (2012) 118–128.
[265] S. Salihovic, H. Nilsson, J. Hagberg, G. Lindstrom, Trends in the analysis of persistent organic pollutants (POPs) in human blood, TrAC 46 (2013) 129–138.
[266] E. Björklund, S. Bøwadt, T. Nilsson, L. Mathiasson, Pressurized fluid extraction of polychlorinated biphenyls in solid environmental samples, J. Chromatogr. A 836 (1999) 285–293.
[267] E. Bjorklund, T. Nilsson, S. Bwadt, Pressurised liquid extraction of persistent organic pollutants in environmental analysis, TrAC 19 (2000) 434–445.
[268] A. Nieto, F. Borrull, R.M. Marcé, E. Pocurull, Pressurized liquid extraction of contaminants from environmental samples, Curr. Anal. Chem. 4 (2008) 157–167.
[269] H. Sun, X. Ge, Y. Lv, A. Wang, Application of accelerated solvent extraction in the analysis of organic contaminants, bioactive and nutritional compounds in food and feed, J. Chromatogr. A 1237 (2012) 1–23.
[270] G. Anitescu, L.L. Tavlarides, Supercritical extraction of contaminants from soils and sediments, J. Supercrit. Fluids 38 (2006) 167–180.
[271] M. Farré, L. Kantiani, M. Petrovic, S. Pérez, D. Barceló, Achievements and future trends in the analysis of emerging organic contaminants in environmental samples by mass spectrometry and bioanalytical techniques, J. Chromatogr. A 1259 (2012) 86–99.
[272] M. Palmiotto, A. Colombo, E. Davoli, A GC/MS-MS versus GC/HRMS dioxin analysis comparison: some critical considerations for low-level environmental samples, Compr. Anal. Chem. 61 (2013) 455–469.

Chapter 2

Bioanalytical Approaches to Understanding Toxicological Implications of Mixtures of Persistent Organic Pollutants in Marine Wildlife

Ling Jin,[1] Caroline Gaus[1] and Beate I. Escher[1,2,3,*]

[1]The University of Queensland, National Research Centre for Environmental Toxicology (Entox), Brisbane, QLD, Australia; [2]UFZ - Helmholtz Centre for Environmental Research, Cell Toxicology, Leipzig, Germany; [3]Eberhard Karls University Tübingen, Environmental Toxicology, Center for Applied Geosciences, Tübingen, Germany

**Corresponding author: E-mail: b.escher@uq.edu.au*

Chapter Outline

Comprehensive Analytical Chemistry, Vol. 67. http://dx.doi.org/10.1016/B978-0-444-63299-9.00002-8

1. MARINE WILDLIFE HEALTH AT RISK

1.1 Environmental Threats

Many coastal environments around the globe have been undergoing substantial modification due to rapid urbanization and industrial development. Given the increasingly shared nature of the coastal zone, human activities along with natural factors inevitably exert pressures on marine wildlife. Extensive records of anthropogenic and natural threats leading to wildlife mortality have been documented in various parts of the world. Direct threats from human activities include hunting, by-catch in fisheries, boat strikes, habitat loss, and pollution. In addition, naturally occurring pressures, such as severe climatic events, infectious diseases, and macroparasites may in some cases be exacerbated by anthropogenic factors [1,2].

Over recent decades, marine wildlife populations, in particular megafauna species, have severely declined worldwide [3]. In Eastern Australian waters, for example, census surveys indicate that loggerhead turtle populations have declined by up to 86% since the 1970s [4]. Similarly, the numbers of dugongs worldwide appear to have declined by at least 20% over the last 90 years (approximately three dugong generations) [5]. Today, these and numerous other marine wildlife species are listed under the Red List of Threatened Species [6] due to their critical population status, ranging from endangered to vulnerable to extinct, and are of high conservation priority.

In the Northern hemisphere, the Deepwater Horizon oil spill in 2010 was highlighted as the largest accidental marine oil spill in the history of the petroleum industry. The spill of approximately 800 million liters of crude oil contaminated vast areas of the Gulf of Mexico, including deep-ocean communities and more than 1600 km of shoreline. A multitude of organisms were affected, including fish, mollusks, crustaceans, marine turtles and mammals, as well as seabirds. Over 20 million hectares of the Gulf of Mexico were closed to fishing [7]. While a massive response ensued to protect the coastal life-supporting ecosystem, dolphins, sea turtles, and other marine life continued to die in record numbers. Symptoms consistent with petroleum exposure were observed suggesting that mortalities were likely related to wildlife oiling during the spill [7,8].

In the Southern hemisphere, a mass mortality event of marine wildlife occurred around a major port along the East Australian coast which has been undergoing intensive development of its port resources since May 2011 [9]. Higher than usual mortality rates of green turtles and dugongs have been recorded since early 2011. One hundred and sixty-three stranded green turtles had been reported by November 2011, in contrast to an average of 32 green turtles per year for the last 5 years [10]. There was also an increase in the number of stranded dugongs, as well as outbreaks of diseases in fish in this region [11]. Suspected triggering factors were the loss of seagrass beds subsequent to increased freshwater and sediment outflow from extensive flooding, as well as increased industrial activity. These events highlight that the exact cause(s) of mortality are difficult to determine, but exemplify that marine wildlife subpopulations can be impacted by expanding human activities, environmental degradation, and climatic factors.

1.2 Increasing Morbidity and Mortality

Among the multiple threats, disease has been identified as an important contributing factor to the decline in marine wildlife population. The incidence of chronic disease, which can lead to death, has been reported to be increasing in marine wildlife populations, in some cases reaching panzootic proportions [12].

High prevalence of parasitic infection in sea turtles has been observed worldwide such as in Costa Rica [13], Florida [14], Taiwan [15], and Australia [16]. A health surveillance [16] of green turtles stranded in Eastern Australia over 2006–2009 revealed that parasitic and generalized infections were among the most frequently occurring causes of death. Particularly, spirorchiidiasis, a disease caused by spirorchid fluke infection, was the predominant diagnosis affecting neurological, gastrointestinal and, in some cases, multisystemic organs. In a recent health assessment [17] of green turtles from Gladstone, North East Australia, spirorchiid flukes were also found to be the most commonly identified infectious agent on complete necropsy. Other diagnosed infectious diseases included fibropapillomatosis (FP) and bacterial gastroenteritis.

FP is another common anecdotal observation of disease in marine turtles worldwide [18]. It is a severely debilitating disease externally and internally characterized by epizootic tumors of soft tissues. Most tumors appear to be benign and can be tolerated for years. However, larger tumors can severely hamper the mechanical ability of turtles to swim, ingest food, and avoid predation [19]. Additionally, internal tumors may interfere with systemic function, and may lead to death [20].

Flint et al. [21] conducted necropsies on two sirenian species, dugong (*Dugong dugon*) from Australia and Florida manatee (*Trichechus manatus latirostris*) stranded between 1997 and 2010. Though geographically separate, the two populations of sirenia displayed similar causes of mortality. Except for a small proportion of mortalities that could be directly attributed to human activities such as boat strikes and entanglement, higher natural-related mortality was not fully resolved. Possible causes include perennial microbiological, parasitic infections, and seasonal cold stress, and a significant number of the mortality cases were left undiagnosed.

Despite the fact that diseases are increasingly known to play a significant role in the population health of wildlife species, their etiology or the underlying causes of rising incidences are poorly understood. The rise in zoonotic diseases is driven by a complex interplay of environmental (global warming, ocean acidification, pollution), ecological (habitat destruction or fragmentation), and epidemiologic (increasing human densities encroaching on decreasing wildlife populations, global movements of plants and animals) factors [2,22]. For example, the cause of FP has not yet been conclusively ascertained and is generally accepted to have a multifactorial etiology. Viral infection due to decreased immune response, possibly via exposure to pollutants, is commonly suggested to contribute in the etiology of this disease [19,23]). Studies from Florida and Hawaii established a link between FP and compromised immune function in green turtles [24,25]. Notably, the higher prevalence of fibropapilloma is observed in more environmentally challenged areas near agricultural and/or industrial lands [19,26], where a variety of chemicals can be released. A recent study by Keller et al. [27] showed that none of the targeted 164 persistent organic pollutants (POPs) increased in concentration with increasing prevalence of FP in Hawaiian green sea turtles affected by FP, implying that POPs are not a major cofactor for the onset of FP. However, the highest concentrations of POPs were found in a group of stranded, severely tumored, emaciated turtles, suggesting possible contribution of these chemicals following mobilization from storage lipids into blood to the progression of the disease during late-stage weight loss.

2. ROLE OF POPs IN MARINE WILDLIFE HEALTH

2.1 POP Exposure: Status and Trend

POPs encompass a group of substances that are not easily degraded, accumulate in organisms, and exhibit a range of acute or chronic toxicity (http://www.chem.unep.ch/pops/). A large proportion of POPs are halogenated organic chemicals—such as, polychlorinated dibenzo-*p*-dioxins/dibenzo furans (PCDD/Fs), organochlorine pesticides (OCPs), polychlorinated biphenyls (PCBs), and polychlorinated naphthalenes. The list is expanding with the recognition of emerging POPs of concern such as brominated flame retardants (e.g., polybrominated diphenyl ethers, PBDEs and hexabromocyclododecanes, HBCDs), perfluorinated compounds (PFCs), and chlorinated paraffins.

The distribution and levels of POPs in the marine environment are governed by source releases and subsequent (predominantly particle-bound) transport processes. POPs often originate from land-based agricultural and industrial activities via wastewater, river, and/or surface runoff. Sediment represents a secondary source of POPs to the low trophic marine wildlife and many POPs have the potential to biomagnify in organisms feeding on higher trophic levels. For example, in the nearshore foraging habitats of marine turtles and dugongs in Eastern Australia, highest concentrations were shown to deposit nearshore, leading to elevated concentrations in the sediment–seagrass–turtle/dugong system. The trends in exposure levels in these seagrass consumers were significantly correlated with the concentrations in the sediment and seagrass from their foraging habitats [28,29].

In general, while there has been a global decline of OCPs and PCBs in marine wildlife in accordance with the international efforts to control their emissions [30], no continuous decreasing trends of PBDEs (reviewed in [31]) and dioxins [30] have been observed despite the regulatory measures, and increasing trends in emerging POPs, such as alternative flame retardants, have been found in marine wildlife worldwide. Remarkable increases in HBCD concentrations have been detected globally over the last decades and such trends may presumably continue as no regulations are currently enforced on HBCDs [32]. In addition, an increasing shift over time from high brominated PBDE congeners to alternate halogenated flame retardants, such as dechlorane plus, *bis*(2,4,6-tribromophenoxy)ethane, and decabromodiphenylethane, were recently, and for the first time, observed in marine cetacean samples [33]. These trends reflect the replacement of PBDEs by these alternatives, which will increase in usage following the phasing out of PBDE commercial mixtures.

Contrary to general trends at the global scale, regional exposure levels in marine wildlife display regional-specific spatial characteristics due to different stages of chemical usage patterns and/or control measures. In the Northern hemisphere, PCBs generally contributed the major proportion to the overall toxic equivalent concentrations (TEQ) of dioxins and dioxin-like chemicals in marine biota [34,35], which is in contrast to the relatively minor contribution of PCB concentrations analyzed from some turtles [36] and dugongs [34] from the Eastern Australia coast. PCDD/F concentrations were particularly elevated in these animals, which were comparable to or higher than those reported in a range of higher trophic marine mammals near relatively polluted areas such as the Mediterranean, Baltic, and Japan [35].

Of the limited available information on PBDE levels in Australian marine biota, PBDE levels measured in green turtles and dugongs and several carnivorous turtles from the Southeast Queensland coast are at the low end of the concentration ranges in marine biota worldwide [37,38], indicating low level input into the marine system of Queensland. In accord with biota results, low PBDE levels have also been measured in Queensland sediments [39]. These findings are in general agreement with global estimates which suggest low PBDE usage in Australia. A previous large scale study, however, has found relatively high PBDE levels in Australian human milk, exceeding those in a European cohort [40]. This discrepancy in contamination trends between terrestrial and marine biota suggests that the transport of PBDEs into the Australian marine system may occur in future.

Similar to PBDEs, OCP concentrations in marine wildlife species from Queensland have been reported to be relatively low compared to those observed in marine biota elsewhere [34]. This is surprising considering the long-term and ubiquitous use of these POPs in the tropical agriculture of North Queensland. However, to date, only a limited number of species samples have been analyzed for OCPs, and the historical distribution and use of these chemicals in Australia is not clearly known.

2.2 Toxicological Implications of POPs

POPs pose serious concerns for human and wildlife health. As demonstrated in numerous laboratory animal-based studies, the effects elicited by these chemicals encompass cancer [41], endocrine disruption [42], reproductive dysfunction [43], behavioral abnormalities [44], disturbance of the immune system [45], damage to the liver [46], and nervous system [47].

While the body burdens of legacy POPs in marine wildlife have been declining substantially on the global scale as a result of worldwide regulation/mitigation measures, their levels remain toxicologically relevant to marine wildlife species in specific parts of the world. It was estimated that 20% of the dugong population [35] and 31% of the green turtle population [36] from the Australian east coast are exposed to dioxin-like compounds that may result in compromised immune function and developmental effects in mammals and birds, respectively [48].

Field evidence also implies the impairment of normal body function, such as immunity and endocrine function by a range of POPs as contributing risk factors. In loggerhead turtles from North Carolina, Keller et al. [49] found suppressed lysozyme activity associated with blood concentrations of 4,4′-dichlorodiphenyldichloroethylene (*p,p′*-DDE) and the sum of chlordanes, and enhanced lymphocyte proliferation responses associated with sum concentrations of PCBs. Hall et al. [50] reported that the elevated levels of circulating thyroid hormones in gray seals (*Halichoerus grypus*) from the UK were associated with their blubber concentrations of PBDEs.

2.3 Need for Mechanistic and Quantitative Understanding of Exposure and Effect

The ultimate goal in marine wildlife toxicology is to establish the quantitative role of chemical pollution in marine wildlife, that is, its contribution among all other environmental threats. To achieve this goal, we need tools to evaluate the combined effect of the complex cocktails of POPs accumulated in marine wildlife. For effective evaluation, we need better mechanistic understanding of chemicals' impact on wildlife health. With respect to exposure assessment, we need more quantitative assessment of chemicals contributing to the overall toxicologically relevant body burdens. This in-depth understanding will inform various stakeholders concerning conservation strategies for marine wildlife.

While our knowledge of POP exposure and effect in marine wildlife is expanding [51], traditionally, exposure and effect assessment is segregated, and data generated are often phenomenological. An integrated approach toward comprehensive understanding of the issue is hindered by our current methodological limitations in the quantitative extraction of whole mixtures of POPs from wildlife tissue/fluid, and mechanistic and quantitative screening of the combined effect of these complex mixtures. In the following sections, we will review the currently used approaches to evaluating mixture effects of POPs in marine wildlife.

3. CURRENT APPROACHES TO EVALUATING MIXTURE EFFECTS OF POPs IN MARINE WILDLIFE

3.1 Exposure-Driven Approaches

Exposure-driven approaches usually start with measuring concentrations of typical groups of POPs that are expected to be present in wildlife tissue and with the known potential to cause adverse health effects. Once the concentrations of selected chemical groups are measured, there are three major approaches (Figure 1) that can direct to the toxicological implications of these chemicals: (1) simple risk assessment by comparing the detected levels with critical body burden; (2) statistical correlation with (impaired) functional parameters to infer causal relationship; and (3) toxicity testing with an environmentally relevant chemical mixture.

3.1.1 Targeted Chemical Analysis and Risk Characterization

A simple risk assessment involves determining the concentrations of target POPs in tissue/fluid of a wildlife species and deriving the safe body burden from laboratory animal experiments for the species of interest. By comparing the measured body burden of the chemical with its safe level, we can calculate

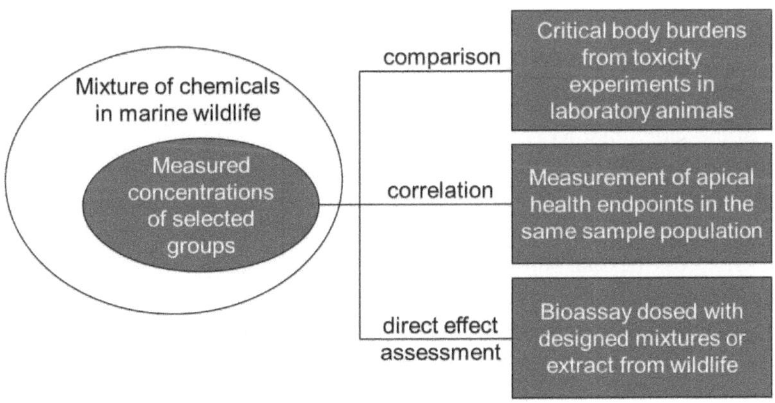

FIGURE 1 Major exposure-driven approaches.

the risk quotient (RQ) as an estimate of the extent to which the body burden of POPs in the species exceed the safe level.

$$RQ = \frac{\text{Measured body burden}}{\text{Acceptable body burden}} \tag{1}$$

If the RQ for a certain chemical exceeds unity, there is a high likelihood that the chemical of concern poses a potential risk and appropriate control or management may be necessary. In contrast, an RQ less than unity means low or negligible risk to the target organisms due to that pollutant.

As it is ethically prohibitive and practically limited to test the effects of POPs directly on marine wildlife species, particularly protected megafauna, much of the knowledge about the potential adverse effects that POPs may elicit in marine wildlife is based on extrapolation from laboratory animal data (such as fish, birds, and rats). Interspecies extrapolation from well-established dose–response relationships in laboratory animals for which critical concentrations have been derived (e.g., no-observable-adverse-effect-level, NOAEL, lowest-observable-adverse-effect-level, LOAEL), represents a useful and simple approach for initial risk assessment of POPs in marine wildlife. In this approach, LOAELs/NOAELs of POPs in test animals are converted to critical/safe body burdens, and a set of uncertainty factors are applied. Using this, combined with probabilistic tools, we can further estimate the percentage of a species population that may be at risk due to chemical exposure (examples are given in Section 2.2). However, sensitivity in responding to chemicals may vary between wildlife species and well-studied laboratory animals. The lack of species-specific LOAEL/NOAEL values may lead to under- or overestimation of the risk posed by these chemicals; the uncertainty of such interspecies variability is commonly considered to lie within approximately one order of magnitude [48].

This approach was initially developed for single chemical safety evaluation. Its adaptation to mixtures of chemicals is not straightforward and depends on the mode of action (MOA), i.e., the molecular mechanism and toxic pathway by which the chemical acts [52]. If each component of a given mixture elicits its effect via a common MOA, a possible approach is to assess the risk of the mixture based on the classical mixture toxicity concept of concentration addition (CA); that is, the risk quotient of a given mixture can be estimated as the sum of the individual RQ of each mixture component [53].

This approach is not often applied in the practical risk assessment of environmental mixtures due to the requirement of detailed data and evaluation for each single substance in a mixture. Often, a total sum of mass-based concentrations is used to express the total burden of a chemical group (e.g., total PBDEs, total DDTs, total chlordanes). While mass-based expression is useful in studies of mass loading and fluxes, it has little meaning for toxicology. Even in a simple mixture of total DDTs, for instance, there are DDT and two major metabolites DDD and DDE, each with a different toxic potency on a defined assessment endpoint. Depending on the composition profile of total DDTs, there can be different degrees of toxicity.

A viable option is to express a complex mixture of chemicals as an equivalent concentration against a reference chemical, the so-called TEQ. The premise is that the chemicals included share a common MOA and thus act concentration additive. The TEQ concept was initially developed for dioxin-like chemicals but has been expanded to many more chemical groups and MOAs. The TEQ approach is based on the fact that PCDD/Fs and dioxin-like PCBs exist in the environment as complex mixtures in which each individual dioxin- and dioxin-like congener has a different potency on the same MOA, that is, via binding to arylhydrocarbon receptor (AhR). By utilizing toxic equivalency factors (TEFs), the toxicity of a mixture is expressed as a single TEQ value. The TEF is the toxic potency of each congener relative to 2,3,7,8-TCDD, which is the most potent congener with a TEF value of 1. The TEQ of a mixture in a biological sample, for instance, can be calculated as the sum of the products of the lipid-normalized concentrations of individual congeners (C_i) in unit of picogram per gram of lipid $\left(\mathrm{pg}\ \mathrm{g}_{\mathrm{lip}}^{-1}\right)$ and the corresponding TEF value (TEF_i) [54].

$$\mathrm{TEQ}\left(\mathrm{pg}\,\mathrm{g}_{\mathrm{lip}}^{-1}\right)=\sum_{i=1}^{n}\left[\mathrm{C}_{\mathrm{i}}\left(\mathrm{pg}\,\mathrm{g}_{\mathrm{lip}}^{-1}\right)\cdot \mathrm{TEF}_{\mathrm{i}}\right] \tag{2}$$

The TEF_i value for each dioxin congener is not derived from a single biological endpoint but is a consensus value that integrates information from a wide range of literature data sources. In regulatory practice, the TEQ approach is currently limited to the risk evaluation of dioxin-like chemicals but for research there is a wider application range.

The concept of bioanalytical equivalent concentrations (BEQ) [55–59], an analog to TEQ, is emerging as an approach to express the combined effect of chemical mixtures in various environmental matrices that are acting according to the same MOA (not necessarily directly linked to toxicity) in cell-based bioassays (discussed in detail in Section 3.2.2).

3.1.2 Correlative Studies Associating Phenotypic Responses with POP Burdens

Measuring health relevant functional parameters in the same species population where target analysis of POP groups is conducted represents a step further from extrapolating toxicity measures in laboratory animals. There is a growing body of literature that explores the correlation/association between tissue burdens of dominant POP groups and compromised functional endpoints (Table 1). The quantification of correlation/association is usually based on regression models, which return a correlation coefficient. If the coefficient is greater (less) than zero, there is a positive (negative) correlation, subject to a statistical test to indicate whether the correlation is significant. The studies summarized in Table 1 mainly focused on legacy POPs such as OCPs, PCBs, and PBDEs, linking their body burden in wildlife species with clinical health parameters, endocrine function, and immunity.

TABLE 1 Summary of Correlative Studies Associating Functional Health Parameters With Body Burdens of Persistent Organic Pollutants (POPs) in Marine Wildlife

Species	Target POPs	Functional Parameters[a]	Type of Correlation[b]	References
Harbor seal	PCBs	Lymphocyte proliferation	↑	[60]
	PCBs	Thyroid hormone receptor gene expression	↑	[61]
	PCBs	TT4	↓	[61]
Gray seal	PBDEs	Circulating thyroid hormone	↑	[50]
	PCBs, DDTs	TT3, FT3	↓	[62]
Hooded seal	HCHs, PCBs, DDTs, CHLs	TT3:FT3 ratio	↑	[63]
California sea lion	PCBs	Serum vitamin A and T3	↓	[64]
Beluga whale	PCBs	AhR and cyp1A1 expression	↑	[65]
Bottlenose dolphin	PCBs	TT4, FT4, T3	↓	[66]
	PCBs	Lymphocyte proliferation	↓	[66]
	PCBs	Indices of innate immunity	↓	[66]
	OH-PCBs	cyp1A1 expression	↑	[67]
Common dolphin (male)	PCBs, DDTs	Blubber retinoids	↑	[68]
Common dolphin (female)	PCBs, DDTs	Blubber retinoids	↓	[68]
Polar bear (male)	DDTs, CHLs, HCB	Baculum bone mineral density	↓	[69]

Polar bear (female)	PCBs, DDTs, CHLs	Ovary length	↓	[69]
	PBDEs	Ovary weight	↓	[69]
	HCB	Uterine horn length	↓	[69]
Polar bear	PCBs	Plasma cortisol	↑	[70]
	OCPs	Plasma cortisol	↓	[70]
Greenland shark	OCPs, PCBs	Plasma retinoids	↓	[71]
Green sea turtle	*p,p′*-DDE	Hematocrit	↑	[72]
	γ-HCH	Total blood proteins	↓	[72]
Loggerhead sea turtle	*p,p′*-DDE	Lymphocyte proliferation	↑	[49]
	PCBs	Lymphocyte proliferation	↑	[49]
	CHLs	Red blood cell count, hemoglobin, hematocrit	↓	[73]
	Mirex, dl-PCBs	Heterophils:lymphocyte ratio	↑	[73]
	Mirex, dl-PCBs	Alkaline phosphatase activity	↓	[73]
	PCBs, DDTs, CHLs, mirex	Aspartate aminotransferase activity	↑	[73]
	PCBs, DDTs, CHLs, Dieldrin, mirex	White blood cell count	↑	[73]
Wild fish (shiner surfperch; Pacific staghorn sculpin)	PCBs	T4	↓	[74]
	PCBs	T3, T3/T4 ratio	↑	[74]

AhR: arylhydrocarbon receptor; CHLs: chlordanes; *p,p′*-DDE: 4,4′-dichlorodiphenyldichloroethylene; DDTs: dichlorodiphenyltrichloroethanes; HCHs: hexachlorocyclohexanes; OCPs: organochlorine pesticides; PCBs: polychlorinated biphenyls; PBDEs: polybrominated diphenyl ethers.
[a]*TT3: total triiodothyrine; FT3: free triiodothyrine; TT4: total thyroxine; FT4: free thyroxine; T3: triiodothyrine; T4: thyroxine; cyp1A1: cytochrome P450A1.*
[b]*↑: positive correlation; ↓: negative correlation (these results were reported in the respective articles).*

Such approaches have the advantage of providing more direct, species-specific evidence that certain POPs are likely linked to the adverse health effect, thus facilitating further causal investigations, should there be a significant statistical correlation between functional health parameters and POP concentrations. Lack of or weak correlation does not necessarily mean toxicological irrelevance of these chemicals, as health parameters can be influenced by intrinsic biological traits and environmental variables such as temperature and food availability. The recovery or exacerbation of health impairment by these factors may confound the causal relationships. Additionally, with the focus on particular chemical groups, others are excluded that are (or may be) present with a minor (or unknown) contribution to the mixture composition, but high toxic potency. This can be effective in event-driven investigations where marine wildlife species are expected to be contaminated by typical known chemicals arising from incidental high volume release or spills [7]. In many cases, however, the habitat of marine wildlife receives an unknown cocktail of chemicals. Mixtures of individual chemical components, even at nondetectable levels, can add up to overall toxic effects if combined effects of chemicals in mixture are taken into account. In this sense, the statistical correlation approach may not be able to reveal the toxicological relevance of these mixtures and individual chemicals therein. Recent advances in nontarget analytical approaches provide tools that may overcome the problem of traditional target analyses (detailed in Section 3.2.4).

3.1.3 Designed Reference Mixture Studies

The above approaches do not inform directly on the toxicological consequences of the measured levels of POPs in wildlife tissue. Therefore, an alternative strategy is to mix the analyzed chemicals at the relevant ratios present in the wildlife and dose the mixture into a bioassay to test its toxic potential. Such an approach has rarely been systematically applied for wildlife toxicology studies. One study was conducted by Keller et al. [49] where Aroclor 1254 (a commercial mixture of PCB congeners) was found to increase in vitro lymphocyte proliferation responses at concentrations similar to those measured in loggerhead sea turtles. Yordy et al. [75] investigated the interactions of four prevalent POPs (*p,p'*-DDE, *trans*-nonachlor, and PCB congeners 138 and 180) that had been detected in cetacean blubber with the estrogen receptor (ER)-specific E-Screen assay. Estrogenic/antiestrogenic activity was determined for the individual chemicals and their binary, tertiary, and quaternary combinations. While the concentrations of these chemicals in dolphin blood were lower than their respective lowest observable effect concentration derived from the E-Screen assay, estrogenic/antiestrogenic activity was observed in contaminant mixtures extracted directly from the blubber of bottlenose dolphins (*Tursiops truncatus*).

These mixture toxicity studies have focused on a few priority components that are likely to result in biological effect. The environmental realism of chemical mixtures, which usually contain compounds with different MOAs at arbitrary concentrations, is often neglected. In addition, the lack of quantitative information regarding how components interact in the assay provides limited predictive value for the real mixtures in the environment.

Over the last decade, there has been a major breakthrough in the development of mixture toxicity concept, enabling mixture effects to be more explicitly described in a quantitative manner [76]. There are typically four classes of mixture effects according to the chemicals' interaction, target site, and MOA. Independent action (IA) and CA/dose addition (DA) are the two most common mechanisms by which mixtures of chemicals elicit their combined effect. When chemicals act on different modes of toxic action or at different target sites, the IA model applies. The expected mixture effect ($effect_{mixture}$) can be calculated according to the joint probability of the effect elicited by each component ($effect_i$) as statistically independent events. The effect is in the scale of 0–1 (0–100%).

$$effect_{mixture} = 1 - \prod_{i=1}^{n} \left(1 - effect_i\right) \tag{3}$$

When chemicals act at the same target site or the same MOA, the CA or DA model applies. In such cases, the fractions of individual component concentrations eliciting the same effect level, which are also termed toxic units, sum up to an overall toxic unit of the mixture, assuming that one component can be replaced totally or in part by an equal toxic unit of another, without changing the overall combined effect:

$$EC_{mixture} = \frac{1}{\sum_{i=1}^{n} \frac{p_i}{EC_i}} \tag{4}$$

where effective concentration (EC) is the concentration of a chemical or a mixture at a certain effect level and p_i is the relative fraction of a component in a given mixture.

For a few well-studied MOAs, such as specific receptor-mediated endpoints, the mixture concept has been well established. Typically, for these MOAs, the concentration-addition model provides good approximations of the combined effect [77–81]. Beyond receptor-specific MOA, the CA model has also demonstrated its applicability to other MOAs (e.g., baseline toxicity [82] and transcriptional factor Nrf2-mediated oxidative stress response [57]).

These mechanistically based quantitative models provide reliable effect predictions of defined mixtures, which in return add tools to dissect complex environmental mixtures of POPs in biological tissue.

3.2 Effect-Driven Approaches

While exposure-driven approaches target select chemical groups, effect-driven approaches aim at quantifying the overall toxic potential of a biological sample extract containing the entire range of chemicals (here we define this whole sample assessment as effect analysis). Further attempts are often made to identify the causative chemicals and quantify the contribution of these identified components to the total toxicologically relevant burden by the virtue of fractionation and instrumental techniques. The overall framework of such effect-driven approaches is termed effect-directed analysis [83]. The nontarget nature of the approach includes parent POPs, their known or unknown transformation products, and other unidentified compounds that are accumulated in wildlife species. This approach also provides a guided way to quantify the exposure levels of individual chemicals.

3.2.1 Toxicity Pathway and MOA-Based Cell Bioassays

The development of the concepts of adverse outcome pathway (AOP) [84] and toxicity pathway [85] has led to a paradigm shift in chemical risk assessment from an over-reliance on in vivo testing to a greater use of in vitro, molecular, and computational tools. An AOP is a conceptual construct that describes a sequential series of causally linked events spanning multiple levels of biological organization that lead from a direct molecular initiating event (MIE) to an ultimate adverse health or ecotoxicological effect (Figure 2) [84]. Implicit in the definition of AOP, a toxicity pathway is defined as a "cellular response pathway that, when sufficiently perturbed, is expected to result in adverse health effect" [85]. While AOP targets a holistic picture at different levels of biological organization, a toxicity pathway focuses almost exclusively on MIEs and proximal cellular responses that can be measured and modeled in vitro. The use of cell-based or animal alternatives replaces conventional approaches involving high-dose in vivo phenotypic responses with the in vitro assessment of cellular signaling pathways that can be perturbed by chemicals [86]. A pathway-based vision has also led to a blurred division between human toxicology and ecotoxicology [87]. These developments open an important alternative testing strategy in wildlife toxicology as it is ethically prohibitive and logistically impractical

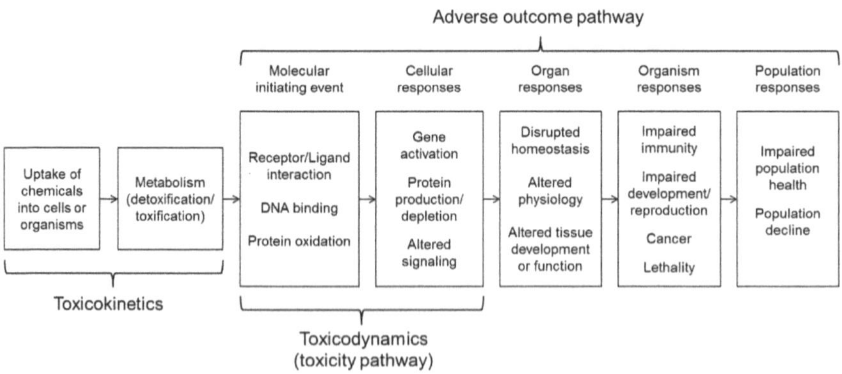

FIGURE 2 A conceptual framework describing the major causally linked events in an adverse outcome pathway from lower tiered molecular responses to higher level population responses. *Adapted with permission to reprint from Ref. [84].*

to expose protected wildlife species. In addition, the current toxicity pathway-based approach has moved beyond testing individual chemicals and demonstrated its applicability to screening environmental samples (typically water [88] and sediment [59,89]) containing mixtures of chemicals. These advances can be transferred to biological samples by designing a test battery of bioassays relevant to wildlife toxicology to assess the toxic potential of mixtures of POPs accumulated in wildlife tissue. A typical panel of assays would desirably encompass endpoints covering major events along the toxicity pathway, including the induction of xenobiotic metabolism and adaptive stress responses as early steps and functional impairment, such as immunosuppression, as later steps. Thus, the manifestation of effect can be related to early toxicokinetic indicators and defense mechanisms as biomarkers of chemical exposure.

In the early steps of a toxicity pathway, chemicals interact with major macro-biomolecules, triggering a signaling cascade down the pathway. Therefore, many bioassays are designed to target these MIEs as typical early warning indicators. Currently, effect-directed analysis for marine wildlife toxicology studies usually focus on receptor-mediated endpoints (Table 2), such as AhR and various hormone receptors, as these receptors have the specificity to a limited number of compounds, which can explain a large proportion of the observed effect in the bioassays. Beyond receptor-mediated endpoints, there are nonspecific MOAs yet to be explored as an early defense mechanism (e.g., cellular stress response pathways [99]) to link mixture chemical exposure to effect indicators.

3.2.2 Principles of Effect Analysis

The quantification of overall mixture effect and components' contribution is achieved by a series of algorithms, forming the major steps of effect-directed analysis (Figure 3). Firstly, extracts from wildlife tissue/fluid samples are prepared and screened in a bioassay indicative of a particular MOA. The total biologically relevant chemical burden is expressed as BEQ_{bio} against a reference compound acting on the biological endpoint. The BEQ_{bio} is defined as the ratio of the EC of the reference compound and that of the sample extract. EC is derived from dose–response modeling via endpoint measurement corresponding to a dilution series of the reference compound or sample extracts. In contrast to a reference compound, the EC of a sample extract containing unknown/unresolved cocktail of chemical mixtures cannot be expressed in commonly defined concentration unit for a single chemical (compound mass per volume of the bioassay). Instead, the EC of a sample extract can be expressed as the relative enrichment or dilution of an original sample transferred into a bioassay (e.g., mass of lipid for fatty tissue per volume of the bioassay). With Eqn (5), the BEQ_{bio} is derived for the sample containing total bioactive chemicals at a lipid-normalized concentration equivalent to the reference compound.

$$BEQ_{bio}\left(ng\ g_{lip}^{-1}\right) = \frac{EC_{reference\ compound}\left(ng\ L_{bioassay}^{-1}\right)}{EC_{sample}\left(g_{lip}\ L_{bioassay}^{-1}\right)} \tag{5}$$

TABLE 2 Summary of Mode of Action (MOA)-Based Bioassays Used in Effect Analysis in the Context of Marine Wildlife Toxicology

Assay Name[a]	Target MOA	References
DR-CALUX	AhR-mediated (ant)agonism	[90,91]
AhR-CAFLUX	AhR activation	[58]
AhR-CALUX	AhR activation	[92]
AR-CALUX	AR-mediated (ant)agonism	[90,93]
ER-CALUX	ER-mediated (ant)agonism	[90,92,94]
E-Screen	ER-mediated agonism	[75,91]
GR-CALUX	GR-mediated (ant)agonism	[90]
PPARγ2-CALUX	PPARγ2-mediated (ant)agonism	[90]
Radiolabeled T4-TTR binding assay	Competitive binding to TTR	[95–97]
H4IIE EROD activity assay	Induction of EROD activity	[98]
Rainbow trout hepatic ER binding assay	ER-mediated agonism	[98]
Goldfish testicular AR binding assay	AR-mediated agonism	[98]
Goldfish SSBP binding assay	Binding to SSBP	[98]

[a]*CALUX: chemically activated luciferase expression; CAFLUX: chemically activated fluorescent expression; AR: androgen receptor; ER: estrogen receptor; GR: glucocorticoid receptor; PPARγ2: peroxisome proliferator-activated receptor gamma isoform two; TTR: transthyretin; EROD: ethoyresorufin-O-deethylase; SSBP: sex steroid binding protein.*

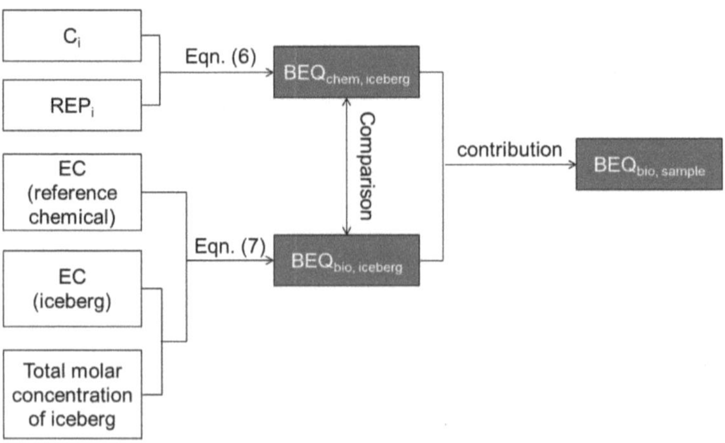

FIGURE 3 The schematic of major steps in effect analysis.

Instrumental analysis then identifies and quantifies possible individual chemicals that are positive for the endpoint. With the available analytical data, the total chemical-based BEQ (BEQ_{chem}) can be calculated as the summation of the instrumentally determined concentration of a chemical (C_i) multiplied by its relative effect potency (REP). The REP is an analog to the TEF in the TEQ approach. In contrast to the TEF, which is a consensus value, the REP is defined as the REP of the chemical that would elicit the same activity as the reference compound, which is derived through profiling individual chemicals in the same bioassay.

$$BEQ_{chem}\left(ng\, g_{lip}^{-1}\right) = \sum_{i=1}^{n}\left[C_i\left(ng\, g_{lip}^{-1}\right)\cdot REP_i\right] \quad (6)$$

$$REP_i = \frac{EC_{reference\ compound}\left(ng\, L_{bioassay}^{-1}\right)}{EC_i\left(ng\, L_{bioassay}^{-1}\right)} \quad (7)$$

For samples with unknown composition of chemicals, the overall effect burden of chemical mixtures present in the biological tissue can be quantitatively estimated from screening in in vitro bioassays. The difference between BEQ_{chem} and BEQ_{bio} is then a measure of the amount of unknown chemicals positive for the endpoint are in the sample in addition to the target chemicals quantified with chemical analysis.

As reviewed in Section 3.1.3, the CA model generally provides robust predictions of mixture effects for a number of well-established specific and nonspecific MOAs, meaning that additivity applies well to BEQ_{chem}. BEQ_{chem} of identified chemicals from a complex mixture approximates their BEQ_{bio}, thus allowing the quantitative attribution of mixture effect to component fractions. To ascertain this for less explored MOAs, a designed mixture toxicity experiment needs to be conducted. This involves profiling individual identified chemicals in a certain bioassay to derive their respective EC and REP values, and mixing the identified chemicals at equimolar, equipotent, or environmentally realistic ratios for dose–response curves, to compare with modeled CA/IA curves. The total BEQ_{chem} of the identified chemicals ($BEQ_{chem,\ iceberg}$) is summed from individuals according to Eqn (6). The BEQ_{bio} of the identified chemicals ($BEQ_{bio,\ iceberg}$) is derived from dose–response curves for the identified chemicals mixed at ratios in a given sample:

$$BEQ_{bio,iceberg}\left(ng\, g_{lip}^{-1}\right) = \frac{\dfrac{EC_{reference\ compound}\left(ng\, L_{bioassay}^{-1}\right)}{EC_{iceberg}\left(mol\, L_{bioassay}^{-1}\right)}}{\text{Total concentration of iceberg in sample}\left(mol\, g_{lip}^{-1}\right)} \times MW_{reference\ compound}\left(g\, mol^{-1}\right) \quad (8)$$

where MW is the molecular weight.

If CA applies, $BEQ_{chem,\ iceberg}$ approximates $BEQ_{bio,\ iceberg}$ [100], thus enabling the calculation of the percentage contribution of identified component fractions to the overall chemical burden that is biologically active (expressed as $BEQ_{bio,\ sample}$).

3.2.3 Application of Effect Analysis

There are a number of studies assessing the total burden of chemicals effective on certain MOAs in marine wildlife species (Table 2). Expressing the total burden as an equivalent concentration to a reference compound against the corresponding MOA (e.g., TCDD equivalent for AhR, estradiol equivalent for ER, etc.) puts exposure of POP mixtures into a toxicological context. While current studies focus on specific receptor-mediated MOAs, particularly related to endocrine disruption and dioxin-like activity, nonspecific MOAs relevant to marine wildlife health need to be included for more comprehensive toxicological profiling of POP mixtures. These may include oxidative stress, DNA damage/repair, inflammation, and immunosuppression. With the expanding knowledge of toxicity pathways and AOP, we may design a more targeted test battery of bioassays to evaluate the overall toxic potential of POP mixtures.

As a useful parameter in initial hazard screening of POP mixtures in marine wildlife species, the BEQ values would become more meaningful in the risk assessment of these mixtures, if there were an effect-based trigger (EBT) value for benchmarking, similar to LOAEL/NOAEL in single chemical safety evaluation. The concept of EBT has been recently developed for mixtures of organic micropollutants in water quality assessment [57,82,101,102]. Its adaptation to POP mixtures in marine wildlife species has yet to be explored.

3.2.4 Identifying the Contributing Chemicals: Bioassay-Directed Fractionation

With the understanding of overall toxicity, there is a growing body of studies aimed at identifying contributing chemical components in the mixture. Biological samples are typically fractionated according to the physicochemical properties of expected compounds (Figure 4). For example, Hewitt et al. [98] fractionated compounds from liver extracts (after lipid removal) of male white sucker exposed to pulp and paper mill effluent, according to their octanol–water partition coefficients (K_{ow}) by reverse-phase high performance liquid chromatography. Different chemical fractions eluted from an analytical C18 column were tested for their potential to induce ethyoxyresorufin-*O*-deethylase (EROD) activity and to bind sex steroid receptors. The $\log K_{ow}$ four to five fraction caused elevated EROD induction. The $\log K_{ow}$ three to four fraction dominated the binding activity to rainbow trout hepatic ER and testicular AR. Binding to goldfish plasma sex steroid binding protein was observed for all $\log K_{ow}>2$ fractions. While characteristic fractions were revealed in this study, possible contributing chemicals were not resolved to account for the observed biological effects.

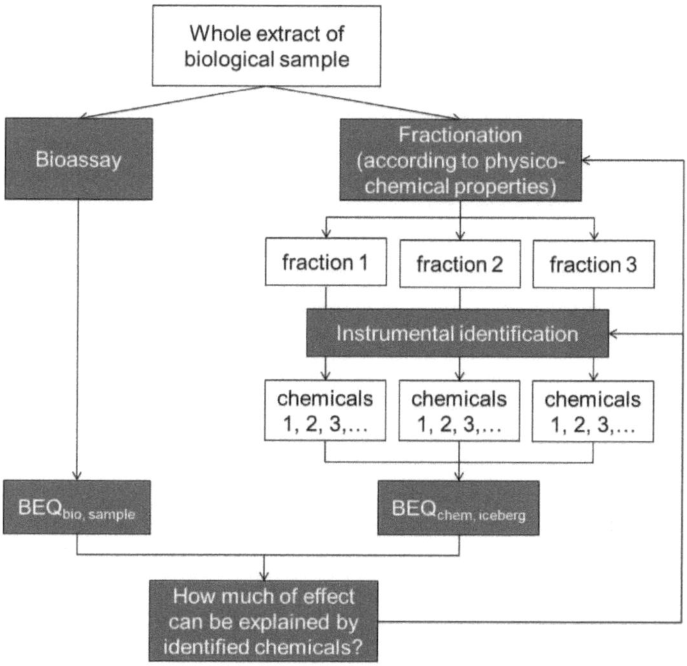

FIGURE 4 The schematic of fractionation and instrumental techniques to identify contributing chemicals to overall mixture effect in a biological sample.

Advances in biological sample preparation methods have allowed more focused target analysis of chemicals contributing to the overall mixture effect. Sophisticated methods are available for removing interfering lipids and natural hormones in biota samples (e.g., fish muscle [93], polar bear plasma [95]) while maintaining satisfactory recoveries for major target chemicals of interest. These advances allowed Bytingsvik et al. [96], for example, to conduct the targeted analysis of known chemical classes with thyroid hormone disrupting properties (e.g., PCBs, OH-PCBs, PBDEs, OH-PBDEs, HBCDs, and PFCs) to account for the transthyretin-binding activity of polar bear plasma extracts, where almost half of the activity could still not be explained by the chemicals examined.

3.2.5 Identifying the Contributing Chemicals: Instrumental Techniques

Recent developments in library-based compound identification software packages coupled to advanced hyphenated instrumental techniques (e.g., two dimensional gas/liquid chromatography and time-of-flight mass spectrometry) have made a step further in nontarget chemical analysis [103,104]. The compound identification can be achieved by searching a standard reference database, by analyzing mass spectra from reference standards, and by de novo interpretation.

Custom software has been developed to organize the identified mass spectra and related information with shared access [103,104]. This reduces the need for the screening of targeted chemicals according to the availability of their standards.

Applying these approaches, Simon et al. [97] identified branched nonylphenols and mono- and dihydroxylated-octaCBs as additional thyroid hormone disrupting compounds in polar bear plasma samples, which were not identified using target analytical screening in a previous study by the same group [96]. These newly identified chemicals together with previously targeted priority ones (e.g., OH-PCBs and PFCs) accounted for 70–80% of total thyroid hormone disrupting activity [97]. Suzuki et al. [90] identified several major dioxin-like compounds in acid-treated tissue extracts of Baikal seals explaining the entire AhR induction effect, and *p,p′*-DDE as an important causative androgen antagonist, accounting for 59% of the total androgenic activity.

It is noted that the MOAs involved in these cases are often chemical specific, and therefore a limited number of dominant and/or potent chemicals can explain a majority of the total effect. This may not be necessarily the case for nonspecific biological endpoints, such as baseline toxicity, oxidative stress response, and genotoxicity, which can target a much wider range of chemicals, as found in environmental water samples [57,82,100,105]. In this sense, $BEQ_{bio,\ sample}$ provides a sum measure of the total toxicologically relevant burden to complement chemical-based monitoring.

3.2.6 Factors Affecting the Interpretation of Bioassay Results

The complexity of biological matrices and assay-specific characteristics may sometimes cause artifacts in biological response when dosing sample extracts into cell-based bioassays, thus affecting the derivation of BEQ_{bio}. First, attention should be paid to matrix interference due to coextracted lipids, proteins, and endogenous bioactive molecules. Lipids from fatty tissue have been found to cause interferences with biological responses within the typically used 24-h exposure period. Faster kinetics of AhR induction, for example, was observed with TCDD spiked into a small amount of lipids compared to the native TCDD standard. This matrix effect was, however, eliminated by a longer exposure period (e.g., 72h) [58]. These observations prompted the development of methods for removing lipid matrices and endogenous compounds for wildlife tissue/blood samples that may interfere with the bioassay interpretation of exogenous POPs [90,93,95]. Second, many currently used bioassays are intrinsically lacking in metabolic competence [106]; however, under in vivo conditions, many parent chemicals can be biotransformed into less active or even more toxic metabolites. This is also relevant to mixture effect screening for environmental sample extracts [89,107,108]. The incorporation of metabolic enzymes in these bioassays is thus necessary to more accurately reflect the in vivo exposure condition.

3.3 Methodological Gaps and Recommendations

The entire realm of environmental mixtures of POPs in wildlife is generally not reflected in current approaches to understand their combined effect. Exposure-driven investigations usually focus on a few priority POPs out of the entire chemical mixture (as shown in Figure 1). In reviewed examples of effect-directed analysis, the mixture was either determined from acid-treated tissue extracts, thus sacrificing nonpersistent chemicals, or extracted entirely with the need for sophisticated matrix removal efforts.

Recent advances in polymer-based passive sampling have opened up the possibility of rapid, quantitative extraction of organic chemicals from biological samples, particularly lipid-rich tissue, based on equilibrium partitioning between the polymer and lipid phases [109–112]. The partition coefficient between lipids and polymer ($K_{lip\text{-}polymer}$) was found to be independent of hydrophobicity and the generally constant $K_{lip\text{-}polymer}$ allows uniform sampling of the entire range of chemicals with varying physicochemical properties from the tissue [58,113]. Our previous work [58] further demonstrated the applicability of passive sampling to in vitro bioanalytical screening of the entire range of bioaccumulative chemicals in fatty tissue without the need for clean-up efforts to remove the lipid matrix. The mixture composition in the tissue samples is maintained in passive sampling extracts, thus allowing mass balance calculation of lipid-normalized total chemical burdens from passive sampling extracts in cell-based bioassays. This method also provides sufficient sensitivity to cover a dynamic detection range of chemical burdens and to allow quantitative links between exposure levels and in vitro effect. These results served as a basis for the ongoing work in our group (unpublished) on effect screening in nonspecific endpoints without resolving individual chemical composition in the sample extracts.

Building on the polymer-lipid partitioning model, passive sampling may apply to even more complex biological matrices, such as blood. Based on the reported partitioning coefficients between polymer and major components in blood, our ongoing work is to design a quantitative polymer-blood system that may allow the enrichment of bioaccumulative chemicals into polymer samplers for in vitro effect screening (unpublished).

A complete understanding of pathways leading to adverse effect is still lacking, particularly with respect to the quantitative relationship from lower to higher levels of biological organization. This is particularly important to the issue of species susceptibility, which is arguably overlooked in model in vitro bioassays as an alternative screening tool. Differences in species sensitivity to dioxin-like chemicals in the AhR-mediated toxicity pathway, for example, may complicate the risk evaluation for wildlife species. Vertebrates are generally sensitive to these chemicals, whereas invertebrates tend to be insensitive. Although it remains unexplored how marine wildlife species, such as mammals and reptiles, vary in their susceptibility to dioxin-like chemicals, large sensitivity differences have been observed across vertebrate classes. Lake trout and zebrafish, for example, exhibit nearly 40-fold difference in sensitivity for early-life-stage mortality caused by TCDD [114]. Domestic chicken can differ from other avian species in their sensitivity to TCDD by an order of magnitude or more [115]. Species susceptibility is presumed to be attributed in a large part to the specific characteristics of their AhR and its capacity to bind these chemicals; however, this crucial issue is not fully explored in terms of quantitative structure–activity relationship to offer a robust predictive model [115–118]. In silico tools are thus desired to model the species-specific AhR ligand binding domain characteristics in relation to the binding affinity of various dioxin-like chemicals to AhR. Such understanding of similar biomolecules involved in the MIEs in a specific toxicity pathway will facilitate interspecies extrapolations based on a few model species for mixture screening, reducing the need for endless species-specific testing.

4. CONCLUSION

An unbiased sampling strategy is desired to extract the entire mixture of POPs from marine wildlife tissue/fluid (Figure 5). Toxicity pathway-oriented high-throughput bioassays indicative of early molecular warning events can serve as lower tier screening tools to evaluate the toxicological/biological relevance of the entire chemical burden. Should the mixture trigger responses, sample

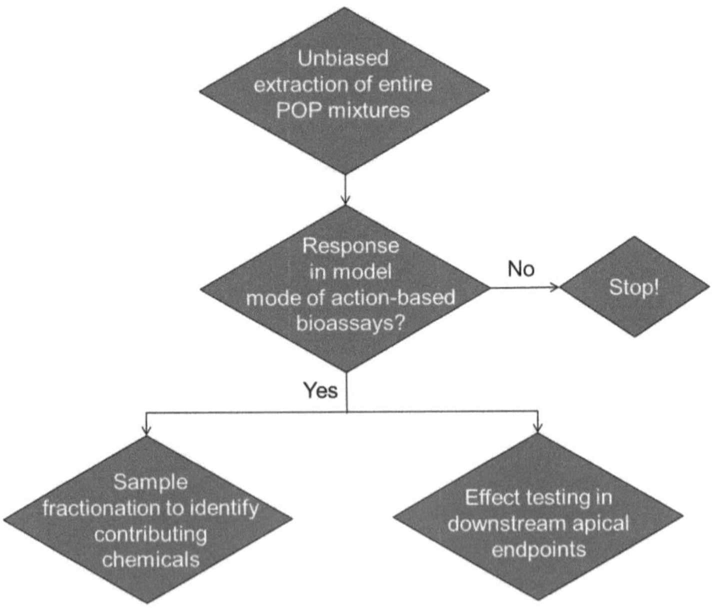

FIGURE 5 Proposed framework for integrated approaches to linking persistent organic pollutant exposure and mixture effect.

extracts can be further screened against apical endpoints downstream the toxicity pathway for insights into the later manifestation of effect. Meanwhile, sample mixtures can be fractionated to quantify and prioritize the contributing POP components with bioassays and nontargeted instrumental analysis [97]. To incorporate species susceptibility, functional assays [119,120] together with quantitative pathway prediction models can be developed for a specific marine wildlife species to address species-specific issue. Thus, integrated exposure and effect screening can be achieved. Only when we have these adequate tools can we initiate more strategic sampling of marine wildlife samples for hypothesis-driven investigations for specific wildlife population. With a mechanistic and quantitative understanding of fate and effect of POPs, we may take a step further toward understanding the overall chemical pollution (including inorganics) and the role it plays in the larger picture of total environmental threats.

ACKNOWLEDGMENTS

The National Research Center for Environmental Toxicology (Entox) is a joint venture between The University of Queensland and Queensland Health Forensic and Scientific Services (QHFSS). Part of the work described in this review was supported by Sea World Research and Rescue Foundation (Ref no.: SWR/1/2013), and the Society of Environmental Toxicology and Chemistry (SETAC) Doctoral Fellowship, sponsored by the Procter & Gamble Company. Ling Jin acknowledges the Prime Minister's Australia Asia Endeavor Award funded by the Australia Government to support his PhD. Sharon Grant is thanked for language editing of the manuscript.

REFERENCES

[1] D. Harvell, R. Aronson, N. Baron, J. Connell, A. Dobson, S. Ellner, L. Gerber, K. Kim, A. Kuris, H. McCallum, K. Lafferty, B. McKay, J. Porter, M. Pascual, G. Smith, K. Sutherland, J. Ward, Front. Ecol. Environ. 2 (2004) 375–382.

[2] M.F. Van Bressem, J.A. Raga, G. Di Guardo, P.D. Jepson, P.J. Duignan, U. Siebert, T. Barrett, M.C.D. Santos, I.B. Moreno, S. Siciliano, A. Aguilar, K. Van Waerebeek, Dis. Aquat. Org. 86 (2009) 143–157.

[3] A.M. Magera, J.E. Mills Flemming, K. Kaschner, L.B. Christensen, H.K. Lotze, PLoS One 8 (2013) e77908.

[4] C.J. Limpus, A Biological Review of Australian Marine Turtles 1. Loggerhead Turtle *Caretta caretta* (Linnaeus), 2008.

[5] I. Lawler, H. Marsh, B. McDonald, T. Stokes, Dugong in the Great Barrier Reef: Current State of Knowledge, CRC Reef Research Centre Ltd, Townsville, Queensland, 2002.

[6] IUCN, IUCN Red List of Threatened Species, International Union for Conservation of Nature, 2008. Cambridge, United Kingdom.

[7] L.H. Schwacke, C.R. Smith, F.I. Townsend, R.S. Wells, L.B. Hart, B.C. Balmer, T.K. Collier, S. De Guise, M.M. Fry, L.J. Guillette, S.V. Lamb, S.M. Lane, W.E. McFee, N.J. Place, M.C. Tumlin, G.M. Ylitalo, E.S. Zolman, T.K. Rowles, Environ. Sci. Technol. 48 (2013) 93–103.

[8] C.H. Peterson, S.D. Rice, J.W. Short, D. Esler, J.L. Bodkin, B.E. Ballachey, D.B. Irons, Science 302 (2003) 2082–2086.

[9] C. Gaus, S. Grant, N.L. Jin, K. Goot, L. Chen, A.C. Villa, F. Neugebauer, L. Qi, C. Limpus, Haematology, Blood Chemistry and Environmental Contaminants in Green Turtle Blood from the Gladstone Port, Report to: Gladstone Ports Corporation Limited and Department of Environment and Resource Management, 2012.
[10] DERM, Strandings Hot Spots, 2012.
[11] DEEDI, Fish Health Sampling Reports. Gladstone Harbour, Department of Employment, Economic Development and Innovation, Queensland, Brisbane, 2011.
[12] A.A. Cunningham, Br. Med. J. 331 (2005) 1214–1215.
[13] M. Santoro, J.A. Morales, B. Rodríguez-Ortíz, Vet. Rec. 161 (2007) 482–486.
[14] B.A. Stacy, A.M. Foley, E. Greiner, L.H. Herbst, A. Bolten, P. Klein, C.A. Manire, E.R. Jacobson, Dis. Aquat. Org. 89 (2010) 237–259.
[15] H. Chen, R.J. Kuo, T.C. Chang, C.K. Hus, Rodney A. Bray, I.J. Cheng, J. Parasitol. 98 (2011) 437–439.
[16] M. Flint, J. Patterson-Kane, C. Limpus, P. Mills, EcoHealth 7 (2010) 135–145.
[17] M. Flint, P.A. Eden, C.J. Limpus, H. Owen, C. Gaus, P.C. Mills, EcoHealth. (2014). DOI: 10.1007/s10393-014-0972-5. In press.
[18] A.M. Foley, B.A. Schroeder, A.E. Redlow, K.J. Fick-Child, W.G. Teas, J. Wildl. Dis. 41 (2005) 29–41.
[19] A.A. Aguirre, P.L. Lutz, EcoHealth 1 (2004) 275–283.
[20] A.A. Aguirre, G.H. Balazs, T.R. Spraker, S.K.K. Murakawa, B. Zimmerman, J. Aquat. Anim. Health 14 (2002) 298–304.
[21] M. Flint, M. de Wit, H. Owen, Geographically separate populations of sirenia display similar causes of mortality, in: 19th Biennial Conference on the Biology of Marine Mammals, Tampa, Florida, 2011.
[22] G.D. Bossart, Vet. Pathol. 48 (2011) 676–690.
[23] L.H. Herbst, P.A. Klein, Environ. Health Perspect. 103 (1995) 27–30.
[24] C. Cray, R. Varella, G.D. Bossart, P. Lutz, J. Zoo Wildl. Med. 32 (2001) 436–440.
[25] T. Work, R. Rameyer, G. Balazs, C. Cray, S. Chang, J. Wildl. Dis. 37 (2001) 574–581.
[26] C.J. Limpus, A Biological Review of Australian Marine Turtles 2. Green Turtle *Chelonia mydas* (Linnaeus), 2008.
[27] J.M. Keller, G.H. Balazs, F. Nilsen, M. Rice, T.M. Work, B.A. Jensen, Environ. Sci. Technol. 48 (2014) 7807–7816.
[28] C. Gaus, O. Papke, W. Blanchard, D. Haynes, D. Connell, J. Muller, Organohalogen Compd. 52 (2001) 95–99.
[29] S. Hermanussen, C.J. Limpus, O. Päpke, W. Blanshard, D. Connell, C. Gaus, Organohalogen Compd. 66 (2004) 1837–1843.
[30] R.J. Law, Mar. Pollut. Bull. 82 (2014) 7–10.
[31] L. Jin, J.C.W. Lam, M.B. Murphy, P.K.S. Lam, Temporal trends of polybrominated diphenyl ethers and hexabromocyclododecanes in marine mammals with special reference to Hong Kong, South China, in: B.G. Loganathan, P.K.S. Lam (Eds.), Global Contamination Trends of Persistent Organic Chemicals, CRC press, 2012, pp. 497–513.
[32] R.J. Law, A. Covaci, S. Harrad, D. Herzke, M.A.E. Abdallah, K. Femie, L.M.L. Toms, H. Takigami, Environ. Int. 65 (2014) 147–158.
[33] B.Q. Zhu, N.L.S. Lai, T.C. Wai, L.L. Chan, J.C.W. Lam, P.K.S. Lam, Environ. Int. 66 (2014) 65–70.
[34] W. Vetter, E. Scholz, C. Gaus, J.F. Müller, D. Haynes, Arch. Environ. Contam. Toxicol. 41 (2001) 221–231.
[35] C. Gaus, M. O'Donohue, D. Connell, J. Mueller, D. Haynes, Olaf Papke, Organohalogen Compd. 66 (2004) 1559–1566.

[36] S.F. Hermanussen, Distribution and Fate of Persistent Organic Pollutants in Nearshore Marine Turtle Habitats of Queensland, Australia, The University of Queensland, Brisbane, 2009.
[37] S. Hermanussen, V. Matthews, O. Päpke, C.J. Limpus, C. Gaus, Mar. Pollut. Bull. 57 (2008) 409–418.
[38] J.P. van de Merwe, M. Hodge, H.A. Olszowy, J.M. Whittier, S.Y. Lee, Mar. Pollut. Bull. 60 (2010) 579–588.
[39] L.-M.L. Toms, M. Mortimer, R.K. Symons, O. Paepke, J.F. Mueller, Environ. Int. 34 (2008) 58–66.
[40] L.-M.L. Toms, F.A. Harden, R.K. Symons, D. Burniston, P. Fürst, J.F. Müller, Chemosphere 68 (2007) 797–803.
[41] P. Cole, D. Trichopoulos, H. Pastides, T. Starr, J.S. Mandel, Regul. Toxicol. Pharmacol. 38 (2003) 378–388.
[42] R.G. Ellis-Hutchings, G.N. Cherr, L.A. Hanna, C.L. Keen, Toxicol. Appl. Pharmacol. 215 (2006) 135–145.
[43] P. Pocar, N. Fiandanese, C. Secchi, A. Berrini, B. Fischer, J.S. Schmidt, K. Schaedlich, S.M. Rhind, Z.L. Zhang, V. Borromeo, Toxicol. Sci. 126 (2012) 213–226.
[44] A. Weinand-Härer, H. Lilienthal, K.-A. Bucholski, G. Winneke, Environ. Toxicol. Pharmacol. 3 (1997) 97–103.
[45] K. Hayashi, T. Fukuyama, A. Ohnuma, Y. Tajima, Y. Kashimoto, T. Yoshida, T. Kosaka, J. Immunotoxicol. 10 (2013) 119–124.
[46] S.H. Seo, Y.M. Ryu, K.N. Kim, H.W. Kim, Y.R. Kim, S.H. Lee, N.H. Won, M.K. Kim, Mol. Cell. Toxicol. 3 (2007) 52.
[47] M.M.L. Dingemans, M. van den Berg, R.H.S. Westerink, Environ. Health Perspect. 119 (2011) 900–907.
[48] WHO., Assessment of the Health Risk of Dioxins: Re-evaluation of the Tolerable Daily Intake (TDI), 1998. Geneva, Switzerland.
[49] J.M. Keller, P.D. McClellan-Green, J.R. Kucklick, D.E. Keil, M.M. Peden-Adams, Environ. Health Perspect. 114 (2006) 70–76.
[50] A.J. Hall, O.I. Kalantzi, G.O. Thomas, Environ. Pollut. 126 (2003) 29–37.
[51] R.J. Letcher, J.O. Bustnes, R. Dietz, B.M. Jenssen, E.H. Jorgensen, C. Sonne, J. Verreault, M.M. Vijayan, G.W. Gabrielsen, Sci. Total Environ. 408 (2010) 2995–3043.
[52] G.M. Rand, Fundamentals of Aquatic Toxicology: Effects, Environmental Fate and Risk Assessment, Taylor and Francis, Washington, DC, 1995.
[53] T. Backhaus, M. Faust, Environ. Sci. Technol. 46 (2012) 2564–2573.
[54] M. Van den Berg, L.S. Birnbaum, M. Denison, M. De Vito, W. Farland, M. Feeley, H. Fiedler, H. Hakansson, A. Hanberg, L. Haws, M. Rose, S. Safe, D. Schrenk, C. Tohyama, A. Tritscher, J. Tuomisto, M. Tysklind, N. Walker, R.E. Peterson, Toxicol. Sci. 93 (2006) 223–241.
[55] H. Kojima, S. Takeuchi, T. Tsutsumi, K. Yamaguchi, K. Anezaki, K. Kubo, M. Iida, T. Takahashi, S. Kobayashi, K. Jin, T. Nagai, Chemosphere 83 (2011) 753–759.
[56] J. Haedrich, L.A.P. Hoogenboom, G. Eppe, L. Goeyens, M. Elskens, P. Malagocki, M.L. Scippo, H. Vanderperren, I. Windal, A. Kotz, M.S. Denison, R. Malisch, Organohalogen Compd. 74 (2012) 193–196.
[57] B.I. Escher, C. van Daele, M. Dutt, J.Y.M. Tang, R. Altenburger, Environ. Sci. Technol. 47 (2013) 7002–7011.
[58] L. Jin, C. Gaus, L. van Mourik, B.I. Escher, Environ. Sci. Technol. 47 (2013) 7982–7988.
[59] J.Y. Li, J.Y.M. Tang, L. Jin, B.I. Escher, Environ. Toxicol. Chem. 32 (2013) 2888–2896.
[60] M. Levin, S. De Guise, P.S. Ross, Environ. Toxicol. Chem. 24 (2005) 1247–1252.

[61] M. Tabuchi, A. Veldhoen, N. Dangerfield, S. Jeffries, C.C. Helbing, P.S. Ross, Environ. Health Perspect. 114 (2006) 1024–1031.
[62] E.G. Sormo, I. Jussi, M. Jussi, M. Braathen, J.U. Skaare, B.M. Jenssen, Environ. Toxicol. Chem. 24 (2005) 610–616.
[63] G.D. Villanger, K.M. Gabrielsen, K.M. Kovacs, C. Lydersen, E. Lie, M. Karimi, E.G. Sormo, B.M. Jenssen, Chemosphere 92 (2013) 828–842.
[64] C. Debier, G.M. Ylitalo, M. Weise, F. Gulland, D.P. Costa, B.J. Le Boeuf, T. de Tillesse, Y. Larondelle, Environ. Pollut. 134 (2005) 323–332.
[65] M. Noel, L.L. Loseto, C.C. Helbing, N. Veldhoen, N.J. Dangerfield, P.S. Ross, Environ. Sci. Technol. 48 (2014) 2942–2951.
[66] L.H. Schwacke, E.S. Zolman, B.C. Balmer, S. De Guise, R.C. George, J. Hoguet, A.A. Hohn, J.R. Kucklick, S. Lamb, M. Levin, J.A. Litz, W.E. McFee, N.J. Place, F.I. Townsend, R.S. Wells, T.K. Rowles, Proc. R. Soc. B 279 (2011) 48–57.
[67] E.W. Montie, P.A. Fair, G.D. Bossart, G.B. Mitchum, M. Houde, D.C.G. Muir, R.J. Letcher, W.E. McFee, V.R. Starczak, J.J. Stegeman, M.E. Hahn, Aquat. Toxicol. 86 (2008) 397–412.
[68] V. Tornero, A. Borrell, A. Aguilar, J. Forcada, C. Lockyer, Environ. Pollut. 140 (2006) 312–321.
[69] C. Sonne, P.S. Leifsson, R. Dietz, E.W. Born, R.J. Letcher, L. Hyldstrup, F.F. Riget, M. Kirkegaard, D.C.G. Muir, Environ. Sci. Technol. 40 (2006) 5668–5674.
[70] I.C. Oskam, E. Ropstad, E. Lie, A.E. Derocher, O. Wiig, E. Dahl, S. Larsen, J.U. Skaare, J. Toxicol. Environ. Health Part A 67 (2004) 959–977.
[71] K. Molde, T.M. Ciesielski, A.T. Fisk, C. Lydersen, K.M. Kovacs, E.G. Sormo, B.M. Jenssen, Sci. Total Environ. 442 (2013) 445–454.
[72] L.M. Komoroske, R.L. Lewison, J.A. Seminoff, D.D. Deheyn, P.H. Dutton, Chemosphere 84 (2011) 544–552.
[73] J.M. Keller, J.R. Kucklick, M.A. Stamper, C.A. Harms, P.D. McClellan-Green, Environ. Health Perspect. 112 (2004) 1074–1079.
[74] N.K. Brar, C. Waggoner, J.A. Reyes, R. Fairey, K.M. Kelley, Aquat. Toxicol. 96 (2010) 203–215.
[75] J.E. Yordy, M.A.M. Mollenhauer, R.M. Wilson, R.S. Wells, A. Hohn, J. Sweeney, L.H. Schwacke, T.K. Rowles, J.R. Kucklick, M.M. Peden-Adams, Environ. Toxicol. Chem. 29 (2010) 2143–2153.
[76] A. Kortenkamp, T. Backhaus, M. Faust, State of the Art Report on Mixture Toxicity, 2009. European Commission 070307/2007/485103/ETU/D.1.
[77] W.S. Baldwin, J.A. Roling, Toxicol. Sci. 107 (2009) 93–105.
[78] G.J. Howard, J.J. Schlezinger, M.E. Hahn, T.F. Webster, Environ. Health Perspect. 118 (2010) 666–672.
[79] S. Ermler, M. Scholze, A. Kortenkamp, Toxicol. Appl. Pharmacol. 257 (2011) 189–197.
[80] R.M. Evans, M. Scholze, A. Kortenkamp, PLoS One 7 (2012) e43606.
[81] G. de Sousa, A. Nawaz, J-P. Cravedi, R. Rahmani, Toxicol. Sci. 141 (2014) 234–243.
[82] J.Y.M. Tang, S. McCarty, E. Glenn, P.A. Neale, M.S.J. Warne, B.I. Escher, Water Res. 47 (2013) 3300–3314.
[83] R.M. Burgess, K.T. Ho, W. Brack, M. Lamoree, Environ. Toxicol. Chem. 32 (2013) 1935–1945.
[84] G.T. Ankley, R.S. Bennett, R.J. Erickson, D.J. Hoff, M.W. Hornung, R.D. Johnson, D.R. Mount, J.W. Nichols, C.L. Russom, P.K. Schmieder, J.A. Serrrano, J.E. Tietge, D.L. Villeneuve, Environ. Toxicol. Chem. 29 (2010) 730–741.
[85] NRC, Toxicity Testing in the 21st Century: A Vision and a Strategy, National Academy of Sciences, Washington, DC., 2007.

[86] S. Bhattacharya, Q. Zhang, P.L. Carmichael, K. Boekelheide, M.E. Andersen, PLoS One 6 (2011) e20887.

[87] E.J. Perkins, G.T. Ankley, K.M. Crofton, N. Garcia-Reyero, C.A. LaLone, M.S. Johnson, J.E. Tietge, D.L. Villeneuve, Environ. Health Perspect. 121 (2013) 1002–1010.

[88] B.I. Escher, M. Allinson, R. Altenburger, P.A. Bain, P. Balaguer, W. Busch, J. Crago, N.D. Denslow, E. Dopp, K. Hilscherova, A.R. Humpage, A. Kumar, M. Grimaldi, B.S. Jayasinghe, B. Jarosova, A. Jia, S. Makarov, K.A. Maruya, A. Medvedev, A.C. Mehinto, J.E. Mendez, A. Poulsen, E. Prochazka, J. Richard, A. Schifferli, D. Schlenk, S. Scholz, F. Shiraish, S. Snyder, G.Y. Su, J.Y.M. Tang, B. van der Burg, S.C. van der Linden, I. Werner, S.D. Westerheide, C.K.C. Wong, M. Yang, B.H.Y. Yeung, X.W. Zhang, F.D.L. Leusch, Environ. Sci. Technol. 48 (2014) 1940–1956.

[89] M. Montano, J. Weiss, L. Hoffmann, A.C. Gutleb, A.J. Murk, Environ. Sci. Technol. 47 (2013) 8878–8886.

[90] G. Suzuki, N.M. Tue, S. van der Linden, A. Brouwer, B. van der Burg, M. van Velzen, M. Lamoree, M. Someya, S. Takahashi, T. Isobe, Y. Tajima, T.K. Yamada, H. Takigami, S. Tanabe, Environ. Sci. Technol. 45 (2011) 10203–10211.

[91] J.L. Balaam, K.V. Thomas, J. Environ. Monit. 9 (2007) 419–423.

[92] S.E. Erdmann, R. Dietz, C. Sonne, T.O. Bechshoft, K. Vorkamp, R.J. Letcher, M.H. Long, E.C. Bonefeld-Jorgensen, Chemosphere 92 (2013) 583–591.

[93] E. Simon, M.H. Lamoree, T. Hamers, J.M. Weiss, J. Balaam, J. de Boer, P.E.G. Leonards, Environ. Sci. Technol. 44 (2010) 8322–8329.

[94] J. Legler, P. Leonards, A. Spenkelink, A.J. Murk, Ecotoxicology 12 (2003) 239–249.

[95] E. Simon, J. Bytingsvik, W. Jonker, P.E.G. Leonards, J. de Boer, B.M. Jenssen, E. Lie, J. Aars, T. Hamers, M.H. Lamoree, Environ. Sci. Technol. 45 (2011) 7936–7944.

[96] J. Bytingsvik, E. Simon, P.E.G. Leonards, M. Lamoree, E. Lie, J. Aars, A.E. Derocher, O. Wiig, B.M. Jenssen, T. Hamers, Environ. Sci. Technol. 47 (2013) 4778–4786.

[97] E. Simon, M.v. Velzen, S.H. Brandsma, E. Lie, K. Løken, J.d. Boer, J. Bytingsvik, B.M. Jenssen, J. Aars, T. Hamers, M. Lamoree, Environ. Sci. Technol. 47 (2013) 8902–8912.

[98] L.M. Hewitt, A.C. Pryce, J.L. Parrott, V. Marlatt, C. Wood, K. Oakes, G.J. Van der Kraak, Environ. Toxicol. Chem. 22 (2003) 2890–2897.

[99] S.O. Simmons, C.-Y. Fan, R. Ramabhadran, Toxicol. Sci. 111 (2009) 202–225.

[100] J.Y.M. Tang, F. Busetti, J.W.A. Charrois, B.I. Escher, Water Res. 60 (2014) 289–299.

[101] W. Brand, C.M. de Jongh, S.C. van der Linden, W. Mennes, L.M. Puijker, C.J. van Leeuwen, A.P. van Wezel, M. Schriks, M.B. Heringa, Environ. Int. 55 (2013) 109–118.

[102] J.Y.M. Tang, B.I. Escher, Environ. Toxicol. Chem. 33 (2014) 1427–1436.

[103] E. Hoh, N.G. Dodder, S.J. Lehotay, K.C. Pangallo, C.M. Reddy, K.A. Maruya, Environ. Sci. Technol. 46 (2012) 8001–8008.

[104] E.L. Schymanski, M. Bataineh, K.U. Goss, W. Brack, TrAC Trends Anal. Chem. 28 (2009) 550–561.

[105] R.Y.L. Yeh, M.J. Farre, D. Stalter, J.Y.M. Tang, J. Molendijk, B.I. Esther, Water Res. 59 (2014) 172–184.

[106] R.R. Tice, C.P. Austin, R.J. Kavlock, J.R. Bucher, Environ. Health Perspect. 121 (2013) 756–765.

[107] A. Natsch, T. Haupt, Toxicol. Sci. 135 (2013) 356–368.

[108] M. Montaño, A.C. Gutleb, A.J. Murk, Environ. Sci. Technol. 47 (2013) 6071–6081.

[109] A. Jahnke, M.S. McLachlan, P. Mayer, Chemosphere 73 (2008) 1575–1581.

[110] A. Jahnke, P. Mayer, D. Broman, M.S. McLachlan, Chemosphere 77 (2009) 764–770.

[111] P. Mayer, L. Toräng, N. Glæsner, J.Å. Jönsson, Anal. Chem. 81 (2009) 1536–1542.

[112] A. Jahnke, P. Mayer, M. Adolfsson-Erici, M.S. McLachlan, Environ. Toxicol. Chem. 30 (2011) 1515–1521.

[113] I.J. Allan, K. Baek, T.O. Haugen, K.L. Hawley, A.S. Hogfeldt, A.D. Lillicrap, Environ. Sci. Technol. 47 (2013) 11660–11667.

[114] G.E. Elonen, R.L. Spehar, G.W. Holcombe, R.D. Johnson, J.D. Fernandez, R.J. Erickson, J.E. Tietge, P.M. Cook, Environ. Toxicol. Chem. 17 (1998) 472–483.

[115] J.A. Head, M.E. Hahn, S.W. Kennedy, Environ. Sci. Technol. 42 (2008) 7535–7541.

[116] M.E. Hahn, Chem.-Biol. Interact. 141 (2002) 131–160.

[117] R. Farmahin, D.M. Wu, D. Crump, J.C. Herve, S.P. Jones, M.E. Hahn, S.I. Karchner, J.P. Giesy, S.J. Bursian, M.J. Zwiernik, S.W. Kennedy, Environ. Sci. Technol. 46 (2012) 2967–2975.

[118] D. Fraccalvieri, A.A. Soshilov, S.I. Karchner, D.G. Franks, A. Pandini, L. Bonati, M.E. Hahn, M.S. Denison, Biochemistry 52 (2013) 714–725.

[119] Y.N. Lu, V.R. Nerurkar, A.A. Aguirre, T.M. Work, G.H. Balazs, R. Yanagihara, In Vitro Cell. Dev. Biol. Anim. 35 (1999) 389–393.

[120] H. Frouin, M. Fortier, M. Fournier, Toxicology 270 (2010) 66–76.

Chapter 3

Fast Analytical Techniques Based on Microextraction

Ruifen Jiang and Gangfeng Ouyang*
School of Chemistry and Chemical Engineering, Sun Yat-sen University, Guangzhou, China
**Corresponding author: E-mail: cesoygf@mail.sysu.edu.cn*

Chapter Outline

1. INTRODUCTION

A typical procedure for the analysis of persistent organic pollutants (POPs) from environmental samples involves sampling, sample preparation, and quantification steps. Each of these analytical steps is critical for gathering precise and informative results, and must be followed in order.

Among the analysis steps, sampling and sample preparation are considered to be the most time- and labor-consuming steps in the entire analytical method. Sampling is a prerequisite step that can be especially tedious when the sampling

Comprehensive Analytical Chemistry, Vol. 67. http://dx.doi.org/10.1016/B978-0-444-63299-9.00003-X

site is far away from the laboratory or a large amount of sample should be collected. For the sample preparation step, classical approaches including liquid–liquid extraction (LLE), Soxhlet extraction, accelerated solvent extraction, solid phase extraction (SPE), and supercritical fluid extraction are the most commonly used and well-established methods. Most of these extraction techniques require the use of a large amount of sample, organic solvent (often toxic and relatively expensive), and long extraction time to maximize recovery. Besides, in many cases, the automation of such procedures is rather problematic and requires special instrumentation.

As alternatives to classical approaches for sample preparation, microextraction techniques [1] have attracted much attention recently. Microextraction, which has been developed in various configurations, addresses the need for a reduction in sample volume, solvent use, and size of extraction device, while it also facilitates rapid and convenient sample preparation. In addition, the miniaturization of the extraction device enables the automation of the techniques. Further, most of the microextraction techniques are nonexhaustive approaches, which cause minimal perturbation to the sample system. In comparison to exhaustive techniques, better characterization and more accurate information about the investigation system or process can be obtained when these microextraction-based strategies are applied.

Microextraction approaches including solid phase microextraction (SPME), liquid phase microextraction (LPME), stir bar sorptive extraction (SBSE), and thin-film microextraction (TFME) have been widely applied to various matrices including biological, environmental, and food samples. This chapter aims at providing an overview of the application of these microextraction techniques for sampling of POPs in environmental samples during 2004–2014, and demonstrating the panorama of opportunity of these microextraction techniques in POP analysis.

2. SOLID PHASE MICROEXTRACTION

SPME was developed by Janusz Pawliszyn in 1990 and has been rapidly developed into several formats and widely used in environmental sample preparation [2]. The originally developed and most commonly used format is the syringe-like fiber geometry in which the extraction phase is coated on a fiber support, such as fused silica fiber or metal wire (Figure 1(a)). The supported fiber is attached to a tube, which can be retracted into a protective needle used to pierce the septa. The syringe-like SPME fiber can be mounted on a commercial holder for easy handling. During sampling, the needle pierces through the septum on the vial cap, and the whole device sits on top of the vial that contains the sample (Figure 1(b)). Next, the fiber coating is exposed directly to the sample or the headspace (HS) of the sample for a period of time. After sampling, the fiber can be directly injected into an analytical instrument for analysis (Figure 1(c)).

SPME is a nonexhaustive extraction method. After exposing the coating into the sample matrix, the analytes partition between the sample matrix and the fiber coating. The extraction process is considered to be complete when the

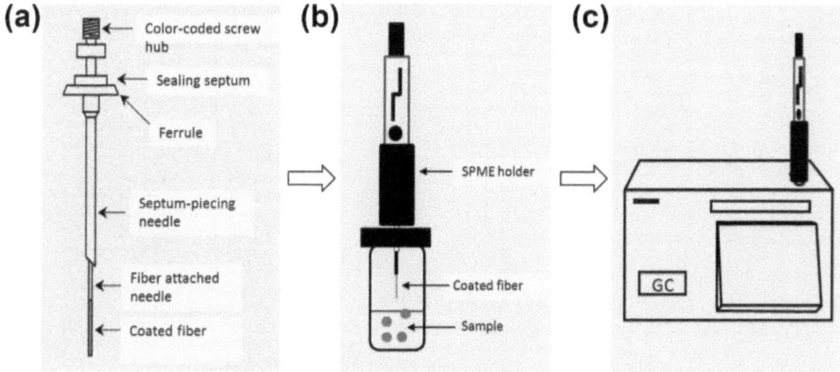

FIGURE 1 Schematic of fiber extraction and injection.

analyte concentration reaches equilibrium between the sample matrix and the coating. For liquid coating direct immersion (DI) extraction, the equilibrium extraction amount can be expressed by Eqn (1). From this equation, one can see that the equilibrium extraction amount (n_e) is proportional to the distribution coefficient (K_{es}) and the volume of the extraction phase (V_e). In addition, when the volume of the sample matrix (V_s) is large enough, such as when conducting in-field sampling, the equilibrium equation can be simplified into $n = K_{es}V_eC_s$:

$$n_e = \frac{K_{es}V_eC_sV_s}{K_{es}V_e + V_s},\ when\ V_s >> K_{es}V_e,\ n = K_{es}V_eC_s \quad (1)$$

The kinetic process of SPME can be described using different sorption models [3–5]. Among them, Ai Jiu proposed a kinetic uptake model based on Fick's diffusion law [5]. The sampling process can be expressed using Eqn (2), in which n_t is the sampling amount during time, t; n_e is the equilibrium extracted amount as described above; a is the extraction rate that is related to the geometry of the sorbent and the conditions of the sample matrix:

$$n_t = n_e[1 - \exp(-at)] \quad (2)$$

The advantages of the SPME procedure with respect to traditional sample preparation techniques are the fact that it uses no (thermal desorption, TD) or little organic solvent (solvent desorption). The SPME approach is relatively simple, fast, and cheap. For TD SPME, it integrates sampling, sample concentration, and sample introduction into one step. Besides, SPME methods are suitable for liquid, solid, and gaseous sample analysis, typically for analysis in which only a small volume of sample is available. Last, the fiber is housed in a commercial holder that is portable, and the whole technique is easily automated.

Table 1 lists the applications of SPME coupled with a typical separation and detection system for sampling of POPs in environmental samples with a focus on 2004–2014 and sampling of organochlorine pesticides (OCPs), polychlorinated

TABLE 1 Application of SPME for Sampling of Persistent Organic Pollutants in Environmental Samples

Target Compounds	Sample Matrices	Sample Preparation	Sorbent Phase	LOD μg/L	Linear Range μg/L	Recovery	Calibration Method	References
Lindane, heptachlor	Water	HS-SPME-GC-ECD	PDMS	0.097 and 0.050	0.1–1	95.6–103%	External calibration	[6]
6 OCPs	Sea, sewage, ground water	DI-SPME-HPLC-UV	PDMS/DVB CW/TPR	0.3–3.6 for PDMS/DVB 0.6–2.7 for CW/TPR	1–200	25–100%	External calibration	[7]
20 Pesticides	Rainwater collected from March 2002 to July 2003	DI-SPME-GC-MS/MS	PDMS	0.005–0.5	0.1–5	NA	External calibration	[8]
OCPs	Drinking, surface, influent, wastewater	In-tube-SPME-GC-MS/MS	PDMS	0.0003–0.001	NA	86–112%	External calibration	[9]
OCPs and OPPs	River water	SPE-DI-SPME-GC-MS	PA	NA	NA	63–104%	External calibration	[10]
THM and OCPs	River water, tap water	DI+HS-SPME-GC-MS	DVB/CAR/PDMS	0.002–2 for OCPs 0.3–0.77 for THM	0.03–100	60–120% for river water 62–120% for tap water	External calibration	[11]

OPPs	Sea water	DI-SPME-GC-MS/MS	PDMS	0.0008–0.0156	0.1–1500	80.9–120%	Standard addition	[12]
Flame retardants	Tap, effluent, influent, waste water	HS-SPME-GC-MS	CAR/PDMS and PDMS	0.0013–0.046 for CAR/PDMS, 0.0023–0.016 for PDMS	0.1–10	17–117%	External calibration	[13]
PBDEs and PBBs	Tap, effluent, influent wastewater	HS-SPME-GC-MS/MS	PDMS	0.0000075–0.00019	0.00012–0.503	74–117%	External calibration	[14]
Brominated flame retardants	Water	HS-SPME-GC-MS	PDMS	0.0006–4.16	Varied	45–146%	External calibration	[15]
17 OCPs	Natural water	DI-SPME-GC-MS	Activated carbon	0.08–0.30	0.2–50	−6.9–11.3%	External calibration	[16]
γ-Hexachloro-cyclohexane	Water	Circulate cool-HS-SPME-GC-ECD	Activated carbon	0.02	0.5–120	NA	External calibration	[17]
OCPs	River water	MA-HS-SPME-GC-ECD	G nanosheets	0.00016–0.00093	0.001–1.5	80.1–101.1%	Standard addition	[18]
5 OCPs	Lake, pond water	HS-SPME-GC-ECD	G	0.00083–0.0115	0.002–1	81–121%	External calibration	[19]

Continued

TABLE 1 Application of SPME for Sampling of Persistent Organic Pollutants in Environmental Samples—cont'd

Target Compounds	Sample Matrices	Sample Preparation	Sorbent Phase	LOD μg/L	Linear Range μg/L	Recovery	Calibration Method	References
OCPs	Rain water	DI-SPME-GC-MS	G	0.00019–0.018		77–120%	External calibration	[20]
PBDEs	Canal water	Plunger-in-needle-DI-SPME-GC-MS	G	0.0002–0.0053	0.005–1	74.8–81.9%	External calibration	[21]
PBDEs	Reservoir, wastewater	HS-SPME-GC-ECD	SWNT	0.00008–0.0008	0.005–0.5	74–109%	External calibration	[22]
PBDEs	Tap water, river water, waste water	DI-SPME-GC-MS	Bamboo charcoal	0.00025–0.00062	0.001–1	79.1–105.1%	External calibration	[23]
OCPs	River water	DI-SPME-GC-ECD	PANI	0.0001–0.0016	0.001–5	93–110%	External calibration	[24]
4 OCPs	Seawater	HS-SPME-GC-ECD	G-PANI	0.0036–0.011	0.2–250	81–112%	External calibration	[25]
OCPs	Lake water	HS-SPME-GC-ECD	PANI-IL	0.00012–0.00031	0.0015–5	88.9–112.9%	External calibration	[26]
PCBs	Surface water	HS-SPME-GC-μECD	Fluorinated PANI	0.00005–0.0001	0.0005–0.1	83–110.7%	External calibration	[27]
PBDEs	River water	HS-SPME-GC-ECD	Perfluorooctanesulfonic acid-doped PANI	0.0001–0.0002	0.001–4	NA	External calibration	[28]
OCPs	Real world water	DI-SPME-ECD	PP-doped-PPy	0.000015–0.00066	0.001–10	88–110%	External calibration	[29]

EDPs	Well water	HS-SPME-GC-ECD	Polythiophene	0.02–0.64	0.03–300	90–110%	External calibration	[30]
16 OCPs	Lake water, river water	HS-SPME-GC-μECD	Benzo-15-crown-5	0.0001–0.0005	0.001–1	85–107%	External calibration	[31]
11 OCPs and derivations	Lake water	HS-SPME-GC-ECD	Polymethylphenylvinylsiloxane	0.000835–0.013	0.00625–2.5	71.5–115.5%	External calibration	[32]
DDT	River water and well water	HS-SPME-GC-μECD	Titania	0.00020–0.00098	0.005–5	67–85%	External calibration	[33]
PBDEs	Lake water	DI-SPME-GC-MS	Etched SSW	0.0002–0.0006	0.001–0.2	77–118%	External calibration	[34]
OCPs	Water	HS-SPME-MS	PDMS/polyvinyl alcohol	0.3–3.9	10–1470	80–110%	External calibration	[35]
PBDE	Wastewater	DI-SPME-GC-MS or ECD	Silicate molecular imprinted polymer	0.0002–0.0036 0.001–0.0088	NA	NA	Standard addition	[36]
PFOS, PFOA	River water	In-tube-SPME-LC-MS	Cp-pora-PLOT-amine	0.0015–0.0032	0.05–5	81–85.4%	External calibration	[37]
PFOS, PFOA	River water	DI-SPME-LC-ESI-MS	DMOD-3-TMOS-AC and 3-TMOX-1-PPA	0.0025 for PFOS 0.0075 for PFOA	0.01-10 for PFOS 0.05–5 for PFOA	88–102%	External calibration	[38]
OCPs	Estuarine sediments	HS-SPME-ECD	PDMS	0.029–0.301	0.2–4.0	71–121%	External calibration	[39]

Continued

TABLE 1 Application of SPME for Sampling of Persistent Organic Pollutants in Environmental Samples—cont'd

Target Compounds	Sample Matrices	Sample Preparation	Sorbent Phase	LOD μg/L	Linear Range μg/L	Recovery	Calibration Method	References
OCPs	Soil	HS-SPME-GC-ECD	PDMS	0.1–0.51	0.2–5.0	NA	External calibration	[40]
11 OCPs	Landfill soil	MA-HS-SPME-GC-MS/MS	PDMS	0.02–3.6	3–54	8–51%	Standard addition	[41]
7 OCPs	Sediments from Portuguese coastal areas, estuarine	MA-HS-SPME-GC-MS/ECD	PDMS	0.005–0.11 ng/g with GC–MS and 0.01–0.26 ng/g with GC-ECD	Up to 1000 ng/g	NA	Standard addition	[42,43]
5 OCPs	Agriculture soils	MA-DI-SPME-HPLC-UV	PDMS/DVB	36–164 ng/g	200–2000	NA	External calibration	[44]
5 OCPs	Soil	MAME-DI-SPME-HPLC-UV	PDMS/DVB	53–96 ng/g	120–2000	NA	NA	[45]
28 OCPs	Marine sediment	PHWE-SPME-GC-MS	PDMS/DVB	0.11–16 ng/g	0.24–500 ng/g	80–115%	External calibration	[46]
PBBs and PBDEs	Soil samples including sandy–marine, river, soil, sludge	HS-SPME-GC-MS/MS	PDMS	0.0052–0.214 ng/g for sediment, 0.014–0.625 ng/g for soil	NA	81.5–104% for sediment 96–105% for soil	Standard addition	[47]

PBDEs	Sediment	HS-SPME-GC-ECD	PDMS	0.5–1.3 ng/g	NA	NA	Standard addition	[48]
PBDEs	Soil	HS-SPME-GC-MS	PM-β-CD/OH-TSO	0.013–0.078 ng/g	0.1–10	78.2–99.3%	External calibration	[49]
Tetra-hexa-brominated diphenyl ethers	Sediments	HS-SPME-GC-MS/MS	PA	0.05–0.5 ng/g	NA	76–111%	Standard addition	[50]
PCBs	Landfill leachates and sediments	MA-HS-SPME-GC-MS/MS	PDMS	0.0001–0.007	0.01–1.2	10–126%	Internal standard	[51]
PCBs	Ash	MA-HS-SPME-GC-ECD/MS	PDMS/DVB	0.2–1.5	NA	83–111%	NA	[52]
Dioxins	Soil	Ultrasonic/heat-cool-SPME-GC-MS/MS	PDMS	2.2–77.2	50–250	NA	NA	[53]

CW/TPR, carbowax/templated resin; DMOD-3-TMOS-AC, dimethyloctadecyl[3-(trimethoxysilyl)propyl]ammonium chloride; DVB, divinylbenzene; ECD, electron capture detector; EDP, endocrine disruptor pesticides; GC, gas chromatography; HPLC, high pressure liquid chromatography; IL, ionic liquid; LOD, limit of detection; MA, microwave assisted; MAME, microwave-assisted micellar extraction; MS, mass spectrometry; PANI, polyaniline; PFOA, perfluoroalkyl acid; PHWE, pressurized hot water extraction; PM-β-CD/OH-TSO, permethylated-beta-cyclodextrin/hydroxyl-termination silicone oil; PP, polyphosphate; PPy, polypyrrole; SSW, stainless steel wire; SWNT, single wall carbon nanotube; THM, trihalomethane; UV, ultraviolet; PA, polyacrylate; PDMS, polydimethylsiloxane; SPME, solid phase microextraction.

biphenyls (PCBs), polybrominated diphenyl ethers (PBDEs), and perfluoroalkyl substances (PFOS) in environmental water, soil, and sediment. Since the allowed concentration of POPs in environmental samples is very low, the main criteria for selection of a specific analytical method are low detection limits and high selectivity. For example, the World Health Organization (WHO) regulated the maximum individual residue concentration of OCPs in drinking water to be 100 ng/L and the total amount of pesticides to be 500 ng/L. In most applications, SPME can achieve adequate sensitivity, accuracy, and precision for current legislation requirements after optimizing the sampling conditions including types of fiber, extraction modes, quantification methods, extraction time, desorption conditions, and sample ionic strength and pH value. In the following sections, the selection of extraction fibers, extraction modes, and quantification methods will be discussed in detail, especially the ones for soil sample analysis.

2.1 Selection of Extraction Fiber

Selection of a suitable fiber coating is the first step in SPME method development. Varied commercially fibers and homemade fibers are available in the literature. Commercially available SPME coatings include PDMS, PA, polyethyleneglycol (PEG), carbopack Z-PDMS, DVB-PDMS, carboxen–PDMS, and DVB-CAR-PDMS, and can be classified as liquid sorbents and solid sorbents. Liquid coatings such as PDMS, PA, and PEG are composed primarily of "liquid-like" polymers, which retain analytes through absorption processes. The analytes partition in and out of the coating, and are attracted to the coating primarily by van der Waals interaction and hydrogen bonds. Solid coatings such as carbopack Z, carboxen, and DVB are porous carbon or polymer, which extracts analytes by surface interaction (adsorption), such as pi–pi bonding, hydrogen bonding, or van der Waals interaction. The retention of the analyte depends upon the contact area, which is in turn dependent on the size of the analyte and the pore diameter.

Selection of an appropriate fiber coating for a particular application depends on the following three criteria: (1) molecular weight and size of an analyte, (2) analyte polarity, and (3) complexity of the matrix. The molecular weight of an analyte determines how rapidly it moves within the coating. Molecules with low weight move faster and are difficult to be retained by liquid coatings, especially ones with low thickness. Generally, for small analytes (MW <125), a solid coating is a better choice over a liquid coating. Carboxen–PDMS fibers were found to be the best choice for this type of compounds. On the other hand, for an analyte with a high molecular weight, such as most POPs, liquid coating is a better option. In addition, DVB sorbent was found to be suitable for many of the large molecules due to their larger pores, and has been widely used for the sampling of OCPs in soil samples [44–46]. Last, it is worth mentioning that carbopack Z is a porous graphitized carbon black, which was originally developed to extract coplanar PCBs and dioxins. The coplanar congeners strongly interact with the carbon surface, and result in being more selective toward planar congeners than nonplanar ones.

The second criterion for fiber selection is the analyte polarity. According to the "like dissolves like" principle, a nonpolar compound tends to dissolve more in a nonpolar coating, while a polar compound prefers to retain in polar coating. PDMS was the most common nonpolar coating that is highly crosslinked and thermal stable. It has been widely reported for the sampling of POPs in environmental samples [6,8,12,13], especially solid samples such as agriculture soils [44], estuarine sediment [39,42,43,48], and landfill soil [47,51,54]. There are only two polar fiber coatings among commercial fibers: 60-μm PEG and 85-μmPA. PEG, the same material as carbowax, and the most polar phase, has been applied for sampling OCPs in water samples. PA fiber, slightly less polar than PEG, is suitable for extraction of the analytes that are poorly extracted by PDMS. PA has been reported for preconcentration of pesticides [10] and flame retardants [50] in river water and sediments, respectively.

The last criterion is the consideration of matrix complexity. For a solid coating that is based on adsorption, competition for space in fiber coating can be challenging. Compounds with low affinity can be displaced by other compounds with high affinity, which results in wrong quantification results. Silva et al. [55] also found that the solid coating was easily contaminated by complex matrices such as protein, sugar, or particle, and shortens the lifetime. An overcoated layer of PDMS on the surface of DVB coating can effectively reduce biofouling and increase the lifetime of the DVB fiber when applied for complex matrix analysis.

In addition to commercial fibers, homemade fibers were widely investigated to overcome the limitations of commercial fibers regarding extraction efficiency and cost [56]. Carbon-based coating was the most introduced. Sun et al. [16] proposed a simple method to prepare an activated carbon fiber for extraction of 17 banned OCPs in water samples. A carbon rod was directly immersed in different concentrations of zinc chloride and then activated in a stove at 750°C under a nitrogen environment. The extraction efficiency of the prepared fiber for targeted OCPs was compared with that of commercial PDMS, PA, and Carboxen PDMS fibers. The results indicated a significant improvement in the extraction amount for activated carbon fiber. To further improve the extraction capacity of the developed carbon fiber, the same group investigated an SPME-circulated cooling device for the determination of two OCPs in water samples. The cooling activated carbon fiber revealed a higher extraction efficiency when compared with the one without cooling [17].

Graphene is an electron-rich and hydrophobic nanomaterial with a large specific area and π–π electrostatic stacking property, and can serve as an extraordinarily good sorbent in sample preparation. Graphene has been used as an SPME extraction phase and for HS and DI sampling of OCPs from aqueous samples [18–20]. Similar LODs (<0.0008 μg/L) were obtained from these reports. The good performance of graphene fiber for OCP sampling is due to a large surface area from its porous, wrinkled, and three-dimensional structure, and strong interaction between the aromatic rings of the OCP compound and the delocalized π–π electron on the coating. A similar interaction was also observed between the graphene coating and PBDE compounds [21].

Other carbon-based material such as polymer-functionalized SWNTs [22] and Fe_3O_4 nanosheet-modified bamboo charcoal [23] were also used as SPME coatings for the DI sampling of PBDEs in water samples. Both coatings showed LODs ranging from 0.00008 to 0.0008 μg/L for selected compounds.

There are several methods that have been applied for SPME coating preparation. They are the dipping technique, sol–gel technology [20], physical deposition [19], and electrochemical modification. Among them, the electrochemical method is one of the most promising coating methods, and has been widely used for SPME coating preparation. PANI and its derivatives are common coatings prepared by the electrochemical method, and have been applied for sampling of OCPs, PCBs, and PBDEs. Li et al. [24] electrodeposited a layer of PANI coating on an SSW and used this for sampling of OCPs in water. An LOD range of 0.0001–0.0016 μg/L was observed for various OCPs and proved to have a better extraction efficiency than the one obtained by the commercial carbowax-DVB coating. On the other hand, the application of PANI coating was limited by poor thermal stability, that is, 220 °C for common PANI. Thermal treatment of PANI at different temperatures destroys crystalline and chemical structures, and influences the extraction efficiency. In Mehdinia's group [25], the researchers combined the PANI material with thermal stable graphene to reinforce the stability of this conducting polymer. The thermal stability of the coating was enhanced, but the obtained LODs were higher than the one without graphene addition. The authors claimed that the NH group in PANI was a specific group interacting with the OCPs, and these groups were masked to some level when graphene was added. In order to simultaneously improve the extraction capacity as well as the thermal stability of the PANI coating, Gao et al. [26] employed a nanostructured PANI-IL coating. The coating was thermally stable up to 350 °C, and can be used for >250 times injections. In addition, the extraction capacity of the PANI-IL was better than that of the pure PANI coating. Last, fluorinated [27] and perfluorooctanesulfonic acid-doped [28] PANI coatings were developed to analyze PCBs and PBDEs in water samples. The results demonstrated an excellent thermal stability and extraction efficiency of the proposed fibers compared to those of pure PANI and commercial fibers.

Besides PANI, PPy is another popular conducting polymer that has been widely used for SPME coating. Mollahosseini [29] introduced a PPy coating doped with PP to enhance the thermal stability of pure PPy coating. The PP-doped PPy coating was then applied for the HS sampling of 11 OCPs from water samples, and results indicated excellent LODs (ranging from 0.00012 to 0.00031 μg/L) for selected OCPs due to strong π–π interaction between the polymer and compounds. Polythiophene is another conducting polymer that received increasing attention due to the higher extraction efficiency toward some analytes. Pelit and Dizdaş [30] fabricated a polythiophene coating by the electropolymerization method, and applied the fiber for determination of EDPs. The coating achieved satisfactory results compared with those of commercial fibers and provided LODs ranging from 0.02 to 0.64 μg/L for the selected compounds.

In addition to the above-introduced homemade fibers, coatings made of benzo-15-Crown-5 [31], polymethylphenylvinylsiloxane [32], and titania particles [33] were also reported for the sampling of OCPs in aqueous samples. All of them provided a remarkable extraction efficiency toward the selected compounds.

2.2 Selection of Extraction Mode

The two common operation modes for SPME are DI and HS sampling [2]. In DI-SPME, the fiber is immersed in the sample matrix, and the analyte gets transported directly from the sample matrix to the extraction phase. The DI mode is suitable for the sampling of less volatile compounds in a simple liquid sample or compounds in a gaseous sample. Agitation methods, such as high air-flow rate, rapid movement of fiber or vial, sonication of samples, are frequently used to accelerate the extraction rate in the DI mode. The DI mode has been applied for sampling of OCPs [7,8,16,20,24,29], PBDEs [23,34,36], and PFOS [38] in environmental aqueous samples.

The application of the DI mode for solid sample analysis is more challenging than for aqueous sampling. DI-SPME has been reported for soil/sediment analysis in two protocols. The first one is to directly expose the coating to the solid matrices [57,58] (Figure 2(a)). Analytes present in soil or sediments distribute between the matrix and fiber coating till they reach equilibrium. Since the soil sample cannot be well agitated and the volume of SPME fiber is very small, only the analytes from the soil close to the fiber coating could be extracted, and the determined concentration may not represent the whole sample. The other approach is to prepare a suspension of the solid sample in water (Figure 2(b)). An appropriate amount of water was spiked into the solid sample to dissolve the analytes presented in the sample before being extracted by the DI-SPME [59]. This procedure is quite effective if the analytes are not too strongly bound to the solid matrix. Research found that the addition of water to soil samples improved the release and transport of volatile and semivolatile compounds from soil samples to the SPME fiber. But directly immersing the SPME fiber into the soil or sediment, or their slurry can easily damage the coating during extraction. As a result, DI-SPME fiber has not been reported for sampling of POPs from a solid sample. The analytes should be extracted from the solid matrix and dissolved in water before using DI-SPME (Figure 2(c)). In this application, techniques such as MA and PHWE [46] were utilized to preextract the analytes into water; then, the extractant was further sampled by DI-SPME. Moreno's research group applied a surfactant to facilitate the desorption of analytes from

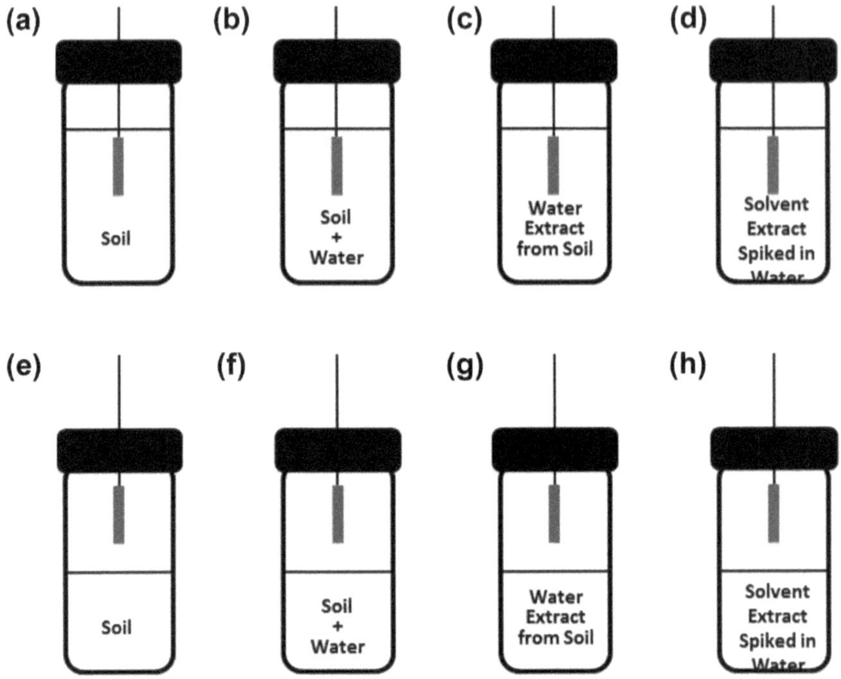

FIGURE 2 DI and HS-SPME from solid sample analysis.

the solid sample, and developed an approach by coupling MAME with the DI-SPME method for determination of OCPs in agricultural soils [44,45]. The last approach was to extract the analytes from a solid sample using an organic solvent, and then spike the extracted organic solvent into water for DI-SPME sampling (Figure 2(d)) [41].

The main challenge of DI-SPME sampling is the potential fiber contamination and inaccurate quantification results due to the matrix effect in complex samples, a process that is also called (bio)fouling. HS sampling was introduced to address the challenges associated with DI sampling but is limited to volatile compounds. For HS sampling, the fiber is inserted into the HS of the closed sample vial, which prevents the fiber from coming into direct contact with a dirty sample matrix and lengthens the lifetime of the fiber. In addition, HS sampling usually has a high sampling rate for volatile compounds due to a high diffusion coefficient in the gas phase. Many research studies have investigated the application of HS-SPME on sampling of OCPs [6,13,19,25,30–33,35], PBDEs [14,22,28], PCBs [15,27] in environmental aqueous samples. In these applications, high temperatures ranging from 70 to 100 °C were usually required to facilitate the release of analytes into the HS. However, it is dangerous to use temperatures >100 °C because of the water boiling issue. Besides, a high temperature decreases the distribution coefficient and results in a low extraction efficiency. To overcome this limitation, Liao et al. [17] developed a circulating cool system, which can cool down the fiber coating temperature during sampling. The extraction efficiency of γ-hexachlorocyclohexane obtained from the cooling device was higher than that without cooling. Pawlisyzn's research group [60] developed another cooled fiber device using liquid CO_2 as the coolant, whose coating temperature can be <0 °C. This cooled SPME can achieve exhaustive sampling of organic pollutants from soil samples. In addition, this device has been coupled with a commercial autosampler and achieved full automation [61–63].

HS sampling is advantageous for solid sample analysis. The sampling can be conducted by directly exposing the fiber on the HS of the solid sample [47] (Figure 2(e)), or the slurry of soil and water mixtures [39,48–50] (Figure 2(f)). Similar to water sampling, a high temperature was commonly required in these applications. However, most analytes such as OCPs have a much stronger interaction with solid samples than with aqueous samples. Higher energies such as microwave [51] and ultrasound [53] were usually applied to assist extraction. Researchers [18] compared the extraction efficiency of the HS sampling of OCPs from water with and without the assistance of microwaves. Results revealed that, with microwave assistance, the extraction efficiency with sample temperature at 36 °C was better than that without microwaves at 100 °C. The only limitation of this approach is the requirement of a modified microwave device that combined the SPME extraction capacity. In the other approaches, similar to DI-SPME for solid sample analysis, organic solvent [41–43] (Figure 2(h)) can be used to preextract the analytes from the soil before proceeding with HS-SPME. No reports were found for sampling POPs in soil using water as a preextraction solvent as shown in Figure 2(g).

2.3 Selection of a Quantification Method

2.3.1 Detection Technique

As aforementioned, the SPME fiber is housed in a commercial holder and the whole device is similar to a microsyringe. Except for special applications, the SPME method uses an existing injector on the GC instrument, and analytes are thermally desorbed into the inlet. For years, chemical techniques for analyzing POPs were still centered on the use of GC coupled to various conventional detectors, typically the ECD. SPME coupled with GC-ECD for sampling POPs, such as OCPs, PCBs, and PBDEs, were widely reported in the literature (Table 1). The LODs of these approaches can be lower than parts per trillion (ppt) and sub-ppt. On the other hand, though the ECD is a very sensitive detection system, unambiguous identification of all the congeners of PCBs and PBDEs is impossible and misinterpretation easily occurs. In recent years, with the advent and popularity of chemical instrumentation, MS has become a sensitive and selective analytical tool for testing these types of compounds. SPME coupled with GC–MS can achieve a comparable sensitivity with that of GC-ECD for analysis of OCPs and PBDEs, but a higher selectivity toward the congeners of PCBs and PBDEs. Tandem MS was also applied for the analysis of complex matrices and showed a further improvement of sensitivity and selectivity.

In addition to GC, HPLC was applied for the analysis of OCPs, PBDEs especially PFOS compounds. For HPLC application, the interface needs to be adjusted when used in combination with SPME [37], or offline desorption using a small volume of an appropriate solvent can be employed [7,38,44,45]. Offline solvent desorption may result in a dilution effect and decrease the sensitivity of the method. SPME coupled with HPLC usually obtained higher LODs than that coupled with GC. Both UV [44,45] and MS [37,38] detectors were reported to integrate with HPLC and SPME sample preparation for analysis of OCPs and PFOS, respectively.

SPME can be integrated directly with MS. A fiber introduction mass spectrometry (FIMS) was introduced by Silva et al. to determine OCPs in herbal infusions of *Passiflora* L. [35]. Although the obtained LODs (0.3–3.9 μg/L) were much higher than for those coupled with GC separation, the simplicity, fast detection (20 s), as well as robustness are the main advantages of the FIMS method.

2.3.2 Calibration Method

SPME is a nonexhaustive extraction technique, for which a calibration method should be developed to obtain the relationship between the extraction amount and the initial sample concentration. Several calibration methods have been developed for SPME sample preparation [64]. Traditional calibration methods including external calibration, standard addition, and internal calibration can be employed for SPME quantification.

The most reported for POP analysis is the external calibration method (Table 1). This method involves preparation of several standard solutions using a matrix-matched blank sample. The standard solutions and unknown samples are subsequently analyzed using the same extraction conditions in order to establish the relationship between the extracted amount and the analyte concentration in the sample. The concentration of the target analytes is then determined from the obtained calibration curve. This is the simplest calibration method for simple sample matrices; however, it may be an inappropriate method when the sample is complex or the extraction conditions are difficult to reproduce. Although this method has been applied for quantification of POPs in aqueous samples, and proved to be accurate in most publications, some researchers found serious matrix effects when using this calibration method for real sample analysis. For example, Torres Padron et al. [7] and Chen [34] both found that the spiked relative recovery of OCPs in real water samples was much <100%. They claimed that this was because the spiked OCPs bound to the complex matrix in real water, and resulted in a lower free concentration. Since the SPME can only extract the unbound compounds, the external calibration curve done by spiking in pure water is not appreciate for complex matrix calibration. In addition, for soil sample analysis, experimental results showed that different soil compositions significantly influenced the slope of the external calibration curve [41,65], which in turn demonstrated that the selection of a matrix-matched sample is very important for the external calibration method.

The standard addition method is accomplished by preparing a series of known concentrations of target analytes using an unknown sample. Compared to external calibration, this approach is able to compensate for matrix effect to some extent. The standard addition method has been frequently applied for quantifying OCPs [42,43], PBDEs [47,48,50] in soil samples, and achieved satisfactory sample recoveries. The limitation of this method is the requirement of a relatively large amount of unknown samples. Besides, the standard addition was shown to have a matrix effect in soil applications [41].

The internal standard approach involves adding a known concentration of an internal standard that is not present in the sample matrix, and can mimic the behavior of an analyte during extraction and instrument quantification. Calibration is then done by comparing the ratio of the analyte signal to that of the internal standard with a known concentration [51]. This method shows a significant advantage in terms of minimizing matrix effects and loss of target analytes during sample preparation. However, it is difficult to find a suitable compound to be an internal standard. Isotopically labeled internal standards are preferable, but they are usually not easy to find and are generally expensive.

Besides traditional calibration methods, equilibrium calibration and exhaustive calibration method were reported for soil sample quantification [59,66]. Equilibrium calibration is based on the relationship between equilibrium extraction amount and sample matrix concentration. When the distribution coefficient and the sample volume are known, the concentration of the sample matrix can be calculated by determining the equilibrium extraction amount. One advantage of equilibrium sampling is to minimize the matrix effect in the calibration process, while the limitation of this method is the long equilibrium time and the difficulty in obtaining an accurate distribution coefficient. Exhaustive extraction occurred when >95% of the analytes present in the sample matrix was accumulated on the fiber coating. For SPME, due to the small volume of the extraction phase, exhaustive extraction rarely happened unless a high distribution coefficient was achieved.

2.4 Method Validation

After optimization of the SPME method, the method should be validated in order to demonstrate the suitability for a particular application. Validation experiments should be conducted to evaluate the method sensitivity, precision, and accuracy. First of all, method sensitivity, expressed as the LOD, should be determined for a developed method to compare with the regulation limit. As shown in Table 1, most methods were found to have detection limits in a range below the concentrations set by regulatory bodies. For example, the WHO has set the maximum individual residue concentration of OCPs in drinking water as 100 ng/L and the total amount of pesticides as 500 ng/L. Second, the precision, which was expressed in terms of relative standard deviation% values should be determined. For the reported data, most of them were below 15% for SPME method (not shown in Table 1). Third, the accuracy of the SPME methods can be evaluated by relative recoveries of spiked samples; comparing with the results obtained from standard methods for that particular application [40,44]; or analysis of certified reference materials [48] with known concentrations. In most reports, SPME was found to have no matrix effect, and recoveries ranged from 70% to 120%. However, in some complex matrices, such as soil [51], sewage [7], and wastewater [14], SPME can have serious matrix effects and low relative recoveries. The developed SPME methods were also compared to standard methods such as Soxhlet extraction for sediment analysis [39,46]. Comparable quantification results were obtained from the developed SPME method and the standard methods.

2.4.1 Summary of SPME for POP Analysis

SPME has been proved to be practical for the analysis of POPs from environmental samples. After careful optimization, SPME can provide precise and accurate quantification results. The optimization procedures including selection of fiber coating, extraction mode, and quantification method should be carefully investigated. In addition, extraction and desorption conditions, such as time and temperature, should be evaluated during the method development process. Last, sample conditions such as ionic strength and pH value should be optimized since they may significantly influence the extraction efficiency.

3. LIQUID PHASE MICROEXTRACTION

Solvent (liquid phase)-based microextraction approaches are being used in various forms and procedures, such as single drop microextraction (SDME) [67], hollow fiber liquid phase microextraction (HF-LPME) [68–70], and dispersive liquid–liquid microextraction (DLLME) [71,72]. These micro-LLE techniques eliminate the disadvantages of conventional liquid extraction methods, such as time-consuming operation and need for specialized apparatus. They are inexpensive and offer considerable freedom in selecting appropriate solvents for the extraction of different analytes. Since very little solvent is used, exposure to a toxic organic solvent is only minimal. At the same time, LPME combines extraction, concentration, and sample introduction in one step. The application of LPME for the sampling of POPs in environmental samples is listed in Table 2. The applications of SDME and HF-LPME for pesticides analysis are well documented [73]; thus, in this section, SDME and HF-LPME will be briefly introduced in terms of operation procedure and extraction condition optimization. More focus will be on the application of DLLME for POP analysis.

TABLE 2 Application of LPME for Sampling of Persistent Organic Pollutants in Environmental Samples

Compounds	Samples	Sample Preparation Methods	Solvent Used	Salt (pH)	LOD μg/L	Linear Range μg/L	EF	Recovery	Others	References
11 OCPs	Sea water	DI-SDME-GC-ECD	4 μL of 2:1 (v/v) *n*-hexane and ethyl acetate	10 mg/L NA	0.001–0.012	0.006–2	272–875	72–91%	15 min extraction with 10 mL water at 900 rpm	[74]
14 OCPs and pyrethroid	Drinking water	DI-SDME-GC-ECD	1.6 μL of hexane	No addition NA	0.003–0.6 *0.01–2*	Varied	NA	71–107%	15 min extraction with 10 mL water at 380 rpm	[75]
18 OCPs	Wastewater	DI-SDME-GC-MS	2 μL toluene	No addition NA	0.022–0.101	0.5–16	5–232	43–103%	37 min extraction with 10 mL water at 300 rpm	[76]
BDE-209	Water	DI-SDME-RP-HPLC-VWR	3 μL toluene	No addition no adjustment	0.7	1–1000	10.6	91.5–102.8%	15 min extraction with 5 mL water in 600 rpm	[77]
5 OPPs	Soil from ecological agriculture farm	UA-HS-SDME-GC-NPD	3 μL octanol	20% SC no adjustment	0.1–2.0 ng/g	5–200 ng/g	1.4–12.7	NA	30 min extraction with 10 mL water ultrasonically	[78]

Continued

TABLE 2 Application of LPME for Sampling of Persistent Organic Pollutants in Environmental Samples—cont'd

Compounds	Samples	Sample Preparation Methods	Solvent Used	Salt (pH)	LOD μg/L	Linear Range μg/L	EF	Recovery	Others	References
PBDE	Tap water, river water, leachate	HF-MMLLE-GC-MS	10 μL *n*-undecane	No addition No adjustment	0.0002–0.0009	0.001–0.1	<5200	86.1–110.4%	60 min extraction with 100 mL water at 1200 rpm	[79]
5 OCPs	Water	HF-LPME-GC-MS	3 μL toluene	No addition NA	0.1	1–500	NA	NA	15 min extraction with 20 mL water at 480 rpm	[80]
DDT	Agriculture field water, seawater, and river water	MA-HS-CT-HF-LLME-GC-ECD	4 μL 1-octanol	No addition pH = 6	0.02–0.03	0.05–3	92–299	94–101.3%	6.5 min HS extraction with 10 mL water and microwave assisted	[81]
40 POPs	Orange juice, tomato pulp, porcine plasma	HF-LPME-GC-MS-MS	20 μL toluene	NA NA	0.0009–0.182	LOQ–10	13–161	65–120%	37 min Extraction with 4 mL water at 1100 rpm	[82]

OPPs	Water	IL-solvent bar-HPLC	Ionic liquid	No addition NA	0.015–0.026	1–200	NA	86.7–103.7%	60 min extraction with 25 mL water at 600 rpm	[83]
PBDE	Soil, dust, spiked lake water	HF-LPME-GC-ICP-MS	4 μL decane	No addition NA	0.015–0.0405	0.2–20	30–84	86.7–110.9%	20 min extraction with 10 mL water at 1000 rpm	[84]
OCPs and PCBs	Marine sediments	MASE-HF-LPME-GC-MS	5 μL toluene	NA NA	0.07–0.7 ng/g	5–500 ng/g	NA	73–111% for OCPs 86–110% for PCBs	1 g + 10 mL water at 80 °C for 20 min, then extraction with HF at 700 rpm	[85]
OCPs	River, tap, sea, and reservoir water	DLLME-GC-MS	13 μL tetrachloroethane and TBME mix	3% SC NA	0.0008–0.0025 *0.0004–0.0013*	0.002–5	1885–2648	54–119.4%	90 s extraction with 10 mL water by manual shaking	[86]
17 Pesticides	River water, sewage water, seawater	DLLME-GC-MS–MS	178 μL trichloroethane 1.9 mL ACN	No addition, no adjustment	0.0005–0.018	0.08–40	46–72	82–118%	3 min extraction with 10 mL water by manual shaking	[87]

Continued

TABLE 2 Application of LPME for Sampling of Persistent Organic Pollutants in Environmental Samples—cont'd

Compounds	Samples	Sample Preparation Methods	Solvent Used	Salt (pH)	LOD µg/L	Linear Range µg/L	EF	Recovery	Others	References
BDE-209	River water	DLLME-HPLC-VWD	22 µL tetrachloroethane 1 mL Tetrahydrofuran	No addition (decrease) no adjustment	0.2	1–500	153	89.9–95.8%	5 s extraction with 5 mL water by manual shaking	[88]
PBDEs	Landfill leachate, water	DLLME-RPHPLC-VWD	20 µL tetrachloroethane 1 mL ACN	No addition NA	0.0124–0.0556	0.05–100	268–305	87–1 19.1%	5 s extraction with 5 mL water by manual shaking	[89]
6 Pesticides	Tap water, river water, lake water	UA-DLLME-GC-FID	15 µL carbon tetrachloride 1 mL ACN	No addition NA	0.09–0.57	1–100	315–1153	90.5–107.7%	5 min Extraction with 5 mL water by ultrasound	[90]
PBDE	River, lake water	UA-DLLME-GC–MS	10 µL tetrachloroethylene 1 mL acetone	No addition NA	0.0004–0.00215	0.001–0.5	845–1050	70.6–105.1%	2 min extraction with 5 mL water by ultrasound	[91]

PBDE	Well, river, sea, leachate, and clover water	SPE-DLLME-GC-ECD	22 μL tetrachloroethylene 1 mL elution sample	No addition NA	0.00003–0.00015	0.0001–0.5	6838–9405	64.5–110%	100 mL extraction with LC-C18, then DLLME	[92]
OCPs	Lake water	Low-density-DLLME-GC-ECD	11 μL 1-nonanol 400 μL methanol	No addition NA	0.0002–0.0094	0.005–5	1309–3629	73–119%	3 min extraction with 10 mL water by manual shaking	[93]
OCPs	Seawater, field fresh water	Low-density-USAEME-GC-ECD	10 μL decanol USA	2% SC NA	0.0006–0.0029	0.005–2.5	1202–4587	75–107% 70–99%	4 min extraction with 10 mL by manual shaking and ultrasound	[94]
6 OCPs	Agricultural surface water samples	Homogeneous LLME flotation-assisted GC-MS	150 μL 1-decanol 2.5 mL ethanol	1.5 mol/L NA	1.4–7	10–150 (0.998)	48.7–845.2	NA	5 min extraction in 22 mL water with air flotation	[95]

Continued

TABLE 2 Application of LPME for Sampling of Persistent Organic Pollutants in Environmental Samples—cont'd

Compounds	Samples	Sample Preparation Methods	Solvent Used	Salt (pH)	LOD μg/L	Linear Range μg/L	EF	Recovery	Others	References
OCPs	Lake water, seawater, stream water	Surfactant assisted-DLLME-GC-ECD	10–12 μL dodecylacetate and 2-dodecanol 120 μL surfactant solution	NA NA	0.0005–0.002	0.01–1	1901–3530 for DA 1881–3108 for 2-D	90.7–111.2%	0.5 min extraction in 8 mL water without shaking	[96]
OCPs	River water, reservoir water	Low-density magnetofluid-DLLME-GC-ECD	15 μL magnetofluid (oleic acid-coated Fe_3O_4)	3% SC NA	0.0018–0.0084	0.05–10	156–214	70.1–98.6%	4 min extraction in 30 mL with vortex agitating	[97]
PCBs, PBDEs	River water	Temperature-assisted IL-DLLME-HPLC	40 μL IL, 1 mL methanol	NA pH = 1	0.1–0.4	0.5–500	278–343	81–125%	3 min extraction with 5 mL water at 40 °C. IL was separated and dissolved	[98]
8 PCB	Tap water and well water	DSLPME-GC-ECD	10 μL octanol	1 mol/L Na_2SO_4 NA	0.01–0.07	0.1–10	578–729	93–105%	10 min extraction in 8 mL water with a stir bar	[99]

PBDE	River water	SFO-DLLME-HPLC-UV	25 μL 2-dodencanol without disperser	No addition NA	0.01–0.04	0.5–75	921–1462	92–118%	25 min extraction in 40 mL water at 60 °C with a stir bar and cooling to separate	[100]
OCPs	Tap, well, and lake water	USAEME-GC-uECD	200 μL chloroform Ultrasonication	NA NA	0.002–0.016	1–10000	NA	80–102%	15 min extraction in 10 mL water with ultrasound	[101]
PBDE	Water sample and soil	Surfactant-assisted USAEM-GC-MS	50 μL isooctane Ultrasonication	No addition, no adjustment	0.001–0.002	0.005–10	250	>96%	10 mL water + 40 surfactant, extract for 7 min at 80 °C ultrasonically	[102]
PFOS	Tap, river, and well water	VALLME-GC-MS	50 μL octanol vortex agitation	No addition NA	0.0016	0.005–0.5	250	90.8–95.5%	2 min extraction in 20 mL water with vortex centrifugation	[103]

Continued

TABLE 2 Application of LPME for Sampling of Persistent Organic Pollutants in Environmental Samples—cont'd

Compounds	Samples	Sample Preparation Methods	Solvent Used	Salt (pH)	LOD μg/L	Linear Range μg/L	EF	Recovery	Others	References
PBDE	Sediments	Ultrasound leaching-SFO-DLLME-GC-MSMS	22 mg 1-dodecanol 0.1 mL MeOH	~1 mol/LSC NA	0.0005–0.0018 ng/g	0.002–1	NA	71–104%	1 g Sediment +1.2 method, 0.4 mL extractant was backextracted for 10 min by DLLME	[104]
PBDE	Sediments	USAL-DSPE-DLLME-GC-MSMS	60 μL carbon tetrachloride 1 mL acetone	NA NA	0.02–0.06	0.07–1000	NA	>80%	0.25 g Sediment + 1.5 acetone–ultrasonic–1 mL extractant into 5 mL water–shake for 10 s kept for 35 °C	[105]

EF, enrichment factor; LOQ, limit of quantification; MASE, microwave-assisted solvent extraction; OPP, organophosphorus pesticide; USA, ultrasound assisted.

4. SINGLE DROP MICROEXTRACTION

SDME was first introduced by Cantwell and Jeannot in 1996 [106] as an alternative SPME concept, in which a single liquid drop was utilized as the extraction phase. This microdrop of the solvent is normally suspended directly at the tip of the needle of a GC microsyringe, and could be placed in the HS or immersed into the bulk of the sample (Figure 3(a)). The analytes diffuse into this droplet in a similar manner as into SPME fiber. After exposure for a predetermined time, the microdrop is retracted into the microsyringe and injected into the analytical instruments such as GC, HPLC coupled with various detectors. SDME became a very popular LPME technique because it combines extraction and concentration in a minimum number of steps. Besides, unlike fibers, solvent drops can be renewed for each extraction, and become an elegant way to overcome the limited available fiber coatings, and possible carryover. Last, the SDME technique has been coupled with an autosampler and achieved full automation [107].

SDMEs were applied in the extraction of OCPs [74,76], OPPs, and BDE-209 [77] in different types of water using DI-SDME. However, SDME in the HS mode has not yet been employed for POP analysis, as is evident from the lack of HS-SDME publications in the open literature. HS sampling, as described in the SPME section, requires a high temperature to release the analytes from the sample matrix, especially for the less volatile POP compounds. However, the high temperature would in turn cause the loss of extraction solvent in the droplet. The reported DI-SDME used n-hexane, octanol, and toluene as the extraction phase with a volume ranging from 1.6 to 3 μL. A larger volume of the extraction phase can be used with assistance of a specially made funnel designed by Qian et al. [74]. The volumes of water sample were about 5–10 mL. Fifteen to fourty minutes was found to be an adequate extraction period with the agitation rate ranging from 380 to 900 rpm. Longer extraction times can lead to solvent loss and decrease the extraction capacity, while a higher agitation rate may result in an unstable droplet. Typical SDME EFs of OCPs and BDE-209 for water samples ranged between 5 and 875.

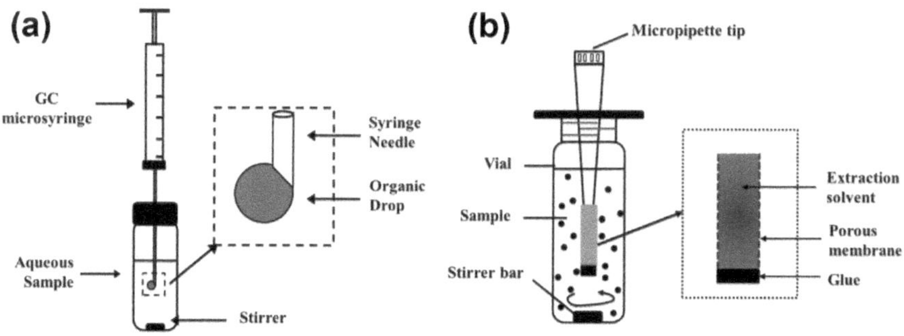

FIGURE 3 Configuration of SDME (a) and HF-LPME (b).

SDME was also applied for the sampling of pesticides in soil. The pesticides were extracted from the soil using water with the assistance of ultrasound. The extractant was further preconcentrated by HS-SDME for 30 min at 60 °C. A more detailed application of SDME for pesticide sampling and other organic compounds can be found in the literature [108].

The limitations of SDME include (1) the volume of the extractant microdrop is small. It is often not >3 μL, which confines the extraction capacity and affects the extraction efficiency; (2) the microdrop is unstable and easily dislodged from the tip of the microsyringe needle due to prolonged extraction time, fast stirring rate, or complex matrix; (3) the choice of the extractant is limited, because it must satisfy certain conditions such as water immiscibility and appropriate density to maintain the suspension; (4) reproducibility is often poor due to significant dissolution of the organic extractant during sampling; (5) the major applications of SDME are limited to liquid matrices, and very few studies have been addressed for solid matrices.

4.1 Hollow Fiber LPME

An innovative modification to SDME, which can further improve the extraction capability of LPME and minimize some of the main shortcomings of SDME, is the use of a polymeric membrane that serves as a support for extracting the solvent and an interface between donor and acceptor phases (Figure 3(b)). HF-LPME uses an HF to stabilize and protect the extraction solvent, which allows an effective agitation of the sample matrix during sampling. In addition, the small pore size on the HF prevents large molecules and particles from entering the acceptor phase, resulting in a clean-up of the sample matrix during sampling. Based on this consideration, HF-LPME involves not only a sample enrichment technique but also a sample clean-up procedure, which makes it applicable to complex matrices. Further, HF-LPME has achieved automation when combined with a GC autosampler [109].

HF-LPME has been reported for preconcentration of OCPs, PCBs, and PBDEs in various types of water samples and solid samples. Undecane [79], octanol [81], and toluene [80,82,85] were used as the extraction solvent. Among them, toluene appears to be the most commonly used, because it provides good immobilization in HF pores, and is immiscible in water and stable enough over the extraction time. In addition, IL [83], which is a group of solvents with much potential in LPME, was reported for extraction of OPPs in water. The ability of ILs to extract more polar pesticides makes them suitable for combination with the LC system. The volume of the acceptor phase in HF-LPME (ranging from 3 to 20 μL) was much higher than in SDME, which enhanced the extraction capacity. Agitation rate for HF-LPME is not limited since the extraction solvent was protected by HF. High agitation rate enhances the extraction rate, and a broad range of stirring rate ranging from 480 to 1200 rpm was reported for POP analysis in water. Two agitation methods can be applied during extraction. One method was to rotate the water sample, while the other method was to refresh the extraction solvent by repeatedly pulling in and pushing out the extraction solvent from the HF using a microsyringe [69]. Temperature was also found to increase the diffusion coefficient and then the extraction rate. However, a higher temperature (>50 °C) was reported to provoke the solvent loss in the hollow membrane during sampling. Compared to SDME, the extraction time for HF-LPME was relatively longer, because the analytes should diffuse through the wall of HF to be extracted by the solvent. Sampling periods ranging from 20 to 60 min were reported for the sampling of OCPs and PBDEs in water and solid sample.

As discussed above, HF-LPME also has a clean-up effect because of the protection of hollow membrane. Therefore, HF-LPME can be directly applied for complex samples. Manso et al. [82] utilized HF for sampling of POPs in orange juice and tomato pulp without any pretreatment. The HF with 20 μL of toluene was directly immersed in fruit slurry diluted with water. EFs ranging from 13 to 161 were obtained for 40 compounds under optimized conditions. The EFs for HF-LPME can be achieved up to 5200 for PBDEs in water samples using *n*-undecane as the extraction solvent [79]. The method sensitivity was also comparable to that of SDME. However, similar to other microextraction techniques, HF-LPME also has some shortcomings. First of all, the operation of HF-LPME requires skillful practice. Second, the extraction time of HF-LPME is relatively long compared to that of other LPME techniques. Third, although semiautomation of HF-LPME has been reported, full automation still needs to be developed. Last, the extraction organic solvent is limited to one that is insoluble in water.

4.2 Dispersive Liquid–Liquid Microextraction

DLLME is an LPME technique introduced by Assadi and coworkers [110] in 2006. Like other microextraction techniques, it has quickly grabbed the attention of researchers and become very popular among analytical chemists. This new technique is based on a ternary component solvent system to form a cloudy point extraction. An appropriate mixture of two solvents—one acting as an extraction solvent and the other as a disperser solvent—was injected into a sample solution with a syringe; the mixture forms a cloudy solution in the aqueous phase; subsequent centrifugation was applied to the sediment as the extraction phase for further separation (Figure 4(a)).

DLLME has been widely reported for the analysis of POPs in environmental samples including water, soil, and sediment. Sampling conditions, such as types of extraction solvent and dispersive solvent, amount of the solvent used, agitation conditions, and pH and salt adjustment need to be optimized to achieve a good extraction performance.

4.2.1 Selection of Extraction Solvent

The type of extraction solvent is a key parameter that influences the EF in a DLLME process. Several criteria should be followed when selecting an extraction solvent: (1) the extraction solvent must have a high capacity for target

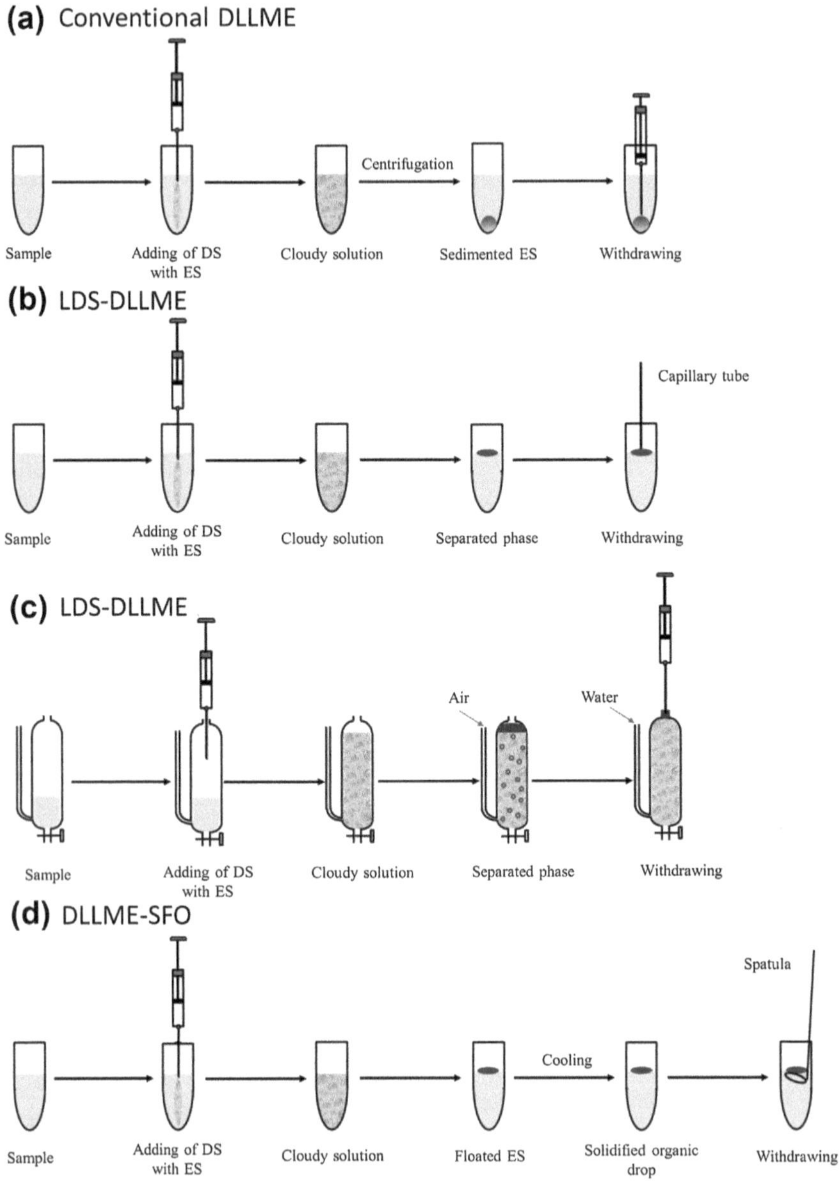

FIGURE 4 Various configurations of DLLME.

analytes; (2) low solubility in the aqueous phase; and (3) be compatible with the analytical instrument being used. In addition to these, (4) the solvent should have a density greater than that of water for easy separation by centrifugation, and form a cloudy solution in the presence of the disperser solvent. Solvents that meet all the requirements are not too many, and usually halogenated hydrocarbons such as tetrachloroethane [86,88,89,91,92], trichloroethane [87], and carbon tetrachloride [90]. These solvents have been used for extraction of OCPs and PBDEs in environmental samples. The amount of extraction solvent ranged from 10 to 178 μL with a sample volume of 5 or 10 mL. Various EFs ranging from 46 to 2648 and 153 to 1050 were achieved using these halogenated hydrocarbons as the extraction phase for sampling of pesticides and PBDEs from water samples, respectively (refer to Table 2).

Although halogenated solvent achieved high EF for the reported target compounds, its toxicity restricts the wide application of this method. To overcome this drawback, researchers have attempted to employ extraction solvent with a lower density than that of water. For example, nonanol [93], decanol [94,95], dodecanol [96], and dodecylacetate [96] have been investigated to extract OCPs from environmental water including river water, tap water, and well water. Unlike heavy solvents that can be separated by centrifugation, the lighter solvent floats on the surface of the sample after extraction. The solvent can be removed from the extraction cell using a capillary tube [94] (Figure 4(b)). However, the volume of the extraction solvent is usually very small ranging from 10 to 150 μL, and forms a very thin layer of solvent film floating on top of the water surface after extraction. It is difficult to separate the organic solvent from the water sample unless some special devices were employed. Hosseini et al. [95] designed a special extraction cell (Figure 4(c)) for sampling of six OCPs in water samples. After extraction, the low-density extraction solvent was separated from the sample matrix with the assistance of airflow. The extraction cell had a narrow opening, and the extraction solvent can be collected by elevating the water level inside the cell. Similar devices were reported elsewhere [111–113], and each design worked on a similar principle: centrifugation of the low-density organic solvent after the extraction procedure to accumulate the solvent on top of the aqueous sample, followed by elevation of the floated film to the narrow part of the device by adding water and withdrawing of the extraction solvent for subsequent analysis.

Another novel approach that was first introduced by Khalili Zanjani et al. to separate the low-density extraction solvent is to solidify the solvent by cooling down the sample matrix [114]. After extraction, centrifugation is applied as in conventional DLLME to cause flotation of the organic solvent on top of the aqueous phase, then followed by cooling in a beaker containing ice. As the extraction solvent solidifies, a spatula can be used to remove the floating solid solvent from the sample matrix, and the solid solvent can be melted in another vial before injection (Figure 4(d)). According to the literature [115], the extraction solvent for solidified floating organic drop microextraction must meet the following requirements: (1) it should have low volatility in order to avoid loss during sampling; (2) it should have the lowest solubility in the water sample; (3) it should have a melting point near room temperature (in the range of 10–30 °C); and (4) it should be compatible with the analytical instrumentation being used for the determination of the target analyte. These requirements regarding extraction solvent limit the application of this technique. Only dodecanol [100,104] has been reported for POP sampling. Regarding the EFs, a lighter solvent provided comparable values ranging from 1202 to 4587.

Other extraction solvents including IL [98], magnetofluid [97] were also reported as extraction solvents. These two extraction materials may become popular in DLLME due to the low toxicity of IL and easy separation of magnetofluid (oleic acid-coated Fe_3O_4 nanoparticles). However, one more step was required to dissolve the extracted IL and magnetofluid using an organic solvent before injection into the analytical instrument. The EF of IL and magnetofluid for sampling of OCPs from water samples were not as high as that when using halogenated solvents or heavy alcohols.

4.2.2 Selection of Disperser Method

The main purpose of a disperser solvent in traditional DLLME is to assist the emulsification of extraction solvent in the sample matrix. It should be able to dissolve in both the aqueous phase and the solvent phase, scatter the extraction droplets in the sample matrix, and form a cloudy solution during extraction. The large contact surface area between the extraction solvent and sample matrix significantly enhances the extraction kinetic and shortens the extraction time. Acetonitrile (ACN) [87,89,90] and light alcohols [92,93,95,98] were usually selected as disperser solvents for the sampling of OCPs. The volume of the disperser solvent directly affects the formation of a cloudy solution, the dispersion degree of extraction solvent in the aqueous phase, and subsequently the extraction efficiency of the method. In addition, variation of disperser solvent volume affects the volume of the sedimented phase. Hence, it is necessary to optimize the volume of the disperser solvent and extraction solvent simultaneously. Experimental results indicated that a volume ratio of 1:5 or 1:10 of the extraction solvent to disperser solvent was commonly applied in DLLME. With the assistance of a disperser solvent, DLLME can significantly enhance the extraction rate. The extraction time in DLLME ranged from 5 s to 5 min, which is much shorter than for other LPME techniques and SPME.

After injecting the mixture of extraction solvent and disperser solvent into the sample matrix, other approaches such as manual shaking [86–89,93,94], ultrasonication [90,91], and temperature change [98] can be employed to further assist dispersion (Table 2). Manual shaking is the most simple but effective. Wei et al. [94] found that manual shaking is a key step to obtain high extraction efficiency. Compared to the method without manual shaking, the extraction amount increased >100% for 10 s manual shaking. The same group also compared the dispersive efficiency of ultrasonication and vortex agitation. Peak areas of analytes done by ultrasound assistance were greater than those by vortex mixing. Homogeneous emulsion was more evident with the ultrasonication-assisted method. The temperature-based method is a well-known technique that improves the dispersion efficiency by increasing the speed of mass transfer between two immiscible phases. The only limitation of using temperature is the potential loss of extraction solvent at high temperatures during sample pretreatment.

The significance of a disperser solvent in the DLLME process is obvious. However, the amount of disperser solvent is usually much larger than extraction solvent. Besides, the available disperser solvent is limited by the extraction solvent and the sample matrix. Hence, a new dispersion method without the addition of a disperser solvent was investigated. Ultrasound [94,101,102] was frequently used to assist the emulsification of extraction solvent in water samples. An ultrasound-assisted emulsification microextraction method proposed by Wei et al. [94] was used to sample OCPs from aqueous samples. In this method, 10 μL of low-toxicity decanol was added to 10 mL of water, and the mixture was manually shaken for 10 s before applying ultrasound for 4 min. After extraction, decanol was separated from the sample with a capillary tube, and injected directly into the GC-ECD for analysis. EFs ranging from 1773 to 4587 were achieved for OCPs in water samples. The obtained LODs were comparable to conventional DLLME [93], which used a large amount of disperser solvent. Vortexing [103] and stirring [100] were also reported to disperse the extraction solvent without disperser solvent addition. In general, the extraction time (10–25 min) for nondisperser solvent DLLME was longer than the one with disperser solvent.

4.2.3 Salt and pH Effect

Salt addition alters the ionic strength of the sample, and has been investigated universally in both SPME and LLE. Generally, the addition of salt can decrease the solubility of analytes in aqueous samples and enhance their partitioning into the organic phase [74,78,86,94,97,104]. This "salting-out" effect, which is compound specific, has been discussed widely, but some contradictory results have been reported [75,88]. Some researchers reported that ionic strength exerts a negative effect. The presence of salt may cause changes in physical properties of the extraction surface, and reduces the rate of analyte diffusion to the organic phase. Therefore, as shown in Table 2, most of the published works have been conducted without salt addition.

While examining the effect of ionic strength of an aqueous sample solution on extraction efficiency, Lambropoulou et al. [116] explained that next to the salting-out effect, a second effect, adverse for the extraction, may also take place [117]. By this effect, the addition of a salt can also change the physicochemical properties of the Nernst diffusion film, thus reducing the diffusion rate of the analyte into the organic phase. Depending on the analyte and the salt content in the aqueous solution, one of these two effects may be dominant, enhancing or restricting the extraction. It is also possible that these two effects cancel each other, and in this case, changes in the ionic strength of the sample solution do not affect extraction efficiency.

The pH value of a solution changes the ionization form of certain analytes and thereby affects their water solubility and extractability. Thus, pH appears to be an important factor that requires to be optimized in an LPME process. There are many reports where, in two-phase or three-phase LPME, pH changes in the donor aqueous solution resulted in a higher analyte preconcentration [73]. However, in most LPME applications for sample preparation of POPs in water samples (Table 2), the relationship between pH and extraction efficiency has not been studied extensively, and the pH of water samples was usually not adjusted. On the other hand, for biological samples, such as urine, which contains a large amount of biomolecular interference, researchers found that acidic conditions aid in removing large molecular proteins, and reduce the biomolecular interference, and enhance extraction efficiency [98].

4.2.4 Summary of DLLME for POP Analysis

DLLME has been widely applied for preconcentration of POPs in environmental aqueous samples and demonstrated to be a fast, simple, and reliable technique. However, DLLME cannot be directly applied for soil and sediment sample preparation. Additional steps should be applied to extract the target compounds from the solid sample before DLLME. Lana et al. [104] and Fontana et al. [105] both reported similar DLLME procedures to extract PBDEs in sediments. The sediment was first treated with methanol with ultrasound assistance, and then the extractant was mixed with extraction solvent (dodecanol or carbotetrachloride) and was injected into pure water to form a cloudy solution. After solidification or centrifugation, the extraction solvent was separated from the water sample and injected into the instrument for quantification.

Before applying to the real sample analysis, the DLLME method should be optimized with regard to the extraction solvent, disperser solvent or method, as well as sample ionic strength and pH value. After sample preparation, the extracted solvent can be analyzed by GC or HPLC coupled with various detectors. The sensitivity of DLLME was proved to be comparable with other microextraction techniques, and sufficient for routing testing. Besides high sensitivity, DLLME also benefits by (1) fast extraction kinetic due to the large surface area between the fine droplets of the extraction solvent and aqueous sample; (2) high EF; (3) a small amount of extraction solvent used. On the other hand, the shortcomings of DLLME regarding the future development include the use of a relatively large amount of disperser solvent and toxic extraction solvent, the potential loss of a low-density solvent, and the difficulty of full automation.

5. OTHER MICROEXTRACTION TECHNIQUES

Besides SPME and LPME, other microextraction methods, such as SBSE, extraction rod, and TFME, were reported for sampling of POPs in environmental samples (Table 3).

SBSE was introduced by Baltussen et al. [118] based on the principle of SPME. A stir bar coated with 50–300 μL of PDMS was employed as extraction

TABLE 3 Other Microextraction Techniques for Persistent Organic Pollutants Analysis

Compounds	Samples	Sample Preparation Methods	LOD μg/L	Linear Range μg/L	Recovery	References
PBDEs	Water	SBSE-TD-GC-MS	0.0002–0.0035	0.05–0.3	82–105.8%	[119]
OCPs, PCB, PBDE	Seawater	SBSE-TD-GC-MS	0.000011–0.0375	0.0005–0.5	86–118%	[120]
15 OCPs	River water	SBSE-TD-GCGC-TOF	0.00001–0.000044	0.00005–0.0001	25–97%	[121]
OCPs	Natural water	Dual SBSE-TD-LTM-GS-MS	0.0014–0.006	0.1–1	30–63%	[124]
Multipesticides	River water	Sequential-SBSE-TD-GC-MS	0.0026–40	0.02–1	82–113%	[125]
OCPs and PCB	Water	SBSE-solvent desorption-GC-MS	NA	0.025–0.25	21.8–92.6%	[122]
25 EDCs	River water, wastewater	SBSE-solvent desorption-GC-MS/MS	0.0002–0.001	0.01–50	>70%	[123]
OCPs	Water	HS-PDMS-SBSE-GC-MS	0.02–1.59	5–17	88.1–100%	[126]
PBDE, PCB	Sediment	Ultrasound extraction-SBBSE-TD-GC-MS	0.10–0.68	0.5–50	62.8–91.5%	[127]

Continued

TABLE 3 Other Microextraction Techniques for Persistent Organic Pollutants Analysis—cont'd

Compounds	Samples	Sample Preparation Methods	LOD μg/L	Linear Range μg/L	Recovery	References
OCPs	Soil	PSWE-SBSE-TD-GC-MS	0.002–4.7 ng/g	0.05–30	4.1–85.2%	[128]
Halogenated flame retardant	Sea, river, rain water, and landfill leachate	PDMS rod-GC-NCI-MS	0.0004–0.01	0.05–0.5	53–130%	[129]
PFOS, PFOA	Water	PDMS rod and PES tube-LC-MS/MS	0.0002–0.02 for PES 0.002–0.25 for PDMS	LOQ-1	85–108%	[130]
15 OCPs	Coastal water	Polymer-coated HF-GC-MS	0.001–0.008	1–750	83.2–109%	[131]
Malathion	River water, tap water	μSPE-GC-MS	0.005–0.02	0.1–100	94–97%	[132]
OCPs	Lake water	DI-PDMS film-GC-ECD	0.00077–0.01025	0.01–5	89.6–99.4%	[134]
8 OCPs	Tap, snow, lake, and reservoir water	TiO_2 nanotube array membrane-GC-ECD	0.062–0.212	0.1–100	78–102.8%	[135]
11 Pesticides	River water, lake water, farmland water	Magnetic PPy-SPE-GC-MS	0.09–0.29	0.2–10	74–129%	[136]
OCPs	Dried and sieved soil	MSPD-LLE-GC-ECD	0.3–1.8 ng/g	15–750 ng/g	59–93.1%	[137]

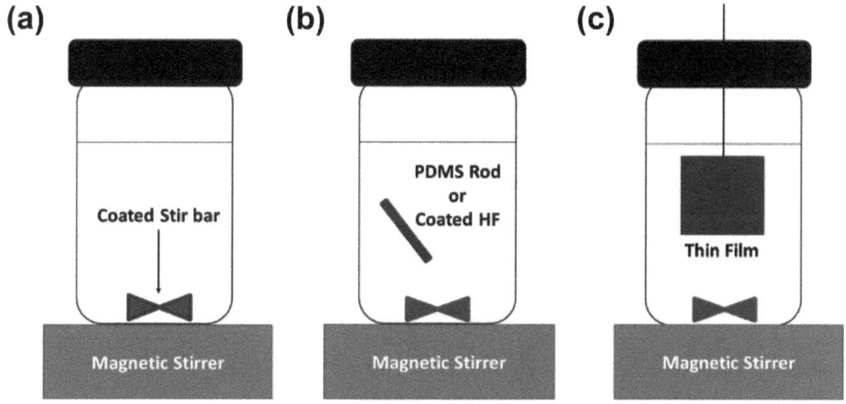

FIGURE 5 Schematic of SBSE, PDMS rod sampling, and thin film.

phase to preconcentrate analytes from aqueous sample. In general, the PDMS coated stir bar is placed inside the water (Figure 5(a)). After stirring for a period of time, the stir bar is removed from the solution and rinsed with deionized water, and dried and placed in an "injector tray." The compounds are thermodesorbed in a programmable temperature vaporization injector [119–121] or solvent desorbed in a vial [122,123] before introducing into the GC instrument. SBSE presents advantages over traditional sample preparation techniques because no/less organic solvent is used and a small volume of sample is required to achieve comparable sensitivity. With respect to SPME, lower detection limits and higher capacity can be obtained due to the larger volume of the extraction phase. SBSE has been successfully applied for sampling of OCPs [120,121], PBDEs [119], and PCBs [120] in aqueous samples including seawater and river water. LODs ranging from 0.01 to 38 ng/L were reported by different publications when coupled with GC–MS analysis. Solvent desorption SBSE theoretically has a lower sensitivity compared to that of TD-SBSE because of solvent dilution effect. On the other hand, solvent desorption provides multiuse of desorbed solvent for other analyses. Similar to SPME, a matrix effect was found in SBSE sampling [119]. The presence of dissolved humic substances (fulvic and humic acids) generally influences the quantification of organic compounds in waters. The extent of this effect depends on both the concentration and the origin of humic substances. In this sense, one should consider the concentration of dissolved humic substances during quantification.

The sampling capacity of the SBSE can be further improved by using two stir bars. Ochiai et al. reported an application of dual stir bars to preconcentrate the pesticides from natural water samples [124,125]. Two stir bars were sampled sequentially, and thermally desorbed simultaneously. By doing this, the extraction capacity was doubled. Although the reported LODs were not higher than the single SBSE sampling in other reports, dual SBSE is still worth considering for trace analysis.

SBSE takes the advantage of a stir bar, and is made for direct aqueous sampling. It is difficult to conduct HS sampling using SBSE. However, if a specially made device is employed, SBSE can also achieve HS sampling. Grossi reported a homemade device for HS-SBSE sampling of OCPs in water sampling [126]. The result indicated a good reproducibility of the device (2.1–14.8%), and LODs ranging from 0.02 to 50 μg/L. This HS SBSE is suitable for direct solid sampling since the stir bar does not come into contact with the sample matrix. Otherwise, SBSE for solid samples requires a preextraction step to release the analytes from the sample before being extracted by the stir bar. Ultrasonic extraction [127] and pressurized subcritical water extraction (PSWE) [128] was employed in the preextraction step, and followed by SBSE sampling. The obtained results were compared with those of reference methods, Soxhlet extraction, and pressurized liquid extraction, and showed a good agreement between methods.

The application of a PDMS rod is similar to SPME coated with PDMS but without fiber support. The sample preparation procedure of a PDMS rod is similar to the SBSE. The rod was exposed directly to the water sample under magnetic stirring for 15 h at room temperature. After extraction, the rod was removed from the vial using tweezers, washed with Milli-Q water, dried with a lint-free tissue, and placed inside an inert contained organic solvent. The insert was sonicated for 15 min, and then, the desorbed solvent was injected into the instrument for quantification. An LOD range of 0.4–10 ng/L was observed for sampling of halogenated flame retardants in natural water using PDMS rod sampling [129]. A PDMS rod was also used to sample PFOS and PFOA in water samples [130], and the extraction efficiency was compared with that of another polymer material, polyethersulfone (PES). The overall results demonstrated that PES led to a better sensitivity than PDMS, particularly for the most polar compounds, and reached detection limit in the range of 0.2–20 ng/L. Furthermore, a polypropylene HF coated with amphiphilic polyhydroxylated polyparaphenylene (PP-PH-PPP) was synthesized by Lee et al. [131] and applied for sampling of 15 OCPs from water samples. The extraction efficiency was compared with those of established methods including the LPME, SPME, and SBSE techniques, PP-PH-PPP extraction rods showed a better selectivity and sensitivity toward target compounds.

Microsolid phase extraction (μ-SPE) is similar to the extraction rod, and is regarded as a miniaturized equilibrium-based microextraction technique. Considering the sorbent nature used in μ-SPE, the sorption process depends on the surface area or the total amount of sorbent. The sample preparation procedure of μ-SPE can refer to the extraction rod sampling. Bagheri et al. reported a μ-SPE method using a PPy coated graphite tube as sorbent for preconcentration of malathion in river and tap water samples [132]. The extraction performance of the developed method was compared to that of the SPME, SBSE, and SPE methods. The result showed that μ-SPE provided the shortest extraction time and lowest LODs toward selected compounds.

Both SBSE and extraction rod increased the extraction capacity as well as extraction time. A sampling period ranging from 5 to 24 h was reported for SBSE and extraction rod sampling. Long extraction time results from a large thickness of both extraction phases. TFME was introduced to overcome this shortcoming by employing an extraction phase with a high surface area-to-volume ratio [133]. The large volume of a thin film increases the extraction capacity, while a large surface area enhances the extraction rate. TFME has been applied to extract OCPs in environmental water samples with a PDMS membrane [134] and TiO_2 nanotube arrays [135]. The films were directly immersed in the water sample with a constant depth, and extracted with magnetic stirring (Figure 5(c)). After equilibrium, the films were taken out of the sample, rinsed with distilled water, dried in air, and desorbed in an organic solvent. The desorbed solvent was further concentrated with nitrogen purge, and 1 μL of desorbed solvent was injected into the GC instrument. The experimental result showed that the equilibrium times for both films were <60 min. The extraction performance of TiO_2 was compared to that of DLLME and traditional SPE method, and showed that the proposed TiO_2 film provided a much higher EF than the target compounds. Furthermore, it is worth mentioning that the extracted film can be directly thermally desorbed into a programmable temperature vaporization (PTV) injector connected with the GC instrument to further improve the method sensitivity [133].

The last reported microextraction technique for POP analysis is matrix solid phase dispersion (MSPD). In this technique, a solid sorbent is used as the extraction phase as well as dispersion material during the sampling process. This technique has been applied for aqueous [136] and soil sample analysis [137]. For aqueous samples, the solid dispersant, magnetic PPy nanowires, was mixed with the water sample, and vortexed vigorously to form a homogeneous dispersion solution. Then the pesticide-absorbed magnetic PPy nanowires were separated from the sample by an external magnet, and redissolved in the organic solvent. For solid sample analysis, the solid sample, dispersing material and clean-up sorbent were mixed in a mortar and transferred into a polymeric extraction tube containing an appropriate volume of extraction solvent. After extraction, the solvent was collected for instrument quantification. The result showed that after optimization of the composition and amount of dispersing material, the obtained sensitivity was considered to be satisfactory.

6. SUMMARY OF MICROEXTRACTION FOR POPs ANALYSIS

The increasing demand for faster, more cost-effective, and environmentally friendly analytical methods is a major incentive to improve the classical procedures used for sample treatment in environmental analysis. As a consequence, the current trend is to simplify, accelerate, and/or miniaturize the pretreatment procedures. SPME, LPME as well as SBSE, extraction rod, TFME are well-designed microextraction techniques that are frequently applied for sampling of POPs in environmental samples. Each microextraction technique reveals advantages and disadvantages in different applications and together can be considered as a toolbox from which the appropriate technique for each application can be selected. With regard to liquid samples, DLLME is probably the first choice for sample preparation, especially for relatively polar compounds. DLLME provides a short extraction time, simple operation procedure, reproducible result, and sufficient sensitivity. It can be easily coupled with most instruments, such as GC and LC, without any modification. However, when considering the toxic organic solvent used in DLLME, polymer-based microextraction, such as SPME, SBSE, extraction rod, and TFME, would be preferable. SPME shows advantages over other solvent-free techniques with respect to various types of fiber, easy integration with the instrument, and full automation. SBSE shows a higher extraction capacity than SPME, but longer extraction time as well. TFME can be used to solve the issue of long extraction time, but the film should be homemade since no analytical thin film was commercialized for GC analysis. Another advantage for solvent-free microextraction techniques is for solid sample preparation. A solid sample is a complex matrix that may result in a quantification error or biofouling of the extraction phase. Generally, POP compounds should be released from the solid matrix by either temperature or water extraction before being further sampled by the extraction phase. HS-SPME was found to be the most suitable method for solid sample pretreatment. Otherwise, direct HF-LPME is the second choice since it overcomes the contamination issue from the complex matrix.

LIST OF ABBREVIATIONS

3-TMOX-1-PPA 3-(trimethoxysilyl)-1-propanamine
μSPE Micro-solid phase extraction
ACN Acetonitrile
CAR Carboxen
CW/TPR Carbowax/templated resin
DI Direct immersion
DMOD-3-TMOS-AC Dimethyloctadecyl[3-(trimethoxysilyl)propyl]ammonium chloride
DVB Divinylbenzene
ECD Electron capture detector
EDP Endocrine disruptor pesticides
EF Enrichment factor
EPA Environment protection agency
G Graphene
GC Gas chromatography
HPLC High pressure liquid chromatography
HS Headspace
IL Ionic liquid
LOD Limit of detection
LOQ Limit of quantification

MA Microwave assistance
MASE Microwave-assisted solvent extraction
MIP Molecular imprinted polymer
MS Mass spectrometry
MSPD Matrix solid phase dispersion
MW Molecular weight
NA Not available
OCPs Organochlorine pesticides
OPPs Organophosphorus pesticides
PA Polyacrylate
PANI Polyaniline
PBB Polybrominated biphenyls
PBDEs Polybrominated diphenyl ethers
PCBs Polychlorinated biphenyls
PDMS Polydimethylsiloxane
PES Polyethersulfone
PES Polyethersulfone
PFOA Perfluoroalkyl acid
PFOS Perfluoroalkyl sulfide
PHWE Pressurized hot water extraction
PM-β-CD/OH-TSO Permethylated-beta-cyclodextrin/hydroxyl-termination silicone oil
POPs Persistent organic pollutants
PP-doped-Ppy Polyphosphate doped polypyrrole
PP-PH-PPP Polyhydroxylated polyparaphenylene
PSWE Pressurized subcritical water extraction
SBSE Stir bar sorptive extraction
SFOD Solidified floating organic drop microextraction
SPME Solid phase microextraction
SSW Stainless steel wire
SWNT Single wall nanotube
TBME Tert-butyl methyl ether
TD Thermal desorption
TFME Thin-film microextraction
THF Tetrahydrofuran
USA Ultrasound assisted
USE Ultrasound extraction
UV Ultraviolet

REFERENCES

[1] L. Ramos, J.J. Ramos, U.A.T. Brinkman, Miniaturization in sample treatment for environmental analysis, Anal. Bioanal. Chem. 381 (2005) 119–140.

[2] J. Pawliszyn, Handbook of Solid Phase Microextraction, Elsevier, 2012.

[3] J. Koziel, M. Jia, J. Pawliszyn, Air sampling with porous solid-phase microextraction fibers, Anal. Chem. 72 (2000) 5178–5186.

[4] Y. Chen, J.A. Koziel, J. Pawliszyn, Calibration for on-site analysis of hydrocarbons in aqueous and gaseous samples using solid-phase microextraction, Anal. Chem. 75 (2003) 6485–6493.

[5] J. Ai, Solid phase microextraction for quantitative analysis in nonequilibrium situations, Anal. Chem. 69 (1997) 1230–1236.
[6] N. Ratola, L. Santos, P. Herbert, A. Alves, Uncertainty associated to the analysis of organochlorine pesticides in water by solid-phase microextraction/gas chromatography-electron capture detection-evaluation using two different approaches, Anal. Chim. Acta 573–574 (2006) 202–208.
[7] M.E. Torres Padrón, Z. Sosa Ferrera, J.J. Santana Rodríguez, Optimisation of solid-phase microextraction coupled to HPLC–UV for the determination of organochlorine pesticides and their metabolites in environmental liquid samples, Anal. Bioanal. Chem. 386 (2006) 332–340.
[8] A. Scheyer, S. Morville, P. Mirabel, M. Millet, Analysis of trace levels of pesticides in rainwater using SPME and GC–tandem mass spectrometry, Anal. Bioanal. Chem. 384 (2006) 475–487.
[9] W. Boonjob, J.B. Quintana, R. Rodil, R. Cela, M. Miró, In-line sequential injection-based hollow-fiber sorptive microextraction as a front-end to gas chromatography–mass spectrometry: a novel fully automatic sample processing technique for residue analysis, Anal. Bioanal. Chem. 405 (2013) 8653–8662.
[10] R.I. Bonansea, M.V. Amé, D.A. Wunderlin, Determination of priority pesticides in water samples combining SPE and SPME coupled to GC–MS. A case study: Suquía River basin (Argentina), Chemosphere 90 (2013) 1860–1869.
[11] J. Merib, V. Simão, A.N. Dias, E. Carasek, Simultaneous determination of trihalomethanes and organochlorine pesticides in water samples by direct immersion-headspace-solid phase microextraction, J. Chromatogr. A 1321 (2013) 30–37.
[12] D. García-Rodríguez, A.M. Carro, R.A. Lorenzo, R. Cela, F. Fernández, Determination of trace levels of aquaculture chemotherapeutants in seawater samples by SPME–GC–MS/MS, J. Sep. Sci. 31 (2008) 2882–2890.
[13] M. Polo, M. Llompart, C. Carcia Jares, G. Gomes Noya, M.-H. Bollain, R. Cela, Development of a solid-phase microextraction method for the analysis of phenolic flame retardants in water samples, J. Chromatogr. A 1124 (2006) 11–21.
[14] M. Polo, G. Gómez Noya, J.B. Quintana, M. Llompart, C. García Jares, R. Cela, Development of a solid-phase microextraction gas chromatography/tandem mass spectrometry method for polybrominated diphenyl ethers and polybrominated biphenyls in water samples, Anal. Chem. 76 (2004) 1054–1062.
[15] P. López, S.A. Brandsma, P.E.G. Leonards, J. De Boer, Methods for the determination of phenolic brominated flame retardants, and by-products, formulation intermediates and decomposition products of brominated flame retardants in water, J. Chromatogr. A 1216 (2009) 334–345.
[16] T. Sun, J. Jia, D. Zhong, Y. Wang, Determination of 17 kinds of banned organochlorine pesticides in water by activated carbon fiber-solid phase microextraction coupled with GC–MS, Anal. Sci. 22 (2006) 293–298.
[17] L. Liao, J. Yang, Y. Wang, T. Sun, J. Jia, Study on a novel circulating cooling solid-phase microextraction method, J. Chromatogr. A 1135 (2006) 1–5.
[18] V.K. Ponnusamy, J.F. Jen, A novel graphene nanosheets coated stainless steel fiber for microwave assisted headspace solid phase microextraction of organochlorine pesticides in aqueous samples followed by gas chromatography with electron capture detection, J. Chromatogr. A 1218 (2011) 6861–6868.
[19] Y.B. Luo, B.F. Yuan, Q.W. Yu, Y.Q. Feng, Substrateless graphene fiber: a sorbent for solid-phase microextraction, J. Chromatogr. A 1268 (2012) 9–15.

[20] Y. Ke, F. Zhu, F. Zeng, T. Luan, C. Su, G. Ouyang, Preparation of graphene-coated solid-phase microextraction fiber and its application on organochlorine pesticides determination, J. Chromatogr. A 1300 (2013) 187–192.

[21] H. Zhang, H.K. Lee, Plunger-in-needle solid-phase microextraction with graphene-based sol–gel coating as sorbent for determination of polybrominated diphenyl ethers, J. Chromatogr. A 1218 (2011) 4509–4516.

[22] W. Zhang, Y. Sun, C. Wu, J. Xing, J. Li, Polymer-functionalized single-walled carbon nanotubes as a novel sol–gel solid-phase micro-extraction coated fiber for determination of polybrominated diphenyl ethers in water samples with gas chromatography-electron capture detection, Anal. Chem. 81 (2009) 2912–2920.

[23] R. Zhao, Y. Liu, X. Chen, J. Yuan, A. Bai, J. Zhou, Preconcentration and determination of polybrominated diphenyl ethers in environmental water samples by solid-phase microextraction with Fe_3O_4-coated bamboo charcoal fibers prior to gas chromatography–mass spectrometry., Anal. Chim. Acta 769 (2013) 65–71.

[24] X. Li, M. Zhong, J. Chen, Electrodeposited polyaniline as a fiber coating for solid-phase microextraction of organochlorine pesticides from water, J. Sep. Sci. 31 (2008) 2839–2845.

[25] A. Mehdinia, H. Khani, S. Mozaffari, Fibers coated with a graphene–polyaniline nanocomposite for the headspace solid-phase microextraction of organochlorine pesticides from seawater samples, Microchim. Acta 181 (2013) 89–95.

[26] Z. Gao, W. Li, B. Liu, F. Liang, H. He, S. Yang, et al., Nano-structured polyaniline-ionic liquid composite film coated steel wire for headspace solid-phase microextraction of organochlorine pesticides in water, J. Chromatogr. A 1218 (2011) 6285–6291.

[27] Y. Wang, Y. Li, J. Zhang, S. Xu, S. Yang, C. Sun, A novel fluorinated polyaniline-based solid-phase microextraction coupled with gas chromatography for quantitative determination of polychlorinated biphenyls in water samples, Anal. Chim. Acta 646 (2009) 78–84.

[28] Y. Wang, J. Zhang, D. Sheng, C. Sun, Preparation and applications of perfluorinated ion doped polyaniline based solid-phase microextraction fiber, J. Chromatogr. A 1217 (2010) 4523–4528.

[29] A. Mollahosseini, E. Noroozian, Polyphosphate-doped polypyrrole coated on steel fiber for the solid-phase microextraction of organochlorine pesticides in water, Anal. Chim. Acta 638 (2009) 169–174.

[30] L. Pelit, T.N. Dizdaş, Preparation and application of a polythiophene solid-phase microextraction fiber for the determination of endocrine-disruptor pesticides in well waters, J. Sep. Sci. 36 (2013) 3234–3241.

[31] D. Wang, J. Peng, J. Xing, C. Wu, Y. Xu, Solid-phase microextraction and gas chromatography-electron capture detection analysis of trace organochlorine pesticides in water using novel benzo-15-crown-5 sol–gel coating, J. Chromatogr. Sci. 42 (2004) 57–61.

[32] C. Dong, Z. Zeng, M. Yang, Determination of organochlorine pesticides and their derivations in water after HS-SPME using polymethylphenylvinylsiloxane-coated fiber by GC-ECD, Water Res. 39 (2005) 4204–4210.

[33] D. Cao, J. Lü, J. Liu, G. Jiang, In situ fabrication of nanostructured titania coating on the surface of titanium wire: a new approach for preparation of solid-phase microextraction fiber, Anal. Chim. Acta 611 (2008) 56–61.

[34] X.F. Chen, C.G. Cheng, X. Wang, R.S. Zhao, Sensitive determination of polybrominated diphenyl ethers in environmental water samples with etched stainless steel wire based on solid-phase microextraction prior to gas chromatography–mass spectrometry, Anal. Methods 4 (2012) 2908–2913.

[35] C. Silva, G. Zuin, J.H. Yariwake, M. Nogueira, F. Augusto, Fiber introduction mass spectrometry: determination of pesticides in herbal infusions using a novel sol–gel PDMS/PVA fiber for solid-phase microextraction, J. Mass Spectrom. 42 (2007) 825–829.

[36] M.K.Y. Li, N.Y. Lei, C. Gong, Y. Yu, K.H. Lam, M.H.W. Lam, et al., An organically modified silicate molecularly imprinted solid-phase microextraction device for the determination of polybrominated diphenyl ethers, Anal. Chim. Acta 633 (2009) 197–203.

[37] K. Saito, E. Uemura, A. Ishizaki, H. Kataoka, Determination of perfluorooctanoic acid and perfluorooctane sulfonate by automated in-tube solid-phase microextraction coupled with liquid chromatography–mass spectroscopy, Anal. Chim. Acta 658 (2010) 141–146.

[38] C. Chen, J. Wang, S. Yang, Z. Yan, Q. Cai, S. Yao, Analysis of perfluorooctane sulfonate and perfluorooctanoic acid with a mixed-mode coating-based solid-phase microextraction fiber, Talanta 114 (2013) 11–16.

[39] S.M. Chang, R.A. Doong, Concentration and fate of persistent organochlorine pesticides in estuarine sediments using headspace solid-phase microextraction, Chemosphere 62 (2006) 1869–1878.

[40] R. Zhao, X. Wang, J. Yuan, T. Jiang, S. Fu, X. Xu, A novel headspace solid-phase microextraction method for the exact determination of organochlorine pesticides in environmental soil samples, Anal. Bioanal. Chem. 384 (2006) 1584–1589.

[41] P. Herbert, S. Morais, P. Paíga, Development and validation of a novel method for the analysis of chlorinated pesticides in soils using microwave-assisted extraction–headspace solid phase microextraction and gas chromatography–tandem mass spectrometry, J. Chromatogr. A 384 (2006) 810–816.

[42] P. Carvalho, P. Rodrigues, Organochlorine pesticides levels in Portuguese coastal areas, Chemosphere 75 (2009) 595–600.

[43] P.N. Carvalho, P.N.R. Rodrigues, F. Alves, R. Evangelista, M.C.P. Basto, M.T.S.D. Vasconcelos, An expeditious method for the determination of organochlorine pesticides residues in estuarine sediments using microwave assisted pre-extraction and automated headspace solid-phase microextraction coupled to gas chromatography–mass spectrometry, Talanta 76 (2008) 1124–1129.

[44] Ä. Guez, G. Vega Moreno, Z. Sosa Ferrera, J.J. Santana Rodrígue z, Sample extraction method combining micellar extraction–SPME and HPLC for the determination of organochlorine pesticides in agricultural soils, J. Agric. Food Chem. 54 (2006) 7747–7752.

[45] D.V. Moreno, Z.S. Ferrera, J.S. Rodr, Microwave assisted micellar extraction coupled with solid phase microextraction for the determination of organochlorine pesticides in soil samples, Anal. Chim. Acta 571 (2006) 51–57.

[46] Y. Lu, Y. Zhu, Determination of chlorobenzenes in textiles by pressurized hot water extraction followed by vortex-assisted liquid-liquid microextraction and gas chromatography–mass spectrometry, J. Chromatogr. A 1319 (2013) 27–34.

[47] C. Salgado-Petinal, M. Garcia-Chao, M. Llompart, C. Garcia-Jares, R. Cela, Headspace solid-phase microextraction gas chromatography tandem mass spectrometry for the determination of brominated flame retardants in environmental solid samples, Anal. Bioanal. Chem. 385 (2006) 637–644.

[48] R. Montes, M. Ramil, I. Rodriguez, Rapid screening of polychlorinated biphenyls in sediments using non-equilibrium solid-phase microextraction and fast gas chromatography with electron-capture detection, J. Chromatogr. A 1124 (2006) 43–50.

[49] J. Zhou, F. Yang, D. Cha, Z. Zeng, Y. Xu, Headspace solid-phase microextraction with novel sol-gel permethylated-beta-cyclodextrin/hydroxyl-termination silicone oil fiber for determination of polybrominated diphenyl ethers by gas chromatography–mass spectrometry in soil, Talanta 73 (2007) 870–877.

[50] R. Montes, I. Rodríguez, R. Cela, Solid-phase microextraction with simultaneous oxidative sample treatment for the sensitive determination of tetra-to hexa-brominated diphenyl ethers in sediments, J. Chromatogr. A 1217 (2010) 14–21.
[51] P. Herbert, A.L. Silva, M.J. João, L. Santos, A. Alves, Determination of semi-volatile priority pollutants in landfill leachates and sediments using microwave-assisted headspace solid-phase microextraction, Anal. Bioanal. Chem. 386 (2006) 324–331.
[52] A. Navalón, A. Prieto, L. Araujo, J.L. Vílchez, Determination of pyrimethanil and kresoxim-methyl in soils by headspace solid-phase microextraction and gas chromatography–mass spectrometry, Anal. Bioanal. Chem. 379 (2004) 1100–1105.
[53] K.-J. Chia, T.-Y. Lee, S.-D. Huang, Simple device for the solid-phase microextraction screening of polychlorodibenzo-p-dioxins and polychlorodibenzofurans in heavily contaminated soil samples, Anal. Chim. Acta 527 (2004) 157–162.
[54] T. Barri, S. Bergstr, A. Hussen, J. Norberg, J. Ake, Extracting syringe for determination of organochlorine pesticides in leachate water and soil-water slurry: a novel technology for environmental analysis, J. Chromatogr. A 1111 (2006) 11–20.
[55] É.A. Souza Silva, J. Pawliszyn, Optimization of fiber coating structure enables direct immersion solid phase microextraction and high-throughput determination of complex samples, Anal. Chem. 84 (2012) 6933–6938.
[56] J. Xu, J. Zheng, J. Tian, F. Zhu, F. Zeng, New materials in solid-phase microextraction, TrAC Trends Anal. Chem. 47 (2013) 68–83.
[57] J.M. Conder, T.W. La Point, Solid-phase microextraction for predicting the bioavailability of 2,4,6-trinitrotoluene and its primary transformation products in sediment and water, Environ. Toxicol. Chem. 24 (2005) 1059–1066.
[58] S.R. Drescher, S.D. Brown, Solid phase microextraction-gas chromatographic-mass spectrometric determination of nitrous oxide evolution to measure denitrification in estuarine soils and sediments, J. Chromatogr. A 1133 (2006) 300–304.
[59] M.T.O. Jonker, S.A. Van der Heijden, J.P. Kreitinger, S.B. Hawthorne, Predicting PAH bioaccumulation and toxicity in earthworms exposed to manufactured gas plant soils with solid-phase microextraction, Environ. Sci. Technol. 41 (2007) 7472–7478.
[60] Z. Zhang, J. Pawliszyn, Quantitative extraction using an internally cooled solid phase microextraction device, Anal. Chem. 67 (1995) 34–43.
[61] Y. Chen, J. Pawliszyn, Miniaturization and Automation of an Internally Cooled Coated Fiber Device, vol. 78, 2006, pp. 5222–5226.
[62] A.R. Ghiasvand, S. Hosseinzadeh, J. Pawliszyn, New cold-fiber headspace solid-phase microextraction device for quantitative extraction of polycyclic aromatic hydrocarbons in sediment, J. Chromatogr. A. 1124 (2006) 35–42.
[63] R. Jiang, E. Carasek, S. Risticevic, E. Cudjoe, J. Warren, J. Pawliszyn, Evaluation of a completely automated cold fiber device using compounds with varying volatility and polarity, Anal. Chim. Acta 742 (2012) 22–29.
[64] G. Ouyang, J. Pawliszyn, A critical review in calibration methods for solid-phase microextraction, Anal. Chim. Acta 627 (2008) 184–197.
[65] R.D. Đurović, T.M. Đorđević, Effects of soil composition on solid phase microextraction determination of triazine and organophosphorus pesticides, J. Environ. Sci. Heal. Part B 47 (2012) 851–857.
[66] M. Bergknut, E. Sehlin, S. Lundstedt, P.L. Andersson, P. Haglund, M. Tysklind, Comparison of techniques for estimating PAH bioavailability: uptake in Eisenia fetida, passive samplers and leaching using various solvents and additives, Environ. Pollut. 145 (2007) 154–160.
[67] L. Xu, C. Basheer, H.K. Lee, Developments in single-drop microextraction, J. Chromatogr. A 1152 (2007) 184–192.

[68] G. Shen, H.K. Lee, Hollow fiber-protected liquid-phase microextraction of triazine herbicides, Anal. Chem. 74 (2002) 648–654.

[69] L. Zhao, H.K. Lee, Liquid-phase microextraction combined with hollow fiber as a sample preparation technique prior to gas chromatography/mass spectrometry, Anal. Chem. 74 (2002) 2486–2492.

[70] X. Jiang, H.K. Lee, Solvent bar microextraction, Anal. Chem. 76 (2004) 5591–5596.

[71] M. Rezaee, Y. Yamini, M. Faraji, Evolution of dispersive liquid-liquid microextraction method, J. Chromatogr. A 1217 (2010) 2342–2357.

[72] L. Kocúrová, I.S. Balogh, Š. Jana, V. Andruch, J. Šandrejová, Recent advances in dispersive liquid–liquid microextraction using organic solvents lighter than water. a review, Microchem. J. 102 (2012) 11–17.

[73] D.A. Lambropoulou, T.A. Albanis, Liquid-phase micro-extraction techniques in pesticide residue analysis, J. Biochem. Biophys. Methods 70 (2007) 195–228.

[74] L. Qian, Y. He, Funnelform single-drop microextraction for gas chromatography-electron-capture detection, J. Chromatogr. A 1134 (2006) 32–37.

[75] E. Carlos, R. Alves, A.A. Neves, Simultaneous determination of the organochlorine and pyrethroid pesticides in drinking water by single drop microextraction and gas chromatography, J. Braz. Chem. Soc. 24 (2013) 1217–1227.

[76] C. Cortada, L. Vidal, S. Tejada, A. Romo, A. Canals, Determination of organochlorine pesticides in complex matrices by single-drop microextraction coupled to gas chromatography–mass spectrometry, Anal. Chim. Acta 638 (2009) 29–35.

[77] X. Wang, Y. Li, G. Wei, Determination of decabromodiphenyl ether in water samples by single-drop microextraction and RP-HPLC, J. Sep. Sci. 30 (2007) 2698–2702.

[78] A. Salemi, R. Rasoolzadeh, M. Mohebbi, M. Vosough, Ultrasonic assisted headspace single drop micro-extraction and gas chromatography with nitrogen–phosphorus detector for determination of organophosphorus pesticides in soil, Anal. Chim. Acta 769 (2013) 121–126.

[79] T. Barri, S. Bergstr, J.A. Jönsson, Determination of polybrominated diphenyl ethers at trace levels in environmental waters using hollow-fiber microporous membrane liquid–liquid extraction and gas chromatography–mass spectrometry, J. Chromatogr. A 1133 (2006) 41–48.

[80] C. Yan, H. Wu, A liquid-phase microextraction method, combining a dual gauge microsyringe with a hollow fiber membrane, for the determination of organochlorine pesticides in aqueous solution by gas chromatography/ion trap mass spectrometry, Rapid Commun. Mass Spectrom. 18 (2004) 3015–3018.

[81] P.V. Kumar, J.F. Jen, Rapid determination of dichlorodiphenyltrichloroethane and its main metabolites in aqueous samples by one-step microwave-assisted headspace controlled-temperature LPME and GC with ECD, Chemsphere 83 (2011) 200–207.

[82] J. Manso, New home-made assembly for hollow-fibre membrane extraction of persistent organic pollutants from real world samples, J. Chromatogr. A 1218 (2011) 7923–7935.

[83] Y. Zhang, R. Wang, P. Su, Y. Yang, Ionic liquid-based solvent bar microextraction for determination of organophosphorus pesticides in water samples, Anal. Methods 5 (2013) 5074–5078.

[84] Q. Xiao, B. Hu, J. Duan, M. He, W. Zu, Analysis of PBDEs in soil, dust, spiked lake water, and human serum samples by hollow fiber-liquid phase microextraction combined with GC–ICP–MS, J. Am. Soc. Mass Spectrom. 18 (2007) 1740–1748.

[85] C. Basheer, J.P. Obbard, H.K. Lee, Analysis of persistent organic pollutants in marine sediments using a novel microwave assisted solvent extraction and liquid-phase microextraction technique, J. Chromatogr. A. 1068 (2005) 221–228.

[86] W.C. Tsai, S. Da Huang, Dispersive liquid–liquid microextraction with little solvent consumption combined with gas chromatography–mass spectrometry for the pretreatment of organochlorine pesticides in aqueous samples, J. Chromatogr. A 1216 (2009) 5171–5175.

[87] A.M. Carro, S.F. Racamonde, D. Garcia-Rodriguez, P. Gonzalez, R.A. Lorenzo, Dispersive liquid–liquid microextraction coupled with programmed temperature vaporization-large volume injection-gas chromatography–tandem mass spectrometry for multiclass pesticides in water, J. Chromatogr. A 1253 (2012) 134–143.

[88] Y. Li, J. Hu, X. Liu, L. Fu, X. Zhang, X. Wang, et al., Dispersive liquid–liquid microextraction followed by reversed phase HPLC for the determination of decabrominated diphenyl ether in natural water. J. Sep. Sci. 31 (2008) 2371–2376.

[89] Y. Li, G. Wei, J. Hu, X. Liu, X. Zhao, X. Wang, Dispersive liquid–liquid microextraction followed by reversed phase-high performance liquid chromatography for the determination of polybrominated diphenyl ethers at trace levels in landfill leachate and environmental water samples. Anal. Chim. Acta 615 (2008) 96–103.

[90] S. Cui, Q. Chen, W. Wang, J. Miao, A. Wang, J. Chen, Ultra-preconcentration and determination of multiple pesticide residues in water samples using ultrasound-assisted dispersive liquid–liquid microextraction and GC-FID, Chromatographia 76 (2013) 671–678.

[91] Q. Zhang, T. Liang, L. Guan, Ultrasound-assisted dispersive liquid-liquid microextraction combined with gas chromatography–mass spectrometry in negative chemical ionization mode for the determination of polybrominated diphenyl ethers, J. Sep. Sci. (2013) 1263–1269.

[92] X. Liu, J. Li, Z. Zhao, W. Zhang, K. Lin, C. Huang, et al., Solid-phase extraction combined with dispersive liquid–liquid microextraction for the determination for polybrominated diphenyl ethers in different environmental matrices, J. Chromatogr. A 1216 (2009) 2220–2226.

[93] C.C. Chang, S.Y. Wei, S. Da Huang, Improved solvent collection system for a dispersive liquid–liquid microextraction of organochlorine pesticides from water using low-density organic solvent, J. Sep. Sci. 34 (2011) 837–843.

[94] S. Wei, M. Leong, Y. Li, S. Huang, Development of liquid phase microextraction based on manual shaking and ultrasound-assisted emulsification method for analysis of organochlorine pesticides in aqueous samples, J. Chromatogr. A 1218 (2011) 9142–9148.

[95] M. Haji Hosseini, P. Asaadi, M. Rezaee, M.R. Rezaei, M.R. Pourjavid, M. Arabieh, et al., Homogeneous liquid–liquid microextraction via flotation assistance (HLLME-FA) method for the pretreatment of organochlorine pesticides in aqueous samples, Chromatographia 76 (2013) 1779–1784.

[96] Y. Li, P. Chen, S. Huang, Water with low concentration of surfactant in dispersed solvent-assisted emulsion dispersive liquid–liquid microextraction for the determination of organochlorine pesticides in aqueous samples, J. Chromatogr. A 1300 (2013) 51–57.

[97] Z. Shen, Z. He, P. Wang, Z. Zhou, M. Sun, J. Li, et al., Low-density magnetofluid dispersive liquid–liquid microextraction for the fast determination of organochlorine pesticides in water samples by GC-ECD, Anal. Chim. Acta 793 (2013) 37–43.

[98] A. Zhao, X. Wang, M. Ma, W. Wang, H. Sun, Z. Yan, et al., Temperature-assisted ionic liquid dispersive liquid–liquid microextraction combined with high performance liquid chromatography for the determination of PCBs and PBDEs in water and urine samples, Microchim. Acta 177 (2012) 229–236.

[99] J. Hassan, M. Shamsipur, Extraction of ultra traces of polychlorinated biphenyls in aqueous samples using suspended liquid-phase microextraction and gas chromatography-electron capture detection, Environ. Monit. Assess. 185 (2013) 3637–3644.

[100] H. Liu, M. Zhang, X. Wang, Y. Zou, W. Wang, M. Ma, Extraction and determination of polybrominated diphenyl ethers in water and urine samples using solidified floating organic drop microextraction along with high performance liquid chromatography, Microchim. Acta (2012) 303–309.

[101] S. Ozcan, A. Tor, M.E. Aydin, Application of ultrasound-assisted emulsification-micro-extraction for the analysis of organochlorine pesticides in waters, Water Res. 43 (2009) 4269–4277.

[102] A.R. Fontana, M.F. Silva, L.D. Martínez, R.G. Wuilloud, J.C. Altamirano, Determination of polybrominated diphenyl ethers in water and soil samples by cloud point extraction-ultrasound-assisted back-extraction-gas chromatography–mass spectrometry, J. Chromatogr. A 1216 (2009) 4339–4346.

[103] A. Papadopoulou, I. Román, A. Canals, Fast screening of perfluorooctane sulfonate in water using vortex-assisted liquid–liquid microextraction coupled to liquid chromatography–mass spectrometry, Anal. Chim. Acta 691 (2011) 56–61.

[104] N.B. Lana, P. Berton, A. Covaci, A.G. Atencio, N.F. Ciocco, J.C. Altamirano, Ultrasound leaching-dispersive liquid–liquid microextraction based on solidification of floating organic droplet for determination of polybrominated diphenyl ethers in sediment samples by gas chromatography–tandem mass spectrometry, J. Chromatogr. A 1285 (2013) 15–21.

[105] A. Fontana, N. Lana, L. Martinez, J. Altamirano, Ultrasound-assisted leaching-dispersive solid-phase extraction followed by liquid–liquid microextraction for the determination of polybrominated diphenyl ethers in sediment samples by gas chromatography-tandem mass spectrometry, Talanta 82 (2010) 359–366.

[106] M.A. Jeannot, F.F. Cantwell, Solvent microextraction into a single drop, Anal. Chem. 68 (1996) 2236–2240.

[107] G. Ouyang, W. Zhao, J. Pawliszyn, Kinetic calibration for automated headspace liquid-phase microextraction, Anal. Chem. 77 (2005) 8122–8128.

[108] Y.B. Pakade, D.K. Tewary, Development and applications of single-drop microextraction for pesticide residue analysis: a review, J. Sep. Sci. 33 (2010) 3683–3691.

[109] G. Ouyang, W. Zhao, J. Pawliszyn, Automation and optimization of liquid-phase microextraction by gas chromatography, J. Chromatogr. A 1138 (2007) 47–54.

[110] S. Berijani, Y. Assadi, M. Anbia, M.-R. Milani Hosseini, E. Aghaee, Dispersive liquid–liquid microextraction combined with gas chromatography–flame photometric detection. Very simple, rapid and sensitive method for the determination of organophosphorus pesticides in water, J. Chromatogr. A 1123 (2006) 1–9.

[111] M.A. Farajzadeh, S.E. Seyedi, M.S. Shalamzari, M. Bamorowat, Dispersive liquid–liquid microextraction using extraction solvent lighter than water, J. Sep. Sci. 32 (2009) 3191–3200.

[112] P. Hashemi, S. Beyranvand, R.S. Mansur, A.R. Ghiasvand, Development of a simple device for dispersive liquid–liquid microextraction with lighter than water organic solvents: isolation and enrichment of glycyrrhizic acid from licorice, Anal. Chim. Acta 655 (2009) 60–65.

[113] A. Saleh, Y. Yamini, M. Faraji, M. Rezaee, M. Ghambarian, Ultrasound-assisted emulsification microextraction method based on applying low density organic solvents followed by gas chromatography analysis for the determination of polycyclic aromatic hydrocarbons in water samples, J. Chromatogr. A 1216 (2009) 6673–6679.

[114] M.R. Khalili Zanjani, Y. Yamini, S. Shariati, J.A. Jönsson, A new liquid-phase microextraction method based on solidification of floating organic drop, Anal. Chim. Acta 585 (2007) 286–293.

[115] M.S. Bidabadi, S. Dadfarnia, A.M.H. Shabani, Solidified floating organic drop microextraction (SFODME) for simultaneous separation/preconcentration and determination of cobalt

and nickel by graphite furnace atomic absorption spectrometry (GFAAS), J. Hazard. Mater. 166 (2009) 291–296.

[116] D.A. Lambropoulou, E. Psillakis, T.A. Albanis, N. Kalogerakis, Single-drop microextraction for the analysis of organophosphorous insecticides in water, Anal. Chim. Acta 516 (2004) 205–211.

[117] E. Psillakis, N. Kalogerakis, Application of solvent microextraction to the analysis of nitroaromatic explosives in water samples., J. Chromatogr. A 907 (2001) 211–219.

[118] E. Baltussen, P. Sandra, F. David, C. Cramers, Stir bar sorptive extraction (SBSE), a novel extraction technique for aqueous samples: Theory and principles, J. Microcolumn Sep. 11 (1999) 737–747.

[119] A. Prieto, O. Zuloaga, A. Usobiaga, Use of experimental design in the optimisation of stir bar sorptive extraction followed by thermal desorption for the determination of brominated flame retardants in water, Anal. Bioanal. Chem. 390 (2008) 739–748.

[120] J. Sánchez-Avila, J. Quintana, F. Ventura, R. Tauler, C.M. Duarte, S. Lacorte, Stir bar sorptive extraction-thermal desorption-gas chromatography–mass spectrometry: an effective tool for determining persistent organic pollutants and nonylphenol in coastal waters in compliance with existing directives, Mar. Pollut. Bull. 60 (2010) 103–112.

[121] N. Ochiai, T. Ieda, K. Sasamoto, Y. Takazawa, S. Hashimoto, A. Fushimi, et al., Stir bar sorptive extraction and comprehensive two-dimensional gas chromatography coupled to high-resolution time-of-flight mass spectrometry for ultra-trace analysis of organochlorine pesticides in river water, J. Chromatogr. A 1218 (2011) 6851–6860.

[122] P. Serôdio, J.M. Nogueira, Multi-residue screening of endocrine disrupters chemicals in water samples by stir bar sorptive extraction-liquid desorption-capillary gas chromatography–mass spectrometry detection, Anal. Chim. Acta 517 (2004) 21–32.

[123] N.S. Chary, S. Herrera, M.J. Gómez, A.R. Fernández-Alba, Parts per trillion level determination of endocrine-disrupting chlorinated compounds in river water and wastewater effluent by stir-bar-sorptive extraction followed by gas chromatography–triple quadrupole mass spectrometry, Anal. Bioanal. Chem. 404 (2012) 1993–2006.

[124] N. Ochiai, K. Sasamoto, H. Kanda, S. Nakamura, Fast screening of pesticide multiresidues in aqueous samples by dual stir bar sorptive extraction-thermal desorption-low thermal mass gas chromatography–mass spectrometry, J. Chromatogr. A 1130 (2006) 83–90.

[125] N. Ochiai, K. Sasamoto, H. Kanda, E. Pfannkoch, Sequential stir bar sorptive extraction for uniform enrichment of trace amounts of organic pollutants in water samples, J. Chromatogr. A 1200 (2008) 72–79.

[126] P. Grossi, I.R.B. Olivares, D.R. de Freitas, F.M. Lancas, A novel HS-SBSE system coupled with gas chromatography and mass spectrometry for the analysis of organochlorine pesticides in water samples, J. Sep. Sci. 31 (2008) 3630–3637.

[127] P. Tölgyessy, B. Vrana, K. Šilhárová, Determination of polychlorinated biphenyls and polybrominated diphenyl ethers in sediment by ultrasonic solvent extraction followed by stir bar sorptive extraction, Chromatographia (2013) 177–185.

[128] R. Rodil, P. Popp, Development of pressurized subcritical water extraction combined with stir bar sorptive extraction for the analysis of organochlorine pesticides and chlorobenzenes in soils, J. Chromatogr. A 1124 (2006) 82–90.

[129] C. Valls-Cantenys, E. Villaverde-de-Sáa, R. Rodil, J.B. Quintana, M. Iglesias, V. Salvadó, et al., Application of polydimethylsiloxane rod extraction to the determination of sixteen halogenated flame retardants in water samples, Anal. Chim. Acta 770 (2013) 85–93.

[130] E. Villaverde-de-Sáa, I. Racamonde, Ion-pair sorptive extraction of perfluorinated compounds from water with low-cost polymeric materials: polyethersulfone vs polydimethylsiloxane, Anal. Chim. Acta 740 (2012) 50–57.

[131] C. Basheer, V. Suresh, R. Renu, H.K. Lee, Development and application of polymer-coated hollow fiber membrane microextraction to the determination of organochlorine pesticides in water, J. Chromatogr. A 1033 (2004) 213–220.

[132] H. Bagheri, A. Aghakhani, Z. Ayazi, M. Khakinezhad, A polypyrrole-based sorptive microextraction coating for preconcentration of malathion from aquatic media, Chromatographia 74 (2011) 731–735.

[133] R. Jiang, J. Pawliszyn, Thin-film microextraction offers another geometry for solid-phase microextraction, TrAC Trends Anal. Chem. 39 (2012) 245–253.

[134] F. Wei, F.F. Zhang, H. Liao, X.Y. Dong, Y.H. Li, H. Chen, Preparation of novel polydimethylsiloxane solid-phase microextraction film and its application in liquid sample pretreatment, J. Sep. Sci. 34 (2011) 331–339.

[135] Q. Zhou, Y. Huang, J. Xiao, G. Xie, Micro-solid phase equilibrium extraction with highly ordered TiO_2 nanotube arrays a new approach for the enrichment and measurement of OCPs at trace level in environmental water samples, Anal. Bioanal. Chem. 400 (2011) 205–212.

[136] Q. Zhao, Q. Lu, Y. Feng, Dispersive microextraction based on magnetic polypyrrole nanowires for the fast determination of pesticide residues in beverage and environmental water samples, Anal. Bioanal. Chem. 405 (2013) 4765–4776.

[137] A. Salemi, E. Shafiei, M. Vosough, Optimization of matrix solid phase dispersion coupled with gas chromatography electron capture detection for determination of chlorinated pesticides in soil, Talanta 101 (2012) 504–509.

Chapter 4

Application of Passive Sampling Techniques in Measurement of HOCs in Aquatic Environments

Lian-Jun Bao[1] and Eddy Y. Zeng[1,2,*]
[1]*State Key Laboratory of Organic Geochemistry, Guangzhou Institute of Geochemistry, Chinese Academy of Sciences, Guangzhou, China;* [2]*School of Environment, Jinan University, Guangzhou, China*
**Corresponding author: E-mail: eddyzeng@gig.ac.cn*

Chapter Outline

1. INTRODUCTION

Freely dissolved concentrations of hydrophobic organic compounds (HOCs), such as polycyclic aromatic hydrocarbons (PAHs), dichlorodiphenyltrichloroethane (DDT) and its metabolites (the sum is designated as DDTs), and polychlorinated biphenyls (PCBs), in open water and sediment porewater have been the topics of numerous studies, because they largely dictate the geochemical

Comprehensive Analytical Chemistry, Vol. 67. http://dx.doi.org/10.1016/B978-0-444-63299-9.00004-1

processes of aquatic HOCs, such as riverine transport, sedimentation, resuspension from sediment, and volatilization [1]. Monitoring of HOCs in aquatic environments across the globe has constantly provided updates on the worldwide occurrence, global geochemical cycling, and ecological risk of HOCs [2]. Conventional sampling methods for sensing HOCs in open waters and sediment porewater follow a general protocol of sample collection with active samplers, i.e., pump and grab, filtration, or extraction (Soxhlet and liquid–liquid extraction), and then purification and instrumental analysis [3]. These active sampling methods require power supply which may not be feasible in remote areas. Thus the use of conventional active sampling methods to monitor HOCs in global waters is extremely difficult. Furthermore, active sampling by water pumping is also pegged with some drawbacks, such as high unit cost and limited sampling time by battery lifespan [4]. As a result, it is time consuming and costly to use active sampling methods for measuring time-weighted average concentrations or grab sampling HOCs in the global aquatic environment.

Conversely, passive sampling techniques are easy to operate and cost effective, which can be the key factor in conducting large-scale monitoring programs across the globe. For instance, Lohmann and Muir [2] called for establishing a global network for monitoring POPs in the world's aquatic environments using passive sampling devices, especially with polyethylene (PE) as the sorption phase. Generally, passive sampling techniques are typically operated based on the phase partitioning of target analytes, driven by the chemical potential gradient between the sampler's sorbent phase and sample matrix [5]. Available passive samplers for HOCs, such as semipermeable membrane devices (SPMD) [6], solid-phase microextraction (SPME) fibers [7], PE devices [8], and polyoxymethlyene (POM) films [9], normally consist of a sorption phase (sorbent) and a protection/support mechanism used to minimize physical/microbial damage and/or facilitate subsequent instrumental analysis.

Early applications of passive sampling techniques mainly involved in-house passive sampling of freely dissolved concentrations of HOCs in water or sediment field-collected with active sampling methods and transported to the laboratory. However, factors impacting the partitioning processes of freely dissolved HOCs among the environmental compartments, e.g., temperature and salinity, may not be accurately accounted for in laboratory measurements [10]. Therefore, in situ passive sampling of HOCs may be an advantageous approach over in-house sampling and is also bound to be the best mode for applications of passive samplers.

To comprehend the current state of passive sampling techniques suitable for field applications, this chapter compiles and analyzes available information about some new developments in typical passive samplers for sensing HOCs in aquatic environments (Table 1). The compiled information is synthesized to characterize the field application of passive sampling techniques, such as monitoring of occurrence and trends, assessment of bioavailability, determination of site-specific partition coefficients, and measurement of intercompartmental flux. In addition, the impacts of environmental factors on field passive sampling of HOCs are also examined.

TABLE 1 Detection Limits of Hydrophobic Organic Chemicals with Typical Passive Samplers Based on Calibration Methods in the Field Application

Sampler	Sorbent Phase	Calibration Method	Detection Limits	Field sites (i.e.)	Reference
Water (pg/L)					
SPMD[a]	Triolein (0.27 g)	Equilibrium	0.05–0.4 (PBDEs[f])	Scheldt estuary[i]	[6]
SPME[b]-based sampler	PDMS[c] fiber (100 μm)		73 (*p,p′*-DDE)	South California Bight	[11]
			43 (*o,p′*-DDE)		[12]
			90 (*o,p′*-DDD)		[11]
Passive water sampler	LDPE[d] (10 g; 50 μm)		0.04–60 (PAHs[e])	Reservoirs (China)	[13]
			0.6–8.3 (DDTs[g])	Hailing Bay (China)	[13]
Sediment Porewater (pg/L)					
In situ sampler	PDMS fibers (length: 95 mm; 30 μm)	Equilibrium	0.005–50 (PAHs)	Elbe River[j]	[14]
Multisection passive sampler	LDPE (0.13 g; 50 μm)	Sampling rate calibrated	810–1880 (DDXs[h])	Hailing Bay (China)	[15]

[a]*SPMD = semipermeable membrane device.*
[b]*SPME = solid-phase microextraction.*
[c]*PDMS = polydimethylsiloxane.*
[d]*LDPE = low-density polyethylene.*
[e]*PAHs = polycyclic aromatic hydrocarbons.*
[f]*PBDEs = polybrominated diphenyl ethers.*
[g]*Sum of p,p′-DDT, p,p′-DDD, p,p′-DDE, o,p′-DDT, o,p′-DDD, and o,p′-DDE.*
[h]*Sum of p,p′-DDT, p,p′-DDD, p,p′-DDE, o,p′-DDT, o,p′-DDD, o,p′-DDE, p,p′-DDMU, p,p′-DDNU, and p,p′-DBP.*
[i]*In the Netherlands.*
[j]*Near Geesthacht/Germany.*

2. PASSIVE SAMPLING DEVICES IN FIELD APPLICATION

2.1 Semipermeable Membrane Device

The SPMD developed by Huckins and his coworkers [16] in 1990 is typically composed of a thin-walled layflat low-density PE (LDPE) tube (50–100 μm in thickness) filled with a pure, neutral, and high molecular weight lipid (e.g., triolein, $m_a > 600$ Da). Freely dissolved target analytes can pass through the PE membrane and concentrate in the lipid. These devices mimic biological systems to measure the likely bioavailable fraction of HOCs in water and sediment porewater [17–19]. For field deployment, the commercially available device comprises five racks of standard SPMDs placed in a stainless steel canister, with holes to allow water to enter, and sealed with a protective cover [20]. Four different types (small, large, well, and disposable) of canisters are available for field deployment (http://www.est-lab.com/spmd.php). The deployment time of SPMDs ranges from days to months depending on the target compounds and typically is 28 days for HOCs in water [17,18].

The SPMD technique has been adopted by many governmental agencies (e.g., the United States Geological Survey, United States Environmental Protection Agency, and United Kingdom Environmental Agency) as a routine monitoring tool or standard method for determining time-weighted average concentrations of freely dissolved HOCs in aqueous and atmospheric media [20]. However, there may be some artifacts with SPMD in field application. For instance, Louch et al. [21] inferred that total suspended solids adhered to the PE tube might contribute to the significant differences between the SPMD-measured concentrations of eight PAH compounds (phenanthrene, 1-methylphenanthrene, 2-methylphenanthrene, 3-methylphenanthrene, 9-methylphenanthrene, fluoranthene, pyrene, and chrysene) at different locations of the same river. The requirements of long equilibrium time and labor-intensive sample processing may also limit the field application of SPMD [22].

2.2 SPME Fiber

Arthur and Pawliszyn [23] introduced the SPME-based passive sampling technique to environmental analysis in 1990 to eliminate the problems of high blank values and interferences from plastic cartridges for solid-phase extraction of HOCs. A typical SPME scheme is composed of a fused silica fiber coated with a layer of polymer, such as polydimethylsiloxane (PDMS) and polyacrylate. The choice of polymer generally hinges on the hydrophobicity of target compound. Because the polymer coating is fragile and easily broken, certain external protective mechanisms may be necessary in field deployment of SPME fibers. Zeng et al. [11] developed an SPME sampler, consisting of a 100-μm PDMS-coated fiber assembly enclosed in a copper casing with a series of 8-mm holes. This sampler was successfully employed to measure the concentrations of DDT and its metabolites at three coastal locations off southern California. Subsequently, a modified protective mechanism (Figure 1(a)), copper casing wrapped by glass fiber filter (GFF) membrane outfitted with 100-mesh copper meshes added to filter fine particles, was adopted in SPME samplers for measuring freely dissolved HOC concentrations in overlying water and sediment porewater [24,25]. Witt et al. [14] also utilized a perforated copper pipe with 5-mm holes to house nine PDMS fibers (each with a fiber length of 92 mm, Figure 1(c)) fixed by a polytetrafluoroethylene-armored clamp for sensing freely dissolved HOC in sediment porewater. A fiber-retracted SPME sampler (Figure 1(b)) was specifically designed for time-weighted average water sampling [7]. This special sampler consists of an external protective copper tube and a commercial PDMS fiber assembly inserted into a deactivated Silcosteel-treated tube used as diffusion path [7].

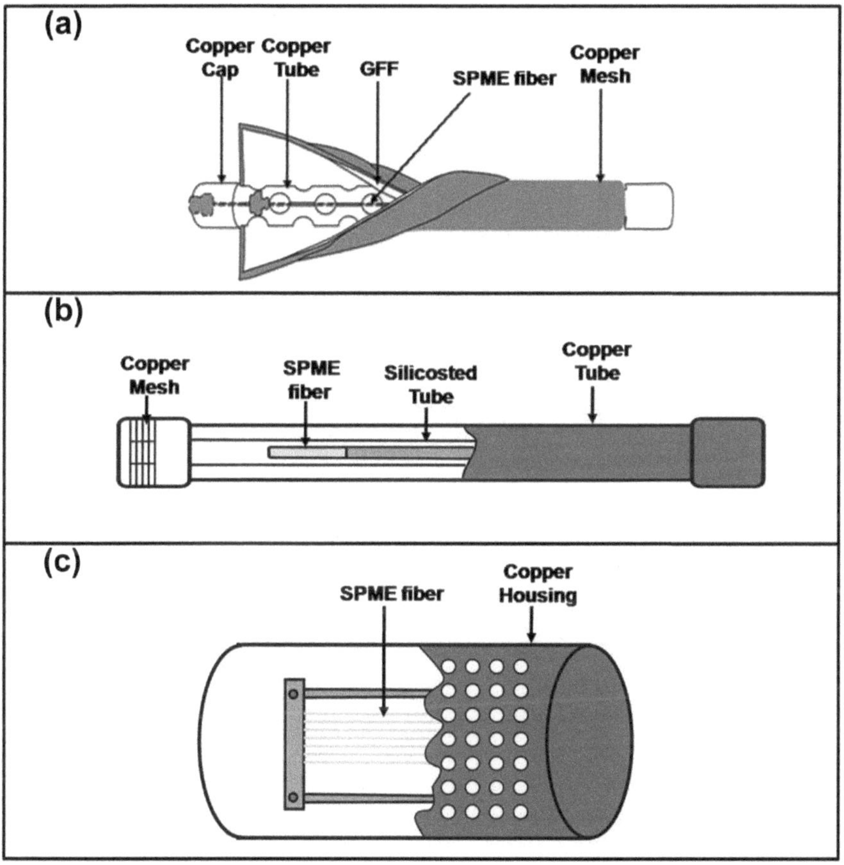

FIGURE 1 Typical passive samplers based on solid-phase microextraction (SPME) fiber: (a) a modified SPME sampler [24,25], GFF = glass fiber filter; (b) SPME fiber-retracted time-weighted average water sampler [7]; and (c) in situ sampler [14].

The SPME fiber sampler, which combines analyte extraction, concentration, and separation into one single step, has been widely used for in situ and ex situ sampling of HOCs in plants, fish, air, water, and sediment [26–28]. On the other hand, SPME fibers are fragile; in addition, the detection limits of SPME fiber methods are generally higher than those of SPMDs and PE-based samplers, because the sorbent phase in SPME is considerably smaller than those of the latter samplers. If large sorbent phase volumes (i.e., thickness of polymer coating >100 μm) are used, the sensitivity of PDMS fiber methods may be improved; however, longer deployment time with larger sorbent volume will be required to attain equilibrium sampling.

2.3 LDPE and POM Membranes

LDPE and POM are both thermoplastic polymers with long chains of hydrocarbons and carbon–oxygen bounds, respectively, which are suitable materials as sorbent phases because they cheap and tough. As mentioned earlier, LDPE tubes were initially used as the layflats in SPMDs to shield particles (in water or sediment) from penetrating into the sorbent phase (e.g., triolein). Later, scientists found that LDPE films could be used directly as sorbents in place of SPMDs and were able to measure low levels of organic compounds in water [29,30]. Obviously, LDPE films are much more cost effective and simpler to handle than SPMDs. As with LDPE, POM films are also suitable sorbent materials in passive samplers and have been gaining acceptance in recent years [31–33]. For example, a POM-based sampler has been used to measure the dissolved concentrations of PCBs in porewater for assessing the effectiveness of activated carbon remediation for contaminated sediment [32].

In some water field applications, LDPE and POM-based devices are quite simple. For instance, naked POM films were attached to a triangular aluminum frame or fastened with plastic wire; the device was anchored to a heavy object (e.g., a brick) and suspended in water by a buoy [31,34]. In other cases, bare LDPE films (Figure 2(a)) were stuck on a stainless steel or copper wire, woven into a circle, and then deployed in seawater by attaching the circle to a buoy with a nylon rope [8,37]. In a field sediment deployment, Tomaszewsky et al. [38] cut LDPE film into round shape and fastened it to a circular frame, which was directly inserted into sediment and horizontally deployed at the depth of 15 cm. Alternatively, LDPE and POM films were attached to sediment-penetrating rods made of stainless steel and then vertically placed in sediment (Figure 2(c)) [32]. Although the effects of particles adhered on the surface of LDPE film have remained unclear on quantitative measurements, preemptive measures have been implemented to minimize such effects. For example, two passive samplers for collecting freely dissolved hydrophobic organics in open water and sediment porewater contained GFF to prevent large particles from entering into LDPE films [13,15]. The open water sampler (Figure 2(d)) consists of a rectangular copper box with two open frames, which are filled with two 80-mesh

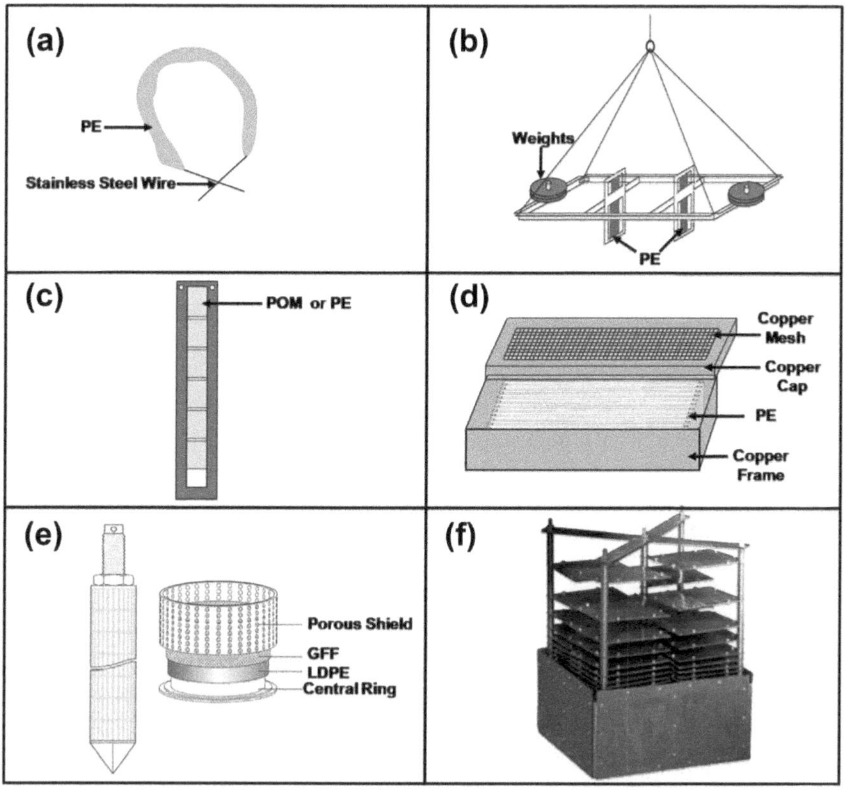

FIGURE 2 Typical passive samplers based on polyethylene (PE) or polyoxymethlyene (POM) film: (a) PE membrane [8]; (b) PE deployment device in sediment [35]; (c) PE and POM mounted on a sediment-penetrating rod [32]; (d) passive water sampler [13]; (e) multisection passive sampler [15], LDPE=low-density polyethylene, GFF=glass fiber filter; and (f) passive sampling device for measuring sediment–water diffusion fluxes of hydrophobic organic chemicals [36].

copper meshes, two stainless steel sieve plates, and GFF [13]. Strips of LDPE in the copper box are fastened in a bracket of comb-like structures [13]. The sediment porewater sampler is composed of a series of sampling cells isolated from each other with seclusion rings [15]. Each sampling cell contains LDPE film wrapped by GFF and outfitted with a porous stainless steel. The sediment porewater sampler (Figure 2(e)) is able to measure concentration profiles of porewater HOCs at every 2-cm sediment depth [15].

3. FIELD APPLICATION OF PASSIVE SAMPLERS

Most passive samplers have been employed to measure freely dissolved HOCs in environmental media. The chemical potential characterized by the difference in the dissolved concentrations of an organic contaminant between two interacting phases (e.g., water–sediment and soil–air) is the driving force behind the partitioning of the contaminant between the two phases and the key factor governing the intercompartmental diffusive flux [39–41]. Freely dissolved HOCs in water or sediment porewater have been considered to be the phase that is bioavailable for benthic organisms, which is a critical parameter for assessing the quality of aquatic ecological systems. Therefore, field applications of passive samplers have mainly focused on monitoring freely dissolved or gaseous HOCs, determining their site-specific partition coefficients and bioavailability, and measuring intercompartmental diffusive fluxes.

3.1 Measurement of Freely Dissolved or Gaseous HOCs

3.1.1 Water

While there have been frequent and comprehensive programs for monitoring atmospheric HOCs with passive air samplers, studies on passive sampling of HOCs in open water have been less common and comprehensive. Concentrations of aqueous HOCs in rivers, lakes, or estuaries have been determined mainly in association with the development and field validation of novel passive samplers or new quantitative methods with available samplers. For example, Zeng et al. [11] developed and field tested an SPME-based sampler using PDMS-coated fiber as the sorbent phase, and later they used it to determine the spatial distribution of *p,p′*-DDE (<0.073–2.6 ng/L) throughout the Southern California Bight [11,12]. Ouyang et al. [7] tested three types of SPME-based passive samplers (fiber-retracted, PDMS-rod, and PDMS membrane) in Hamilton Harbour (Canada) and obtained average concentrations of PAHs at 113.1 ± 14.6, 33.1 ± 2, and 53.2 ± 7.2 ng/L, respectively, at depths of 1, 11, and 21 m in the water column with the fiber-retracted sampler. In addition, the aqueous concentrations of DDT and its major metabolites obtained by the SPME-based sampling device and modified SPME sampler were 0.60–5.94 and 0.89–3.51 ng/L in the coast off southern California and Hailing Bay of Guangdong Province, China [11,24]. In other studies, SPMDs deployed at two sites of Scheldt estuary and four sites of the North Sea along the Dutch Coast and 10 freshwater sites in Amsterdam obtained time-weighted concentrations of BDE congeners at 0.0001–0.005 ng/L and PCBs at 0.04–1.3 ng/L [6,42]. In contrast to the high cost of SPME fiber and complex postsampling processing procedures with SPMD, PE- and POM-based samplers are much cheaper and relatively simple to handle, and therefore have found more recent applications in field monitoring. For example, PE-based samplers were deployed in Boston Harbor (Massachusetts) [8], Auke Lake (Alaska) [29], Narragansett Bay (Rhode Island) [41], four reservoirs and Hailing Bay of Guangdong Province, China [13], and the Palos Verdes Shelf of California, USA [43]. In open Baltic Sea, the median concentrations of polychlorinated dibenzo-*p*-dioxins and dibenzofurans and PCBs deduced with POM samplers were 2.3 pg toxic equivalents/m^3 and 15 pg/L [9], while the dissolved concentrations of PCBs in overlying water of the Grasse River varied between 10 and 30 ng/L from 2006 to 2009 [34]. Overall, the use of passive sampling methods for measuring aqueous HOCs in open water has been gradually extended to widespread areas around the world. However, such application has yet to gain large momentum.

Despite the recommendation of Lohmann and Muir [2] in 2010 for establishing a global network for monitoring aqueous HOCs in waters with passive sampling techniques, little progress has been made toward this proposed goal. Because monitoring of HOCs, including POPs and other persistent organic chemicals, in open waters of background or remote regions would provide critical information for evaluating the effectiveness of Stockholm Convention on POPs, it is significant to initiate and accelerate on building a global network based on passive sampling methods.

3.1.2 Sediment Porewater

Besides passive sampling, three other methods, i.e., traditional active sampling, modeling approach, and biomarkers, have also been used for measuring dissolved HOC concentrations in sediment porewater. However, these three methods have some significant limitations in field application. For instance, the prerequisite for a large volume of porewater to lower detection limits for an active sampling method is difficult to satisfy in field sampling [10]. Possible sequestration of HOCs in black carbon embedded in sediment and uncertainty in partition coefficients may affect the accuracy of modeling results [35,44]. In addition, natural biological variability (e.g., lipid content and composition) of benthic species as biomarkers would also impact the estimates of dissolved concentrations of HOCs in sediment porewater. Therefore, passive sampling can be a better alternative for determining dissolved HOCs in porewater.

Similar to passive sampling of aqueous HOCs, in situ measurement of HOCs in sediment porewater has been conducted only at a limited number of sites. In fact, most data have been acquired in laboratory by passive sampling of field-collected sediment. In situ exposure of passive samplers has been conducted only in a few sites, i.e., South San Francisco Bay [38], Boston Harbor [35], Stockholm Harbor of Baltic Sea [45], a town canal (Entenfleet, Germany) [14], Elbe River [14], and Hailing Bay (China) [15]. Under the field conditions, in situ PE and PDMS samplers were placed within the top layer of sediment to measure single-point porewater concentrations of PCBs and PAHs [14,38]. Two other approaches have been developed which can measure chemical concentration profiles in sediment porewater. Oen et al. [32] used PE or POM-mounted sediment-penetrating rods (Figure 2(c)) to obtain sediment porewater concentrations of PCBs at 0–40 cm depths with 5-cm intervals. In addition, Liu et al. [15] deployed a multisection passive sampler (Figure 2(d)) in Hailing Bay of South China and obtained vertical sediment porewater concentration profiles of DDT and its metabolites within a 40 cm depth with 2-cm intervals. Moreover the results were consistent with the metabolism pathways of DDT. Apparently, the capability of measuring vertical profiles of sediment porewater HOCs has created a new dimension in assessing the mobility of buried contaminants and the effectiveness of contaminated sediment remediation efforts. Subsequently, it also has great implications for protecting aquatic environments and ultimately human health.

Freely dissolved concentrations of HOCs in sediment porewater are usually greater than those in overlying water. Consequently, monitoring HOCs in sediment porewater with passive samplers is expected to draw more attention in coming years due to its importance for minimizing human health risk through aquatic foodweb transfer. In field applications, one cost-effective approach would be to conduct passive sampling of both sediment porewater and overlying HOCs with the same sampler.

3.2 Measurement of Site-Specific Partition Coefficients

The partition coefficient of an HOC between organic carbon (OC)/dissolved OC (DOC) and water (K_{oc} or K_{doc}) is a critical parameter for fugacity-based modeling of its geochemical fate. Although K_{oc} or K_{doc} can be estimated, accumulated field data suggest that measured and predicted values are often substantially different [46], e.g., the measured log K_{oc} values of phenanthrene (6.07–7.03) in Boston Harbor of Massachusetts, USA were two orders of magnitude greater than the predicted value Eqn (4.12) [47]. To account for the impact of organic matter on K_{doc}, humic acid (Aldrich) was used in the determination of K_{doc}, but laboratory-determined K_{doc} values of benzo[*a*]pyrene were still substantially different from their site-specific values. One of the reasons for the distinction may probably be attributed to the difference between the properties of commercial humic acid and naturally occurring DOC [48]. Passive sampling techniques have two advantages over liquid–liquid extraction for determining the site-specific K_{oc} or K_{doc}. First, only the freely dissolved portions of target compounds are sensed by passive samplers, whereas dissolved and DOM-bound target compounds may be retained by liquid–liquid extraction, thereby underestimating K_{oc} or K_{doc} [49]. Second, nondepletive extraction of HOCs with passive samplers does not exhaust target compounds in water or sediment porewater as compared to liquid–liquid extraction. Vaes et al. [50] suggested using <5% of the change in the concentration of a target compound before and after extraction as the criterion of nondepletive extraction, i.e., equilibration of the target compound between two interacting phases is not impacted by extraction.

Polymer-coated SPME fibers with 7–100 μm coating thickness often take up only small portions of analytes being sampled and are strong sorbents of HOCs. As a result, they are preferable choices for nondepletive extraction and have been often used in determination of site-specific K_{oc} or K_{doc} [51]. For example, the log K_{oc} values of five PAH compounds (phenanthrene: 4.65±0.03; pyrene: 5.26±0.02; benz[*a*]anthracene: 6.21±0.01; benzo[*b*]fluoranthene: 6.72±0.01; and benzo[*ghi*]perylene: 7.18±0.06) in the Danish soils were estimated from the aqueous PAH concentrations obtained by an SPME fiber in the negligible depletion mode and the fraction of OC divided by the total concentration in soil [52]. Similarly, the log K_{doc} values of eight BDE congeners (BDE-28, -47,

-66, -99, -100, -153, -154, and -183) in sediments collected from four sites of USA (Greasy Creek, Benton County, Oregon; Jordan Lake Reservoir, Chatham County, North Carolina; Leaf Lake, El Dorado County, California; and San Diego Creek, Orange County, California) ranged from 5.10 to 8.02 [53]. These data have definitely filled in a large knowledge gap in site-specific K_{oc} or K_{doc} and are more preferable than predicted values in phase-partition modeling. In addition, a previous study [54] has demonstrated that K_{oc} values of phenanthrene varied with different compositions of organic materials in sediment, i.e., lignite, charcoal, activated carbon, and lignite coke. Therefore, site-specific K_{oc} or K_{doc} derived with passive sampling methods are vital for improving the accuracy of fugacity modeling under different environmental conditions and further our understanding of the mechanism for HOCs' sorption to DOC.

3.3 Bioavailability Assessment

Passive samplers, which measure freely dissolved concentrations of HOCs, can hence be used to assess the bioavailability in sediment or soil. Numerous studies have suggested that the log-transformed amounts of HOCs sorbed on passive sampling phases, such as PE, SPME fiber, and POM, are significantly correlated with those (normalized in lipid) in benthic organisms (*Nereis virens*, *Lumbriculus variegates*, *Hexagenia* sp., and *Eisenia fetida*) exposed to the same sediment or soil in laboratory [55–57]. However, the inhabiting environmental factors of benthic organisms, such as temperature, salinity, and food variability, are expected to have some effects on the internal bioaccumulation of HOCs in organisms under field conditions.

Van der Heijden and Jonker [58] found that in situ SPME fiber was a better predicting tool for accumulation of PAHs in *L. variegates* than ex situ SPME. In addition, the concentrations of petroleum hydrocarbons determined by in situ SPME tended to underestimate the amounts in worms [59]. Nevertheless, these two studies demonstrated that the laboratory-based POM samplers were the same or better in predicting the amounts of PAHs and petroleum hydrocarbon in worms than in situ SPME fiber, and may be an alternative choice for in situ assessing the bioavailability of HOCs in sediment. On the other hand, incidents of sediment and soil contamination by petroleum residues from oil leaking are on the rise [60]. Related toxicity tests showed that low content (0.01%) of unresolved complex mixture in sediment contaminated with petroleum residues increased the bioavailability of PAHs for *L. variegates*, whereas a high level (5%) posed opposite effects [61]. Apparently, further efforts are needed to determine whether in situ or ex situ passive samplers (e.g., POM) are better predictors of the bioavailability of sediment-associated HOCs and petroleum residues.

Another application of passive sampling techniques is toxicity assessment for sediment HOCs in aquatic organisms. Li et al. [62] found that the coefficient (r^2) of correlation between the mortality to *Hyalella azteca* (expressed in probit units) and toxic units of cypermethrin was elevated from 0.26 (p=0.06) to 0.52 (p=0.004), with the toxic unit estimated from matrix-SPME instead of

OC-normalized method. This is probably because freely dissolved HOCs derived with matrix-SPME fiber may be able to penetrate through cell membranes and subsequently cause toxic effects on organisms. Moreover, Tenax-extractable concentrations of permethrin in sediments were significantly correlated with various endpoints of chronic toxicity to *Chironomus dilutes* [63]. However, the median lethal and effects concentrations, estimated from OC-normalized, SPME fiber, and Tenax, of bifenthrin on *C. dilutes* and *H. azteca* in different sediments were inconsistent, probably resulting from biological degradation or ingestion of sediment particles by the worms [64].

3.4 Intercompartmental Flux Measurement

The molecular diffusion flux of an HOC across the sediment–water or air–water interfaces indicates not only the magnitude of intercompartmental transfer but also the directions of such transfer; hence it is an important process in the geochemical cycling of HOCs. Because passive samplers can minimize the disturbance of HOCs distributing between two interacting compartments, they may be a powerful tool for measuring intercompartmental fluxes of HOCs.

However, to our knowledge, only a few successful measurements of fluxes of HOCs across the air–water interface with passive samplers have been conducted. Lohmann et al. [41] determined a net air–water exchange gradient with PE sampler in Narragansett Bay of Rhode Island, USA and observed a trend of volatilization for most PAHs, but they did not calculate the air–water fluxes of PAHs, probably due to the lack of wind speed and rainfall data. Wind speed is a key parameter for calculating the intercompartmental flux [65], whereas rainfall occurring during the sampling period may increase the dissolved concentrations of volatile PAHs in water [41].

Furthermore, two passive samplers, an infinite-sink benthic flux chamber and an integrated passive sampling device, have been developed to measure the diffusive fluxes of HOCs across the sediment–water interface [36,66]. The benthic chamber with SPMD as the sorbent phase was deployed in Oslo Harbor of Norway and escaping fluxes of pyrene and PCB-52 from contaminated sediment to overlying water were determined at 0.3–1.6 μg m^{-2} d^{-1} and 2–8 ng m^{-2} d^{-1}, respectively [66]. However, there are three intrinsic deficits with the benthic chamber approach. First, a chamber is normally so deployed that its interior space is isolated from overlying water, thus only the escaping fraction of HOCs from sediment is measured by the chamber. Therefore, the chamber approach is unable to measure depositing fluxes from overlying water to sediment. Second, the benthic chamber approach is based on the assumption that the lengths and slopes of the diffusion boundary layers at the sorbent phase–water and sediment–water interfaces are essentially identical. Apparently, this assumption may not always be satisfied in field deployment; large measurement uncertainties can be produced as a result. Third, benthic chamber deployment would interfere with current flows near the seafloor, and consequently the boundary layer at the sediment–water interface. Thus, large uncertainties in measured fluxes of HOCs may be induced from benthic chambers.

An integrated passive sampling device with LDPE as the sorbent phase (Figure 2(f)) has been developed to mitigate the above-mentioned deficits [36]. This device is able to measure concentration profiles of HOCs near the sediment–water interface, from which sediment–water diffusion fluxes of HOCs can be calculated with a Fickian diffusion-based mathematical model. This device was field tested in Hailing Bay of South China where sediment was heavily contaminated with DDTs, and diffusion fluxes (5.9–150 ng m^{-2} d^{-1}) of individual DDT metabolites were obtained; a custom-made benthic chamber (with LDPE as the sorbent phase) deployed in parallel with the passive sampling device obtained comparable results (5.5–85 ng m^{-2} d^{-1}) [36]. More significantly, the passive sampling device was able to detect the diffusion fluxes of *o,p′*-DDT and *p,p′*-DDT as being depositing from overlying water to sediment, which was otherwise not distinguished with the benthic chamber approach.

3.5 Quantitation Methods in Field Passive Sampling

The principle of passive sampling is characterized as the process of a dissolved target compound being enriched in the sampler's sorbent phase, i.e., "like dissolves like." The fundamental mechanism governing the partitioning of the analyte of interest between the sorbent phase and water is the theory of thermodynamics and mass-transfer kinetics [67]. Basically, the amount of a target HOC extracted by a sorbent phase is proportional to the concentration of the analyte in water or porewater and the exposure time. Two regimes, i.e., kinetic and near equilibrium regimes, were defined to describe the extraction profile of a target analyte (Figure 3). Once the equilibrium partitioning of an analyte between the sorbent phase and water is reached, the amount of the analyte retained by the

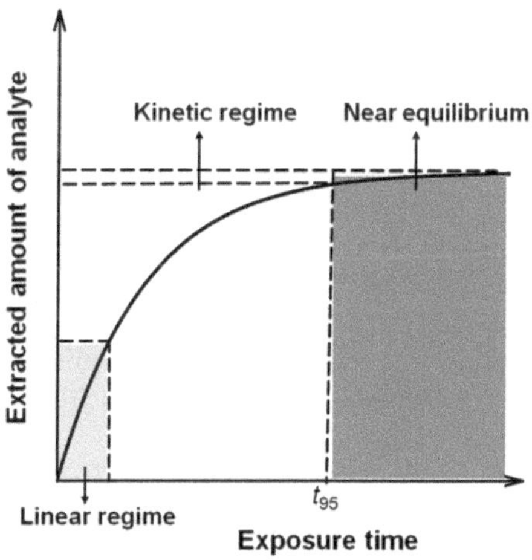

FIGURE 3 Typical extraction profile of a target analyte in a passive sampler's sorbent phase [67].

sampler becomes constant, if the concentration of the analyte in water remains unchanged. Before the equilibrium regime is reached, the mass transfer of an analyte from water into the sorbent phase is kinetically controlled. Within the kinetic regime, there is a linear range (Figure 3), where the amount of the analyte sequestered by the sorbent phase increases linearly with exposure time. If a passive sampler is designed properly, e.g., fiber-retracted SPME device [7], the time-weighted average concentration of an analyte can be obtained during an optimized sampling period.

Currently, there are two main quantitation methods in field passive sampling, on the basis of equilibrium partitioning and kinetic diffusion-controlled process. Specifically, when partitioning of a target analyte between the sorbent phase and water or porewater is equilibrated, the dissolved concentration (C_{free}) of the analyte can be calculated by

$$C_{\text{free}} = \frac{C_{\text{sorbent}}}{K_{\text{sw}}} \quad (1)$$

where C_{sorbent} is the analyte concentration in the sorbent phase at equilibrium and K_{sw} is the equilibrium partition coefficient of the target compound. The value of K_{sw} is somewhat dependent on some environmental factors, such as temperature, salinity, and pressure [68,69]. Additionally, K_{sw} for a large number of HOCs in different sorbent phases such as PDMS and PE have been determined and can be readily used in method development. Quantitation methods based on equilibrium partitioning are generally easy to handle and capable of sensing low levels of HOCs in water or porewater. For example, if the PE–water partition coefficient for a target compound is $10^5\,L_w/kg_{PE}$ and the lowest detectable analyte amount in instrumental analysis is 10 pg, then the detection limit is 10 pg/L for 1 g of the sorbent phase (PE) and a final extract volume of 100 μL [8]. On the other hand, the time to equilibrium can be quite long for very hydrophobic compounds, perhaps weeks or months (Table 2). For instance, the time to equilibrium for field sampling of aqueous pentachlorobenzene and PCBs with log K_{ow} at 5.16–7.26 varied between 23 and 52 days with a 15-μm PDMS fiber [70]. The time to equilibrium was more than 119 days for sampling freely dissolved PAHs in Oslo Harbor of Norway with 100-μm LDPE and 500-μm POM film [31]. In prolonged field deployment, loss or damage of the samplers is inevitable, thereby greatly undermining the practical value of equilibrium partitioning-based quantitation methods. Long equilibrium times may also restrict the utility of passive samplers in extremely remote regions if frequent back and forth traveling and extended stays on-site are inconvenient.

To address the above-mentioned issue with equilibrium partitioning, a kinetically diffusion-controlled quantitation method has been used to determine C_{free}, i.e.,

$$C_{\text{free}} == \frac{C_{\text{sorbent}}(t)}{\left(1 - e^{-k_{\text{abs}}t}\right) \times K_{\text{sw}}} \quad (2)$$

TABLE 2 Summaries of Partition Equilibrium Time (t_{eq}) between Target Compound and Sorbent of Passive Sampling Devices under Static Condition

Passive Sampler	t_{eq} (days)	Log K_{ow} Range	Target Compound	Reference
15-μm PDMS fiber	23–51	5.16–7.20	Hexachlorobenzene, *p,p′*-DDE 5 PCB congeners and fluoranthene	[70]
74-μm LDPE	60	3.37–7.23	38 PAH compounds	[55]
35-μm PDMS fiber	28	4.46–6.72	Hexachlorobenzene, *p,p′*-DDE, permethrin chloropyrifos, and phenanthrene	[71]
51-μm LDPE	239	4.46–8.18	107 PCB congeners	[38]
28-μm PDMS fiber and 17-μm POM	30	3.37–7.23	14 PAH compounds	[9]
100-μm LDPE and 500-μm POM	>119	3.37–7.23	14 PAH compounds	[31]
28.5-μm PDMS fiber	30	4.46–7.23	13 PAH compounds	[58]
100-μm PDMS fiber	14–110	4.46–7.36	3 PAH compounds and 3 PCB congeners; 6 insecticides	[25]

PDMS = polydimethylsiloxane; LDPE = low-density polyethylene; POM = polyoxymethylene; PAH = polycyclic aromatic hydrocarbons; PCB = polychlorinated biphenyl.

where k_{abs} is the absorption rate coefficient, which relates to the property of the analyte under consideration and environmental factors, such as hydrological conditions. One approach to estimating k_{abs} is the use of performance reference compounds (PRCs) which are preloaded into the sorbent phase prior to deployment (Figure 4(a)). If the PRC is an isotope-labeled counterpart of a target analyte, k_{abs} of the target analyte is the same as the desorption rate (k_{des}) of the preloaded PRC (Figure 4(b)), which can easily be determined by the retained faction (f) of the PRC after exposure, i.e.,

$$k_{des} = -\frac{\ln f}{t} = -\frac{\ln\left(\frac{q}{q_0}\right)}{t} \tag{3}$$

where q_0 and q are the dosed and remaining amounts of the PRC on the sorbent phase and t is the exposure time. Thus, Eqn (2) can be revised to

$$C_{free} == \frac{C_{sorbent}(t)}{(1-f)\times K_{sw}} \tag{4}$$

It is generally not necessary to correct C_{free} for site-specific characteristics (e.g., temperature, flow, and biofouling), as the simultaneous use of PRCs automatically offset the impacts of environmental factors on the uptake rates of the target analytes. Ideally, the preloaded PRCs should be isotopically labeled analogues of the target analytes under investigation. In addition, the optimal f value should range from 20% to 80% so as to minimize the analytical uncertainties for k_{abs}. Nevertheless, there are still two limitations of the PRC-calibrated method.

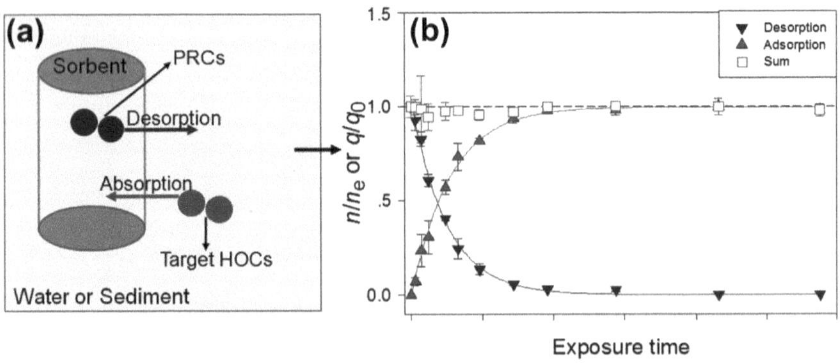

FIGURE 4 Schematic showing (a) calibration method of performance reference compounds (PRCs) and (b) absorption kinetics of p,p'-DDE and desorption kinetics of corresponding isotope-labeled PRC (p,p'-DDE-d_8) on the PDMS fiber in sediment under mixing conditions [72] (q_0 is the initial amount of PRC on the fiber; q is the amount of PRC remaining on the fiber at time t; n and n_e are the amounts of target compound on the fiber at time t and at equilibrium, respectively; sum $= q/q_0 + n/n_e$). HOCs = hydrophobic organic compounds.

First, isotopically labeled PRCs are costly and may not be always available for all target compounds. Alternatively, unlabeled chemicals, with physiochemical properties similar to those of the target analytes and rarely found in the environment, have been used as PRCs. The k_{abs} value of a target analyte can be adjusted by the alternative PRC through calibration, e.g., molar volume adjustment [73], one-dimensional diffusion model [74], and exposure adjustment factors [75]. Nevertheless, such calibrated methods still cannot completely eliminate the variability in k_{abs} resulting from the difference in physiochemical property between the alternative PRC and target analyte. For example, Fernandez et al. [35] reported that the concentrations of pyrene and chrysene in sediment porewater calibrated by phenanthrene-d_{10} through the molar volume adjustment were 40% and 60% of those calibrated by pyrene-d_{10} and chrysene-d_{12}, respectively.

In addition, k_{des} can be expressed as a function of the diffusion coefficient in aqueous layer (D_w) and within the layer (D_s) of the sorbent phase, K_{sw}, the diffusion layer thickness in the aqueous layer (δ_w) and in the sorbent phase (δ_s), the transport surface area (A), and volume (V_s) of sorbent, i.e.,

$$\frac{1}{k_{des}} = \frac{V_s \delta_w K_{sw}}{A D_w} + \frac{V_s \delta_s}{A D_s} \tag{5}$$

When the resistance to mass transfer of PRCs occurred in the aqueous boundary layer, Eqn (5) can be revised to yield

$$\frac{1}{k_{des}} = \frac{V_f \delta_w K_{sw}}{A D_w} \tag{6}$$

The sampling rate (R_s) is defined as

$$R_s = \frac{A D_w}{\delta_w} \tag{7}$$

Combing Eqns (6) and (7) yields the following the equation:

$$k_{des} = \frac{R_s}{K_{sw} V_s} \tag{8}$$

Therefore, a nonlinear least-squares regression method was also explored to estimate R_s by fitting all f values covering the full range of 0–100% as a function of K_{sw} [76], i.e.,

$$f = \exp\left(-\frac{R_s t}{K_{sw} V_s}\right) \tag{9}$$

However, large uncertainties in R_s can be deduced by f values within <20% and >80%, unless a large number of PRCs or passive samplers (deployed at the same location) are used to generate sufficient data points within these ranges

[77]. Obviously, it is quite costly and therefore may not be feasible to eliminate most uncertainties in R_s by the nonlinear least-squares regression approach in field sampling.

Second, the PRC-calibrated method may not be suitable for all passive samplers under field conditions. In other words, PRCs probably behave differently with different sorbent phases. For instance, Perron et al. [78] found that the C_{free} values of PAHs and PCBs estimated with POM-based sampler were three times greater than those obtained with PE-based sampler using the same PRC-calibrated method. This difference between POM and PE samplers was attributed to the different structures of the sorbent phases. POM, also known as acetal, is a functional polymer containing oxygenic groups, while PE is a thermoplastic polymer consisting of long hydrocarbon chains only. Therefore, as indicated by limited information, the electron donor–acceptor interactions between PAHs or PCBs and oxygen groups in POM are stronger than those between the target compounds and PE, which consists of van der Waals dispersive forces only, and may retard the release of PRCs from POM [78].

4. FACTORS IMPACTING PASSIVE SAMPLING

4.1 Temperature

Significant changes in temperature may have some effects on the solubilities of target compounds in water or sorbent phases, and thereby play an important role in field passive sampling if environmental conditions are substantially distinct from those used to determine quantitation parameters, such as K_{sw} and R_s, in the laboratory [68,69]. Specifically, a modified van't Hoff equation was used to describe the impacts of temperature on K_{sw} in PDMS fiber and PE [68,69]:

$$\ln K_{sw}^{T_2} = \ln K_{sw}^{T_1} + \left(\frac{-\Delta H_{sw}}{R}\right)\left(\frac{1}{T_2} - \frac{1}{T_1}\right) \quad (10)$$

where $K_{sw}^{T_2}$ and $K_{sw}^{T_1}$ are partition coefficients at temperatures T_2 and T_1 (K), respectively; ΔH_{sw} is the change in the partial molar excess enthalpy for a target compound in the sorbent phase and water; and R is the universal ideal gas constant (8.13 J/mol K). Equation (10) indicates that the K_{sw} values of higher molecular weight compounds can be more strongly impacted by temperatures, as ΔH_{sw} increases with increasing molecular size [79]. Muijs and Jonker [80] obtained ΔH_{sw} for several PAHs (phenanthrene, anthracene, fluoranthene, pyrene, benzo[*a*]anthracene, chrysene, benzo[*e*]pyrene, benzo[*b*]fluoroanthene, and benzo[*k*]fluoranthene) in the range of −21 to −46 kJ/mol between 5 and 30 °C, equivalent to the difference of 42–65% for K_{sw} values in PDMS fiber within 5–30 °C.

The effects of temperature on measured K_{sw} values in PE-based sampler have been identified for a few HOCs [81]. For example, Adams et al. [8] observed significantly positive ($r^2>0.71$ and $p<0.01$) correlations between ln K_{sw} and 1/(RT) for PCB-52, pyrene, and phenanthrene. Booij et al. [30] also reported that K_{sw} for partitioning of PCB-28, PCB-52, and five PAHs compounds (acenaphthene, phenanthrene, fluoranthene, pyrene, and benzo[*a*]anthracene) between PE and water at 2 °C were one to three times greater than those at 30 °C, but a similar trend was not observed for chrysene, benzo[*e*]pyrene, and benzo[*a*]pyrene. In addition, most K_{sw} values for HOCs between PDMS-coated fiber and water have been determined at 25 °C, with only a few at other temperatures [69]. Overall, temperature may significantly influence in situ determination of HOC concentrations by impacting K_{sw}. It is interesting to note that a flow-through system, which was developed to determine the K_{sw} values of PAHs between PDMS fiber and water with environmental variables (e.g., temperature and turbulence) [82,83], may provide a new opportunity to measure K_{sw} accurately for HOCs for all passive samplers.

4.2 Salinity

Currently, passive samplers have been mainly deployed in overlying water or sediments of estuaries, harbors, and oceans, where dissolved salts may have some impact on the partitioning of HOCs between the sampler's sorbent phase and ambient water. In general, dissolved salts can decrease the aqueous solubilities of HOCs, thereby shifting the sorption equilibrium toward the sorbent phase [69]. The empirical Setschenow equation has been used to describe the effects of salinity on solubility [84] or K_{sw}

$$\log\left(\frac{C_w^{salt}}{C_w^{fresh}}\right) = -K^s[\text{salt}]_{tot} \tag{11}$$

and

$$\log\left(\frac{K_{sw}^{salt}}{K_{sw}^{fresh}}\right) = K^s[\text{salt}]_{tot} \tag{12}$$

where C_w^{salt} and C_w^{fresh} are the solubilities of a target compound in freshwater and salt water, respectively; K_{sw}^{salt} and K_{sw}^{fresh} are the partition coefficients of the target compound in freshwater and salt water, respectively; K^s is the Setschenow constant (L/mol) for the analyte in the salt; and $[\text{salt}]_{tot}$ is the total salt content (mol/L). As shown in Eqn (12), an increase in salt concentration would elevate the K_{sw} of HOCs, which has been verified with PDMS fiber and PE [8,85].

Similarly, increased salinity also has a positive influence on water viscosity and decreases the diffusivities of HOCs in water, and therefore slows the process of equilibrating between the sorbent phase and water. Reitsma et al. [81] found that the amount of benzo[*a*]pyrene in PE was reduced from 56% to 11% of the expected equilibrium mass as salinity increased from 0 to 100‰ at 28 d. Overall, salinity is an important factor in determining the C_{free} of HOCs with field passive sampling, especially when equilibrium extraction is used as the quantitation method.

4.3 Other Factors

Biofouling is the result of accumulation of microorganisms, plants, algae, or animals on the surface of a sampler in field application. Several studies have shown that periphyton and macrofauna attached on the SPMD surface may reduce sampling rates [75,86,87]. On the other hand, Booij et al. [88] compared the sampling rates of PAHs with SPMDs in shielded and unshielded exposures, calibrated with the dissipation rates of PRCs (acenapthene-d_{10}, phenanthrene-d_{10}, fluoranthene-d_{10}, and chrysene-d_{12}), but did not observe any difference between the two exposure modes (unshielded and shielded with stainless steel mesh screens). The effects of biofouling on the determination of dissolved HOCs with SPME fiber, PE-, and POM-based samplers have not been reported.

In addition, water flow velocity may also be an important factor in determining the sampling rate of HOCs in field passive sampling. Apparently, high water flow rate can increase diffusivities of target compounds and therefore enhance sampling rates if the diffusion kinetics are boundary layer controlled. Booij et al. [88] found that an enforced flow rate up to 90cm/s increased the sampling rate of PAHs with SPMDs by a factor of 1.2–2.2.

5. CONCLUSIONS

Passive samplers have been used in situ to monitor freely dissolved HOCs in open water and sediment porewater. Measurement of fluxes of HOCs across the air–water interface has been far from successful with passive sampling methods. By comparison, the escaping/depositing fluxes of DDTs at the sediment–water interface under field conditions have been successfully measured with the development of a passive sampling device. Equilibrium extraction is easy to handle and has relatively high sensitivity in sensing aqueous HOCs in aquatic environment. However, its field application is impacted by several environmental factors, such as temperature and salinity. On the other hand, PRC calibration methods can circumvent the environmental impacts on passive sampling of HOCs, but determination of the sampling rates for target HOCs based on the lost fractions of alternative analogues instead of isotope-labeled compounds as PRCs should merit more research.

REFERENCES

[1] E. Jurado, J.-M. Zaldívar, D. Marinov, J. Dachs, Fate of persistent organic pollutants in the water column: does turbulent mixing matter? Mar. Pollut. Bull. 54 (2007) 441–451.

[2] R. Lohmann, D. Muir, Global aquatic passive sampling (AQUA-GAPS): using passive samplers to monitor POPs in the waters of the world, Environ. Sci. Technol. 44 (2010) 860–864.

[3] P. Konieczka, L. Wolska, J. Namieśnik, Quality problems in determination of organic compounds in environmental samples, such as PAHs and PCBs, Trends Anal. Chem. 29 (2009) 706–717.

[4] Y. Chen, J. Pawliszyn, Time-weighted average passive sampling with solid-phase microextraction device, Anal. Chem. 75 (2003) 2004–2010.

[5] S. Seethapathy, T. Górecki, X. Li, Passive sampling in environmental analysis, J. Chromatogr. A 1184 (2008) 234–253.

[6] K. Booij, B.N. Zegers, J.P. Boon, Levels of some polybrominated diphenyl ether (PBDE) flame retardants along the Dutch coast as derived from their accumulation in SPMDs and blue mussels (*Mytilus edulis*), Chemosphere 46 (2002) 683–688.

[7] G. Ouyang, W. Zhao, L. Bragg, Z. Qin, M. Alaee, J. Pawliszyn, Time-weighted average water sampling in Lake Ontario with solid-phase microextraction passive samplers, Environ. Sci. Technol. 41 (2007) 4026–4031.

[8] R.G. Adams, R. Lohmann, L.A. Fernandez, J.K. Macfarlane, P.M. Gschwend, Polyethylene devices: passive samplers for measuring dissolved hydrophobic organic compounds in aquatic environments, Environ. Sci. Technol. 41 (2007) 1317–1323.

[9] G. Cornelissen, K. Wiberg, D. Broman, H.P.H. Arp, Y. Persson, K. Sundqvist, P. Jonsson, Freely dissolved concentrations and sediment-water activity ratios of PCDD/Fs and PCBs in the open Baltic Sea, Environ. Sci. Technol. 42 (2008) 8733–8739.

[10] L.J. Bao, E.Y. Zeng, Passive sampling techniques for sensing freely dissolved hydrophobic organic chemicals in sediment porewaters, Trends Anal. Chem. 30 (2011) 1422–1428.

[11] E.Y. Zeng, D. Tsukada, D.W. Diehl, Development of a solid-phase microextraction-based method for sampling of persistent chlorinated hydrocarbons in an urbanized coastal environment, Environ. Sci. Technol. 38 (2004) 5737–5743.

[12] E.Y. Zeng, D. Tsukada, D.W. Diehl, J. Peng, K. Schiff, J.A. Noblet, K.A. Maruya, Distribution and mass inventory of total dichlorodiphenyldichloroethylene in the water column of the Southern California Bight, Environ. Sci. Technol. 39 (2005) 8170–8176.

[13] L.-J. Bao, S.-P. Xu, E.Y. Zeng, Development and field validation of a low-density polyethylene-containing passive sampler for measuring dissolved hydrophobic organic compounds in open waters, Environ. Toxicol. Chem. 31 (2012) 1012–1018.

[14] G. Witt, S.-C. Lang, D. Ullmann, G. Schaffrath, D. Schulz-Bull, P. Mayer, Passive equilibrium sampler for in situ measurements of freely dissolved concentrations of hydrophobic organic chemicals in sediments, Environ. Sci. Technol. 47 (2013) 7830–7839.

[15] H.-H. Liu, L.-J. Bao, W.-H. Feng, S.-P. Xu, F.-C. Wu, E.Y. Zeng, A multisection passive sampler for measuring sediment porewater profile of dichlorodiphenyltrichloroethane and its metabolites, Anal. Chem. 85 (2013) 7117–7124.

[16] J.N. Huckins, M.W. Tubergen, G.K. Manuweera, Semipermeable membrane devices containing model lipid: a new approach to monitoring the bioavailability of lipophilic contaminants and estimating their bioconcentration potential, Chemosphere 20 (1990) 535–552.

[17] J.A. Lebo, R.W. Gale, J.D. Petty, D.E. Tillitt, J.N. Huckins, J.C. Meadows, C.E. Orazio, K.R. Echols, D.J. Schroeder, Use of the semipermeable membrane device as an in situ sampler of waterborne bioavailable PCDD and PCDF residues at sub-parts-per-quadrillion concentrations, Environ. Sci. Technol. 29 (1995) 2886–2892.

[18] Z. Wang, Y. Wang, M. Ma, Y. Lu, J. Huckins, Use of triolein-semipermeable membrane devices to assess the bioconcentration and sediment sorption of hydrophobic organic contaminates in the Huaihe River, China, Environ. Toxicol. Chem. 21 (2002) 2378–2384.

[19] K.S. Williamson, J.D. Petty, J.N. Hunkins, J.A. Lebo, E.M. Kaiser, Sequestration of priority pollutant PAHs from sediment pore water employing semipermeable membrane devices, Chemosphere 49 (2002) 717–729.

[20] J. Huckins, D. Alvarez, Semipermeable Membrane Device, http://www.cerc.usgs.gov/pubs/center/pdfdocs/spmd.pdf (accessed July, 2013).

[21] J. Louch, G. Allen, C. Erickson, G. Wilson, D. Schmedding, Interpreting results from field deployments of semipermeable membrane devices, Environ. Sci. Technol. 37 (2003) 1202–1207.

[22] F.A. Esteve-Turrillas, A. Pastor, V. Yusa, M. de la Guardia, Using semi-permeable membrane devices as passive samplers, Trends Anal. Chem. 26 (2007) 703–712.

[23] C.L. Arthur, J. Pawliszyn, Solid phase microextraction with thermal desorption using fused silica optical fibers, Anal. Chem. 62 (1990) 2145–2148.

[24] Y.-N. Xing, Y. Guo, M. Xie, R.-L. Shen, E.Y. Zeng, Detection of DDT and its metabolites in two estuaries of South China using a SPME-based device: first report of p,p′-DDMU in water column, Environ. Pollut. 157 (2009) 1382–1387.

[25] K.A. Maruya, E.Y. Zeng, D. Tsukada, S.M. Bay, A passive sampler based solid-phase microextraction for quantifying hydrophobic organic contaminants in sediment pore water, Environ. Toxicol. Chem. 28 (2009) 733–740.

[26] M.D.F. Alpendurada, Solid-phase microextraction: a promising technique for sample preparation in environmental analysis, J. Chromatogr. A 889 (2000) 3–14.

[27] E.M. Sheehan, M.A. Limmer, P. Mayer, U.G. Karlson, J.G. Burken, Time-weighted average SPME analysis for *in planta* determination of cVOCs, Environ. Sci. Technol. 46 (2012) 3319–3325.

[28] G. Ouyang, K.D. Oakes, L. Bragg, S. Wang, H. Liu, S. Cui, M.R. Servos, D.G. Dixon, J. Pawliszyn, Sampling-rate calibration for rapid and nonlethal monitoring of organic contaminants in fish muscle by solid-phase microextraction, Environ. Sci. Technol. 45 (2012) 7792–7798.

[29] M.G. Carls, L.G. Holland, J.W. Short, R.A. Heintz, S.D. Rice, Monitoring polynuclear aromatic hydrocarbons in aqueous environments with passive low-density polyethylene, Environ. Toxicol. Chem. 23 (2004) 1416–1424.

[30] K. Booij, H.E. Hofmans, C.V. Fischer, E.M. Van Weerlee, Temperature-dependent uptake rates of nonpolar organic compounds by semipermeable membrane devices and low-density polyethylene membranes, Environ. Sci. Technol. 37 (2003) 361–366.

[31] G. Cornelissen, A. Pettersen, D. Broman, P. Mayer, G.D. Breedveld, Field testing of equilibrium passive samplers to determine freely dissolved native polycyclic aromatic hydrocarbon concentrations, Environ. Toxicol. Chem. 27 (2008) 499–508.

[32] A.M.P. Oen, E.M.L. Janssen, G. Cornelissen, G.D. Breedveld, E. Eek, R.G. Luthy, In situ measurement of PCB pore water concentration profiles in activated carbon-amended sediment using passive samplers, Environ. Sci. Technol. 45 (2011) 4053–4059.

[33] S.B. Hawthorne, M.T.O. Jonker, S.A. Van der Heijden, C.B. Grabanski, N.A. Azzolina, D.J. Miller, Measuring picogram per liter concentrations of freely dissolved parent and alkyl PAHs (PAH-34), using passive sampling with polyoxymethylene, Anal. Chem. 83 (2011) 6754–6761.

[34] B. Beckinghan, U. Ghosh, Polyoxymethylene passive samplers to monitor changes in bioavailability and flux of PCBs after activated carbon amendment to sediment in the field, Chemosphere 91 (2013) 1401–1407.

[35] L.A. Fernandez, J.K. Macfarlane, A.P. Tcaciuc, P.M. Gschwend, Measurement of freely dissolved PAH concentrations in sediment beds using passive sampling with low-density polyethylene strips, Environ. Sci. Technol. 43 (2009) 1430–1436.

[36] H.-H. Liu, L.-J. Bao, K. Zhang, S.-P. Xu, F.-C. Wu, E.Y. Zeng, Novel passive sampling device for measuring sediment-water diffusion fluxes of hydrophobic organic chemicals, Environ. Sci. Technol. 47 (2013) 9866–9873.

[37] V.P. Sacks, R. Lohmann, Development and use of polyethylene passive samplers to detect triclosans and alkylphenols in an urban estuary, Environ. Sci. Technol. 45 (2011) 2270–2277.

[38] J.E. Tomaszewsky, R.G. Luthy, Field deployment of polyethylene devices to measure PCB concentrations in pore water of contaminated sediment, Environ. Sci. Technol. 42 (2008) 6086–6091.
[39] W. Lick, The sediment-water flux of HOCs due to "diffusion" or is there a well-mixed layer? If there is, does it matter? Environ. Sci. Technol. 40 (2006) 5610–5617.
[40] A. Cabrerizo, J. Dachs, D. Barceló, Development of a soil fugacity sampler for determination of air-soil partitioning of persistent organic pollutants under field controlled conditions, Environ. Sci. Technol. 43 (2009) 8257–8263.
[41] R. Lohmann, M. Dapsis, E.J. Morgan, V. Dekany, P. Luey, Determining air-water exchange, spatial and temporal trends of freely dissolved PAHs in an urban estuary using passive polyethylene samplers, Environ. Sci. Technol. 45 (2011) 2655–2662.
[42] F. Verweij, K. Booij, K. Satumalay, N. van der Molen, R. van der Oost, Assessment of bioavailable PAH, PCB and OCP concentrations in water, using semipermeable membrane devices (SPMDs), sediments and caged carp, Chemosphere 54 (2004) 1675–1689.
[43] L.A. Fernandez, W. Lao, K.A. Maruya, C. White, R.M. Burgess, Passive sampling to measure baseline dissolved persistent organic pollutant concentrations in the water column of the Palos Verdes Shelf superfund site, Environ. Sci. Technol. 46 (2012) 11937–11947.
[44] A. Accardi-Dey, P.M. Gschwend, Reinterpreting literature sorption data considering both absorption into organic carbon and adsorption onto black carbon, Environ. Sci. Technol. 37 (2003) 99–106.
[45] A. Jahnke, P. Mayer, M.S. McLachlan, Sensitive equilibrium sampling to study polychlorinated biphenyl disposition in Baltic Sea sediment, Environ. Sci. Technol. 46 (2012) 10114–10122.
[46] T.D. Bucheli, Ö. Gustafsson, Ubiquitous observations of enhanced solid affinities for aromatic organochlorines in field situations: are in situ dissolved exposures overestimated by existing partitioning models? Environ. Toxicol. Chem. 20 (2001) 1450–1456.
[47] S.E. Mcgroddy, J.W. Farrington, Sediment porewater partitioning of polycyclic aromatic hydrocarbons in three cores from Boston Harbor, Massachusetts, Environ. Sci. Technol. 29 (1995) 1542–1550.
[48] L.P. Burkhard, Estimating dissolved organic carbon partition coefficients for nonionic organic chemicals, Environ. Sci. Technol. 34 (2000) 4663–4668.
[49] S. Lee, J. Gan, W.P. Liu, A.S. Anderson, Evaluation of K_d underestimation using solid phase microextraction, Environ. Sci. Technol. 37 (2003) 5597–5602.
[50] W.H.J. Vaes, E.U. Ramos, H.J.M. Verhaar, W. Seinen, J.L.M. Hermens, Measurement of the free concentration using solid-phase microextraction: binding to protein, Anal. Chem. 68 (1996) 4463–4467.
[51] S.B. Hawthorne, C.B. Grabanski, D.J. Miller, J.P. Kreitinger, Solid-phase microextraction measurement of parent and alkyl polycyclic aromatic hydrocarbons in milliliter sediment pore water samples and determination of K_{DOC} values, Environ. Sci. Technol. 39 (2005) 2795–2803.
[52] T.L. Ter Laak, S.O. Agbo, A. Barendregt, J.L.M. Hermens, Freely dissolved concentrations of PAHs in soil pore water: measurements via solid-phase extraction and consequences for soil tests, Environ. Sci. Technol. 40 (2006) 1307–1313.
[53] W. Wang, L. Moreno-Moreno, Q. Ye, J. Gan, Improved measurements of partition coefficients for polybrominated diphenyl ethers, Environ. Sci. Technol. 45 (2011) 1521–1527.
[54] S. Kleineidam, C. Schüth, P. Grathwohl, Solubility-normalized combined adsorption-partitioning sorption isotherms for organic pollutants, Environ. Sci. Technol. 36 (2002) 4689–4697.
[55] A.E. Vinturella, R.M. Burgess, B.A. Coull, K.M. Thompson, J.P. Shine, Use of passive samplers to mimic uptake of polycyclic aromatic hydrocarbons by benthic polychaetes, Environ. Sci. Technol. 38 (2004) 1154–1160.

[56] A.D. Harwood, P.F. Landrum, M.J. Lydy, Can SPME fiber and tenax methods predict the bioavailability of biotransformed insecticides? Environ. Sci. Technol. 46 (2012) 2413–2419.

[57] J.L. Gomez-Eyles, M.T.O. Jonker, M.E. Hodson, C.D. Collins, Passive samplers provide a better prediction of PAH bioaccumulation in earthworms and plant roots than exhaustive, mild solvent, and cyclodextrin extractions, Environ. Sci. Technol. 46 (2012) 962–969.

[58] S.A. Van der Heijden, M.T.O. Jonker, PAH bioavailability in field sediments: comparing different methods for predicting in situ bioaccumulation, Environ. Sci. Technol. 43 (2009) 3757–3763.

[59] M. Barry, M.T.O. Jonker, Does equilibrium passive sampling reflect actual in situ bioaccumulation of PAHs and petroleum hydrocarbon mixtures in aquatic worms? Environ. Sci. Technol. 46 (2012) 937–944.

[60] J.H. Paul, D. Hollander, P. Coble, K.L. Daly, S. Murasko, D. English, J. Basso, J. Delaney, L. McDaniel, C.W. Kovach, Toxicity and mutagenicity of Gulf of Mexico waters during and after the deepwater horizon oil spill, Environ. Sci. Technol. 47 (2013) 9651–9659.

[61] J. Du, W.T. Mehler, M.J. Lydy, J. You, Toxicity of sediment-associated unresolved complex mixture and its impact on bioavailability of polycyclic aromatic hydrocarbons, J. Hazard Mater. 203 (2012) 169–175.

[62] H. Li, B. Sun, X. Chen, M.J. Lydy, J. You, Addition of contaminant bioavailability and species susceptibility to a sediment toxicity assessment: application in an urban stream in China, Environ. Pollut. 178 (2013) 135–141.

[63] J. Du, J. Pang, J. You, Bioavailability-based chronic toxicity measurements of permethrin to *Chironomus dilutus*, Environ. Toxicol. Chem. 32 (2013) 1403–1411.

[64] A.D. Harwood, P.F. Landrum, M.J. Lydy, Bioavailability-based toxicity endpoints of bifenthrin for *Hyalella azteca* and *Chironomus dilutus*, Chemosphere 90 (2013) 1117–1122.

[65] A. Palm, I. Cousins, Ö. Gustafsson, J. Axelman, K. Grunder, D. Broman, E. Brorström-Lundén, Evaluation of sequentially-coupled POP fluxes estimated from simultaneous measurements in multiple compartments of an air–water–sediment system, Environ. Pollut. 128 (2004) 85–97.

[66] E. Eek, G. Cornelissen, G.D. Breedveld, Field measurement of diffusional mass transfer of HOCs at the sediment-water interface, Environ. Sci. Technol. 44 (2010) 6752–6759.

[67] G. Ouyang, J. Pawliszyn, A critical review in calibration methods for solid-phase microextraction, Anal. Chim. Acta 627 (2008) 184–197.

[68] R. Lohmann, Critical review of low-density polyethylene's partitioning and diffusion coefficients for trace organic contaminants and implications for its use as a passive sampler, Environ. Sci. Technol. 46 (2012) 606–618.

[69] E.L. DiFilippo, R.P. Eganhouse, Assessment of PDMS-water partition coefficients: implications for passive environmental sampling of hydrophobic organic compounds, Environ. Sci. Technol. 44 (2010) 6917–6925.

[70] P. Mayer, W.H.J. Vaes, F. Wijnker, K.C.H.M. Legierse, R.H. Kraaij, J. Tolls, J.L.M. Hermens, Sensing dissolved sediment porewater concentrations of persistent and bioaccumulative pollutants using disposable solid-phase microextraction fibers, Environ. Sci. Technol. 34 (2000) 5177–5183.

[71] J. You, P.F. Landrum, M.J. Lydy, Comparison of chemical approaches for assessing bioavailability of sediment-associated contaminants, Environ. Sci. Technol. 40 (2006) 6348–6353.

[72] L.-J. Bao, F. Jia, J. Crago, E.Y. Zeng, D. Schlenk, J. Gan, Assessing bioavailability of DDT and metabolites in marine sediments using solid-phase microextraction with performance reference compounds, Environ. Toxicol. Chem. 32 (2013) 1946–1953.

[73] J.N. Huckins, J.D. Petty, K. Booij, Monitors of Organic Chemicals in the Environment, Springer, New York, USA, 2006.

[74] L.A. Fernandez, C.F. Harvey, P.M. Gschwend, Using performance reference compounds in polyethylene passive samplers to deduce sediment porewater concentrations for numerous target chemicals, Environ. Sci. Technol. 43 (2009) 8888–8894.

[75] J.N. Huckins, J.D. Petty, J.A. Lebo, F.V. Almeida, K. Booij, D.A. Alvarez, W.L. Cranor, R.C. Clark, B.B. Mogensen, Development of the permeability/performance reference compound approach for in situ calibration of semipermeable membrane devices, Environ. Sci. Technol. 36 (2002) 85–91.

[76] K. Booij, F. Smedes, An improved method for estimating in situ sampling rates of nonpolar passive samplers, Environ. Sci. Technol. 44 (2010) 6789–6794.

[77] E.Y. Zeng, C.S. Wong, Response to "letter to the editor concerning the viewpoint; recognizing the limitations of performance reference compound (PRC)-calibration technique in passive water sampling", Environ. Sci. Technol. 48 (2014) 1369.

[78] M.M. Perron, R.M. Burgess, E.M. Suuberg, M.G. Cantwell, K.G. Pennell, Performance of passive samplers for monitoring estuarine water column concentrations: 1. contaminants of concern, Environ. Toxicol. Chem. 32 (2013) 2182–2189.

[79] R.P. Schwarzenbach, P.M. Gschwend, D.M. Imboden, Environmental Organic Chemistry, John Wiley & Sons, Inc., Hoboken, NJ, 2003.

[80] B. Muijs, M.T.O. Jonker, Temperature-dependent bioaccumulation of polycyclic aromatic hydrocarbons, Environ. Sci. Technol. 43 (2009) 4517–4523.

[81] P.J. Reitsma, D. Adelman, R. Lohmann, Challenges of using polyethylene passive samplers to determine dissolved concentrations of parent and alkylated PAHs under cold and saline conditions, Environ. Sci. Technol. 47 (2013) 10429–10437.

[82] G. Ouyang, Y. Chen, J. Pawliszyn, Flow-through system for the generation of standard aqueous solution of polycyclic aromatic hydrocarbons, J. Chromatogr. A 1105 (2006) 176–179.

[83] G. Ouyang, S. Cui, Z. Qin, J. Pawliszyn, One-calibrant kinetic calibration for on-site water sampling with solid-phase microextraction, Anal. Chem. 81 (2009) 5629–5636.

[84] R.P. Schwarzenbach, P.M. Gschwend, D.M. Imboden, Environmental Organic Chemistry, John Wiley & Sons, Inc., New York, 1993.

[85] E.Y. Zeng, D. Tsukada, J.A. Noblet, J. Peng, Determination of polydimethylsiloxane-seawater distribution coefficients for polychlorinated biphenyls and chlorinated pesticides by solid-phase microextraction and gas chromatography-mass spectrometry, J. Chromatogr. A 1066 (2005) 165–175.

[86] B.J. Richardson, P.K.S. Lam, G.J. Zheng, K.E. McClellan, S.B. De Lucaabbott, Biofouling confounds the uptake of trace organic contaminants by semi-permeable membrane devices (SPMDs), Mar. Pollut. Bull. 44 (2002) 1372–1379.

[87] G.S. Ellis, C.E. Rostad, J.N. Huckins, C.J. Schmitt, J.D. Petty, P. Maccarthy, Evaluation of lipid-containing semipermeable membrane devices for monitoring organochlorine contaminants in the Upper Mississippi River, Environ. Toxicol. Chem. 14 (1995) 1875–1884.

[88] K. Booij, R. van Bommel, A. Mets, R. Dekker, Little effect of excessive biofouling on the uptake of organic contaminants by semipermeable membrane devices, Chemosphere 65 (2006) 2485–2492.

Chapter 5

Assessment of Sediment Toxicity with SPME-Based Approaches

Jing You,[1,*] Huizhen Li[1] and Michael J. Lydy[2,*]
[1]State Key Laboratory of Organic Geochemistry, Guangzhou Institute of Geochemistry, Chinese Academy of Sciences, Guangzhou, China; [2]Center for Fisheries, Aquaculture and Aquatic Sciences and Department of Zoology, Southern Illinois University, Carbondale, IL, United States
**Corresponding authors: E-mail: youjing@gig.ac.cn, mlydy@siu.edu*

Chapter Outline

1. INTRODUCTION

Legacy hydrophobic organic contaminants (HOCs), such as polycyclic aromatic hydrocarbons (PAHs), polychlorobiphenyls (PCBs), and organochlorine pesticides, are still ubiquitous in the environment, and thus there is a continuing need to monitor their occurrence, distribution, and relative risk to nontarget species. Meanwhile, emerging HOCs, including halogenated flame retardants and current-use pesticides, have been detected at elevated concentrations in various

Comprehensive Analytical Chemistry, Vol. 67. http://dx.doi.org/10.1016/B978-0-444-63299-9.00005-3

regions of the world, suggesting an urgent challenge to better understand their environmental fate and toxicological effects. After being released into the environment, HOCs have a strong affinity for sediment organic carbon (OC), because of their high hydrophobicity. Over time, sediment becomes an important sink for legacy and emerging HOCs; therefore, the assessment of sediment quality is one of the vital tasks for understanding the fate and risk of HOCs.

A variety of approaches have been applied to assess the risk of sediment-associated HOCs [1]. The simplest and perhaps the most commonly used sediment risk assessment approach is the deterministic hazard quotient method, which provides a conservative estimate for screening-level risk assessments. Hazard quotients are calculated by dividing the sediment concentrations of contaminants measured in the field by the established benchmark values, such as the empirical sediment quality criteria [2,3] and the mechanistically developed sediment benchmarks on the basis of equilibrium partitioning (EqP) theory [4–10].

Total concentrations of HOCs in sediment quantified by exhaustive extraction have been generally used to assess sediment quality when the empirical sediment quality criteria are applied. Nevertheless, a number of factors can affect the bioavailability of HOCs in sediment, such as the geochemical composition of the sediment [11–13], physicochemical properties of the target chemicals [14,15], HOC–sediment contact time [13,16], and various biological factors [17]. A growing body of evidence has suggested that the bioavailability of HOCs in sediment may vary significantly from site to site and the total HOC concentrations in sediment may fail to predict sediment toxicity [1,14,18]. As a consequence, there is an urgent need for incorporating bioavailability measurements into sediment risk assessments [1,14,18–20]. Several mechanistic approaches, including normalizing chemical concentrations to sediment OC and measuring HOC concentrations in sediment pore water using passive sampling methods (PSMs), have been proposed for evaluating sediment quality by incorporating bioavailability of sediment-associated HOCs [4–10,21].

The objectives of the present review are to provide a summary of the mechanistic approaches for incorporating bioavailability into sediment quality assessments and to review the applications and the scientific merit of using PSMs to evaluate the toxicity of HOCs in sediment to benthic organisms with a focus on solid-phase microextraction (SPME). Potential future uses of PSMs for evaluating sediment toxicity are also outlined. Other PSMs exist that allow measurement of bioavailability of metals and polar contaminants in sediment, including peepers, diffusive gradients in thin films, Chemcatcher, and polar organic chemical integrative samplers; however discussion of metals and polar compounds is beyond the scope of the present paper, and these topics have been previously reviewed [22,23].

2. BIOAVAILABILITY IN SEDIMENT QUALITY ASSESSMENTS

Generally speaking, bioavailability is defined as the fraction of contaminants that can be taken up by organisms within a given amount of time [24]. Since the term bioavailability has been used in multiple disciplines to describe different concepts, the National Research Council (NRC) [25] suggested the use of the phrase "bioavailability process" as a replacement for the term bioavailability. The NRC also separated the bioavailability process into four stages including desorption, transport (bound and released chemicals), movement across cell membranes, and initiation of toxic effects. Similarly, the International Standardization Organization [26] defined bioavailability as a dynamic process that contained three phases, namely environmental availability, environmental bioavailability, and toxicological bioavailability. In order to better understand the mechanisms involved in bioavailability of a chemical in an environmental matrix, Semple et al. [27] separated bioavailability into two concepts (a bioaccessible fraction and a bioavailable fraction) and demonstrated that the bioaccessible fraction contained both bioavailable and potential bioavailable components. Reichenberg and Mayer [24] further explained the two sides of bioavailability, bioaccessibility and chemical activity (or fugacity) from the points of mass quantity and energy state, respectively. Bioaccessibility is usually operationally determined, while chemical activity can be quantified using mechanistic approaches and was closely related to the freely dissolved concentration (C_{free}) of a chemical [24]. Based on its definition, chemical activity of a compound remains constant throughout a system when thermodynamic equilibrium is reached, e.g., a sediment-pore water-(passive sampler)-organism system, suggesting that the measured chemical activity in one phase is applicable in the other phases. This characteristic makes chemical activity convenient in use in multimedia systems. In addition, both chemical activity and C_{free} are based on the same basic strategy of diffusion and partitioning, so chemical activity can be determined from C_{free} through normalizing by solubility of the compound [24]. To date, various PSMs have been developed to measure C_{free} and subsequently chemical activity of HOCs in sediment pore water.

In addition to understanding the definition of bioavailability, it is also important to obtain accurate measurements of bioavailability in order to incorporate this concept into sediment risk assessments. A number of studies in the past 20 years have been devoted to improving methods to predict bioavailability of HOCs in sediment. The most commonly used methods can be classified into three categories, i.e., bioassays (e.g., bioaccumulation and toxicity tests), theoretical estimations by modeling (e.g., equilibrium partition theory), and chemical measurements (e.g., PSMs).

Body residues of HOCs in benthic organisms quantified from bioaccumulation testing or field-collected organisms have been conventionally used to assess bioavailability and subsequently to predict toxicity of HOCs [28–30]. However, toxicity of a chemical may be influenced by a number of factors, which are difficult to control during the testing, including the occurrence and degree of transformation of the chemical in the exposure media and in the test organisms [23,31]. Furthermore, it is laborious to analyze HOCs in tissues, especially at sites with low concentrations of HOCs [32]. Finally, it is sometimes difficult, if not impossible, to collect enough organisms for tissue analysis at sites with high sediment concentrations of HOCs [33].

Compared to bioassays and the associated tissue analyses, chemical approaches for quantifying bioavailability, including sediment extraction and biomimetic extraction, are more easily controlled and less expensive. As discussed above, bioaccumulation is not related to the total concentrations of HOCs in the sediment measured by exhaustive extraction, but instead is proportional to the chemical activity of HOCs in the sediment [34]. Therefore, mechanistic approaches to predict the bioavailability, bioaccumulation, and toxicity of sediment-associated HOCs have been developed on the basis of estimating their chemical activity [1,14,18]. Di Toro et al. [35] demonstrated that the chemical activity of contaminants of concern is the same in all phases (sediment OC, sediment pore water, and organism lipid) when equilibrium is reached within a system, regardless of uptake routes of exposure. Therefore, OC-normalized sediment concentrations have been proposed to estimate the bioaccumulation of HOCs in sediment and C_{free} in sediment pore water and the bioaccumulation potential (biota-sediment accumulation factor, BSAF) are estimated using Eqns (1) and (2), respectively.

$$C_{free} = \frac{C_s}{f_{oc} \times K_{oc}} \quad (1)$$

$$\text{BSAF} = \frac{C_b \times f_{OC}}{C_s \times f_{lip}} \quad (2)$$

Where, C_s and C_b are the chemical concentrations in sediment and organisms, respectively, and f_{oc} and f_{lip} are the fractions of OC in the sediment and lipids in the organisms, respectively. The K_{oc} is the partition coefficient of HOCs between the sediment OC and water.

Recent studies have shown that OC content of a sediment alone cannot explain the large variation observed in bioavailability found for HOCs in sediment, but instead that the composition of OC also played an important role in sequestration of HOCs by OC [11–13]. For example, the presence of black carbon (BC) in sediment reduced the partitioning of PAHs into pore water and limited their bioaccumulation potential to benthic organisms [11,36]. The BC content of a sediment is generally determined using a chemothermal oxidation method proposed by Gustafsson et al. [37] and the partitioning of PAHs between BC and water (K_{BC}) can be taken into consideration when estimating C_{free} (Eqn (3)).

$$C_{free} = \frac{C_s}{(f_{NSOC} \times K_{oc}) + (f_{BC} \times K_{BC} \times C_{free}^{n-1})} \quad (3)$$

Where, f_{NSOC} and f_{BC} are the fractions of natural sedimentary OC and BC in a sediment, respectively, and "n" is the Freundlich exponent. The n accounts for nonlinear sorptive behavior and an n value of 0.6 has been used for PAHs [38].

While more accurate C_{free} estimates are achieved by taking K_{BC} into account compared with using K_{oc} alone, the K_{BC} correction method has only been used for PAHs [39]. Meanwhile, other types of sedimentary OC, such as residue oils in sediment, may also serve as super-sorbents for HOCs; thereby, limiting the bioaccumulation of these types of compounds [40,41]. Thus, a direct measurement of C_{free} has been proposed to estimate the bioavailability of HOCs in sediment, and subsequently to assess sediment quality [21,42]. Although measuring C_{free} has historically been challenging, recent progress in PSM development has provided an effective means for directly estimating C_{free} of HOCs in sediment [43]. Several studies have demonstrated that C_{free} predicted well the body residues of HOCs in benthic organisms [14,15,34]. Currently, the concept that chemical activity controls the bioavailability of sediment-associated contaminants is widely accepted, and PSM is considered a practical and reliable method for measuring C_{free}, which is a good indicator for chemical activity [1,21,42,44]. Recently, a series of reviews [14,18,45,46] summarized the application of PSMs for measuring C_{free} in contaminated sediments from different aspects, including a general summary of the applications, the scientific rationale for using PSMs, the practical considerations in employing PSMs ex situ and in situ, and the management application of PSMs for decision making.

3. CALIBRATION OF PASSIVE SAMPLING METHODS

Passive sampling techniques are defined as "any sampling techniques based on free flow of analyte molecules from the sampled medium to a collecting medium, as a result of a difference in chemical potential of the analytes between the two media" [47]. When using PSMs, the receiving phase is used to trap and retain the analytes [23]. So far, various receiving phases have been employed to analyze C_{free} of HOCs in sediment, such as polydimethylsiloxane (PDMS), which is used in the form of SPME fiber or thin film [19,20,43,48], polyacrylate (PA) fiber [49–54], low-density polyethylene (LDPE) membrane [55,56], and polyoxymethylene (POM) membrane [57,58]. Rusina et al. [59] analyzed several critical properties of commonly used polymers (e.g., PDMS, LDPE, and POM), including the release of oligomers, swelling of the polymer in solvents, diffusion coefficients and transport resistances of PAHs, and the partition coefficients of PAHs between polymer and water, as well as provided general guidelines for selecting the receiving phase of PSMs. The polymer–water partition coefficients of PAHs varied by two orders of magnitude across polymers, with an order of POM < PDMS < LDPE. The total mass transfer resistance (the sum of resistances in water boundary layer, biofilm, and polymers) of PAHs for PDMS sampling was generally lower than that for LDPE sampling, suggesting less time being required for PAHs to reach equilibrium in PDMS compared to LDPE [59].

Regardless of the polymer material used, the exchange kinetic rates between environmental media and the passive sampler can be modeled using a first-order one-compartment model (Eqn (4)).

$$C_{\text{sampler}}(t) = C_{\text{medium}} \frac{k_1}{k_2}(1 - e^{-k_2 t}) \tag{4}$$

Where, C_{sampler} is the analyte concentration on the passive sampler at exposure time t, C_{medium} is the analyte concentration in the environmental media, and k_1 and k_2 are the rate constants describing uptake onto and elimination off the sampler, respectively.

The passive sampling process can be terminated before or after equilibrium is reached between the two media for the target chemicals, based on the type of calibration method used [47]. As suggested by Mayer et al. [60] and Ouyang and Pawliszyn [61,62], the uptake process for equilibrium sampling can be divided into three types, including kinetic, intermediate and near-equilibrium regimes (Figure 1). Finally, passive sampling can be conducted in two modes, namely equilibrium sampling and kinetically controlled sampling.

3.1 Equilibrium Sampling

As discussed earlier, EqP theory assumes that chemical activity and fugacity are the same in individual medium when the equilibrium of a chemical is reached among the lipids of biota, the OC of sediment, and pore water [35]. This relationship is similar for PSMs when the system is at equilibrium. After a passive sampler is introduced into the sediment for some length of time, a thermodynamic equilibrium is established for the target compounds among the sampler, sediment OC, and pore water. The sampler–medium partition coefficient ($K_{\text{sampler–medium}}$) is defined as the ratio of the concentration of HOC on the sampler (C_{sampler}) and that in a given environmental medium (C_{medium}, e.g., C_{free}) at equilibrium (Eqn (5)). In addition, $K_{\text{sampler–medium}}$ can be estimated from the uptake and elimination rate constants (Eqn (6)).

$$K_{\text{sampler-medium}} = \frac{C_{\text{sampler}}}{C_{\text{medium}}} \tag{5}$$

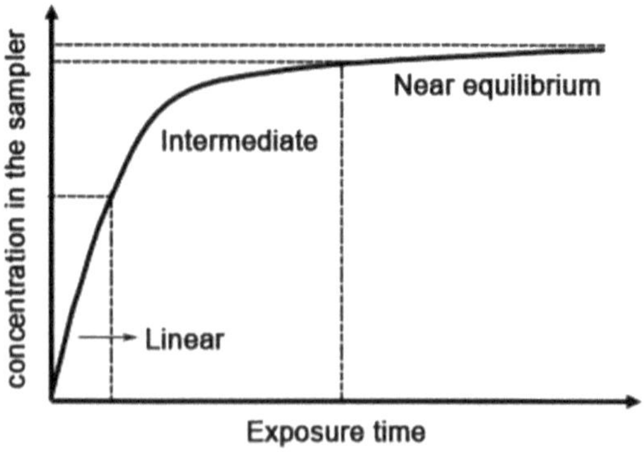

FIGURE 1 Uptake profile of passive sampling. *Adapted from [60].*

$$K_{\text{sampler-medium}} = \frac{k_1}{k_2} \tag{6}$$

If 90% of equilibrium is assumed to represent the equilibrium state, then the time required to reach this state (t_{90}) can be calculated using Eqn (7) [60–62].

$$t_{90} = \frac{\ln 10}{k_2} = K_{\text{sampler-medium}} \times \frac{\ln 10}{k_1} \tag{7}$$

Since equilibrium calibration is less impacted by the fluctuation of environmental conditions (e.g., temperature and hydrodynamics) and can generate more robust data than kinetically controlled sampling, it is preferable to determine C_{free} by equilibrating the sampler with its environment [18]. Mayer et al. [43] reported that the analytical precision of PSMs at equilibrium was significantly better compared to that in the kinetic regime, with relative standard deviations increasing from 20% to 30% in the kinetic regime compared to <10% when the system was near equilibrium. Series measurements at different sampling times or parallel analyses using PSMs with different surface area/volume (A/V) ratios are generally conducted to confirm the equilibrium condition. Nevertheless, the time for the high octanol–water partition coefficient HOCs (log $K_{ow}>6$) to attain equilibrium might be too long to be useful in bioassays, especially in the case of field applications [63]. Therefore, kinetically controlled sampling has been developed for PSM applications without the need to reach equilibrium.

3.2 Kinetically Controlled Sampling

To shorten the sampling time, kinetically controlled calibration was applied to quantify C_{free} in the kinetic regimes of the uptake curve (Figure 1). To conduct kinetically controlled calibration, the rate of mass transfer was assumed to be linearly related to the difference in fugacity between the phases [23,47]; and thus the partitioning process followed Fisk's first law (Eqn (8)):

$$J = -D \times \left(\frac{dc}{dz}\right) \tag{8}$$

Where, J and D are the flux and diffusion coefficient of a chemical, respectively, and $\frac{dc}{dz}$ stands for the concentration gradient of the chemical. At the beginning stage of the sampling, there is little to no chemical desorption from the receiving phase to the medium and the sampler is operated in the kinetic regime. In this case, Eqn (4) can be rewritten as Eqn (9).

$$C_{\text{sampler}}(t) = C_{\text{medium}} \times k_1 \times t \tag{9}$$

The uptake and elimination rate constants k_1 and k_2 control the exposure kinetics. Generally, when $K_{\text{sampler–medium}}$ is small, mass transfer is limited by diffusion inside the sampler, and k_1 will increase with $K_{\text{sampler–medium}}$. Conversely, for chemicals with larger $K_{\text{sampler–medium}}$, which is the case for HOCs, mass transfer is limited by aqueous diffusion, a uniform k_1 will be used for HOCs with different $K_{\text{sampler–medium}}$ and k_2 will rapidly decrease with increasing $K_{\text{sampler–medium}}$ [64]. Based on the first-order diffusion model, both chemical properties and sampler designs will affect k_1 and k_2 (Eqns (10) and (11), respectively) [64].

$$k_1 = \frac{A}{V} \times \frac{1}{\frac{\delta_{\text{m}}}{D_{\text{m}}} \times \frac{r}{\delta_{\text{m}} + r} + \frac{1}{m_{\text{t}} \times K_{\text{sampler-medium}}}} \tag{10}$$

$$k_2 = \frac{A}{V \times K_{\text{sampler-medium}}} \times \frac{1}{\frac{\delta_{\text{m}}}{D_{\text{m}}} \times \frac{r}{\delta_{\text{m}} + r} + \frac{1}{m_{\text{t}} \times K_{\text{sampler-medium}}}} \tag{11}$$

Where, A is the surface area interface between the sampler and the medium, V is the volume of the sampler, δ_{m} is the diffusion layer thickness in the medium, r is the radius of curvature of the sampler surface, D_{m} is the diffusion coefficient of the analyte in the medium, and m_{t} is the mass transfer coefficient of the sampler.

The parameters δ_{m}, D_{m}, and m_{t} are affected by the physicochemical properties of the contaminants and deployment conditions, like temperature, biofouling [64,65] and turbulence [23]. In addition, the sampler design can also influence the uptake kinetics, including the shape of the samplers [66]; A/V ratio [64] and surface area/diffusion distance (A/L) ratio [23]. Consequently, strict calibration of kinetic coefficients in the laboratory is usually required before using kinetically controlled sampling in the field. In addition, performance reference compounds (PRCs) have been employed to estimate sampling rates of target HOCs in sediment compared to the dissipation rates of the PRCs from the samplers [67]. The PRC method assumes that the rates of sorption and desorption of the PRCs and the target contaminants are the same, thus isotopically labeled compounds are preferred to be used as the PRCs [18].

Overall, a good calibration method is critical for accurately quantifying HOCs in sediment. The uptake of sediment-associated HOCs onto a passive sampler involves a variety of processes, including chemical desorption from the sediment, mass transfer of the chemical within the environmental media, and partitioning of the chemicals onto the samplers. Uptake kinetics and calibration methods for PSMs will not be covered in the present paper, because they have been detailed in previous reviews [18,23,24,60,61,62,64].

Our improved understanding of passive sampling processes has promoted the application of PSMs in analyzing C_{free} of contaminants in various environmental media, including sediment [14]. While PSMs have been extensively used to quantify C_{free} of HOCs in sediment and to subsequently estimate the accumulated HOCs in organisms, few studies have directly linked PSM measurements to sediment toxicity [1,14,20]. The applications of PSMs, with a special focus on SPME, on evaluating the toxicity of sediment-associated HOCs to benthic organisms are discussed below.

4. INTRODUCTION OF SOLID PHASE MICROEXTRACTION (SPME)

4.1 Development of SPME

The SPME method was originally introduced by Authur and Pawliszyn [68] as a solvent-free replacement for traditional extraction techniques to analyze bulk chemical concentrations in water, such as liquid–liquid extraction (LLE) and solid phase extraction (SPE). By direct desorption of all the analytes from SPME fibers into chromatographic systems, high analytical sensitivity and automation can be achieved [69]. In addition, SPME performs sampling, isolation, and cleanup in a single step [70]; which can greatly reduce the time and labor required for sample preparation. Finally, SPME also has the advantages of being easy to operate, being small in size, and requiring relatively short equilibration times and the fibers can often be reused.

The most commonly used SPME configuration is a fused silica core coated with a thin layer of polymer, e.g., PDMS or PA. The commercialized fibers usually contain a 100–200 μm silica core with a 7 to 100 μm thick polymer coating [70]. After the fiber is introduced into a sampling matrix, the analytes transfer from the matrix onto the fiber until a thermodynamic equilibrium is achieved. Chemicals are quantified in aqueous media with SPMEs under three operating modes, including headspace SPMEs, direct immersing SPMEs, and membrane-protected SPMEs. In addition to the traditional calibration methods (e.g., including external and internal standard calibrations and standard addition calibration), diffusion-based calibration methods have been proposed for onsite sampling to achieve time-weighted average concentrations [61,62]. Additional calibration methods based on Fick's first law of diffusion have also been proposed, e.g., interface [71] and flow cross models [72]. In order to improve the sensitivity and selectivity of SPME methods, extraction conditions, e.g., the thickness of the fiber, agitation method and intensity and extraction time, need to be precisely controlled. Therefore, it is important to select appropriate operating modes, calibration methods, and extraction conditions for SPME according to the objectives of the analysis.

4.2 Types of SPME

To date, both ex situ and in situ SPME methods have been widely used to monitor bulk contaminant concentrations in a variety of environmental media [23,70]. Alternatively, SPME has also been used as a robust passive sampling technique to quantify chemical activity and C_{free} of HOCs [43,73]. Unlike LLE and SPE, SPME is a nonexhaustive extraction technique. Under nondepletive conditions, SPME does not disturb the initial equilibrium of chemicals between phases, so it is a good candidate for C_{free} measurements [43,73].

Two types of SPMEs have been proposed to measure C_{free} of HOCs in sediment, namely pore water SPME [74,75] and matrix SPME [15,43]. Before conducting pore water SPME, pore water is first acquired from sediment by centrifugation and the fibers are placed into the pore water with or without flocculation [74,75]. Conversely, in matrix SPME, the fibers are directly inserted into sediment, and therefore, the sediment is less disturbed [43].

In order to achieve nondepletive conditions and maintain the original chemical equilibrium within the system, the amount of chemical sorbed by the SPME fibers should be less than 5% of the total chemical mass within that matrix [73]. This nondepletive condition is difficult to achieve for HOCs in an aqueous solution due to their low solubility, and thus a large volume of water sample is usually required. When matrix-like sediment is used, HOCs extracted by SPME fibers only account for a small fraction of the total HOCs in the matrix, thus the nondepletive condition can be attained if the loss of chemical to the fibers can be quickly replenished from the matrix. In other words, the entire sediment matrix is used as a reservoir for the chemicals and matrix SPME can be operated as an equilibrium and nondepletive extraction [43].

Because the partition coefficients for HOCs between PDMS and water (K_{PDMSw}) are generally high, matrix SPME with a PDMS coating is highly sensitive at measuring C_{free} of HOCs in sediment, with detection limits being in the pg/L to ng/L levels [43]. Due to its simplicity and high sensitivity, PDMS-coated SPMEs have shown great potential to be used as in situ techniques to measure C_{free} of HOCs in aquatic environments [20]. Since K_{PDMSw} values are essential to accurately quantify C_{free}, a great deal of effort has been made to estimate the K_{PDMSw} values. Di Filippo and Eganhouse [76] provided an extensive review of K_{PDMSw} values for various classes of HOCs and developed a multiparameter linear solvation energy relationship for predicting K_{PDMSw} values based on the physicochemical parameters of the target HOCs. Chemical activity can also be quantified by matrix-SPME as the fugacity (chemical activity) of the chemical on the SPME fibers and in the sediment pore water is the same at equilibrium.

As previously discussed, matrix SPME may require long exposure times for highly hydrophobic contaminants to reach equilibrium, especially in sediments with high viscosity and in field-collected sediments [43]. The time for a chemical to reach equilibrium between the sediment pore water and SPME fiber is directly related to the chemical's hydrophobicity and the thickness of the unstirred boundary layer around the fiber, which is inversely related to the *A*/*V* ratio of the fiber [18]. Therefore, the use of SPME fibers with large *A*/*V* ratios or thin PDMS films and elevated agitation intensity during the sampling process have been proposed to shorten equilibrium times [19,20,45,48,77]. Agitation decreases the thickness of the unstirred boundary layer around the fiber and subsequently increases the uptake rate of HOCs onto the fibers up to a certain threshold level, where the uptake rate no longer increases with increasing intensity. In addition, vigorous agitation may disturb the sediment–pore water equilibrium; thereby, introducing error into the C_{free} measurement. Thus, a gentle agitation is normally preferred in laboratory SPME testing [43,77]. On the other hand, it is difficult for chemicals to reach equilibrium under field conditions, and various kinetic calibration methods have been used for in situ SPME testing. Compared to other passive samplers, such as PE and POM, the use of the PRC approach for SPME is underdeveloped. Recent studies, however, have successfully quantified C_{free} for sediment-associated HOCs using SPME fibers, which were preloaded with isotopically labeled PRCs [78,79].

Chemical activity governs spontaneous processes, such as diffusion, thus C_{free} directly represents the effective concentration for sediment-associated HOCs being taken up by small organisms. Kraaij et al. [34] showed that the measured body residues of PAHs and PCBs in benthic organisms after sediment exposure matched well with those estimated by multiplying SPME-measured C_{free} in sediment and bioconcentration factors. Since that paper was published, a number of other studies have supported the ability of SPME to predict the bioaccumulation potential of various HOCs in sediment to benthic organisms [14,19,80,81].

5. APPLICATION OF SPME IN SEDIMENT TOXICITY ASSESSMENT

5.1 Overview

In addition to understanding the bioaccumulation potential of sediment-associated HOCs, it is also important to know their toxicity to benthic organisms when assessing the overall sediment quality. Total sediment concentrations, which are usually determined using exhaustive extraction techniques, have traditionally been used as the dose metrics for sediment toxicity testing and environmental risk assessments [28]. Recent studies have suggested that the use of internal body residues, instead of external media concentrations, is a more effective approach for assessing environmental hazard, because internal doses take into account the bioavailability of toxicants [28–30,82,83]. As a result, there is a growing interest in using the bioavailable fraction of a compound in sediment management; however, this application is still limited due to the absence of reliable and cost-effective techniques for assessing bioavailability [25].

Extensive research has shown that PSMs have great potential to predict the bioavailability of sediment-associated HOCs. As discussed earlier, PSMs, e.g., SPME, are effective in quantifying C_{free} of HOCs in sediment and predicting the accumulated HOCs in benthic organisms, which are exposed to contaminated sediment [14,19,20,34,84]. Additional research is attempting to use PSM measurements to describe the exposure of HOCs in sediment and link these doses to the toxicity in benthic organisms [18,20,21,42]. For instance, SPME has been used to evaluate sediment toxicity caused by PAHs [17,85–89], TNTs [52,53], and pesticides [33,90–94]. As shown in Table 1, most studies on the application of SPMEs to measure sediment toxicity to the benthos deal with PAH-contaminated sediments; however, the toxicity related to pesticides in sediment has been an important concern as well.

TABLE 1 Application of Solid-phase Microextraction (SPME) on Evaluating the Toxicity of Hydrophobic Organic Contaminants (HOC) in Sediment, Soil, and Water Samples

Chemical Class	Experimental Conditions	Chemical	PSM Type	Species	Toxic Endpoint	Dose Metric	Reference
PAHs	45 sediments from six manufactured gas plant (MGP) sites	34 parent and alkylated PAHs	Pore water SPME (centrifugation, flocculation, addition of d-PAH internal standards, SPME, and GC/MS analysis)	Benthic organisms	Sediment benchmark [9]	Toxic units derived from the total C_{free} of 34 and 16 PAHs	[85]
	97 sediments from six former MGP and two aluminum smelter sites	34 parent and alkylated PAHs	Pore water SPME	*Hyalella azteca*	Predicted 28-day survival based on equilibrium partitioning and hydrocarbon narcosis models	C_{free} for classifying toxic and nontoxic sediment samples	[86]
	34 sediment samples from four MGP sites	34 PAHs	Pore water SPME, supercritical fluid extraction	*H. azteca*	Survival and growth of *H. azteca* in 28-day bioassays	Rapidly released concentrations as determined by mild supercritical fluid extraction or the C_{free} measured by SPME	[87]

	Laboratory-spiked sediment	Three pairs of isomers, including two homocyclic compounds (anthracene and phenanthrene), two azaarenes (acridine and phenanthridine), and two main transformation products of azaarenes (acridone and phenanthridone)	SPME fiber with polydimethylsiloxane (PDMS) at equilibrium	*Lumbriculus variegatus*	28-day reproduction	EC50 and EC10 based on bulk sediment concentration or C_{free} measured by SPME	[89]
	192 sediment samples from 12 field sites	34 PAHs	Pore water SPME	*H. azteca*	Toxicity predicted by hydrocarbon narcosis model	Bulk sediment concentrations and C_{free} values predicted by K_{OC} model, $K_{OC}K_{BC}$ model, or SPME measurement	[88]
	Field sediment from electronic waste recycling site	16 PAHs	Co-exposed matrix SPME and 24-h Tenax extraction	*L. variegatus* and *H. azteca*	Bioaccumulation potential in *L. variegatus* and survival of *H. azteca* in 28-day bioassays	Organic carbon normalized sediment concentration, C_{free}, and 24-h Tenax extractable concentration	[17]

Continued

TABLE 1 Application of Solid-phase Microextraction (SPME) on Evaluating the Toxicity of Hydrophobic Organic Contaminants (HOC) in Sediment, Soil, and Water Samples—cont'd

Chemical Class	Experimental Conditions	Chemical	PSM Type	Species	Toxic Endpoint	Dose Metric	Reference
	In situ water and sediment	PAHs and metals	The Sediment Ecotoxicity Assessment (SEA) Ring: (SPME for PAHs while diffusive gradient in thin film (DGT) for metals	In situ toxicity test durations included 2-day (amphipod, polychaete, mysid and bivalve) and 10-day (amphipod) exposures	Measured endpoints were both sublethal and lethal effects as well as bioaccumulation	In situ co-exposure of passive sampling devices with organisms	[109]
	In situ exposure in shallow estuarine wetland, lab toxicity testing	PAHs and metals	SEA Ring: (SPME for PAHs while DGT for metals	Amphipod *Leptocheirus plumulosus*, Polychaete *Neanthes arenaceodentata*, Mysid *Americamysis bahia*	Weight of evidence: passive sampling measurements, bioaccumulation, toxicity, lab and in situ SEA Ring	Bulk sediment concentrations and C_{free}	[110]

	335 sediments from 19 sites impacted by MGP, aluminum smelters, and other pyrogenic sources	34 parent and alkylated PAHs	Pore water SPME	Amphipod *H. azteca* and *L. plumulosus*	28-day survival	Toxic units derived from the total C_{free} of 34 and 16 PAHs	[111]
	Spiked water	Complex hydrocarbon mixture, including a gasoline, kerosene, a crude oil, and related products	SPME fiber	Rainbow trout	96-h Lethality	Bioavailable petroleum hydrocarbons (BPHs): the total moles of hydrocarbons that partition to the SPME fiber and be quantified using gas chromatography/flame ionization detector	[98]
	Laboratory-spiked water	Two halogenated aromatic compounds	SPME fibers with polyacrylate (PA) coating	*Chironomus riparius*	96-h Survival	Total body residues estimated with SPME	[49]

Continued

TABLE 1 Application of Solid-phase Microextraction (SPME) on Evaluating the Toxicity of Hydrophobic Organic Contaminants (HOC) in Sediment, Soil, and Water Samples—cont'd

Chemical Class	Experimental Conditions	Chemical	PSM Type	Species	Toxic Endpoint	Dose Metric	Reference
	Laboratory-spiked water	Chlorobenzene mixture	Co-exposure SPME with PDMS fibers	*C. riparius and L. variegatus*	96-h Survival	Total body residues estimated with SPME	[99]
	Laboratory-spiked water	1,2,3,4-tetrachlorobenzene	Co-exposure SPME with PDMS fibers	*C. riparius*	28-day chronic toxicity (lethality and growth)	Body residues estimated with SPME	[51]
	Water of leachable constituents of four solid and three liquid resins	Complex hydrocarbon-based mixtures	SPME fibers	*Daphnia magna*	Lethality	Total water concentration, and bioavailable concentrations determined using SPME fiber measurements	[112]
	Laboratory dosing system	PAH mixtures	Passive dosing with the silicone PDMS	Benthic marine amphipods, *Orchomonella pinguis* and *Corophium volutato*	Baseline toxicity	Chemical activity	[104]

	Laboratory dosing system	Anthracene, chrysene, and benzo(a)pyrene	Passive dosing with the silicone PDMS	*D. magna*	48-h lethality	Sum chemical activity–response curve based on the effective chemical activity and the effective concentration on a lipid basis	[105]
	Laboratory dosing system	Six PAHs: naphthalene, anthracene, and pyrene	Passive dosing with the silicone PDMS	The terrestrial springtail *Folsomia candida*	Effective lethal chemical activities; effective lethal body burdens, lethal membrane burden for baseline toxicity	Dose-response curves	[113]
	15 soil samples from MGP sites	34 PAHs	Pore water SPME	Earthworms	K_{oc}, body residues and worm mortality estimated from literature lethal body burden or critical body residue	C_{free} measured by SPME method or predicted from one-parameter LFER model	[114]
	Soil of various organic matter, aging time	Pyrene	SPME fibers with PDMS	*F. candida*	EC50 values	Soil concentration and C_{free}	[115]

Continued

TABLE 1 Application of Solid-phase Microextraction (SPME) on Evaluating the Toxicity of Hydrophobic Organic Contaminants (HOC) in Sediment, Soil, and Water Samples—cont'd

Chemical Class	Experimental Conditions	Chemical	PSM Type	Species	Toxic Endpoint	Dose Metric	Reference
TNT	Laboratory-spiked water and sediment	2,4,6-Trinitrotoluene (TNT) and its degradation products	SPME fiber with PA coating	*Tubifex tubifex*	Lethality	External matrix (water, sediment) and internal (organism) concentrations, SPME fiber concentrations	[52]
	Laboratory-spiked water and sediment	TNT and its degradation products	SPME fiber with PA coating	*T. tubifex, Chironomus tentans,* and *Ceriodaphnia dubia*	Lethality	External matrix (water, sediment) and internal (organism) concentrations, SPME fiber concentrations	[53]

Pesticide	Laboratory-spiked sediments	Pyrethroids	Equilibrium SPME using PDMS fibers	*C. tentans*	10-day acute mortality	The LC50s were expressed using HOC concentrations on a bulk sediment basis (C_S), organic carbon-normalized sediment basis ($C_{S\text{-}OC}$), porewater basis (C_{pw}), dissolved organic carbon-normalized porewater basis ($C_{pw\text{-}DOC}$), and freely dissolved pore water basis (C_{free})	[90]
	Laboratory-spiked sediment aged for 7, 28 and 90 days	Permethrin and dichlorodiphenyl trichloroethane (DDT)	Disposable PDMS fibers	*Chironomus dilutus* and *H. azteca*	10-day mortality	Concentrations in sediment, organism and PDMS fiber at equilibrium	[33]

Continued

TABLE 1 Application of Solid-phase Microextraction (SPME) on Evaluating the Toxicity of Hydrophobic Organic Contaminants (HOC) in Sediment, Soil, and Water Samples—cont'd

Chemical Class	Experimental Conditions	Chemical	PSM Type	Species	Toxic Endpoint	Dose Metric	Reference
	Three laboratory-spiked sediments	Bifenthrin	Equilibrium SPME using PDMS fibers and Tenax extraction at 6 and 24 h	*H. azteca* and *C. dilutus*	10-day mortality and immobilization	Bioavailability-based LC50 and EC50 using four metrics, including SPME fiber concentration, SPME-derived C_{free}, and 6- and 24-h Tenax extractable concentrations	[91]
	Field-collected sediments and in situ exposure	Pyrethroids and chlorpyrifos	Field-deployed SPME fibers coated with PDMS, laboratory-exposed SPME fibers, and a 24-h Tenax extraction	*H. azteca*, *C.dilutus*, and *Hexagenia* sp.	10-day mortality	There is a significant relationship between Tenax extractable bifenthrin concentrations and mortality. Due to sensitivity limitation, no SPME data are available	[92]

	Field-collected sediments	Cypermethrin	Equilibrium SPME using PDMS fibers and 24-h Tenax extraction	*H. azteca*	10-day acute mortality	Bioavailability toxic units (TUs) from SPME and Tenax measurements. Traditional TU from organic carbon normalized sediment concentration	[94]
	Laboratory-spiked waters with suspended sediment	Bifenthrin, Esfenvalerate, *cis*-Permethrin, and cypermethrin	SPME using PDMS fibers	*C. dubia*	96-h Static water toxicity assays	C_{free}, relationship between LC50 and concentrations of suspended sediment in water	[116]
	Four laboratory-spiked waters, including a laboratory reconstituted water, two natural waters, and modified natural water	Permethrin	48 h and equilibrium SPME using PDMS fibers	*D. magna*	48-h water-only mortality	48-h and equilibrium-based SPME fiber concentrations	[97]
	Laboratory-spiked water	DDT, DDD, DDE, permethrin, bifenthrin, tefluthrin, and chlorpyrifos	Disposable PDMS fibers	*C. dilutus* and *H. azteca*	4-day and 10-day mortality and immobilization	Concentration in water and organism and PDMS fiber at equilibrium	[95]

5.2 Measurement Endpoints

In general, methods that incorporate bioavailability-based measurements into sediment toxicity assessment can be classified into four categories based on the endpoints chosen for the SPME measurements, including SPME-measured freely dissolved concentration in sediment pore water (C_{free}), concentrations in the organisms (C_b) estimated by SPME, concentrations on SPME coatings, (e.g., C_{PDMS}), and chemical activity. While the four terms could be calculated from each other, they are often used for different purposes. Ding et al. [95,96] proposed a conceptual model using C_{PDMS} at equilibrium as a dose metric to reflect sediment toxicity of pesticides to benthic organisms (Figure 2). The C_{PDMS} at equilibrium reflected the C_{free} or chemical activity in the sediment pore water, which directly drives the absorbed dose in the organisms and subsequently toxicity, so C_{PDMS} should also reflect toxicity. It should be noted that equilibrium is generally attainable for a narcotic chemicals (e.g., PAHs and PCBs) among sediment, pore water, and SPME fiber, and steady state concentrations are achievable for narcotic HOCs, which are usually less toxic in an organism. Conversely, equilibrium and steady state are most likely not achievable for HOCs with specific modes of action (e.g., some pesticides). For HOCs with specific modes of action, toxicity (e.g., lethal or sublethal effects) may occur prior to their body residues reaching steady state levels in the organisms.

5.2.1 Freely Dissolved Concentration

Freely dissolved concentration of HOC in sediment pore water, which is determined by using SPME, can be directly used as a dose metric for toxicity assessments. Since C_{free} takes bioavailability directly into account, it is a better dose metric than bulk sediment concentrations for sediment toxicity. Harwood et al. [97] suggested that the relationship between observed permethrin toxicity to *Daphnia magna* and water concentrations was improved by replacing the total

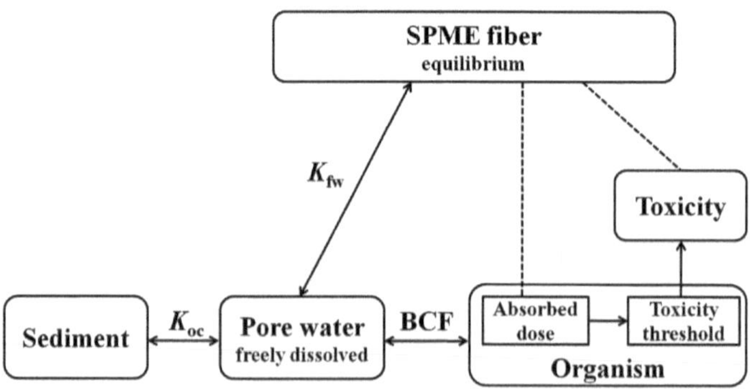

FIGURE 2 Conceptual model using solid-phase microextraction (SPME) fiber concentration as a dose metric to estimate the toxicity of hydrophobic organic contaminants (HOCs) in sediment. *Adapted from [95,96].*

permethrin concentration in water by C_{free} of permethrin. Another finding of this study was that the median lethal concentration (LC50) expressed using C_{free} was matrix-independent, that is, C_{free} based LC50 values might be used in either water or sediment matrices. In sediment toxicity testing with pyrethroids, Xu et al. [90] compared five parameters to express the lethality of pyrethroids in sediment to *Chironomus tentans*, including total sediment concentration (C_s), OC-normalized concentration ($C_{s\text{-}oc}$), pore water concentration (C_{pw}), dissolved organic carbon (DOC)-normalized pore water concentration ($C_{pw\text{-}DOC}$), and C_{free}, and concluded that the SPME-measured C_{free} was the best dose metric for expressing the toxicity across sediments. In addition, the measurement of C_{free} provides a means of assessing sediment risk directly by comparing C_{free} to water quality criteria [21].

5.2.2 Concentration in Organisms

SPME has been widely used to estimate bioaccumulation and predict concentrations of HOCs in organisms [14,24]. Therefore, the relative toxicity of a sample can be calculated by comparing HOC concentrations in organisms estimated from SPMEs to the values of critical body residues, which have been previously determined in laboratory toxicity tests [49,98,99]. Parkerton et al. [98] noted bioavailable petroleum hydrocarbons measured using SPMEs at equilibrium were correlated to the observed acute toxicity to rainbow trout (*Oncorhynchus mykiss*). Recent research linking internal exposure to the toxicity of narcotic chemicals has developed various models to predict toxicity, such as the lipid target model [28,30]. Hence, body residues estimated by SPMEs can be combined with the lipid target model or toxicokinetic models to predict toxicity.

5.2.3 Concentration on the SPME Fibers

Moreover, concentrations of HOCs on the polymers, e.g., SPME fibers, can be directly used as dose metrics in sediment toxicity assessments (Figure 2, [95,96]). This approach acknowledges that the absorbed dose (internal concentration) is a better dose metric than chemical concentration in the external environment, particularly when there are substantial bioavailability limitations, e.g., significant binding ability of planar compounds to BC. Since SPME fiber concentrations are proportional to body residues, SPMEs can serve as a reliable estimate of exposure for toxicity evaluation. Based on this conceptual model, the toxicity of dichlorodiphenyltrichloroethane and permethrin in sediment to *Chironomus dilutus* and *Hyalella azteca* was evaluated by establishing a dose–response relationship using SPME fiber concentrations and body residues as dose metrics [33]. To establish a toxicity threshold of critical bioavailable petroleum hydrocarbons Parkerton et al. [98] converted critical body residues for narcotic chemicals to SPME concentrations. Therefore, toxicity is expected in the target organisms, if the measured bioavailable petroleum hydrocarbon concentrations are greater than the threshold values.

In addition, SPME measurements can also be directly incorporated into sediment risk assessments and management decisions by developing sediment quality criteria, which are based on fiber concentrations. When $K_{polymer-water}$ values, which are the ratios of chemical concentrations on the polymer and in the water, are available for specific compounds, water quality criteria for specific chemicals can be converted into polymer-based effect concentrations. Consequently, the PSM-measured concentrations of target HOCs can be compared to the polymer-based effect concentrations to assess sediment risk. Although this method is straightforward and easy to conduct, specific polymers and standardized methods are required for the developed polymer-based sediment quality criteria to be used. Recently, bioavailability-based toxic endpoints have been developed based on concentrations on the PDMS fibers and Tenax sorbents [91]. In this study, the LC50 and median effect concentrations (EC50) of bifenthrin to benthic invertebrates in 10-day sediment toxicity tests were established using four bioavailability-based dose metrics, including SPME fiber concentrations, SPME measured C_{free}, and 6- and 24-h Tenax extractable concentrations. Results showed that these bioavailability-based endpoints are interrelated and more accurate in toxicity prediction than sediment-based ones across sediment types. Greater precision (less relative standard deviations) has also been reported for the bioavailability-based dose metric compared to bulk sediment concentrations [100,101].

5.2.4 Chemical Activity

When PSMs are used, chemical activity is another important measurement and is useful to predict the spontaneous processes, including the partitioning of organic substances into lipid membranes [24]. Therefore, risk of particular sediment can be assessed by relating the measured chemical activity to the chemical activity range that is needed to initiate baseline toxicity. Critical membrane concentrations for baseline toxicity generally require chemical activities ranging from 0.01 to 0.1 [24,102].

Witt et al. [103] measured chemical activity and C_{free} of PAHs in sediment core samples using SPME and found the sum chemical activity more directly reflected baseline toxicity (narcotic action) of individual PAHs compared to the toxicity contribution of individual PAHs expressed as C_{free}. For example, phenanthrene dominated the composition of C_{free} (48%), but it only contributed moderately to the overall exposure when expressed as chemical activity (16%). On the other hand, benzo[a]pyrene contributed only 1.1% to the sum C_{free} of PAHs, while it contributed 16% to the chemical activity. The difference in contributions of individual PAHs to the sum C_{free} and chemical activity implied the good application of chemical activity as dose metrics in assessing baseline toxicity across chemicals.

5.3 Toxicity of Chemical Mixtures

To date, most studies using SPMEs to assess toxicity have focused on narcotic compounds, particularly PAHs, and SPMEs have been noted as promising techniques for assessing the aquatic toxicity of complex mixtures of narcotic compounds [49,98,99,104,105]. Concentration addition was the most effective model to describe mixture toxicity for narcotic compounds, and thus biomimetic methods can be applied to predict baseline toxicity of complex mixtures using chemical activity as a dose metric. To assess baseline toxicity of a sediment, chemical activity of individual HOCs in that sediment can be directly summed and compared to the threshold values [104,105].

The SPMEs cannot only determine C_{free} and bioaccumulation potential of a HOC in sediment, but also the toxicity to benthic organisms. By taking advantage of the concentration additive properties of narcotic compounds, baseline toxicity of a complex mixture can be assessed, and chemical activity measured by PSMs is the preferred dose metric for this analysis. However, many contaminants are vulnerable to biotransformation in organisms [93]. It should be noted that passive sampling cannot quantify the biotransformation of contaminants, thus it overestimates sediment toxicity induced by compounds that are easily biotransformed. Several studies have shown that SPMEs successfully estimated body residues and toxicity of contaminants that were biotransformed, including pyrethroid insecticides and TNT [33,52,54,93,95–97]. Yet, the relationship between SPME measurements and chemical toxicity in these studies was only presented for individual chemicals across sediments, and it is still not clear how the biotransformation would affect the toxicity when sediment is contaminated by a mixture of contaminants [106]. Furthermore, the instability of these contaminants in environmental media also affects the measurements of C_{free} by SPMEs and thus parameters that take instability into account need to be considered when designing SPME calibration methods [95,96,107]. Overall, additional studies are required to better understand the role of degradation and biotransformation of chemicals in estimating sediment toxicity with SPMEs.

An additional complication to the use of SPME is its application in assessing sediment toxicity of pesticides. The previously mentioned toxicity prediction models for narcotic compounds cannot be used to predict sediment toxicity caused by pesticides, because most pesticides cause toxicity by specific modes of action. To address this issue, the toxic unit (TU), which is equal to the ratio of OC-normalized sediment concentration and LC50 or EC50, has been applied to calculate the toxicity of individual compounds and TU addition has been used to predict mixture toxicity for chemicals with similar modes of action. Taking bioavailability into account, a concept coined "bioavailable TU" was introduced by replacing the OC-normalized sediment concentration with the bioavailable concentration measured using bioavailability-based extraction methods, such as SPME [108]. A recent study of an urban stream in China demonstrated the benefit of using bioavailable TUs for estimating sediment toxicity of pyrethroid insecticides [94]. The traditional TU approach based on OC-normalized chemical concentrations suggested that the pyrethroid, cypermethrin, was the major contributor to sediment toxicity to *H. azteca* in this urban stream, while the traditional TUs overestimated toxicity. Instead, bioavailable TUs estimated from C_{free}

of cypermethrin measured by SPME significantly improved the prediction of sediment toxicity to *H. azteca* compared to traditional TUs [94].

6. SCIENTIFIC MERITS OF USING PASSIVE SAMPLING FOR EVALUATING BIOAVAILABILITY AND TOXICITY

For HOC-contaminated sediments, accurate estimates of contaminant bioavailability and toxicity are important for proper assessment of sediment risk. While bulk sediment concentrations cannot accurately estimate bioavailability, C_{free} shows great promise in improving the accuracy of the bioavailability prediction. Passive samplers, like SPME and PEs, have been widely used to quantify the bioavailable fraction of HOCs in sediment [14].

Compared to the traditional extraction techniques, passive samplers are more cost-effective and simpler at measuring chemical activity. In addition, most passive samplers only absorb the freely dissolved portion of contaminants, which have desorbed from the binding phases [18]. The PSM-measured concentrations reflect the bioavailability of contaminants in sediment, or in other words, passive samplers are a good indicator for HOC bioavailability and sediment toxicity. Recent developments in PSMs have significantly improved the feasibility of measuring C_{free} accurately and more effectively. Although equilibrium measurements are the preferred method for passive sampling, newly developed calibration methods, such as the use of PRCs, make the assessment of C_{free} possible prior to the system reaching equilibrium.

In conclusion, as a more relevant dose metric compared to bulk sediment concentrations, C_{free} better represents exposure as well as the bioaccumulation potential of contaminants to organisms. By incorporating C_{free} into the dose–response relationship, exposure can be linked to adverse effects in organisms. Bioavailability may be determined using chemical activity, the deployment of passive samplers in sediments during toxicity testing, the application of internal effect concentrations, and target lipid models, as well as adjusting water quality criteria based on polymer concentrations.

7. CONCLUSIONS AND PERSPECTIVES

Overall, intensive research in the past 20 years has provided us with a better understanding of how bioavailability of HOCs in sediment affects sediment quality assessment and how to estimate bioavailability through the use of PSMs. The C_{free} (chemical activity) derived from the PSMs is a better dose metric of sediment-associated HOCs than the OC-normalized HOC concentration. Incorporation of bioavailability-based endpoints into toxicity evaluations will significantly improve sediment quality assessments.

A variety of PSMs, such as SPME, have been developed and applied for research purposes, but there is still a gap between the scientific understanding of these devices and the regulatory application of bioavailability concepts during risk assessments. Therefore, before their implementation, PSMs should be better validated and standardized, including the establishment of solid quality assurance/quality control strategies to ensure the performance of these techniques and the development of partition coefficient databases for the selected polymers. Furthermore, although conducting C_{free} studies at equilibrium is the preferred method, the long equilibrium times required for highly hydrophobic compounds requires effective kinetic calibration methods, particularly in the case of in situ PSMs. Finally, most studies using PSMs have focused on narcotic compounds, such as PAHs, in sediment. More research is needed to investigate the relationship between bioavailability-based doses and toxic effects of contaminants exhibiting specific modes of action and those with a tendency of being biotransformed.

ACKNOWLEDGMENTS

Financial support by the National Natural Science Foundation of China (No. 41222024 and 41473106) is thankfully acknowledged.

REFERENCES

[1] K.A. Maruya, P.F. Landrum, R.M. Burgess, J.P. Shine, Incorporating contaminant bioavailability into sediment quality assessment frameworks, Integr. Environ. Assess. Manag. 8 (2012) 659–673.

[2] E.R. Long, D.D. MacDonald, Recommended uses of empirically derived, sediment quality guidelines for marine and estuarine ecosystems, Hum. Ecol. Risk Assess. 4 (1998) 1019–1039.

[3] E.R. Long, C.G. Ingersoll, D.D. MacDonald, Calculation and uses of mean sediment quality guideline quotients: a critical review, Environ. Sci. Technol. 40 (2006) 1726–1736.

[4] D.M. Di Toro, J.A. McGrath, D.J. Hanse, Technical basis for narcotic chemicals and polycyclic aromatic hydrocarbon criteria. I. Water and tissue, Environ. Toxicol. Chem. 19 (2000) 1951–1970.

[5] D.M. Di Toro, J.A. McGrath, Technical basis for narcotic chemicals and polycyclic aromatic hydrocarbon criteria. II. Mixtures and sediments, Environ. Toxicol. Chem. 19 (2000) 1971–1982.

[6] USEPA (United States Environmental Protection Agency), Technical Basis for the Derivation of Equilibrium Partitioning Sediment Benchmarks (ESBs) for the Protection of Benthic Organisms: Nonionic Organics, Office of Research and Development, Washington, DC, USA, 2003. EPA-600-R-02-014.

[7] USEPA (United States Environmental Protection Agency), Procedures for the Derivation of Equilibrium Partitioning Sediment Benchmarks (ESBs) for the Protection of Benthic Organisms: Endrin, EPA-600-R-02-009, Office of Research and Development, Washington, DC, USA, 2003.

[8] USEPA (United States Environmental Protection Agency), Procedures for the Derivation of Equilibrium Partitioning Sediment Benchmarks (ESBs) for the Protection of Benthic Organisms: Dieldrin, EPA-600-R-02-010, Office of Research and Development, Washington, DC, USA, 2003.

[9] USEPA (United States Environmental Protection Agency), Procedures for the Derivation of Equilibrium Partitioning Sediment Benchmarks (ESBs) for the Protection of Benthic Organisms: PAH Mixtures, EPA-600-R-02-013, Office of Research and Development, Washington, DC, USA, 2003.

[10] USEPA (United States Environmental Protection Agency), Procedures for the Derivation of Equilibrium Partitioning Sediment Benchmarks (ESBs) for the Protection of Benthic Organisms: Compendium of Tier 2 Values for Nonionic Organics, EPA-600-R-02-016, Office of Research and Development, Washington, DC, USA, 2008.
[11] G. Cornelissen, Ö. Gustafsson, T.D. Bucheli, M.T.O. Jonker, A.A. Koelmans, P.C.M. van Noort, Extensive sorption of organic compounds to black carbon, coal, and kerogen in sediments and soils: mechanisms and consequences for distribution, bioaccumulation, and biodegradation, Environ. Sci. Technol. 39 (2005) 6881–6895.
[12] G.A.C. Ehlers, A.P. Loibner, Linking organic pollutant (bio)availability with geosorbent properties and biomimetic methodology: a review of geosorbent characterisation and (bio) availability prediction, Environ. Pollut. 141 (2006) 494–512.
[13] W.J. Weber Jr., W. Huang, E.J. LeBoeuf, Geosorbent organic matter and its relationship to the binding and sequestration of organic contaminants, Colloids Surf. A: Physicochem. Eng. Aspects 151 (1999) 167–179.
[14] M.J. Lydy, P.F. Landrum, A.M.P. Oen, M. Allinson, F. Smedes, A.D. Harwood, H. Li, K.A. Maruya, J. Liu, Passive sampling methods for contaminated sediments: state of the science for organic contaminants, Integr. Environ. Assess. Manag. 10 (2014) 167–178.
[15] J. You, S. Pehkonen, P.F. Landrum, M.J. Lydy, Desorption of hydrophobic compounds from laboratory-spiked sediments measured by Tenax absorbent and matrix solid-phase microextraction, Environ. Sci. Technol. 41 (2007) 5672–5678.
[16] R.G. Luthy, G.R. Aiken, M.L. Brusseau, S.D. Cunningham, P.M. Gschwend, J.J. Pignatello, M. Reinhard, S.J. Traina, W.J. Weber Jr., J.C. Westall, Sequestration of hydrophobic organic contaminants by geosorbents, Environ. Sci. Technol. 31 (1997) 3341–3347.
[17] J. Pang, B. Sun, H. Li, W.T. Mehler, J. You, Influence of bioturbation on bioavailability and toxicity of PAHs in sediment from an electronic waste recycling site in South China, Ecotox. Environ. Saf. 84 (2012) 227–233.
[18] P. Mayer, T.F. Parkerton, R.G. Adams, J.G. Cargill, J. Gan, T. Gouin, P.M. Gschwend, S.B. Hawthorne, P. Helm, G. Witt, J. You, B.I. Escher, Passive sampling methods for contaminated sediments: scientific rationale supporting use of freely dissolved concentrations, Integr. Environ. Assess. Manag. 10 (2014) 197–209.
[19] X. Cui, P. Mayer, J. Gan, Methods to assess bioavailability of hydrophobic organic contaminants: principles, operations, and limitations, Environ. Pollut. 172 (2013) 223–234.
[20] J. You, A.D. Harwood, H. Li, M.J. Lydy, Techniques for assessing bioavailability of sediment-associated contaminants: SPME versus Tenax extraction, J. Environ. Monit. 13 (2011) 792–800.
[21] USEPA (United States Environmental Protection Agency), Equilibrium Partitioning Sediment Benchmarks (ESBs) for the Protection of Benthic Organisms: Procedures for the Determination of the Freely Dissolved Interstitial Water Concentrations of Nonionic Organics, EPA/600/R-02/012, Office of Research and Development, Washington, DC, USA, 2012.
[22] W.J.G.M. Peijnenburg, P.R. Teasdale, D. Reible, J. Mondon, W.W. Bennett, P.G.C. Campbell, Passive sampling methods for contaminated sediments: State of the science for metals, Integr. Environ. Assess. Manag. 10 (2014) 179–196.
[23] B. Vrana, I.J. Allan, R. Greenwood, Passive sampling techniques for monitoring pollutants in water, Trends Anal. Chem. 24 (2005) 845–868.
[24] F. Reichenberg, P. Mayer, Two complementary sides of bioavailability: accessibility and chemical activity of organic contaminants in sediments and soils, Environ. Toxicol. Chem. 25 (2006) 1239–1245.
[25] NRC (National Research Council), Bioavailability of Contaminants in Soils and Sediments: Processes, Tools and Applications, National Academies Press, Washington, DC, 2003.

[26] ISO (International Standardization Organization), Soil Quality—Requirements and Guidance for the Selection and Application of Methods for the Assessment of Bioavailability of Contaminants in Soil and Soil Materials, ISO 17402:2008, 2008.

[27] K.T. Semple, K.J. Docie, K.C. Jones, P. Burauel, A. Craven, H. Harms, Defining bioavailability and bioaccessibility of contaminated soil and sediment is complicated, Environ. Sci. Technol. 38 (2004) 228A–231A.

[28] B.I. Escher, J.L.M. Hermens, Internal exposure: linking bioavailability to effects, Environ. Sci. Technol. 38 (2004) 455A–462A.

[29] B.I. Escher, R. Ashauer, S. Dyer, J.L.M. Hermens, J.-H. Lee, H.A. Leslie, P. Mayer, J.P. Meador, M.S.J. Warne, Crucial role of mechanisms and modes of toxic action for understanding tissue residue toxicity and internal effect concentrations of organic chemicals, Integr. Environ. Assess. Manag. 7 (2011) 28–49.

[30] J. Meador, Rationale and procedure for using the tissue-residue approach for toxicity assessment and determination of tissue, water, and sediment quality guideline for aquatic organisms, Hum. Ecol. Risk Assess. 12 (2006) 1018–1073.

[31] A.E. Vinturella, R.M. Burgess, B.A. Coull, K.M. Thompson, J.P. Shine, Use of passive samplers to mimic uptake of polycyclic aromatic hydrocarbons by benthic polychaetes. Use of passive samplers to mimic uptake of polycyclic aromatic hydrocarbons by benthic polychaetes, Environ. Sci. Technol. 38 (2004) 1154–1160.

[32] A.A. Fay, B.J. Brownawell, A.A. Elskus, A.E. McElroy, Critical body residues in the marine amphipod *Ampelisca abdita*: sediment exposures with non-ionic organic contaminants, Environ. Toxicol. Chem. 19 (2000) 1028–1035.

[33] Y. Ding, P.F. Landrum, J. You, M.J. Lydy, Assessing bioavailability and toxicity of permethrin and DDT in sediment using matrix solid phase microextraction, Ecotoxicology 22 (2013) 109–117.

[34] R. Kraaij, P. Mayer, F.J.M. Busser, M. van het Bolscher, W. Seinen, J. Tolls, A. Belfroid, Measured pore water concentrations make equilibrium partitioning work—A data analysis, Environ. Sci. Technol. 37 (2003) 268–274.

[35] D.M. Di Toro, C.S. Zarba, D.J. Hansen, W.J. Berry, R.C. Swartz, C.E. Cowan, S.P. Pavlou, H.E. Allen, N.A. Thomas, P.R. Paquin, Technical basis for establishing sediment quality criteria for nonionic organic chemicals by using equilibrium partitioning, Environ. Toxicol. Chem. 10 (1991) 1541–1583.

[36] S. Pehkonen, J. You, J. Akkanen, J.V.K. Kukkonen, M.J. Lydy, Influence of black carbon and chemical planarity on bioavailability of sediment-associated contaminants, Environ. Toxicol. Chem. 29 (2010) 1976–1983.

[37] Ö. Gustafsson, F. Haghseta, C. Chan, J. MacFarlane, P.M. Gschwend, Quantification of the dilute sedimentary soot phase: Implications for PAH speciation and bioavailability, Environ. Sci. Technol. 31 (1997) 203–209.

[38] A. Accardi-Dey, P.M. Gschwend, M. Ralph, Assessing the combined roles of natural organic, matter and black carbon as sorbents in sediments, Environ. Sci. Technol. 36 (2002) 21–29.

[39] S.B.K. Driscoll, C.B. Amos, M.E. McArdle, C.A. Menzie, A. Coleman, Predicting sediment toxicity at former manufactured gas plants using equilibrium partitioning benchmarks for PAH mixtures, Soil Sediment Contam. 18 (2009) 307–319.

[40] J. Du, W.T. Mehler, M.J. Lydy, J. You, Toxicity of sediment-associated unresolved complex mixture and its impact on bioavailability of polycyclic aromatic hydrocarbons, J. Hazard. Mater. 203-204 (2012) 169–175.

[41] M.T.O. Jonker, A. Barendregt, Oil is a sedimentary supersorbent for polychlorinated biphenyls, Environ. Sci. Technol. 40 (2006) 3829–3835.

[42] USEPA (United States Environmental Protection Agency), Guidelines for Using Passive Samplers to Monitor Organic Contaminants at Superfund Sediment Sites, 9200.1–110 FS, Office of Superfund Remediation and Technology Innovation and Office of Research and Development, Washington, DC, USA, 2012.

[43] P. Mayer, W.H.J. Vaes, F. Wijnker, R.H. Kraaij, J. Tolls, J.L.M. Hermens, Sensing dissolved sediment pore water concentrations of persistent and bioaccumulative pollutants using disposable solid-phase microextraction fibers, Environ. Sci. Technol. 34 (2000) 5177–5183.

[44] ITRC (Interstate Technology & Regulatory Council), Incorporating Bioavailability Considerations into the Evaluation of Contaminated Sediment Sites, CS-1, Interstate Technology & Regulatory Council, Washington, D.C, 2011. Contaminated Sediments Team. www.itrcweb.org.

[45] U. Ghosh, S.K. Driscoll, R.M. Burgess, M.T.O. Jonker, D. Reible, F. Gobas, Y. Choi, S.E. Apitz, K.A. Maruya, W.R. Gala, M. Mortimer, C. Beegan, Passive sampling methods for contaminated sediments: practical guidance for selection, calibration, and implementation, Integr. Environ. Assess. Manag. 10 (2014) 210–223.

[46] M.S. Greenberg, P.M. Chapman, I.J. Allan, K.A. Anderson, S.E. Apitz, C. Beegan, T.S. Bridges, S.S. Brown, J.G. Cargill IV, M.C. McCulloch, C.A. Menzie, J.P. Shine, T.F. Parkerton, Passive sampling methods for contaminated sediments: risk assessment and management, Integr. Environ. Assess. Manag. 10 (2014) 224–236.

[47] T. Gorecki, J. Namiesnik, Passive sampling, Trends Anal. Chem. 21 (2002) 276–291.

[48] R. Jiang, J. Pawliszyn, Thin-film microextraction offers another geometry for solid-phase microextraction, Trends Anal. Chem. 39 (2012) 245–253.

[49] H.A. Leslie, A.J.P. Oosthoek, F.J.M. Busser, M.H.S. Kraak, J.L.M. Hermens, Biomimetic solid-phase microextraction to predict body residues and toxicity of chemicals that act by narcosis, Environ. Toxicol. Chem. 21 (2002) 229–234.

[50] H.A. Leslie, T.L. ter Laak, F.J.M. Busser, M.H.S. Kraak, J.L.M. Hermens, Bioconcentration of organic chemicals: Is a solid-phase microextraction fiber a good surrogate for biota? Environ. Sci. Technol. 36 (2002) 5399–5404.

[51] H.A. Leslie, M.H.S. Kraak, J.L.M. Hermens, Chronic toxicity and body residues of the nonpolar narcotic 1,2,3,4-tetrachlorobenzene in *Chironomus riparius*, Environ. Toxicol. Chem. 23 (2004) 2022–2028.

[52] J.M. Conder, T.W. La Point, J.A. Steevens, G.R. Lotufo, Recommendations for assessment of TNT toxicity in sediments, Environ. Toxicol. Chem. 23 (2004) 141–149.

[53] J.M. Conder, G.R. Lotufo, A.T. Bowen, P.K. Turner, T.W. La Point, J.A. Steevens, Solid phase microextraction fibers for estimating the toxicity of nitroaromatic compounds, Aquat. Ecosys. Health Manag. 7 (2004) 387–397.

[54] J.M. Conder, T.W. La Point, Solid-phase microextraction for predicting the bioavailability of 2,4,6-trinitrotoluene and its primary transformation products in sediment and water, Environ. Toxicol. Chem. 24 (2005) 1059–1066.

[55] R.G. Adams, R. Lohmann, L.A. Fernandez, J.K. MacFarlane, P.M. Gschwend, Polyethylene devices: passive samplers for measuring dissolved hydrophobic organic compounds in aquatic environments, Environ. Sci. Technol. 41 (2007) 1317–1323.

[56] R. Lohmann, Critical review of low-density polyethylene's partitioning and diffusion coefficients for trace organic contaminants and implications for its use as a passive sampler, Environ. Sci. Technol. 46 (2012) 606–618.

[57] S. Endo, S.E. Hale, K.-U. Goss, H.P.H. Arp, Equilibrium partition coefficients of diverse polar and nonpolar organic compounds to polyoxymethylene (POM) passive sampling devices, Environ. Sci. Technol. 45 (2011) 10124–10132.

[58] M.T.O. Jonker, A.A. Koelmans, Polyoxymethylene solid phase extraction as a partitioning method for hydrophobic organic chemicals in sediment and soot, Environ. Sci. Technol. 35 (2001) 3742–3748.
[59] T.P. Rusina, F. Smedes, J. Klanova, K. Booij, I. Holoubek, Polymer selection for passive sampling: a comparison of critical properties, Chemosphere 68 (2007) 1344–1351.
[60] P. Mayer, J. Tolls, J.L.M. Hermens, D. Mackay, Equilibrium sampling devices, Environ. Sci. Technol. 37 (2003) 184–191.
[61] G. Ouyang, J. Pawliszyn, Configuration and calibration methods for passive sampling techniques, J. Chromatogr. A 1168 (2007) 226–235.
[62] G. Ouyang, J. Pawliszyn, A critical review in calibration methods for solid-phase microextraction, Anal. Chim. Acta 627 (2008) 184–197.
[63] G. Cornelissen, A. Pettersen, D. Broman, P. Mayer, G.D. Breedveld, Field testing of equilibrium passive samplers to determine freely dissolved native polycyclic aromatic hydrocarbon concentrations, Environ. Toxicol. Chem. 27 (2008) 499–508.
[64] S. Bayen, T.L. ter Laak, J. Buffle, J.L.M. Hermens, Dynamic exposure of organisms and passive samplers to hydrophobic chemicals, Environ. Sci. Technol. 43 (2009) 2206–2215.
[65] S. Seethapathy, T. Gorecki, X. Li, Passive sampling in environmental analysis, J. Chromatogr. A 1184 (2008) 234–253.
[66] J. Namieśnik, B. Zabiegala, A. Kot-Wasik, M. Partyka, A. Wasik, Passive sampling and/or extraction techniques in environmental analysis: a review, Anal. Bioanal. Chem. 381 (2005) 279–301.
[67] K. Booij, H.M. Sleiderink, F. Smedes, Calibrating the uptake kinetics of semipermeable membrane devices using exposure standards, Environ. Toxicol. Chem. 17 (1998) 1236–1245.
[68] C.L. Arthur, J. Pawliszyn, Solid phase microextraction with thermal desorption using fused silica optical fibers, Anal. Chem. 62 (1990) 2145–2148.
[69] C.L. Arthur, L.M. Killam, K.D. Buchholz, J. Pawliszyn, J.R. Berg, Automation and optimization of solid-phase microextraction, Anal. Chem. 64 (1992) 1960–1966.
[70] J. Pawliszyn, B. Pawliszyn, M. Pawliszyn, Solid phase microextraction (SPME), Chem. Educ. 2 (1997) 1–7.
[71] J.A. Koziel, I. Novak, Sampling and sample-preparation strategies based on solid phase microextraction for analysis of indoor air, Trends Anal. Chem. 21 (2002) 840–850.
[72] Y. Chen, J.A. Koziel, J. Pawliszyn, Calibration for on-site analysis of hydrocarbons in aqueous and gaseous samples using solid phase microextraction, Anal. Chem. 75 (2003) 6485–6493.
[73] W.H.J. Vaes, E.U. Ramos, H.J.M. Verhaar, W. Seinen, J.L.M. Hermens, Measurement of the free concentration using solid-phase microextraction: binding to protein, Anal. Chem. 68 (1996) 4463–4467.
[74] S. Bondrenko, F. Spurlock, J. Gan, Analysis of pyrethroids in sediment pore water by solid-phase microextraction, Environ. Toxicol. Chem. 26 (2007) 2587–2593.
[75] S.B. Hawthorne, C.B. Grabanski, D.J. Miller, J.P. Kreitinger, Solid-phase microextraction measurement of parent and alkyl polycyclic aromatic hydrocarbons in milliliter sediment pore water samples and determination of K_{DOC} values, Environ. Sci. Technol. 39 (2005) 2795–2803.
[76] E.L. Di Filippo, R.P. Eganhouse, Assessment of PDMS-water partition coefficients: Implications for passive environmental sampling of hydrophobic organic compounds, Environ. Sci. Technol. 44 (2010) 6917–6925.
[77] A.D. Harwood, P.F. Landrum, M.J. Lydy, A comparison of exposure methods for SPME-based bioavailability estimates, Chemosphere 86 (2012) 506–511.

[78] L. Bao, F. Jia, J. Crago, E.Y. Zeng, D. Schlenk, J. Gan, Assessing bioavailability of DDT and metabolites in marine sediments using solid-phase microextraction with performance reference compounds, Environ. Toxicol. Chem. 32 (2013) 1946–1953.

[79] X. Cui, L. Bao, J. Gan, Solid-phase microextraction (SPME) with stable isotope calibration for measuring bioavailability of hydrophobic organic contaminants, Environ. Sci. Technol. 47 (2013) 9833–9840.

[80] T.A. Trimble, J. You, M.J. Lydy, Bioavailability of PCBs from field-collected sediments: application of Tenax extraction and matrix-SPME techniques, Chemosphere 71 (2008) 337–344.

[81] J. You, P.F. Landrum, M.J. Lydy, Comparison of chemical approaches for assessing bioavailability of sediment-associated contaminants, Environ. Sci. Technol. 40 (2006) 6348–6353.

[82] L.J. Schuler, P.F. Landrum, M.J. Lydy, Time-dependent toxicity of fluoranthene to freshwater invertebrates and the role of biotransformation on lethal body residues, Environ. Sci. Technol. 38 (2004) 6247–6255.

[83] L.J. Schuler, P.F. Landrum, M.J. Lydy, Response spectrum of fluoranthene and pentachlorobenzene for the fathead minnow (*Pimephales promelas*), Environ. Toxicol. Chem. 26 (2007) 1248–1257.

[84] J. You, P.F. Landrum, T.A. Trimble, M.J. Lydy, Availability of polychlorinated biphenyls in field-contaminated sediment, Environ. Toxicol. Chem. 26 (2007) 1940–1948.

[85] S.B. Hawthorne, D.J. Miller, J.P. Kreitinger, Measurement of total polycyclic aromatic hydrocarbon concentrations in sediments and toxic units used for estimating risk to benthic invertebrates at manufactured gas plant sites, Environ. Toxicol. Chem. 25 (2006) 287–296.

[86] S.B. Hawthorne, A.A. Nicholas, E.F. Neuhauser, J.P. Kreitinger, Predicting bioavailability of sediment polycyclic aromatic hydrocarbons to *Hyalella azteca* using equilibrium partitioning, supercritical fluid extraction, and pore water concentrations, Environ. Sci. Technol. 41 (2007) 6297–6304.

[87] J.P. Kreitinger, E.F. Neuhauser, F.G. Doherty, S.B. Hawthorne, Greatly reduced bioavailability and toxicity of polycyclic aromatic hydrocarbons to *Hyalella azteca* in sediments from manufactured-gas plant sites, Environ. Toxicol. Chem. 26 (2007) 1146–1157.

[88] K.M. McDonough, N.A. Azzolina, S.B. Hawthorne, D.V. Nakles, E.F. Neuhauser, An evaluation of the ability of chemical measurements to predict polycyclic aromatic hydrocarbon-contaminated sediment toxicity to *Hyalella azteca*, Environ. Toxicol. Chem. 29 (2010) 1545–1550.

[89] M.L. Paumen, P. Stol, T.L. ter Laak, M.H.S. Kraak, C.A.M. van Gestel, W. Admiraal, Chronic exposure of the oligochaete *Lumbriculus variegatus* to polycyclic aromatic compounds (PACs): bioavailability and effects on reproduction, Environ. Sci. Technol. 42 (2008) 3434–3440.

[90] Y. Xu, F. Spurlock, Z. Wang, J. Gan, Comparison of five methods for measuring sediment toxicity of hydrophobic contaminants, Environ. Sci. Technol. 41 (2007) 8394–8399.

[91] A.D. Harwood, P.F. Landrum, M.J. Lydy, Bioavailability-based toxicity endpoints of bifenthrin for *Hyalella azteca* and *Chironomus dilutus*, Chemosphere 90 (2013) 1117–1122.

[92] A.D. Harwood, P.F. Landrum, D. Weston, M.J. Lydy, Using SPME fibers and Tenax to predict the bioavailability of pyrethroids and chlorpyrifos in field sediments, Environ. Pollut. 173 (2013) 47–51.

[93] A.D. Harwood, P.F. Landrum, M.J. Lydy, Can SPME fiber and Tenax methods predict the bioavailability of biotransformed insecticides? Environ. Sci. Technol. 46 (2012) 2413–2419.

[94] H. Li, B. Sun, X. Chen, M.J. Lydy, J. You, Addition of contaminant bioavailability and species susceptibility to a sediment toxicity assessment: application in an urban stream in China, Environ. Pollut. 178 (2013) 135–141.

[95] Y. Ding, P.F. Landrum, J. You, A.D. Harwood, M.J. Lydy, Use of solid phase microextraction to estimate toxicity: relating fiber concentrations to toxicity-part I, Environ. Toxicol. Chem. 31 (2012) 2159–2167.

[96] Y. Ding, P.F. Landrum, J. You, A.D. Harwood, M.J. Lydy, Use of solid phase microextraction to estimate toxicity: relating fiber concentrations to body residues-part II, Environ. Toxicol. Chem. 31 (2012) 2168–2174.

[97] A.D. Harwood, A.R. Bunch, D.L. Flickinger, J. You, M.J. Lydy, Predicting the toxicity of permethrin to *Daphnia magna* in water using SPME fibers, Arch. Environ. Contam. Toxicol. 62 (2012) 438–444.

[98] T.F. Parkerton, M.A. Stone, D.J. Letinski, Assessing the aquatic toxicity of complex hydrocarbon mixtures using solid phase microextraction, Toxicol. Lett. 112–113 (2000) 273–282.

[99] H.A. Leslie, J.L.M. Hermens, M.H.S. Kraak, Baseline toxicity of a chlorobenzene mixture and total body residues measured and estimated with solid-phase microextraction, Environ. Toxicol. Chem. 23 (2004) 2017–2021.

[100] J. Du, J. Pang, J. You, Bioavailability-based chronic toxicity measurements of permethrin to *Chironomus dilutus*, Environ. Toxicol. Chem. 32 (2013) 1403–1411.

[101] J. Du, Y. Li, Z.C. Huang, J. You, Chronic toxicity thresholds for sediment-associated benzo[a] pyrene in the midge (*Chironomus dilutus*), Arch. Environ. Contam. Toxicol. 66 (2014) 370–378.

[102] D. Mackay, J.A. Arnot, F. Wania, R.E. Bailey, Chemical activity as an integrating concept in environmental assessment and management of contaminants, Integr. Environ. Assess. Manag. 7 (2011) 248–255.

[103] G. Witt, G.A. Liehr, D. Borck, P. Mayer, Matrix solid-phase microextraction for measuring freely dissolved concentrations and chemical activity of PAHs in sediment cores from the western Baltic Sea, Chemosphere 74 (2009) 522–529.

[104] M. Engraff, C. Solere, K.E.C. Smith, P. Mayer, I. Dahllof, Aquatic toxicity of PAHs and PAH mixtures at saturation to benthic amphipods: linking toxic effects to chemical activity, Aquat. Toxicol. 102 (2011) 142–149.

[105] K.E.C. Smith, S.N. Schmidt, N. Dom, R. Blust, M. Holmstrup, P. Mayer, Baseline toxic mixtures of non-toxic chemicals: "Solubility addition" increases exposure for solid hydrophobic chemicals, Environ. Sci. Technol. 47 (2013) 2026–2033.

[106] M.J. Lydy, J.B. Belden, C.E. Wheelock, B.D. Hammock, D.L. Denton, Challenges in regulating pesticide mixtures, Ecol. Soc. 9 (2004) 1–15.

[107] W. Lao, K.A. Maruya, D. Tsukada, A two-component mass balance model for calibration of solid-phase microextraction fibers for pyrethroids in seawater, Anal. Chem. 84 (2012) 9362–9369.

[108] J. You, S. Pehkonen, D.P. Weston, M.J. Lydy, Chemical availability and sediment toxicity of pyrethroid insecticides to *Hyalella azteca*: application to field sediment with unexpectedly low toxicity, Environ. Toxicol. Chem. 27 (2008) 2124–2130.

[109] G.A. Burton, G. Rosen, D.B. Chadwick, M.S. Greenberg, W.K. Taulbee, G.R. Lotufo, D.D. Reible, A sediment ecotoxicity assessment platform for in situ measures of chemistry, bioaccumulation and toxicity. Part 1: system description and proof of concept, Environ. Pollut. 162 (2012) 449–456.

[110] G. Rosen, D.B. Chadwick, G.A. Burton, W.K. Taulbee, M.S. Greenberg, G.R. Lotufo, D.D. Reible, A sediment ecotoxicity assessment platform for in situ measures of chemistry, bioaccumulation and toxicity. Part 2: Integrated application to a shallow estuary, Environ. Pollut. 162 (2012) 457–465.

[111] H.P.H. Arp, N.A. Azzolina, G. Cornelissen, S.B. Hawthorne, Predicting pore water EPA-34 PAH concentrations and toxicity in pyrogenic-impacted sediments using pyrene content, Environ. Sci. Technol. 45 (2011) 5139–5146.

[112] R.W. Woods, D.J. Letinski, E.J. Febbo, C.L. Dzamba, M.J. Connelly, T.F. Parkerton, Assessing the aquatic hazard of commercial hydrocarbon resins, Ecotox. Environ. Saf. 66 (2007) 159–168.

[113] S.N. Schmidt, K.E.C. Smith, M. Holmstrup, P. Mayer, Uptake and toxicity of polycyclic aromatic hydrocarbons in terrestrial springtails—studying bioconcentration kinetics and linking toxicity to chemical activity, Environ. Toxicol. Chem. 32 (2013) 361–369.

[114] M.T.O. Jonker, S.A. van der Heijden, J.P. Kreitinger, S.B. Hawthorne, Predicting PAH bioaccumulation and toxicity in earthworms exposed to manufactured gas plant soils with solid-phase microextraction, Environ. Sci. Technol. 41 (2007) 7472–7478.

[115] B. Styrishave, M. Mortensen, P.H. Krogh, O. Andersen, J. Jensen, Solid-phase microextraction (SPME) as a tool to predict the bioavailability and toxicity of pyrene to the Springtail, *Folsomia candida*, under various soil conditions, Environ. Sci. Technol. 42 (2008) 1332–1336.

[116] W. Yang, F. Spurlock, W. Liu, J. Gan, Inhibition of aquatic toxicity of pyrethroid insecticides by suspended sediment, Environ. Toxicol. Chem. 25 (2006) 1913–1919.

Chapter 6

Pharmaceuticals and Personal Care Products (PPCPs) in the Environment and Their Removal from Wastewater through Constructed Wetlands

Cristina Ávila and Joan García*

GEMMA–Group of Environmental Engineering and Microbiology, Department of Hydraulic, Maritime and Environmental Engineering, Universitat Politècnica de Catalunya-BarcelonaTech, Barcelona, Spain

**Corresponding author: E-mail: joan.garcia@upc.edu*

Chapter Outline

Comprehensive Analytical Chemistry, Vol. 67. http://dx.doi.org/10.1016/B978-0-444-63299-9.00006-5

1. INTRODUCTION

Water scarcity forces people to rely on unsafe sources of drinking water. The use of poor water quality can increase the risk of waterborne diseases (e.g., cholera and typhoid fever). Every year almost 2 million child deaths are related to unclean water and poor sanitation [156].

Although Millennium Development Goal (MDG) number 7 aims to halve by 2015 the proportion of people without sustainable access to safe drinking water and basic sanitation, the situation is getting worse. Water scarcity is an increasing problem, as needs for water rise along with population growth, urbanization, and higher industrial and domestic water demand. On national and local scales, there is a need for appropriately funded infrastructure and robust governance mechanisms in order to protect water resources and ensure a sustainable development and distribution of water-derived benefits. Initiatives are emerging worldwide to attain this objective. For that reason, wastewater treatment, reuse, and reclamation represent key practices in that approach [158].

More than one-third of the planet's accessible renewable freshwater is used for agricultural, industrial, and municipal purposes. Most of the activities eventually lead to the contamination of freshwater systems worldwide with thousands of industrial and natural chemical compounds. Some 2 million tons of waste, derived from all industrial and human activities (e.g., synthetic compounds used in industrial and consumer products; diffuse agricultural pollution from the application of fertilizers and pesticides; and oil and gasoline components through accidental spills) are released into receiving waters everyday with incomplete or no treatment at all (Figure 1) [139,157].

During the past decade, the occurrence of organic micropollutants in the environment has attracted great interest since a generalized concern arise about the possible undesirable effects of many of these pollutants in the environment

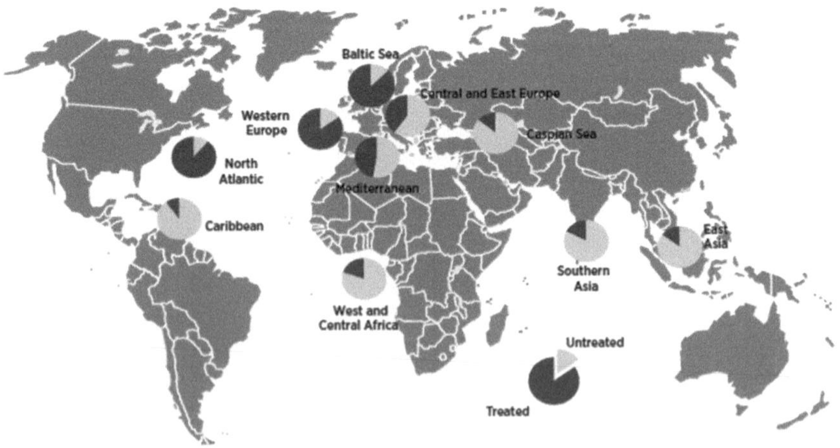

FIGURE 1 Ratio of treated to untreated wastewater discharged into water bodies. *Adaptation of [158].*

and on living organisms [34,91]. The trace pollutants referred to as "emerging organic contaminants" (EOCs) mainly comprise a group of man-made compounds such as pharmaceutical and personal care products (PPCPs), pesticides, plasticizers, and antiseptics that are continuously discharged into the environment as a result of consumer activities, waste disposal, accidental releases, and purposeful introduction [35].

One of the main sources of EOCs into the environment is the discharge of effluents from wastewater treatment plants (WWTPs), where their removal is usually incomplete [66]. Although a whole array of advanced treatment technologies have lately appeared aiming at the elimination of EOCs, such as ozonation, chemical advanced oxidation, or ultraviolet (UV) radiation [83,94,136] and have shown to enhance their elimination, the cost of these treatments is oftentimes too expensive and not justified under the current concept and standards of sanitation and wastewater treatment. Moreover, they could lead to the generation of transformation products being in some cases more persistent or toxic than the parental compound [50].

To this regard, decentralized, extensive low-cost treatment technologies, such as constructed wetlands, emerge as a great alternative for wastewater treatment and reuse, which require almost no maintenance and energy consumption. Constructed wetlands (CWs) are natural wastewater treatment systems that emphasize the processes happening in natural wetlands in order to improve their treatment capacity [78]. They constitute an alternative cost-effective technology to conventional WWTPs, in particular in the context of small communities with less than 2000 people equivalent (PE) [129]. Apart from the removal of conventional wastewater pollutants (e.g., biochemical oxygen demand, total suspended solids, nitrogen, and phosphorus) and heavy metals, CWs have proven to have a great potential for the removal of EOCs. The existing literature includes mostly studies on the occurrence and behavior of pharmaceuticals at horizontal subsurface flow (HF) wetlands at micro- and mesoscale [7,8,41,69,70,104,131,178,179], as well as pilot-scale [103,106]. Moreover, several studies have been carried out in vertical subsurface flow (VF) systems [100,101,143] and others at full-scale tertiary treatment systems, which include the use of a free water surface (FWS) wetland [68,95,107,108]. A couple of other examples comprise the use of a CW after a previous treatment system like a conventional WWTP or an anaerobic digester [132,160], and the use of a three-stage hybrid constructed wetland systems based on different wetland types [4,5]. As an example, an aerial view of a surface CW system, acting as a tertiary treatment unit after a conventional activated-sludge unit in Spain, which was evaluated by [107] can be seen in Figure 2.

With the purpose of providing further insight about the occurrence and possible undesirable effects of PPCPs in the environment, as well as the technologies that could avoid their release into the aquatic environment, in this chapter we describe the current knowledge on their source, fate, and removal in conventional WWTPs and tertiary treatment units.

Finally, this chapter reviews the efficiency of CWs at removing PPCPs from domestic wastewater.

FIGURE 2 Aerial view of the activated sludge wastewater treatment plant (WWTP) and tertiary treatment constructed wetland (CW) system at the Natural Park of Aiguamolls de l'Empordà, Girona, Spain. *Photo: Consorci Costa Brava.*

1.1 Emerging Organic Contaminants

The EOCs are, in most cases, unregulated contaminants whose emissions have emerged as an environmental problem. They comprise a diverse collection of thousands of chemical substances, such as pharmaceuticals, personal care products (PCPs), pesticides, hormones, surfactants, flame retardants, plasticizers, industrial additives, and agents, as well as their transformation products [49]. The steady increase in the manufacture and consumption of this type of compounds [23], together with the development of new analytical techniques that have allowed their detection [48,61], and attracted the attention and concern of the scientific community around a decade ago [63]. Since then, a number of authors have done extensive research about the occurrence, fate, and ecotoxicological effects of these pollutants in the environment [12,21,34,89].

Europe stands as a pioneer in this field and many programs/projects are carried out, which investigate the occurrence, fate, and ecotoxicology of these pollutants in all the different compartments of the aquatic environment, as well as potential solutions concerning the elimination of these in wastewater. Some examples of finalized projects are:

- POSEIDON (2001–2004): Assessment of Technologies for the Removal of Pharmaceuticals and Personal Care Products in Sewage and Drinking Water Facilities to Improve Indirect Potable Water Reuse
- P-THREE (2002–2005): Removal of Persistent Polar Pollutants through Improved Treatment of Wastewater Effluents

- EMCO (2004–2007): Reduction of Environmental Risks, Posed by Emerging Contaminants, through Advanced Treatment of Municipal and Industrial Wastes
- ERAPharm (2004–2007): Environmental Risk Assessment of Pharmaceuticals
- KNAPPE (2007–2008): Knowledge and Need Assessment on Pharmaceutical Products in Environmental Waters
- PILLS (2009–2012): Pharmaceutical Input and Elimination from Local Sources

In September 2005 the NORMAN network (network of reference laboratories for monitoring of emerging environmental pollutants) was created with the financial support of the European Commission, with the purpose of being a self-sustaining network of reference laboratories, research centers, and related organizations around Europe for the monitoring and biomonitoring of EOCs. The aim is to enhance the exchange of information, to harmonize methodologies, and to stimulate coordinated, interdisciplinary projects on problem-oriented research to address identified needs.

Likewise, the Environmental Protection Agency (EPA, USA) has also made great efforts to raise attention about the potential problems derived from the ubiquity of pharmaceuticals in the environment, not only asking scientist to increase the research on the field, but also calling the attention of authorities, healthcare providers, pharmaceutical companies, and patients [36,38]. The ultimate goal is to develop mitigation practices to tackle this environmental problem and to avoid the risk associated with these pollutants [11].

Although major strides have been made in the past years in increasing knowledge about the matter, one of the main conclusions at this point is that knowledge about emerging contaminants' fate and impact in the environment and living organisms is fragmentary, and often dealing with only a part of the problem, therefore much remains to be understood [87].

Of the rich array of EOCs, the ones receiving the greatest attention up-to-date are the following: antibiotics, due to the increase in antibiotic-resistance genes in aquatic microbiota, and potentially human pathogens [39,86,124,135]; X-ray contrast media, which have shown high persistence in the environment; cytostatic drugs, which, being majorly used in the chemotherapy of oncological patients, are designed to prevent the growth and proliferation of cells, and thus are considered highly hazardous compounds due to their genotoxic properties [175]; and finally, endocrine disrupting compounds (EDCs) such as estrogens, polychlorinated biphenyls (PCBs), and organochlorine pesticides (e.g., DDT), among others, which have been found to induce disruption of the reproductive system, female and male reproductive health in humans and wildlife, sex ratio (in wild fish and mollusks), neurodevelopment in children and wildlife, as well as to cause hormone-related cancers, bone disorders, metabolic disorders, and even immune function diseases and disorders in humans and wildlife [15].

Daughton [36] conducted a great discussion about the universe of chemicals that occur in the environment, both known (regulated or not) and unknown, so as to put efforts in widening the understanding of the exposure and risk of chemicals in ecosystems and human health. A large portion of the chemicals occurring in analyzed water samples are not (or cannot be) identified, due to limitations in the repertoire of available analysis tools. Every day, and for decades now, a whole array of new drugs are released into the environment with mechanisms of action never before encountered by biological systems, and only a very small percentage of commercially used contaminants are being investigated. Definitively, industrial and technological advances with respect to the production of chemicals pose substantial challenges to the evolution and design of regulatory practices.

The limited information on the occurrence, fate, and effects of EOCs makes it difficult for authorities to regulate their disposal and limit their concentrations. Some groups of compounds have received regulation, such as pesticides and other priority pollutants [43]. Moreover, the European Commission is working on the preparation of a candidate list of 553 EDCs so as to prioritize further detailed review of these. In this list, the industrial chemical bisphenol A (monomer of hard plastics and epoxy resins) and the estrogenic steroid hormone 17α-ethinylestradiol are included, among others, together with many new pesticides and other industrial chemicals [46]. This list is subjected to changes in response to developments in scientific knowledge or changes in patterns of chemical usage. On the other hand, in the USA the EPA holds a list of compounds called the Contaminant Candidate List (CCL), which comprises a list of contaminants that although are not subject to any drinking water regulation, are being monitored with the aim of possibly being included in future regulations. The last one includes several EDCs, and some flame retardants [183].

However, there are no laws, acts, or regulations of any kind, which include guidelines for EOCs concentrations in WWTPs, drinking water facilities, or the environment. The lack of regulations allows unlimited and widespread discharge of EOCs in the environment, which have unknown consequences for the environment and public health. Efforts should be harmonized and integrated between the different actors involved in EOCs' life cycle, as well as researchers and public authorities, so as to obtain more productive outcomes and understanding of the problem, and to take action to limit their presence in the environment [36,87,139].

1.1.1 Pharmaceuticals and Personal Care Products

There are a large number of EOCs that are continuously released to the environment due to their wide use in society. A subclass underneath this class is PPCPs, which in turn comprise an extraordinarily diverse group of chemicals used in veterinary medicine, agricultural practice, human health, and cosmetic care.

PPCPs include prescription and non-prescription medications, nutritional supplements, diagnostic agents, as well as other consumer products such as disinfectants, fragrances, sunscreen, and cosmetics. Even though most of them are polar, have a short half-lifetime in water, and are found in trace concentrations, they are considered pseudo-persistent pollutants [12]. This is because their universal, frequent usage by multitude of individual and animals causes a continuous discharge into the environment oftentimes sourcing from either untreated or insufficient wastewater treatment.

The molecular structures of PPCPs are typically large and complex, and differ a lot among substances, containing functional groups such as hydroxyl, carboxyl, ketone, and amine. It is important to note that the classification of pharmaceuticals by their active substances within subgroups of pharmaceuticals does not imply that they follow a certain chemical behavior. Small changes in the chemical structure may have significant effects on solubility and polarity, and other properties, which will in turn determine their environmental distribution in air, water, sediments, soils, and animals [91].

One of the physicochemical properties of organic compounds most applied in the field of environmental chemistry is the *n*-octanol–water partition coefficient (K_{ow}), which is defined as the ratio of the compound's concentration in a known volume of *n*-octanol to its concentration in a known volume of water at equilibrium and at constant temperature. K_{ow} values are constant for a given compound and reflect the lipophilicity of a compound. Hence, the higher the K_{ow} of a chemical, the higher its lipophilicity. K_{ow} values are generally directly proportional to molecular weight.

K_{ow} values can be measured in the laboratory with shake-flask systems, or can be estimated by models. K_{ow} values have been measured for a wide variety of chemicals and range from about 0.001 to over 10,000,000. Since the range of K_{ow} values encompass 10 orders of magnitude, log K_{ow} values are used instead, and range −3 to 7. A substance is usually considered lipophilic at log K_{ow} values ranging from 3-4 to 6.5 ($K_{ow} > 100$) [29].

However, K_{ow} assays are done adjusting the pH of the aqueous phase so that the predominant form of the compound is unionized. In this way, although some substances could be expected to be fairly lipophilic when having log K_{ow} values ranging 3-4, the occurrence of carboxylic groups or other polar groups in their structure are prone to be ionized and thus, increasing their hydrophilicity. Since many pharmaceuticals are ionizable at environmental pH conditions, log K_{ow} might not always be an appropriate predictor of the hydrophobicity of a compound. In particular, most nonsteroidal anti-inflammatory drugs (NSAIDs) are ionized at environmental pH, and behave as highly polar compounds. Hence, for weak acids the hydrophobicity also depends on its pK_a and the occurring pH of the media where they are contained (Eqn (1)).

The ionization constant (K_a) is a measure of (weak) acid strength as expressed by the concentration of ionized molecules divided by the concentration of unionized molecules. In this way, the lower the pK_a, the higher the K_a and the stronger the acid:

$$\log [A^-]/[AH] = pH - pKa \quad (1)$$

Hence, for a given pK_a of a compound, the higher the pH, the higher the ionization of the compound.

To this regard, the octanol–water distribution coefficient (D_{ow}), which is defined as the ratio of the sum of all forms of the compound (ionized + unionized) in each of the two phases, constitutes a better descriptor for ionizable compounds. In this case the pH of the aqueous phase is buffered to a specific value such that the pH is not significantly perturbed by the introduction of the compound. The log D_{ow} is pH dependent, and hence one must specify the pH at which the parameter was measured. For unionizable compounds, log K_{ow} = log D_{ow} at any pH, however for many pharmaceuticals such as NSAIDs (e.g., ibuprofen and diclofenac), the molecule occurs mainly in its ionized form and thus holds a hydrophilic behavior at environmental pHs (log D_{ow} = 1.1 for both substances at pH = 8).

In the case of PCPs, fragrances are generally semi-volatile organic compounds, fairly hydrophobic substances, which have been for many centuries used for multiple purposes (e.g., perfumes, deodorants, washing and cleaning agents, and cosmetics). Typical concentrations of fragrances range from 0.3% in face creams to as much as 20% in perfume extracts [22]. Most of the fragrances used nowadays are synthesized by industry, and nitro musks and polycyclic musks are the two main categories. The latter are the major musks used today, accounting for almost two-thirds of worldwide production, with an estimated worldwide production volume of 6000 t/year [133]. Tonalide (AHTN, 6-acetyl-1,1,2,4,4,7-hexamethyltetraline) is one of the most widely used polycyclic musks around the world and is used in cosmetics and detergents. It is designed to be very lipophilic in order to sorb to organic materials such as the skin, and in consequence is poorly soluble in water, as indicated by its high octanol–water coefficient (log K_{ow} = 5.9). Moreover, oxybenzone is an organic UV filter used in sunscreen and other cosmetics. Another commonly found PCP is triclosan, an antibacterial and antifungal agent present in soaps, deodorants, toothpastes, shaving creams, and in a number of consumer products [109]. PCPs differ from pharmaceuticals in that large amounts can be directly introduced in the environment, by direct release into recreational waters or volatilized into the air (e.g., musks).

1.1.2 Transformation Products

The compounds resulting from the structural change of pharmaceuticals within the human body are called metabolites. As with metabolism, when a substance is introduced to the environment or a treatment plant, it can undergo different structural changes by a variety of biotic (e.g., biodegradation and biotransformation) and non-biotic processes (e.g., phototransformation and hydrolysis) resulting in partial or complete transformation of the original compound. This results in changes in their physicochemical properties, which in turn affects their behavior and toxicity. Although the transformation of parent compounds could be expected to reduce toxicity, in some cases transformation products could have higher persistence and/or more harmful effects than the parent substance. This has particularly been found during natural and technical photolytic processes and advanced oxidation processes (AOPs) [14,50,91].

Little is known about the degradation pathways and products of most PPCPs in the environment, and less is known about their intermediates of degradation, which have just started to be explored [49,111]. Gómez et al. [60] have recently found transformation products of various PPCPs at the Henares River basin (central Spain) at concentrations much higher than those of their parent compounds. Various transformation products of ibuprofen and other pharmaceuticals were identified by [130] in municipal wastewater treated by a membrane bioreactor.

These findings highlight the importance of including degradation products in monitoring and in research programs so as to have a better understanding of the behavior and potential toxicology of PPCPs in the environment. However, it is important to have in mind that given the universe of chemicals that are released every day into the environment and, the difficulty and expense involved in their analysis, it would be impossible (if not infinite) to take them all into consideration for research programs or rutinary monitoring. Instead, well-balanced efforts must be put based on predicted risks associated to each specific contaminant in order to find a compromise solution.

1.2 Sources and Fate of PPCPs in the Environment

Contaminants enter the environment through multitude of sources, whose components are depicted in Figure 3.

Consumption of PPCPs is high worldwide. As an example of human consumption, it is worth noting how the consumption of NSAIDs in Spain raised from 26.30 DHD (defined daily doses/1000 inhabitants-day) to 45.81 DHD from 1992 to 2006 [1]. Of the NSAIDs, one of the compounds attaining the greatest increase was ibuprofen, whose consumption increased from 0.29 DHD in 1992 up to 21.30 in 2006. The NSAID DCF (commonly known as voltaren) however showed rather stable values of consumption overtime, averaging 7 DHD. According to IMS Health data 942 tons of the anticonvulsant carbamazepine and 877 tons of DCF were sold in 2007 in 76 major countries, which are believed to account for 96% of the global pharmaceutical market. Antibiotic consumption did not fluctuate a lot during the last decade in Spain as is shown in the report of [2]. Values ranged 0.1 DHD for most studied quinolones up to a maximum value of 6.6 DHD found for the penicillin amoxicilin. As can be seen in the report by the OECD (Organization for Economic Cooperation and Development), most countries in the world have experimented a significant increase

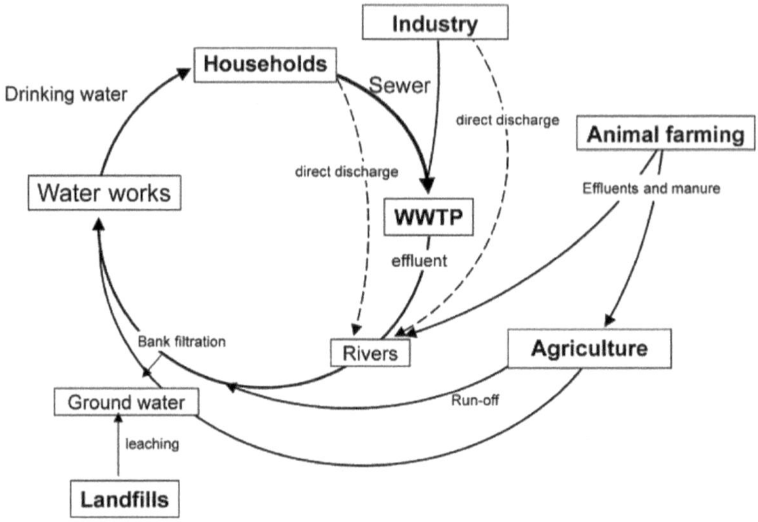

FIGURE 3 Sources and fate of contaminants in the (partially) closed water cycle with indirect potable reuse. *From Ref. [127].*

in consumption of pharmaceuticals in the last decade [122]. To this regard, Daughton and Ruhoy [37] have called to note the importance of prescribing the "lowest effective dose" of pharmaceuticals on patients as a strategy to diminish PPCP concentrations in the environment and eventually avoid potential adverse environmental impacts.

Once pharmaceuticals are ingested and a fraction metabolized, they are excreted and reach the urban sewer network up to the WWTPs. The current treatment processes applied for municipal wastewater treatment often fail to completely remove PPCPs, which leads to their subsequent release in the terrestrial and aquatic ecosystems through disposal and reuse applications.

On the other hand, the second major source of PPCPs in the environment after sewage effluent discharge is terrestrial run-off derived from the agri-food (and especially meat) industry. To have an idea of the levels of consumption of antibiotics on livestock farms, while human antibiotic use in the USA is maintained below 8 billion pounds per year, livestock farms have been increasing their use, constituting to date nearly four-fifths of the total amount of antibiotics used in the USA [52].

A significant source of concern for PPCPs in groundwater is the leaching from municipal landfills. Improper disposal of unused or expired drugs (either through the toilet or landfill) constitute another source of these contaminants in the environment. Depending on the physicochemical characteristics of the organic contaminant (and their transformation products) and the characteristics of the soil, PPCPs can either reach groundwater and pollute aquifers or remain retained within the soil and eventually accumulate, thus affecting the ecosystem and human beings through the food web. As a matter of fact, PPCPs have been detected at trace concentrations in the freshwater environment for decades [63,76,89,96,97]. As a particular example, Kim et al. [82] documented the frequent detection of many pharmaceuticals, hormones, antibiotics, as well as flame retardants in surface waters in South Korea. Various EDCs, as well as associated toxicological effects over caged male fathead minnows, were found in 90% of sediment and water samples of 11 studied lakes across Minnesota, USA [168]. Stackelberg et al. [145] found pharmaceuticals compounds and other organic wastewater contaminants in a conventional drinking-water-treatment plant before and after treatment. Pharmaceuticals were also found at the ng/L level in the tap water of Berlin [66]. The authors attribute this phenomenon to be characteristic of conurbations like Berlin, with high municipal wastewater discharges and low surface water flows, when groundwater recharge is used in drinking water production.

Some other scattered examples, which show the ubiquity of the occurrence of PPCPs in various compartments of the aquatic environment around the world, are displayed in Table 1.

Concentrations of PPCPs in surface waters (primarily due to their incomplete elimination in WWTPs) as well as groundwater (due to filtration through the soil) are in the range of ng/L to μg/L. The different pathways that PPCPs can undertake from resource up to drinking water level are explained in detail in the study by [63]. In general, the polar nature and low volatility of PPCPs prevents their escape from the aquatic realm, being primarily distributed through aqueous transport and food chain dispersal [38]. Although the half-life in aquatic ecosystem for most PPCPs is relatively low [169], some substances are fairly recalcitrant and may accumulate in soil, reaching concentrations of g/kg. As an example, clofibric acid (a compound used for reducing cholesterol levels in blood) has been found to have a half-life of 21 years in soil [67]. In relation to this, the application of sewage sludge could eventually contribute to soil, surface, and groundwater contamination, either by accumulation of these contaminants in soil in the long term, or by possible leaching and migration of contaminants through the soil zone [26,147]. Another major pathway of groundwater contamination is due to groundwater–surface water interaction through infiltration from losing reaches of rivers, especially near industrial premises and sewage works [147]. The contamination of the groundwater could pose serious challenges to safety regarding drinking water supply. A recent study by the GAO [55] showed that most drinking water samples grabbed in USA were contaminated by pharmaceuticals, mostly estrogen-based hormones and antibiotics.

Moreover, there is a clear need to include irrigation as an additional exposure route for PPCPs in terrestrial ecosystems, so as to assess the potential risks derived. Driven by the imposing water scarcity, irrigation through water reclamation is being widely implemented in Europe for multiple purposes, mainly agriculture, due to the advantages related to nutrient recovery and socioeconomic implications. The practice of wastewater reuse without proper

TABLE 1 Minimum, Maximum, and Average Concentrations (in ng/L) of Various PPCPs in Natural Water Bodies around the Globe

	Min	Max	Avg	Sample Source	Reference
Diclofenac	10	50	29	River Elbe and at the mouth of its tributaries (Germany)	[166]
Ibuprofen	<20	70	35		
Acetaminophen	<5	65	16		
Bisphenol A	16	100	40		
Ibuprofen	11	38	28	Han, Nakdong and Youngsan rivers (South Korea)	[82]
Acetaminophen	4	73	33		
Diclofenac	1	6.8	3.0		
Oxybenzone	1	3	2		
Acetaminophen	<LOD	712	–	Ebro River basin (Spain)	[98]
Triclosan	<LOD	8	n.a.	Surface waters of Greenwich Bay (USA)	[80]
Acetaminophen	2	17	5	Near-shore habitats of Lake Michigan—river discharges into the lake from predominantly urban watersheds (USA)	[53]
Ibuprofen	2	84	8		

Carbamazepine	2	118	40	Aquifers at the delta area of the Llobregat River (NE Spain)	[149]
Ibuprofen	<LOD	185	185		
Diclofenac	35	477	256		
Acetaminophen	–	1890	180	Groundwater used for drinking-water supply in California (USA)	[54]
Diclofenac	<LOD	1.2	0.2	Urban groundwater in the district of Poble Sec, Barcelona (Spain)	[98]
Ibuprofen	<LOD	2	0.2		
Acetaminophen	<LOD	<LOD	<LOD		
Erythromycin	<LOD	<LOD	<LOD		
Enrofloxacin	12	264	75		
Ibuprofen	<LOD	200	–	Berlin groundwater wells (Germany)	[66]
Diclofenac	<LOD	380	–		
Clofibric acid	<LOD	7300	–		

management poses potential risks to human health and the environment due to the potential occurrence of pathogens and chemical pollutants [50,116].

1.3 Toxicity and Evaluation of Risks Associated to PPCPs

The ubiquity of PPCPs in the environment has created a great concern among scientists due to the possible toxicological effects that these substances may pose on ecosystems and public health. In the last years numerous studies are evaluating the possible toxicological effects related to the occurrence of PPCPs and other emerging organic contaminants, mainly on aquatic organisms [184], as well as the management of risk [203] with the purpose of giving support to decision taking by stakeholders and eventually policy making.

Although occurring concentrations of PPCPs in the environment are oftentimes very low (ranging from ng/L to μg/L) and may not seem to pose an appreciable risk to human health, these are continuously released to the environment, constituting so-called "pseudo-persistent" contaminants. In many cases the possible consequences of their occurrence in the environment are not well understood, but in others the effects seem evident and alarming. As an example, the reduction of 90% of the population of vultures in India, Pakistan, and Nepal since the 1990s is related to the use of the anti-inflammatory drug diclofenac on cattle in these countries, and therefore its use as a veterinary drug was banned in 2005 in India [10,121].

What is more, musk fragrances are found to be ubiquitous, persistent, bioaccumulative pollutants in aquatic organisms that are sometimes highly toxic [38,175]. Amino musk transformation products are toxicologically significant. Polycyclic musks fragrances have been found to accumulate in several aquatic organisms, such as fish and mussels, as also in high trophic level aquatic organisms and human breast milk [185,186].

It is important to note that pharmaceuticals are designed to be biologically active, by modulating endocrine and immune systems and cellular signal transduction, in order to serve their therapeutic purposes. This in consequence will potentially affect the physiological and biochemical functions of biological systems and ecosystems. Moreover, some of them are estrogen hormones (such as ethynilestradiol), which belong to the category of EDCs, showing toxicity at ng/L levels [73,187]. However, PPCPs are quite diverse among them, and therefore they may have very different effects over ecosystems [63]. Although PPCPs have received less scrutiny than pesticides (many are EDCs of concern), these differ from the latter in that the disposal of PPCPs is widespread, long-term, and in lower doses. Thus, aquatic organisms receive a more chronic exposure.

Although little is known about the occurrence and effects of pharmaceuticals in the environment, more data exist for antibiotics than for any other therapeutic class. This is a result of their extensive use in human therapy and animal husbandry, as well as plant agriculture and aquaculture. It has become clear that the intense use and misuse of antibiotics has increased and spread the occurrence of highly resistant pathogenic bacteria. There is real concern about the effects on aquatic microbiota and consequent alteration of the structure of the microbial community, as well as on the development of resistance in potential human pathogens. Although occurrent bacteria in freshwater sources are commonly non-pathogenic, there could be exchange of genetic material between pathogenic and non-pathogenic bacteria, conferring on them increased antibiotic resistance [188–193].

Research efforts aiming at elucidating the potential toxicological significance of PPCPs in ecosystems and eventually in public health requires documentation of contaminant uptake, modes of action, and biological end points, as well as exposure. Most of the published literature is based on acute effects on aquatic organisms, and in hazard and risk assessments. Assessment factors are applied to the acute effects data for extrapolation to chronic (long-term) effects. This approach is oftentimes nonrealistic, given that the ratio between acute and chronic toxicity shows to differ depending on the PPCP [34,184]. In general, PPCPs do not show high acute toxicity, but chronic toxicity and potential subtle effects are marginally known.

As an example, Cleuvers (2003) found how the NSAIDs, ibuprofen and diclofenac, and the anticonvulsant carbamazepine were toxic for algae, and discuss that although acute effects of single substances in the aquatic environment are very unlikely due to the small concentrations, a combination effect of several compounds and/or their transformation products can occur. [194] evaluated acute and chronic toxicity data available for PCPs, including the antiseptic triclosan and the musk fragrance tonalide (AHTN). Only TCS and triclocarban presented any hazard. Moreover, [195] assessed the ecotoxicity of various micropollutants on macroalgae and crustaceans in a FWS wetland in Sweden and reported a good quality of the effluent water, comparable to that treated under advanced tertiary treatment processes.

Conversely, [184] found how few substances such as DCF were the compounds with the highest acute toxicity within the class of NSAIDs, and their average concentrations found at the effluent of conventional WWTPs were in the range of those being toxic for fish. What is more, Lai et al. [196] using a food-web model found that natural and synthetic estrogens, such as the hormone ethynilestradiol, could be bioaccumulated in fish in river systems, although to a lesser extent than other EDCs such as DDT [196]. Likewise, AHTN and other fragrances, which are made hydrophobic so as to adsorb to tissue might accumulate in aquatic biota [133].

Although acute toxicity in the aquatic systems due to the occurrence of PPCPs is unlikely to occur, there is a lack of chronic toxicity data, and effects from long-term, low-level environmental exposures should be further addressed, so as to have a better understanding of the environmental risk of these compounds. What is more, since pharmaceuticals are not present alone in the environment, but as multicomponent mixtures, an accurate prediction of the chronic mixture toxicity is indispensable for an environmental risk assessment. Significant synergistic toxic effects may occur through additive exposures; therefore they should not be handled in isolation. The effects and interaction of a cocktail of commonly used pharmaceuticals, including CBZ, IB, and the antibiotic sulfamethoxazole were studied using in vitro tests on human and zebrafish cells by Pomati et al. [197], and they found that the mixture of drugs at ng/L can inhibit cell proliferation on aquatic life. However, the research in this new field of ecotoxicology is just starting and much remains to be learned [91,147].

Another issue of very recent concern is the inhibition of multixenobiotic resistance (MXR) in aquatic organisms, which has especially been found in environmental samples from polluted locations. This mechanism is used as a "first line of defense" against endogenous and exogenous potentially toxic xenobiotics, which are expelled out of the cell by various transmembrane transport proteins. It is still unclear whether the so-called MXR inhibitors or chemosensitizers [198] are natural or man-made substances. It is also unknown whether these are constituted by a few powerful MXR inhibitors, or if the inhibition is caused by the mere presence of a large number of chemicals in polluted water. It is also discussed whether these substances could play key roles in potentiating the effects of other xenobiotics [38]. Research on this topic has also just been initiated, such as the study of the inhibition of MXR by musk fragrances [199], and still multiple aspects need to be addressed on this topic [200].

Even though ecological effects caused by PPCPs may not be manifest, it does not mean that they do not exist. Potential subtle effects of these substances over time may occur [38]. In order to have a better understanding of mechanisms of action of these contaminants in the environment, the target- or biomolecule-oriented, or mode-of-action-based investigation activities are preferred over traditional standard ecotoxicological assays. In general, assessing the impact of PPCPs in aquatic systems remains a major challenge, requiring improved analytical and modeling tools, the development of methods to classify existing and new chemicals on the basis of their potential to harm humans and ecosystems, and the development of attenuation technologies and other strategies that minimize their introduction into the environment [139].

1.4 Occurrence and Fate of PPCPs in Conventional WWTPs

The occurrence and fate of PPCPs have been extensively assessed in many conventional wastewater treatment technologies, such as activated sludge systems, which are oftentimes unable to degrade them, and thus constitute their major source point into the environment. The activated sludge process is the most common type of secondary treatment used in municipal WWTPs worldwide. It consists of a two-stage suspended growth biological treatment process designed to majorly remove organic matter. The first stage is comprised of an aerated reactor in which the organic matter is eliminated by a mixed microbial population, while the second stage consists of a settling tank or clarifier that separates solids (activated sludge) from water. A portion of the activated sludge is wasted while the rest is returned to the reactor. While hydraulic retention time of the water is of about 8 h, for conventional activated sludge, the average solids retention time is 5–10 days. Conventional WWTPs are an intensive wastewater treatment technology. They are able to treat great amounts of water, occupying a relatively small surface area [148].

A compilation of results of various studies including influent and effluent concentration of PPCPs at conventional WWTPs based on activated sludge is displayed in Table 2.

NSAIDs take part of a therapeutical group widely used nowadays, and consequently acetaminophen and ibuprofen are the ones being found at higher concentration levels in raw wastewater. In the lower concentration levels of PPCPs in the environment have been reported the detection of steroid hormones, generally at low ng/L and oftentimes below detection limits [25,38]. However, it should be noted that in spite of the low concentrations, estrogens could induce serious chronic effects, such as endocrine disruption [73]. The occurrence of antibiotics in WWTPs has also been reported in various studies [170].

Activated sludge WWTPs are not specifically designed to eliminate PPCPs and thus removal rates are very variable and compound dependent. A recent report by the EPA includes a review of recent literature on wastewater treatment technologies and their ability to remove 16 emerging organic contaminants, including many PPCPs. Average removal efficiencies were calculated from a collection of as many as 41 full-scale WWTPs based on activated sludge, and removal rates for various PPCPs are displayed in Table 3. Note that the treatment systems used to obtain the presented data on the table did not include activated sludge systems that reported design modifications from the conventional activated sludge WWTPs (e.g., those that aim to remove nutrients). Although there are many variations of this process, further division of activated sludge categories was impractical [47].

Treatment performance was found to vary among the conventional WWTPs under study, majorly as a function of the characteristics of inflowing wastewater. Average removal efficiencies were very variable and compound dependent.

PPCPs may be removed from wastewater during activated sludge treatment by mechanisms of biodegradation and/or adsorption to the particulate matter. Some compounds undergo rapid biodegradation (e.g., acetaminophen) in conventional WWTPs [90,112,136,173]. Likewise, ibuprofen and triclosan are consistently removed in most treatment plants [27,93]. However, some compounds exhibit variable removal (e.g., diclofenac), while others, such as carbamazepine, have shown to be very recalcitrant [90]. In fact, in some occasions higher concentrations have been found in the effluent than

TABLE 2 Minimum, Maximum, and (Average) Concentrations of Various PPCPs (in ng/L) and Percentage of Detection in Influent and Effluents of Various Conventional WWTPs around the World

Compound	WWTPs Influent	% Detection	WWTP Effluent	% Detection	Country	Ref.
Ibuprofen	1200–2679 (2021)	n.a.	<LOD–2400 (489)	n.a.	Austria	[27]
	968–2986 (1681)	100	131–424 (263)	100	UK	[79]
	2800–25,400 (12,500)	n.a.	500–2600 (1500)	n.a.	Greece	[90]
	Max: 16,500 (8450)	n.a.	Max: 773 (384)	n.a.	Canada	[93]
	170–83,500 (14,600)	97	2–24,600 (1960)	93	n.a.	[112]
	3730–353,000 (69,700)	n.a.	<LOD–26,500 (4130)	n.a.	Spain	[137]
	(1900)	n.a.	(250)	n.a.	USA	[173]
Diclofenac	905–4114 (2572)	n.a.	780–1680 (1366)	n.a.	Austria	[27]
	57–1161 (260)	100	6–496 (179)	100	UK	[79]
	<LOD–3900 (2000)	n.a.	<LOD–2600 (1300)	n.a.	Greece	[90]
	Max: 1010 (204)	n.a.	Max: 748 (194)	n.a.	Canada	[93]
	105–4110 (1340)	81	35–1950 (680)	85	n.a.	[112]

	<LOD	0	<LOD	0	Spain	[137]
Acetaminophen	68,107–482,687 (211,380)	100	1826–24,525 (11,733)	100	UK	[79]
	108,383–246,641 (178,116)	100	>80–1575 (353)	86	UK	[79]
	4700–52,500 (20,600)	n.a.	500–1700 (900)	n.a.	Greece	[90]
	1571–37,458 (23,202)	n.a.	<LOQ	n.a.	Spain	[136]
	(960)	n.a.	<LOD	n.a.	USA	[173]
Tonalide	210–1106 (760)	n.a.	144–170 (158)	n.a.	Austria	[27]
	Max: 2000 (804)	n.a.	Max: 600 (274)	n.a.	Canada	[93]
	210–1690 (990)	100	144–200 (162)	100	n.a.	[112]
Oxybenzone	n.a.	n.a.	1–30 (11)	71	South Korea	[82]
Triclosan	33–463 (228)	100	13–82 (57)	100	UK	[79]
	<LOD–1000 (800)	n.a.	<LOD	n.a.	Greece	[90]
	4010 (1930)	n.a.	Max: 324 (108)	n.a.	Canada	[93]

Continued

TABLE 2 Minimum, Maximum, and (Average) Concentrations of Various PPCPs (in ng/L) and Percentage of Detection in Influent and Effluents of Various Conventional WWTPs around the World—cont'd

Compound	WWTPs Influent	% Detection	WWTP Effluent	% Detection	Country	Ref.
	1520–4430	n.a.	(810)	n.a.	USA	[80]
	<LOQ–2417 (860)	n.a.	<LOQ–512 (219)	n.a.	Spain	[136]
Ethynilestradiol	0.4–70 (4.2)	91	0.2–5 (0.9)	59	n.a.	[112]
	<LOD	0	<LOD	0	Sweden	[181]
	n.a.	n.a.	1.3	14	South Korea	[82]

TABLE 3 Minimum, Maximum, and Average Removal Efficiencies of Various PPCPs at Full-Scale Activated Sludge WWTPs Treating Municipal Wastewater

	Removal Efficiency (%)			
Compound	**Min**	**Max**	**Avg**	**# of WWTPs Used to Calculate Removal**
Ibuprofen	43	100	90	32
Acetaminophen	>90	100	97	4
Diclofenac	7	>99	44	23
Tonalide	13	97	67	20
Oxybenzone	>8	>96	76	6
Triclosan	>67	100	89	22
Ethynilestradiol	>1	>99	66	13

Adapted from Ref. [47].

in the influent, which is attributed to desorption and/or deconjugation processes [79,181]. The more hydrophobic substances may attach to suspended solids or activated sludge. In this respect, Carballa et al. [24] found how compounds with high sorption properties such as tonalide and diclofenac were significantly removed during coagulation–flocculation assays encompassing different doses and types of coagulants at different temperatures, whereas hydrophilic compounds (e.g., ibuprofen and carbamazepine) were not affected. Conversely, Jelić et al. [75] found no accumulation of diclofenac in sludge samples from three conventional WWTPs.

Treated effluent of WWTPs can either be disposed in surface waters or be reused for multiple applications. Additionally, the waste sludge, after its drying and treatment, could be applied to the land as fertilizers. In this way, remaining concentrations of PPCPs may transfer to soil and surface waters.

1.5 Removal of PPCPs through Tertiary and Advanced Treatment Technologies

Since conventional activated-sludge WWTPs have shown to be unable to completely eliminate PPCPs, as well as other specific and priority substances included in the Water Framework Directive [45], a number of advanced (or tertiary) treatment technologies, which can provide further removal of these substances are being explored and developed lately in this sense. Already, a whole array of oftentimes fairly standardized unit processes is available so as to remove PPCPs by transformation or removal by physical and chemical methods, including adsorption or filtration. These include activated carbon [142,150], ozonation [136], photo-Fenton [134], UV radiation [83], ultrasonic irradiation [117], chemical AOPs (involving hydroxyl radicals), which generally use a combination of oxidization agents (e.g., H_2O_2, O_3, and TiO_2), irradiation (UV or ultrasound), and catalysts [85,94]; chlorine dioxide oxidation [72]; membrane technologies [142,172]; and membrane bioreactors [84,130,140]. As an example the anti-inflammatory drug diclofenac has consistently shown to be well photodegraded under UV treatment [3,83,176]. This substance also gets well mineralized under ozonation [136] and chlorine disinfection [90].

Average removal efficiencies of various PPCPs achieved under different tertiary treatment technologies are displayed in Table 4 [47].

Although conventional tertiary or advanced treatment technologies have shown to enhance the elimination of these contaminants, they are very unlikely to be implemented in the context of wastewater treatment of small communities—even if the technology is improved—due to energy and economic constraints. To this regard, decentralized, extensive low-cost treatment technologies, such as constructed wetlands, emerge as a great alternative for wastewater treatment and reuse, which require almost no maintenance and energy consumption. Their successful implementation at full scale has shown their capacity to degrade all kind of contaminants, from conventional parameters to heavy metals or pathogens. Just in recent years their potential on the removal of PPCPs is being explored. The characteristics, advantages and disadvantages, and performance of CWs in the treatment of municipal wastewater are shown in the following section.

2. CWs: A DECENTRALIZED WASTEWATER TREATMENT ECOTECHNOLOGY

CWs are natural wastewater treatment systems that consist of a properly designed shallow basin, which contains a substrate that is planted with aquatic vegetation. Other components such as microorganisms and aquatic invertebrates develop naturally. These systems are constructed to mimic the microbiological, biological, physical, and chemical processes that occur in a natural wetland but in a more controlled environment to treat wastewater [78].

These natural treatment systems constitute a cost-effective and technically feasible approach for the treatment of wastewater for several reasons:

- Can be less expensive to build than other treatment options.
- Low operational and maintenance (O&M) expenses (energy and supplies). Easy to maintain (do not require highly qualified staff).
- Able to tolerate fluctuations in flow and load.

TABLE 4 Minimum, Maximum, and Average Removal Efficiencies of Various PPCPs under Different Advanced Treatment Technologies for Municipal Wastewater Treatment at Full Scale [47]

Compound	Technology	Min	Max	Avg (%)	# Systems Used to Calculate Removal
Ibuprofen	Chlorine disinfection	>43	100	78	5
	Ozonation	>90	100	95	2
	Reverse osmosis	–	–	72	1
	UV disinfection	>81	100	90	6
Acetaminophen	Chlorine disinfection	>90	>99	95	2
	Reverse osmosis	–	–	90	1
	UV disinfection	–	–	90	1
Diclofenac	Chlorine disinfection	>18	>90	66	3
	Reverse osmosis	–	–	90	1
	UV disinfection	>86	>91	89	3
Tonalide	Chlorine disinfection	64	93	79	2
	UV disinfection	–	–	52	1
Oxybenzone	Chlorine disinfection	>8	>95	51	2
	Reverse osmosis	–	–	95	1
	UV disinfection	>89	>96	92	3
Triclosan	Chlorine disinfection	>67	>99	83	4
	Ozonation	99	100	99	2
	Reverse osmosis	–	–	67	1
	UV disinfection	>71	>99	90	5
Ethynilestradiol	Chlorine disinfection	1	72	42	4

- Promote biodiversity. As an example, restored wetlands are becoming common in agricultural landscapes of northern Europe so as to increase nutrient retention as well as increasing species diversity [102,151].
- Can be easily integrated into the natural and rural landscape, and provide aesthetic, commercial, and habitat value [95].
- Possible commercial revenue from flower harvesting and/or aquaculture [74].

However, they have some limitations:

- Generally high land area requirement: they may be more economical than other technologies, but only if land is available and affordable. While activated sludge technology requires approximately 0.06 m^2/PE [32], vertical subsurface, horizontal subsurface, and FWS CWs require 2, 5, and 8 m^2/PE, respectively, to attain the same contaminant removal efficiencies [32,77,162].
- Poor maintenance or design can lead to excessive clogging, which can affect hydraulics and treatment performance [88].

Since the implementation of the Water Framework Directive 2000/60/EC [45], as well as the Directive 91/271/CEE concerning urban wastewater treatment [44], there is a need to comply with stringent regulations in many countries around Europe, especially for the treatment of small communities. This has attracted the focus to decentralized wastewater treatment ecotechnologies, given an increasing demand for economical, aesthetic, and ecologically sustainable treatment systems.

To this regard, CWs and other biological treatment technologies such as waste stabilization ponds, require very low to no energy consumption, and are easy to operate. Thus, they constitute a good alternative for wastewater treatment in small communities. Indeed, CWs have been extensively used at full scale across the world since the 1980s, for the treatment of not only municipal wastewater but all kinds of wastewater including stormwater, landfill leachate, tannery wastewater, and other industrial wastewaters [13,119,138,163]. They have been widely implemented in countries such as France [113], UK [32,165], Denmark [19], or Czech Republic [161], and their implementation is increasing in countries such as Spain [159] or Italy [99]. Numerous studies have shown their capability to maintain hydraulic, technical, economic, environmental, and ecological benefits [40,161,180].

There are several types of constructed wetlands. They can mainly be divided into subsurface flow (horizontal or vertical), surface flow constructed wetlands, and hybrid systems, which are a combination of different types.

2.1 Horizontal Subsurface Flow Constructed Wetlands

This type of wetlands consists of gravel beds planted with wetland vegetation, where the wastewater is intended to remain below the top of the substrate and to flow horizontally in and around the roots and rhizomes of the plants. These wetlands are always flooded and water depth usually remains between

0.3 and 0.9 m (between 0.05 and 0.1 m beneath the surface of the bed). The recommended organic loading rate(OLR) is around 6 g of biochemical oxygen demand (BOD_5)/m^2 d [57]. The biofilm that grows on the granular media and the roots and rhizomes of the vegetation plays a major role in the removal of contaminants from wastewater.

HF CWs are typically composed by: inlet piping, an impervious liner, filter media, emergent vegetation, and outlet piping. A diagram of a conventional HF wetland is depicted in Figure 4.

They are good for total suspended solids (TSS), BOD_5, and bacteria removal and denitrification. However, they are poor at nitrification because of limited oxygen transfer capacity (OTC) [30,155]. Since the water is not exposed, the risk associated with exposure to pathogenic organisms is minimized, and they do not provide suitable habitat for mosquitoes [78].

2.2 Vertical Subsurface Flow Constructed Wetlands

In this type of wetlands water is distributed across the surface of a sand or gravel bed planted with emergent vegetation, and the effluent is collected from the bottom of the media, where the water is freely draining. The treatment of the water occurs as it percolates through the filter media and the plant root zone.

There exist several variations of VF wetlands. In the most common design the bed is usually comprised of several layers of different grain size and contains a network of draining perforated pipes. They are usually fed intermittently, and frequently loading and resting periods are applied. This is the case of many systems in France, where two or three beds work alternatively. In the French VF systems, the wastewater is fed without a previous primary treatment [115], causing the accumulation of a layer of solids on the top of the bed, which in turn acts as a filter. The alternation of cycles of feed and rest promote mineralization of the solid deposits during resting phases [113]. In VF wetlands in general—usually comprised of a single bed—it is the special operating conditions

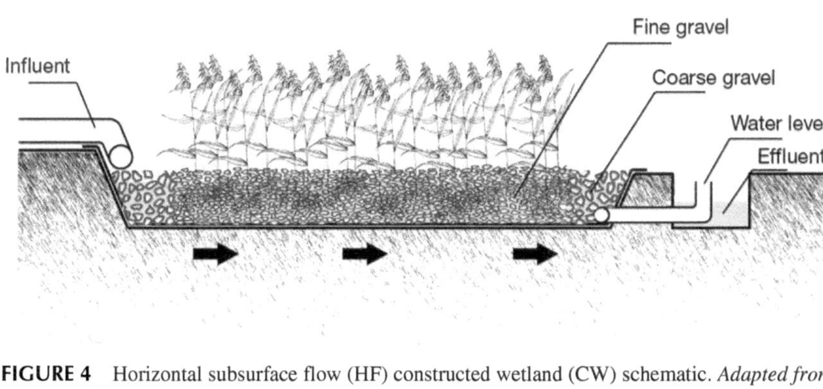

FIGURE 4 Horizontal subsurface flow (HF) constructed wetland (CW) schematic. *Adapted from Ref. [125].*

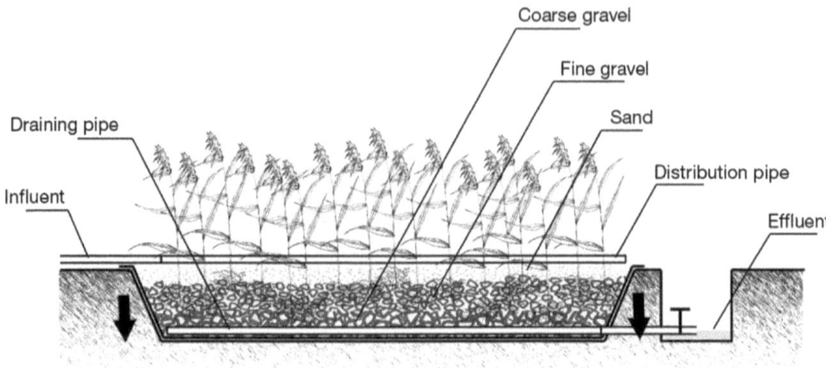

FIGURE 5 Basic elements of a vertical subsurface flow (VF) constructed wetland (CW). *Adapted from Ref. [125].*

(i.e., intermittent dosing and unsaturated bed) that allow a higher OLR to be applied in comparison to HF beds. Applied OLR values vary significantly from place to place and range from 20 g BOD_5/m^2d in Denmark [19] up to 180 g BOD_5/m^2d in French VF systems (just screened wastewater). The components of a typical VF wetland are shown in Figure 5.

The configuration of these systems confers very different properties to those of HF wetlands. They have a much greater OTC than HF beds, which makes them good for nitrification. Oxygen transfer is achieved by means of diluted oxygen present in wastewater, convection while intermittent loading, and diffusion processes occurring between doses [153]. The high OTC also leads to good removal of BOD_5 and chemical oxygen demand (COD); they are also able to remove some bacteria [65] and are also considerably smaller (about 2–3 m^2/PE) than HF systems (about 5 m^2/PE) [32,162]. Nevertheless, they are less good for suspended solids removal and, more specially for single VF systems were no resting periods are allowed, clogging of the media may occur rapidly if the granular material selection and the hydraulic loading rates (HLRs) are not correct [30,81,128].

2.3 Free Water Surface Wetlands

In this type of wetlands, the wastewater flows above the top of the media (with some kind of impervious layer underneath), which can be of many types (e.g., clay, soil, and gravel), since the wastewater is not intended to be filtered through it. These wetlands contain areas of open water, floating vegetation, and emerging plants, either by design or as an unavoidable consequence of the design configuration. A schematic of a conventional FWS wetland is depicted in Figure 6.

Because the water is exposed during the treatment process, it is common that these types of wetlands attract a variety of wildlife (i.e., insects, mollusks, fish, amphibians, reptiles, birds, and mammals) (Kadlec and Knight, 1996).

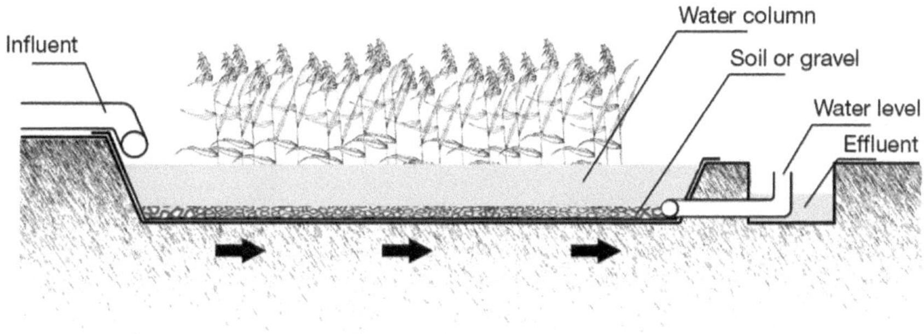

FIGURE 6 Free water surface (FWS) constructed wetland (CW) schematic. *Adapted from Ref. [125].*

Moreover, since they represent a risk for public health due to human exposure to pathogens, they are commonly applied for tertiary treatment of secondary or tertiary treatment effluents (e.g., activated sludge systems, waste stabilization ponds, and SSF wetlands). They are especially appropriate for restoration of deteriorated areas or attenuation of agricultural runoff at tertiary treatment level [102,108,151].

2.4 Hybrid Systems

Various types of CWs may be combined in order to achieve higher treatment efficiency, especially for nitrogen and pathogens. These hybrid systems are normally comprised of VF and HF systems arranged in different possible manners. While in HF wetlands nitrification is not achieved due to a lack of oxygen, VF wetlands can provide good conditions for nitrification but no denitrification occurs in these systems. Thenceforward, the strengths and weaknesses of each type of system balance each other out, and in consequence it is possible to obtain an effluent low in BOD and in total nitrogen (TN) concentrations [162]. Different combinations are possible, including HF followed by VF wetlands, VF followed by HF wetlands, and other stages of filters including water recirculation from one stage to another [19].

3. REMOVAL AND BEHAVIOR OF PPCPs IN CWs FOR URBAN WASTEWATER TREATMENT

The elimination of PPCPs in CWs takes place as a result of the simultaneous occurrence of various removal mechanisms, including biodegradation, adsorption, photodegradation, plant uptake, or hydrolysis. Some removal mechanisms will predominate in some wetland configurations in respect to others (e.g., photodegradation will only take place in FWS wetlands).

Average removal efficiencies of target PPCPs at several pilot and full-scale CW systems of different configurations are shown in Table 5.

TABLE 5 Average Removal Efficiencies (R.E.) of Various PPCPs at Different Pilot and Full-Scale CW Systems of Various Configurations

Compound	Range	Avg R.E. (%)	Type of Treatment	Ref.
Ibuprofen	95–96	96	Restoration FWS wetland (receives secondary effluent from a WWTP)	[95]
	48 (shallow)-81 (deep)	n.a.	Pilot-scale HF wetlands (different depths) with injection of contaminants	[106]
	n.a.	99	Experimental pilot-scale VF wetland (5 m^2)	[100]
	n.a.	<15	Restored wetland system fed with natural waters impacted by urban and agricultural runoff	[102]
	n.a.	>95	Pilot-scale VF wetland (5 m^2)	[6]
Diclofenac	n.a.	73	Experimental pilot-scale VF wetland (5 m^2)	[100]
	n.a.	<15	Restored wetland system fed with natural waters impacted by urban and agricultural runoff	[102]
	n.a.	65	Pilot-scale VF wetland (5 m^2)	[6]
	65–87	77	Three hybrid treatment systems (ponds and CWs)	[68]
	73–96	n.a.	FWS wetland (receives secondary effluent from WWTP)	[107]
	n.a.	21	Household HF wetland	[101]

TABLE 5 Average Removal Efficiencies (R.E.) of Various PPCPs at Different Pilot and Full-Scale CW Systems of Various Configurations—cont'd

Compound	Range	Avg R.E. (%)	Type of Treatment	Ref.
Acetaminophen	>90	>90	Pilot-scale VF wetland (5 m^2)	[6]
Tonalide	88–90	89	Restoration FWS wetland (receives secondary effluent from a WWTP)	[95]
	n.a.	82	Experimental pilot-scale VF wetland (5 m^2)	[100]
	n.a.	15–40	Restored wetland system fed with natural waters impacted by urban and agricultural run-off	[102]
	n.a.	70	Pilot-scale VF wetland (5 m^2)	[6]
Oxybenzone	n.a.	>95	Pilot-scale VF wetland (5 m^2)	[6]
	n.a.	98	Household HF wetland	[101]
	n.a.	77	Tertiary treatment by pond + FWS wetland	[108]
Triclosan	n.a.	>40	Restored wetland system fed with natural waters impacted by urban and agricultural runoff	[102]
	n.a.	>80	Pilot-scale VF wetland (5 m^2)	[6]
	n.a.	86	Tertiary treatment by pond + FWS wetland	[108]

The anti-inflammatory drug ibuprofen has shown to be readily removed in the three types of CWs (HF, VF, and FWS). Results by [106] and [103] show high removal efficiencies in HF wetlands (60–80%), observing that the elimination rates are higher in shallower (0.27 m) than deeper beds (0.5 m). Matamoros et al. [104] found a prevalence of OH-IB over Ca-IB metabolite in an HF wetland, which would be explained by its predominant aerobic biodegradation. A comparison between different seasons by [69] in HF beds shows a seasonal variability, with lower removal efficiencies in winter (<40%) than in summer (>90%). Moreover, two injection experiments carried out at that same experimental plant in Barcelona, Spain, showed how its removal highly depended on the existence of a high redox status, which would promote aerobic biodegradation of this drug [7,8]. The removal of IB is also high in FWS wetlands, as documented by [95] and [107]; where efficiencies of over 90% are detected. High elimination rates were also reported for VF beds, with efficiencies of around 99% for unsaturated vertical wetlands [100].

Conversely, although diclofenac has been reported to be recalcitrant in activated sludge WWTPs [66], this substance shows a variable removal in CWs, ranging from values as low as 21% in HF beds [101,103] up to values of 99% at experimental-scale HF wetland under an injection experiment [7] or a value of 73% in a VF wetland in Denmark [100]. In FWS wetlands average removal efficiencies have been reported to be as high as 96% and 73–96% in studies carried out by [95] and [107]; respectively. This could be attributed to its large removal through photodegradation, as reported by [3]. This substance is fairly hydrophilic and was not detected in the particulate matter associated to the gravel matrix of CWs, as reported by [103]. A high variability on the elimination of diclofenac was also observed by [69] in various experimental mesoscale HF wetlands, who showed that its degradation was favored by high temperatures (summer VS winter) and found variable removal efficiencies in Barcelona (around 70%) and in León (<12%) wetlands. Although reductive dehalogenation has been claimed as a possible important removal mechanism for this compound [7], other studies show how its removal seems to be enhanced by more oxidizing conditions [8].

With regard to the removal of the estrogen ethynilestradiol in CWs, Gray and Sedlak [62] found a 41% removal after spiking it in concentrations slightly above background concentrations in a full-scale FWS wetland (750 m^2) with a depth of 0.5 m and a hydraulic retention time of 84 h, and reported sorption as a major mechanism for its removal in the aqueous phase. Song et al.[143] evaluated the influence of saturated versus unsaturated conditions in VF units with regard to the removal of estrogens, and found out that unsaturated conditions (and thus more oxidized conditions) were better for the removal of this type of pollutants.

What is more, the fate of the musk fragrance tonalide in CWs has been previously investigated by several authors, such as [107] who reported elimination rates of 88–90% in a full-scale FWS wetland, which did not show a dependence on the season, presumably due to their low biodegradability properties (i.e., major removal mechanism through interaction with organic matter and sediment). In fact, in another study, this time in two pilot HF wetlands at Les Franqueses del Vallès (Barcelona), Matamoros and Bayona [103] found this fragrance together with galaxolide to be the most abundant on the gravel bed and the suspended particulate matter, attributing the removal of these musks to hydrophobic interactions with the organic matter and the biofilm. Moreover, no water depth effect dependence was found for this compound. On the contrary, Hijosa-Valsero et al. [69] reported a possible dependence of this compound on temperature, since the removal efficiencies were significantly higher in summer than in wintertime, presumably due to the release of hydrophobic compounds in winter (i.e., galaxolide and tonalide), under a lower biofilm and plant activity. Removal efficiencies in VF wetlands were also reported by [100] to be moderately removed with elimination rates ranging from 75% to 82%.

The behavior of the sunscreen agent oxybenzone in CWs has been reported by [100] in a VF wetland, where it was very efficiently removed (>95%), and by [101], where the removal was also >95% in HF wetlands and 90% in VF wetlands. The authors suggest that it could be partially removed by sorption onto the particulate matter, although it is majorly removed by biodegradation [5,8].

With regard to the elimination of triclosan, Park et al. [123] reported removal efficiencies in FWS wetlands in South Korea of above 60%. Matamoros et al. [102] found triclosan to be the substance achieving the highest removal efficiency (>40%) among all studied compounds in a restored wetland in Denmark.

3.1 Design and Operational Factors Affecting the Removal of PPCPs in CWs

The removal of contaminants in CWs occurs as a result of complex physicochemical and microbial interactions. The rates of these processes depend on a variety of design and operational factors such as depth of the bed, type and size of media, hydraulic and OLRs, feeding strategy, and artificial/external aeration, among others. The design of CWs is often carried out using the black box concept, and reduced treatment efficiency may occur when wetlands are constructed without considering the influence of these parameters. Most of the available research concerning design and operational parameters' influence on treatment performance focuses on conventional water-quality parameters. Only in the last decade, the effect of these parameters on the removal of PPCPs has been investigated. Those include very few studies evaluating the influence of water depth [103,106]; primary treatment [8]; operation under cycles of saturation/unsaturation [8]; type of organic matter [104]; type of granular media [41,42]; and other different design parameters and modes of operation [5,69,70,177,178]. Therefore, the optimization of the performance of CWs in terms of PPCP removal stands as a necessity that can be achieved through the identification of the optimal design and operational factors of these treatment systems.

3.1.1 Behavior of PPCPs in HF CWs

HF wetlands display seasonal effects for PPCP removal, with lower removal efficiencies during the cold season at lower water temperatures. Seasonal patterns have been observed at various CW systems at mesoscale in Barcelona, NE Spain [8,69]; and in León, NW Spain [70,131], removal efficiencies being consistently better in summer than in winter for almost all of the PPCPs studied.

It is generally known that microbial activity is highly influenced by temperature. Microorganisms in wetlands usually reach their optimal activity at warm temperatures (15–25 °C), especially nitrifying and proteolytic bacteria [154]. The same trend has been found to occur for denitrifying bacteria [16,144]. The elimination of PPCPs has been found to depend on water temperature, being especially important for biodegradable compounds (e.g., ibuprofen and diclofenac). Moreover, the presence of vegetation during the summer season could also contribute to the removal of PPCPs, either by plant assimilation, and to a higher extent by the transfer of oxygen to the rhizosphere [18,155]. This should allow the creation of microenvironments with different redox conditions, which in turn would promote the development of microbial biofilms with functionally different respiration processes [8,78,167].

What is more, it is also widely known that removal of most pollutants, including PPCPs, in CWs occur primarily due to microbial activity [59]. In fact, the removal efficiency of PPCPs in CWs is oftentimes higher than that achieved in conventional WWTPs, which could presumably be due to their higher microbial diversity occurring in wetland ecosystems [78]. CWs can be designed to favor a wide range of redox conditions, therefore enhancing a variety of biological processes and removal of multiple contaminants in the same CW bed. In turn, CW technology can be optimized if the factors influencing their performance are identified.

For instance, operating HF wetlands in batch (or cycles of fill and drain) seems to be a very important factor promoting the elimination of most contaminants. The alternation of phases of saturation and unsaturation promotes the existence of a higher redox status as compared to functioning under saturated conditions, which in turn enhances significantly the elimination of PPCPs [8,178]. Substances whose major removal mechanism is thought to be biodegradation under aerobic conditions (e.g., ibuprofen) are those that show the largest differences [8]. In general, the operation under cycles of saturation and unsaturation seems to be a good strategy to enhance the treatment performance of HF CWs by promoting the fluctuation of redox conditions where robust aerobic facultative biofilms can operate [51,146].

Other design factors for HF wetlands are water depth, applied HLRs, grain size, and type of media. In particular, water depth seems to be a determining factor in this type of wetlands, which affects the redox conditions, oxygen supply, and thus the treatment performance of the wetlands. Shallower wetlands appear to have higher redox potential, and thus promote more variate and energetically favorable reactions, which in turn enhance the degradation of PPCPs [106]. Other authors have shown experimental evidence, which supports this hypothesis [56,58,64,71,103].

With regard to the granular media of HF beds, in general a wide range of materials and sizes have been applied around the world, and its selection is in many cases dictated by the availability, price, and local practices of a certain region. Some studies indicate the importance of using hard, durable, and homogeneous materials that do not contain fine grains, which might clog the media [57,126]. In general, finer materials (D_{60}=3.5 vs 10mm) have been recommended in HF beds when low organic loads are applied since they provide a greater surface area for microbial biofilms. Finer gravel promotes a higher growth of the vegetation and in conjunction a higher removal of pollutants such as ammonia [56]. However, it is important to have in mind that the finer the material, the greater the risk of clogging and hydraulic problems. Moreover, although usually granite gravel is used in HF wetlands, other alternative sorbents have emerged so as to increase the adsorption capacity of these wetlands, such as light expanded clay aggregates [41,42]. This is expected to be especially important for hydrophobic compounds such as fragrances. The influence of granular size in HF wetlands' PPCPs removal has not to date been assessed, although it should majorly show effects on hydrophobic substances that get attached to particulate matter.

Finally, it is well known that in HF wetlands the oxygen demand exerted by the degradation of pollutants in wastewater exceeds the amount of oxygen available within the wetland bed. In general, subsurface oxygen limitation is one of the main rate-limiting factors for traditional HF wetlands [20,78,155]. For that reason, other variants such as the use of active aeration are being developed and implemented worldwide and have shown to enhance the treatment performance of this type of CWs [118,164]. However, this practice comes at a heavy O&M cost, which is only justified if the improvement in the performance of the wetland offsets the cost of operation equipment and energy needs [78,120]. To the best of our knowledge there are still no studies, which assess the removal of PPCPs within actively aerated HF wetlands.

In conclusion, seasonal variations in environmental conditions are relevant factors to take into consideration for HF CW design, and O&M, especially at places showing high seasonality. In general, it has been observed that the treatment performance of HF CWs on PPCP removal is enhanced by the adoption of strategies that promote a higher redox status of the system. Operating HF wetlands in cycles of saturation and unsaturation stands as a very relevant strategy, which greatly enhances PPCP removal by promoting a higher redox status of the wetland bed. Water depth is also a critical factor influencing performance, shallower wetlands of about 0.25–0.30cm being more efficient in PPCP removal than deeper HF CWs.

3.1.2 Behavior of PPCPs in VF CWs

In general, it has been observed that VF wetlands are more efficient on the removal of PPCPs than HF wetlands, which agrees with other laboratory and experimental studies that suggest that aerobic pathways are in general more efficient in the degradation of these substances than anaerobic ones [8,28,100,182]. However, the relative importance of a particular degradation process in VF CWs can vary as a function of environmental factors, organic contaminant being treated, wetland configuration, and specific design and operational parameters. For instance, in unsaturated intermittently loaded VF wetlands (typical), diffusion and convection processes depend on the loading strategy, being affected by the number of pulses, the volume of the pulse, and the duration of surface filter dewatering [114,153]. The choice of the filter media (i.e., grain size, grain material, and depth of layers) is also crucial, and should simultaneously satisfy treatment needs, while avoiding clogging of the bed and maintaining oxygen renewal in the media [31,81].

With regard to the loading frequency, a higher volume of batch dosed at lower feeding frequencies should favor oxygen transfer into the media due to diffusion and mass convection. However, if the volume is too high, it could lead to decreased removal efficiency due to reduced contact time between the water and the biomass. On the other hand, if the loading is too frequent, the capacity of oxygen renewal occurring between pulses could be limited, thus affecting removal processes such as nitrification. To this respect, differences could be more notable in the long term [114,153]. Further studies in this direction are needed.

In general, the loading frequency is an important operational parameter to guarantee good treatment performance of this type of treatment systems, where sufficient oxygen renewal and mineralization of the organic matter should be allowed. Applied HLRs and potential accumulation of solid deposits on the bed surface should be carefully supervised and controlled, especially in non-rested VF wetlands, so as to avoid filter clogging and surface ponding, which could eventually decrease the efficiency and lifetime of the treatment system [92,128].

Moreover, grain size is again an important factor to take into account. A recent study has evaluated the influence of this factor on PPCP removal and has shown a greater performance of sand-based (1–3 mm) than gravel-based (4–8 mm) VF wetlands [6]. Results suggest that in gravel-based VF wetlands, lower contact together with lower solids retention capacity due to higher pores between grains would result in reduced treatment efficiency. In fact, it has been observed that a rapid flow downward does not favor ammonia sorption onto the gravel bed, which might eventually decrease nitrification activity [114,153]. The lower surface area per unit volume in gravel-based as compared to sand-based VF beds, provided by a larger grain size, reduces oxygen diffusion, and entrapment of solids, and may provide insufficient contact time between the biofilm and pollutants. On the other hand, in sand-based VFs, the higher surface area for biofilm attachment and higher oxygen availability may promote the existence or larger biomass and more diverse microbial communities. In fact, higher microbial biomass and activities have been found to take place in the upper part (about first 10 cm of a 50 cm VF) of a VF wetland, where a higher redox status prevails [152].

Although higher elimination rates have been found at sand-based systems [6], their long-term operation may make them more vulnerable to clogging. In practice, grain size as well as loading frequency should be carefully selected, as well as HLRs controlled [100], in order to ensure desired removal of PPCPs, as well as enough oxygen renewal and mineralization of organic matter so that no clogging develops in the filter beds. Nevertheless, in French-style VF wetlands, where various beds alternate cycles of feed and rest, it has been observed that very high hydraulic overloads can be applied without observing a decrease in the infiltration capacity or treatment performance of the system (approx. 0.4–1.8 m/d). Nevertheless, in particular case scenarios the grain and type of media used should be adapted to available local materials, and the use of gravel could constitute an acceptable solution.

3.2 Behavior of PPCPs in Hybrid CW Systems

With the purpose of evaluating the treatment capacity and possible performance enhancement when various CW of different types are combined, a hybrid CW system consisting of three stages of VF, HF, and FWS wetland (in this study followed by an open-air water tank indicated for water reuse) was designed, constructed, and monitored at experimental and full scale for the removal of conventional water-quality parameters and PPCPs [4,5,9].

Results obtained in a full-scale hybrid system of the cited characteristics (surface areas of 317, 219, and 240 m^2 for the VF, HF, and FWS wetlands, respectively) showed to perform remarkably well in the removal of PPCPs, achieving very high overall removal efficiencies for the majority of the studied compounds (above 80% removal for all compounds) [4]. The accumulated removal efficiencies in each treatment unit are displayed in Figure 7.

Final effluent concentrations of target PPCPs were very low, being below the limit of detection for acetaminophen. The rest were in the ng/L order (20–100), which is in the range of those found in the environment, such as those reported by [105] in small ponds or lagoons. These concentrations are also in the range of those obtained in advanced treatment technologies applied at full scale, such as ozonation or membrane filtration [136,142].

The high removal efficiencies can be explained by differing existing physicochemical conditions at different CW configurations, which would allow for the combination and synergy of various physicochemical and biological removal mechanisms to occur (e.g., biodegradation, sorption, volatilization, hydrolysis, and photodegradation) and thus achieve improved treatment efficiency of most pollutants [51,201]. In this sense, while aerobic metabolic pathways and solids retention are enhanced in VF wetlands, other removal mechanisms such as anaerobic biodegradation and sorption would predominate in HF beds. At last, the FWS wetland would be responsible for potential photodegradation of compounds, and less importantly through adsorption onto organic matter and uptake of plant material [108].

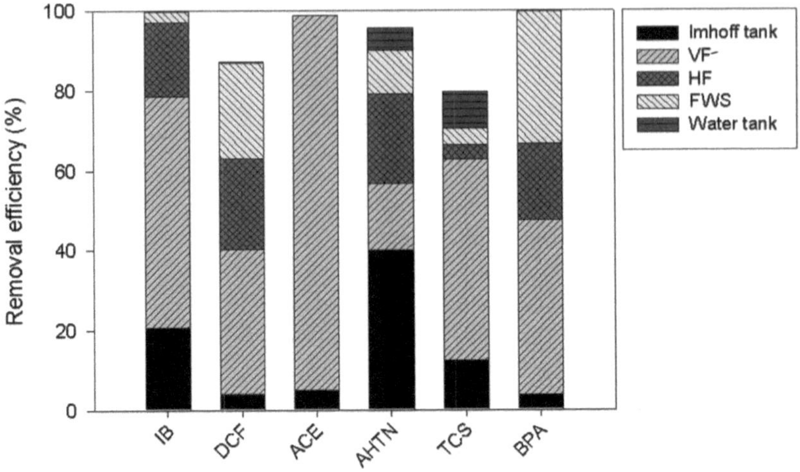

FIGURE 7 Accumulated removal efficiencies of various PPCPs at the different units of the pilot hybrid treatment system [4]. Note: ibuprofen: IB; diclofenac: DCF; acetaminophen: ACE; tonalide: AHTN; triclosan: TCS; the industrial chemical compound bisphenol A: BPA.

The superior treatment performance of the VF over the other treatment units could be owed to energetically favorable aerobic microbial reactions, as well as hydrolysis reactions, taking place within this wetland type and provided by its design and operation strategy, which results in high effluent redox potentials and dissolved oxygen concentrations. However, the synergic nitrification–denitrification activity observed in the VF bed within the cited [4] and a previous study at the same treatment plant [9] suggest the coexistence of both aerobic and anaerobic microenvironments within this wetland bed, which would allow the occurrence of both processes to take place. This could be partially explained by the occurrence of a saturated layer at the bottom of the wetland bed, below draining pipes. Similarly, this finding indicates that although aerobic biodegradation and sorption onto organic matter may be the major removal mechanisms contributing to PPCP reduction in VF wetlands, alternative processes based on anaerobic metabolism could simultaneously be occurring at anoxic microsites or micropores within the wetland bed (i.e., in lower layers) [7,33], which contribute to its elimination.

While substances like the musk fragrance tonalide show a great reduction within the HF presumably due to sorption processes, results for triclosan would exhibit little sorption capacity. The behavior of triclosan within the three different wetland configurations indicate aerobic biodegradation as the major removal mechanism involved in the elimination of this compound [141,171]. Although triclosan has been detected in plant and sediments of an HF CW and its concentration generally decreased from inflow to outflow [174], sorption and plant uptake do not appear to constitute a principal mechanism of triclosan removal in CWs. The similarly high removal efficiencies achieved for diclofenac in the current study in the HF bed when compared to the VF wetland suggest that various alternative mechanisms may determine the elimination of this compound, and to that respect, anaerobic biodegradation through reductive dehalogenation could constitute a predominant degradation pathway of diclofenac when anaerobic conditions prevail [7,69,123].

The FWS wetland performed especially well for diclofenac. These results support photodegradation as a principal removal mechanism involved in diclofenac attenuation in water bodies [3,17,108,202], together with less predominant mechanisms (i.e., aerobic/anaerobic biodegradation and plant uptake). Although the removal of triclosan was negligible, some more reduction within the water reuse tank seemed to occur, indicating that photooxidation processes may constitute a small contribution to its removal [5,108,110,202]. Although the reduction of tonalide concentrations in the FWS could be attributed to sorption onto particulate matter and sediment, further reduction was achieved at the water reuse tank, suggesting photodegradation through sunlight exposure as one of the principal mechanisms of the fragrance tonalide removal within this type of wetland configuration. Similarly, high removal efficiencies were obtained at other FWS wetlands operating as a tertiary treatment step [95,107,108].

In conclusion, hybrid CW systems consisting of a combination of CW configurations, including a VF CW, an HF wetland, and an FWS wetland operating in series, have shown to be a very robust ecotechnology for wastewater treatment and reuse in small communities [4,5]. The combination of different wetland configurations has shown to optimize a number of important treatment processes, achieving an excellent overall removal efficiency of PPCPs (above 80% for all studied substances), conventional water-quality parameters (>90%), and a high disinfection capacity [9]. This has been possible thanks to the occurrence of complementary abiotic/biotic removal pathways under differing physicochemical conditions existing at wetlands of different configuration.

3.3 Concluding Remarks

This chapter shows how PPCPs are a subgroup within the so-called "emerging organic contaminants" which comprise an extraordinarily diverse array of chemicals used in veterinary medicine, agricultural practice, human health, and cosmetic care. PPCPs include prescription and non-prescription medications, nutritional supplements, diagnostic agents, as well as other consumer products such as disinfectants, fragrances, sunscreen agents, and cosmetics. Even though most of them are polar, have a short half-lifetime in water, and are found in trace concentrations, they are considered pseudo-persistent pollutants. This is because of their universal, frequent usage by multitude of individuals and animals, which causes a continuous discharge into the environment oftentimes sourcing from either untreated or insufficient wastewater treatment. Little is known about the degradation pathways and products of most PPCPs in the environment, and less is known about their intermediates of degradation, which have just started to be explored. Conventional activated sludge WWTPs often provide insufficient removal of these substances, and in this context CWs constitute a good option for the removal of pharmaceuticals and care products from wastewater.

CWs emphasize the transformation processes taking place in natural wetlands to treat wastewater. Wetlands are complex ecosystems and comprise a wide range of microenvironments that aim at pollutant removal. Biodegradation, sorption, and photodegradation seem to be the major mechanisms involved in PPCP removal in CWs. While aerobic biodegradation and sorption predominate in VF wetlands, HF wetlands should be especially important at anoxic biodegradation and sorption of contaminants. In surface flow wetlands photodegradation plays a major role in PPCP removal. Other mechanisms such as hydrolysis, plant uptake, sedimentation, contaminant accumulation, metabolic transformation, or phytovolatilization also contribute at a lower rate to PPCP removal.

CWs can be optimized so as to improve treatment performance. Design and operational factors such as water depth, HLR, grain size, or use of active aeration can have a major impact on the removal efficiency of PPCP removal. In general, VF wetlands have shown to be more efficient on the removal of PPCPs than HF wetlands due to the aerobic environment promoted in the intermittently-fed, unsaturated typical VF beds.

The use of hybrid CW systems, which combine different wetland types, has shown to enhance the removal of PPCPs by combining different treatment processes occurring at various (physicochemical and microbiological) microsites of the different wetland units. A combination of VF, HF, and FWS wetlands in series has been proven to achieve very high removal rates (above 80%) for most studied PPCPs, and constitutes a good alternative for the treatment of domestic wastewater in small or rural communities.

REFERENCES

[1] AEMPS, Utilización de antiinflamatorios no esteroides (AINE) en España 1992–2006, Agencia Española de Medicamentos y Productos Sanitarios. Spanish Ministry of Health and Social Politics, 2006. Available: http://www.aemps.gob.es/medicamentosUsoHumano/observatorio/docs/AINE.pdf (accessed 02.06.13).

[2] AEMPS, Uso de antibióticos en España (1992–2009), Agencia Española de Medicamentos y Productos Sanitarios. Spanish Ministry of Health and Social Politics, 2009. Available: http://www.aemps.gob.es/medicamentosUsoHumano/observatorio/docs/antibioticos.pdf (accessed 02.06.13).

[3] R. Andreozzi, R. Marotta, N. Paxéus, Pharmaceuticals in STP effluents and their sol1ar photodegradation in aquatic environment, Chemosphere 50 (2003) 1319–1330.

[4] C. Ávila, J.M. Bayona, I. Martín, J.J. Salas, J. García, Emerging organic contaminant removal in a full-scale hybrid constructed wetland system for wastewater treatment and reuse, Ecol. Eng. (2014). http://dx.doi.org/10.1016/j.ecoleng.2014.07.056.

[5] C. Ávila, V. Matamoros, C. Reyes-Contreras, B. Piña, M. Casado, L. Mita, C. Rivetti, C. Barata, J. García, J.M. Bayona, Attenuation of emerging contaminants in a hybrid constructed wetland system under different hydraulic loading rates and their associated toxicological effects in wastewater, Sci. Total Environ. 470–471 (2014) 1272–1280.

[6] C. Ávila, J. Nivala, L. Olsson, K. Kassa, T. Headley, R. Mueller, J.M. Bayona, J. García, Emerging organic contaminant removal in vertical subsurface flow constructed wetlands: influence of media size, loading frequency and use of active aeration, Sci. Total Environ. 494–495 (2014) 211–217.

[7] C. Ávila, A. Pedescoll, V. Matamoros, J.M. Bayona, J. García, Capacity of a horizontal subsurface flow constructed wetland system for the removal of emerging pollutants: an injection experiment, Chemosphere 81 (2010) 1137–1142.

[8] C. Ávila, C. Reyes, J.M. Bayona, J. García, Emerging organic contaminant removal depending on primary treatment and operational strategy in horizontal subsurface flow constructed wetlands: influence of redox, Water Res. 47 (2013) 315–325.

[9] C. Ávila, J.J. Salas, I. Martín, C. Aragón, J. García, Integrated treatment of combined sewer wastewater and stormwater in a hybrid constructed wetland system in southern Spain and its further reuse, Ecol. Eng. 50 (2013) 13–20.

[10] A. Balmford, Pollution, politics and vultures, Science 339 (6120) (2013) 653–654.

[11] D. Barceló, Emerging pollutants in water analysis, Trends Anal. Chem. 22 (2003) 14–16.

[12] D. Barceló, M. Petrović, Pharmaceuticals and personal care products (PPCPs) in the environment, Anal. Bioanal. Chem. 387 (2007) 1141–1142.

[13] L.L. Behrends, E. Balley, P. Jansen, L. Houke, S. Smith, Integrated constructed wetland systems: design, operation, and performance of low-cost decentralized wastewater treatment systems, Water Sci. Technol. 55 (2007) 155–161.

[14] F.J. Benitez, J.L. Acero, F.J. Real, G. Roldan, E. Rodriguez, Photolysis of model emerging contaminants in ultra-pure water: kinetics, by-products formation and degradation pathways, Water Res. 47 (2013) 870–880.

[15] Å. Bergman, J.J. Heindel, S. Jobling, K.A. Kidd, R.T. Zoeller (Eds.), State of the Science of Endocrine Disrupting Chemicals—2012, United Nations Environment Programme and the World Health Organization, Geneva, Switzerland, 2013. Available: http://www.who.int/ceh/publications/endocrine/en/index.html (accessed 04.03.13).

[16] S. Boulêtreau, E. Salvo, E. Lyautey, S. Mastrorillo, F. Garabetian, Temperature dependence of denitrification in phototrophic river biofilms, Sci. Total Environ. 416 (2012) 323–328.

[17] H.R. Buser, T. Poiger, M.D. Müller, Occurrence and fate of the pharmaceutical drug diclofenac in surface waters: rapid photodegradation in a lake, Environ. Sci. Technol. 32 (1998) 3449–3456.

[18] H. Brix, Macrophyte-mediated oxygen transfer in wetlands: transport mechanisms and rates, in: G.A. Moshiri (Ed.), Constructed Welands for Water Quality Improvement, Lewis Publishers, Boca Raton, Florida, 1993, pp. 391–398.

[19] H. Brix, C. Arias, The use of vertical flow constructed wetland for on-site treatment of domestic wastewater: new Danish guidelines, Ecol. Eng. 25 (2005) 491–500.

[20] H. Brix, H.H. Schierup, Soil oxygenation in constructed reed beds: the role of macrophyte and soil-atmosphere interface oxygen transport, in: P.F. Cooper, B.C. Findlater (Eds.), Constructed Wetlands in Water Pollution Control, Pergamon Press, Oxford, United Kingdom, 1990, pp. 53–66.

[21] B.W. Brooks, J.P. Berninger, L.A. Kristofco, A.J. Ramirez, J.K. Stanley, T.W. Valenti, Chapter eight—pharmaceuticals in the environment: lessons learned for reducing uncertainties in environmental risk assessment, Prog. Mol. Biol. Transl. Sci. 112 (2012) 231–258.

[22] P.H. Cadby, W.R. Troy, M.G.H. Vey, Consumer exposure to fragrance ingredients: providing estimates for safety evaluation, Regul. Toxicol. Pharmacol. 36 (2002) 246–252.

[23] J. Campbell, Drugs on Tap: Pharmaceuticals in Our Drinking Water. Pharmacy Practice, 2007. Available: http://www.cielap.org/pdf/news.DrugsOnTap.pdf (accessed 17.05.10).

[24] M. Carballa, F. Omil, J.M. Lema, Removal of cosmetic ingredients and pharmaceuticals in sewage primary treatment, Water Res. 39 (2005) 4790–4796.

[25] M. Carballa, F. Omil, J.M. Lema, M. Llompart, C. García-Jares, I. Rodríguez, M. Gómez, T. Ternes, Behavior of pharmaceuticals, cosmetics and hormones in a sewage treatment plant, Water Res. 38 (2004) 2918–2926.

[26] F. Chen, G.G. Ying, L.X. Kong, L. Wang, J.L. Zhao, L.J. Zhou, L.J. Zhang, Distribution and accumulation of endocrine-disrupting chemicals and pharmaceuticals in wastewater irrigated soils in Hebei, China, Environ. Pollut. 159 (2011) 1490–1498.

[27] M. Clara, B. Strenn, O. Gans, E. Martinez, N. Kreuzinger, H. Kroiss, Removal of selected pharmaceuticals, fragrances and endocrine disrupting compounds in a membrane bioreactor and conventional treatment plants, Water Res. 39 (2005) 4797–4807.

[28] J.L. Conkle, J. Gan, M.A. Anderson, Degradation and sorption of commonly detected PPCPs in wetland sediments under aerobic and anaerobic conditions, J. Soils Sediments 12 (2012) 1164–1173.

[29] D.W. Connell, Basic Concepts of Environmental Chemistry, Second ed., Taylor & Francis, Boca Raton, FL, USA, 2005, p. 465.

[30] P. Cooper, A review of the design and performance of vertical-flow and hybrid reed bed treatment systems, Water Sci. Technol. 40 (1999) 1–9.

[31] P.F. Cooper, Sizing vertical flow and hybrid constructed wetland systems, in: Proceedings of the 1st International Seminar on the Use of Aquatic Macrophytes for Wastewater Treatment in Constructed Wetlands. Lisbon, Portugal. May 8–10, 2003.

[32] P. Cooper, The performance of vertical flow constructed wetland systems with special reference to the significance of oxygen transfer and hydraulic loading rates, Water Sci. Technol. 51 (2005) 81–90.

[33] P.F. Cooper, G.D. Job, M.B. Green, R.B.E. Shutes, Reed beds and constructed wetlands for wastewater treatment, WRc, Swindon, UK, 1996. 184 pp.

[34] V.L. Cunningham, M. Buzby, T. Hutchinson, F. Mastrocco, N. Parke, N. Roden, Effects of human pharmaceuticals on aquatic life: next steps, Environ. Sci. Technol. 40 (2006) 3456–3462.

[35] C.G. Daughton, Non-regulated water contaminants: emerging research, Environ. Impact Asses. 24 (2004) 711–732.

[36] C.G. Daughton, PPCPs in the environment: future research—beginning with the end always in mind (Chapter 33), in: K. Kümmerer (Ed.), Pharmaceuticals in the Environment, Second ed., Springer, 2004, pp. 463–495. Available: http://www.epa.gov/ppcp/pdf/daughton-needs.pdf (accessed 10.11.09).

[37] C.G. Daughton, I.S. Ruhoy, Lower-dose prescribing: minimizing "side effects" of pharmaceuticals on society and the environment, Sci. Total Environ. 443 (2013) 324–337.

[38] C.G. Daughton, T.A. Ternes, Pharmaceuticals and personal care products in the environment: agents of subtle change? Environ. Health Persp. 107 (1999) 907–938.

[39] M.S. Díaz-Cruz, M.J. López de Alda, D. Barceló, Environmental behavior and analysis of veterinary and human drugs in soils, sediments and sludge, Trends Anal. Chem. 22 (2003) 340–351.

[40] A. Dixon, M. Simon, T. Burkitt, Assessing the environmental impact of two options for small-scale wastewater treatment: comparing a reed bed and an aerated biological filter using a life cycle approach, Ecol. Eng. 20 (2003) 297–308.

[41] A. Dordio, A.J.P. Carvalho, D.M. Teixeira, C.B. Dias, A.P. Pinto, Removal of pharmaceuticals in microcosm constructed wetlands using Typha spp. and LECA, Bioresour. Technol. 101 (2010) 886–892.

[42] A.V. Dordio, A.J. Estevao Candeias, A.P. Pinto, C. Teixeria da Costa, A.J. Palace Carvalho, Preliminary media screening for application in the removal of clofibric acid, carbamazepine and ibuprofen by SSF-constructed wetlands, Ecol. Eng. 35 (2009) 290–302.

[43] EC, Directive 76/769/ECC: use of certain dangerous substances and preparations, Off. J L262 (1976) Final.

[44] EC, Directive 91/271/EEC of 21 May 1991 concerning urban wastewater treatment, Off. J. L135 (1991) 40–52.

[45] EC, Directive 2000/60/EC of the European Parliament and of the Council of 23 October 2000 establishing a framework for Community action in the field of water policy, Off. J. L327 (2000) 1–73.

[46] EC, Endocrine Disrupters Website. How the European Comission Uses the Precautionary Principle to Tackle Endocrine Eisrupters, 2000. Available: http://ec.europa.eu/environment/endocrine/strategy/substances_en.htm (accessed 10.07.13).

[47] EPA, Treating Contaminants of Emerging Concern. A Literature Review Database, United States Environmental Protection Agency, 2010. EPA-820-R-10–002. Available: http://water.epa.gov/scitech/swguidance/ppcp/upload/cecliterature.pdf (accessed 23.05.13).

[48] M. Farré, L. Kantiani, M. Petrović, S. Pérez, D. Barceló, Achievements and future trends in the analysis of emerging organic contaminants in environmental samples by mass spectrometry and bioanalytical techniques, J. Chromatogr. A 1259 (2012) 86–99.

[49] M. Farré, S. Perez, L. Kantiani, D. Barceló, Fate and toxicity of emerging pollutants, their metabolites and transformafarretion products in the aquatic environment, Trends Anal. Chem. 27 (2008) 991–1007.

[50] D. Fatta-Kassinos, M.I. Vasquez, K. Kümmerer, Transformation products of pharmaceuticals in surface waters and wastewater formed during photolysis and advanced oxidation processes—degradation, elucidation of by products and assessment of their biological potency, Chemosphere 85 (2011) 693–709.

[51] J.L. Faulwetter, V. Gagnon, C. Sundberg, F. Chazarenc, M.D. Burr, J. Brisson, A.K. Camper, O.R. Stein, Microbial processes influencing performance of treatment wetlands: a review, Ecol. Eng. 35 (2009) 987–1004.

[52] FDA, Food and Drug Administration, Department of Health and Human Services, USA. Summary Report on Antimicrobials Sold or Distributed for Use in Food-producing Animals, 2011. Available: http://www.fda.gov/downloads/ForIndustry/UserFees/AnimalDrugUserFeeActADUFA/UCM338170.pdf (accessed 01.04.13).

[53] P.J. Ferguson, M.J. Bernot, J.C. Doll, T.E. Lauer, Detection of pharmaceuticals and personal care products (PPCPs) in near-shore habitats of souther Lake Michigan, Sci. Total Environ. 458-460 (2013) 187–196.

[54] M.S. Fram, K. Belitz, Occurrence and concentrations of pharmaceutical compounds in groundwater used for public drinking-water supply in California, Sci. Total Environ. 409 (2011) 3409–3417.

[55] GAO, United States Government Accountability Office. Action Needed to Sustain Agencies' Collaboration on Pharmaceuticals in Drinking Water, 2011. Report to Congressional Requesters. Available: http://www.gao.gov/new.items/d11346.pdf (accessed 24.06.13).

[56] J. García, P. Aguirre, J. Barragán, R. Mujeriego, V. Matamoros, J.M. Bayona, Effect of key design parameters on the efficiency of horizontal subsurface flow constructed wetlands: long-term performance pilot study, Ecol. Eng. 25 (2005) 405–418.

[57] J. García, A. Corzo, Water Treatment with Constructed Wetlands: Handbook of Design, Construction and Operation of Subsurface Flow Wetlands Systems, Universitat Politècnica de Catalunya, Barcelona, Spain, 2008. Available: http://upcommons.upc.edu/e-prints/bitstream/2117/2474/1/JGarcia_and_ACorzo.pdf (accessed 15.05.10) (in Spanish).

[58] J. García, E. Ojeda, E. Sales, F. Chico, T. Píriz, P. Aguirre, R. Mujeriego, Spatial variations of temperature, redox potential and contaminants in horizontal flow reed beds, Ecol. Eng. 21 (2003) 129–142.

[59] J. García, D.P.L. Rousseau, J. Morató, E. Lesage, V. Matamoros, J.M. Bayona, Contaminant removal processes in subsurface-flow constructed wetlands: a review, Crit. Rev. Env. Sci. Technol. 40 (2010) 561–661.

[60] M.J. Gómez, S. Herrera, D. Solé, E. García-Calvo, A.R. Fernández-Alba, Spatio-temporal evaluation of organic contaminants and their transformation products along a river basin affected by urban, agricultural and industrial pollution, Sci. Total Environ. 420 (2012) 134–145.

[61] S. González, M. Petrović, D. Barceló, Advanced liquid chromatography/mass spectrometry (LC-MS) methods applied to wastewater removal and the fate of surfactants in the environment, Trends Anal. Chem. 26 (2007) 116–124.

[62] J.L. Gray, D.L. Sedlak, The fate of estrogenic hormones in an engineered treatment wetland with dense macrophytes, Water Environ. Res. 77 (2005) 24–31.

[63] B. Halling-Sørensen, S. Nors Nielsen, P.F. Lanzky, F. Ingerslev, H.C. Holten Lützhøft, E. Jørgensen, Occurrence, fate and effects of pharmaceutical substances in the environment—A review, Chemosphere 36 (1998) 357–393.

[64] E.H. Headley, L. Davison, Treatment at different depths and vertical mixing within a 1-m deep horizontal subsurface-flow wetland, Ecol. Eng. 25 (2005) 567–582.

[65] T. Headley, J. Nivala, K. Kassa, L. Olsson, S. Wallace, H. Brix, M. van Afferden, R. Müller, Escherichia coli removal and internal dynamics in subsurface flow ecotechnologies: effects of design and plants. Ecol. Eng. 61 (2013) 564–574.

[66] T. Heberer, Occurrence, fate, and removal of pharmaceutical residues in the aquatic environment: a review of recent research data, Toxicol. Lett. 31 (2002) 5–17.

[67] M.D. Hernando, M. Mezcua, A.R. Fernández-Alba, D. Barceló, Environmental risk assessment of pharmaceutical residues in wastewater effluents, surface water and sediments, Talanta 69 (2006) 334–342.

[68] M. Hijosa-Valsero, V. Matamoros, J. Martín-Villacorta, E. Bécares, J.M. Bayona, Assessment of full-scale natural systems for the removal of PPCPs from wastewater in small communities, Water Res. 44 (2010) 1429–1439.

[69] M. Hijosa-Valsero, V. Matamoros, A. Pedescoll, J. Martín-Villacorta, E. Bécares, J. García, J.M. Bayona, Evaluation of primary treatment and loading regimes in the removal of pharmaceuticals and personal care products from urban wastewaters by subsurface-flow constructed wetlands, Int. J. Environ. Anal. Chem. 91 (2011) 632–653.

[70] M. Hijosa-Valsero, V. Matamoros, R. Sidrach-Cardona, J. Martín-Villacorta, E. Bécares, J.M. Bayona, Comprehensive assessment of the design configuration of constructed wetlands for the removal of pharmaceuticals and personal care products from urban wastewaters, Water Res. 44 (2010) 3669–3678.

[71] Y. Huang, L. Ortiz, P. Aguirre, J. García, R. Mujeriego, J.M. Bayona, Effect of design parameters in horizontal flow constructed wetland on the behaviour of volatile fatty acids and volatile alkylsulfides, Chemosphere 59 (2005) 769–777.

[72] M.M. Huber, S. Korhonen, T.A. Ternes, U. von Gunten, Oxidation of pharmaceuticals during water treatment with chlorine dioxide, Water Res. 39 (2005) 3607–3617.

[73] F. Ingerslev, E. Vaclavik, B. Halling-Sorensen, Pharmaceuticals and personal care products: a source of endocrine disruption in the environment, Pure Appl. Chem. 75 (2003) 1881–1893.

[74] B.B. Jana, Sewage-fed aquaculture: the Calcutta model, Ecol. Eng. 11 (1998) 73–85.

[75] A. Jelić, M. Gros, A. Ginebreda, R. Cespedes-Sánchez, F. Ventura, M. Petrović, D. Barceló, Occurrence, partition and removal of pharmaceuticals in sewage water and sludge during wastewater treatment, Water Res. 45 (2011) 1165–1176.

[76] A. Jurado, E. Vàzquez-Suñé, J. Carrera, M. López de Alda, E. Pujades, D. Barceló, Emerging organic contaminants in groundwater in Spain: a review of sources, recent occurrence and fate in a European context, Sci. Total Environ. 440 (2012) 82–94.
[77] R.H. Kadlec, Comparison of free water and horizontal subsurface treatment wetlands, Ecol. Eng. 35 (2009) 159–174.
[78] R.H. Kadlec, S.D. Wallace, Treatment Wetlands, second ed., CRC Press, Boca Raton, Florida, 2009, p. 1016.
[79] B. Kasprzyk-Hordern, R.M. Dinsdale, A.J. Guwy, The removal of pharmaceuticals, personal care products, endocrine disruptors and illicit drugs during wastewater treatment and its impact on the quality of receiving waters, Water Res. 43 (2009) 363–380.
[80] K. Katz, T. Inoue, H. Ietsugu, T. Koba, H. Sasaki, N. Miyaji, K. Kitagawa, P.K. Sharma, T. Nagasawa, Performance of six multi-stage hybrid wetland systems for treating high-content wastewater in the cold climate of Hokkaido, Japan, Ecol. Eng. 51 (2013) 256–263.
[81] K. Kayser, S. Kunst, Processes in vertical-flow reed beds: nitrification, oxygen transfer, and soil clogging, Water Sci.Technol. 51 (2005) 177–184.
[82] S.D. Kim, J. Cho, I.S. Kim, B.J. Vanderford, S.A. Snyder, Occurrence and removal of pharmaceuticals and endocrine disruptors in South Korean surface, drinking and waste waters, Water Res. 41 (2007) 1013–1021.
[83] I. Kim, N. Yamashita, H. Tanaka, Performance of UV and UV/H2O2 processes for the removal of pharmaceuticals detected in secondary effluent of a sewage treatment plant in Japan, J. Hazard. Mat. 166 (2009) 1134–1140.
[84] K. Kimura, H. Hara, Y. Watanabe, Removal of pharmaceutical compounds by submerged membrane bioreactors (MBRs), Desalination 178 (2005) 135–140.
[85] M. Klavarioti, D. Mantzavinos, D. Kassinos, Removal of residual pharmaceuticals from aqueous systems by advanced oxidation processes, Environ. Int. 35 (2009) 402–417.
[86] C.W. Knapp, J. Dolfing, P.A. Ehlert, D.W. Graham, Evidence of increasing antibiotic resistance gene abundances in archived soils since 1940, Environ.Sci. Technol. 44 (2010) 580–587.
[87] KNAPPE, Knowledge and Need Assessment on Pharmaceutical Products in Environmental Waters, 2008. FP7 Project. Final report. Available: http://environmentalhealthcollaborative.org/images/KNAPPE_REPORT_FINAL.pdf.
[88] P. Knowles, G. Dotro, J. Nivala, J. García, Clogging in subsurface-flow treatment wetlands: occurrence and contributing factors, Ecol. Eng. 37 (2011) 99–112.
[89] D.W. Kolpin, E.T. Furlong, M.T. Meyer, E.M. Thurman, S.D. Zaugg, L.B. Barber, H.T. Buxton, Pharmaceuticals, hormones, and other organic wastewater contaminants in U.S. Streams, 1999–2000: a national Reconnaissance, Environ. Sci. Technol. 36 (2002) 1202–1211.
[90] C.I. Kosma, D.A. Lambropoulou, T.A. Albanis, Occurrence and removal of PPCPs in municipal and hospital wastewaters in Greece, J. Hazard. Mat. 179 (2010) 804–817.
[91] K. Kümmerer, The presence of pharmaceuticals in the environment due to human use—present knowledge and future challenges, J. Environ. Manage. 90 (2009) 2354–2366.
[92] G. Langergraber, R. Haberl, J. Laber, A. Pressl, Evaluation of substrate clogging processes in vertical flow constructed wetlands, Water Sci. Technol. 48 (2003) 25–34.
[93] L. Lishman, S.A. Smyth, K. Sarafin, S. Kleywegt, J. Toito, T. Peart, B. Lee, M. Servos, M. Beland, P. Seto, Occurrence and reductions of pharmaceuticals and personal care proucts and estrogens by municipal wastewater treatment plants in Ontario, Canada, Sci. Total Environ. 367 (2006) 544–558.
[94] Z. Liu, Y. Kanjo, S. Mizutani, Removal mechanisms for endocrine disrupting compounds (EDCs) in wastewater treatment—physical means, biodegradation, and chemical advanced oxidation: a review, Sci. Total Environ. 407 (2009) 731–748.

[95] E. Llorens, V. Matamoros, V. Domingo, J.M. Bayona, J. García, Water quality improvement in a full-scale tertiary constructed wetland: effects on conventional and specific organic contaminants, Sci. Total Environ. 407 (2009) 2517–2524.

[96] R. Loos, G. Locoro, S. Comero, S. Contini, D. Schwesig, F. Werres, Pan-European survey on the occurrence of selected polar organic persistent pollutants in ground water, Water Res. 44 (2010) 4115–4126.

[97] R. Loos, B.M. Gawlik, G. Locoro, E. Rimaviciute, S. Contini, G. Bidoglio, EU-wide survey of polar organic persistent pollutants in European river waters, Environ. Pollut. 157 (2009) 561–568.

[98] R. López-Serna, M. Petrović, D. Barceló, Occurrence and distribution of multi-class pharmaceuticals and their active metabolites and transformation products in the Ebro River basin (NE Spain), Sci. Total Environ. 440 (2012) 280–289.

[99] F. Masi, N. Martinuzzi, R. Bresciani, L. Giovannelli, G. Conte, Tolerance to hydraulic and organic load fluctuations in constructed wetlands, Water Sci. Technol. 56 (2007) 39–48.

[100] V. Matamoros, C. Arias, H. Brix, J.M. Bayona, Removal of pharmaceuticals and personal care products (PPCPs) from urban wastewater in a pilot vertical flow constructed wetland and a sand filter, Environ. Sci. Technol. 41 (2007) 8171–8177.

[101] V. Matamoros, C. Arias, H. Brix, J.M. Bayona, Preliminary screening of small-scale domestic wastewater treatment systems for removal of pharmaceutical and personal care products, Water Res. 43 (2009) 55–62.

[102] V. Matamoros, C.A. Arias, L.X. Nguyen, V. Salvadó, H. Brix, Occurrence and behavior of emerging contaminants in surface water and a restored wetland, Chemosphere 88 (2012) 1083–1089.

[103] V. Matamoros, J.M. Bayona, Elimination of pharmaceuticals and personal care products in subsurface flow constructed wetlands, Environ. Sci. Technol. 40 (2006) 5811–5816.

[104] V. Matamoros, A. Caselles-Osorio, J. García, J.M. Bayona, Behaviour of pharmaceutical products and biodegradation intermediates in horizontal subsurface flow constructed wetland. A microcosm experiment, Sci. Total Environ. 394 (2008) 171–176.

[105] V. Matamoros, A. Duhec, J. Albaigés, J.M. Bayona, Photodegradation of carbamazepine, ibuprofen, ketoprofen and 17α-ethinylestradiol in fresh and seawater, Water Air Soil Pollut. 196 (2009) 161–168.

[106] V. Matamoros, J. García, J.M. Bayona, Behavior of selected pharmaceuticals in subsurface flow constructed wetlands: a pilot-scale study, Environ. Sci. Technol. 39 (2005) 5449–5454.

[107] V. Matamoros, J. García, J.M. Bayona, Organic micropollutant removal in a full-scale surface flow constructed wetland fed with secondary effluent, Water Res. 42 (2008) 653–660.

[108] V. Matamoros, V. Salvadó, Evaluation of the seasonal performance of a water reclamation pond-constructed wetland system for removing emerging contaminants, Chemosphere 86 (2012) 111–117.

[109] MedlinePlus, U.S. National Library of Medicine, 2010. Available: http://www.nlm.nih.gov/medlineplus/ (accessed 28.08.12).

[110] M. Mezcua, M.J. Gómez, I. Ferrer, A. Aguera, M.D. Hernando, A.R. Fernández-Alba, Evidence of 2,7/2,8-dibenzodichloro-p-dioxin as a photodegradation product of triclosan in water and wastewater samples, Anal. Chim. Acta 524 (2004) 241–247.

[111] X.-S. Miao, C.D. Metcalfe, Determination of carbamazepine and its metabolites in aqueous samples using liquid chromatography-electrospray tandem mass spectrometry, Anal. Chem. 75 (2003) 3731–3738.

[112] C. Miège, J.M. Choubert, L. Ribeiro, M. Eusèbe, M. Coquery, Fate of pharmaceuticals and personal care products in wastewater treatment plants—conception of a database and first results, Environ. Pollut. 157 (2009) 1721–1726.

[113] P. Molle, A. Liénard, C. Boutin, G. Merlin, A. Iwema, How to treat raw sewage with constructed wetlands: an overview of the French systems, Water Sci. Technol. 51 (2005) 11–21.

[114] P. Molle, A. Liénard, A. Grasmick, A. Iwema, Effect of reeds and feeding operations on hydraulic behavior of vertical flow constructed wetlands under hydraulic overloads, Water Res. 40 (2006) 606–612.

[115] P. Molle, S. Prost-Boucle, A. Liénard, Potential for total nitrogen removal by combining vertical flow and horizontal flow constructed wetlands: a full-scale experiment study, Ecol. Eng. 34 (2008) 23–29.

[116] I. Muñoz, M.J. Gómez-Ramos, A. Agüera, J.F. García-Reyes, A. Molina-Díaz, A.R. Fernández-Alba, Chemical evaluation of contaminants in wastewater effluents and the environmental risk of reusing effluents in agriculture, Trends Anal. Chem. 28 (2009) 676–694.

[117] V. Naddeo, S. Meriç, D. Kassinos, V. Belgiorno, M. Guida, Fate of pharmaceuticals in contaminated urban wastewater effluent under ultrasonic irradiation, Water Res. 42 (2009) 4019–4027.

[118] J. Nivala, T. Headley, S. Wallace, K. Bernhard, H. Brix, M. van Afferden, R.A. Müller, Comparative analysis of constructed wetlands: the design and construction of the ecotechnology research facility in Langenreichenbach, Germany, Ecol. Eng. 61 (2013) 527–543.

[119] J. Nivala, M.B. Hoos, C. Cross, S. Wallace, G. Parkin, Treatment of landfill leachate using an aerated, horizontal subsurface-flow constructed wetland, Sci. Total Environ. 380 (2007) 19–27.

[120] J. Nivala, S. Wallace, T. Headley, K. Kassa, H. Brix, M. van Afferden, R. Müller, Oxygen transfer and consumption in subsurface flow treatment wetlands, Ecol. Eng. 61 (2013) 544–554.

[121] J.L. Oaks, M. Gilbert, M.Z. Virani, R.T. Watson, C.U. Meteyer, B.A. Rideout, H.L. Shivaprasad, S. Ahmed, M.J.I. Chudhry, M. Arshad, S. Mahmood, A. Ali, A.A. Khan, Diclofenac residues as the cause of vulture population decline in Pakistan, Nature 427 (2004) 630–633.

[122] OECD, Organisation for Economic Co-operation and Development, Health Policy Studies. OECD Health Data: Health Expenditure and Financing: OECD Health Statistics (Database). Pharmaceutical Pricing Policies in a Global Market, 2008, 2013. Available: http://www.oecd-ilibrary.org/social-issues-migration-health/pharmaceutical-expenditure-per-capita_pharmexpcap-table-en (accessed 24.03.13).

[123] N. Park, B.J. Vanderford, S.A. Snyder, S. Sarp, S.D. Kim, J. Cho, Effective controls of micropollutants included in wastewater effluent using constructed wetlands under anoxic conditions, Ecol. Eng. 35 (2009) 418–423.

[124] S. Patra, T.K. Das, C. Ávila, V. Cabello, F. Castillo, D. Sarkar, S. Lahiri, B.B. Jana, Cadmium tolerance and antibiotic resistance in *Escherichia coli* isolated from waste stabilization ponds, Indian J. Exp. Biol. 50 (2012) 300–307.

[125] A. Pedescoll, Clogging in Horizontal Subsurface Flow Constructed Wetlands (Doctoral thesis), Universitat Politècnica de Catalunya, Spain, 2010.

[126] A. Pedescoll, E. Uggetti, E. Llorens, F. Granés, D. García, J. García, Practical method based on saturated hydraulic conductivity used to assess clogging in subsurface flow constructed wetlands, Ecol. Eng. 35 (2009) 1216–1224.

[127] M. Petrovic, S. Gonzalez, D. Barceló, Analysis and removal of emerging contaminants in wastewater and drinking water, Trends Anal. Chem. 22 (2003) 685–696.

[128] C. Platzer, K. Mauch, Soil clogging in vertical flow reed beds: mechanisms, parameters, consequences and solutions? Water Sci. Technol. 35 (1997) 175–182.

[129] J. Puigagut, J. Villaseñor, J.J. Salas, E. Bécares, J. García, Subsurface-flow constructed wetlands in Spain for the sanitation of small communities: a comparative study, Ecol. Eng. 30 (2007) 312–319.

[130] J.B. Quintana, S. Weiss, T. Reemtsma, Pathways and metabolites of microbial degradation of selected acidic pharmaceutical and their occurrence in municipal wastewater treated by a membrane bioreactor, Water Res. 39 (2005) 2654–2664.

[131] C. Reyes-Contreras, M. Hijosa-Valsero, R. Sidrach-Cardona, J.M. Bayona, E. Bécares, Temporal evolution in PPCP removal from urban wastewater by constructed wetlands of different configuration: a medium-term study, Chemosphere 88 (2012) 161–167.

[132] C. Reyes-Contreras, V. Matamoros, I. Ruiz, M. Soto, J.M. Bayona, Evaluation of PPCPs removal in a combined anaerobic digester-constructed wetland pilot plant treating urban wastewater, Chemosphere 84 (2011) 1200–1207.

[133] G.G. Rimkus, Polycyclic musk fragrances in the aquatic environment, Toxicol. Lett. 111 (1999) 37–56.

[134] J.L. Rodríguez-Gil, M. Catalá, S. González, R. Romo, Y. Valcárcel, Y. Segura, R. Molina, J.A. Melero, F. Martínez, Heterogeneous photo-Fenton treatment for the reduction of pharmaceutical contamination in Madrid rivers and ecotoxicological evaluation by a miniaturized fern spores bioassay, Chemosphere 80 (2010) 381–388.

[135] A. Rodríguez-Rojas, J. Rodríguez-Beltrán, A. Couce, J. Blázquez, Antibiotics and antibiotic resistance: a bitter fight against evolution, Int. J. Med. Microbiol. 303 (2013) 293–297.

[136] R. Rosal, A. Rodríguez, J.A. Perdigón-Melón, A. Petre, E. García-Calvo, M.J. Gómez, A. Agüera, A.R. Fernández-Alba, Occurrence of emerging pollutants in urban wastewater and their removal through biological treatment followed by ozonation, Water Res. 44 (2010) 578–588.

[137] J.L. Santos, I. Aparicio, M. Callejón, E. Alonso, Occurrence of pharmaceutically active compounds during 1-year period in wastewaters from four wastewater treatment plants in Seville (Spain), J. Hazard. Mat. 164 (2009) 1509–1516.

[138] R. Schulz, S.K.C. Peall, C. Hugo, V. Krause, Concentration, load and toxicity of spraydrift-borne azinphos-methyl at the inlet and outlet of a constructed wetland, Ecol. Eng. 18 (2001) 239–245.

[139] R.P. Schwarzenbach, B.I. Escher, K. Fenner, T.B. Hofstetter, C.A. Johnson, U. von Gunten, B. Wehrli, The challenge of micropollutants in aquatic systems, Science 313 (2006) 1072–1077.

[140] F.P. Shariati, M.R. Mehrnia, B.M. Salmasi, M. Heran, C. Wisniewski, M.H. Sarrafzadeh, Membrane bioreactor for treatment of pharmaceutical wastewater containing acetaminophen, Desalination 250 (2010) 798–800.

[141] H. Singer, S. Muller, C. Tixier, L. Pillonel, Triclosan: occurrence and fate of a widely used biocide in the aquatic environment: field measurements in wastewater treatment plants, surface waters, and lake sediments, Environ. Sci. Technol. 36 (2002) 4998–5004.

[142] S.A. Snyder, S. Adham, A.M. Redding, F.S. Cannon, J. DeCarolis, J. Oppenheimer, E.C. Wert, Y. Yoon, Role of membranes and activated carbon in the removal of endocrine disruptors and pharmaceuticals, Desalination 202 (2007) 156–181.

[143] H.-L. Song, K. Nakano, T. Taniguchi, M. Nomura, O. Nishimura, Estrogen removal from treated municipal effluent in small-scale constructed wetland with different depth, Bioresour. Technol. 100 (2009) 2945–2951.

[144] D.J. Spieles, W.J. Mitsch, The effects of season and hydrologic and chemical loading on nitrate retention in constructed wetlands: a comparison of low- and high-nutrient riverine systems, Ecol. Eng. 14 (2000) 77–91.

[145] P.E. Stackelberg, E.T. Furlong, M.T. Meyer, S.D. Zaugg, A.K. Henderson, D.B. Reissman, Persistence of pharmaceutical compounds and other organic wastewater contaminants in a conventional drinking-water-treatment plant, Sci. Total Environ. 329 (2004) 99–113.

[146] O.R. Stein, P.B. Hook, J.A. Biederman, W.C. Allen, D.J. Borden, Does batch operation enhance oxidation in subsurface constructed wetlands? Water Sci. Technol. 48 (2003) 149–156.

[147] M. Stuart, D. Lapworth, E. Crane, A. Hart, Review of risk from potential emerging contaminants in UK groundwater, Sci. Total Environ. 416 (2012) 1–21.

[148] G. Tchobanoglous, F.L. Burton, in: Metcalf, Eddy (Eds.), Wastewater Engineering: Treatment, Disposal and Reuse, third ed., McGraw-Hill, New York, 1991.

[149] G. Teijón, L. Candela, K. Tamoh, A. Molina-Díaz, A.R. Fernández-Alba, Occurrence of emerging contaminants, priority substances (2008/105/CE) and heavy metals in treated wastewater and groundwater at Depurbaix facility (Barcelona, Spain), Sci. Total Environ. 408 (2010) 3584–3595.

[150] T.A. Ternes, M. Meisenheimer, D. McDowell, F. Sacher, H.-J. Brauch, B. Haist-Gulde, G. Preuss, U. Wilme, N. Zullei-Seibert, Removal of pharmaceuticals during drinking water treatment, Environ. Sci. Technol. 36 (2002) 3855–3863.

[151] G. Thiere, S. Milenkovski, P.-E. Lindgren, G. Sahlén, O. Berglund, S.E.B. Weisner, Wetland creation in agricultural landscapes: biodiversity benefits on local and regional scales, Biol. Conserv. 142 (2009) 964–973.

[152] A. Tietz, A. Kirschner, G. Langergraber, K. Sleytr, R. Haberl, Characterisation of microbial biocoenosis in vertical subsurface flow constructed wetlands, Sci. Total Environ. 380 (2007) 163–172.

[153] A. Torrens, P. Molle, C. Boutin, M. Salgot, Impact of design and operation variables on the performance of vertical-flow constructed wetlands and intermittent sand filters treating pond effluent, Water Res. 43 (2009) 1851–1858.

[154] M. Truu, J. Huhanson, J. Truu, Microbial biomass, activity and community composition in constructed wetlands, Sci. Total Environ. 407 (2009) 3958–3971.

[155] L. Tyroller, D.P.L. Rousseau, S. Santa, J. García, Application of the gas tracer method for measuring oxygen transfer rates in subsurface flow constructed wetlands, Water Res. 44 (2010) 4217–4225.

[156] UNDP, Human Development Report 2006. Beyond Scarcity: Power, Poverty and the Global Water Crisis. United Nations Development Programme, 2006. New York. Available: http://hdr.undp.org/en/media/HDR06-complete.pdf (accessed 20.07.13).

[157] UNESCO, United Nations Educational, Scientific, and Cultural Organization. World Water Assessment Programme. Water for People, Water for Life—the United Nations World Water Development Report, Berghahn Books, Barcelona, 2003. Available: http://sustainabledevelopment.un.org/content/documents/WWDR_english_129556e.pdf (accessed 03.05.13).

[158] UNESCO, United Nations Educational, Scientific, and Cultural Organization, World Water Assessment Programme (WWAP), 2012. The United Nations World Water Development Report 4: Managing Water under Uncertainty and Risk. Available http://www.unesco.org/new/fileadmin/MULTIMEDIA/HQ/SC/pdf/WWDR4%20Volume%201-Managing%20Water%20under%20Uncertainty%20and%20Risk.pdf (accessed 20.05.13).

[159] I. Vera, J. García, K. Sáez, L. Moragas, G. Vidal, Performance evaluation of eight years experience of constructed wetland systems in Catalonia as alternative treatment for small communities, Ecol. Eng. 37 (2011) 364–371.

[160] P. Verlicchi, A. Galletti, M. Petrovic, D. Barceló, M. Al Aukidy, E. Zambello, Removal of selected pharmaceuticals from domestic wastewater in an activated sludge system followed by a horizontal subsurface flow bed – analysis of their respective contributions, Sci. Total Environ. 454–455 (2013) 411–425.

[161] J. Vymazal, The use of sub-surface constructed wetlands for wastewater treatment in the Czech Republic: 10 years experience, Ecol. Eng. 18 (2002) 633–646.

[162] J. Vymazal, Horizontal sub-surface flow and hybrid constructed wetlands systems for wastewater treatment, Ecol. Eng. 25 (2005) 478–490.

[163] J. Vymazal, The use of constructed wetlands with horizontal sub-surface flow for various types of wastewater, Ecol. Eng. 35 (2009) 1–17.

[164] S.D. Wallace, 2001. Patent: System for Removing Pollutants from Water. United States: US 6,200,469 B1.

[165] C.M. Weedon, Compact vertical flow constructed wetland systems–first two years' performance, Water Sci. Technol. 48 (2003) 15–23.

[166] S. Wiegel, A. Aulinger, R. Brockmeyer, H. Harms, J. Löffler, H. Reincke, R. Schmidt, B. Stachel, W. von Tümpling, A. Wanke, Pharmaceuticals in the river Elbe and its tributaries, Chemosphere 57 (2004) 107–126.

[167] A. Wiessner, U. Kappelmeyer, P. Kuschk, M. Kästner, Influence of the redox condition dynamics on the removal efficiency of a laboratory-scale constructed wetlands, Water Res 39 (2005) 248–256.

[168] J.H. Writer, L.B. Barber, G.K. Brown, H.E. Taylor, R.L. Kiesling, M.L. Ferrey, N.D. Jahns, S.E. Bartell, H.L. Schoenfuss, Anthropogenica tracers, endocrine disrupting chemicals, and endocrine disruption in Minnesota lakes, Sci. Total Environ. 409 (2010) 100–111.

[169] H. Yamamoto, Y. Nakamura, S. Moriguchi, Y. Nakamura, Y. Honda, I. Tamura, Y. Hirata, A. Hayashi, J. Sekizawa, Persistence and partitioning of eight selected pharmaceuticals in the aquatic environment: laboratory photolysis, biodegradation, and sorption experiments, Water Res. 43 (2009) 351–362.

[170] S. Yang, K. Carlson, Routine monitoring of antibiotics in water and wastewater with a radioimmunoassay technique, Water Res. 38 (2004) 3155–3166.

[171] G.-G. Ying, X.-Y. Yu, R.S. Kookana, Biological degradation of triclocarban and triclosan in a soil under aerobic and anaerobic conditions and comparison with environmental fate modelling, Environ. Pollut. 150 (2007) 300–305.

[172] Y. Yoon, P. Westerhoff, S.A. Snyder, E.C. Wert, Nanofiltration and ultrafiltration of endocrine disrupting compounds, pharmaceuticals and personal care products, J. Membr. Sci. 270 (2006) 88–100.

[173] J.T. Yu, E.J. Bouwer, M. Coelhan, Occurrence and biodegradability studies of selected pharmaceuticals and personal care products in sewage effluent, Agric. Water Manage. 86 (2006) 72–80.

[174] F.M. Zarate Jr., S.E. Schulwitz, K.J. Stevens, B.J. Venables, Bioconcentration of triclosan, methyl-triclosan, and triclocarban in the plants and sediments of a constructed wetland, Chemosphere 88 (2012) 323–329.

[175] J. Zhang, V.W.C. Chang, A. Giannis, J.-Y. Wang, Removal of cytostatic drugs from aquatic environment: a review, Sci. Total Environ. 445–446 (2013) 281–298.

[176] Y. Zhang, S.U. Geissen, C. Gal, Carbamazepine and diclofenac: removal in wastewater treatment plants and occurrence in water bodies, Chemosphere 73 (2008) 1151–1161.

[177] D.Q. Zhang, R.M. Gersberg, T. Hua, J. Zhu, N.A. Tuan, S.K. Tan, Pharmaceutical removal in tropical subsurface flow constructed wetlands at varying hydraulic loading rates, Chemosphere 87 (2012) 273–277.

[178] D.Q. Zhang, R.M. Gersberg, J. Zhu, T. Hua, K.B.S.N. Jinadasa, S.K. Tan, Batch versus continuous feeding strategies for pharmaceutical removal by subsurface flow constructed wetland, Environ. Pollut. 167 (2012) 124–131.

[179] D.Q. Zhang, S.K. Tan, R.M. Gersberg, S. Sadreddini, J. Zhu, N.A. Tuan, Removal of pharmaceutical compounds in tropical constructed wetlands, Ecol. Eng. 37 (2011) 460–464.

[180] J.B. Zhou, M.M. Jiang, B. Chen, G.Q. Chen, Emergy evaluations for constructed wetland and conventional wastewater treatments, Commun. Nonlinear Sci. Numer. Simul. 14 (2009) 1781–1789.

[181] S. Zorita, L. Mårtensson, L. Mathiasson, Occurrence and removal of pharmaceuticals in a municipal sewage treatment system in the south of Sweden, Sci. Total Environ. 407 (2009) 2760–2770.

[182] C. Zwiener, F.H. Frimmel, Short-term tests with a pilot sewage plant and biofilm reactors for the biological degradation of the pharmaceutical compounds clofibric acid, ibuprofen, and diclofenac, Sci. Total Environ. 309 (2003) 201–211.

[183] CCL3, Contaminant candidate list 3, United States Environmental Protection Agency, 2008. Available http://www2.epa.gov/ccl/ccl-3-support-documents-final-ccl-3-and-other-documents (accessed 10 .01.14).

[184] K. Fent, A.A. Weston, D. Caminada, Ecotoxicology of human pharmaceuticals, Aquat. Toxicol. 76 (2006) 122–159.

[185] J. Yin, H. Wang, J. Zhang, N. Zhou, F. Gao, Y. Wu, J. Xiang, B. Shao, The occurrence of synthetic musks in human breast milk in Sichuan, China, Chemosphere 87 (2012) 1018–1023.

[186] K. Kannan, J.L. Reiner, S.H. Yun, E.E. Perrotta, L. Tao, B. Johnson-Restrepo, B.D. Rodan, Polycyclic musk compounds in higher trophic level aquatic organisms and humans from the United States, Chemosphere 61 (2005) 693–700.

[187] D.G.J. Larsson, M. Adolfsson-Erici, J. Parkkonen, M. Pettersson, A.H. Berg, P.-E. Olsson, L. Förlin, Ethinyloestradiol – an undesired fish contraceptive? Aquat. Toxicol. 45 (1999) 91–97.

[188] W. Witte, Medical consequences of antibiotic use in agriculture, Science 279 (1998) 996–997.

[189] T.M. Barbosa, S.B. Levy, The impact of antibiotic use on resistance development and persistence, Drug Resist. Update 3 (2000) 303–311.

[190] A. Schlüter, R. Szczepanowski, N. Kurz, S. Schneiker, I. Krahn, A. Pühler, Erythromycin resistance-conferring plasmid pRSB105, isolated from a sewage treatment plant, harbors a new macrolide resistance determinant, an integron-containing Tn402-like element, and a large region of unknown function, Appl. Environ. Microb. 73 (2007) 1952–1960.

[191] B.G. Kelly, A. Vespermann, D.J. Bolton, Gene transfer events and their occurrence in selected environments, Food Chem. Toxicol. 47 (2009) 978–983.

[192] Y. Zhang, C.F. Marrs, C. Simon, C. Xi, Wastewater treatment contributes to selective increase of antibiotic resistance among Acinetobacter spp, Sci. Total Environ. 407 (2009) 3702–3706.

[193] J.H. Middleton, J.D. Salierno, Antibiotic resistance in triclosan tolerant fecal coliforms isolated from surface waters near wastewater treatment plant outflows (Morris County, NJ, USA), Ecotox. Environ. Safe. 88 (2013) 79–88.

[194] J.M. Brausch, G.M. Rand, A review of personal care products in the aquatic environment: environmental concentrations and toxicity, Chemosphere 82 (2011) 1518–1532.

[195] M. Breitholtz, M. Näslund, D. Sträe, H. Borg, R. Grabic, J. Fick, An evaluation of free water surface wetlands as tertiary sewage water treatment of micro-pollutants, Ecotox. Environ. Safe. 78 (2012) 63–71.

[196] K.M. Lai, M.D. Scrimshaw, J.N. Lester, Prediction of the bioaccumulation factors and body burden of natural and synthetic estrogens in aquatic organisms in the river systems, Sci. Total Environ. 289 (2002) 159–168.

[197] F. Pomati, C. Orlandi, M. Clerici, F. Luciani, E. Zuccato, Effects and interactions in an environmentally relevant mixture of pharmaceuticals, Toxicol. Sci. 102 (2008) 129–137.

[198] T. Smital, B. Kurelec, The chemosensitizers of multixenobiotic resistance mechanism in aquatic invertebrates: a new class of pollutants, Mutat. Res. 399 (1998) 43–53.
[199] T. Luckenbach, I. Corsi, D. Epel, Fatal attraction: synthetic musk fragrances compromise multixenobiotic defense systems in mussels, Mar. Environ. Res. 56 (2004) 215–219.
[200] T. Smital, T. Luckenbach, R. Sauerborn, A.M. Hamdoun, R.L. Vega, D. Epel, Emerging contaminants –pesticides, PPCPs, microbial degradation products and natural substances as inhibitors of multixenobiotic defense in aquatic organisms, Mutat. Res. 552 (2004) 101–117.
[201] G. Imfeld, M. Braeckevelt, P. Kuschk, H.H. Richnow, Monitoring and assessing processes of organic chemicals removal in constructed wetlands, Chemosphere 74 (2009) 349–362.
[202] V. Matamoros, V. Salvadó, Evaluation of a coagulation/flocculation-lamellar clarifier and filtration-UV-chlorination reactor for removing emerging contaminants at full-scale wastewater treatment plants in Spain, J. Enviro. Manag. 117 (2013) 96–102.
[203] O.V. Enick, M.M. Moore, Assessing the assessments: Pharmaceuticals in the environment, Environ. Impact Asses. 27 (2007) 707–729.

Chapter 7

Occurrence and Fate of Pharmaceuticals and Personal Care Products in Wastewater

Trine Eggen[1,*] and Christian Vogelsang[2]
[1]Norwegian Institute of Agriculture & Environment Research, Ås, Norway; [2]Norwegian Institute for Water Research, Oslo, Norway
**Corresponding autor: E-mail: Trine.Eggen@bioforsk.no*

Chapter Outline

Comprehensive Analytical Chemistry, Vol. 67. http://dx.doi.org/10.1016/B978-0-444-63299-9.00007-7

1. INTRODUCTION

Pharmaceuticals and personal care products (PPCPs) enter the environment via regular domestic use [1]. Thousands of different substances are used in medicines such as painkillers, antibiotics, contraceptives, beta-blockers, lipid regulators, tranquilizers, and impotence drugs. During and following administration, humans excrete a combination of intact and metabolized pharmaceuticals that enter wastewater. In addition, there are also thousands of chemicals that are used in personal care products in high consumption volume products, such as skin care products, dental care products, soaps, sunscreen agents, and hairstyle products etc. [2]. Unlike nontopical pharmaceuticals, personal care products, and topical medicines do not have to pass through the human body since they enter the wastewater after their regular use during showering or bathing [3].

In this chapter we will take a closer look at the factors that contribute to determining the fate of PPCPs in conventional wastewater treatment. Processes have been designed and optimized for the removal of suspended solids and compounds responsible for eutrophication (e.g., biodegradable organic matter, nitrogen, and phosphorous), but they fall short for the removal of many more recalcitrant compounds. This results in the inefficient removal of a major number of PPCPs. Numerous studies have not only reported PPCPs in surface waters such as rivers, lakes, and fjords, but also their occurrence in groundwater which are subsequently detected in drinking water [4–8]. It is now well established that wastewater effluents are the main entry routes for PPCPs into the environment through their release into surface waters or via terrestrial systems when sewage effluent is used for irrigation or where sewage sludge is applied as fertilizer or soil conditioner to agricultural land [4,9]. Posttreatment of the secondary effluent by methods such as advanced oxidation, membrane filtration, and adsorption to reduce the environmental risk of selected PPCPs are found to be appropriate, especially for water reuse purposes, aquifer recharge and for protecting drinking water sources, and are also considered here.

This chapter focuses on how structure and physiochemical properties of the individual PPCPs influence their fate in different wastewater treatment processes. Typical concentrations in wastewater influent and effluent and biosolids of selected PPCPs are presented in Table 1. PPCPs which are frequently detected or measured in micrograms per liter or milligrams per kilogram dry weight (dw) concentrations in wastewater or biosolids should be of particular concern. Structure, properties, typical observed percent removal, and rate kinetics if available, for secondary and tertiary treatment processes are presented for some of the PPCPs (Tables 2 and 3). Due to the large variation of observed removal efficiencies in conventional treatment processes, both average and removal range, depending on available data are presented.

For a more thorough overview of the fate of PPCPs in wastewater treatment processes, other review papers are recommended [4,10,11].

2. FACTORS THAT INFLUENCE THE INFLUENT CONCENTRATIONS OF PPCPs TO WASTEWATER TREATMENT PLANTS

The amount of individual PPCPs in the untreated sewage that enters a wastewater treatment plant (WWTP) is influenced by factors such as discharges from local PPCP production facilities, improper disposal of unused PPCP products, usage patterns and total population in the area, connectivity of the population to WWTPs, and overflows and leakages in sewer pipelines [12]. Many pharmaceuticals are partially excreted as metabolites, and the ratio and mixture of metabolites may change during transport to the WWTPs due to biotic and abiotic transformations and degradation occurring within the wastewater. Therefore, metabolites excreted by humans should be accounted for when the mass flux of a compound during wastewater treatment is described [2]. The use of PPCPs may be country and culture specific, and may vary drastically with season. Seasonal changes are particularly prominent in typical tourist areas. Extreme peaks may be expected related to large social events in less densely populated areas (e.g., review by Ref. [10]). Pharmaceuticals that tend to sorb to organic matter and particles, for instance antibiotics, may show enhanced loading to WWTPs during stormwater events due to flush-out of settled matter in the sewer system. Last, but not least, biased pharmaceutical load data due to improper sample collection, preparation, and analysis [13] may lead to erroneous conclusions regarding the presence of pharmaceuticals in the influent. The impact of these factors on the influent concentration of PPCPs to WWTPs is considered below.

TABLE 1 Information of Therapeutic Class, CAS# and Measured Residues in Influent and Effluent Wastewater after Conventional Treatment, and Biosolids of Selected PPCPs

Class	Name CAS#	Wastewater (µg/L)		Ref.	Biosolids (mg/kg dw)	Ref.
		Influent	Effluent			
Analgesic and anti-inflammatories	Diclofenac CAS# 15307-86-5	(<0.001–94.2); (0.105–4.11); avg 0.34/0.25	(<0.001–0.69)/ (0.035–1.95)/avg 0.20/0.31	[4,137,140]	(0.002–0.029)	[77]
	Ibuprofen CAS# 15687-27-1	(<0.001–603)/ (0.17–83.5)/(2.5–45) avg 8.6 Fq:100	(nd–55)/(0.002–95)/ (0.016–4.7) avg 0.51, Fq:92	[4,63,137]	(0.003–0.2)/ (0.053–0.49) avg 0.15, Fq: 100	[63,77]
	Ketoprofen CAS# 22071-15-4	(<0.004–8.56)/ (0.08–5.7)	(0.003–3.92)/ (0.04–1.62)	[4,137]		
	Mefenamic acid CAS# 22071-15-4	(<0.017–1.27)/ (0.14–3.2)/(1.6–3.2)	(<0.005–0.39)/ (0.09–2.40)/(0.8–2.3)	[4,137]		
	Naproxen CAS# 22204-53-1	(<0.002–52.9)/ (1.79–611)/(1.7–25.0) avg 6.28, Fq:100	(<0.005–5.09)/ (0.17–33.9)/(0.02–3.5) avg 0.58 Fq:100	[4,63,137]	(0.0003–0.64)/ (0.003–0.15) avg 0.024 Fq:75	[63,77]

Antibiotics	Azithromycin CAS# 83905-01-5	(0.061–2.5) avg 0.29 Fq:100	(0.057–1.3) avg 0.21 Fq:100	[63]	(0.081–0.85) 0.36 Fq:100	[63]
	Ciprofloxacin CAS# 85721-33-1	(0.017–2.5) avg 0.6 Fq:100	(0.022–0.62) avg 0.2 Fq:81	[63]	(1.78–16.0) avg 6.5 Fq:100	[63]
	Norfloxacin CAS# 70458-96-7	Avg <0.055 Fq:0	Avg <0.03, Fq:0	[63]		
	Ofloxacin CAS# 82419-36-1	(0.006–2.8) avg 0.12 Fq:100	(0.006–0.21) avg 0.045 Fq:100	[63]	(0.15–3.2) avg 0.69 Fq:100/ avg 7.8	[63,74]
	Roxithromycin CAS# 80214-83-1	(0.018–17)/(0.001–0.004) avg 0.002 Fq:33	(0.01–5)/(0.0006–0.009) avg 0.004 Fq:50	[10,63]		
	Sulfamethoxazole CAS# 723-46-6	(<0.003–0.98)/ (0.059–3.1) avg 0.57 Fq:89	(<0.003–1.15)/ (0.033–1.8) avg 0.160 Fq:67	[4,63]	(0.0015–0.051) avg 0.010 Fq:25	[63]
	Tetracycline CAS# 60-54-8	(0.024–0.12) avg 0.053 Fq:81	(0.007–0.036) avg 0.009 Fq:25	[63]	(0.044–0.97) avg 0.24 Fq:100	[63]
	Trimethoprim CAS# 738-70-5	(<0.06–6.8)/(0.079–0.81) avg 0.24 Fq:100	(<0.01–3.05)/ (0.018–0.58) avg 0.17 Fq:100	[4,63]	(0.001–0.14) avg 0.018 Fq:50	[63]
Beta-blockers	Atenolol CAS# 29122-68-7	(0.1–33.1)/(5.1–11.2) avg 7.8	(0.13–7.6)/(0.26–5.9) avg 2.8	[4,136]	(0.00–1.65)	[77]
	Metoprolol CAS# 37350-58-6	(0.002–1.52)/avg 2.4/2.24	(0.003–0.25)/avg 1.39/1.60	[4,140]	(0.015–0.549)	[77]

Continued

TABLE 1 Information of Therapeutic Class, CAS# and Measured Residues in Influent and Effluent Wastewater after Conventional Treatment, and Biosolids of Selected PPCPs—cont'd

Class	Name CAS#	Wastewater (μg/L)		Ref.	Biosolids (mg/kg dw)	Ref.
		Influent	Effluent			
Contrast media	Iopromide CAS# 73334-07-3	(0.01–6.6) avg 2.2	(0.01–9.3) avg 2.5	[10]		
Diabetic	Hydrochlorothiazide CAS# 58-93-5	(2.3–4.8)/(0.61–10)/ avg 1.65/1.46	(1.8–11)/0.67–1.7)/ avg 1.27/1.01	[10,140]	(0.00–1.686)	[77]
	Metformin CAS# 657-24-9	Avg 73.7/84.4/>10/ 101–129	Avg 1.8/1.2/>4/ 2.2–21	[82,140,143]	(0.149–7.8)	[77]
Lipid regulators	Bezafibrate CAS# 41859-67-0	(0.05–1.39)	(0.03–0.67)	[4]		
	Clofibric acid CAS# 882-09-7	(0–0.74)	(nd–0.33)	[4]		
	Gemfibrozil CAS# 25812-30-0	(0.10–17.1)	(<0.0025–5.24)	[4]		

Psychiatric drugs/ Anticonvulsant	Carbamazepine CAS# 298-46-4	(<0.04–3.78)/avg 0.22/0.56	(<0.005–4.60)/avg 0.20/0.54	[4,140]	(0–0.13) Fq:15/16	[138]
	Diazepam CAS# 439-14-5	(<0.001–94.2)/ (0.105–4.11)	(<0.001–0.69)/ (0.035–1.95)	[4,137]	(0.002–0.005) Fq: 2/16	[138]
	Primidone CAS# 125-33-7		max 0.12/max 0.8	[139,145]		
Musk substance	Galaxolide CAS# 1222-05-5	(0.03–25)/avg 14.40/ avg 1.9	(<0.06–2.77)/avg 0.69	[4,141,142]	Avg: 15.61/ avg 3.0	[141,142]
	Tonalide CAS# 1506-02-1	(<0.05–1.93)/avg 2.84/avg 0.58	(<0.05–0.32)/avg 0.21	[4,141,142]	Avg: 2.32/avg 1.5	[141,142]
Antimicrobial agent	Triclosan CAS# 3380-34-5	(0.22–7.0) avg 1.9/ (0.87–1.83)	(0.005–2.5) avg 0.32/(0.05–0.36)	[10,144]	(0.6–46.4), avg (0.63–38.6)	[144]

Data are presented as (Min–max) and average=avg. frequency (fq.) of detection in samples in study by Guerra et al. [63] is shown; influent (n=36), effluent (n=36), and sludge (n=24).

TABLE 2 Structure and Properties Relevant for the Removal Efficiency (Charge at pH 7, pK_a, Log K_{ow}, Log D_{ow}, Sorption) for Selected PPCP

		Properties [10,82]							
Name	Structure	Charge at pH 7.0	MW	pK_a	Log K_{ow}	Log D_{ow}	Log K_d (L/kg)	K_{biol} (L/g*d)	Percentage Sorption Sludge
Atenolol	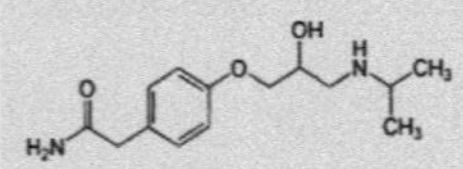	+1	266.3	14.08	0.16	−1.3			<5
Azithromycin	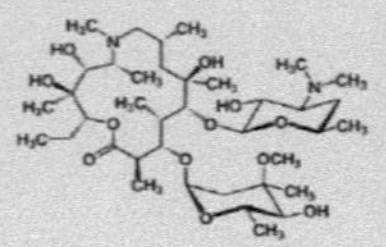	+2	749	8.7/9.5	4.02	2.8	2.5–2.7	0.1–1.2	<10
Bezafibrate	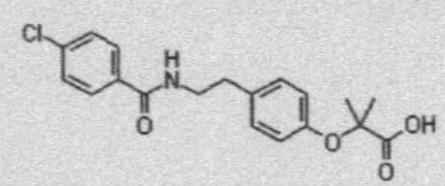	−1	361.8	3.7; 13.6	4.25	2.7	0.77–4.5		
Carbamazepine	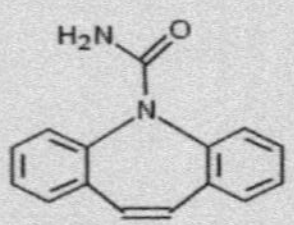	0	236.3	13.9	2.45	2.5	0.1	<0.005–0.1	<5/3/5
Ciprofloxacin	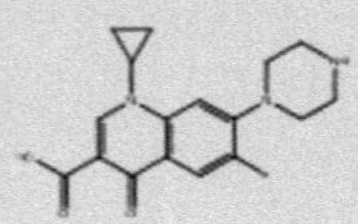	1; zwitterion ; 0;−1	331.4	6.38	0.28	0.3	4.3		70–80/77

Clofibric acid	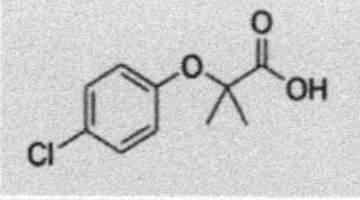	–1	214.7	–3.18	2.57	1.0		0.09–0.8	
Diazepam	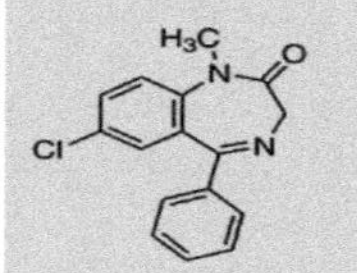	0	284.7	3.4	2.82		1.3	<0.16–0.4	42/64
Diclofenac	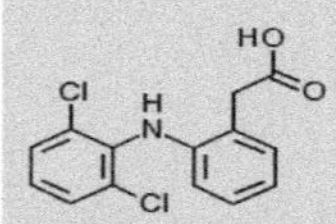	–1	296.2	4.15	4.51	3.0	1.2	0.002–1.2	<5
Galaxolide	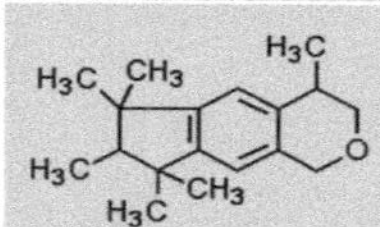	0	258.4				3.7/3.3	5.9	
Hydrochlorothiazide	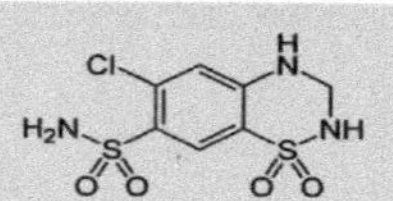	+2	297.7	9.98/8.76	–0.58				100/100
Ibuprofen	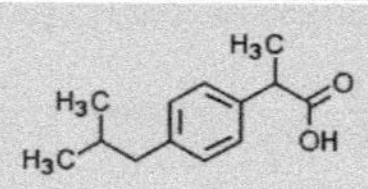	–1	206.3	4.51	3.97/0.45		0.9	1.3–35	<5

Continued

TABLE 2 Structure and Properties Relevant for the Removal Efficiency (Charge at pH 7, pK_a, Log K_{ow}, Log D_{ow}, Sorption) for Selected PPCP—cont'd

Name	Structure	Charge at pH 7.0	MW	pK_a	Log K_{ow}	Log D_{ow}	Log K_d (L/kg)	K_{biol} (L/g*d)	Percentage Sorption Sludge
	Properties [10,82]								
Iopromide	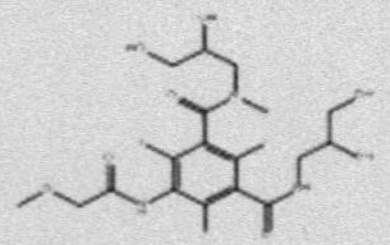	0	791.1		−2.49		1	0.026–2.5	<5
Ketoprofen	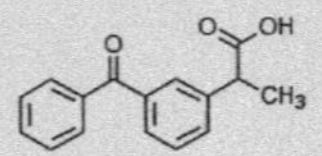	−1	254.3	4.45	3.12	1.7	1.2		0
Mefenamic acid	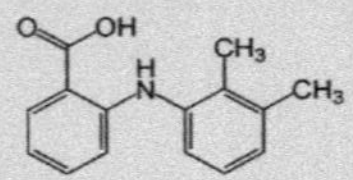	−1	241.3	4.2	5.12	3.7	2.6	0.34–0.40	7
Metformin	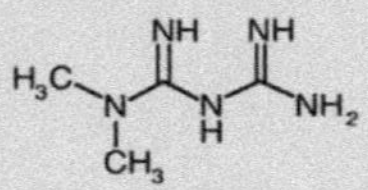	+2	129.17			−3.3			
Metoprolol	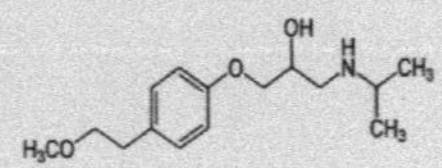	+1	267.4	9.6	1.88	0.4		0.35–0.40	0

Naproxen	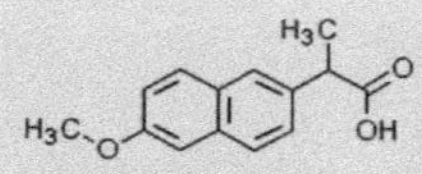	−1	230.3	4.2	3.18	1.7	1.1	0.08–2.9	<5
Ofloxacin	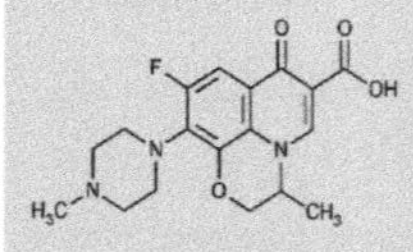	1; zwitterion;0;−1	361.4	5.97	−0.39	−0.4	4.2		60
Primidone	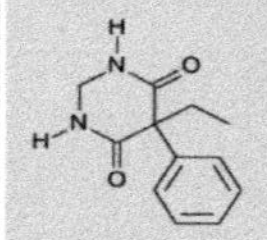	0	218.3		0.91	0.9			
Roxithromycin	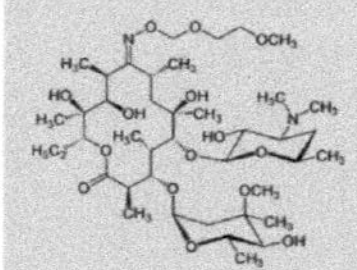	+1	837.1	8.8	2.75	1.7	2.2–2.7	0.022–0.2	<5
Sulfamethoxazole	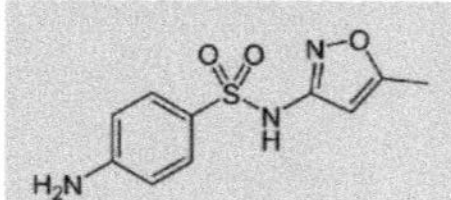	−1	253.3	5.7	0.89	−0.2	2.1–2.7	0.3	0/20
Tetracycline	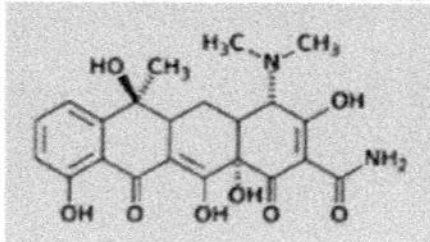	1; zwitterion;0;−1	444.4	3.3/7.7/9.7	−0.18		3.9		

Continued

TABLE 2 Structure and Properties Relevant for the Removal Efficiency (Charge at pH 7, pK_a, Log K_{ow}, Log D_{ow}, Sorption) for Selected PPCP—cont'd

		Properties [10,82]							
Name	Structure	Charge at pH 7.0	MW	pK_a	Log K_{ow}	Log D_{ow}	Log K_d (L/kg)	K_{biol} (L/g*d)	Percentage Sorption Sludge
Trimethoprim	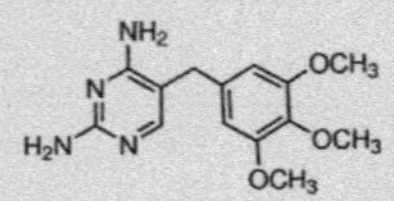	1; 0	290.3	7.2	0.91	0.4	2.2–2.6	0.15	<5
Tonalide	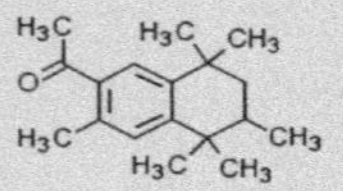	0	258.4				3.7/3.4	5.7	
Triclosan	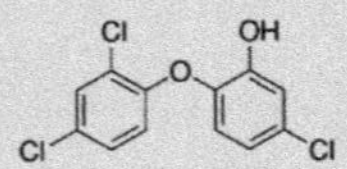	0;–1	289.5						

2.1 Use Patterns and Pathways to Wastewater

Use pattern of PPCPs: In general, organic contaminants, including PPCPs, enter the environment in flux. This is the result of the steady development of new products demanded by industry and agriculture, and also social and demographic changes request new substances such as diagnostic agents or therapeutic drugs for health care and treating diseases more efficiently [14]. The entry of new chemicals to the market is increasingly evaluated for potential environmental hazards by the authorities (REACH program). An environmental risk assessment (ERA) is required for all new marketing authorization applications for medical products in Europe and in the United States, and also for products with an increase in potential for environmental exposure. The European Medicines Evaluation Agency and US Food and Drug Administration have published guidelines for ERA [15,16]. The introduction of new pharmaceuticals might vary between countries and over time [17,18]. The European consumption of pharmaceutical classes, such as antidepressants and antidiabetics, increased by 79% and 82%, respectively, from 2000 to 2010 [19].

Pharmaceutical concentrations in influent wastewaters may be generally predicted from national wholesale and usage statistics, if such are available [20]. However, a range of institutional, social, and cultural factors influence the pharmaceutical use patterns, and the concentrations will vary with place and time. Specific sources such as hospital effluents have shown elevated per capita loads for certain pharmaceuticals; e.g., antibiotics, analgesics, antiflammatories, lipid regulators, and antitumor drugs in addition to pharmaceuticals related to diagnostic and laboratory activity and disinfectants [21–25]. Antibiotics and iodinated contrast media are frequently found in high concentration; e.g., identified in all samples (n = 17) and measured average concentrations of ciprofloxacin and iopamidol were 32 μg/L and 2600 μg/L (before treatment), respectively [26]. Also the corrosion inhibitor benzotriazole related to NMR analysis was found in all samples with an average concentration of 23.6 μg/L [26]. Ciprofloxacin and furosemide have been measured at concentrations of up to 125 μg/L and 392 μg/L, respectively, in hospital effluents [24,25]. Despite the high concentrations of certain compounds, the hospital effluent load of pharmaceuticals typically constitute <10% of the total load to the sewer [22,24,25,27]. Some measured concentration ranges for selected PPCPs in typical influent and effluent wastewater after conventional treatment, and in biosolids are shown in Table 1.

Temporal and spatial concentration variation: Data from surveys of raw sewage from European cities have shown distinct temporal and spatial patterns in illicit drug use across Europe [28,29]. Elevated levels of cocaine and ecstasy were found in sewage samples collected during weekends as compared to weekdays [29]. Also seasonal variations are observed; the antibiotic chloramphenicol typically used to treat infections in the respiratory tract was found at significantly higher concentrations during winter than during summer, while the

TABLE 3 Removal of PPCPs, Presented as Percent, in Selected Secondary and Tertiary Treatment Processes

Name	Rate Constants [10,120,147]					
	Biological degradation	Ozonation	$^{\bullet}OH$		Photochemical degradation	
	k_{biol} [10,147] (L/g*d)	$k_{O3\ pH7/8}$ (M/s)	$k_{^{\bullet}OH}$[a] (10^9M/s)	Ref. O_3 $^{\bullet}OH$	ϕ_{254}[b] [120] (10^{-2}) [mol/ ein-stein]	ε_{254}[b] [120] (10^{-3}) [L/ mol cm]
Atenolol	1.1–1.9	6.3×10^5/ 1.6×10^4	8.0	[98,151]	nd	nd
Azithromycin	0.1–1.2	1.1×10^5/nd	2.9	[117]	nd	nd
Bezafibrate	0.77–4.5	590/nd	7.4	[116]	nd	nd
Carbamazepine	<0.005–0.1	3.0×10^5/ 3.0×10^5	8.8	[116,158]	0.06	6.07
Ciprofloxacin		1.9×10^4/nd	4.1/5.94	[117,120]	1.18	17.2
Clofibric acid	0.09–0.8	nd	nd		nd	nd
Diazepam	<0.16–0.4	0.75/nd	7.2	[116]	nd	nd
Diclofenac	0.002–1.2	1.0×10^6/nd	7.5	[116]	29.2	27.5
Gemfibrozil	1.28	5.0×10^4/nd	10	[150,158]	nd	nd
Ibuprofen	1.3–35	9,6/nd	7.4	[116]	19.2	0.256
Iopromide	0.026–2.5	<0.8/nd	3,3	[116]	nd	nd

Typical Observed Removal Percentage									
Conventional Treatment				Tertiary Treatment					
ommonly onventional*	MBR	CAS	Ref	PAC [82]	Ozone	Ref. ozon	UV_{254}	UV_{254}/H_2O_2	Ref. UV and UV/ H_2O_2
–85.1) avg 3.4/avg 64.5	Avg 65.5	Avg: (<10–76)/0	[4,10, 136,146]	>90	>50-90, <d.l.[c]	[82,149, 155]	nd	nd	
		Avg: (39–74)/avg (<0–55)	[10,59]	70–90	>50–90	[82,117]	nd	nd	
.1–70.5) avg 1.2	Avg 95.6	Avg: (36–99.9)/ avg 48.4	[4,10, 146]	70–90	<50, >50–90	[82,148, 154]	nd	nd	
–62.3) avg: 2.7/avg 23.1	Avg 13.4/avg (0–13)/ avg 3	Avg: (<0–35)/ avg 10	[4,10, 47,136, 140]	>90	>90	[82,148, 149,154, 155]	<50	>50–90, >90	[125, 156]
		Avg: (18–93)	[10]	<70	<50, >50–90	[82,117]	<50	>50–90, <d.l.	[125]
–93.6) avg: 1.4	Avg 71.8	Avg: (28–84)/avg 27.7	[4,10, 146]	nd	nd		nd	nd	
		Avg 8 (SRT 80d)	[10]	nd	>50–90	[154]	nd	nd	
–81.4) avg 1.4/avg 81.4	Avg 17.3 (0–87.4)/ avg 87.4/ avg: <0	Avg (7–60)/ avg 50.1/ avg 41	[4,10, 47,136, 140,146]	70–90	>90	[82,148, 154,155]	>90	>90	[125, 156]
–92.3), avg 3.3/avg 92.3	Avg 98.95, (32.5– 90)/avg 89.6	Avg (5–68)/ avg 38.8	[4,10, 47,136, 146]	70–90	>90, <d.l.[c]	[82,149]	nd	nd	
2–100), avg 1.4/avg 98.2	Avg:96.6, <90	(60–70)	[4,47, 69,136]	70–90	nd		<50	>50–90	[125]
		Avg: (<0–50)/0	[10,69]	<70	<50, >50–90	[82,154]	nd	nd	

Continued

TABLE 3 Removal of PPCPs, Presented as Percent, in Selected Secondary and Tertiary Treatment Processes—cont'd

Name	Rate Constants [10,120,147]					
	Biological degradation	Ozonation	•OH		Photochemical degradation	
	k_{biol} [10,147] (L/g*d)	$k_{O3\ pH7/8}$ (M/s)	$k_{•OH}$[a] (10^9M/s)	Ref. O_3 •OH	ϕ_{254}[b] [120] (10^{-2}) [mol/ein-stein]	ε_{254}[b] [120] (10^{-3}) [L/mol cm]
Ketoprofen	nd	nd	6.89	[120]	29.8	15.3
Mefenamic acid	0.35–0.40	nd	nd		nd	nd
Metoprolol	0.35–0.40	nd	7.84	[120]	3.47	0.565
Naproxen	0.08–2.9/9	nd	8.61	[157]	2.78	4.00
Norfloxacin	nd	nd	1	[120]	0.34	15.4
Ofloxacin	nd	nd	nd		nd	nd
Primidone	nd	nd	6.70	[120]	8.2	0.22
Roxithro-mycin	0.022–0.2/9	nd	nd		nd	nd
Sulfamethox-azole	0.3	2.6×10^6	5.5	[116]	3.79	13.2
Tetracycline		1.9×10^6/nd	7.7	[117,120]	0.38	8.82
Trimethoprim	0.15	2.7×10^5	6.9	[117]	0.118	2.94

(Min–max), avg = average, *not specified conventional processes From Luo et al. [4] and Behera et al. [136]; nd = no data found.
[a]*Rate constant $k_{•OH}$ for Effluent organic matter (EffOM) 0.35–0.79 [152,153]; NHO_2/NO_2^- 10.0 [121]; Carbonate 0.39 [121]; Bicarbonate 0.0085 [121]; Bromide ion 10.6.*
[b]*Average reported value is presented.*
[c]*Residual concentration <LOD, but the low influent concentration did not allow to conclude whether the removal was >90% or not.*

Typical Observed Removal Percentage									
Conventional Treatment			Tertiary Treatment						
ommonly onventional*	MBR	CAS	Ref	PAC [82]	Ozone	Ref. ozon	UV_{254}	UV_{254}/ H_2O_2	Ref. UV and UV/ H_2O_2
0.8–100), avg: 1.7/avg: 94.2	Avg: 70.5, (43.9–95)/ avg 91.9	Avg: (30–92)/ avg 51.5	[4,10, 47,136, 146]	70–90	nd		>90	>90	[125, 156]
<0–70.2) avg: 6.0/avg <0	Avg 74.8	Avg 29.4	[4,136, 146]	>90	nd		nd	nd	
–56.4) avg: 7.6/avg: 23	Avg 58.7/avg 29	Avg: (<10–31)/ avg 42/<0	[4,10, 136,140, 146]	>90	nd		<50	>50–90, >90	[125, 156]
3.3–98.6), avg: 5.5/avg: 95.7	Avg: 40.1, (36–91.6)	Avg: (35–93)/ (40–50)	[4,10, 47,69, 136]	70–90	nd		<50, >50–90	>50 –90, >90	[125, 156]
		Avg (<0–96)	[10]	>90	nd		<50	>50–90	[125]
	Avg 94.0	Avg: (24–84)/ avg 23.8	[10,146]	>90	nd		nd	nd	
	Avg: 12.4		[47]	<70	nd		<50	>50–90, >90	[125]
		Avg: (<0–46)/avg: (<0–38)	[10,59]	nd	nd		nd	nd	
–88.9), avg: 4.6/avg: 51.9	avg: 91.9, (52–80.8)/ avg 60.5	avg: (4.5–100)/ avg 55.6/ avg (<0–60) (3–< 0,9,60)	[4,10, 47,59, 136,146]	<70	>90	[82,149, 154,155]	>50–90, >90	>90	[125, 156]
		Avg: (<0–72)	[10]	nd	>50–90	[117]	<50, >50–90	>90	[156]
<0–81.6), Avg 6.8/avg: 69	Avg: 16.6, (<0–90)	Avg: 16.6, (0–90)/avg (<0–20)	[4,10, 47,59, 136]	>90	>90	[82,148, 154,155]	<50	>50 –90, >90	[125, 156]

insect repellent *N,N*-diethyl toluamide, biocides, and sunscreen agents such as octocrylen or benzophenone-3 have been found in higher concentrations during summer [30,31]. The antiviral agent Oseltamivir™, used for treatment of influenza type A and B [32] exhibited elevated concentration in surface water during times of influenza events because this agent is hardly removed in WWTPs [33]. Also daily and seasonal variations of antibiotics have been measured [34–36].

Pharmaceutical metabolites: Pharmaceutical drugs such as antibiotics are biotransformed by metabolic processes in the human body to facilitate excretion [37]. The drugs may undergo three different processes, also called detoxification pathways, to render them readily extractable [38]. The pharmaceutical is oxidized, reduced, or hydrolyzed and the metabolites may be more active/toxic than the parent compound (phase I). However, many compounds need to undergo conjugation of carboxyl, hydroxyl, amino, and sulfhydryl groups with charged species such as glutathione, sulfate, glycine, or glucuronic acid to be excreted (phase II). Some metabolites may be further modified by e.g., acetylation (phase III). These metabolites are in general more soluble and more susceptible to biotransformation. The anti-inflammatory diclofenac is eliminated via both urinary and biliary excretion. Little or no free unchanged diclofenac is eliminated via urine, approximately 65% of the dose, while the rest is excreted via the bile and feces and include both metabolites and conjugates of unchanged diclofenac, particularly as glucuronide and sulfate conjugates [39].

2.2 Changes and Losses during Transport to the WWTPs

When the PPCPs end up in a WWTP via a sewer pipeline system, a range of factors alter the composition and concentrations of compounds in transit to the WWTP. The wastewater can spend from hours to days in transit, depending on the dimensions of the pipeline network, the water volumes and where the locations of the discharge. Since the pipelines typically are covered with more or less active biofilm and the wastewater itself is rich in active microorganisms, and the retention time in the pipeline can be at least as long as in the bioreactor at the WWTP itself, a range of biotic and abiotic processes may occur. The conjugated unchanged pharmaceuticals are prone to deconjugate in the sewer system and/or in the WWTP sometimes causing a measured increased concentration of the parent compounds [10,39]. Conjugated pharmaceutical metabolites may be deconjugated back to their parent compounds [10,39] or further degraded. Moreover, the parent compounds may be biotransformed [40], and compounds may sorb to organic matter and particles. However, the inherent heterogeneity and complexity of the sewer pipeline network makes a proper and controlled assessment of these processes difficult. The mechanisms controlling these processes should be the same as those determining the fate of PPCPs in WWTP processes and these mechanisms are described in the Sections 3 and 4.

Pharmaceuticals that tend to sorb to organic matter and particles (e.g., some antibiotics) may accumulate in biofilm structures and sewer sediments.

During stormwater events the first-flush effect may remove biofilm and resuspend large amounts of sediments, thereby increasing the wastewater content of these compounds. Stormwater events may also lead to large overflows and losses of untreated wastewater to the receiving aquatic environment, particularly in combined sewer systems.

2.3 Sampling and Analysis for Pharmaceuticals

A proper sampling methodology is needed in order to collect a representative sample from an influent of WWTP. Ort et al. [13] showed that sampling intervals of 5 min or shorter may be required to properly account for temporal PPCP variations. If percent removals of PPCPs over the whole WWTP are sought, the sampling program should take into account the hydraulic retention time (HRT) in the plant by scheduling the sampling of the influent and effluent according to the HRT and/or collecting composite samples over a period that is longer than the HRT [41]. In addition, biased pharmaceutical load data due to improper sample preparation and analysis have been reported [42]. Commonly, only the dissolved fraction of the water sample is analyzed which might result in a significant underestimation of compounds with high sorption constants (K_{OC}). In the last decade, new and improved analytical methodologies have drastically reduced the detection levels (from μg/L to ng/L levels) regarding the quantification of PPCPs in environmental samples. However, authentic reference standards for method development for many pharmaceutical metabolites are still not available and there are numerous analytical challenges still being faced [23].

3. FACTORS INFLUENCING THE FATE OF PPCPs IN PRIMARY AND SECONDARY WASTEWATER TREATMENT PROCESSES

Despite the fact that conventional WWTPs are not designed to remove trace contaminants such as PPCPs, commonly applied treatment units at domestic WWTPs can contribute to the removal of such compounds. The fate of a given compound in a WWTP is influenced by several factors; the wastewater composition, the types, sequence order and physical arrangements of the different treatment processes that are applied, the operational conditions during treatment, and the compound's physicochemical properties. There is great variation in applied configurations for WWTPs around the world. More literature about different treatment systems is found elsewhere (e.g., Metcalf & Eddy [43]).

There are many published papers related to PPCPs and municipal wastewater treatment and observed removal efficiency (See Table 3). The following description of the fate of selected PPCPs in WWTPs is based on a number of recently published comprehensive reviews and selected case studies.

In terms of degradation, organic compounds are either fully degraded (mineralized to water and carbon dioxide) or transformed to metabolites. The use of the term degradation in this chapter refers to transformation of the parent substance and not necessarily complete mineralization.

3.1 Primary and Secondary Wastewater Treatment

The treatment processes in WWTPs are usually divided into pretreatment, primary treatment, and secondary treatment (see Figure 1). The pretreatment is applied primarily to remove larger objects, sand, and grease from the wastewater and to some extent equalize variations in the hydraulic load prior to the subsequent treatment steps. For primary treatment the wastewater is usually pumped to a primary clarifier to remove suspended solids by settling (sometimes enhanced by chemical precipitation) while grease and oil is skimmed off at the surface.

PPCPs that tend to sorb to the suspended solid particles or are soluble in grease and oil, might be partially removed during primary treatment. However, according to Ternes et al. [2] the removal of trace pollutants in a WWTP essentially depends on the biological treatment during secondary treatment. This treatment step is designed primarily to remove biodegradable organic matter, nitrogen, and/or phosphorous. To maintain the necessary microbial activity in the bioreactor, a sufficient part of the biomass needs to be retained. This can be done by a conventional activated sludge (CAS) system in which the sludge is separated in a secondary clarifier and recirculated back to the bioreactor. The removal of phosphorous might be improved by chemical precipitation in a separate treatment step or by coprecipitation in the CAS system. The latter will also improve the settling properties of the sludge. In a biofilm reactor, the biofilm grows on a stationary solid support (e.g., trickling filters, rotating contactors) or on a suspended solid support (e.g., moving bed biological reactor, MBBR) making a secondary clarifier unnecessary. The granular sludge bioreactor is a hybrid system of the CAS and the MBBR as the granulation of the sludge creates a solid support itself (a few millimeters in size) that makes it possible to retain it within the bioreactor using appropriate reactor configurations. Another hybrid system is the membrane biological

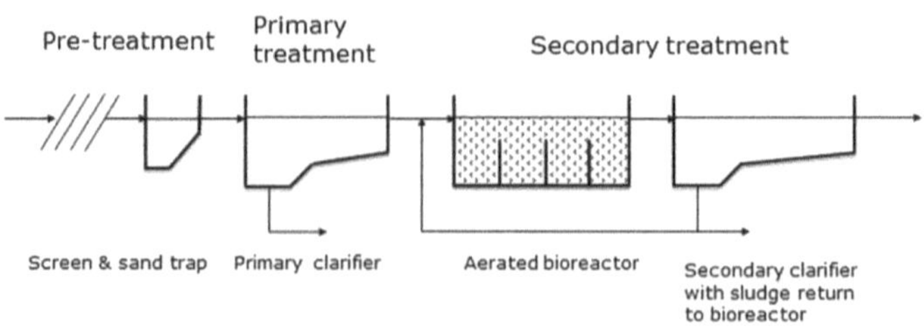

FIGURE 1 An illustration of a typical conventional WWTP containing of a pretreatment, primary treatment, and secondary treatment step.

reactor (MBR) in which the biomass is retained by membrane filtration (usually microfiltration (MF) or ultrafiltration (UF)) either within the bioreactor itself or outside the bioreactor. These different systems of retaining the biomass in the reactor influence the actual sludge age or the average sludge retention time (SRT) and thus, also the microbial community in the reactor. In WWTPs that need to remove nitrogen, the SRT is typically >10 days in order to adequately retain a slow-growing nitrifying community [44]. In the CAS and MBR systems, the SRT is regulated by the amount of the produced biomass that is removed from the systems. In biofilm and granular sludge systems, the sludge age is partially decoupled from the biomass growth and the actual sludge age is literally infinite.

3.2 Removal of PPCPs by Biodegradation

3.2.1 General Considerations Regarding the Biodegradation of Xenobiotics

Xenobiotics (compounds "foreign" to nature) such as PPCPs might be transformed or degraded in the presence of microorganisms that can use them as growth substrates. A prerequisite for transformation is that the xenobiotic is available at an appropriate concentration in the water (given by the microorganisms' affinity for the compound), the environmental conditions are generally suitable, and that compound's structure is degradable by the microorganisms' enzymes. One of the reasons for persistence of xenobiotics is that their unfamiliar structure and functional groups are not readily degraded by common microorganisms. The commonly very low influent concentrations of PPCPs (<1 mg/L) are also a challenge and two scenarios for the biological transformation or degradation of the trace pollutants are likely [2]. If a primary substrate is available for the bacteria to grow on, cometabolism may occur, in which case bacteria break down or partially convert the trace pollutant but do not use it as a carbon source. In another likely scenario, mixed-substrate growth takes place and the bacteria use the trace pollutant as a carbon and energy source and may mineralize it completely.

Organic compounds are commonly classified according to their biodegradability according to their kinetic reaction rate constant (k_{biol}), expressed as liter per gram suspended solid per day ($L/g_{ss}{\cdot}d$), as determined in lab-scale tests, e.g., [45]:

- Very highly biodegradable $k_{biol} > 5\,L/g_{ss} \cdot d$
- Highly biodegradable $1 < k_{biol} < 5\,L/g_{ss} \cdot d$
- Moderately biodegradable $0.5 < k_{biol} < 1\,L/g_{ss} \cdot d$
- Hardly biodegradable $0.5 < k_{biol}\,L/g_{ss} \cdot d$

The kinetic reaction rate constants for several PPCPs are presented in Table 3 and how operational conditions influence degradation is discussed in this section.

3.2.2 Stereochemical Properties Influencing the Biodegradation of PPCPs

A compound's biodegradability depends on its molecular structure, physicochemical properties and available enzymes to transform the structure. In general, the most easily degradable substances are nonbranched linear and unsaturated aliphatic compounds. Polycyclic structures, structures with halogen or other electron-attracting functional groups, or highly branched structures, are less biodegradable and many of these substances are rather persistent in conventional enhanced treatment processes [46,47].

Two groups of compounds which need especial attention are (1) compounds containing electron withdrawing functional groups (the highly oxidized compounds) and (2) compounds with electron donor groups (less oxidized, more reduced compounds). Compounds with electron withdrawing functional groups (e.g., halogens, amides) are less susceptible to oxidation and therefore there is little degradation under aerobic conditions; examples among PPCPs are diclofenac, clofibric acid, diazepam, carbamazepine, ofloxacin, and norfloxacin. On the other hand, electron donor groups (e.g., amines, alcohols, ethers) are generally easier degraded by oxidation; for example, ibuprofen, ketoprofen, gemfibrozil, and atenolol.

Neither the BIOWIN (Biodegradation Probability Program for Windows based only on biodegradation) nor the submodels within BIOWIN showed good prediction for biodegradability of 18 PPCPs in a batch study by Yu et al. [48]. Based on a laboratory scale MBR study with 8 halogenated organic compounds (including 4 pharmaceuticals), the BIOWIN index and the halogen content (more specific index for halogenated compounds) showed both to be insufficient to predict removal of halogenated compounds in MBR (including both biodegradation and sorption processes) [49]. However, the ratio of the halogen content and the pH-dependent partition coefficient, log D_{ow} were better suited to predict removal of the halogenated compounds.

3.2.3 The Fate of Human Pharmaceutical Metabolites

Certain pharmaceutical metabolites return to the parent substance during primary and secondary treatment. Examples of negative removal related to redissolution from bile and feces during wastewater are antibiotics, specifically macrolides such as roxithromycin, erythromycin, clarithromycin, and azithromycin which are mainly excreted with bile and feces [4,10]. Deconjugation or resorption from particles is typically found for fluoroquinolones (ciprofloxacin, ofloxacin, norfloxacin), lincomycin, tetracycline, trimethoprim, and sulfamethoxazole. Other pharmaceuticals that show negative removal during secondary treatment are ibuprofen (hydroxyl- and carboxy-metabolites might be hydrolyzed back to parent compounds [41], diclofenac (due to deconjugation of glucuronidated or sulfated diclofenac or desorption from particles [50], carbamazepine (enzymatic cleavage of glucuronic conjugate [51,52], and iopromide. For iopromide the reason for release is unknown, but the contact media is designed for persistence and the observed negative removal might be related to analytical challenges [10]. Estrone (E1) shows mostly high removal but negative removal has also been observed and is assumed to be related to oxidation of estradiol or deconjugation of other estrogens present [10].

3.2.4 *Plant Operational Factors that Influence the Biotransformation of PPCPs in Bioreactors*

As seen from Table 3, many PPCPs are regarded as readily biodegradable ($k_{biol} > 1\,L/g_{ss}\cdot d$), but the observed removal rates vary greatly both between the different biodegradable compounds and the different conventional WWTPs. Even compounds with general low removal efficiencies might show high removal in some studies. Factors that influence fate processes of PPCPs and might explain some of the observed variation are discussed below.

3.2.4.1 Effect of Biomass Concentration, Redox Conditions, Bioreactor Configuration, and SRT

The SRT, the average time the activated sludge solids are in the system, is one of the most important parameters in biological wastewater processes. In CAS systems that are not required to remove nitrogen, the bioreactors are typically operated with SRTs of just a few days. In CAS systems that require nitrogen removal, the SRTs are usually in the order of 10–20 days—even as high as 50–80 days are used—to retain slow-growing nitrifying bacteria in sufficient amounts in the sludge. The conventional biological nitrogen removal includes both an aerobic stage, in which ammonia is oxidized to nitrite by nitrifying bacteria, and an anoxic stage, where nitrite is reduced to nitrogen gas by denitrifiers. The nitrifying bacteria may play a key role in the biotransformation of pharmaceuticals in WWTPs. It has been shown that their enzyme ammonium monooxygenase, which is necessary for the oxidation of ammonium, is able to co-oxidize pharmaceuticals such as iopromide and trimethoprim [53]. In addition, the high SRTs needed for the nitrifiers to proliferate also improve the proliferation of other slow-growing microorganisms resulting in a more heterogenic microbial community. This increases the probability that some of the microorganisms present in the sludge in appreciable numbers have the ability to transform the PPCPs. Improved removals with increasing SRT have been observed with, e.g., natural estrogens, the synthetic estrogen ethinylestradiol (EE2), naproxene, ibuprofen, bezafibrate, iopromide, trimethoprim, diclofenac, galaxolide, tonalide, and triclosan [45,53–56].

Based on studies by several authors [4,10,45,57,58] the following classification of PPCPs can be found: (1) highly biodegradable compounds under both aerobic and anoxic conditions, e.g., fluoxetine, estrone (E1), and 17β-estradiol (E2) and the musk fragrances galaxolide, tonalide, and celestolide; (2) highly biodegradable only under aerobic conditions, e.g., ibuprofen, naproxen, diclofenac,

17α-ethinylestradiol (EE2), roxithromycin, and erythromycin; and (3) compounds resistant to biodegradation, e.g., carbamazepine, iopromide, diazepam, clofibric acid, sulfamethoxazole, and trimethoprim. It is important to mention that iopromide and trimethoprim have shown improved removal efficiency at SRT 50 days [53], indicating an enhanced removal for certain resistant PPCPs with long SRT.

In contrast to CAS, MBRs also retain microorganisms that are not associated with sludge flocs as long as their doubling time (i.e., growth rate) is shorter than the SRT. This has been shown to have an impact on the heterogeneity of the microbial community in the MBR system as compared to the CAS system [59]. The large fraction of planktonic microorganisms and smaller flocs found in MBR compared with in CAS systems (10–100 μm vs 100–500 μm) also increases the viable fraction of the biomass in MBR due to improved accessibility to the substrates [60]. Furthermore, the SRT in MBR systems is usually kept even higher than in CAS systems with nitrogen removal (typically 15–80 days), which leads to higher biomass concentrations. The higher biomass concentration decreases the food to microorganism ratio, which makes it more likely that the microorganisms will target less energy-rich substrates [61].

3.2.4.2 Effect of the HRT

The HRT is the average length of time that a soluble compound remains in the bioreactor (defined as the volume of the aeration tank divided by the influent flow rate) and controls the time for sorption and biodegradation in water. Commonly, HRT in CAS and MBR is in the range of 2–24 h and 7–15 h, respectively [10]. High variation is observed, generally, compounds that are moderately biodegradable (high k_{biol}) and have low sorption potential (low K_d; see below) show increased removal when the HRT is increased [62]. Enhanced PPCP removal with increased HRT have been reported in a full-scale study that included five WWTPs and a suggested removal with HRT > 16 h [63].

3.2.4.3 Temperature

Biological reactions are greatly influenced by temperature, and lower efficiencies have been observed during winter seasons in colder climates [64]. Higher removal efficiencies of PPCPs in summer than in winter have been reported in Italy [65]. It is expected that countries with average ambient temperature in the range of 15–20 °C will have less seasonal differences than countries in colder climate zones. Less removal efficiency in the winter period might also be combined with less sunlight—and less photodegradation—and a general higher use of pharmaceuticals such as, painkillers and antibiotics (as discussed in Section 2.1). Sunlight will also influence on the removal of photosensitive PPCPs. This is supported by a study where it has been seen highest removal of PPCPs like diclofenac, ketoprofen, naproxen, and triclosan in the sixth or seven lagoon (10 lagoons in series) and not in the first lagoon where the sedimentation was highest and the removal by adsorption is dominating [66].

Optimum temperature in activated sludge systems is typically between 30 and 38 °C corresponding to mesophilic microbial populations, which are the most common. Since few microorganisms grow in the temperature range of 40–45 °C, removal efficiency is less optimal in this range. Thermophilic conditions, temperatures ranging from 49 to 57 °C or up to 70 °C, and processes operating in this temperature range might contribute to biodegradation. In general, lower temperatures resulted in removal rates that were yielded less than optimal condition [67,68].

3.3 Removal of PPCPs by Sorption

All WWTPs produce sludge which is separated from the wastewater and treated separately. Hence, any PPCPs associated with this sludge are also removed from the wastewater. During pretreatment, coarse fragments and grease are separated from the wastewater, and during primary treatment, suspended solids that settle in the primary clarifier are removed. Therefore, PPCPs can be removed during these treatment processes. For example, up to 40% removal of the highly hydrophobic musk fragrances galaxolide and tonalide have been reported [69]. For several PPCPs, the sorption and removal through secondary sludge are appreciable (Table 2).

Sorption of PPCPs may occur though (1) *absorption* due to hydrophobic interactions of the aliphatic and aromatic groups of a compound with the lipophilic cell membrane of the microorganisms or the lipid fractions of the suspended solids and (2) *adsorption* due to electrostatic interactions of positively charged groups of chemicals with the negatively charged surfaces of the microorganisms [10]. Factors such as pH, redox potential, stereochemical structures, and nature of both the sorbent and the sorbed molecule influence the sorption [27].

3.3.1 Factors Influencing the Sorption of PPCPs

The quantity of a compound that will be sorbed to suspended solids in the wastewater at equilibrium (C_{sorbed} [μg/L]), is assumed to be proportional to the concentration of the compound in solution ($C_{dissolved}$ [μg/L]) [70]:

$$C_{sorbed} = K_d \cdot SS \cdot C_{dissolved} \tag{1}$$

K_d is the solid–liquid partition coefficient of the compound (L/kg SS), defined as the partition of the compound between the sludge and the water phase. SS is the concentration of suspended solids in the raw wastewater. Hence, the expected sorption of a compound increases with increasing concentrations of the compound itself, the amount of suspended solids and the compound's partition coefficient.

3.3.1.1 Partition Coefficients and the Influence of pH

Based on the behavior of the well-known and now regulated hydrophobic persistent organic pollutants, it was generally accepted that log K_{ow} values <2.5 imply low sorption to sludge, while compounds with log K_{ow} > 4 tend to sorb strongly to sludge [4]. This was also supported by Ternes et al. [70], who claimed that the removal by sorption in municipal WWTPs is negligible in comparison to the total mass balance for K_d values of less than 500 L/kg·SS (i.e., log K_d < 2.7).

However, in the same period it was observed that the partition, and thereby the K_d value, of certain PPCPs were highly dependent on the pH of the wastewater and was linked to changes in the ionic properties of the compounds. Acidic functional groups such as COOH- and OH- are negatively charged above their pK_a and neutral below their pK_a, while basic functional groups such as NH_2-are positively charged below their pK_a and neutral above their pK_a. Thus, it was suggested that D_{ow}, the pH-dependent derivation of the common octanol–water partition coefficient K_{ow}, is more appropriate partition coefficient than K_{ow} for dissociable PPCPs [70]:

$$D_{ow} = \frac{K_{ow}}{1+10^{pH-pK_a}} \tag{2}$$

For instance, the observed higher sorption of the acidic diclofenac (pK_a 4.2) to primary sludge at pH 6.6 than to secondary sludge at pH 7.5 was explained, at least partially by a higher portion of protonated and neutral diclofenac at pH 6.6 than at pH 7.5 [70]. This could reasonably well describe the eight times higher D_{ow} value (114.4) for diclofenac at pH 6.6 than at pH 7.5 (14.4), hence, implying that the increased sorption at least partially, was caused by increased protonation of diclofenac.

The electrostatic interactions are important for the sorption of positively charged compounds to negatively charged sludge and particles in wastewater treatment, and probably to explain the observed unexpected high concentration in biosolids, in milligrams per kilogram level, for even very hydrophilic but cationic pharmaceuticals such as metformin, atenolol, and hydrochlorothiazide (Table 1 (residue levels) and Table 2 (properties)). Negatively charged compounds will tend to be repulsed by the negatively charged sludge. However, also negatively charged compounds may be partially removed if their K_d values (or D_{ow} values) are high enough to counter the electrostatic repulsion.

The phenomena with observed higher sorption to sludge for certain positively charged PPCPs with low-medium low log D_{ow} than neutral or negatively charge PPCPs with comparable log D_{ow} has been discussed thoroughly by Stevens-Garmon et al. [71]. In their study, five positively charged pharmaceuticals (D_{ow} 0.86–2.89) showed much higher measured K_d values than other PPCPs with comparable D_{ow}. For example, the positive amitriptyline with D_{ow} 2.12, showed a log K_d 3.4–3.7 compared with the neutral carbamazepine and negative atorvastatin (both D_{ow} 2.77) which showed a much lower log K_d in the range of 1.6–2.3 [71]. For compounds with both acidic and basic functional groups, Stevens-Garmon et al. [71] suggested that not only do the charged part, but also the hydrophobic part interfere with and influence the strength of sorption.

The removal of sulfamethoxazole, diclofenac, ibuprofen, and ketoprofen, all with pK_a values in the range 4.2–5.7, were found to be highly influenced by pH [47]. Operating the MBR under acidic conditions instead of neutral has shown to improve the removal of acidic pharmaceuticals [72]. Kimura et al. [73] also reported significant removal of acidic pharmaceuticals (clofibric acid, ibuprofen, ketoprofen, naproxen, and mefenamic acid) with modest pH variation. They suggested both protonation and change from negative to neutral charge and activation of enzymes after decrease of pH as explanations for the high removal efficiency. The nondissociating carbamazepine was not affected by pH [47].

Since some compounds have more than one functional group (zwitterions) with quite different values of pK_a, their net charge can vary greatly over a large pH range. Zwitterions such as ciprofloxacin, ofloxacin tetracycline, and norfloxacin have shown high sorption (up to 70–90%) in sludge [10,63,74].

Since more than 70% of the more than 50,000 compounds in the World Drug Index are dissociable compounds [75], it is important to take into account pH-dependent partitioning for many pharmaceuticals.

3.3.1.2 Effect from Chemical Precipitation

Positively charged coagulants such as iron and aluminum salts are frequently used to improve the removal of phosphorous from wastewater. It is also used in CAS systems (coprecipitation) to improve the settling properties of the activated sludge. The coagulants precipitate certain dissolved compounds (e.g., phosphate) and build larger aggregates of colloidal and suspended solids in the water leading to increased sludge production. Due to the generally low concentrations of PPCPs in wastewater, any precipitation of these compounds found in the dissolved state might be limited. Suárez et al. [76] investigated the efficiency of coagulation–flocculation as a pretreatment of a hospital effluent before discharge to the sewer. As expected, the best removals were observed for the hydrophobic musk compounds galaxolide, tonalide, and celestolide (>77% average removal). The less lipophilic compounds diclofenac, naproxen, and ibuprofen showed somewhat lower removals (12–42%) and nil or poor removal was measured for carbamazepine, sulfamethoxazole, roxithromycin, trimethoprim, erythromicyn, iopromide, and diazepam [76]. There is a paucity of studies on the effect of coagulants, however, it has been demonstrated that type of coagulant and reaction time are shown to influence the removal efficiency [76].

3.3.2 Transfer of PPCPs to Biosolids

As seen from Table 1 the dominating groups of PPCPs that are often found in high concentrations in biosolids (>1 mg/kg dw) can be divided into three groups; (1) hydrophobic compounds with intermediate to high log K_{ow} such as galaxolide and tonalide (log K_{ow}>4), (2) certain hydrophilic dissociable zwitterionic antibiotics with pK_a values ranging from 6 to 8 such as ciprofloxacin, norfloxacin, and ofloxacin, and (3) basic and positively charged hydrophilic compounds in wastewater such as metformin (pK_a 12.25) and atenolol (pK_a 9.6). The musk compounds and triclosan have been detected in concentrations up to 7–30 mg/kg dw and the antibiotics ciprofloxacin, norfloxacin, and ofloxacin have been reported with average values around 6–7 mg/kg dw (Table 1). Negatively charged PPCPs have usually not been found in significant levels in sludge.

Environmental data for the basic pharmaceuticals, e.g., atenolol, hydrochlorothiazide, and metformin (up to 1.65 and 7.8 mg/kg dw in biosolids) [77] are still scarce, but there is a reason to believe there might be significant concentration in the sludge.

3.4 Evaporation of PPCPs

Since pharmaceuticals in general have low fugacity properties toward air, evaporation is not a relevant process for their removal [69]. In fact, the removal was nearly negligible for the intentionally volatile fragrance musk compounds galaxolide and tonalide while it was significant removal (16%) for celestolide [45].

4. FACTORS INFLUENCING THE FATE OF PPCPs IN TERTIARY WASTEWATER TREATMENT PROCESSES

Sand filtration is a commonly applied tertiary treatment processes in WWTPs and is used to remove residual suspended solids and colloidal matter (including particulate BOD) from the secondary clarifier. It is also used as a pretreatment step to subsequent tertiary treatment processes. However, sand filters are not efficient in the removal of residuals of PPCPs [78]. The inadequate removal of some pharmaceuticals of environmental concern by conventional treatment methods (e.g., those marked with an asterisk in Table 1), may therefore advocate more advanced tertiary treatment to reduce their discharge to the environment [10]. Moreover, a growing awareness of water scarcity and the concomitant need to reuse wastewater for high water quality purposes or groundwater recharge has also drawn attention to treatment technologies such as advanced oxidation processes (AOPs), membrane filtration, and adsorption as a means to improve and control the quality of the reclaimed water [4,79]. However, there are rather limited full-scale experiences with these treatment processes for the removal of PPCPs from secondary effluents [80,81]. Often two or more of these processes are combined, for instance, membrane filtration prior to AOP. This would result in the improved removal of effluent organic matter (Efom), thereby reducing the need for oxidant dosing and the potential formation of by-products. Another combination is adsorption by powdered activated carbon (PAC) prior to membrane UF, thus using PAC to adsorb the PPCPs and UF to separate the used PAC and organics/microorganisms from the water [82].

4.1 Membrane Filtration (MF, UF, Nanofiltration, and Reverse Osmosis)

4.1.1 Factors Affecting the Rejection of PPCPs by Membrane Filtration

The use of MF and UF membranes in MBR processes may increase the removal of PPCPs by improved retention of the suspended solids in which the more hydrophobic/neutral PPCPs may be adsorbed. However, the typical pore sizes of MF (100–5000 nm) and UF membranes (10–100 nm) make these membranes generally less efficient for the removal of PPCPs [83,84].

Nanofiltration (NF; 1–10 nm) membranes and reverse osmosis (RO; 0.1–1 nm) are shown to be more suitable [4,84,85]. In such membrane processes the rejection of PPCPs can generally be achieved by size exclusion/steric hindrance, adsorption onto membrane, and/or charge repulsion [85]. The predominating mechanisms and the degree of their relative importance in a given case is dependent on properties of the target compound (molecular weight (MW) and effective diameter (MWd), pK_a, hydrophobicity, and diffusion coefficient); membrane properties (molecular weight cutoff (MWCO), pore size, surface charge, hydrophobicity/hydrophilicity, and surface roughness); operating conditions (pH, ionic strength, hardness, and organic matter); and membrane fouling [85,86]. In a comprehensive literature review by Bellona et al. [85] these factors were included in a rejection diagram that qualitatively allows the prediction of the rejection of a given compound in high-pressure membranes if certain properties of the compound and the membrane are known. In general, if the MW of the compound is larger than the MWCO of the membrane and the compound is charged, the rejection can be expected to be very high due to both steric and electrostatic exclusion. If MW > MWCO, but the compound is noncharged thereby decreasing the potential rejection from electrostatic exclusion, then the compound's log K_{ow} should also be considered. If log $K_{ow} > 2$, the rejection can be expected to be moderate to high, largely depending on partitioning and potential adsorption on the membrane [87]. At Log $K_{ow} < 2$, partitioning has limited effects and the rejection is governed by the MWd of the compound as compared to the pore size of the membrane. If the MWd is smaller than the pore size, the rejection can be expected to be low or moderate. It should be noted that for more rod-like than spherical shaped polar compounds (i.e., high dipole moments; Debye > 3), the tip of the dipole will be directed toward the pore due to electrostatic attraction (e.g., sulfamethoxazole and carbamazepine), which will significantly reduce the rejection as compared to compounds with similar MW that are more spherically shaped (e.g., ibuprofen) [88]. Electrostatic repulsion is a particularly important rejection mechanism for charged compounds for which the MW < MWCO and where the membrane surface is charged. For membranes with low surface charges, the same compounds are expected to be only moderately rejected. For neutral compounds, electrostatic repulsion is less important and the degree of rejection is primarily determined by steric exclusion [85].

Xu et al. [89] found that membrane fouling significantly affected the rejection of organic compounds by cellulose triacetate RO, commercial polyamide NF, and ultra-low pressure RO membranes, but it was less important for thin-film composite RO membranes. Foulant precipitation and cake-layer formation changed the hydrophobicity, surface charge, and surface morphology of the membranes. Due to a greater negatively charged membrane surface, more ionic organic micropollutants were rejected. Moreover, the membrane fouling resulted in an increased adsorption and rejection of hydrophobic nonionic compounds. However, potential membrane swelling due to the increased surface charge could decrease the rejection of hydrophilic nonionic compounds [89].

4.1.2 Observed Rejection of PPCPs by NF and RO

Ng et al. [84] examined the rejection of selected PPCPs by MF, UF, and by RO after different types of pretreatments (combinations of UF, MF, chemical flocculation, and a fiber filter). The PPCPs had a wide variation in MW, pK_a, and log K_{ow}. All compounds (atenolol, carbamazepine, clofibric acid, erythromycin, gemfibrozil, ibuprofen, paracetamol, and sulfamethoxazole) had an effective diameter (0.59–1.17 nm) larger than the nominal pore size of the RO membrane (0.1 nm). Hence, size exclusion was probably an important rejection mechanism and the cause for the high rejection results (94 to >99%) for all compounds and pretreatment methods [84]. However, electrostatic exclusion probably did contribute to the rejection of the negatively charged PPCPs (carbamazepine, clofibric acid, gemfibrozil, ibuprofen, and sulfamethoxazole; as given by their pK_a) by the negatively charged polyamide RO membrane surface at the operating pH [84]. From pilot scale studies that utilized an MF-RO unit, a UF-RO unit, and two MBR-RO units, no rejection to a significant degree of the selected PPCPs by the MF and UF units was reported, while the RO membrane removed all of the 21–29 compounds in all cases down to or below the detection limit of the analyses [83]. Liu et al. [90] reported rejections >98% for four different antibiotics (norfloxacin, ofloxacin, azithromycin, and roxithromycin) by an NF membrane (NFX, polyamine membrane operated at 0.2 MPa). The MWCO of the membrane was ~150–300 Da, and hence, all four compounds had MWs larger than this, and steric exclusion probably played an important role in the rejection.

While RO and tight NF membranes can remove nearly all organic and inorganic contaminants, these rejected compounds are merely concentrated in brine streams which must be disposed off properly. The brine from an RO/NF process would be expected to have much greater toxicity than the influent water [83].

4.2 Adsorption Processes (GAC and PAC)

Adsorption by activated carbon is a well-known process for the removal of various organic contaminants, particularly in drinking water treatment [83]. It is most commonly applied as a powdered feed (PAC; 5–50 μm diameters) or in a granular form (GAC; 100–2400 μm diameters) in packed bed filters. The PAC needs to be separated from the water afterward, for instance, by UF or sand filtration [82]. An advantage of employing PAC is that it can provide fresh carbon continuously or can be used seasonally or occasionally when risk of trace organics is present at a high level [83]. Filter bed volume (GAC), PAC dosage, contact time, and GAC regeneration/replacement rate are the main process control parameters [83]. Used PAC must be disposed through land filling or other solids handling while spent GAC must also be disposed or regenerated.

The adsorption of a compound to activated carbon is dependent on the affinity between the compound and the activated carbon and the availability of sites for adsorption. Adsorption may occur on the outer surface in macropores and mesopores of the activated carbon, but the amount of material adsorbed is regarded as negligible compared to what may be adsorbed within micropores and submicropores [91]. When used in secondary effluents the competition from EfOM can be very high due to direct competition for available adsorption sites and blocking/constriction of the pores [92]. Hence, the different types or levels of treatment prior to the GAC/PAC treatment can significantly influence the adsorption efficiency [83]. It has been observed that wastewater treated only with chemical coagulation prior to PAC treatment showed strong reduction in the PAC adsorption capacity due to the high concentrations of low MW organics which are impossible to remove by precipitation [82].

The fate of more than 40 pharmaceuticals in a large pilot scale PAC-UF process in Switzerland was monitored by 25 sampling programs (24-h and 72-h composite samples) spanning more than a year of operation [82]. Observed affinity of selected pharmaceuticals toward PAC based on the measured removal at a PAC dose of 10–20 mg/L are shown in Table 3. The very high removals (median >90%) found for more or less all the positively charged compounds, independent of their hydrophobicity (log K_{ow} or log D_{ow}) (e.g., atenolol log D_{ow} −1.3 to propranolol D_{ow} 3.48) (Table 2), strongly suggested that the affinity toward PAC for these compounds was governed by electrostatic interactions. It was argued that the adsorbed EfOM would lower the net surface pH to negative of the otherwise neutral or slightly positively charged PAC [94], thereby attracting the positively charged compounds and repulsing the negatively charged compounds. The compounds with medium affinity to PAC, showing 70–90% removal, were predominantly negatively charged substances with hydrophobic properties (e.g., acidic lipid regulators and analgesic and anti-inflammatories, log D_{ow} 1.7–3.4, Table 2) indicating that the electrostatic repulsion could be offset by hydrophobic partitioning, though other mechanisms also contribute to the attraction [93]. It has been reported that for compounds with log D_{ow} > 3.7, hydrophobic partitioning is the dominating mechanism leading to PAC adsorption [93], which explains high removal (>90%), for instance, for mefenamic acid (log D_{ow} 5.12). The somewhat unexpected poor removal (70–90%) of the positively charged azithromycin (log D_{ow} 2.8) was suggested to be caused by exclusion in the micropores of the PAC due to its large size [82]. The rather low and variable removal observed for the compounds with the lowest affinity for PAC (sulfamethoxazole, ciprofloxacin, primidone, and iopromide) was possibly due to different EfOM contents and composition through the one year period of operation [82]. When the PAC dose was increased to 60 mg/L, the removal of these compounds increased to >90%. Higher doses of PAC lead however to higher costs and larger amounts of sludge produced. Contradictable removal has been observed in a full-scale tertiary GAC-plant treating the secondary effluent from a biological WWTP (no nitrification/denitrification) (United Kingdom) over a 3-year period during which the GAC plant was commissioned [95]. For the five selected pharmaceuticals that were found in the influent to the GAC (propranolol, mebeverine, carbamazepine, indomethacine, and diclofenac), they observed removals ranging from 17% for propranolol to >98% for diclofenac. It is known from breakthrough curves that compounds with greater hydrophilicity break through activated carbon faster than hydrophobic compounds [83], indicating that regeneration/replacement rate of the GAC was possibly suboptimal.

Reungoat et al. [78] evaluated and compared the performance of biological activated carbon (BAC) filters during long-term operation. The results demonstrated that BAC had a great potential for removal of PPCP such as diclofenac, carbamazepine, sulfamethoxazole, and gemfibrozil with removal that exceeded 90%. The main mechanism for PPCP removal by the BAC filter appeared to be biodegradation rather than adsorption.

4.3 Advanced Oxidation Processes (Ozonation and UV/H_2O_2)

The term AOP is used for oxidation processes where $^{\bullet}OH$ radicals play an important role. $^{\bullet}OH$ radicals are much more reactive and nonselective than all other oxidizing species that are used in oxidative pollution abatement in water [96]. This also makes AOPs superior to conventional treatment processes. However, when AOPs are applied to secondary effluents to remove PPCPs, the aim is usually to only partially oxidize the pollutants to harmless transformation products since complete mineralization would require excessive oxidant doses.

4.3.1 The Generation of $^{\bullet}OH$ Radicals

Typically $^{\bullet}OH$ is generated in situ by applying oxidants (ozone and/or hydrogen peroxide), a radiation source (UV radiation or gamma rays), ultrasound waves, or catalysts (metal salts, metal oxides, or semiconductors). In most cases two or more of these are combined to accelerate the generation of $^{\bullet}OH$ [97].

When ozone (O_3) is used alone (ozonation) in the treatment of secondary effluents, the generation of $^{\bullet}OH$ radicals will be very dependent on the EfOM. O_3 is a selective oxidant that primarily reacts with electron-rich organic moieties (ERM) such as activated aromatic compounds, deprotonated amines, olefins, and sulfides. Since secondary effluents usually are rather rich in compounds with such functional groups, considerable amounts of •OH radicals may be generate in situ [98–101]. The protonation/deprotonation of amines is dependent on their pK_a values; hence, an increase in pH tends to increase the degree of deprotonation, and therefore, subsequently increase the generation of •OH radicals [99,102–104]. A high degree of hydroxylation makes the moieties more electron rich, which also positively impacts the •OH radical generation [103]. However, since the •OH radical generation itself reduces the amount of the moieties from which they are formed, the rate of generation decreased exponentially with time [99,100,102]. A direct relationship exists between the reduction in the secondary effluents' UV absorption at 254 nm (UV_{254}) and the production of •OH radicals. This could potentially be of high practical value for control strategies in full-scale applications [81]. Significant efforts have been performed in identifying appropriate surrogate parameters for the reactions with the oxidants [102,105–107].

When ozone is added to secondary effluents at doses typically used for micropollutant removal in such waters (1–12 mg O_3/L; [108]) all ozone will usually be decomposed within the first 20 s [99]. However, when the access to ERM is limited and higher ozone doses are applied, ozone decomposition enters a second phase in which •OH radicals are generated through a complex chain of reactions involving multiple radical species [99,109]. The generation of •OH radicals by this autocatalytic decomposition of ozone is not influenced or limited as much by the EfOM or ERM as the one occurring in the first phase, and more stable, but low •OH radical generation rates would be observed [110].

If hydrogen peroxide (H_2O_2) is combined with O_3 (the peroxone process), the generation of •OH radicals is accelerated by direct reaction between the two oxidants [111]. The generation is further accelerated at a higher pH as the dissociated peroxide anion (HO_2^-; pK_a value of 11.6) reacts much faster with O_3 [112]. •OH radicals can also be generated by UV irradiation of the oxidant. For instance, UV radiation of H_2O_2 leads to photolysis of the hydrogen peroxide and the generation of two •OH radicals [112]:

$$H_2O_2 + h\nu \rightarrow {}^{\bullet}OH + {}^{\bullet}OH \tag{3}$$

The efficiency of the formation of •OH radicals is largely determined by the amount of the UV light that reaches the hydrogen peroxide molecules and how well these molecules absorb the UV light. In most UV/H_2O_2 systems, low pressure UV lamps (LP-UV) with a monochromatic light source (253.7 nm) are the most commonly used UV sources [97], but also medium pressure UV lamps (MP-UV) with a polychromatic light source emitting in the whole UV region and partially in the visible region are used [81,113]. Unfortunately, the limited spectral overlap between the absorption spectra of the hydrogen peroxide molecule and the emitted UV radiation from both LP-UV and MP-UV significantly limits the generation of •OH radicals [112]. This is particularly the case if the inner filter effects caused by, for example, aromatic compounds and shielding due to particulate matter are prominent. However, since the dissociated peroxide anion (HO_2^-; pK_a value of 11.6) absorbs $UV_{254\,nm}$ more than 10 times better than the hydrogen peroxide, the •OH radicals generation can be increased by running the oxidation under alkaline conditions [112]. To improve the generation efficiency of •OH radicals further, alternative light sources such as Xe-doped Hg arcs [112] and UV LED [114] with strong emissions in the low UV region, where H_2O_2 has a higher molar absorption [112], have been proposed. Also extensive pretreatment such as membrane filtration (from MF to RO) prior to the UV/H_2O_2 process, as is applied in some reuse applications in the USA, will improve the generation efficiency of •OH radicals [81,90].

The absorption spectre of O_3 has a much larger absorption cross-section at 254 nm than H_2O_2, and inner filter effects by, for example, aromatics could therefore be less problematic [112]. However, it has been shown that H_2O_2 is the primary product of ozone photolysis, and the generation efficiency is therefore not significantly improved [112,115]. A wide range of different UV AOP systems have been developed, however, in the treatment of secondary effluents and water reuse applications UV/H_2O_2 has been the preferred technology in full-scale installations and also has been the most studied [81].

4.3.2 Transformation of PPCPs during Ozonation

During ozonation, PPCPs may be directly oxidized by the ozone or may be attacked by the •OH radicals. Only compounds that contain electron-rich functional groups such as amines, phenol, and double bonds, which are reactive with ozone (see Table 4), will be directly oxidized [116,117]. More recalcitrant compounds are characterized by the presence of electron-withdrawing functional groups such as fluoro, nitro, chloro, amide, and carboxyl (Table 4). Hence, this means that many PPCPs will not be readily oxidized by ozone, as indicated by second-order rate constants $\left(k''_{O_3}\right)$ spreading over more than 10 orders of magnitude (10^{-3}–10^{10}/M s) [101,118], as shown for selected PPCPs in Table 3. However, the •OH radicals are much less selective and extremely potent, capable of oxidizing a wide range of organic and inorganic compounds [112]. The •OH radicals attack the organic compounds by three types of reactions as summarized in Table 4: (1) radical addition to double bonds, but not to C=O, (2) hydrogen abstraction, and (3) electron transfer [111,112]. The strong and nonselective oxidative nature of the •OH radicals is revealed by generally very high second-order rate constants $\left(k''_{\bullet OH}\right)$. For more than a thousand organic and inorganic compounds for which these constants have been determined, they vary by only four orders of magnitude and with most second-order constants being in the order of 10^9/M s [99,118]. Second-order rate constants for the same selected PPCPs are shown in Table 3.

Table 3 also shows typical removals of the selected PPCPs reported in literature. As can be seen, all the readily removed compounds have $k''_{O_3} > 10^4$/M s. However, due to the •OH radicals, compounds with much lower second-order rate constants may also be transformed, as shown for diazepam, iopromide,

TABLE 4 Important Reaction Mechanisms of Active Compounds in AOPs and the Chemical Groups They are Reactive with

Active Compound	Important Reaction Mechanisms	Reactive Groups	Important Groups with Limited Reactivity	Ref[a]
•OH radicals	Radical addition to double bonds	C=C[a], C=N, S=O (sulfoxides)	C=O, O–H, C–Cl	[11,112]
	Hydrogen abstraction related to the R–H bond dissociation energy	S–H >> tertiary C–H > secondary C–H > primary C–H		
Ozone	Direct oxidation, cycloaddition, electrophilic substitution	Electron-rich functional groups: C=C, deprotonated N-R_3, aniline, phenol, methoxy groups amines (NH-R_2, NH_2-R)	Electron-withdrawing functional groups: Fluoro, nitro, chloro, amide, carboxyl	[104,109, 116,117, 131,158, 159]
	Decomposition to generate •OH radicals through electron transfer to electron-rich organic moieties in EfOM	Activated aromatic compounds, deprotonated amines (N-R_3, NH-R_2, NH_2-R), C=C of olefins, sulfides	–	

[a]*The radical addition to C=C is promoted by electron-donating and electron-withdrawing substituents [111].*

and bezafibrate. When ozonating secondary effluents it was found that •OH-based oxidation was significant for all transformed compounds that had $k''_{O_3} < 10^4$/M s [99]. This is, however, not the case for the azithromycin, ciprofloxacin, and tetracycline which have reported k''_{O_3} of $1.1 \cdot 10^5$, $1.9 \cdot 10^4$ and $1.9 \cdot 10^6$, respectively [116]. As suggested by Margot et al. [82] for azithromycin, this may be caused by possible sorption of these substances to colloidal particles (submicron particles) that may protect them against ozone attack. Therefore, the exposures to both ozone and •OH radicals should be taken into consideration when predicting the transformation efficiency of selected PPCPs by ozonation [116]. Possible shielding from ozone due to sorption to particles should also be considered. However, the exposure to •OH radicals may be significantly impacted by •OH radical scavengers in secondary effluents. Despite the somewhat lower $k''_{\bullet OH}$ of 0.4–0.8 · 10^9/M s of typical EfOM as compared to the selected PPCPs (3–14 · 10^9/M s), the high concentrations (in the mg/L range) of these compounds as compared to the very low concentrations (in the ng/L to μg/L range) of the targeted PPCPs imply that scavenging of hydroxyl radicals may be significant in secondary effluents and poor transformation efficiencies of the targeted PPCPs. The dissociated forms of the inorganic compounds nitrite (10 · 10^9/M s) and carbonate (0.4 · 10^9/M s L) are also important •OH radical scavengers [119–122]. Therefore, in poorly nitrified or denitrified wastewater, the transformation efficiencies of targeted micropollutants can be low. During normal ozone dosages there will, hence, be extensive and more or less unaltered competition between •OH scavengers and attacks at the target compounds throughout the oxidation process. For the selective ozone, on the other hand, the competition disappears rapidly after the electron-rich moieties present in EfOM are consumed [98].

Dissolved organic carbon (DOC) has been found to be a key water quality parameter that may be used to normalize ozone doses when comparing results obtained with waters of various origins [101]. An ozone dose per DOC (g O_3/g C) to reach 90% transformation of 42 active pharmaceutical ingredients in six separate secondary effluents have been determined [123]. Compounds (e.g., trimethoprim, carbamazepine, naproxen, diclofenac) that were transformed through the application of doses of only 0.4–0.7 g O_3/g DOC were regarded as readily transformable, while moderately transformable compounds (e.g., sulfamethoxazole, citalopram, metoprolol) required 0.7–1.2 g O_3/g DOC, and compounds more or less recalcitrant to transformation by ozone (e.g., ketoprofen, ibuprofen) required 1.5–2.4 g O_3/g DOC. The electron-withdrawing carboxyl groups of ketoprofen and ibuprofen make these compounds more or less recalcitrant to ozonation [123].

4.3.3 Transformation of PPCPs during UV and UV/H_2O_2 Treatment

During UV/H_2O_2 treatment the PPCPs will be subjected to both UV irradiation and the generated •OH radicals. Hence, when predicting the transformation of target compounds by UV/H_2O_2 systems both direct photolysis by UV irradiation and attack by the generated •OH radicals have to be considered. The efficiency of the transformation of a target compound caused directly by the UV irradiation itself is largely dependent on [46,112,120]:

- the fraction of UV light that reaches the compound,
- how well the compound absorbs the light (photons) at the given wavelengths, defined by the compound's molar absorption ε_M (m^2/mol), and
- how many photons that are needed for the transformation, given by the compound's quantum yield Φ_M, which is defined as the fraction of photons that decompose the compound over the total number of photons absorbed by the compound (mol/Einstein).

In Table 3 values for ε_M and Φ_M are shown for some selected pharmaceuticals together with reported removals by direct $UV_{254\,nm}$ photolysis by LP-UV systems of secondary effluents. Some of the pharmaceuticals (diclofenac, ketoprofen, and sometimes sulfamethoxazole) show more than 90% removal, while other such as the antibiotics ciprofloxacin, norfloxacin, and trimethoprim and the psychiatric drugs carbamazepine and primidone showed removals <50%. Due to the broader UV spectrum and higher power input of the MP-UV lamps [124], direct photolysis of trace pollutants may potentially be more significant [112].

In general, the compounds with the highest ε_M and Φ_M values show the highest percentage removal (e.g., diclofenac and ketoprofen, see Table 3). If the applied UV dose is the same for all target compounds, then the expected degradation by UV radiation alone should be more or less proportional to the logarithmic product of ε_M and Φ_M [120]:

$$\ln\left(\frac{[M]_T}{[M]_0}\right) \propto \Phi_M \cdot \varepsilon_M \quad (4)$$

Figure 2 shows percent removal data of selected pharmaceuticals from De la Cruz et al. [125] plotted against the values of the product of ε_M and Φ_M on a log axis. There is a reasonably good fit with the logarithmic curve ($R^2 = 0.83$–0.87) supporting the proposed proportionality.

However, the primary role of the UV irradiation in UV/H_2O_2 is to generate $^{\bullet}OH$ radicals by reaction with H_2O_2 (Eqn (3). The combined removal of the same selected pharmaceuticals is also shown in Table 3. For the pharmaceuticals for which the UV irradiation alone was relatively limited (e.g., the

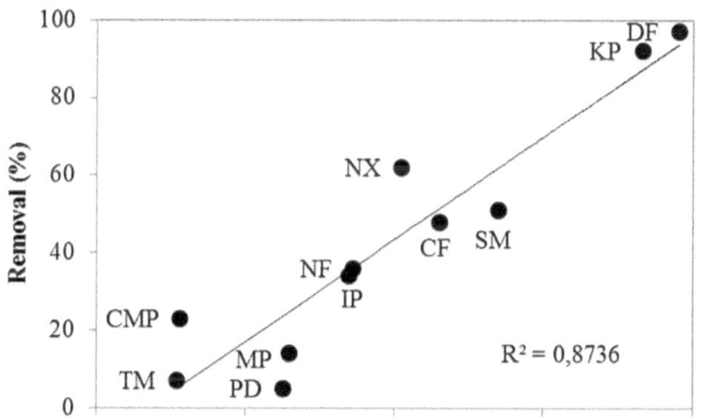

FIGURE 2 Reported removals of selected pharmaceuticals by De la Cruz et al. [125] plotted against the product of reported values for the compounds' quantum yield Φ_M and molar absorption ε_M. The curve shows the logarithmic fit to the data set ($R^2 = 0.87$). CF, ciprofloxacin; CMP, carbamazepine; DF, diclofenac; IP, ibuprofen; KP, ketoprofen; MP, metoprolol; NF, norfloxacin; NX, naproxen; PD, primidone; SM, sulfamethoxazole; TM, trimethoprim.

antibiotics and the psychiatric drugs mentioned above), the percentage removal was significantly improved by the combined UV irradiation and attack by the generated •OH radicals.

4.3.4 Effects on Toxicity and Formation of By-Products

The purpose of the tertiary treatment is to reduce the discharges of PPCPs that can pose harm to the recipient aquatic environment. In membrane filtration and adsorption processes the toxicity is not targeted as such, but just removed from the water. The AOPs need to target the moieties that are essential to the toxic activity in order to be able to influence the specific toxic effect of a given compound [117]. Hence, the transformation that occurs with the targeted compounds should render them at least less toxic, if not completely harmless. In general, the by-products of the transformation processes have higher polarity and are more water soluble than the parent compounds [124,126], and therefore, they are also usually more readily biodegradable. However, ozonation has been reported to both decrease the non-specific toxicity and estrogenicity [78,127] and to lead to increased toxicity and genotoxicity [128,129]. The observed toxicity can originate from bulk organic and inorganic compounds of the EfOM (e.g., *N*-nitrosodimethylamine (NDMA), bromate, or formaldehyde) or from the PPCPs themselves (e.g., hydroxylamines from secondary and tertiary amines, 2,8-dichlorodibenzo-*p*-dioxin (2,8-DCDD) and 2,4-dichlorophenol (2,4-DCP) from triclosan and acridine from carbamazepine [116,130]. However, it has been shown that the observed toxicity can be significantly reduced by a post-biological treatment step [128,129,131] and ozonation should, therefore, normally be followed by a polishing step, e.g., sand filter [131,132]. To be able to cover the effects of by-products and antagonistic or synergistic interactions, chemical analysis (used to determine the fate of the target compounds) should be complemented by the application of bioassays that provide an integrated measure of toxicity.

4.3.5 Feasibility of AOP to Remove PPCPs from Secondary Effluents in Full-scale Treatment

Reported experiences from full-scale treatment of secondary effluents have shown that both ozonation and UV/H_2O_2 have been proven to be effective and practical for the removal of selected PPCPs [80,108,133]. Though, ozonation is generally a more energy efficient process than the UV/H_2O_2 process in producing •OH radicals the low footprint of the UV/H_2O_2 process is an advantage which makes it easier to retrofit in existing treatment plants [81,108]. To control the UV/H_2O_2 process, the lamp power, the number of lamps in operation, and the H_2O_2 dosage are the main process parameters to be varied. However, often the UV and H_2O_2 doses are set based on the worst case scenario, leading to cost-intensive operation. Also in ozonation applications, similar suboptimal control strategies are more the rule than the exception [81]. Hence, the AOPs need to be specifically operated in accordance with its water matrix and targeted contaminant removal [120]. However, proper process control is also limited by the current lack of ability to properly predict how the EfOM influences the oxidant production, scavenging capacity, and the formation of transformation products, particularly for the ozonation process [81]. The use of suitable surrogates for EfOM and mathematical models for effective process control based on real-time assessment of the matrix composition has been discussed in detail by Audenaert [81]. The optimal location of the AOP is after the biological step as the oxidant requirement for DOC oxidation becomes drastically reduced by the biological step [81].

With further developments and improvements, other AOPs may be more favorable than these two AOPs (ozonation and the UV/H_2O_2 process). For instance, the major drawback of the otherwise promising Fenton and photo-Fenton processes [125] has been the large quantities of ferric hydroxide sludge that precipitates at pH>4, posing adverse environmental effects and waste disposal issues [134]. The introduction of heterogeneous catalysts may expand the pH optimum of the processes to neutrality [135].

5. THE PPCPs PROPERTIES AND THE INFLUENCE OF REMOVAL DURING WWTPS

Different WWTPs that apply similar treatment processes show a great variation in observed removal efficiency. As seen in the preceding sections, not only do the properties of each individual PPCP determine their fate in the applied treatment processes, but also the composition of the wastewater and operational conditions will influence all types of treatment processes. For conventional treatment processes factors such as SRT, HRT, temperature, and redox conditions will influence treatment efficiencies, while for tertiary treatment processes, e.g., type and size of filters, amount of active carbon, the regeneration frequency of the active carbon, dose of ozone, light (UV) intensity exposure influence the removal efficiency. In addition, the sampling and analysis methodology as well as deconjugation of pharmaceuticals back to the parent compounds may have a great impact on the observed removal efficiency. In spite of the complexity of influencing factors, some general deductions can be made as discussed below.

In general, sorption plays the dominant role in the removal of individual PPCPs during primary treatment, while biotransformation and/or sorption are the dominating removal mechanisms during secondary wastewater treatment. The advantage of biotransformation is that the compound is actually degraded and not only transferred to the sludge. Since the biotransformation probably seldom ends in complete mineralization, there is still a possibility that the transformation products may pose a risk to the environment upon discharge and/or that it may regenerate back to the parent compound.

The personal care products galaxolide, tonalide, and triclosan as well as the painkiller ibuprofen are commonly classified as readily biodegradable ($k_{biol} > 5\,L/g\,d$) while pharmaceuticals classified as barely degradable ($k_{biol} < 0.5\,L/g\,d$) include carbamazepine, iopromide, diazepam, and clofibric acid. The influences of stereochemical properties such linear or branched side chain as well as the presence of electron-attracting and/or electron-withdrawing groups describe the intrinsic degradation properties. However, optimizing the plant operational parameters might enhance the biodegradation for the more persistent compounds. In particular, increasing the SRT and the HRT to sustain the growth of slow-growing nitrifying bacteria has shown to improve the biotransformation efficiency of low to moderately biodegradable compounds. Strong correlation ($r > 0.8$) between degree of nitrification and removal of certain pharmaceuticals such as several of the antibiotics (norfloxacin, ofloxacin, trimethoprim) in addition, atenolol, bezafibrate, gemfibrozil, and ketoprofen has been reported [82]. Biotransformation may also be improved by retaining the microorganisms that are not associated with flocs, which is the case with the MBR process, thereby increasing the microbial diversity further.

The sorption of PPCPs to particles or biomass and further removal from wastewater via sludge/biosolids is the dominating mechanism for the nondissociating hydrophobic compounds; for instance, galaxolide and tonalide (log $K_{ow} > 4$) which are commonly found in milligrams per kilogram dry weight levels in biosolids.

For the dissociating PPCPs the sorption processes are more complicated. While log K_{ow} is a suitable parameter for the hydrophobic compounds in terms of tendency to sorb, log D_{ow} accounts for electrostatic properties and is thus more appropriate for ionic PPCPs. However, as illustrated in a laboratory scale MBR study by Tadkaew et al. [47] log D_{ow} alone does not explain removal efficiency. Compounds above log $D_{ow} > 3.2$ showed a high removal while for compounds with log $D_{ow} < 3.2$ the removal efficiency varied from less than 20% to nearly complete removal. The compounds containing strong electron-withdrawing groups showed very low removal (<20%) [47]. It has also been reported that the removal of certain acidic pharmaceuticals such as ibuprofen, ketoprofen, sulfamethoxazole, and diclofenac (all pK_a round 4.2–5.7) are highly influenced by pH. For instance, enhanced removal was demonstrated in MBRs under acidic conditions compared to neutral conditions [47,72,73]. A suggested explanation for this pH effect is protonation of these compounds (higher sorption of the neutral and more hydrophobic compounds). However, the activation of enzymes after a change in pH might also be possible. The nondissociating carbamazepine was not affected by pH [47].

BIOWIN did not demonstrate any meaningful correlation for the removal of pharmaceuticals [47–49]. This was not unexpected since removal in MBR includes both biotransformation and sorption processes and not just the biodegradability potential alone, and secondly, only 3 of the 40 test compounds were included in the database used for the development of BIOWIN [47].

Another group of compounds that should be further studied is the strong basic PPCPs that in contrast to other PPCPs with similar log D_{ow} have a generally higher potential for adsorption to negatively charged microbial mass and particles. Stevens-Garmon and coworkers [71] observed that five positively charged pharmaceuticals (D_{ow} 0.86–2.89 at pH 7) showed much higher measured K_d values than other PPCPs with comparable D_{ow}. For instance, the positively charged amitriptyline (D_{ow} 2.12) showed a log K_d in the range of 3.4–3.7 compared with the neutral carbamazepine and negative atorvastatin (both D_{ow} 2.77), which have log K_d in the range of 1.6–2.3 [71].

The highly soluble and strong base metformin (pK_a 12.25, log $K_{ow} < -0.5$) and atenolol (pK_a 9.6, log K_{ow} 0.16) have been measured in sludge from the Nordic countries Faroe Island, Greenland, and Iceland in concentrations as high as 7.8, 1.65 mg/kg, respectively [77]. Other bases such as azithromycin, clarithromycin, and roxithromycin ($pK_a > 8.5$), which are also positively charged at pH around 7, have not shown high transfer to sludge. The mechanisms behind the difference are still not known but it might be related to their high degree of complexity and large molecular size.

For PPCPs that show limited removal due to either low sorption—and further removal via sludge—or low biodegradation during secondary treatment, some of the advanced tertiary treatment processes may be applicable.

In membrane filtration the rejection of a compound by size exclusion is governed by the effective diameter of the compound as compared to the pore size of the membrane, however, the ionic nature of charged compounds may also improve their rejection due to electrostatic repulsion if the membrane is also charged. The rejection of more hydrophobic compounds (log $K_{ow} > 2$) is determined by partitioning and potential adsorption on the membrane if its size does not reject it from penetrating the membrane. Nanofiltration can be used to improve the removal of pharmaceuticals such as azithromycin, carbamazepine, clofibric acid, gemfibrozil, norfloxacin, ofloxacin, roxithromycin, and sulfamethoxazole in secondary effluents.

In GAC and PAC treatment, the adsorption of a compound is dependent on the affinity between the compound and the activated carbon. Positively charged compounds such as atenolol, carbamazepine, metoprolol, ofloxacin, and propranolol have been found to be particularly attracted to the activated carbon due to electrostatic interactions with the net negatively charged surface (due to the adsorbed negatively charged EfOM). Negatively charged compounds (e.g., mefenamic acid) may also adsorb if their hydrophobic properties are strong enough to offset the electrostatic repulsion caused by the negatively charged surface. Since most of the available sites for adsorption are within micropores and submicropores, larger molecules may have fewer sites available and may therefore be less efficiently removed. Also norfloxacin and ofloxacin have shown high removal in spite of the fact that they are highly hydrophilic and not positively charged.

Membrane filtration and GAC/PAC treatment are based on the physical removal of the PPCPs from the wastewater and concentrate them in a brine stream (NF/RO) or on activated carbon, both of which need to be treated separately.

In the AOPs compounds are oxidized by •OH radicals and/or ozone during ozonation and oxidized by •OH radicals and/or transformed by UV irradiation during UV/H_2O_2 treatment. Both the generation of •OH radicals and the efficiency of the oxidation of PPCPs by •OH radicals and ozone and transformation by UV irradiation are highly dependent on the EfOM; high concentrations of EfOM with electron-rich moieties lead to high production of •OH radicals. However, a drawback might be the low selectivity of the •OH radicals in regard to attacking PPCPs typically found in very low concentrations (in ng/L to µg/L) as compared to EfOM concentrations (in mg/L). Ozone is much more selective and primarily attacks compounds with electron-rich functional groups such as carbon double bonds, deprotonated amines and phenols (Tables 4 and 5). Ozonation can be used to improve the removal of carbamazepine and diclofenac which have low removal rates in conventional treatment processes. Ozonation can, e.g., be applied for removal of trimethoprim and sulfamethoxazole. UV irradiation alone is also a viable option if proper pretreatment is applied to improve the UV transmittance of the wastewater to be treated. Hence, UV irradiation can be used to improve the removal of diclofenac and ketoprofen from pretreated secondary effluents. The combination of UV and H_2O_2 will generate•OH radicals that will improve the removal of a large range of compounds in comparison to UV irradiation alone; e.g., carbamazepine, diclofenac, metoprolol, norfloxacin, sulfamethoxazole, tetracycline, and trimethoprim.

As discussed by several authors such as Stevens-Garmon et al. [71] and Tadkaew et al. [46,47], there is no doubt that due to the complexity of the treatment system with its inherent variability in wastewater quantities and composition, sequential treatment processes and associated operating conditions and how these influence the physicochemical properties of the PPCPs, the removal mechanisms are not yet fully understood. An improved understanding of this complexity will make us able to better tailor the removal of selected PPCPs without limiting the removal (or harvesting) of prioritized compounds, possibly also for a reasonable costs.

A simplified overview of some of the PPCPs with low to moderate removal during conventional treatment processes are summarized in Table 5. To illustrate the comparative removal fate of property-like compounds, molecular descriptors such as pK_a, log D_{ow}, in addition to the structure and assumed reactive groups for ozonation are also shown.

TABLE 5 Comparison between Structures, Charge, pK_a, Log D_{ow}, and General Expressed Removal Efficiency for Selected PPCPs

		Descriptors			General Observed Removal					
Name	Structure	Charge at pH 7	pK_a	$logD_{ow}$	Secondary Treatment	Nanofiltration	PAC	Ozonation	UV254 nm	H_2O_2/UV
Atenolol	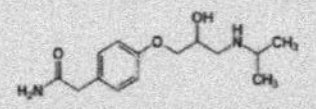	+1	14.08	−1.3	Low/moderate	High	High	Moderate	n.d.	n.d.
Metoprolol	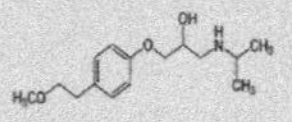	+1	9.6	0.4	Low	n.d.	High	n.d.	Low	Moderate
Azithromycin	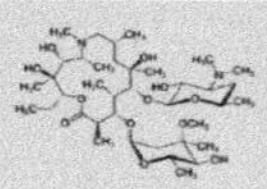	+2	8.7/9.5	2.8	Low/moderate	High	Moderate	Moderate	n.d.	n.d.
Roxithromycin	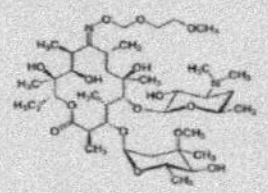	+1	8.8	1.7	Low	High	n.d.	n.d.	n.d.	n.d.
Carbamazepine	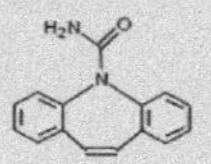	0	13.9	2.5	Low	High	High	High	n.d.	Moderate
Primidone	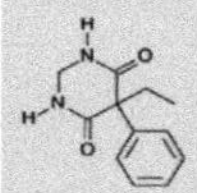	0	11.5	0.9	Low	n.d.	Some	n.d.	Low	Moderate
Ibuprofen	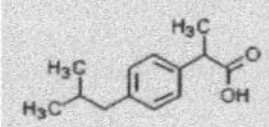	−1	4.9	2.6	High	High	Moderate	n.d.	Some	Moderate

Continued

TABLE 5 Comparison between Structures, Charge, pK_a, Log D_{ow}, and General Expressed Removal Efficiency for Selected PPCPs—cont'd

		Descriptors			General Observed Removal					
Name	Structure	Charge at pH 7	pK_a	$logD_{ow}$	Secondary Treatment	Nanofiltration	PAC	Ozonation	UV254 nm	H_2O_2/UV
Ketoprofen	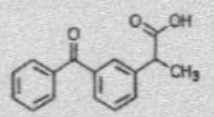	–1	4.5	1.7	Moderate/high	n.d.	n.d.	Moderate	High	High
Clofibric acid	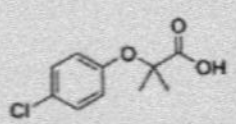	–1	3.5	1.0	Low/moderate	High	n.d.	n.d.	n.d.	n.d.
Diclofenac	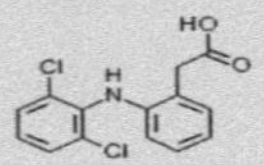	–1	4.1	3.0	Low	n.d.	n.d.	High	High	High
Iopromide	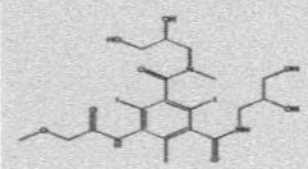	0	11.4	–2.1	Low	n.d.	n.d.	Some	Some	n.d.
Ciprofloxacin	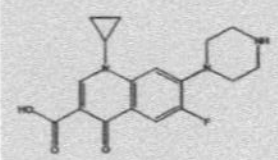	1; Z;0;–1	6.1/8.8	0.3	Moderate	n.d.	n.d.	Some	Some	Moderate
Ofloxacin	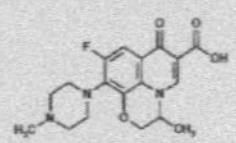	1; Z;0;–1	5.7/7.1	–0.4	Moderate/high	High	High	High	n.d.	n.d.
Tetracycline	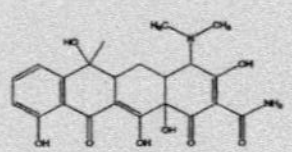	1; Z;0;–1	3.3/7.7/ 9.7		Medium (high variat.)	n.d.	n.d.	n.d.	Moderate	n.d.
Sulfamethox azole	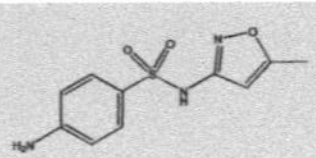	–1	1.8/5.8	–0.2	Medium (high variat.)	High	n.d.	High	Moderate	High

6. CONCLUDING REMARKS

Clean water is crucial in every sense; for the ecosystem and for human beings. After decades with a high focus on sanitation and hygiene worldwide, there is a growing awareness regarding emerging chemical pollutants. Water pollutants disturb the aquatic ecosystem, reduce food security and safety, for example, by restriction of aquatic organisms for food use in contaminated lakes and fjords, or they are transferred to agriculture via irrigation or the application of sewage biosolids. Among scientists, regulators, and the public there is a growing concern related to the thousands of known and unknown chemicals present in our daily life. We do not know which chemicals are present in the environment or in our food, their levels of contamination nor the potential adverse effects and risk that these compounds pose. So far most attention has been given to additives present in all kinds of daily products. However, when additives subsequently enter the environment they become a problem. Even when attention has been drawn to both additives and pharmaceuticals with regard to their potential environmental threats, there is another difficult dimension with pharmaceuticals. This is the unknown response to nontarget organisms in the ecosystem or population in general to bioactive substances outside their designed physiological response. The industry claims that increased costs related to designing eco-friendly pharmaceuticals must be borne by the consumers, meaning more costly medicine and indirectly perhaps, fewer treated patients. So the principle of "polluter pays" is not an obvious and easy way to go. Can emerging compounds to which we are exposed be linked to an increase in new types of health problems?

Wastewater is the major pathway of a wide range of contaminants including PPCPs into the environment. As shown in this chapter, there is still a huge knowledge gap regarding how to remove PPCPs during wastewater treatment processes: which PPCPs would cause a threat to the ecosystem or could transfer to the human food chain, and how to predict PPCPs fate and removal mechanisms during different processes. Avoiding or delimiting the transfer of emerging contaminants via WWTPs must be given the necessary attention and resources to secure both the ecosystem and human health.

REFERENCES

[1] A.B.A. Boxall, L.A. Fogg, P.A. Blackwell, P. Kay, E.J. Pemberton, A. Croxford, Rev. Environ. Contam. Toxicol. 180 (2004) 1–91.

[2] T.A. Ternes, A. Joss, H. Siegrist, Environ. Sci. Technol. 38 (2004) 393A–399A.

[3] C.G. Daughton, I.S. Ruhoy, Environ. Toxicol. Chem. 28 (2009) 2495–2521.

[4] Y. Luo, W. Guoa, H.H. Ngo, L.D. Nghiemb, F.I. Hai, J. Zhang, S. Liang, X.C. Wang, Sci. Total Environ. 473–474 (2014) 619–641.

[5] K.K. Barnes, D.W. Kolpin, E.T. Furlong, S.D. Zaugg, M.T. Meyer, L.B. Barber, Sci. Total Environ. 402 (2008) 192–200.

[6] Z.Q. Ye, H.S. Weinberg, M.T. Meyer, Anal. Chem. 79 (2007) 1135–1144.

[7] M. Benotti, R. Trenholm, B. Vanderford, J. Holady, B. Stanford, S. Snyder, Environ. Sci. Technol. 43 (2009) 597603.

[8] E. Vulliet, C. Cren-Olivé, Environ. Pollut. 159 (2011) 2929.
[9] A.B.A. Boxall, EMBO J. Rep. 5 (2004) 1110–1116.
[10] P. Verlicchi, M. Al Aukidy, E. Zambello, Sci. Total Environ. 429 (2012) 123–155.
[11] K.M. Onesios, J.T. Yu, E.J. Bouwer, Biodegradation 20 (2009) 441–466.
[12] A.B.A. Boxall, Environ. Health Perspect. 120 (2012) 1221–1229.
[13] C. Ort, M.G. Lawrence, J. Reungoat, J.F. Mueller, Environ. Sci. Technol. 44 (2010) 6289–6296.
[14] N.G. Van der Aa, G.J. Kommer, J.E. Van Montfoort, J.F. Versteegh, Water Sci. Technol. 63 (2011) 825–831.
[15] Committee for Medicinal Products for Human Use (Eds.), Guideline on the Environmental Risk Assessment of Medicinal Products for Human Use. EMEA/CHMP/SWP/4447/00 corr 1., London, England, (2006).
[16] US Food and Drug Administration (Eds.), Guidance for Industry for Submission of an Environmental Assessment in Human Drug Applications and Supplements. CMC6, Revision 1, Rockville, MD, USA, (1998).
[17] H. Goossens, M. Ferech, S. Coenen, P. Stephens, Clin. Infect. Dis. 44 (2007) 1091–1095.
[18] R. Alexy, A. Sommer, F.T. Lange, K. Kummerer, Acta Hydrochim. Hydrobiol. 34 (2006) 587–592.
[19] OECD, Health at a Glance: Europe 2012, 2012. http://dx.doi.org/10.1787/9789264183896-38-en.
[20] M. Grung, T. Källqvist, S. Sakshaug, S. Skurtveit, K.V. Thomas, Ecotoxicol. Environ. Saf. 71 (2008) 328–340.
[21] K. Kümmerer (Ed.), Pharmaceuticals in the Environment. Sources, Fate, Effects and Risks, 2001. Germany, Heidelberg.
[22] K.V. Thomas, K. Langford, M. Grung, M. Schlabach, C. Dye, Norwegian Environment Agency, , 2007. TA-2246, Oslo, Norway.
[23] K. Langford, K.V. Thomas, J.Environ. Monit. 13 (2011) 416–421.
[24] A. Hartmann, E.M. Golet, S. Gartiser, A.C. Alder, T. Koller, R.M. Widmer, Arch. Environ. Contam. Toxicol. 36 (1999) 115–119.
[25] K.H. Langford, K.V. Thomas, Environ. Int. 35 (2009) 766–770.
[26] L. Kovalova, H. Siegrist, H. Singer, A. Wittmer, C.S. McArdell, Environ. Sci. Technol. 46 (2012) 1536–1545.
[27] K. Kümmerer, J. Environ. Manage. 90 (2009) 2354–2366.
[28] C. Ort, A.L.N. van Nuijs, J.-D. Berset, et al., Soc. Study Addict. 109 (2014) 1338–1352.
[29] K.V. Thomas, L. Bijlsma, S. Castiglioni, A. Covaci, E. Emke, R. Grabic, F. Hernández, S. Karolak, B. Kasprzyk-Hordern, R.H. Lindberg, M. Lopez de Alda, A. Meierjohann, C. Ort, Y. Pico, J.B. Quintana, M. Reid, J. Rieckermann, S. Terzic, A.L.N. van Nuijs, P. de Voogt, Sci. Total Environ. 432 (2012) 432–439.
[30] Q. Sui, J. Huang, S. Deng, W. Chen, G. Yu, Environ. Sci. Technol. 45 (2011) 3341–3348.
[31] G.A.P. Loraine, M. E, Environ. Sci. Technol. 40 (2006) 687–695.
[32] J.O. Straub, Ecotoxicol. Environ. Saf. 72 (2009) 1625–1634.
[33] J. Fick, R.H. Lindberg, M. Tysklind, P.D. Haemig, J. Waldenstrom, A. Wallensten, B. Olsen, PLoS ONE 10 (2007) 1–5.
[34] A. Göbel, A. Thomsen, C.S. Mcardell, A. Joss, W. Giger, Environ. Sci. Technol. 39 (2005) 3981–3989.
[35] A. Joss, E. Keller, A.C. Alder, A. Gobel, C.S. McArdell, T. Ternes, H. Siegrist, Water Res. 39 (2005) 3139–3152.
[36] B.G. Plósz, H. Leknes, H. Liltved, K.V. Thomas, Sci. Total Environ. 408 (2010) 1915–1924.

[37] K. Kümmerer, A. Henninger, Eur. J. Clin. Microbiol. 9 (2003) 1203–1214.
[38] K. Kümmerer, Annu. Rev. Environ. Resour. 35 (2010) 57–75.
[39] B. Kasprzyk-Hordern, R.M. Dinsdale, A.J. Guwy, Water Res. 42 (2008) 3498–3518.
[40] B.G. Plósz, M.J. Reid, M. Borup, K.H. Langford, K.V. Thomas, Water Res. 47 (2013) 2129–2140.
[41] P.H. Roberts, K.V. Thomas, Sci. Total Environ. 356 (2006) 143–153.
[42] R.H. Lindberg, P. Wennberg, M.I. Johansson, M. Tysklind, B.A. Andersson, Environ. Sci. Technol. 39 (2005) 3421–3429.
[43] G. Tchobanoglous, F.L. Burton, H.D. Stensel (Eds.), Wastewater Engineering: Treatment and Reuse, 2003. New York, USA.
[44] M. Henze, M.C.M. van Loosdrecht, G.A. Ekema, D. Brdjanovic (Eds.), Biological Wastewater Treatment–Principles, Modelling and Design, 2008. London, England.
[45] S. Suárez, J.M. Lema, F. Omil, Water Res. 44 (2010) 3214–3224.
[46] R.P. Schwarzenbach, P.M. Gschwend, D.M. Imboden (Eds.), Environmental Organic Chemistry, Second ed., , 2003. New Jersey, USA.
[47] N. Tadkaew, F.I. Hai, J.A. McDonald, S.J. Khan, L.D. Nghiem, Water Res. 45 (2011) 2439–2451.
[48] J.T. Yu, E.J. Bouwer, M. Coelhan, Agric. Water Manage. 86 (2006) 72–80.
[49] F.I. Hai, K. Tessmer, L.N. Nguyen, J. Kang, W.E. Price, L.D. Nghiem, J. Membr. Sci. 383 (2011) 144–151.
[50] K. Kimura, H. Hara, Y. Watanabe, Desalination 178 (2005) 135–140.
[51] N.M. Vieno, H. Härkki, T. Tuhkanen, L. Kronberg, Environ. Sci. Technol. 41 (2007) 5077–5084.
[52] J. Radjenovic, M. Petrovic, D. Barceló, Anal. Bioanal. Chem. 387 (2007) 1365–1377.
[53] A.L. Batt, S. Kim, D.S. Aga, Environ. Sci. Technol. 40 (2006) 7367–7373.
[54] M. Clara, B. Strenn, O. Gans, H. Kroiss, Water Res. 39 (2005) 97–106.
[55] A.S. Stasinakis, C.I. Kordoutis, V.C. Tsiouma, G. Gatidoum, N.S. Thomaidis, Bioresour. Technol. 101 (2010) 2090–2095.
[56] A. Nikolaou, S. Meric, D. Fatta, Anal. Bioanal. Chem. 387 (2007) 1225–1234.
[57] A. Joss, S. Zabczynski, A. Göbel, B. Hoffmann, D. Löffler, C.S. McArdell, T.A. Ternes, A. Thomsen, H. Siegrist, Water Res. 2006 (2006) 1686–1696.
[58] A. Joss, H. Andersen, T. Ternes, P.R. Richle, H. Siegrist, Environ. Sci. Technol. 38 (2004) 3047–3055.
[59] A. Göbel, C.S. McArdell, A. Joss, H. Siegrist, W. Giger, Sci. Total Environ. 372 (2007) 361–371.
[60] N. Cicek, J.P. Franci, M.T. Suidan, U. Vincent, J.J. Manem, Water Environ. Res. 71 (1999) 64–70.
[61] S. Weiss, T. Reemtsma, Water Res. 42 (2008) 3837.
[62] M. Gros, M. Petrovic, A. Ginebreda, D. Barceló, Environ. Int. 36 (2010) 15–26.
[63] P. Guerra, M. Kim, A. Shah, M. Alaee, S.A. Smyth, Sci. Total Environ. 473–474 (2014) 235–243.
[64] N.M. Vieno, T. Tuhkanen, L. Kronberg, Environ. Sci. Technol. 39 (2005) 8220–8226.
[65] S. Castiglioni, R. Bagnati, R. Fanelli, F. Pomati, D. Calamari, E. Zuccato, Environ. Sci. Technol. 40 (2006) 357–363.
[66] G.G. Ying, R.S. Kookanab, D.W. Kolpinc, J. Environ. Monit. 11 (2009) 1498–1505.
[67] Z. Qiang, H. Dong, B. Zhu, J. Qu, Y. Nie, Chemosphere 92 (2013) 986–992.
[68] Y. Nie, Z. Qiang, H. Zhang, W. Ben, Sep. Purif. Technol. 84 (2012) 9–15.
[69] M. Carballa, F. Omil, J.M. Lema, M. Llompart, C. Garcia-Jares, I. Rodriguez, M. Gómez, T. Ternes, Water Res. 38 (2004) 2918–2926.

[70] T.A. Ternes, N. Herrmann, M. Bonerz, T. Knacker, H. Siegrist, A. Joss, Water Res. 38 (2004) 4075–4084.

[71] J. Stevens-Garmon, J. Drewes, S. Khan, J. McDonald, E. Dickenson, Water Res. 45 (2011) 3417–3426.

[72] N. Tadkaew, M. Sivakumar, S.J. Khan, J.A. McDonald, L.D. Nghiem, Bioresour. Technol. 101 (2010) 1494–1500.

[73] K. Kimura, H. Hara, Y. Watanabe, Water Sci. Technol. 62 (2010) 1084–1089.

[74] A. Jia, Y. Wan, Y. Xiao, J. Hu, Water Res. 46 (2012) 387–394.

[75] D.T. Manallack, Perspect. Med. Chem. 1 (2007) 25–38.

[76] S. Suárez, J.M. Lema, F. Omil, Bioresour. Technol. 100 (2009) 2138–2146.

[77] S. Huber, M. Remberger, A. Goetsch, K. Davanger, L. Kaj, D. Herzke, M. Schlabach, H.Õ. Jörundsdóttir, J. Vester, M. Arnórsson, I. Mortensen, R. Schwartson, M. Dam, Nordic Council of Ministers, 2013. TemaNord 2013:541, Copenhagen, Denmark.

[78] J. Reungoat, B.I. Escher, M. Macova, F.X. Argaud, W. Gernjak, J. Keller, Water Res. 46 (2012) 863–872.

[79] I. Oller, S. Malato, J.A. Sánchez-Pérez, Sci. Total Environ. 409 (2011) 4141–4166.

[80] B.L. Loeb, C.M. Thompson, J. Drago, H. Takahara, S. Baig, Ozone: Sci. Eng 34 (2012) 64–77.

[81] W. Audenaert, (PhD Thesis), Ghent University, Belgium, 2012.

[82] J. Margot, C. Kienle, A. Magnet, M. Weil, L. Rossi, L.F. de Alencastro, C. Abegglen, D. Thonney, N. Chèvre, M. Schärer, D.A. Barry, Sci. Total Environ. 461–462 (2013) 480–498.

[83] S.A. Snyder, S. Adham, A.M. Redding, F.S. Cannon, J. DeCarolis, J. Oppenheimer, E.C. Wert, Y. Yoon, Desalination 202 (2007) 156–181.

[84] K.K. Ng, A.Y.C. Lin, T.H. Yu, C.F. Lin, Sustainable Environ. Res. 21 (2011) 173–180.

[85] C. Bellona, J.E. Drewes, P. Xu, G. Amy, Water Res. 38 (2004) 2795–2809.

[86] A.I. Schäfer, I. Akanyeti, A.J.C. Semião, Adv. Colloid Interface Sci. 164 (2011) 100–117.

[87] Y. Yoon, P. Westerhoff, S.A. Snyder, E.C. Wert, J.R. Yoon, Desalination 202 (2007) 16–23.

[88] L.D. Nghiem, A.I. Schäfer, M. Elimelech, Environ. Sci. Technol. 39 (2005) 7698–7705.

[89] X. Xu, J. Drewes, T. Kim, C. Bellona, G. Amy, J. Memb. Sci. 279 (2006) 165–175.

[90] P. Liu, H. Zhang, Y. Feng, F. Yang, J. Zhang, Chem. Eng. J. 240 (2014) 211–220.

[91] T. Asano, F.L. Burton, H.-L. Leverenz, R. Tsuchihashi, G. Tchobanoglous (Eds.), Water Reuse. Issues, Technologies, and Applications, McGrawHill, New York, USA, 2007.

[92] L.F. Delgado, P. Charles, K. Glucina, C. Morlay, Sci. Total Environ. 435–436 (2012) 509–525.

[93] D.J. de Ridder, A.R.D. Verliefde, S.G.J. Heijman, J.Q.J.C. Verberk, L.C. Rietveld, L.T. van der Aa, J.C. van Dijk, Water Res. 63 (2011) 416–423.

[94] G. Newcombe, J. Colloid Interface Sci. 164 (1994) 452–462.

[95] D.P. Grover, J.L. Zhou, P.E. Frickers, J.W. Readman, J. Hazard. Mater. 185 (2011) 1005–1011.

[96] W.H. Glaze, J.W. Kang, D.H. Chapin, Ozone Sci. Eng. 9 (1987) 335–352.

[97] S.A. Parsons (Ed.), Advanced Oxidation Processes for Water and Wastewater Treatment, IWA Publishing, London, UK, 2004.

[98] Y. Lee, U. von Gunten, Water Res. 44 (2010) 555–566.

[99] M.O. Buffle, J. Schumacher, S. Meylan, M. Jekel, U. von Gunten, Ozone Sci. Eng. 28 (2006) 247–259.

[100] M.O. Buffle, J. Schumacher, E. Salhi, M. Jekel, U. von Gunten, Water Res. 40 (2006) 1884–1894.

[101] M.O. Buffle, U. von Gunten, Environ. Sci. Technol. 40 (2006) 3057–3063.

[102] W.T.M. Audenaert, D. Vandierendonck, S.W.H. Van Hulle, L. Nopens, Water Res. 47 (2013) 2387–2398.
[103] J. Hoigne, in: R.G. Rice, A. Netzer (Eds.), Handbook of Ozone Technology and Applications, Ann Arbor Science, Ann Arbor, 1982.
[104] J. Hoigne, H. Bader, Water Res. 17 (1983) 185–194.
[105] V. Nanaboina, G.V. Korshin, Environ. Sci. Technol. 44 (2010) 6130–6137.
[106] T. Ratpukdi, S. Siripattanakul, E. Khan, Water Res. 44 (2010) 3531–3543.
[107] J.L. Gong, Y.D. Liu, X.B. Sun, Water Res. 42 (2008) 1238–1244.
[108] A. Ried, J. Mielcke, A. Wieland, Ozone Sci. Eng. 31 (2009) 415–421.
[109] F.J. Beltrán (Ed.), Ozone Reaction Kinetics for Water and Wastewater Systems, CRC Press, Florida, USA, 2004.
[110] M.S. Elovitz, U. von Gunten, Ozone Sci. Eng. 21 (1999) 239–260.
[111] C. von Sonntag, Water Sci. Technol. 58 (2008) 1015–1021.
[112] O. Legrini, E. Oliveros, A.M. Braun, Chem. Rev. 93 (1993) 671–698.
[113] W.J. Masschelein (Ed.), Ultraviolet Light in Water and Wastewater Sanitation, CRC Press, Boca Raton Florida, USA, 2002.
[114] S. Vilhunen, M. Sillanpaa, Rev. Environ. Sci. Biotech. 9 (2010) 323–330.
[115] G.R. Peyton, W.H. Glaze, Environ. Sci. Technol. 22 (1988) 761–767.
[116] M.M. Huber, S. Canonica, G.W. Park, U. von Gunten, Environ. Sci. Technol. 37 (2003) 1016–1024.
[117] M.C. Dodd, M. Buffle, U. von Gunten, Environ. Sci. Technol. 40 (2006) 1969–1977.
[118] National Institute of Standards and Technology (Ed.), NDRL/NIST Solution Kinetics Database on the Web. http://kinetics.nist.gov/solution/.
[119] C. Wu, K.G. Linden, Water Res. 44 (2010) 3585–3594.
[120] B.A. Wols, C.H.M. Hofman-Caris, Water Res. 46 (2012) 2815–2827.
[121] G.V. Buxton, C.L. Greenstock, W.P. Helman, A.B. Ross, J. Phys. Chem. Ref. Data 17 (1988) 513–886.
[122] P. Mazellier, E. Leroy, J. De Laat, B. Legube, New J.Chem 26 (2002) 1784–1790.
[123] M.G. Antoniou, G. Hey, S.R. Vega, A. Spiliotopoulou, J. Fick, M. Tysklind, J. la Cour Jansen, H.R. Andersen, Sci. Total Environ. 456–457 (2013) 42–49.
[124] E.J. Rosenfeldt, K.G. Linden, Environ. Sci. Technol. 38 (2004) 5476–5483.
[125] N. De la Cruz, J. Giménez, S. Esplugas, D. Grandjean, L.F. de Alencastro, C. Pulgarín, Water Res. 46 (2012) 1947–1957.
[126] B. Chen, Y. Li, Y. Guo, L. Zhu, J.L. Schnoor, Environ. Sci. Technol. 42 (2008) 1517–1523.
[127] R. Andreozzi, R. Marotta, N. Paxéus, Chemosphere 50 (2003) 1319–1330.
[128] M. Petala, P. Samaras, A. Zouboulis, A. Kungolos, G.P. Sakellaropoulos, Water Res. 42 (2008) 4929–4940.
[129] D. Stalter, A. Magdeburg, J. Oehlmann, Water Res. 44 (2010) 2610–2620.
[130] D.E. Latch, J.L. Packer, B.L. Stender, J. VanOverbeke, W.A. Arnold, K. McNeill, Environ. Toxicol. Chem. 24 (2005) 517–525.
[131] J. Hollender, S.G. Zimmermann, S. Koepke, M. Krauss, C.S. McArdell, C. Ort, H. Singer, U. von Gunten, H. Siegrist, Environ. Sci. Technol. 43 (2009) 7862–7869.
[132] S.G. Zimmermann, M. Wittenwiler, J. Hollender, M. Krauss, C. Ort, H. Siegrist, U. von Gunten, Water Res. 45 (2011) 605–617.
[133] M. Leach, A. Festger, A. Royce, C. Williamson, IUVA News 8 (2006) 11–16.
[134] J.J. Pignatello, E. Oliveros, A. MacKay, Crit. Rev. Environ. Sci. Technol. 36 (2006) 1–84.
[135] S.R. Pouran, A.R.A. Aziz, W.M.A.W. Daud, J. Ind. Eng. Chem. 21 (2014) 53–69.
[136] S.K. Behera, H.W. Kim, J.-E. Oh, H.-S. Park, Sci. Total Environ. 409 (2011) 4351–4360.

[137] A. Ziylan, N.H. Ince, J. Hazard. Mater. 187 (2011) 24–37.

[138] B. Subedi, S. Lee, H.-B. Moon, K. Kannan, Environ. Sci. Technol. 47 (2013) 13321–13329.

[139] X. Yang, R.C. Flowers, H.S. Weinberg, P.C. Singer, Water Res. 45 (2011) 5218–5228.

[140] M. Oosterhuis, F. Sacher, F. ter Laak, Sci. Total Environ. 442 (2013) 380–388 2013.

[141] T. Kupper, C. Plagellat, R.C. Brändi, L.F. de Alencastro, D. Grandjean, J. Tarradellas, Water Res. 40 (2006) 2603–2612.

[142] K. Bester, Chemosphere 57 (2004) 863–870.

[143] M. Scheurer, F. Sacher, H.-J. Brauch, J. Environ. Monitor. 11 (2009) 1608–1613.

[144] Health Canada Environment Canada, Preliminary Assessment, March 2012. http://www.ec.gc.ca/ese. www.ec.gc.ca/ese.

[145] U. Hass, U. Duennbier, G. Massmann, Water Res. 46 (2012) 6013–6022.

[146] D. Barcelo, M. Petrovic, J. Radjenovic, Treating emerging contaminants (pharmaceuticals) in wastewater and drinking water treatment plants, in: M. El Moujabber, L. Mandi, G. Trisorio-Liussi, I. Martin, A. Rabi, R. Rodriguez (Eds.), Technological perspectives for rational use of water resources in the Mediterranean region, CIHEAM, Bari, 2009, pp. 133–140.

[147] S. Suárez, J.M. Lema, F. Omil, Water Res. 44 (2010) 3214–3224.

[148] Q. Sui, J. Huang, S. Deng, G. Yu, Q. Fan, Water Res. 44 (2010) 417–426.

[149] D. Gerrity, S. Gamage, J.C. Holady, D.B. Mawhinney, O. Quiñones, R.A. Trenholm, S.A. Snyder, Water Res. 45 (2011) 2155–2165.

[150] B. Razavi, W. Song, W.J. Cooper, J. Greaves, J. Jeong, J. Phys. Chem. A 113 (2009) 1287–1294.

[151] J. Benner, E. Salhi, T. Ternes, U. von Gunten, Water Res. 42 (2008) 3003–3012.

[152] G. McKay, M.M. Dong, J.L. Kleinman, S.P. Mezyk, F.L. Rosario-Ortiz, Environ. Sci. Technol. 45 (2011) 6932–6937.

[153] P.M. Nagarnaik, B. Boulanger, Chemosphere 85 (2011) 854–860.

[154] H. Schaar, M. Clara, O. Gans, N. Kreuzinger, Environ. Pollut. 158 (2010) 1399–1404.

[155] T.A. Ternes, J. Stüber, N. Herrmann, D. McDowell, A. Ried, M. Kampmann, B. Teiser, Water Res. 37 (2003) 1976–1982.

[156] I. Kim, N. Yamashita, H. Tanaka, J. Hazard. Mater. 166 (2009) 1134–1140.

[157] J.L. Packer, J.J. Werner, D.E. Latch, K. McNeill, W.A. Arnold, Aquat. Sci. 65 (2003) 342–351.

[158] N. Nakada, T. Tanishima, H. Shinohara, K. Kiri, H. Takada, Water Res. 40 (2006) 3297–3303.

[159] Y. Lee, U. von Gunten, Water Res. 46 (2012) 6177–6195.

Chapter 8

Atmospheric Deposition of POPs: Implications for the Chemical Pollution of Aquatic Environments

Javier Castro Jiménez,[1,*] Jordi Dachs[1] and Steven J. Eisenreich[2]
[1]Department of Environmental Chemistry, Institute for Environmental Assessment and Water Research (IDAEA-CSIC), Barcelona, Spain; [2]Department of Hydrology and Hydraulics, and IUPWARE Programme, Vrije Universiteit Brussel (VUB), Brussels, Belgium
**Corresponding author: E-mail: jvcastrojm@gmail.com*

Chapter Outline

1. INTRODUCTION

Persistent organic pollutants (POPs) are chemical substances that persist in the environment, may bioaccumulate through the food web, and pose a risk of toxic effects to human health and especially aquatic ecosystems. Once in the atmosphere, POPs are widely distributed around the globe by a so-called "grasshopper effect" [1] and are variably removed through deposition processes and reactions. Therefore, POPs undergo long-range atmospheric transport (LRAT) depending on properties

Comprehensive Analytical Chemistry, Vol. 67. http://dx.doi.org/10.1016/B978-0-444-63299-9.00008-9

and environmental variables to areas where they are/were not used or produced. The international community has called for global actions to reduce and eliminate releases of these chemicals (UNEP, http://chm.pops.int).

The fact that around 70% of the earth surface is covered by water [2] increases the potential impact of those chemicals in the marine environment (and other major water bodies) as important sinks of POPs. The atmospheric loading of POPs to aquatic environments is required for an accurate assessment of contaminant input pathways and impacts to inland and marine waters. In marine environments, atmospheric deposition occurs at both coastal sites and open sea. At coastal sites, even if atmospheric loadings may be significant, riverine inputs and/or run-off can predominate for some POP families inducing a net volatilization of POPs [3,4]. However, atmospheric deposition is the main source of POPs to the open sea [5,6] and large inland lakes [7].

The importance of the atmospheric pathway for the major POPs families (e.g., polychlorinated biphenyls (PCBs), polychlorinated dibenzo-*p*-dioxins and dibenzofurans (PCDDFs), organochlorine pesticides (OCPs), and polycyclic aromatic hydrocarbons (PAHs)) has been confirmed in the oceans and seas worldwide far from coasts. Comprehensive studies exist for the Atlantic Ocean [8–11], including close to the Arctic and the Southern Oceans [9,12–16], and in the Mediterranean and Black Seas [17–19]. The POPs atmospheric pathway has also proved to be a relevant player in major inland water bodies, such as the Great Lakes in North America [7,20,21], in the large sub-alpine lakes, such as Lake Maggiore (Italy and Switzerland) [22–24], and other high mountain lakes [25,26]. There is considerable evidence of the key role of the atmosphere in delivering POPs to aquatic environments and supporting ecosystem occurrence of POPs. Atmospheric deposition is thought to support the accumulation of POPs in aquatic planktonic organisms, the base of the food web [14,27–29].

Most of these studies investigated the occurrence, deposition, processes, and impacts of POPs with a major research focus. However, when it comes to a regulatory framework for organic chemical pollution management and control in aquatic environments, the question arises: To what extent is atmospheric deposition considered as a source of POPs? There are a number of regional, international, and global regulations dealing with the occurrence and control of POPs in inland and marine waters. Within these programs, key POP families are targeted in various matrices with the objective of identifying hot spots, to assure that current levels are not harmful for biota and humans and to monitoring the effectiveness of control regulations. Potential matrices for POPs monitoring in aquatic environments are water (both dissolved and particle phases) and sediments. Biota samples (e.g., mussels) are often collected and analyzed for POPs too and are used in monitoring programs [30]. The final choice will depend on the pollutant in question and the type of aquatic system. The atmospheric compartment has been identified as key in few monitoring programs aiming at protecting the aquatic environment. A successful pioneering effort has been the Integrated Atmospheric Deposition Network (IADN) in the North American Great Lakes region, which may serve as a model of international involvement.

However, in most cases, atmospheric deposition is not considered nor requested in other legislation such as the Water and Marine Strategy Framework Directives (WFD, MSFD, respectively) in the European Union (EU).

The overall objective of this chapter is to evaluate to what extend the atmosphere is key in controlling the water column concentrations of POPs; to assess the need for the main international regulations on water chemical pollution for considering atmospheric deposition as a relevant input pathway of POPs to aquatic environment, with a particular focus on marine waters. Most important atmospheric deposition processes of POPs are briefly reviewed and discussed in the view of last knowledge available. Moreover, the influence of atmospheric deposition in the POPs bioavailable fraction and background POPs levels in marine waters is discussed.

2. ATMOSPHERIC DEPOSITION OF POPs AND RELEVANCE

2.1 Main Processes

The atmospheric deposition of POPs to aquatic environments can be estimated considering dry deposition, wet deposition, and air–water diffusive fluxes (Figure 1). The relative importance of one process over the other will depend on the physicochemical properties of the pollutant targeted (e.g., solubility, vapor pressure, octanol–water partition coefficient (K_{ow})), and the environmental conditions (e.g., temperature, wind speed, precipitation rates, etc.). For instance, dry deposition may drive the atmospheric loading for those contaminants highly associated with the atmospheric particle phase (e.g., higher molecular weight PCDD/Fs and PAHs), while diffusive fluxes may dominate for those contaminants with a large fraction in the atmospheric gas phase (e.g., PCBs and the low-molecular weight PAHs). In addition, wet deposition may become more relevant in regions with higher precipitation rates or high rain seasons. Deposition fluxes can be

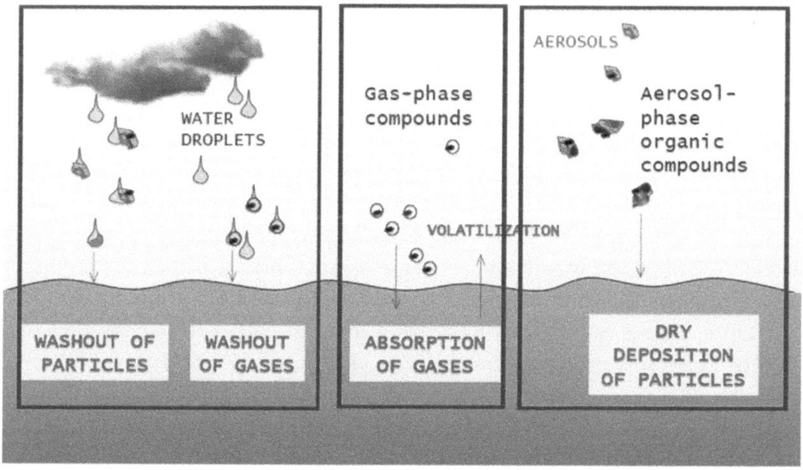

FIGURE 1 Main atmospheric deposition processes: loading of POPs to surface waters.

experimentally measured or estimated by using deposition models. Here, a brief summary of the most common approaches employed to measure and estimate those fluxes is presented. A recent review on sampling procedures for the assessment of atmospheric deposition of POPs, among other contaminants, offering a more complete overview on the methodological approaches can be consulted [31].

2.1.1 Dry Deposition

Atmospheric dry deposition mostly refers to the gravitational settling of atmospheric particles. Due to their physicochemical properties, many POPs can absorb/adsorb to atmospheric particles and then follow their downward way up to the surface. Dry deposition of POPs can be experimentally measured by using deposition samplers [31–34]. However, it is very common to estimate the POPs dry deposition fluxes (F_{DD}, ng/m^{2}d) using the volumetric concentrations of POP in the atmospheric particle phase (C_a, ng/m^3) and the particle deposition velocity (v_d, cm/s) as:

$$F_{DD} = v_d \, C_a \tag{1}$$

Sampling of the atmospheric particle phase is normally achieved by using high volume samplers operating at a various flow rates and durations depending on the target POPs and sampling region. Atmospheric particle phase is collected by using quartz fiber filters (QFFs). Higher volumes will be required for POPs normally present a low levels, such as PCDD/Fs (volumes of ≥1000 m^3 are normally employed) and lower volumes (200–400 m^3) are normally used for more abundant POPs such as some PCB congeners and PAHs. These volumes are only estimative since they may vary depending on the sampling site. Remote sites will generally demand higher sample volumes for a proper quantification of target compounds.

The most important source of uncertainty using this last approach comes from the use of an "unrealistic" or not representative enough v_d value for the study area and compound. Particle deposition velocity depends on many factors such as wind speed, the size distribution of aerosols, atmospheric stability, and relative humidity. In addition, for marine environments, the occurrence of a surface microlayer with reduced surface tension and higher hydrophobicity due to organic compounds and lipids floating at surface also increases the values of v_d [32]. There are very few data on experimentally measured v_d. Most of the measurements have been performed at coastal sites covering a limited geographic area [32,35–38]. Estimates of v_d are even more limited far from the coast, key data for the assessment of POPs dry atmospheric deposition fluxes based on cruise measurements. The first study reporting experimentally measured atmospheric v_d at open sea was from Del Vento and coworkers, which was conducted in the Northeast Atlantic Ocean [33]. They estimated a v_d ranging from 0.08 to 0.3 cm/s. These estimations have been used to calculate the dry deposition fluxes of POPs in various oceanic regions [17,18,39–42]. Additional field measurements of v_d have been recently reported collecting deposition samples over the major oceans of the world during a 7-month circumnavigation campaign [34]. Estimations of v_d (for a set of PAHs) ranging from 0.01 to 10 cm/s were reported. The authors also proposed an empirical parameterization that can be used to estimate the dry deposition velocities of POPs when aerosol samples are taken by using high volume samples (as total suspended matter, TSP), which is the most common approach. This new model has already been used to estimate the dry deposition fluxes of PCDD/Fs and organophosphorus flame retardants (OPEs) to the global oceans, reducing considerably the uncertainty on the calculated dry deposition fluxes [43,44]. The work of González-Gaya and coauthors [34] provides important insight on the variability of dry deposition velocities and main contributing factors at open sea. A better characterization of the size distribution of aerosols and the chemical composition of the surface microlayer and aerosols are highlighted for further research.

2.1.2 Wet Deposition

Wet deposition refers to the POPs depositing with precipitation or snow. First, POPs which are adsorbed to the atmospheric particles will be scavenged by precipitation. Second, POPs will partition between the atmospheric gas phase and the water droplets and eventually find their downward way till the surface [6,45,46] (Figure 1). Similarly to dry deposition, wet deposition fluxes can be experimentally measured by using wet deposition samplers. The wet-only sampler (automatically opens only in the event of precipitation) is the only one exclusively collecting the precipitation. Bulk deposition samplers, gathering both rain and dry deposition are often used [31]. Direct analysis of collected precipitation samples will provide with experimental values. However, it is common to estimate the wet deposition fluxes from the gas and aerosol phase POP concentrations. As explained above, high volume sampling is the most common approach for the collection of air samples. Retention of contaminants in the gas phase can be achieved by using polyurethane foams plugs (PUFs), whereas compound in the particle phase will be trapped by QFFs. A parameterization example successfully implemented for the calculation of wet deposition fluxes of PCDD/Fs and PCBs in the Lake Maggiore area (Italy) [23] is offered here:

The wet deposition flux (F_{WD}, pg/m^{2}d) can be calculated as the product of the precipitation rate P_o (m^3/m^{2}d) and the contaminant concentration in rain water C_r (pg/m^3) that includes both the atmospheric gas and particle phase contributions as explained above:

$$F_{WD} = C_r P_o \tag{2}$$

Contaminant concentrations in precipitation as the result of the scavenging of the atmospheric gas phase by rain drops (C_{rw}) can be calculated using the dimensionless Henry's law constant ($H' = H/RT$) and the measured gas phase concentrations. Henry's law constants (H) values for PCB [47] can be corrected by temperature using Eqn (3):

$$Ln\left(\frac{H_T}{H}\right) = \frac{\Delta H}{R}\left(\frac{1}{T} - \frac{1}{T_T}\right) \tag{3}$$

Where H_T and H are the Henry's law constants (Pa m^3/mol) at the sampling average temperature (T_T) and at standard temperature (T, 298.15 K), ΔH (kJ mol) is the phase transition enthalpy of the compound and R (kJ/K mol) is the gas constant. The correlation of H with temperature proposed by Passivirta and coworkers [48] was used for PCDD/Fs. The scavenging of the particle phase (aerosol) by rain was calculated by multiplying the aerosol concentrations (C_a) by the particle phase washout ratio (W_P), so Eqn (2) becomes,

$$F_{WD} = \left(\frac{C_g}{H'} + W_P\, C_a \right) P_o \tag{4}$$

The value of $W_P = 2 \times 10^5$ was selected for these calculations, consistent with previously proposed W_P values for the same contaminants [6,49,50].

In this case, there is also a source of uncertainty associated to the selection of representative parameters for the study area, such as the washout ratio.

2.1.3 Diffusive Air–Water Exchange

Diffusive deposition refers to the transfer of POPs from the atmospheric gas phase to the water dissolved phase. Contrary to the dry and wet deposition fluxes, the diffusive fluxes cannot be completely experimentally measured and its calculation involves the use of partition and mass transfer coefficients. The most common parameterization for POPs studies is presented hereafter. Indeed, the diffusive deposition refers to the absorption term of the diffusive air–water exchange model (two-film diffusion model), which includes also a volatilization term in order to calculate the contaminant net diffusive flux (Figure 1) Therefore, the diffusive air–water exchange can be calculated by determining the POP absorption (F_{AWAbs}, ng/m^2 d) and volatilization (F_{AWVol}, ng/m^2 d) fluxes as:

$$F_{AWAbs} = -\,10^{-3} k_{AW}\, C_g/H' \tag{5}$$

$$F_{AWVol} = 10^{-3} k_{AW}\, C_w \tag{6}$$

Where k_{AW} is the air–water mass transfer coefficient (m/d), C_g (pg/m^3) and C_w (pg/m^3) are the pollutant concentration in the gas and dissolved phases, respectively, and H' the dimensionless Henry's law constant for the targeted POPs.

The estimation of the diffusive air–water exchange of PAHs in open Mediterranean waters can be considered as a practical example [18]. For these calculations, H' values and their temperature dependence as well as k_{AW} values were estimated as explained elsewhere [5,51]. The nonlinear influence of wind speed was taken into account by correcting the water-side mass transfer coefficient considering a Weibull distribution of wind speeds [5,52]. It has been proved that for POPs predominantly in the atmospheric gas phase, such as PAHs, the diffusive fluxes can be various order of magnitude higher than the dry deposition fluxes [18].

2.2 Relevance in Aquatic Environments

In order to illustrate the relevance of the atmospheric deposition of POPs in aquatic environments, we will present here a number of case studies. We have selected representative aquatic environments, inland and marine waters, where extensive field work has been conducted on the elucidation and quantification of the atmospheric inputs of key POPs families.

2.2.1 Case Study 1: The North American Great Lakes

Measurements and modeling of organic contaminants, especially PCBs, in the Great Lakes system in the 1970s–1980s supported the conclusion that atmospheric deposition represented the principal source to the upper lakes and a significant source to the lower lakes [7,53]. Considering the magnitude of the environmental reservoir available, the general susceptibility of the Great Lakes to deleterious effects and the resistance of many organics to chemical and biological degradation, PCBs and other trace organics present serious environmental and health problem in the Great Lakes Basin for the decades following.

This part of the chapter will address the dominant role of atmospheric concentrations and deposition of POPs in the Great Lakes as a clear indicator that the atmospheric pathway is important and should be considered in international POPs protocols. In the Great Lakes, the International Joint Commission (IJC), Environment Canada and the US Environmental Protection Agency (US EPA) lead the international effort to ascertain the significance of the atmospheric pathway in contaminating the Great Lakes' ecosystems, identify sources, and lead the way toward their mitigation. Especially important in the revisions of the Clean Air Act of 1990 is the establishment of the Great Waters Program, and also establishment of the binational IADN cooperated between Canada and the United States. IADN will be profiled here as a paradigm of atmospheric deposition programs. Also presented are a few specific examples of the relative importance of the atmospheric pathway of POPs through construction of whole lake mass balances.

2.2.1.1 The North American Great Lakes

The North American Great Lakes are major physical, cultural, transport, and economic features of North America and serve a population of ~40 M people within its watershed. Four of the five Great Lakes form an international boundary between Canada and the United States of ~1200 km. Together, the Great Lakes represent ~21% of the world's supply of freshwater, and ~84% of North America's freshwater. The five NA Great Lakes–Superior, Michigan, Huron, Erie, and Ontario–represent the largest freshwater system in the world, and Lake Superior is the largest single freshwater lake by surface area and second by volume to Lake Baikal, Russia (http://www.epa.gov/greatlakes/basicinfo.html).

Individual Great Lakes' surface areas represents up to two-third of the total watershed area (Table 1). The upper Great Lakes are deep and cold with average

TABLE 1 Physical Characteristics of the North American Great Lakes

Lake	Surface Area (km^2)	Avg./Max Depth (m)	Water Retention Time (year)	Length/ Breadth (km)
Superior	82,100	147/406	191	563/257
Michigan	57,800	85/282	99	494/190
Huron	59,600	59/229	22	332/245
Erie	25,700	19/64	2.6	388/92
Ontario	18,960	86/244	6	193/85

Ref.: http://coastwatch.glerl.noaa.gov/statistic/physical.html.

TABLE 2 Susceptibility of Great Lakes to Input of Airborne Organic Contaminants [7]

1. Near and downwind of major pollution sources
2. Large surface areas and surface area/basin area ratio
3. Macrometeorological patterns
4. Long water *and* chemical residence times
5. Slow rate of removal from the water column
   a. Slow sedimentation
   b. Chemically persistent
6. Sensitive biota
7. Dynamic redistribution through circulation and mixing
8. Bio concentration/magnification

depths of 147 (maximum depth is 404 m) and 85 m, respectively, for Superior and Michigan. Water retention times range from 191 years for Superior to 2.6 years for the shallower Lake Erie. The Great Lakes' watershed is dominated by forestry, agriculture, industry, transport, con-urbanizations, tourism/recreation, and commercial/recreational fishing, and is a source of drinking water for tens of millions of people.

With lake surface area comprising from 27% (Ontario) to 64% (Superior) of the total basin area, and lake surface areas ranging from 19,000 (Ontario) to 82,000 km^2 (Superior), it is not surprising that atmospheric deposition of pollutants may play a significant role compared to total input. When coupled with the long water retention times of the lakes (Superior, 190 years; Michigan, 99 years), the relative importance of trace organic contaminant loadings compared to all sources are important or even dominate, and the recovery perhaps long. The susceptibility of the Great Lakes to input of airborne organic pollutants [7] (Table 2) is related specifically to the atmospheric burden of organics in the basin resulting transport from regional and local sources, the inefficiency of effective in-lake removal processes, and the sensitivity of the biota. The Great Lakes are near and downwind of major pollution areas, and meteorological patterns accompanying eastward-moving low- or high-pressure systems distribute emitted pollutants to sensitive areas. Even though in-lake processes of sedimentation and burial and volatilization can be effective physical and chemical natural cleansing processes for many organics, chemical persistence and biotic uptake into the food web are still important.

2.2.1.2 Integrated Atmospheric Deposition Network

In July 1988, the IJC submitted a report to the Water Quality Board outlining a binational plan for addressing the problem of airborne toxic substances in the Great Lakes Basin. A key element of this plan was the establishment of the IADN for the Great Lakes Basin. Annex 15 of the Great Lakes Water Quality Agreement (GLWQA; http://www.epa.gov/greatlakes/glwqa/) sets out actions to be taken by the Parties with the general intent of reducing atmospheric deposition of toxic substances in the Great Lakes Basin. This network was designed to consist of several Master (research grade) Stations augmented by a number of Satellite (routine) Stations. The objectives of this network were to acquire sufficient, quality assured data to estimate with a specified degree of confidence the loading to the Great Lakes Basin of selected toxic substances and the relative importance of the atmospheric pathway from which could be ascertained appropriate control strategies. The IJC report also identified many critical research issues that must be addressed in order to *understand* and *quantify* the importance of the atmospheric pathway for toxic substances. IADN began in 1990 in a pilot phase and has been fully implemented since 1992. The US activities are funded by the US EPA and conducted/coordinated at Indiana University in the United States, and by Environment Canada on the Canadian side.

IADN is a highly successful binational program and a paradigm for worldwide application. Viewed from a global perspective, IADN is a leading international effort in the assessment of the role of the atmospheric impacts of persistent, toxic substances on aquatic systems. It has successfully satisfied its mission to identify and assess the extent of atmospheric deposition of toxic pollutants to the Great Waters (IADN Information Resources, http://www.epa.gov/greatlakes/monitoring/air2/iadn/resources.html; 2002 and 2008 IADN Peer Review Reports accessible from the link).

IADN today (Figure 2) consists of five master monitoring and research stations providing temporal and regional, lake-by-lake assessments, and a set of satellite stations on the Great Lakes positioned to address specific regional issues (e.g., contributions of atmospheric inputs near urban/industrial centers (e.g., Chicago, Cleveland, Toronto) to whole lake loading.

IADN collects gas, particle, and precipitation samples for analysis of selected toxic chemicals at each of its master stations, one on each of the Great Lakes. Details on collection, analysis, and quality assurance of the analysis of

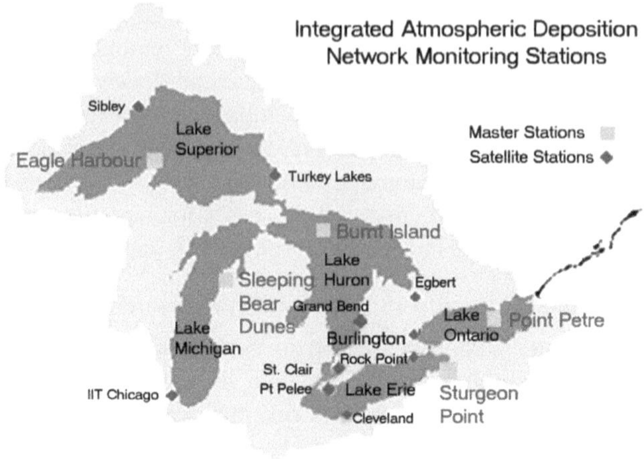

FIGURE 2 Location of IADN sites on the Great Lakes on the Canadian and US sides of the border. *Ref. [54].*

air and precipitation samples can be found at http://www.epa.gov/greatlakes/monitoring/air2/iadn/resources.html. This archive is an up to date collection of documents on the implementing US and Canadian and binational legislation governing IADN, the design of IADN, annual reports, chemical loading reports, standard operating procedures, quality assurance procedures, and peer reviews conducted by external experts about every 5 years.

The list of target organic analytes includes 23 organochlorine pesticides (OCPs), most of them included in the Stockholm Convention of POPs, a standardized suite of 84 PCB congeners and 16 PAHs. Detail list of compounds can be found at the above mentioned Website. In addition, other target compounds have been added to the list in the last 10 years of IADN operation including 43 polybrominated diphenyl ether (PBDE) congeners and six other flame retardants (OPEs–three chlorinated OPE, three alkyl phosphates, and six aryl phosphates) that were analyzed in atmospheric gaseous and particle phases at the US IADN sites.

As an example of the type of valuable information available from IADN, Table 3 lists the annual loading rates (wet + dry deposition, ng/m^2 year) to the five Great Lakes for selected monitored compounds for the year 2005. Loading rates of all organochlorine compounds, PAHs and PCBs monitored for the years 2001–2005 can be found elsewhere [54].

For sum PCBs, these concentrations correspond to annual loads of 71 kg/year for Superior, 48 kg/year for Michigan, 45 kg/year for Huron, 18 kg/year for Erie, and 34 kg/year for Ontario. Again, net air–water exchange loads are estimated as well but with a high level of uncertainty.

For PBDEs and selected alternative flame retardants, air (gas + particle) and precipitation samples were collected every 12 days at the 5 IADN master

TABLE 3 Loading Rates (Wet + Dry Deposition) of Selected IADN Compounds to the Great Lakes for 2005

Compound ($ng/m^2 year$)	Superior	Michigan	Huron	Erie	Ontario
a-HCH	0.5	0.3	0.3	0.6	0.3
g-HCH	0.4	0.4	0.9	0.6	0.9
Dieldrin	0.3	0.5	0.5	0.4	0.4
HCB	0.0	0.5	0.1	0.1	0.1
p,p′-DDE	0.0	0.5	0.1	0.1	0.2
p,p′-DDT	0.1	0.1	0.1	0.2	0.2
Sum PCBs	2.4	2.3	8.0	2.1	13.0
B(a)P	3.0	11.0	7.0	17.0	13.0
Phenanthrene	9.0	21.0	20.0	37.0	35.0
Pyrene	5.0	23.0	14.0	38.0	28.0

Note: IADN also provides estimates of net air–water exchange fluxes (not presented here).

sites from 2001 to 2005 [21]. Target analytes were pentabromoethyl benzene (PBEB), hexabromobenzene (HBB), 1,2-bis(2,4,6-tribromophenoxy)ethane (TBE or BTBPE), decabromodiphenylethane (DBDPE), 2-ethylhexyl-2,3,4,5-tetrabromobenzotae (TBB), and bis(2-hexylethyl)bromo-phthalate (TBPH). Table 4 presents mean concentrations of these polybrominated flame retardants in the atmosphere for 2001–2005 at an urban site (Chicago, Lake Michigan, LM) and a rural/background site (Eagle Harbor, Lake Superior, LS).

The authors concluded that the concentrations of most all the flame retardants were related to population density within a radius of 25 km of the IADN sites, and trends did not suggest that the replacement with the alternative flame retardants had yet become evident in the Great Lakes atmosphere.

2.2.1.3 Mass Balance of PCBs in Lake Superior, 1986—The Importance of Atmospheric Deposition and Air–Water Exchange

In the 1980s–1990s, assessment of the input of PCBs to Lake Superior from all sources indicated that ~90% were derived from atmospheric deposition [7]. Based on multiyear water column measurements, PCB concentrations decreased at a rate of ~0.20 per year in the surface waters of Lake Superior from 1980 to 1992 [55], and are close to equilibrium with the gas phase in recent years [56]. Due to the remote location of Lake Superior and the absence of significant point source loading to the lake, this decrease is representative of a continental decrease in the atmospheric loadings of PCBs.

TABLE 4 Mean Concentrations (2001–2005) of Brominated Flame Retardants at IADN Urban and Rural Sites

Brominated Flame Retardant (pg/m^3)	Urban Site (Chicago/LM)	Rural Site (LS)
BDE-47	6.3	3.2
BDE-99	6.7	4.8
BDE-209	15.6	1.7
Total BDE	56.0	8.9
PBEB	0.8	2.9
HBB	1.0	0.3
TBE	1.6	0.4
DBDPE	2.2	1.2
TBB	4.3	0.8
TBSPH	5.3	0.9

Using three different approaches, it was concluded that volatilization from the water column was the dominant loss process of PCBs from Lake Superior (Figure 3).

The first approach showed that the total amount of PCBs lost from the water column of Lake Superior between 1980 and 1992 was ~26,500 kg, which is significantly greater than the ~4900 kg accumulated in the sediments of Lake Superior since ~1930 [56]. This indicates that only a small fraction of the PCBs lost from the water column are lost due to sediment accumulation. Second, based on a mass balance constructed about the water column in 1986, the dominant loss process of PCBs was concluded again to be volatilization with ~1900 kg of PCBs predicted to have volatilized in 1986, while ~110 kg/year were predicted to have been lost to the sediments. The third approach utilized a pseudo first-order rate constant model, which predicted volatilization to have a rate constant of −0.24 per year and sedimentation to have a rate constant of −0.004 per year for PCB, again supporting the hypothesis that volatilization is the dominant loss process of PCBs from Lake Superior. This study indicated that large oligotrophic lakes in the midlatitudes do not always act as sinks for many hydrophobic organic compounds, but actually are sources in present times acting to increase the residence times of these chemicals in the atmosphere and buffering their concentrations. The net effect may be that the higher latitudes act as the ultimate sink for these chemicals due to the effect of lower temperatures on the air–water and air–terrestrial exchange processes.

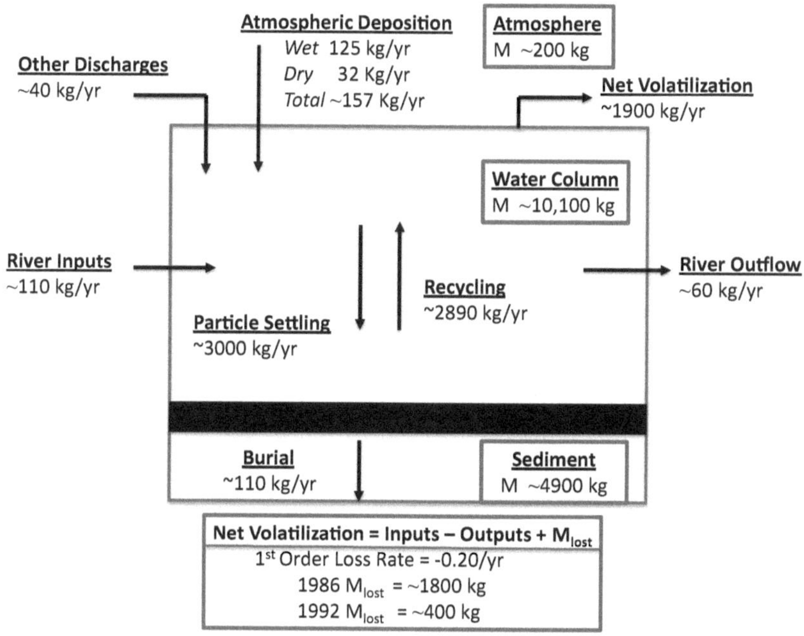

FIGURE 3 Mass balance of PCBs in Lake Superior, 1986. *Ref. [55].*

2.2.1.4 Mass Balance of PCBs in Lake Michigan, 1994—Impact of Urban Areas

The impact of urban air toxics on the proximate water bodies was investigated as part of the AEOLOS Project (Atmospheric Exchange Over Lakes and Oceans) which hypothesis was that emissions of hazardous air pollutants into the coastal urban atmosphere increased atmospheric depositional fluxes to proximate Great Waters [57]. Areas of major field and modeling campaigns focused on Southern Lake Michigan near Chicago, IL. The impact on Lake Michigan of PCB emissions from the urban area of Chicago was evaluated through field experiments to determine atmospheric concentrations, surface water concentrations, wet and dry deposition, and bidirectional gas exchange. These values were used to construct a PCB mass budget for the Lake Michigan ecosystem, which includes the urban influence. PCB emissions in the Chicago atmosphere lead to dramatically increased atmospheric concentrations of PCBs off shore and significantly increased wet, dry particle, and air–water exchange fluxes in Southern Lake Michigan compared to the regional signal. High PCB concentrations occur over the lake only when the wind is from the direction of the urban/industrial complex inclusive of the shoreline from Gary, IN to Evanston, IL toward the lake. Although atmospheric loading of PCBs was much higher in the Southern basin of Lake Michigan, water column concentrations have significantly decreased from 1980 to 1994 through the rapid uptake by settling particles,

	1994 w/o Urban Influence	1994 w/ Urban Influence
Total PCB Mass in Water Column (kg)	965	965
Inputs (kg/year)		
Riverine	45	45
from Green Bay	115	115
Atmospheric Deposition		
Wet	50	100
Dry Aerosol	1090	1180
Gas Absorption	280	880
Total Loadings	**1580**	**2320**
Outflows (kg/year)		
to Lake Huron	10	10
Burial	730	730
Net Loss	440	440
Volatilization	640	1100
Total Losses	**1820**	**2280**
Loadings − Losses =	−240	40
	−25% of inventory	4% of inventory

FIGURE 4 Results from the mass budgets for PCBs in Lake Michigan in 1994 without and with urban-related atmospheric loadings. *Ref. [57].*

and have continued to decrease at a rate of 0.17 per year. In contrast to earlier PCB mass budgets that missed significant inputs, outputs and/or process terms, inclusion of the urban-influenced air–water exchange processes and a reevaluation of new lake data effectively closed the PCB budget for Lake Michigan (Figure 4). Atmospheric exchange processes dominated the whole lake budget. The Chicago area, with high industrial density, emits large quantities of hazardous organic pollutants and has important impacts on the downwind coastal atmosphere and proximate surface waters.

The Lake Michigan Mass Balance Project (LMMB-PCBs) (http://www.epa.gov/glnpo/lmmb/) conducted concurrently reached similar conclusions. More recent data from IADN and modeling suggested that the Chicago area resulted in ~10–20% increase in atmospheric deposition of PCBs to the whole lake budget.

These are examples of in-depth assessments of the occurrence of legacy POPs in the Great Lakes region. However, recent studies show that not only legacy POPs are present at high levels in the atmosphere of the Great Lake's, but also new emerging contaminants, such as OPEs. Atmospheric concentrations of these contaminants have been reported to be about 2–3 orders of magnitude higher than the concentrations of PBDEs in similar samples [58]. OPEs do certainly reach the Great Lake's waters from the atmosphere, although more data are needed to assess their current inputs and predominant pathways in these aquatic systems. OPEs have been added to the group of priority contaminants to be analyzed by the IADN [58].

2.2.2 Case Study 2: The Mediterranean Sea

The Mediterranean Sea (~2.5 million km^2 of surface) is a semienclosed marine environment located in the center of the Mediterranean Region, comprising two main basins, the Western and the Eastern, connected by the Strait of Sicily (150 km wide). It is surrounded by 21 bordering countries representing ~46,000 km of coastline including the islands. The Mediterranean coastal areas have been the base for the settling and development of some of the largest cities in the region (e.g., Barcelona, Rome, Naples, Athens, Marseille, Alexandria, Istanbul, etc.) and many industrial facilities. Around 30% of the 460 million inhabitants of the Mediterranean countries live in coastal areas [59]. This stable population may be significantly increased in certain seasons of the year due to intense tourist activity, for example, 275 million visitors traveled the Mediterranean region in 2008 [59]. This marine environment holds an intensive industrial activity, with more than 200 petrochemical and energy installations, chemical industries, and chlorine plants located along its coasts [60]. The Mediterranean Sea is under intense anthropogenic pressure (e.g., urbanization, industry, energy generation and consumption, transportation systems, tourism and recreation, agriculture, fisheries, forestry and mining). Most of these activities are responsible for the emission of large amounts of contaminants from land-based sources, including POPs, in the region resulting in local impacts in coastal areas. In addition, a considerable amount of those contaminants can reach open waters (far from the coast line) by means of a combined process implying pollutant emission–atmospheric transport–deposition. Moreover, some specific pressures have a direct effect at open waters, like the intense maritime traffic in the region. Among the world's oceans, the traffic density of merchant vessels in the Mediterranean Sea is particularly important. About 30% of international sea-borne volume originates from or is directed to the 300 Mediterranean ports, or passes through the Mediterranean Sea [61]. Around 2000 active ferry lines, 1500 cargo vessels, and 2000 commercial vessels operate in the Mediterranean resulting in about 200,000 vessels crossing the Mediterranean Sea annually [62].

The atmospheric deposition of POPs and related contaminants in the Mediterranean Sea, both at coastal sites and open Sea has been investigated during the last two decades. A recent review discusses and compiles the deposition fluxes in this marine environment [63]. In this section, we used these data to provide an overall view and add the last findings since then. The most studied atmospheric deposition process of POPs in the Mediterranean Sea is dry deposition. Only few studies have reported wet and diffusive fluxes. However, the magnitude of diffusive fluxes for compounds occurring in the atmospheric gas and water dissolved phases, such as PAHs, is much higher compared to the other mechanisms of atmospheric deposition [18]. Table 5 shows a compilation of reported atmospheric deposition fluxes of POPs and related compounds at open sea and coastal background/low impacted sites in the Mediterranean. The reported fluxes by Castro-Jiménez and coworkers [63] have been updated with the most recent available data, in particular in emerging contaminants. Recent analyses of OPE flame retardants and plasticizers in atmospheric samples collected in two oceanographic campaigns across the Mediterranean Sea and in the Southwest Black Sea demonstrate their overall distribution in these marine environments [64]. The dry deposition fluxes for OPEs (26–320 ug/m^2 year) are in the range or higher than those of PAHs (70–80 ug/m^2 year) for the same set of samples at open waters. OPEs are currently used in many cases as substitutes of PBDEs since their recent ban by the Stockholm Convention in 2009. Some OPEs have been reported to be persistent in the environment (e.g., Tris [2-chloro-1-(chloromethyl)ethyl]phosphate, TDCP) and to bioaccumulate in the food chain [65,66]. In addition, some halogenated OPEs have been found to exhibit various toxic effects. For instance, Tris-(1-chloro-2-propyl)phosphate (TCPP) is considered to be potentially carcinogenic and Tris-(2-chloroethyl) phosphate (TCEP) is toxic to aquatic organisms and carcinogenic for animals [65–69].

These compounds unlike many other POPs and related chemicals contain phosphorous, which is an important nutrient for phytoplankton and microbes. In many cases, Phosphorus is a limiting nutrient, such as the case of the Mediterranean Sea and other oligotrophic aquatic environments. This is a complementary and relevant aspect of the atmospheric deposition of organic chemicals. In the case of OPEs, not only their inherent potential toxic effects can be considered an anthropogenic impact but also their potential interactions with the natural Phosphorus cycle, as recently evaluated by Castro-Jiménez and coworkers in the open Mediterranean Sea [64].

In order to provide with an overall figure of the total loading of toxic organic pollutants to the Mediterranean waters, we followed the approach developed by Castro-Jiménez and coworkers reported elsewhere [63]. Briefly, since most of the existing investigations do not report data for the three atmospheric deposition processes (Figure 1) for the same study; therefore, data from more than one study in similar areas have been combined in order to provide with an estimate of the atmospheric loading of contaminants including the three above mentioned components. Lower and upper ranges of deposition fluxes from Table 5 have been used to estimate an overall range of deposition. A surface of 2.5×10^{12} m^2 was used for the Mediterranean Sea in order to estimate the total atmospheric loading (ton/year).

The total atmospheric input of organic contaminants (sum of all studied families, Table 5) to open Mediterranean Sea waters is estimated to varied from 2100 to 4600 ton/year. PAHs are the predominant contaminants inputted to Mediterranean waters (2000–4100 ton/year), followed by OPEs (13–260 ton/

TABLE 5 Atmospheric Deposition Fluxes of Organic Contaminats Reported in the Last Two Decades for the Mediterranean Basin (Open Sea, Background, and Coastal Sites)

	Deposition Flux					
Contaminant	Dry	Wet	Diffusive	Location	Year	References
PCDD/Fs	(ng/m^2 year)	(ng/m^2 year)	(ng/m^2 year)			
$\sum$2,3,7,8-PCDD/Fs[a]	2–60	n.r	3–25	All basins (open MED)	2006, 2007	[17]
$\sum$2,3,7,8-PCDD/Fs[a,c]	23	19	n.r	Western MED (coastal)	2007–2008	[70]
PCBs	(ng/m^2 year)	(ng/m^2 year)	(ng/m^2 year)			
$\sum_{18}$PCBs[a]	136	125	n.r	Western MED (coastal)	2007–2008	[70]
$\sum_{54}$PCBs[b]	36–400	820[c]	n.r	Eastern MED (background)	2001	[71,72]
$\sum_{41}$PCBs[a]	50–2250	n.r	187–4585	All basins (open MED)	2006, 2007	[19]
PAHs	(ug/m^2 year)	(ug/m^2 year)	(ug/m^2 year)			
$\sum_{30}$PAHs[c]	70–80	n.r	600–1400	All basins (open MED)	2006, 2007	[18]
$\sum_{35}$PAHs[b,c]	60	165	700	Eastern MED (background)	2001, 2002	[73]
$\sum_{15}$PAHs	n.r	n.r	335–510	Western MED (coastal)	2002	[74]
$\sum_{11}$PAHs[a]	15–35	10–40	n.r	Western MED (coastal)	1990	[75]

Continued

TABLE 5 Atmospheric Deposition Fluxes of Organic Contaminats Reported in the Last Two Decades for the Mediterranean Basin (Open Sea, Background, and Coastal Sites)—cont'd

	Deposition Flux					
Contaminant	**Dry**	**Wet**	**Diffusive**	**Location**	**Year**	**References**
OCPs	(ng/m^2 year)		(ug/m^2 year)			
HCB[a]	4–18	n.r	0.03–1.4	All basins (open MED)	2006, 2007	[19]
HCHs[a]	n.r	n.r	1–90	All basins (open MED)	2006, 2007	[19]
PBDEs	(ng/m^2 year)					
$\sum_8$PBDEs[a]	230	n.r	n.r	Western MED (coastal)	2007–2008	Est. from data in [70]
$\sum_5$PBDEs[b]	4–90	n.r	n.r	Western MED (coastal)	2001–2003	[76]
OPEs	(ug/m^2 year)					
$\sum_{14}$OPEs[a]	26–320	n.r	n.r	All basins (open MED)	2006, 2007	[64]

n.r = not reported.
[a]*Samples collected using high volume samplers (gas and aerosol phases).*
[b]*Samples collected using deposition samplers.*
[c]*Mean values.*

year), OCPs (3–230 ton/year), PCBs (~3–20 ton/year), PBDEs (~1 ton/year), and 2,3,7,8-PCDD/Fs (60–260 kg/year). This contaminant cocktail is expected to be deposited every year in Mediterranean open waters from the atmosphere. Higher deposition fluxes are expected in the proximity of the coast in particular for those areas close to medium and large coastal Mediterranean cities or/and in the proximity of industrial facilities.

However, the Mediterranean Sea is not a net sink for all contaminants in all regions. Net volatilization of some PCDD congeners [17] and low MW PAHs [18] has been reported in some sub-basins. Moreover, PCBs were reported close to equilibrium conditions between air and water in a recent work [19].

2.2.3 Case Study 3: The Atlantic Ocean

The Atlantic Ocean, defined as the region delimited by the Arctic and Antarctica, and the American, European and Africa continents is extremely variable in terms of sources of pollutants, climatically, and comprises very different biogeochemical provinces. In addition, the water and atmospheric compartments from the northern hemisphere are relatively isolated from the southern hemisphere. With the exception of pioneering studies, such that of Iwata and coworkers [77], it has not been until the last decade that the atmospheric deposition of POPs has been quantified for this region. Previous attempts also covered some coastal regions such as that influenced by the metropolitan New York/New Jersey region in the Mid Atlantic North America [3,46,78,79], confirming the influence of urban centers as a source of pollutants to adjacent water bodies.

For the open ocean, a major breakthrough came from the measurements of atmospheric concentrations of several POP families in north–south transects [8,12,39,80–82]. These field campaigns allowed to determine the latitudinal atmospheric distribution of legacy POPs, such as PCBs, hexachlorocyclohexanes (HCHs) and PCDD/Fs, emerging POPs such as PBDEs, and for PAHs as well. Generally, it has been observed that POP concentrations are higher in the northern hemisphere than in the southern hemisphere, consistent with larger historical emissions of PCBs and other POPs from North America and Europe.

The data sets generated in the Atlantic Ocean for PCBs [8,80], allowed the first comprehensive estimation of the relative importance of dry, wet, and diffusive exchange as a source of POPs to this marine ecosystem [5,6]. These estimations were obtained by combining the field measurements of atmospheric concentrations of dioxins (PCDD/Fs) and PCBs, the models described above for estimating the magnitude of the different atmospheric deposition processes, and remote sensing derived information for the environmental variables (wind speed, temperature, precipitation). Figure 5 shows the comparison of the three deposition processes for PCBs and PCDD/Fs in the Atlantic Ocean. In the case of PCBs, representative of chemicals found mainly in the

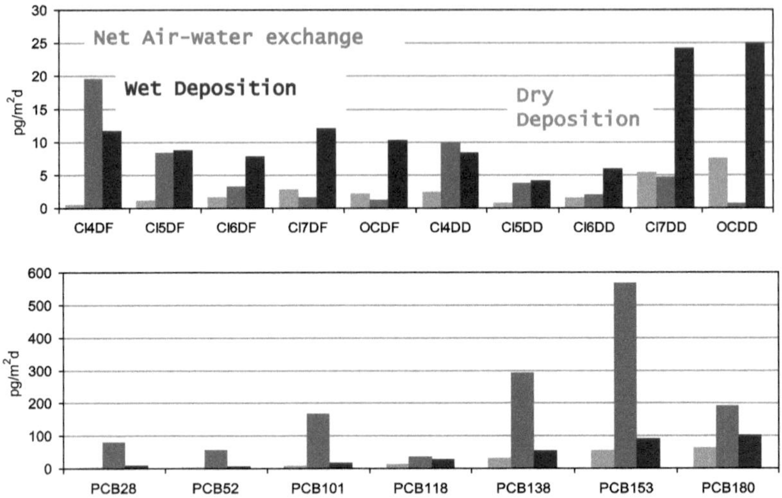

FIGURE 5 Comparison of the average net air–water exchange, dry deposition and wet deposition fluxes of dioxins (upper panel) and PCBs (lower panel) for the Atlantic Ocean. *Refs [5,6].*

gas phase, the net diffusive air–water exchange is significantly of larger magnitude than the dry and wet deposition. This points out to the important role of diffusive exchanges controlling the water column concentrations in the open ocean, since atmospheric deposition is the main input of pollutants to the open ocean due to the relatively lower importance of oceanic currents transporting POPs. In the case of dioxins, which have a much stronger affinity to the aerosol phase due to adsorption to soot carbon [83], there is a greater importance of dry and wet deposition. Indeed, wet deposition is the dominant depositional process for the hexa- to octa-chlorinated dioxins and furans, and diffusive air–water exchange is only the main depositional process for the tetra-chlorinated PCDD/F.

Even though wet deposition appears as an important process, its relevance is highly variable temporally and spatially. Obviously, wet deposition is only relevant during the rain episodes. In addition, the precipitation spatial patterns are mainly localized in midhigh latitudes of the north and southern Atlantic, and in the intertropical convergence zone near the equator [6]. In the case of dioxins, the dry deposition will be the dominating process in those oceanic regions where there are scarce raining events, as is usually the case for most of the tropical and subtropical oceans.

Dry deposition of semivolatile organic pollutants has seldom being measured for the oceanic regions. There are only two studies reporting experimental measurements, one performed in the NE Atlantic [33], and the other in the major oceanic regions of the world during a circumnavigation campaign [34]. These studies report v_d values for PAHs ranging from 0.01 to 10 cm/s, even though in most cases v_d values are below 0.3 cm/s. An empirical equation derived from those global measurements has been recently suggested to predict the v_d values [34]:

$$Log(v_d) = -0.261(\pm 0.038)\ Log\ (P_L) + 0.387(\pm 0.074)\ U_{10} * Chl_s - 3.082(\pm 0.197) \quad (7)$$

Where U_{10} (m/s) is the wind speed at 10m over sea surface, Chl_s (mg/m^3) is the chlorophyll *a* concentration in surface waters, and P_L is the chemical vapor pressure. This equation predicts higher depositional fluxes for the less-volatile (more hydrophobic) chemicals, and higher depositional fluxes in those oceanic regions with higher phytoplankton biomass due to the formation of a surface microlayer enhancing the deposition of hydrophobic fine aerosols [32]. Unfortunately, there are no field measurements of the dry deposition velocity for organic pollutants other than PAHs, and thus it has not been possible to validate this empirical model for other POPs.

Especial attention should be given to the diffusive air–water exchange. This is the main depositional process for many POP, such as PCBs, low MW PAHs, HCHs, HCBs, etc. Diffusive exchange depends on the chemical properties, temperature, and wind speed as before explained, which all affect the magnitude of the air–water mass transfer coefficient. In the Atlantic Ocean, the intertropical regions are those showing higher temperatures and lower wind speeds, while at higher latitudes, the temperature decreases and wind speed increases. However, the diffusive air–water exchange also depends on the air–water concentrations gradient, as given by ($C_g/H'-C_w$). Therefore, any process affecting the chemical concentrations in air and water will also affect the air–water diffusive flux.

During the last decade, there have been a number of studies addressing the biogeochemical processes that occur in the water column that can affect the air–water diffusive exchange. The biological pump is now accepted as a key process affecting this air–water gradient in concentrations. The uptake of POPs by plankton and other organic matter pools, and the subsequent sinking of the organic matter bound POPs to deep waters depletes the water concentrations of POPs, driving and air to water disequilibrium in the POP concentrations [27,28]. The field validation of this process for the marine environment has been recently described in the high productive waters of the North Atlantic and Southern Ocean, where a net air to water exchange of PCBs was found [14,84]. The more hydrophobic POPs are those bioconcentrating to a greater extend in plankton, and therefore, the influence of the biological pump enhancing the air–water exchange is greater for the more hydrophobic chemicals. For example, it has been shown that the air–water disequilibrium in the productive waters of the Southern ocean was correlated with the contaminant K_{ow} [15] (Figure 6). In addition, a recent study in the North Atlantic has shown that these biological pump driven air–water exchange fluxes reduce the potential for atmospheric transport of the more hydrophobic PCBs [14].

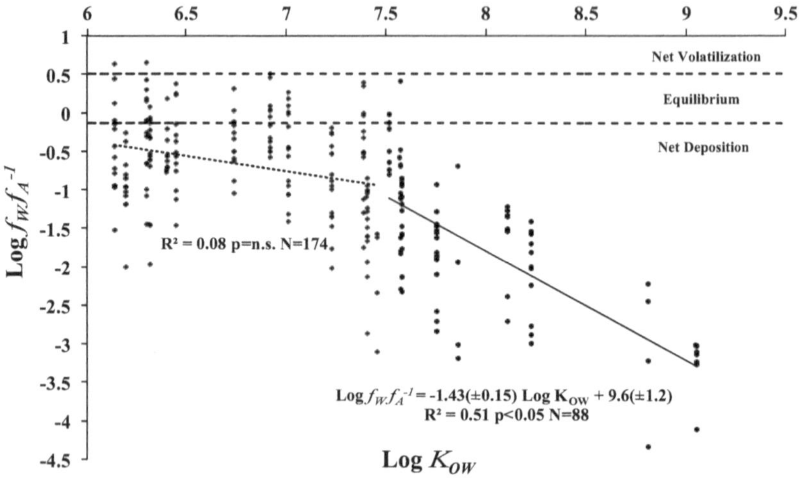

FIGURE 6 Air–water disequilibrium of the fugacity in water (f_W) and fugacity in air (f_A) for PCBs in the southern Ocean surrounding the Antarctica Peninsula. *Ref. [15].*

The oceanic high productivity regions, where a net air to water transfer of gas phase POPs is expected, are located at high latitudes of the North and South Atlantic oceans, and in the upwelling regions offshore Sahara, Namibia, Greenland, and Argentina [28]. However, most of the global oceans are oligotrophic. There are few studies performed in the oligotrophic subtropical and tropical oceans. Net volatilization of PCBs has been described in the southern Atlantic and Pacific Oceans [11,85], while PCBs were reported close to equilibrium conditions between air and water in a north–south Atlantic Ocean transect [81]. The settling fluxes of organic matter in oligotrophic marine regions are small, and thus the biological pump is not a significant process enhancing the atmospheric deposition of POPs. In fact, the subtropical gyres would tend to concentrate POPs in the central region of the gyres, with an associated increase of water temperatures, thus facilitating the POP volatilization from these vast oceanic regions.

For chemicals that degrade in the water column (mainly microbial degradation), air–water disequilibrium may be produced, also increasing the atmospheric deposition of some POPs, a process named "degradative pump." This scenario has been described recently for HCHs in the north Atlantic and the Southern Oceans [15,86]. Therefore, the knowledge gained during the last decade on the atmospheric deposition of organic pollutants to the Atlantic Ocean shows that the relative importance of the three processes depend on the chemical properties, as they affect their gas-particle partitioning. In addition, and on the other hand, atmospheric deposition fluxes are highly variable spatially due to concurrent variability of environmental conditions (wind speed, temperature, precipitation events) and biogeochemical cycles in the oceanic water column.

2.3 Influence on POPs Bioavailable Reservoirs and Exposure

Coastal regions adjacent to urban centers receive important loadings of POPs by run-off, but also by atmospheric deposition. For example, studies performed in Lake Michigan showed that there was a net deposition of PCBs to waters adjacent to Chicago when the air masses were from the urban region, conversely, there was a net volatilization when the air masses were from the cleaner Northern lake Michigan [87,88]. A similar situation has been described in marine regions adjacent to New York (NE Atlantic) and Barcelona (NE Mediterranean) [3,4]. Therefore, coastal run-off can enhance the levels of pollutants in the coastal region. However, few kilometers offshore, atmospheric deposition is the main input route for organic pollutants to the water column.

The mechanism of atmospheric deposition also has an important influence on the levels and bioavailability of organic pollutants in the water column. In the case of high MW PAHs that are strongly associated with aerosol soot carbon in the atmosphere [89], their bioavailability will be low once the soot particles enter the water column. However, diffusive air–water exchange is the main entry route for many POPs to the water column. Diffusion of POPs from the gas phase to the dissolved phase, support reservoirs of truly dissolved POPs in the water column, and thus these are fully bioavailable for being accumulated in biota (entering the food chain), or degraded by microorganisms in the case they are not persistent.

In the case of some new emerging contaminants such as OPEs (mostly associated with the atmospheric particle phase), even though they are generally hydrophobic compounds, they will undergo particle-dissolve water phase partitioning once introduced in the water column (they have a general higher solubility in water compared to POPs). The dissolved phase compounds will be bioavailable, and if degraded, further OPEs will partition from the aerosol particles to the dissolved phase due to a particle to water fugacity gradient. However, there is no experimental confirmation of their degradation in the water column up today. The phosphate ester groups, the "basic skeleton" of OPEs, are common in biological organic molecules (e.g., phospholipids, DNA, RNA) and thus there are specific enzymes able to degrade organic phosphate ester compounds, suggesting that degradation of OPEs could likely occur.

There are a number of studies that have demonstrated the importance of atmospheric deposition, and specifically, the interactions of atmospheric deposition and biogeochemical cycling in the water column controlling the concentrations of POPs in biota. It has been shown that atmospheric deposition alone could account for the PCB levels measured in plankton in the Experimental Lakes area of Canada [27]. Dachs and coworkers [90] also showed that the levels of POPs in plankton from Lake Ontario would decrease in eutrophied conditions due to the depletion of water column levels in the water column. Recent work in the Mediterranean Sea, have described that the concentrations of PCBs in plankton decrease when the water column biomass is higher, and that this decrease in concentrations is of larger magnitude for the more hydrophobic congeners [91]. This scenario is consistent with the role of the biological pump decreasing the concentrations in the water column. A similar trend in the concentrations of PCBs in plankton has been described in the NW Pacific (close to Vancouver) [92] and for the Southern Ocean [15]. In all these cases, the concentrations of PCBs in plankton show a dependency with the biomass, especially for the more hydrophobic congeners. This shows that the levels of POPs in plankton depend on water column biogeochemical processes, and how these interact with atmospheric inputs. More importantly, these studies demonstrate that the levels in plankton do not depend on the distance to coastal sources. Plankton is the first step in the food web, and the enrichment of POP concentrations in plankton when compared to the levels in the dissolved phase is approximately equal to K_{ow}, an enrichment larger than that observed when going to upper levels in the food web. Therefore, it is not possible to predict the levels of POPs in marine biota without a quantification of their atmospheric inputs, and how these interact with the water column processes.

Wet deposition scavenges both POPs present in the gas and aerosol phase, and thus also has the potential to support the bioavailable fraction of POPs in surface waters.

3. FINAL REMARKS

The predominant atmospheric deposition process for a given contaminant will be determined by the inherent physicochemical properties of the chemical (e.g., solubility, vapor pressure, K_{ow}) and the specific conditions of the aquatic environment (e.g., biogeochemistry of the water column) and region (e.g., climatology). However, there is little doubt today that atmospheric transport and deposition via dry and wet deposition and air–water transfer are major drivers of POPs loading and inventories in both large lakes and costal/open marine systems.

From the data and scientific evidence presented in the previous sections, it seems clear that the comprehensive evaluation of the pollution status of a given aquatic environment by POPs cannot be undertaken without addressing the atmospheric pathway in its full dimension. This is particularly true far from the coast line, where atmospheric deposition processes are major drivers of surface water levels, even more than land-based sources is some cases.

It is not clear to what extent this knowledge has been implemented in the main international regulations aiming at protecting the aquatic environments from pollution by organic contaminants and to preserve a good quality status of our waters, such as the EU WFD and the MSFD as well as the regional protocols such as OSPAR, HELCOM, and the Barcelona Convention, among others.

Awareness should be raised in the fact of not considering the atmospheric pathway for the assessment of the exposure and pollution of aquatic environment by POPs and related contaminants.

REFERENCES

[1] D. Mackay, F. Wania, Environ. Sci. Tech. 30 (1966) 390–396.

[2] J.H. Lehr, J. Keeley, Water Encyclopedia: Oceanography, Meteorology, Physics and Chemistry, Water Law, and Water History, Art, and Culture, vol. 4, John Wiley & Sons, Inc.: Hoboken, NJ, 2005, p. 21.

[3] S. Yan, L.A. Rodenburg, J. Dachs, S.J. Eisenreich, Environ. Pollut. 152 (2008) 443–452.

[4] N. García-Flor, J. Dachs, J.M. Bayona, J. Albaigés, Chemosphere 75 (2009) 1144–1152.

[5] E. Jurado, F.M. Jaward, R. Lohmann, K.C. Jones, R. Simó, J. Dachs, Environ. Sci. Technol. 38 (2004) 5505–5513.

[6] E. Jurado, F.M. Jaward, R. Lohmann, K.C. Jones, R. Simó, J. Dachs, Environ. Sci. Technol. 39 (2005) 2426–2435.

[7] S.J. Eisenreich, B.B. Looney, J.D. Thornton, Environ. Sci. Technol. 15 (1981) 30–38.

[8] F.M. Jaward, J. L Barber, K. Booij, K.C. Jones, Environ. Pollut. 132 (2004) 173–181.

[9] R. Lohmann, R. Gioia, K.C. Jones, L. Nizzetto, C. Temme, Z. Xie, D. Schulz-Bull, I. Hand, E. Morgan, L. Jantunen, Environ. Sci. Technol. 43 (2009) 5633–5639.

[10] R. Lohmann, J. Klanova, P. Kukucka, S. Yonis, K. Bollinger, Environ. Sci. Technol. 47 (2013) 13967–13975.

[11] L. Nizzetto, R. Lohmann, R. Gioia, J. Dachs, K.C. Jones, Environ. Sci. Technol. 44 (2010) 6978–6984.

[12] R. Gioia, R. Lohmann, J. Dachs, C. Temme, S. Lakaschus, D. Schulz-Bull, I. Hand, K.C. Jones, J. Geophys. Res. Atmos. 113 (2008) D19302.

[13] R. Gioia, L. Nozetto, R. Lohmann, J. Dachs, C. Temme, K.C. Jones, Environ. Sci. Technol. 42 (2008) 1416–1422.

[14] C. Galbán-Malagón, N. Berrojalbiz, M.J. Ojeda, J. Dachs, Nat. Commun. 3 (2012) 862.

[15] C. Galbán-Malagón, S. Del Vento, A. Cabrerizo, J. Dachs, J. Atmos. Chem. Phys. 13 (2013) 12029–12041.

[16] C. Galbán-Malagón, G. Caballero, A. Cabrerizo, J. Dachs, J Atmos. Environ. 80 (2013) 41–49.

[17] J. Castro-Jiménez, S.J. Eisenreich, M. Ghiani, G. Mariani, H. Skejo, G. Umlauf, J. Wollgast, J.M. Zaldívar, N. Berrojalbiz, H.I. Reuter, J. Dachs, Environ. Sci. Technol. 44 (2010) 5456–5463.

[18] J. Castro-Jiménez, N. Berrojalbiz, J. Wollgast, J. Dachs, Environ. Pollut. 166 (2012) 40–47.

[19] N. Berrojalbiz, J. Castro-Jiménez, G. Mariani, J. Wollgast, G. Hanke, J. Dachs, Atmos. Chem. Phys. 14 (2014) 8947–8959.

[20] B.R. Hillery, M.F. Simcik, I. Basu, R.M. Hoff, W.M.J. Strachan, D. Burniston, C.H. Chan, K.A. Brice, C.W. Sweet, R.A. Hites, Environ. Sci. Technol. 32 (1998) 2216–2221.

[21] Y. Ma, A. Salamova, M. Venier, R.A. Hites, Environ. Sci. Technol. 47 (2013) 11457–11464.

[22] J. Castro-Jiménez, G. Mariani, S.J. Eisenreich, E.H. Christoph, G. Hanke, E. Canuti, H. Skejo, G. Umlauf, Chemosphere 73 (2008) S122–S130.

[23] J. Castro-Jiménez, S.J. Eisenreich, G. Mariani, H. Skejo, G. Umlauf, Atmos. Environ. 56 (2012) 194–202.

[24] G. Mariani, E. Canuti, J. Castro-Jiménez, E.H. Christoph, S.J. Eisenreich, G. Hanke, H. Skejo, Chemosphere 73 (2008) S114–S121.

[25] S.N. Meijer, J. Dachs, P. Fernandez, L. Camarero, J. Catalan, S. Del Vento, B. Van Drooge, E. Jurado, J.O. Grimalt, Environ. Pollut. 140 (2006) 546–560.

[26] S.N. Meijer, J.O. Grimalt, P. Fernández, J. Dachs, Environ. Polltut 157 (2009) 1815–1822.

[27] J. Dachs, S.J. Eisenreich, J.E. Baker, F.C. Ko, J.D. Jeremiason, Environ. Sci. Technol. 33 (1999) 3653–3660.

[28] J. Dachs, R. Lohmann, W.A. Ockenden, L. Méjanelle, S.J. Eisenreich, K.C. Jones, Environ. Sci. Technol. 36 (2002) 4229–4237.
[29] L. Nizzetto, R. Gioia, J. Li, K. Borgå, F. Pomati, R. Bettinetti, J. Dachs, K.C. Jones, Environ. Sci. Technol. 46 (2012) 3204–3211.
[30] J.W. Farrington, H. Takada, Oceanography 27 (2014) 196–213.
[31] M. Amodio, S. Catino, P.R. Dambruoso, G. de Gennaro, D. Di Gilio, P. Giungato, E. Laiola, M. Marzocca, A. Mazzone, A. Sardaro, M. Tutino, Adv. Meteorol. (2014). Article ID 161730.
[32] S. Del Vento, J. Dachs, Atmos. Environ. 41 (2007) 4920–4930.
[33] S. Del Vento, J. Dachs, Environ. Sci. Technol. 41 (2007) 5608–5613.
[34] B. González-Gaya, J. Zuñiga, M. José Ojeda, B. Jiménez, J. Dachs, Environ. Sci. Technol. 48 (2014) 5583–5592.
[35] E. Demircioglu, A. Sofuoglu, M. Odabasi, J. Hazard Mater. 186 (2011) 328–335.
[36] A. Birgül, Y. Tasdemir, S.S. Cindoruk, Atmos. Res. 101 (2011) 341–353.
[37] M. Günindi, Y. Tasdemir, Water Air Soil Pollut. 215 (2011) 427–439.
[38] F. Esen, Y. Tasdemir, S.S. Cindoruk, Atmos. Res. 95 (2010) 379–385.
[39] L. Nizzzetto, R. Lohmann, R. Gioia, A. Jahnke, C. Temme, J. Dachs, P. Herckes, A. Di Guardo, K.C. Jones, Environ. Sci. Technol. 42 (2008) 1580–1585.
[40] A. Möller, Z. Xie, R. Sturm, R. Ebinghaus, Environ. Sci. Technol. 44 (2010) 8977–8982.
[41] A. Möller, Z. Xie, R. Sturm, R. Ebinghaus, Environ. Pollut. 159 (2011) 1577–1583.
[42] A. Möller, R. Sturm, Z. Xie, M. Cai, J. He, R. Ebinghaus, Environ. Sci. Technol. 46 (2012) 3127–3134.
[43] L. Morales, J. Dachs, B. González-Gaya, G. Hernán, M. Ábalos, E. Abad, Environ. Sci. Technol. 48 (2014) 10198–10207.
[44] J. Castro-Jiménez, P. Casal, B. González Gaya, M. Pizarro, J. Dachs, Oral Commun. (September 2014) 34th International Symposium on Halogenated Persistent Organic Pollutants, Madrid.
[45] T.P. Franz, S.J. Eisenreich, Environ. Sci. Technol. 32 (1998) 1771–1778.
[46] D.A. Van Ry, C.L. Gigliotti, T.R. Glenn, E.D. Nelson, L.A. Totten, S.J. Eisenreich, Environ. Sci. Technol. 36 (2002) 3201–3209.
[47] H.A. Bamford, D.L. Poster, R.E. Huie, J.E. Baker, Environ. Sci. Technol. 36 (2002) 4395–4402.
[48] J. Paasivirta, S. Sinkkonen, P. Mikkelson, T. Rantio, F. Wania, Chemosphere 39 (1999) 811–832.
[49] M.S. McLachlan, U. Sellström, Atmos. Environ. 43 (2009) 6084–6090.
[50] S. Dueri, D. Marinov, A. Fiandrino, J. Tronczyński, J.-M. José-Manuel Zaldívar, Int. J. Environ. Res. Public Health 7 (2010) 1467–1485.
[51] H.A. Bamford, D.L. Poster, J.E. Baker, Environ. Toxicol. Chem. 18 (1999) 1905–1912.
[52] P.D. Nightingale, P.S. Liss, P. Schlosser, Geophys. Res. Lett. 27 (2000) 2117–2120.
[53] W.M. J Strachan, S.J. Eisenreich, in: Oceans '87, vol. 19, 1987, pp. 1765–1770.
[54] P. Blanchard, C.V. Audette, M.L. Hulting, I. Base, K.A. Brice, S.M. Bakus, H. Dryfhout-Clark, F. Froude, R.A. Hites, M. Neilson, R. Wu, Environment Canada and the United States EPA, 2005. http://www.epa.gov/glnpo/monitoring/air2/iadn/reports/IADN_Toxics_Deposition_Thru_2005.pdf.
[55] J.D. Jeremiason, K.C. Hornbuckle, S.J. Eisenreich, Environ. Sci. Technol. 28 (1994) 903–914.
[56] J.E. Baker, S.J. Eisenreich, B.J. Eadie, Environ. Sci. Technol. 25 (1991) 500–509.
[57] J.H. Offenberg, M.F. Simcik, J.E. Baker, S.J. Eisenreich, Aquat. Sci. 67 (2005) 79–85.
[58] A. Salamova, Y. Ma, M. Venier, R.A. Hites. Environ. Sci. Technol. Lett. 1 (2014) 8–14.
[59] UNEP/MAP-Plan Bleu, State of the Environment and Development in the Mediterranean, United Nations Environment Programme/Mediterranean Action Plan–Plan Bleu, Athens, 2009.

[60] MED POL Programme, United Nations Environmental Programme. Barcelona Convention (MAP). http://www.unepmap.org/index.php?module=content2&;catid=001017003.
[61] Action Plan for the Protection of the Marine Environment and the Sustainable Development of the Coastal Areas of the Mediterranean (MAP Phase II). Appendix I. http://195.97.36.231/dbases/webdocs/BCP/MAPPhaseII_eng.pdf.
[62] REMPEC (Regional Marine Pollution Emergency Response Center for the Mediterranean Sea). An Overview of Maritime Transport in the Mediterranean, Document REMPEC/WG.10/Inf.25; Meeting of Focal Points of the Regional Marine Pollution Emergency Response Centre for the Mediterranean Sea, Malta, 4–8 October 1994, pp. 1–21.
[63] J. Castro-Jiménez, N. Berrojalbiz, L. Mejanelle, J. Dachs, in: L. McConnell, J. Dachs, C.J. Hapeman (Eds.), Occurrence, Fate and Impact of Atmospheric Pollutants on Environmental and Human Health, ACS Symposium Series, vol. 1149, American Chemical Society, Washington, DC, 2013, pp. 231–260 (Chapter 11).
[64] J. Castro-Jiménez, N. Berrojalbiz, M. Pizarro, J. Environ. Sci. Technol. 48 (2014) 3203–3209.
[65] I. van der Veen, J. de Boer, Chemosphere. 88 (2012) 1119–1153.
[66] European Chemical Agency (ECHA), Tris[2-Chloro-1-(Chloromethyl)ethyl] Phosphate (TDCP) CAS No.: 13674-87-8. EINECS No.: 237-159-2. Risk Assessment Report, 2008.
[67] A.F. Lehner, F. Samsing, W.K. Rumbeiha, J. Med. Toxicol. 6 (2010) 448–458.
[68] European Chemical Agency (ECHA), Member State Committee Support Document for the Identification of Tris (2-chloroethyl) Phosphate as a Substance of Very High Concern Because of its CMR Properties, 2009.
[69] X. Liu, K. Ji, K. Choi, Aquat. Toxicol. 114–115 (2012) 173–181.
[70] J. Castro-Jiménez, G. Mariani, I. Vives, H. Skejo, G. Umlauf, J.M. Zaldivar, S. Dueri, G. Messiaen, T. Laugier, Environ. Pollut. 159 (2011) 1948–1956.
[71] M. Mandalakis, E.G. Stephanou, Environ. Sci. Technol. 38 (2004) 3011–3018.
[72] M. Mandalakis, M. Apostolaki, E.G. Stephanou, Global Biogeochem. Cy. 19 (2005) GB3018.
[73] M. Tsapakis, M. Apostolaki, S.J. Eisenreich, E.G. Stephanou, Environ. Sci. Technol. 40 (2006) 4922–4927.
[74] C. Guitart, N. García-Flor, J.C. Miquel, S.W. Fowler, J. Albaigés, J. Mar. Syst. 79 (2010) 210–217.
[75] E. Lipiatou, J. Albaigés, Mar. Chem. 46 (1994) 153–164.
[76] L. Mejanelle, S. Lacorte, J. Dachs. (Unpublished results).
[77] H. Iwata, S. Tanabe, N. Sakal, R. Tatsukawa, Environ. Sci. Technol. 27 (1993) 1080–1098.
[78] C.L. Gigliotti, P.A. Brunciak, J. Dachs, T.R. Glenn Iv, E.D. Nelson, L.A. Totten, S.J. Eisenreich, Environ. Toxicol. Chem. 21 (2002) 235–244.
[79] R. Lohmann, E.D. Nelson, S.J. Eisenreich, K.C. Jones, Environ. Sci. Technol. 34 (2000) 3086–3093.
[80] R. Lohmann, W.A. Ockenden, J. Shears, K.C. Jones, Environ. Sci. Technol. 35 (2001) 4046–4053.
[81] S. Lakaschus, K. Weber, F. Wania, R. Bruhn, O. Schrems, Environ. Sci. Technol. 36 (2002) 138–145.
[82] F.M. Jaward, J. L Barber, K. Booij, J. Dachs, R. Lohmann, K.C. Jones, Environ. Sci. Technol. 38 (2004) 2617–2625.
[83] R. Lohmann, R. Gioia, S.J. Eisenreich, K.C. Jones, Atmos. Environ. 41 (2007) 7767–7777.
[84] C.J. Galbán-Malagón, S. Del Vento, N. Berrojalbiz, M.J. Ojeda, J. Dachs, Environ. Sci. Technol. 47 (2013) 5578–5587.
[85] L. Zhang, R. Lohmann, Environ. Sci. Technol. 44 (2010) 3832–3838.
[86] C.J. Galbán-Malagón, N. Berrojalbiz, R. Gioia, J. Dachs, Environ. Sci. Technol. 47 (2013) 7195–7203.

[87] H. Zhang, S.J. Eisenreich, T.R. Franz, J.E. Baker, J.H. Offenberg, Environ. Sci. Technol. 33 (1999) 2129–2137.
[88] P. Sun, I. Basu, P. Blanchard, K.A. Brice, R.A. Hites, Environ. Sci. Technol. 41 (2007) 1131–1136.
[89] J. Dachs, S.J. Eisenreich, Environ. Sci. Technol. 34 (2000) 3690–3697.
[90] J. Dachs, S.J. Eisenreich, R.M. Hoff, Environ. Sci. Technol. 34 (2000) 1095–1102.
[91] N. Berrojalbiz, J. Dachs, S. Del Vento, M.J. Ojeda, M.C. Valle, J. Casreo-Jiménez, G. Mariani, J. Wollgast, G. Hanke, Environ. Sci. Technol. 45 (2011) 4315–4322.
[92] H. Frouin, N. Dangerfield, R.W. Macdonald, M. Galbraith, N. Crewe, P. Shaw, D. Mackas, P.S. Ross, Prog. Oceanogr. 115 (2013) 65–75.

Chapter 9

Electronic Waste: A New Source of Halogenated Organic Contaminants

Hong-Gang Ni[1,*] and Eddy Y. Zeng[2]
[1]Peking University Shenzhen Graduate School, Shenzhen, China; [2]School of Environment, Jinan University, Guangzhou, China
**Corresponding author: E-mail: nihg@pkusz.edu.cn*

Chapter Outline

1. INTRODUCTION

Electronic waste (e-waste) and waste electrical and electronic equipment (WEEE) are the two most frequently used terms to define discarded appliances that use electricity. Generally, e-waste refers to waste electronic products, such as computers, televisions, and mobile phones, while WEEE also includes nonelectronic products such as refrigerators and washing machines [1]. However, the distinction between the two is becoming murky, because programmable microprocessors have been increasingly used in refrigerators, washing machines, and ovens which have traditionally been considered as nonelectronics [2]. With all these considerations, the definition of e-waste in this chapter covers all European Union (EU) WEEE Directive categories and products, within

Comprehensive Analytical Chemistry, Vol. 67. http://dx.doi.org/10.1016/B978-0-444-63299-9.00009-0

the context of Article 3(1) of Directive 2008/98/EC, including all components, subassemblies, and consumables which are parts of a product at the time of discarding (http://ec.europa.eu/environment/waste/weee/pdf/faq.pdf).

Rapid advances and improvements in equipment features and capabilities, combined with intense marketing strategies and increasing consumer demand, have led to an accelerating turnover rate for electrical and electronic equipment (EEE) globally. As a result, large amounts of e-waste are generated worldwide every day [3–18]. The rapidly increasing volume of e-waste streams and the hazardous substances contained therein pose serious challenges for processing and final disposal of e-waste [9,15,16,19–28]. Although regular e-waste recycling corporations focus intensively on eliminating the environmental impact with advanced technologies and applications under governmental guidelines, primitive handlings of e-waste, such as manual disassembling, open incineration, and acid dipping, have remained popular in some parts of the world [29]. These inappropriate e-waste recycling and disposal methods lead to the release of heavy metals and organic pollutants into the surrounding environment, which may exert potentially adverse environmental effects. In particular, informal processing of e-waste mostly occurring in developing countries may cause severe environmental pollution, though these countries are also most likely to reuse repaired electronics. Even in developed countries, recycling and disposal of e-waste may induce significant risk to workers and communities because all electronic scrap components contain various contaminants [27]. Therefore, great care must be taken to minimize unsafe exposure in recycling operations.

As the first step toward containing e-waste crisis, quantifying the magnitude of the crisis is critical. However, most published studies on e-waste were mainly restricted to evaluating the generation, exportation, and importation of e-waste or focused on the effects of e-waste management practices; [30–32] very few attempted to estimate the mass emissions of a variety of halogenated organic contaminants (HOCs) from globally generated e-waste. Our previous study [33] investigated environmental and human exposure to halogenated compounds derived from e-waste, but the estimates were restricted to China. In the last few years, a large number of studies have been published on the environmental occurrence of HOCs, including polybrominated diphenyl ethers (PBDEs) [34–36], polychlorinated biphenyls (PCBs) [34,35,37], polychlorinated dibenzo-*p*-dioxins and dibenzofurans (PCDD/Fs) [38–40], polybrominated biphenyls (PBBs) [34,35], tetrabromobisphenol A (TBBPA) [41], polybrominated dibenzo-*p*-dioxins and dibenzofurans (PBDD/Fs) [38,42] and chlorinated polycyclic aromatic hydrocarbons (ClPAHs) [41], originating from e-waste, thus providing sufficient data for an integrated assessment of the e-waste issue. This chapter will present the mass emissions of a variety of HOCs estimated from available data on the amounts of e-waste generated and processed, and contents of the target compounds.

2. AMOUNTS OF GLOBALLY GENERATED E-WASTE

Although the worldwide e-waste crisis has inspired numerous studies, few of them have quantified the figures of e-waste generated globally. In recent years, some methods, such as material flow analysis [4,9], multi-scale integrated analysis of societal and ecosystem metabolism [43], multivariate input–output analysis [44], etc., have been proposed to estimate the amount of e-waste generated in a certain region or country. Even so, as Chung et al. [5] wrote "Large discrepancies in the predicted amounts of e-waste generated were found no matter whether the same or different methods of estimation are used. Moreover, even when agreements between studies were found, the agreed figures were not necessarily the correct figures." Considering the unreliability of the predicted amounts of e-waste, the actual statistical data from the map of the world's e-waste, which was created by a partnership known as the "Solving the E-Waste Problem (StEP) Initiative" based on data gathered by the United Nations agencies, governments, and nongovernmental and scientific organizations, were adopted to discuss the global e-waste generation in this chapter (http://www.step-initiative.org/index.php/WorldMap.html). The interactive e-waste map provides data from 184 countries on the amounts of e-waste generated in 2012 (the latest available data so far). Unless otherwise noted, the amounts of e-waste generated worldwide and other relative data (such as EEE put in the market, population, and purchasing power) in this chapter are all from the map of the world's e-waste created by the StEP.

The map of the world's e-waste indicates that 48.9 million metric tons of e-waste were generated worldwide in 2012, which is equivalent to approximately 7 kg per capita for a global population of 7 billion. Currently, China leads the world in production of EEE. In 2012, China produced about 11.1 million metric tons of EEE, followed by the United States (US) with about 10 million metric tons. However, the amount of e-waste generated in the US in 2012 (9.4 million metric tons) was greater than that in China (7.3 million metric tons), because more products previously put in the market in the US were likely to become obsolete. In terms of the amounts of per capita e-waste generated in 2012 among all countries and regions in the world, Qatar ranked the first (63.1 kg per capita), followed by Luxembourg (48.9 kg per capita), Singapore (36.6 kg per capita), Norway (33.1 kg per capita), Hong Kong (30.3 kg per capita), Brunei Darussalam (30.2 kg per capita), the US (29.8 kg per capita), the United Arab Emirates (29.3 kg per capita), Switzerland (27.0 kg per capita), and Kuwait (26.1 kg per capita), among the top 10.

There are comprehensive reports on the transboundary movement of e-waste globally. These reports [1,12,45] provide important information for understanding the export/import of e-waste in recent years, which is not covered in this chapter. Instead, the relationships between the amount of e-waste generated and population, ratio of e-waste generated to electronic products put in the market and purchasing power are discussed to reveal the causes of e-waste generation and identify effective approaches to contain the global e-waste crisis.

A positive linear correlation is observed between the amounts of e-waste generated in individual countries and regions in 2012 and the corresponding populations (Figure 1). This is reasonable because waste (not only confined to e-waste) generation is related to the number of inhabitants. However, a country with a large population does not necessarily produce a proportional

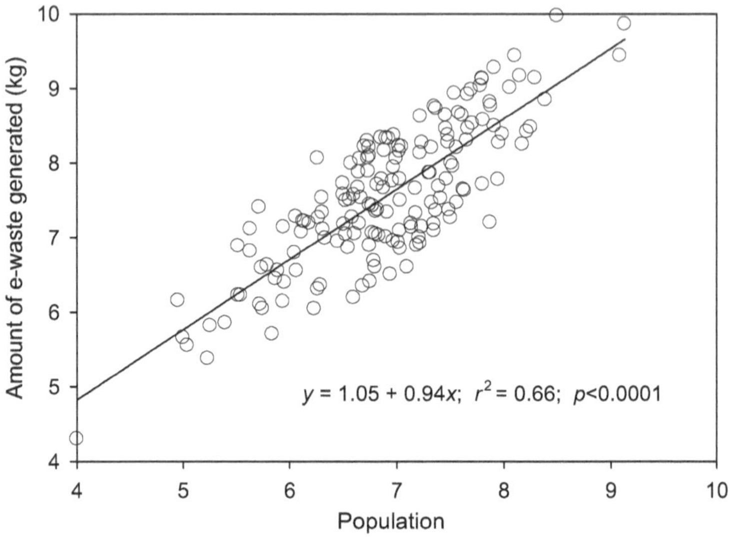

FIGURE 1 Relationship between the logarithmic transformed amounts of e-waste generated and population for 166 countries or regions in 2012.

amount of e-waste. For instance, the US eclipsed China in e-waste generation in 2012 (9.4 vs 7.3 million metric tons), although China's population (~1353 million) is much larger than that of the US (~314 million). This indicates that population is not always the key determinant of e-waste generation. A further investigation indicates that the amounts of EEE put in the market and e-waste generated in individual countries or regions in 2012 are significantly positively correlated (Figure 2), suggesting EEE in the market is an important driving force for e-waste generation. In addition, the slope in Figure 2 is the ratio of e-waste generated to electronic products put in the market, which can be defined as e-waste to EEE ratio. This index is a more appropriate representative of e-waste generation than the absolute amounts of e-waste generated. Similarly, a positive correlation between the e-waste to EEE ratio and purchasing power (USD per capita) is also observed (Figure 3). We have thus reached the following conclusions: (1) The e-waste to EEE ratio can be used as a better index than the absolute amount of e-waste generated to indicate e-waste generation and (2) purchasing power is a key determinant of e-waste generation.

Apparently, continuous increase in purchasing power is definitively a long-term trend with a steadily growing global economy. In this context, the amount of e-waste generated globally will most probably continue to rise in the foreseeing future. Therefore, extended use of EEE, clean production of EEE, and adoption of environmentally sound e-waste recycling methods are the key strategies in dealing with the global e-waste crisis. Apparently, effective global cooperation is a crucial step for implementing these strategies.

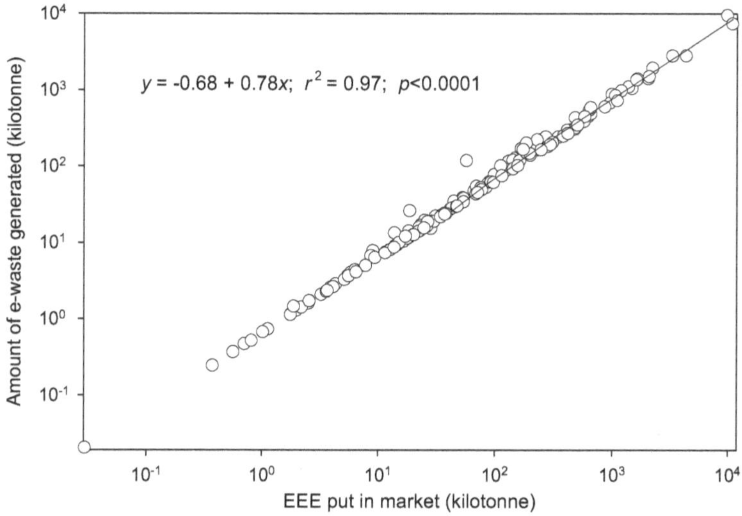

FIGURE 2 Correlation of electrical and electronic equipment put in the market and e-waste generated in 2012 for 166 countries or regions in 2012.

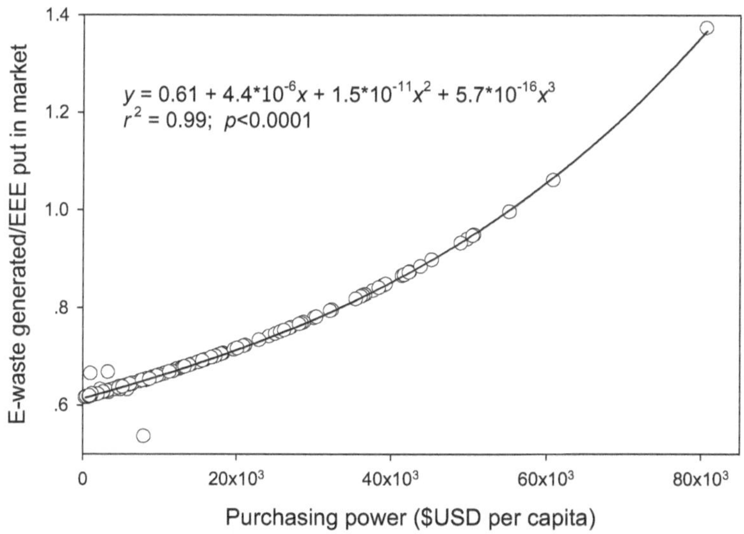

FIGURE 3 Relationship between the ratios of e-waste generated to electrical and electronic equipment put in the market and purchasing power (USD per capita) for 166 countries or regions in 2012.

3. HALOGENATED ORGANIC CONTAMINANTS FROM E-WASTE

It should be emphasized that e-waste is not the only source of HOCs in the environment. For instance, PCDD/Fs can be produced from various sources besides e-waste recycling operations [39,40,46–55]. ClPAHs are mainly emitted from incineration of municipal solid wastes and certain industrial processes [41,54]. Also, only a small portion of brominated flame retardants (BFRs) detected in the environment is released from e-waste, because only 3% of the global production of flame retardants is applied in electronic products [56]. Nevertheless, environmental problems associated with e-waste still deserve particular attention as e-waste is a huge hazard to the eco-environment and human health. Given the large number of studies on environmental and human exposure to toxic chemicals from e-waste recycling operations, only a few classes of HOCs associated with e-waste are reviewed in this chapter.

3.1 Brominated Flame Retardants

BFRs are anthropogenic chemicals that exert inhibitory effects on the ignition of combustible organic materials. They are effective in plastics and textiles, which is intended to reduce the flammability of the products. Among the various types of BFRs, PBDEs have been extensively used in a variety of commercial products, such as furniture, building material, and electronic components. The penta-BDEs and octa-BDEs were included in the target list of the Stockholm Convention on Persistent Organic Pollutants in 2009. Deca-BDEs have also been restricted in Europe since 2008, and were supposedly phased out in the US by the end of 2013 (http://www.epa.gov/oppt/existingchemicals/pubs/actionplans/deccadbe.html). With the ban of PBDEs, some non-PBDE BFRs have been proposed as alternatives of PBDEs in some applications [57], including hexabromocyclododecanes (HBCDs), 1,2-bis(2,4,6-dibromophenoxy) ethane (BTBPE), and decabromodiphenylethane (DBDPE). Because these alternative BFRs possess similar physiochemical properties as PBDEs, they are expected to behave similarly as PBDEs in the environment.

BFRs have been detected in almost all environmental media around e-waste recycling sites, apparently because large amounts of BFRs are released during primitive recycling processes. For example, a study reported that the air PBDE concentration in Guiyu (21.5 ng/m^3), a typical e-waste recycling site in China, was approximately 140 times higher than that in Hong Kong (0.15 ng/m^3) and 70 times higher than that in Guangzhou (0.29 ng/m^3) [58]. Similar results were also reported in another prominent e-waste recycling site, Taizhou in China [59]. Soil and sediments impacted by e-waste recycling processes also contained higher PBDE levels than those from reference areas. Especially, the levels of PBDEs were significantly higher in combusted residues (33,000–97,400 ng/g dry sample weight) than those in acid leaching impacted soil (2720–4250 ng/g dry sample weight) [60].

E-waste-derived BFRs have been detected in humans, fishes, and birds via various exposure routes. Luo et al. [61] confirmed that e-waste recycling operations were the major source of PBDEs in birds collected from the e-waste recycling region by examining the correlation between the levels of PBDEs in birds and e-waste recycling operations. Other studies reported that levels of PBDEs in human tissues (serum, hair, etc.) in e-waste recycling sites were higher than those in control regions [20,62]. Birds from an e-waste recycling site in South China contained high levels of HBCDs (<3.0–200 ng/g lipid weight) [63]. This value was at the high end of the worldwide range, indicating possible contamination of HBCDs in the e-waste recycling site. Occurrence of HBCDs was also detected in aquatic species from several e-waste recycling sites in China [64,65]. It is notable that aquatic species from these e-waste recycling sites contained HBCD levels comparable to those from contaminated sites in Europe [66], suggesting that e-waste recycling site is another HBCD hot spot, in addition to manufacturing facilities for HBCDs or HBCD-containing materials. In addition, other non-PBDE BFRs, such as BTBPE, DBDPE, hexabromobenzene (HBB), pentabromotoluene (PBT), and pentabromoethylbenzene (PBEB), were also detected in environmental media from e-waste recycling sites with higher levels than those in non-e-waste recycling sites [6]. All these results reflect that e-waste recycling operations have been posing adverse impacts on the surrounding environments. More details regarding the environmental occurrence of e-waste-derived BFRs can be found in various published references [13,17,21,28,33,67–71].

3.2 Dioxins and Dioxin-like Compounds

Dioxins and dioxin-like compounds are by-products of various industrial processes, and are commonly regarded as highly toxic chemicals that are environmentally persistent organic pollutants. They include PCDDs (derivatives of dibenzo-*p*-dioxin), PCDFs (derivatives of dibenzofuran), and PCBs (not dioxins, but 12 of them have dioxin-like properties). PCDD/Fs have been synthesized in small quantities for scientific research only [60]. Historically, municipal and medical waste incineration has been the most important source of PCDD/Fs. Another important source is uncontrolled combustion, particularly open burning of waste, accidental fires, and wildfires. For PCBs, they were synthesized for various technical applications and are still used in closed systems. They have entered the environment via accidents such as fires or leaks from transformers or heat exchangers, or from PCB-containing products in landfills or during incineration.

Similar to PBDEs, levels of PCDD/Fs and PCBs in ambient air of e-waste recycling sites were higher than those in other urban sites [10,38,49,72–75]. To date, the highest atmospheric concentration of PCDD/Fs (2765 pg/m^3) worldwide was observed in Guiyu, China [37]. Meanwhile, higher levels of atmospheric PBDD/Fs were also observed at the same site [37]. Obviously, the e-waste recycling operations in the region were the main sources of PCDD/F in the atmosphere. Moreover, the levels of atmospheric PCDD/Fs and PBDD/Fs approximately 50 km away from an e-waste recycling site in South China were at the high end of the global range for urban areas [38]. This indicated that dioxin and dioxin-like compounds from e-waste recycling activities may also impact adjacent areas.

Through comparative analysis, previous studies [54,60] found that concentrations of PCDD/Fs in the soil from e-waste recycling sites were significantly higher than those in a reference site. The difference between the reference site and e-waste recycling sites suggested the likelihood for these chemicals to transport atmospherically from where they were generated to adjacent areas. Comparing to other e-waste disposal activities, combustion of e-waste is the predominant formation mechanism for PCDD/Fs in the ambient environment [60]. A significant positive correlation between the levels of PCDD/Fs in dust and soil samples from an e-waste recycling site indicates that dioxin and dioxin-like compounds initially derived from burning of e-waste can enter ambient air and dust, and finally deposit in the soil [33].

Human exposure to dioxins and dioxin-like compounds derived from e-waste has been documented by several previous studies [35,60,74]. Levels of PCBs and PBBs in human tissues (e.g., hair, human milk, and serum) collected from e-waste recycling sites were normally higher than those from control sites [35,76]. As PCBs and PBBs were mainly used in old electrical and electronic products, their impacts on the environment are expected to gradually diminish. However, the considerably high levels of PCBs and PBBs in environmental compartments near e-waste recycling sites suggested it is not the case [34,77]. All these results implicate that PCBs and PBBs have remained the important groups of e-waste-related organic pollutants, and therefore need to be examined adequately as part of the e-waste problem [34,35,37,77].

3.3 Halogenated Polycyclic Aromatic Hydrocarbons

Halogenated polycyclic aromatic hydrocarbons (HPAHs) are a class of compounds with one or more halogen substituents attached to the aromatic rings of corresponding parent PAHs [78], which include ClPAHs and brominated polycyclic aromatic hydrocarbons. Currently, automobile emission and waste incineration have been considered as the two major emission sources of HPAHs in the environment [79–87]. Besides, primitive e-waste recycling procedures, such as roasting and open burning at low temperatures, are another proven emission source of ClPAHs [86]. However, very few studies on HPAHs associated with e-waste have been conducted and the formation mechanisms of HPAHs have remained unclear. During e-waste recycling processes, ClPAHs may be produced by chlorination of PAHs [86]. A previous investigation found that the mean concentration of ClPAHs in e-waste debris was lower than that in dust from an e-waste recycling workshop [86]. A possible explanation is that the e-waste recycling process, rather than electronic devices themselves, was the major source of ClPAHs [86]. This is similar to the formation of PCDD/F which is mainly originated from e-waste open burning [86]. Overall, there is limited information about the occurrence of e-waste-derived HPAHs in other environmental media.

4. OTHER TOXICS IN E-WASTE

Besides HOCs, there are many other categories of e-waste related toxic chemicals, such as heavy metals and other organic contaminants. Numerous previous studies have documented that most environmental media (air, soil, sediment, dust, etc.) around e-waste sites have been severely contaminated by heavy metals [15,28,73,88–96]. Because environmental and human exposure to heavy metals derived from e-waste have been detailed in our previous publication [97], only latest information (mainly published in 2014) about heavy metals from e-waste is updated herein. Other organic pollutants in addition to HOCs are also discussed in this section.

4.1 Heavy Metals

E-waste may contain various metals, for example, waste printed circuit boards contain Cu at 25–28 mg/g, Fe at 1.3–1.7 mg/g, Sn at 0.8–1.0 mg/g, and Mg at 0.4–1.0 mg/g, which are the main metals in raw materials and pyrolytic residues [98]. E-waste recycling activities are also releasing various other heavy metals such as Hg, Cd, Cr, Cu, Ni, Pb, and Zn [99]. For example, Pb was leached from mobile phone plastics by washing with Milli-Q water at a concentration (5.33 mg/L) exceeding the regulatory limit (5.0 mg/L), which is used to characterize a waste as being hazardous [100]. Similarly, a previous study also obtained the concentration of Ni leached from printed wire boards (21.83 mg/L) which exceeded the regulatory limit of 20 mg/L [100]. These results implicate mobile phone parts as hazardous wastes. In fact, all e-waste may possess the potential of causing contamination risk. An analysis of soil samples collected near an abandoned e-waste recycling plant in Taizhou, China showed that levels of Pb, Cd, and Cu reached 6083, 42, and 2364 mg/kg dry weight, and hazard quotient estimates suggested that Pb, Cd, Hg, and Cu posed high health risk for both adults and children [94]. Furthermore, accumulation of heavy metals in topsoil of e-waste recycling sites significantly increases the likelihood of shallow groundwater contamination [15,93].

It is notable that toxic heavy metals from e-waste recycling activities may contribute to an increase in the levels of heavy metals in human tissues. A recent study has examined levels of Hg in human hair from Guiyu of South China and potential risk factors. The results showed that hair Hg concentrations in volunteers of Guiyu (median: 0.99 μg/g) were significantly higher than those of a control site (median: 0.59 μg/g) [90]. After multiple stepwise regression analyses, hair Hg concentrations in the Guiyu samples were positively associated with the residence time in Guiyu ($p < 0.001$), and negatively related to the distance between homes and waste incineration sites ($p = 0.015$). All these results indicated that residing and working in e-waste sites for an extended period of time may primarily contribute to the high hair Hg concentrations [90]. Evaluating the effects of heavy metal coexposure on DNA oxidative damage in neonates from the same e-waste site suggested that parents' residence in this site and parents' work related to e-waste recycling operations were the risk factors associated with neonate's umbilical cord blood Pd and Cd levels. Exposures to Cd, Cr, and Ni were associated with increased oxidative DNA damage in neonates [73].

4.2 Organic Pollutants

Disposal of e-waste releases not only HOCs, but also other toxic organic compounds. For instance, nonmetal fractions of waste printed circuit boards were thermally treated at 300–500 °C, releasing phenol, 3-bromophenol, 2-methylphenol, and 4-propan-2-ylphenol in liquid products as the major species, which account for >50% of organic pollutants analyzed [98]. This result indicates that various e-waste recycling operations may produce different types of pollutants. As a result, only the most studied organic pollutants associated with e-waste are reviewed herein.

Investigations into the occurrence of PAHs in soils from an e-waste recycling center in South China confirmed that PAHs were mainly produced from combustion of e-waste [101]. Open burning and dismantling sites were the main sources of PAHs. Another recent study conducted in Africa [102] also corroborated that primitive recycling of e-waste is an emerging source of PAHs. They measured the concentrations of the hydroxylate PAH metabolites (OH-PAH), 1-hydroxyphenanthrene (1-OH-phenanthrene), the sum of 2- and 9-hydroxyphenanthrene (2-/9-OH-phenanthrene), 3-hydroxyphenanthrene (3-OH-phenanthrene), 4-hydroxyphenanthrene (4-OH-phenanthrene) and 1-hydroxypyrene (1-OH-pyrene), cotinine, and creatinine in urine samples. In the exposed group, median urinary concentrations were 0.85, 0.54, 0.99, 0.22, and 1.33 μg/g creatinine for 1-OH-phenanthrene, 2-/9-OH-phenanthrene, 3-OH-phenanthrene, 4-OH-phenanthrene, and 1-OH-pyrene, respectively, all significantly higher than those for the control group (0.55, 0.37, 0.63, 0.11, and 0.54 μg/g creatinine, respectively). A multivariate linear regression analysis suggested that e-waste recycling was the key determinant for PAH exposure [102].

Bisphenol A (BPA) was also found in soil samples collected from a typical electronic waste recycling center, South China [103]. The areas around the open burning and dismantling sites, as well as soils far from pollution sources, have been contaminated by dumping of waste residues and fly ash deposition. Concentrations of BPA were higher than those of TBBPA in all soil samples, and they were correlated significantly with each other. A possible explanation is that TBBPA can be transformed to BPA and that BPA can accumulate in soils. The transition between TBBPA and BPA was verified by incubation experiments [103].

Polybromobenzenes, including HBB, PBT, PBEB, pentabromobenzene, tetrabromobenzenes, and tribromotoluene were detected in air samples collected in North China [104]. The median concentrations of individual pollutants ranged from 0.07 to 1.16 pg/m^3, with relatively higher concentrations found near e-waste recycling sites. This confirmed e-waste may be one of the important sources of polybromobenzenes [104].

Occurrences of phthalic acid esters in soil and plants near an e-waste recycling site in East China have been observed. These compounds appeared to pose a great threat to the ambient environment media and human health [105].

Perfluorooctanoic acid (PFOA) is used widely in numerous industrial and consumer products. A recent study evaluated the exposure of mothers to PFOA and the potential hazards to neonates in an e-waste recycling site, South China [106]. The PFOA level was higher in maternal serum samples from the e-waste site than that from a control site (please refer to [106] for details). Mothers from the e-waste site were exposed to higher levels of PFOA than those from the control site. Prenatal exposure to PFOA was associated with decreased neonatal physical development and adverse birth outcome [106]. Furthermore, several other organic chemicals may also be associated with e-waste. For example, e-waste recycling has been implicated as one of the sources of polychlorinated naphthalenes [107]. In addition, a recent study has documented that e-waste recycling is an important source of Dechlorane Plus to the environment [108].

5. INVENTORIES OF HOCs IN E-WASTE

Needless to say, estimating the inventories of HOCs in e-waste is quite a challenge as data are insufficient. Herein, we adopt a prudent approach with meticulous scientific procedures to perform the estimation. As mentioned above, the reliable database of global e-waste amounts can be obtained from the "Solving the E-Waste Problem (StEP) Initiative." In addition, a large number of studies on the environmental occurrence of HOCs associated with e-waste have been published. Therefore, it is somewhat feasible to estimate the global inventories of HOCs in e-waste.

Based on the concentrations of target HOCs in e-waste and volumes of e-waste generated globally each year, the annual inventories of HOCs in e-waste generated worldwide (or within a certain country or region) can be estimated simply by the formula: annual inventory of HOCs in e-waste = annual total amount of e-waste generated × concentration of the target compound in e-waste. Plastics constitute a major portion of flame retardant consumption, and account for approximately 20% of the total weight in electronic products [109]. Therefore, the gross amount of plastics in e-waste and concentrations of BFRs in plastics can also be used to estimate the inventories of BFRs in e-waste. In this chapter, the inventories of HOCs in e-waste are estimated for the year of 2012, but inventories of HOCs in e-waste for other years can also be estimated using the same approach if data are available.

The StEP database indicates that the gross amounts of e-waste generated in the US, the EU, China, and the entire world were 9.359, 9.918, 7.253, and 48.89 million metric tons, respectively, in 2012. With the concentrations of BFRs in plastics [109] and PCDD/Fs [54], PBDD/Fs [42], and ClPAHs [41] in e-waste debris, the annual inventories of PBDEs, TBBPA, PBBs, PBPs, PCDD/Fs, PBDD/Fs, and ClPAHs in e-waste generated in the US, the EU, China, and the entire world were estimated (Table 1). Globally, the annual inventory of PBDEs in e-waste reached approximately 2.5×10^8 kg in 2012, with deca- and none-BDE congeners dominating the PBDE composition. This is consistent with the fact that heavily brominated BDEs are major constituents in electronic products [110]. In addition, the annual inventories of PBDEs in e-waste generated in the US, the EU, and China were comparable. As the world's three largest economic entities, they collectively contributed more than half of the global e-waste generated in 2012. As a consequence, the total annual inventory of PBDEs in e-waste generated in these three entities was also more than 50% of the global figure. Similar trend is also observed for other e-waste-derived organic pollutants.

A more recent study suggested that 2.7 million tons of plastics in e-waste containing flame retardants are discarded globally each year [111]. With the same procedure, the global annual inventories of BFRs in e-waste are also estimated (Figure 4). Overall, the estimated inventories of BFRs based on the annual global generation of e-waste are three times of those based on the annual global disposal of plastics in e-waste. The difference apparently reflects the variability in the estimated relative abundances of plastics in e-waste and concentrations of BFRs in plastics.

The estimated inventories of PCDD/Fs in e-waste in 2012 were quite low (Table 1), similar to the results of a previous study [33]. For example, the annual inventories of the most toxic component, 2,3,7,8-Tetrachlorodibenzo-*p*-dioxin (2,3,7,8-TCDD) in e-waste were only 7, 7, 5, and 35 g/year for the US, the EU, China, and the whole world. The fact is, almost no PCDD/Fs are inherited in e-waste and they are mostly produced from recycling processes, such as combustion or melting of plastics in e-waste [54]. Similarly, PBDD/Fs and HPAHs are also mainly produced from e-waste recycling processes, rather than being embedded in e-waste [41]. Therefore, inventories of PCDD/Fs, PBDD/Fs, and HPAHs in e-waste may under represent the actual environmental exposure of these chemicals from e-waste. In reality, mass loadings of PCDD/Fs, PBDD/Fs, and HPAHs from e-waste should be greater than their inventories in e-waste, as they are mainly originated from recycling operations. Of particular note is that the global inventory of ClPAHs in e-waste (2890 kg/year) is considerably greater than that for PCDD/Fs (104 kg/year) (Table 1). If ClPAHs released from e-waste recycling processes are taken into account, the amount of ClPAHs accumulated in the global environment will be even greater. Amid the high toxicity of HPAHs, which is similar to that of dioxins [41], HPAHs associated with e-waste recycling processes merit the further and comprehensive investigations.

Accumulated evidence has confirmed primitive recycling processes as the main culprit for emissions of toxic organic pollutants from e-waste. It should be noted that no recycling technique is absolutely safe. A recent study [27] showed that PBDEs may be released from electronic products to the environment at any time of their whole life cycles. PBDEs were detected in all air samples collected from the areas of dismantling and crushing, as well as open yards housing piles of e-waste in recycling regions. Assuming all e-waste

TABLE 1 Estimated Annual Inventories of E-Waste-Derived Polybrominated Diphenyl Ethers (PBDEs), Tetrabromobisphenol A (TBBPA), Polybrominated Biphenyls (PBBs) and Polybrominated Phenols (PBPs), Chlorinated Polycyclic Aromatic Hydrocarbons (ClPAHs), Polychlorinated Dibenzo-*p*-dioxins and Dibenzofurans (PCDD/Fs), and Polybrominated Dibenzofurans and Doxin (PBDD/Fs) in the Environments of the United States (US), the European Union (UN), China, and the Entire World

		Annual Inventory in E-Waste (kg/year)			
Compounds	Concentration (ng/g)	USA	EU	China	World
PBDEs					
penta-BDE	170.0	318.2	337.2	246.6	1662
hexa-BDE	2200	4118	4364	3191	2.2×10^4
hepta-BDE	2500	4680	4959	3627	2.4×10^4
octa-BDE	12,000	2.2×10^4	2.4×10^4	1.7×10^4	1.2×10^5
nona-BDE	4,200,000	7.8×10^6	8.3×10^6	6.1×10^6	41×10^7
deca-BDE	21,000,000	3.9×10^7	4.1×10^7	3.0×10^7	2.1×10^8
Sum	25,216,870	4.7×10^7	5.0×10^7	3.7×10^7	2.5×10^8
TBBPA	8100	1.5×10^4	1.6×10^4	1.2×10^4	7.9×10^4
PBBs					
tera-BB	5.00	9.360	9.918	7.253	48.89
penta-BB	45.00	84.24	89.26	65.28	440.0
hexa-BB	200.0	374.4	396.7	290.1	1956
Sum	250.0	468.0	495.9	362.7	2445

Continued

TABLE 1 Estimated Annual Inventories of E-Waste-Derived Polybrominated Diphenyl Ethers (PBDEs), Tetrabromobisphenol A (TBBPA), Polybrominated Biphenyls (PBBs) and Polybrominated Phenols (PBPs), Chlorinated Polycyclic Aromatic Hydrocarbons (ClPAHs), Polychlorinated Dibenzo-*p*-dioxins and Dibenzofurans (PCDD/Fs), and Polybrominated Dibenzofurans and Doxin (PBDD/Fs) in the Environments of the United States (US), the European Union (UN), China, and the Entire World—cont'd

Compounds	Concentration (ng/g)	Annual Inventory in E-Waste (kg/year)			
		USA	EU	China	World
PBPs					
mono-BP	70.00	131.0	138.9	101.5	684.5
di-BP	4400	8237	8728	6383	43,027
tri-BP	56.00	104.8	111.1	81.2	547.6
tetra-BP	35.00	65.52	69.43	50.77	342.3
penta-BP	170.0	318.2	337.2	246.6	1662
Sum	4731	8856	9384	6863	46,264
ClPAHs					
3,9-Cl_2Phe	0.940	8.798	9.323	6.818	45.96
3-ClFlu	0.520	4.867	5.157	3.772	25.42
8-ClFlu	13.20	123.5	130.9	95.74	645.4
1-ClPr	14.90	139.5	147.8	108.1	728.5
3,9,10-Cl_3Phe	5.430	50.82	53.85	39.38	265.5

7-ClBaA	10.60	99.21	105.1	76.88	518.3
6-ClBaP	13.50	126.4	133.9	97.92	660.1
Sum	59.10	553.2	586.2	428.7	2890
PCDDs					
2378-TCDD	0.001	0.007	0.007	0.005	0.035
12378-PeCDD	0.003	0.025	0.027	0.020	0.133
123478-HxCDD	0.003	0.033	0.035	0.025	0.171
123678-HxCDD	0.010	0.095	0.101	0.074	0.499
123789-HxCDD	0.012	0.112	0.119	0.087	0.587
1234678-HpCDD	0.083	0.776	0.822	0.601	4.053
OCDD	0.138	1.292	1.369	1.001	6.747
Sum	0.250	2.340	2.480	1.813	12.22
PCDFs					
2378-TCDF	0.099	0.928	0.984	0.719	4.850
12378-PeCDF	0.025	0.232	0.246	0.180	1.213
23478-PeCDF	0.037	0.344	0.364	0.266	1.794
123478-HxCDF	0.090	0.839	0.889	0.650	4.381
123678-HxCDF	0.030	0.280	0.297	0.217	1.462

Continued

TABLE 1 Estimated Annual Inventories of E-Waste-Derived Polybrominated Diphenyl Ethers (PBDEs), Tetrabromobisphenol A (TBBPA), Polybrominated Biphenyls (PBBs) and Polybrominated Phenols (PBPs), Chlorinated Polycyclic Aromatic Hydrocarbons (ClPAHs), Polychlorinated Dibenzo-*p*-dioxins and Dibenzofurans (PCDD/Fs), and Polybrominated Dibenzofurans and Doxin (PBDD/Fs) in the Environments of the United States (US), the European Union (UN), China, and the Entire World—cont'd

		Annual Inventory in E-Waste (kg/year)			
Compounds	**Concentration (ng/g)**	**USA**	**EU**	**China**	**World**
234678-HxCDF	0.033	0.304	0.322	0.236	1.589
123789-HxCDF	0.003	0.024	0.026	0.019	0.126
1234678-HxCDF	1.210	11.325	12.001	8.776	59.16
1234789-HxCDF	0.295	2.761	2.926	2.140	14.42
OCDF	0.069	0.642	0.680	0.498	3.354
Sum	1.889	17.68	18.74	13.70	92.36
PBDDs					
TriBDD	0.270	2.527	2.678	1.958	13.20
TeBDD	120.0	1123	1190	870.4	5867
PeBDD	1.900	17.78	18.84	13.78	92.90
HxBDD	0.920	8.611	9.125	6.673	44.98
2378TeBDD	0.020	0.187	0.198	0.145	0.978
12378PeBDD	0.070	0.655	0.694	0.508	3.423
123(6)78HxBDD	0.250	2.340	2.480	1.813	12.22
Sum	123.4	1155	1224	895.2	6035

PBDFs					
MoBDF	0.040	0.374	0.397	0.290	1.956
DiBDF	0.600	5.616	5.951	4.352	29.34
TriBDF	0.400	3.744	3.967	2.901	19.56
TeBDF	2.500	23.40	24.80	18.13	122.2
PeBDF	22.00	205.9	218.2	159.6	1076
HxBDF	85.00	795.6	843.0	616.5	4156
HpBDF	140.0	1310	1389	1015	6845
OBDF	160.0	1498	1587	1160	7823
Sum	410.5	3843	4072	2978	20,073

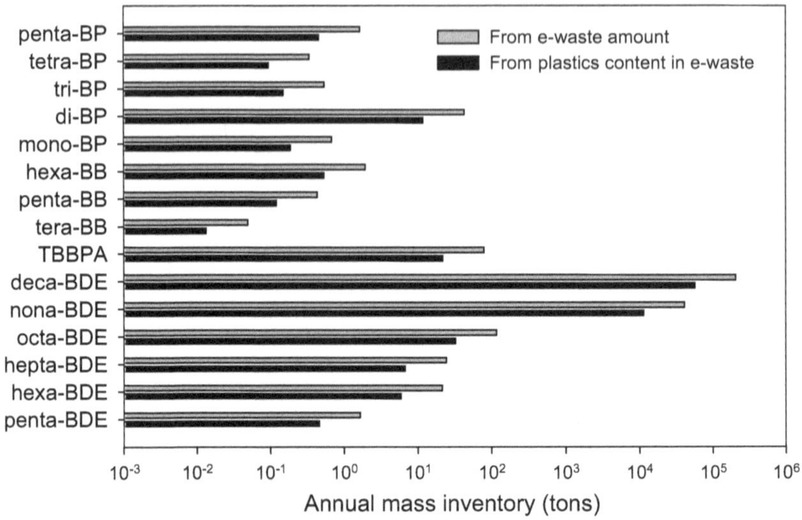

FIGURE 4 Comparison of the global annual inventories of brominated flame retardants in e-waste generated in 2012, estimated from the contents of plastics in e-waste, and the amount of e-waste generated, respectively.

generated globally is recycled in the same manner as in Korea, the emissions of PBDEs during recycling processes in the US, the EU, China, and the entire world are estimated using the emission factors obtained in the study of Park et al. [27] (Table 2).

Because of the large variability in the emission factors of PBDEs in recycling processes, i.e., the lowest and highest emission factors differ by 2–4 orders of magnitude, the estimated emissions of PBDEs during recycling operations also carry large uncertainties [27]. For instance, the estimated emissions of the sum of BDE-203, -206, and -209 for the entire world were 58–9730, 4.6–14,375, and 13–45,520 kg/year in 2012 from dismantling houses, crushing rooms, and open yards, respectively, based on the data provided by StEP. The combined emission of PBDEs from the US, the EU, and China reached more than half of the global emission (Table 2). Compared with the annual inventory in e-waste discussed above, the global annual emission of PBDEs during recycling processes is four orders of magnitude lower than the annual inventory in e-waste. This is obviously because annual mass inventory is the amount of HOCs embedded in e-waste but not the actual emission.

In summary, the global e-waste generation rate is increasing rapidly; [112] therefore, the annual mass inventories of e-waste-derived organic pollutants in the global environment are expected to maintain an upward trend in the future. HOCs in electronic products can enter the environment at any time during their lifespan. Recycling processes, however they may be, would continue to drive HOCs out of e-waste and into the environment.

TABLE 2 Estimated Emission (kg/year) of Polybrominated Diphenyl Ethers (Sum of BDE-203, -206, and -209) from E-Waste Recycling Processes in 2012

		Mass Emission			
	E-Waste[a]	Dismantling	Crushing	Open-Air	Total
US	9.360	11-1862	0.89-2752	2.5-8714	14-13328
EU	9.918	12-1974	0.94-2916	2.7-9234	15-14123
China	7.253	8.6-1443	0.69-2132	1.9-6752	11-10328
World	48.89	58-9730	4.6-14375	13-45520	75-69625

[a]*E-waste generated in million metric tons.*

6. UNCERTAINTIES AND LIMITATIONS

The above discussed procedure is deliberately simplistic, so as to be able to estimate the annual inventories of HOCs in e-waste generated in a certain country or region. Due to the lack of accurate and reliable data, the contents of HOCs in e-waste may vary widely and lead to large uncertainties in the estimates presented in this chapter. Unfortunately, no previous studies have estimated inventories of HOCs in e-waste generated or emissions of HOCs during e-waste recycling processes. Therefore we are unable to validate the estimates presented in this chapter against any previous independent examinations. In spite of this, the uncertainties in the estimated inventories of BFRs in e-waste can be evaluated through comparison with the estimated stock of BFRs in plastics in e-waste. For other HOCs (including HPAHs, PCDD/Fs, and PBDD/Fs) than BFRs associated with e-waste, no available data in the literature can be referenced for conducting uncertainty analysis. Because HPAHs, PCDD/Fs, and PBDD/Fs are mainly emitted from primitive recycling processes, their mass emissions are difficult to estimate. To our best knowledge, no inventories or mass emissions of these contaminants associated with e-waste have been reported to date. As a result, the estimates presented in this chapter may need to be improved via field measurements of the concentrations of HOCs in e-waste or emission factors for specific scenarios.

Whenever feasible, our estimates are presented as numerical ranges with maximum and minimum values, rather than as discrete and definitive numbers (e.g., Table 2). Because the present chapter merely attempts to present a static budget for the annual inventories of selected organic pollutants in e-waste for the year of 2012, certain dynamic features for the emission process of HOCs from e-waste may have been overlooked, e.g., potential temporal variability in the annually accumulated inventories of HOCs in e-waste and governing mechanisms for emission processes of HOCs from e-waste. Nevertheless, even if the estimates on the mass inventories and emissions of e-waste-derived HOCs by this chapter are substantially different from the reality, they still can serve as important sources of information for examining the environmental exposure to HOCs derived from e-waste and assessing the magnitude of the global e-waste crisis.

ACKNOWLEDGMENTS

This project was supported by the Shenzhen Science and Technology Research and Development fund (JCYJ20120614145332607) and the National Natural Science Foundation of China (Nos. 41329002 and 41390240).

REFERENCES

[1] B.H. Robinson, Sci. Total Environ. 408 (2009) 183–191.
[2] L.M. Hilty, Environ. Impact Assess. Rev. 25 (2005) 431–435.
[3] O.A. Ogunseitan, J.M. Schoenung, J.-D.M. Saphores, A.A. Shapiro, Science 326 (2009) 670–671.
[4] P. Andarani, N. Goto, J. Mater. Cycles Waste Manag. 16 (2014) 306–320.
[5] S.S. Chung, Waste Manage. Res. 30 (2012) 1130–1137.
[6] J. Wu, Y. Zhang, X. Luo, Y. She, L. Yu, S. Chen, B. Mai, J. Environ. Sci. 24 (2012) 183–194.
[7] M. Dwivedy, R.K. Mittal, J. Clean. Prod. 37 (2012) 229–242.
[8] J.A. Estrada-Ayub, R. Kahhat, Resour. Conserv. Recy. 86 (2014) 93–106.
[9] Habuer, J. Nakatani, Y. Moriguchi, Waste Manage. 34 (2014) 489–497.
[10] S. Herat, P. Agamuthu, Waste Manage. Res. 30 (2012) 1113–1129.
[11] N.G. Leigh, T. Choi, N.Z. Hoelzel, J. Ind. Ecol. 16 (2012) 940–950.
[12] J. Li, B.N.N. Lopez, L. Liu, N. Zhao, K. Yu, L. Zheng, Waste Manage. 33 (2013) 923–934.
[13] J. Li, P. Shi, H. Shan, Y. Xie, Environ. Technol. 33 (2012) 2547–2551.
[14] N. Milovantseva, J.-D. Saphores, Waste Manage. 33 (2013) 519–529.
[15] J.K. Pradhan, S. Kumar, Environ. Sci. Pollut. Res. 21 (2014) 7913–7928.
[16] M. Rahmani, R. Nabizadeh, K. Yaghmaeian, A.H. Mahvi, M. Yunesian, Resour. Conserv. Recy. 87 (2014) 21–29.
[17] L. Rocchetti, F. Veglio, B. Kopacek, F. Beolchini, Environ. Sci. Technol. 47 (2013) 1581–1588.
[18] Q. Song, Z. Wang, J. Li, X. Zeng, J. Mater. Cycles Waste Manag. 15 (2013) 469–475.
[19] Y.-J. Ben, X.-H. Li, Y.-L. Yang, L. Li, J.-P. Di, W.-Y. Wang, R.-F. Zhou, K. Xiao, M.-Y. Zheng, Y. Tian, X.-B. Xu, Environ. Pollut. 173 (2013) 176–181.
[20] X.H. Bi, G.O. Thomas, K.C. Jones, W.Y. Qu, G.Y. Sheng, F.L. Martin, J.M. Fu, Environ. Sci. Technol. 41 (2007) 5647–5653.
[21] M.N. Danon-Schaffer, A. Mahecha-Botero, J.R. Grace, M. Ikonomou, Sci. Total Environ. 447 (2013) 458–471.
[22] W.J. Deng, P.K.K. Louie, W.K. Liu, X.H. Bi, J.M. Fu, M.H. Wong, Atmos. Environ. 40 (2006) 6945–6955.
[23] I. Labunsk, S. Harrad, D. Santillo, P. Johnston, K. Brigden, Environ. Sci. Process. Impacts 15 (2013) 503–511.
[24] I. Labunska, S. Harrad, M. Wang, D. Santillo, P. Johnston, Environ. Sci. Technol. 48 (2014) 5555–5564.
[25] C.W. Lam, S.-R. Lim, J.M. Schoenung, J. Ind. Ecol. 17 (2013) 299–309.
[26] D. Pant, Int. J. Life Cycle Assess. 18 (2013) 1465–1469.

[27] J.-E. Park, Y.-Y. Kang, W.-I. Kim, T.-W. Jeon, S.-K. Shin, M.-J. Jeong, J.-G. Kim, Sci. Total Environ. 470 (2014) 1414–1421.
[28] Q. Song, J. Li, Environ. Intern. 68 (2014) 82–93.
[29] Y. Li, Y.-P. Duan, F. Huang, J. Yang, N. Xiang, X.-Z. Meng, L. Chen, Waste Manage. 34 (2014) 1059–1065.
[30] R. Widmer, H. Oswald-Krapf, D. Sinha-Khetriwal, M. Schnellmann, H. BÖni, Environ. Impact Assess. Rev. 25 (2005) 436–458.
[31] C.S.C. Wong, S.C. Wu, N.S. Duzgoren-Aydin, A. Aydin, M.H. Wong, Environ. Pollut. 145 (2007) 434–442.
[32] D.S. Khetriwal, P. Kraeuchi, R. Widmer, J. Environ. Manage. 90 (2009) 153–165.
[33] H.-G. Ni, H. Zeng, S. Tao, E.Y. Zeng, Environ. Toxicol. Chem. 29 (2010) 1237–1247.
[34] G. Zhao, H. Zhou, D. Wang, J. Zha, Y. Xu, K. Rao, M. Ma, S. Huang, Z. Wang, Sci. Total Environ. 407 (2009) 2565–2575.
[35] G.F. Zhao, Z.J. Wang, M.H. Dong, K.F. Rao, J.P. Luo, D.H. Wang, J.M. Zha, S.B. Huang, Y.P. Xu, M. Ma, Sci. Total Environ. 397 (2008) 46–57.
[36] Z.Z. Yang, X.R. Zhao, Q. Zhao, Z.F. Qin, X.F. Qin, X.B. Xu, Z.X. Jin, C.X. Xu, Bull. Environ. Contam. Toxicol. 80 (2008) 340–344.
[37] G. Zhao, Y. Xu, W. Li, G. Han, B. Ling, Sci. Total Environ. 377 (2007) 179–191.
[38] H.R. Li, L.P. Yu, G.Y. Sheng, J.M. Fu, P.A. Peng, Environ. Sci. Technol. 41 (2007) 5641–5646.
[39] J.X. Zhu, Y. Hirai, S. Sakai, M.H. Zheng, Chemosphere 73 (2008) S72–S77.
[40] J.X. Zhu, Y. Hirai, G. Yu, S. Sakai, Chemosphere 70 (2008) 703–711.
[41] T. Shi, S.J. Chen, X.J. Luo, X.L. Zhang, C.M. Tang, Y. Luo, Y.J. Ma, J.P. Wu, X.Z. Peng, B.X. Mai, Chemosphere 74 (2009) 910–916.
[42] M. Zennegg, M. Schluep, M. Streicher-Porte, P. Lienemann, R. Haag, A.C. Gerecke, Chemosphere 116 (2014) 34–39.
[43] G. D'Alisa, M. Federica Di Nola, M. Giampietro, J. Clean. Prod. 35 (2012) 59–70.
[44] F. Wang, J. Huisman, A. Stevels, C.P. Balde, Waste Manage. 33 (2013) 2397–2407.
[45] F.O. Ongondo, I.D. Williams, T.J. Cherrett, Waste Manage. 31 (2011) 714–730.
[46] M. Cohen, R. Draxler, R. Artz, Environ. Sci. Technol. 36 (2002) 4831–4835.
[47] F. Karasek, O. Hutzinger, Anal. Chem. 58 (1986) 633–640.
[48] C. Chen, Chemosphere 54 (2004) 1413–1420.
[49] G.J. Zheng, A.O.W. Leung, L.P. Jiao, M.H. Wong, Environ. Intern. 34 (2008) 1050–1061.
[50] H.R. Chao, S.L. Wang, C.C. Lee, H.Y. Yu, Y.K. Lu, O. Päke, Food Chem. Toxicol. 42 (2004) 1299–1308.
[51] J.L. Domingo, M. Schuhmacher, M.C. Agramunt, J.M. Llobet, J. Rivera, L. Muller, Environ. Intern. 28 (2002) 19–27.
[52] R. Ishaq, C. Näf, Y. Zebühr, D. Broman, U. Järnberg, Chemosphere 50 (2003) 1131–1150.
[53] B. Jimenez, R. Rodriguez-Estrella, R. Merino, G. Gomez, L. Rivera, M.J. Gonzalez, E. Abad, J. Rivera, Environ. Pollut. 133 (2005) 139–146.
[54] J. Ma, K. Kannan, J. Cheng, Y. Horii, Q. Wu, W. Wang, Environ. Sci. Technol. 42 (2008) 8252–8259.
[55] S. Wen, F. Yang, J.G. Li, Y. Gong, X.L. Zhang, Y. Hui, Y.N. Wu, Y.F. Zhao, Y. Xu, Chemosphere 74 (2009) 981–987.
[56] P.O. Darnerud, G.S. Eriksen, T. Johannesson, P.B. Larsen, M. Viluksela, Environ. Health Perspect. 109 (2001) 49–68.
[57] A. Covaci, S. Harrad, M.A.E. Abdallah, N. Ali, R.J. Law, D. Herzke, C.A. de Wit, Environ. Intern. 37 (2011) 532–556.
[58] W.J. Deng, J.S. Zheng, X.H. Bi, J.M. Fu, M.H. Wong, Environ. Intern. 33 (2007) 1063–1069.

[59] Y.M. Li, G.B. Jiang, Y.W. Wang, P. Wang, Q.H. Zhang, Chin. Sci. Bull. 53 (2008) 521–528.
[60] A.O.W. Leung, W.J. Luksemburg, A.S. Wong, M.H. Wong, Environ. Sci. Technol. 41 (2007) 2730–2737.
[61] X.J. Luo, X.L. Zhang, J. Liu, J.P. Wu, Y. Luo, S.J. Chen, B.X. Mai, Z.Y. Yang, Environ. Sci. Technol. 43 (2009) 306–311.
[62] J. Yuan, L. Chen, D. Chen, H. Guo, X. Bi, Y. Ju, P. Jiang, J. Shi, Z. Yu, J. Yang, L. Li, Q. Jiang, G. Sheng, J. Fu, T. Wu, X. Chen, Environ. Sci. Technol. 42 (2008) 2195–2200.
[63] M.J. He, X.J. Luo, L.H. Yu, X.L. Zhang, J. Liu, S.J. Chen, D. Chen, B.X. Mai, Environ. Sci. Technol. 44 (2010) 5748–5754.
[64] X. Zhang, F. Yang, C. Luo, S. Wen, X. Zhang, Y. Xu, Chemosphere 76 (2009) 1572–1578.
[65] J.-P. Wu, Y.-T. Guan, Y. Zhang, X.-J. Luo, H. Zhi, S.-J. Chen, B.-X. Mai, Environ. Sci. Technol. 44 (2010) 5490–5495.
[66] A. Covaci, A.C. Gerecke, R.J. Law, S. Voorspoels, M. Kohler, N.V. Heeb, H. Leslie, C.R. Allchin, J. de Boer, Environ. Sci. Technol. 40 (2006) 3679–3688.
[67] K. Betts, Environ. Sci. Technol. 41 (2007) 5576.
[68] Q. Luo, M.H. Wong, Z. Wang, Z. Cai, Environ. Earth Sci. 69 (2013) 2633–2641.
[69] S. Needhidasan, M. Samuel, R. Chidambaram, J. Environ. Health Sci. Eng. 12 (2014).
[70] H.G. Ni, E.Y. Zeng, Environ. Sci. Technol. 43 (2009) 3991–3994.
[71] M. Oteng-Ababio, E.F. Amankwaa, M.A. Chama, Habitat. Intl. 43 (2014) 163–171.
[72] R. Chatterjee, Environ. Sci. Technol. 41 (2007) 5577.
[73] W. Ni, Y. Huang, X. Wang, J. Zhang, K. Wu, Sci. Total Environ. 472 (2014) 354–362.
[74] T. Zhang, Y.-R. Huang, S.-J. Chen, A.-M. Liu, P.-J. Xu, N. Li, L. Qi, Y. Ren, Z.-G. Zhou, B.-X. Mai, J. Environ. Monit. 14 (2012) 3156–3163.
[75] G. Zheng, X. Xu, B. Li, K. Wu, T.A. Yekeen, X. Huo, J. Expo. Sci. Environ. Epidemiol. 23 (2013) 67–72.
[76] G.H. Xing, J.K.Y. Chan, A.O.W. Leung, S.C. Wu, M.H. Wong, Environ. Intern. 35 (2009) 76–82.
[77] A. Leung, Z.W. Cai, M.H. Wong, J. Mater. Cycles Waste Manag. 8 (2006) 21–33.
[78] P.P. Fu, L.S. Von Tungeln, L.H. Chiu, Z.Y. Own, J. Environ. Health C 17 (1999) 71–109.
[79] P. Haglund, T. Alsberg, A. Bergman, B. Jansson, Chemosphere 16 (1987) 2441–2450.
[80] U.L. Nilsson, C.E. Oestman, Environ. Sci. Technol. 27 (1993) 1826–1831.
[81] T. Ohura, T. Amagai, M. Makino, Chemosphere 70 (2008) 2110–2117.
[82] T. Ohura, S. Fujima, T. Amagai, M. Shinomiya, Environ. Sci. Technol. 42 (2008) 3296–3302.
[83] T. Ohura, A. Kitazawa, T. Amagai, M. Makino, Environ. Sci. Technol. 39 (2005) 85–91.
[84] T. Ohura, M. Morita, M. Makino, T. Amagai, K. Shimoi, Chem. Res. Toxicol. 20 (2007) 1237–1241.
[85] T. Ohura, K.-I. Sawada, T. Amagai, M. Shinomiya, Environ. Sci. Technol. 43 (2009) 2269–2275.
[86] J. Ma, Y. Horii, J. Cheng, W. Wang, Q. Wu, T. Ohura, K. Kannan, Environ. Sci. Technol. 43 (2009) 643–649.
[87] Y. Horii, G. Ok, T. Ohura, K. Kannan, Environ. Sci. Technol. 42 (2008) 1904–1909.
[88] J. Li, H. Duan, P. Shi, Waste Manage. Res. 29 (2011) 727–738.
[89] C. Luo, C. Liu, Y. Wang, X. Liu, F. Li, G. Zhang, X. Li, J. Hazard. Mater. 186 (2011) 481–490.
[90] W. Ni, Y. Chenb, Y. Huang, X. Wang, G. Zhang, J. Luo, K. Wu, Environ. Res. 128 (2014) 84–91.
[91] X. Tang, M.Z. Hashmi, D. Long, L. Chen, M.I. Khan, C. Shen, Int. J. Environ. Res. Public Health 11 (2014) 3118–3131.

[92] Q. Wang, A.M. He, B. Gao, L. Chen, Q.Z. Yu, H. Guo, B.J. Shi, P. Jiang, Z.Y. Zhang, P.L. Li, Y.G. Sheng, M.J. Fu, C.T. Wu, M.X. Chen, J. Yuan, J. Environ. Sci. Health Part A 46 (2011) 669–676.

[93] C. Wu, Y. Luo, S. Deng, Y. Teng, J. Song, Sci. Total Environ. 472 (2014) 556–561.

[94] Q. Zhang, J. Ye, J. Chen, H. Xu, C. Wang, M. Zhao, Environ. Pollut. 185 (2014) 258–265.

[95] K. Zhao, W. Fu, X. Liu, D. Huang, C. Zhang, Z. Ye, J. Xu, Environ. Sci. Pollut. Res. 21 (2014) 7165–7176.

[96] J. Zheng, X.-J. Luo, J.-G. Yuan, L.-Y. He, Y.-H. Zhou, Y. Luo, S.-J. Chen, B.-X. Mai, Z.-Y. Yang, Arch. Environ. Contam. Toxicol. 61 (2011) 696–703.

[97] H.-G. Ni, E.Y. Zeng, in: B. Bilitewski (Ed.), Global Risk-based Management of Chemical Additives II: Risk-based Assessment and Management Strategies, Hdb Env Chem, Springer-Verlag, Berlin Heidelberg, 2013.

[98] K.-H. Lin, H.-L. Chiang, J. Hazard. Mater. 271 (2014) 258–265.

[99] A.O.W. Leung, N.S. Duzgoren-Aydin, K.C. Cheung, M.H. Wong, Environ. Sci. Technol. 42 (2008) 2674–2680.

[100] S. Yadav, S. Yadav, P. Kumar, Waste Manage. 34 (2014) 1274–1278.

[101] D.-Y. Huang, C.-P. Liu, F.-B. Li, T.-X. Liu, C.-S. Liu, L. Tao, Y. Wang, Environ. Monitor. Assess. 186 (2014) 3351–3364.

[102] T. Feldt, J.N. Fobil, J. Wittsiepe, M. Wilhelm, H. Till, A. Zoufaly, G. Burchard, T. Goeen, Sci. Total Environ. 466 (2014) 369–376.

[103] D.-Y. Huang, H.-Q. Zhao, C.-P. Liu, C.-X. Sun, Environ. Sci. Pollut. Res. 21 (2014) 5818–5826.

[104] Y. Lin, X. Qiu, Y. Zhao, J. Ma, Q. Yang, T. Zhu, Environ. Sci. Technol. 47 (2013) 12761–12767.

[105] T.T. Ma, P. Christie, Y.M. Luo, Y. Teng, Environ. Geochem. Health 35 (2013) 465–476.

[106] K. Wu, X. Xu, L. Peng, J. LiuA, Y. Guo, X. Huo, Environ. Intern. 48 (2012) 1–8.

[107] Q. Li, Y. Xu, J. Li, X. Pan, X. Liu, G. Zhang, Atmos. Environ. 56 (2012) 228–235.

[108] S.-J. Chen, M. Tian, J. Wang, T. Shi, Y. Luo, X.-J. Luo, B.-X. Mai, Environ. Pollut. 159 (2011) 1290–1296.

[109] K.-I. Choi, S.-H. Lee, M. Osako, Chemosphere 74 (2009) 460–466.

[110] P. Lindberg, U. Sellstrom, L. Haggberg, C.A. de Wit, Environ. Sci. Technol. 38 (2004) 93–96.

[111] J.R. Peeters, P. Vanegas, L. Tange, J. Van Houwelingen, J.R. Duflou, Resour. Conserv. Recy. 84 (2014) 35–43.

[112] T. Wang, Environ. Sci. Technol. 41 (2007) 2655–2656.

Chapter 10

Occurrence and Human Health Risk of Emerging Organic Contaminants in E-Waste

David O. Carpenter
Institute for Health and the Environment, University at Albany, Rensselaer, NY, USA
E-mail: dcarpenter@albany.edu

Chapter Outline

1. INTRODUCTION

E-waste is the term used to describe discarded electrical appliances or electronic devices. As such it includes everything from refrigerators, washing machines, microwave ovens, fluorescent light fixtures to old cell phones, computers, television sets, and any device that has a printed circuit board or electrical wiring. E-waste also includes batteries, transformers, capacitors, circuit boards, and activated glass. E-waste is growing faster than any other component of the municipal waste stream and in 2005, constituted 8% of municipal solid waste, equivalent to some 20–50 million tons per year [1,2]. The lifetime of appliances and especially of electronic devices has gotten shorter in recent years because

Comprehensive Analytical Chemistry, Vol. 67. http://dx.doi.org/10.1016/B978-0-444-63299-9.00010-7

of the rapid development of technology which makes computers and cell phones obsolete within short periods of time. In all societies, there is a great urge to purchase the latest and newest devices that have more and better features.

While it has never been very profitable to reclaim materials from old appliances, the situation has changed recently as the value of copper and other metals has risen. This has resulted in a widespread industry, primarily in poor and developing countries, where metals are extracted and sold. The price of copper, a major component of electric wires and cables, has risen enormously in recent years (Figure 1) and is the major metal of value in electrical wire. However, e-waste contains many other metals that can be sold if removed and separated, including nickel, lead, aluminum, iron, tin, and arsenic, precious metals (gold, silver, platinum) and concentrations of rare elements that are in short supply worldwide (palladium, tantalum, indium, ruthenium, tellurium) and are used in computers, cell phones, cathode ray tubes, and other devices.

It is reported that 80% of the e-waste from developed countries is shipped to developing countries because there is not enough value in most e-waste to cover the cost of responsible recycling [3]. Most e-waste ends up in informal recycling facilities which operate without regulation in Asia and Africa. Larger household appliances constitute about 50%, information and communications equipment about 30%, and consumer electronics about 10% of the e-waste. In the past, most e-waste went to China, where there are several locations that have massive areas of informal recycling. The largest is at Guiyu in Guangdong province, where nearly 60–80% of families are engaged in unregulated e-waste recycling [4]. Guiyu (population 150,000) has 300 companies and 3000 individual workshops spread in over 20 villages, and processes about 50% of the estimated 2.5×10^6 annual tons of e-waste going to China [5]. The methods used involve sorting, incineration, acid and alkaline baths, open burning of wires and cables, and strong acid leaching. There are other large e-waste industries at Taizhou and Qingyuan. However, China is beginning to regulate the industry [6], and this has

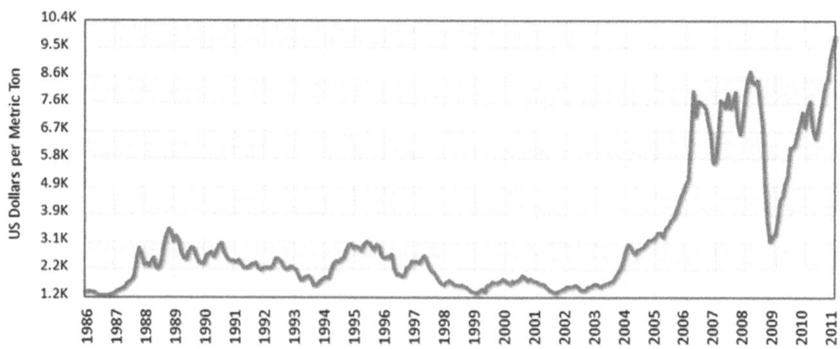

FIGURE 1 Price of copper between 1986 and 2011. *Data originally extracted from* http://www.lme.co.uk *by* en.wikipedia *where it is freely accessible and usable in the public domain.*

had the effect of moving some of the e-waste recycling operations to other countries, especially India, Pakistan, Ghana, and Nigeria and also the Philippines, Sri Lanka, Thailand, Malaysia, and Vietnam [3]. Nigeria is reported to have imported about 40 million personal computers and monitors over the period 2005–2010 [7]. In many of these countries, unlike China, there are few or no regulations to protect either human health or the environment from exposure to chemicals that have adverse effects. However, much of what we know of the hazards of e-waste come from studies in China, where there is a great deal of recent monitoring of concentrations of contaminants and a beginning of study of the adverse health effects on workers and local residents.

Almost all of the value in the e-waste stream comes from the metals. While there are serious adverse health effects from exposure to the metals associated with e-waste recycling, this chapter is focused on the health effects coming from a variety of the organic compounds also present in e-waste. Most of the organic compounds do not have commercial value, and human exposure and environmental contamination are by-products of the methods used to extract metals.

2. ORGANIC COMPOUNDS IN E-WASTE

There are a great variety of organic compounds in e-waste. Some of the most important are the brominated flame retardants, which are components of almost all electronic devices. There are a number of different flame-retardant compounds, including polybrominated diphenyl ethers (PBDEs), decabromodiphenyl oxide (used in electronic equipment like TVs), tetrabromobisphenol A (used in printed circuit boards), polybrominated biphenyls (PBBs), and polybrominated benzenes. Polychlorinated biphenyls (PCBs) are found in old capacitors, transformers, and in the ballasts in discarded fluorescent light fixtures. Burning of plastic produces chlorinated and brominated dibenzofurans and dioxins, as well as polyaromatic hydrocarbons (PAHs). Burning causes a release of particulates of various compositions, often containing PAHs as well as metals. Volatile organic compounds (VOCs) will also be released when plastic is heated. Phthalates are added to plastics to make them soft and can be released when heated. One study [8] reported higher concentrations of perchlorate in the blood of Vietnamese living near an e-waste recycling site as compared to a control. Perchlorate is used in rocket fuels, explosives and fireworks and also in the manufacture of polyvinyl chloride. The latter may be the major source of perchlorate from e-waste.

3. ROUTES OF EXPOSURE TO ORGANIC COMPOUNDS IN E-WASTE

Exposure to organic chemicals can result from ingestion, inhalation, or dermal absorption, and all the three routes are important for e-waste. Inhalation is a particularly important and immediate route. Since the major goal of handling e-waste is usually to extract the metals, it is a common practice to burn the plastic and insulation around electric wires and cables. Most if not all of these contain PBDEs or other flame retardants, and burning will volatilize them as well as generate PBDE-containing particulates that will be inhaled along with PAHs and VOCs. Some plastics will also contain phthalates, which will volatilize. PBDEs in particulates near Guiyu were found to be 58–691 times higher than those from other urban sites in China [9,10]. There is also the problem that organics going into air can travel to sites distant from that where the e-waste recycling is occurring. This is a particular problem for the more persistent organics, such as PCBs and PBDEs, as well as the very toxic products of combustion, furans and dioxins.

In addition to old transformers and capacitors, some older cables contained PCBs as insulators because they were relatively resistant to fire, and these will be volatilized. Elevated concentrations of PCBs have been found in soil, water, air, local fish, and human breast milk in local residents and e-waste workers in Guiyu [9,11]. Chen et al. [12] measured PCBs in the air near an e-waste site and found concentrations to be 12 times higher than those at a rural site. Xing et al. [10] concluded that inhalation of PCBs was responsible for 27% of the exposure of local residents and 93% of the exposure of the workers, with the remainder being due to ingestion.

Unfortunately, low-temperature combustion of PCBs and brominated flame retardants will create chlorinated or brominated dibenzofurans and dioxins, and these also can be inhaled both in a vapor phase and more often bound to particulates. Li et al. [13] measured chlorinated and brominated dioxins and furans in air at Guiyu, and found values that were higher than those measured in any other location in the world. They found that the lower chlorinated dioxin and furan congeners partitioned primarily in the gas phase, while the more highly chlorinated congeners were particulate-bound in the summer. In the winter, however, almost all were particulate-bound. Ren et al. [14] monitored particulates in a typical workshop at Guiyu, and found levels of total PBDEs to be 948 ng/m^3, chlorinated dioxins/furans to be 317 pg/m^3, and brominated dioxins/furans to be 481 pg/m^3. Xiao et al. [15] measured air concentrations of chlorinated dioxins/furans at the Longtang e-waste site, and found them to be 16–17 times higher than those in neighboring areas, and 2–3 orders of magnitude higher than those measured in most urban environments. Clearly inhalation is a major route of exposure to these very toxic organics.

PAHs are formed from combustion, and PAH metabolites have been found to be elevated in the urine of e-waste workers in Ghana [16]. While PAHs will be formed from burning plastic, it is not uncommon in Africa to make the fires for melting the metals by burning old tires or other combustible sources of PAHs. Huang et al. [17] monitored PAHs near the Longtang e-waste site in China, and found very high concentrations in soils nearest to the actual site, decreasing with distance away from the site. In a separate study, Huang et al. [18] monitored tetrabromobisphenol A and bisphenol A in soils at various distances from the Longtang e-waste site, and found the same pattern, with high concentrations near the site and significant concentrations even at some distance away, indicating that, like the PAHs, these compounds were attached to particulates that travel by air before depositing. Bisphenol A is a degradation product of tetrabromobisphenol A, and was found at higher concentrations than its parent compound in the soils.

Most organic pollutants are lipophilic, and therefore, if persistent, accumulate in animal fat. The term persistent organic pollutants (POPs) is commonly used to indicate PCBs, dioxins/furans, halogenated pesticides, persistent brominated compounds and PAHs. This makes ingestion an important route of exposure as well, especially when these chemicals are present in waters containing fish or contaminated food fed to domestic animals. Domestic animals will also consume a certain amount of soil, which can be an indirect route of exposure to humans. Song and Li [6] investigated dietary intake of organics at three e-waste sites (Guiyu, Taizhou, Quinyuan) relative to control sites, and looked at freshwater fish, marine fish, shellfish, meat and meat products, viscera, eggs, cereals, vegetables, and fruits. At Guiyu, the order of the highest PBDE concentrations was freshwater fish > meat > poultry > eggs. At both Guiyu and Taizhou the main dietary source was seafood, while at Quinyuan it was eggs. Eggs were particularly high in PBBs at Quinyuan, probably reflecting some differences in the types of e-waste at the different sites.

E-waste sites have been found to cause extensive contamination of local soils, river sediments, river water, air, and local foodstuffs with PBDEs and other brominated flame retardants, PCBs, and of course metals [9,19]. PBDE concentrations in fish from the Nanyang River near Guiyu were found to be 1000 times higher than those found in other regions [20], while PCB levels in fish were at least an order of magnitude higher than those from control sites [11]. These substances can also be ingested from hand contamination, leading to contamination of foodstuffs. There can be some exposure from consuming plants growing in the area as a result of inadequate surface washing, and even some minor uptake through an interior compartment that is poorly understood [21]. There is clear evidence of PBDE contamination of fish from local streams and domestic duck eggs that are consumed [22,23].

When these volatile compounds land on the skin, they can be absorbed, making dermal exposure also a possible route of exposure. Ma et al. [24] reported that dermal absorption of dioxins and furans were two orders of magnitude higher at an e-waste site than at a comparison industrial site, although the absolute exposure from soil ingestion was four times that of dermal absorption.

A particularly concerning route of exposure is transplacental and/or via breast milk. The developing fetus is especially sensitive to a variety of chemicals during development. Xu et al. [25] have found that there were elevated concentrations of PBDEs and PAHs in umbilical cord blood and placenta in infants born to mothers in Guiyu as compared to a control site. Song and Li [6] have reviewed the studies of levels of the various organics as well as metals in umbilical cord blood, breast milk, human blood, hair, and urine, and report that the weight of evidence strongly supports the conclusion that nearby residents and workers all have elevated concentrations of contaminants.

4. HEALTH EFFECTS OF E-WASTE

The study of adverse health effects among individuals exposed to e-waste is not well advanced, but we know a great deal about the adverse health effects of many of the chemicals found in the air, soil, water, and other media around e-waste sites. Therefore, we will approach the subject of human health effects by discussion of both what is known of workers and residents near e-waste sites, but also the more general literature on health effects of the various chemicals found at e-waste sites. Grant et al. [26] have previously reviewed the earlier literature on health effects of e-waste.

4.1 Cancer

Many of the organic compounds associated with e-waste are known human carcinogens. The evidence is strongest for dioxins/furans and PCBs, all of which are listed by the International Agency for Research on Cancer as Group 1, known human carcinogens [27,28]. These carcinogens are present in air, both in a vapor phase and bound to particulates. Zhang et al. [29] measured atmospheric concentrations of chlorinated and brominated dioxins and furans at Taizhou, and calculated that the lifetime excess cancer risk from these exposures ranged from 0.7×10^{-5} to 5.4×10^{-5}. Wang et al. [30] have calculated the risk of cancer from inhalation of PAHs at Qingyuan e-waste sites as compared to those in Guangzhou, considered as a background level in urban environments, and found that the risk was 1.6-fold higher at the e-waste site. Benzo(b)fluoranthene, dibenz(ah)anthracene, and dibenzo(a)pyrene contributed the most to cancer risk. There is strong evidence for elevations in cancer risk from fine particle air pollution from studies independent of e-waste exposures [31]. PAHs are known to be human carcinogens [32], and that both particulate-bound and gas-phase PAHs contribute to cancer risk [33]. While childhood cancer is also a concern, many of the cancers that result from exposure to children will not manifest until later in life [34].

4.2 Thyroid Dysfunction

Residents of a village close to an e-waste site had both higher serum levels of PBDEs and thyroid-stimulating hormone (TSH) than did residents of a more remote village [35]. However, a study by Wang et al. [36] reported that brominated flame retardants were associated with reduced TSH among residents and workers at Taizhou e-waste sites. These seemingly incompatible results may be due to differences in effects of PBDEs, which have been reported to be associated with lower TSH [37] while PCBs are known to be associated with increased TSH [38].

Giera and Zoeller [39] have recently reviewed the actions of various organics on thyroid function, and report perturbation of normal function by PCBs, PBDEs, bisphenol A and phthalates, all of which are present at e-waste sites. Perchlorate is an antagonist of iodide uptake into the thyroid gland. Because hypothyroidism during pregnancy is a serious risk factor for cognitive impairment of the child [40], the demonstration that perchlorate levels in Thai women are associated with elevated TSH and reduced thyroxine is of concern [41], and suggests the possibility that a similar association may occur in pregnant women working or living near e-waste sites.

4.3 Cognitive Function

PCB exposure is known to cause reductions in IQ [42,43]. Even adults exposed to PCBs show reduced ability to learn and remember [44–46] and an increased frequency of depression [47]. These effects on nervous system function are of particular concern for exposure to children because impaired cognitive function during the period of life when learning is most important will lead to lifelong reduction in productivity. There has been less study on the neurobehavioral effects of brominated flame retardants, but Roze et al. [48] reported that prenatal exposure resulted in poorer motor ability and attention, but better behavior. There is strong evidence that childhood PAH exposure also results in reduced IQ [49,50]. Thus even without clear documentation of studies done on both children and adults near e-waste sites, it is clear that there will be cognitive deficits and behavioral abnormalities among exposed persons.

4.4 Behavioral Effects

There is strong evidence that exposure to PCBs, like exposure to lead, not only causes a reduction in IQ but also results in behavioral changes, including shortened attention span and subjective emotional complaints [51]. Exposure alters behavior by reducing attention span and increasing antisocial behaviors and attention deficit hyperactivity disorder [52,53]. Stewart et al. [54] reported that children exposed to PCBs prenatally show reduced response inhibition, which may be the basis for the impulsive behavior seen in such children, and suggested that this might be even more important than the shortened attention span. Boucher et al. [55] reported particular deficits in executive functioning in relation to prenatal PCB exposure. Perera et al. [56] also reported that PAH exposures were associated with symptoms of anxiety, depression, and attention problems in children.

PCBs and dioxins also have endocrine disrupting actions, and Vreugdenhil et al. [57] investigated the effects of perinatal exposure on play behavior of Dutch children. They found that PCB exposure was associated with less masculinized play behavior in boys, but more masculinized play behavior in girls. In contrast, perinatal dioxin exposure was associated with more feminized play behavior in both boys and girls.

4.5 Endocrine Disruption

Zhou et al. [58] studied pregnant women at Guiyu, and reported significantly elevated levels of estradiol and progesterone in exposed subjects, but a lower concentration of testosterone. Furthermore, they reported large elevations in mRNA for both estrogen receptor alpha and beta, but a reduced mRNA for progesterone. Xu et al. [59] investigated male reproductive function among informal e-waste workers at Guiyu. They found significant increases in rates of epididymitis, impotence, urethritis, azoospermia, asthenospermia, and sexual dysfunction among workers as compared to controls. There is other evidence that elevated serum PCB levels results in reduced testosterone [60] and that elevated PCBs and furans are associated with reduced sperm motility and increased abnormal morphology [61]. Elzanaty et al. [62] studied Inuit and European men in relation to their exposure to POPs, and concluded that alterations in epididymal function related to exposure may be one mechanism for the reduced male reproduction.

There is also strong evidence that exposure to POPs alters female reproductive function. Blanck et al. [63] reported that girls with PBB exposure had earlier puberty and Tanner stage development. Buck et al. [64] showed that consumption of PCB-contaminated fish was associated with impaired female fecundity as measured by time-to-pregnancy. PAHs, however, are known to have antiestrogenic activity [65].

Bisphenol A is not persistent, but is a potent endocrine disrupting substance [66] and it is likely that e-waste workers and nearby residents are exposed to it almost continuously. In addition to having estrogenic actions at low concentrations [67], bisphenol A stimulates release of inflammatory cytokines and has actions on adipose tissues, which may indicate involvement in the development of obesity and the metabolic syndrome [68].

4.6 Immune Function

Studies from the Netherlands [69], Belgium [70], and Canada [71] provide clear evidence that exposure to PCBs and dioxins suppresses immune system function, making children more vulnerable to middle ear infections, respiratory and gastrointestinal infections. In addition, the immune suppression results in a reduced response to immunization for communicable diseases. Both children and adults who live near hazardous waste sites containing POPs are more frequently hospitalized for respiratory infections than those who are away from such sites [72]. PAHs are also immunosuppressive substances, especially for humoral immunity [73]. Immune suppression will not only increase the frequency and severity of infections, but also increase risk of cancer, since the immune system is the first line of defense against cancer progression.

4.7 Respiratory Symptoms

A high proportion of young male e-waste workers in Ghana complain of cough and chest pain as well as dizziness or vertigo, and work without any respiratory protection [16]. Similar symptoms have been found related to dioxin-like toxic equivalents [69]. While this and other studies have reported that dioxins and PCBs reduce risk of allergic disease, there is growing evidence that PCBs but not dioxins increase risk of asthma [70,74]. While the immune suppression increases risks of all kinds of infections, respiratory infections are particularly important because of the combination of local irritation and altered immune system function. Ostro et al. [75] found that concentration of particulates (≤2.5 μm) was associated with significantly increased childhood hospitalizations for several respiratory infections (pneumonia, bronchitis) and asthma.

4.8 Fetal Growth and Development

Xu et al. [76] have reported that in Guiyu, as compared to a control site, there were significantly elevated rates of stillbirth (4.72% vs 1.03%), term low birth weight (3168 vs 3258 g), and lower Apgar scores (9.6 vs 9.9). While in this study they could not distinguish effects of metals from those of organics, in a later study Xu et al. [25] found that PAHs and PBDEs were elevated in cord blood and placentas in infants born in Guiyu, and that these changes were associated with lower birth length and Apgar scores. They also found that insulin growth factor (IGF) 1 was elevated in placenta in relation to PBDE concentrations, while its binding protein (IGFBP-3) levels were elevated in relation to PAH levels. They suggest that these alterations might be responsible for the effects on growth and development. There has long been evidence that prenatal PCB exposure results in a higher risk of low birth weight [77], and that low birth weight is a risk factor for a variety of other adverse health effects, particularly of cognitive function [78].

4.9 Diabetes

Diabetes is not usually considered to be an environmentally induced disease, but in fact it is. There is strong evidence that exposure to POPs increases risk of diabetes [79–81]. While the mechanisms are still unclear, it appears that these substances promote insulin resistance [82], perhaps via gene induction [83].

4.10 Cardiovascular Disease

There is a significant body of evidence that dioxin exposure increases risk of cardiovascular disease (see Ref. [84]). Results from the National Health and Nutrition Examination Survey [85] as well as studies of elderly Swedes [86] and Native Americans [87] show significant associations between serum concentrations of PCBs and chlorinated pesticides and indicators of cardiovascular disease, especially levels of serum lipids and consequent atherosclerosis. Animal studies have shown that individual PCB congeners can promote weight gain, hypertrophy of adipocytes, dyslipidemia and augmented atherosclerosis [88].

5. THE IMPORTANCE OF EXPOSURE TO CHEMICAL MIXTURES

There is strong evidence of harm from each of the organic substances discussed above. However in reality, none of the e-waste workers or nearby residents are exposed to only single organic pollutants. Unfortunately, there has not been adequate study of the interactions when individuals are exposed to multiple chemicals simultaneously. As is clear from the discussions above, many of the organics present at e-waste sites have similar actions, whether the outcome is cancer, cognitive and behavioral effects, or endocrine disruption. In addition, many of these organics also have multiple actions and are known to increase risk of several diseases.

There are a variety of ways in which chemicals can interact, usually falling in the categories of additive, less than additive, antagonistic, or synergistic [89]. The most simple is when two or more chemicals have a common binding site. Then effects may be additive at low concentrations or less than additive, if the receptor site becomes full or saturated. For carcinogens, the United States Environmental Protection Agency (USEPA) [90] has developed a series of formulae for the purpose of providing advice on fish consumption when the fish contains either multiple carcinogens or noncarcinogens. The assumption behind the formulae for carcinogenic effects is that cancer risks are additive and that there is no threshold concentration of a carcinogen that does not increase risk of cancer. Thus, if exposure to chemical 1 increases cancer risk by an amount x, and chemical 2 increases cancer risk by an amount y, the mixture of chemicals 1 and 2 will increase risk of cancer by x plus y. For noncancer risk, USEPA also assumes additive actions when two or more chemicals act at a common site, but assume that there is a threshold concentration below which there are no adverse health effects. These assumptions are subject to debate, but form the basis for USEPA risk assessment.

It is also important to recognize that a single chemical may have actions at more than one receptor. Synergistic effects of mixtures have not often been demonstrated, but have been reported between cigarette smoking and asbestosis exposure [91], and between exposure to dioxins and mercury at an industrial site [92].

Many of the POPs discussed above are in fact not single chemicals but rather mixtures. Dioxins and furans come in 75 congeners and PCBs and PBDEs come as 209 congeners, each varying in numbers of halogen atoms and their positions around the rings. Often the actions of different congeners vary, some showing additive, some antagonistic, and some synergistic actions with other congeners [93].

Many chemicals act at multiple binding sites, and show much more complex responses, often on multiple physiologic systems and altering multiple diseases. "Endocrine disruptors" are exogenous chemicals that interfere with and/or alter the actions of hormones. They may do so by binding to hormone receptors (e.g., estrogen, androgen, or thyroid receptors). Often the chemical affinity and potency at these receptors are low relative to that of the endogenous agonist, but because the chemical, unlike the endogenous agonist, is often persistent there may be significant long-term effects. Alternatively, the chemicals may antagonize these receptors, or may alter the metabolism by inducing degradative enzymes or by competitive binding to proteins that transport the endogenous hormone, which can increase the rate of degradation of the natural hormone [89]. Such actions may cause significant alternation of normal hormonal responses. Most hormone receptors are "nuclear" receptors, in that they bind to an intracellular receptor which activates DNA to upregulate or downregulate gene transcription. Hormone receptors usually include multiple genes, often genes affecting many different physiologic functions.

There is sufficient evidence so as to conclude that most of the organic compounds discussed above act at least in part as endocrine disruptors, and alter gene expression. Some act directly on known hormone receptors, and some act via other nuclear receptors, such as the aryl hydrocarbon receptor (AhR), constitutive androstane receptor (CAR), the peroxisome proliferator-activated receptors (PPARs α, β, and γ), and others. Like estrogen, androgen, and thyroid receptors, activation of each of these leads to gene induction, often of very many genes. Figure 2 (from Ref. [94]) shows how complex these interactions can be from activation of only the AhR. To add to the complexity, Maier et al. [95] reported that the pattern of gene expression upon AhR activation differs in different cell types. Some PCB congeners activate the AhR, others activate the CAR, and these actions result in induction of many but different gene profiles [96,97].

If this does not make things complicated enough already, it is now becoming apparent that some POPs can induce epigenetic changes in DNA which

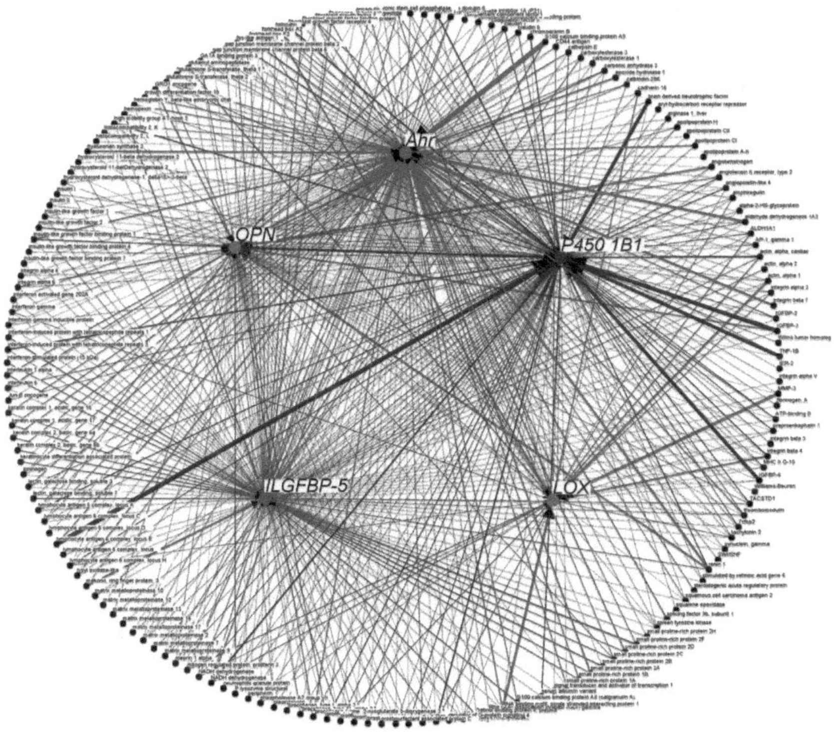

FIGURE 2 Gene–gene interaction networks activated by ligands of the aryl hydrocarbon receptor. The thickness of the line denotes the selection frequency for individual gene for each target. P450 1B1, cytochrome P450 1B1; OPN, osteopontin; LOX, lysyl oxidase; ILGFBP-5, insulin-like growth factor-binding protein-5. *From Ref. [94].*

will influence the development of various diseases [98]. A variety of POPs have been reported to result in DNA hypomethylation in healthy Koreans [99] and Greenland Inuit [100]. Thus there are multiple pathways whereby organic chemicals found at e-waste sites can lead to human disease.

6. CONCLUSIONS

E-waste contains many valuable substances, especially metals, that we as a society cannot afford to simply discard. However, e-waste recycling as it is currently practiced in many developing countries exposes workers and local residents to many dangerous chemicals, both metals and organics. There is clear scientific evidence for the toxicity of many of these chemicals, but altogether too little information on how mixtures of these chemicals interact. While informal e-waste recycling does provide income for many poor people, it is imperative that governments take responsibility for protecting human health. China is making significant progress in doing this, but this has resulted in some of the recycling moving to other countries that lack any significant regulation. E-waste involves a great variety of toxic substances that have the potential to have adverse effects on almost all physiologic functions, and these toxic substances contaminate air, water, food, and soil in the vicinity of e-waste recycling sites that operate without safety controls. While there is great merit in obtaining valuable metals and other useful compounds from obsolete electric and electronic equipment, we must find ways of doing this safely.

REFERENCES

[1] R. Widmer, H. Oswald-Krapf, D. Sinha-Khetriwal, M. Schellmann, H. Böni, Environ. Impact Assess. Rev. 25 (2005) 436–485.

[2] M. Schluep, C. Hagelueken, R. Kuehr, F. Magalini, C. Maurer, C. Meskers, E. Mueller, F. Wang, Sustainable Innovation and Technology Transfer Industrial Sector Studies, UNEP and STeP, Nairobi and Bonn, 2009.

[3] K. Lundgren, The Global Impact of E-Waste: Addressing the Challenge, Switzerland, Geneva, 2012.

[4] L. Zheng, K. Wu, Y.L.Z. Qi, D. Han, B. Zhang, C. Gu, G. Chen, J. Liu, S. Chen, X. Xu, X. Huo, Environ. Res. 108 (2008) 15–20.

[5] N. Suciu, E. Capri, M. Trevisan, T. Tanaka, H. Tien, S. Heise, M. Schuhmacher, M. Nadal, J. Rovira, X. Seguí, J. Casal, R.M. Darbra, Hdb. Env. Chem. 23 (2013) 349–384.

[6] Q. Song, J. Li, Environ. Int. 68 (2014) 82–93.

[7] I.C. Nnorom, O. Osibanjo, Waste Manag. 28 (2008) 1472–1479.

[8] A. Eguchi, T. Kunisue, Q. Wu, P.T.K. Trang, P.H. Viet, K. Kannan, S. Tanabe, Arch. Environ. Cont. Toxicol. 67 (2014) 29–41.

[9] W.J. Deng, J.S. Zheng, X.H. Bi, J.M. Fu, M.H. Wong, Environ. Int. 33 (2007) 1063–1069.

[10] A. Sepúlveda, M. Schluep, F.G. Renaud, M. Streicher, R. Kuehr, C. Hagelüken, A.C. Gerecke, Environ. Impact Assess. Rev. 30 (2010) 28–41.

[11] G.H. Xing, J.K. Yan Chan, A.O.W. Leung, S.C. Wu, M.H. Wong, Environ. Int. 35 (2008) 76–82.

[12] S.J. Chen, M. Tian, J. Zheng, Z.C. Zhu, Y. Luo, X.J. Luo, B.X. Mai, Environ. Sci. Technol. 48 (2014) 3847–3855.
[13] H. Li, L. Yu, G. Sheng, J. Fu, P. Peng, Environ. Sci. Technol. 41 (2007) 5641–5646.
[14] Z. Ren, X. Xiao, D. Chen, X. Bi, B. Huang, M. Liu, J. Hu, P. Peng, G. Sheng, J. Fu, Chemosphere (2014) 143–150.
[15] X. Xiao, J. Hu, P. Chen, D. Chen, W. Huang, P. Peng, M. Ren, Environ. Toxicol. Chem. 33 (2013) 500–507.
[16] T. Feldt, J.N. Fobil, J. Wittsiepe, M. Wilhelm, H. Till, A. Zoufaly, G. Burchard, T. Göen, Sci. Total Environ. 466–467 (2014) 369–376.
[17] D.Y. Huang, C.P. Liu, F.B. Li, T.X. Liu, C.S. Liu, L. Tao, Y. Wang, Environ. Monit. Assess. 186 (2014) 3351–3364.
[18] D.Y. Huang, H.Q. Zhao, C.P. Liu, C.X. Sun, Environ. Sci. Pollut. Res. 21 (2014) 5815–5826.
[19] G. Zhao, Y. Xu, G. Han, B. Ling, Environ. Geochem. Health 28 (2006) 341–351.
[20] Q. Luo, M. Wong, Z. Cai, Talanta 72 (2007) 1644–1649.
[21] H. Hung, G.O. Thomas, K.C. Jones, D. Mackay, Environ. Sci. Technol. 35 (2001) 4066–4073.
[22] Q. Luo, Z.W. Cai, M.H. Wong, Sci. Total Environ. 383 (2007) 115–127.
[23] I. Labunska, S. Harrad, D. Santillo, P. Johnston, L. Yun, Environ. Sci. Technol. 47 (2013) 9258–9266.
[24] J. Ma, K. Kannan, J. Cheng, Y. Horii, Q. Wu, W. Wang, Environ. Sci. Technol. 42 (2008) 8252–8259.
[25] X. Xu, T.A. Yekeen, Q. Xiao, Y. Wang, F. Lu, X. Huo, Environ. Pollut. 182 (2013) 63–69.
[26] K. Grant, F.C. Goldizen, P.D. Sly, M.N. Brune, M. Neira, M. van den Berg, R.E. Norman, Lancet Global Health 1 (2013) e350–e361.
[27] K. Steenland, P. Bertazzi, A. Baccarelli, M. Kogevinas, Environ. Health Perspect. 112 (2004) 1265–1268.
[28] B. Lauby-Secretan, D. Loomis, Y. Grosse, F. El Ghissassi, V. Bouvard, L. Benbrahim-Tallaa, N. Guha, Lancet Oncol. 14 (2013) 287–288.
[29] T. Zhang, Y.R. Huang, S.J. Chen, A.M. Liu, P.J. Xu, N. Li, L. Qi, Y. Ren, Z.G. Zhou, B.X. Mai, J. Environ. Monit. 14 (2012) 3156–3163.
[30] J. Wang, S. Chen, M. Tian, X. Zheng, L. Gonzales, T. Ohura, B. Mai, S.L.M. Simonich, Environ. Sci. Technol. 46 (2012) 9745–9752.
[31] C.A. Pope III, R.T. Burnett, M.J. Thun, E.E. Calle, D. Krewski, K. Ito, G.D. Thurston, JAMA 287 (2002) 1132–1141.
[32] G. Mastrangelo, E. Fadda, V. Marzia, Environ. Health Perspect. 104 (1996) 1166–1170.
[33] N. Ramírez, A. Cuadras, E. Rovira, R.M. Marcé, F. Borrull, Environ. Health Perspect. 119 (2011) 1110–1116.
[34] D.O. Carpenter, S. Bushkin-Bedient, J. Adolesc. Health 52 (2013) 521–529.
[35] J. Yuan, L. Chen, D. Chen, H. Guo, X. Bi, Y. Ju, P. Jiang, J. Shi, Z. Yu, J. Yang, L. Li, Q. Jiang, G. Sheng, J. Fu, T. Wu, X. Chen, Environ. Sci. Technol. 42 (2008) 2195–2200.
[36] H. Wang, Y. Zhang, Q. Liu, F. Wang, J. Nie, Y. Qian, Int. J. Hyg. Environ. Health 213 (2010) 369–380.
[37] J. Chevrier, K.G. Harley, A. Bradman, M. Gharbi, A. Sjödin, B. Eskenazi, Environ. Health Perspect. 118 (2010) 1444–1449.
[38] L.M. Schell, M.V. Gallo, M. Denham, J. Ravenscroft, A.P. DeCaprio, Environ. Health Perspect. 116 (2008) 806–813.
[39] S. Giera, T. Zoeller, in: D.O. Carpenter (Ed.), Effects of Persistent and Bioactive Organic Pollutants on Human Health, Hoboken, New Jersey, 2013.

[40] J.E. Haddow, G.E. Palomaki, W.C. Allan, J.R. Williams, G.J. Knight, J. Gagnon, C.E. O'Heir, M.L. Mitchell, R.J. Hermos, S.E. Waisbren, J.D. Faix, R.Z. Klein, N. Eng. J. Med. 341 (1999) 549–555.
[41] N. Charatcharoenwitthaya, B. Ongphiphadhanakul, E.N. Pearce, C. Somprasit, A. Chanthasenanont, X. He, L. Chailurkit, L.E. Braverman, J. Clin. Endocrinol. Metab. 99 (2014) 2365–2371.
[42] J.L. Jacobson, S.W. Jacobson, N. Engl. J. Med. 335 (1996) 783–789.
[43] P.W. Stewart, E. Lonky, J. Reihman, J. Pagano, B.B. Gump, T. Darvill, Environ. Health Perspect. 116 (2008) 1416–1422.
[44] S.L. Schantz, D.M. Gasior, E. Polverejan, R.J. McCaffrey, A.M. Sweeney, H.E.B. Humphrey, J.C. Gardiner, Environ. Health Perspect. 109 (2001) 605–611.
[45] K.C. Lin, N.W. Guo, P.C. Tsai, C.Y. Yang, Y.L.L. Guo, Environ. Health Perspect. 116 (2008) 184–189.
[46] R.F. Haase, R.J. McCaffrey, A.L. Santiago-Rivera, G.S. Morse, A. Tarbell, Environ. Res. 109 (2009) 73–85.
[47] E.F. Fitzgerald, E.E. Belanger, M.I. Gomez, M. Cayo, R.J. McCaffrey, R.F. Seegal, R.L. Jansing, S. Hwang, H.E. Hicks, Environ. Health Perspect. 116 (2008) 209–215.
[48] E. Roze, L. Meijer, A. Bakker, K.N.J.A. Van Braeckel, P.J.J. Sauer, A.F. Bos, Environ. Health Perspect. 117 (2009) 1953–1958.
[49] F.P. Perera, Z. Li, R. Whyatt, L. Hoepner, S. Wang, D. Camann, V. Rauh, Pediatrics 124 (2009) e195–e202.
[50] S.C. Edwards, W. Jedrychowski, M. Butscher, D. Camann, A. Kieltyka, E. Mroz, E. Flak, Z. Li, S. Wang, V. Rauh, F. Perera, Environ. Health Perspect. 118 (2010) 1326–1331.
[51] M. Peper, M. Klett, R. Morgenstern, Environ. Health Glob. Access Sci. Source 4 (2005) 22, http://dx.doi.org/10.1186/1476-069X-4-22.
[52] S.K. Sagiv, S.W. Thurston, D.C. Bellinger, P.E. Tolbert, L.M. Altshul, S.A. Korrick, Am. J. Epidemiol. 171 (2010) 593–601.
[53] P.A. Eubig, A. Aguiar, S.L. Schantz, Environ. Health Perspect. 118 (2010) 1654–1667.
[54] P. Stewart, J. Reihman, B. Gump, E. Lonky, T. Darvill, J. Pagano, Neurotoxicol. Teratol. 27 (2005) 771–780.
[55] O. Boucher, G. Muckle, C.H. Bastien, Environ. Health Perspect. 117 (2009) 7–16.
[56] F.P. Perera, D. Tang, S. Wang, J. Vishnevetsky, B. Zhang, D. Diaz, D. Camann, V. Rauh, Environ. Health Perspect. 120 (2012) 921–926.
[57] H.J.I. Vreugdenhil, F.M.E. Slijper, P.G.H. Mulder, N. Weislas-Kuperus, Environ. Health Perspect. 110 (2002) A593–A598.
[58] X. Zhou, Y. Ju, Z. Wu, K. Yang, Int. J. Occup. Environ. Health 19 (2013) 22–28.
[59] X. Xu, Y. Zhang, T.A. Yekeen, Y. Li, B. Zhuang, X. Huo, Environ. Sci. Pollut. Res. 21 (2014) 3540–3545.
[60] Goncharov, R. Rej, S. Negoita, M. Schymura, A. Santiago-Rivera, G. Morse, The Akwesasne Task Force on the Environment, D.O. Carpenter, Environ. Health Perspect. 117 (2009) 1454–1460.
[61] Y.L. Guo, P.C. Hsu, C.C. Hsu, G.H. Lambert, Lancet 356 (2000) 1240–1241.
[62] S. Elzanaty, A. Rignell-Hydbom, B.A.G. Jönsson, H.S. Pedersen, J.K. Ludwicki, M. Shevets, V. Zvyezday, G. Toft, J.P. Bonde, L. Rylander, L. Hagmar, E. Bonefeld-Jorgensen, M. Spano, D. Bizzaro, G.C. Manicardi, A. Giwercman, Rep. Toxicol. 22 (2006) 765–773.
[63] H.M. Blanck, M. Marcus, P.E. Tolbert, C. Rubin, A.K. Henderson, V.S. Hertzberg, R.H. Zhang, L. Cameron, Epidemiology 11 (2000) 641–647.

[64] G.M. Buck, J.E. Vena, H.B. Greizerstein, J.M. Weiner, B. McGuinness, P. Mendola, P.J. Kostyniak, M. Swanson, M.S. Bloom, J.R. Olson, Environ. Toxicol. Pharmacol. 12 (2002) 83–92.
[65] K.F. Arcaro, P.W. O'Keefe, Y. Yang, W. Clayton, J.F. Gierthy, Toxicology 133 (1999) 115–127.
[66] W.T. Tsai, J. Environ, Sci. Health Part C 24 (2006) 225–255.
[67] F.S. vom Saal, C. Hughes, Environ. Health Perspect. 113 (2005) 926–933.
[68] N. Ben-Jonathan, E.R. Hugo, T.D. Brandebourg, Mol. Cell Endocrinol. 304 (2009) 49–54.
[69] N. Wiesglas-Kuperus, S. Patandin, G.A.M. Berbers, T.C.J. Sas, P.G.H. Mulder, P.J.J. Sauer, H. Hooijkaas, Environ. Health Persect. 108 (2000) 1203–1207.
[70] R.L. Van Den Heuvel, G. Koppen, J.A. Staessen, E. Den Hond, G. Verheyen, T.S. Nawrot, H.A. Roels, R. Vlietink, G.E.R. Schoeters, Environ. Health Perspect. 110 (2002) 595–600.
[71] F. Dallaire, E. Dewailly, G. Muckle, C. Vézina, S.W. Jacobson, J.L. Jacobson, P. Ayotte, Environ. Health Perspect. 112 (2004) 1359–1364.
[72] J. Ma, M. Kouznetsova, L. Lessner, D.O. Carpenter, Paediatr. Respir. Rev. 8 (2007) 292–298.
[73] A. Szczeklik, J. Szczeklik, Z. Galuszka, J. Musial, E. Kolarzyk, D. Targosz, Environ. Health Perspect. 102 (1994) 302–304.
[74] D.O. Carpenter, J. Ma, L. Lessner, Ann. N.Y. Acad. Sci. 1140 (2008) 201–208.
[75] B. Ostro, L. Roth, B. Malig, M. Marty, Environ. Health Perspect. 117 (2009) 475–480.
[76] X. Xu, H. Yang, A. Chen, Y. Zhou, K. Wu, J. Liu, Y. Zhang, X. Huo, Reproduct. Toxicol. 33 (2012) 94–98.
[77] A. Baibergenova, R. Kudyakov, M. Zdeb, D.O. Carpenter, Environ. Health Perspect. 111 (2003) 1352–1357.
[78] N. Marlow, D. Wolke, M.A. Bracewell, M. Samara, N. Eng. J. Med. 352 (2005) 9–19.
[79] N. Codru, M.J. Schymura, S. Negoita, The Akwesasne Task Force on the Environment, R. Rej, D.O. Carpenter, Environ. Health Perspect. 115 (2007) 1442–1447.
[80] D.K. Lee, M.W. Steffes, A. Sjödin, R.S. Jones, L.L. Needham, D.R. Jacobs Jr., Environ, Health Perspect. 118 (2010) 1235–1242.
[81] D.H. Lee, P.M. Lind, D.R. Jacobs Jr., S. Salihovic, B. Van Bavel, L. Lind, Diabetes Care 34 (2011) 1778–1784.
[82] T.L.M. Hectors, C. Vanparys, L.F. Van Gaal, P.G. Jorens, A. Covaci, R. Blust, Environ. Health Perspect. 121 (2013) 1273–1281.
[83] D.O. Carpenter, Rev. Environ. Health 23 (2008) 59–74.
[84] O. Humblet, L. Birnbaum, E. Rimm, M.A. Mittleman, R. Hauser, Environ. Health Perspect. 116 (2008) 1443–1448.
[85] M.H. Ha, D.H. Lee, D.R. Jacobs Jr., Envrion. Health Perspect. 115 (2007) 1204–1209.
[86] P.M. Lind, B. van Bavel, S. Salihovic, L. Lind, Environ. Health Perspect. 120 (2012) 38–43.
[87] A. Goncharov, R.F. Haase, A. Santiago-Rivera, G. Morse, Akwesasne Task Force on the Environment, R.J. McCaffrey, R. Rej, D.O. Carpenter, Environ. Res. 106 (2008) 226–239.
[88] V. Arsenescu, R.I. Arsenescu, V. King, H. Swanson, L.A. Cassis, Environ. Health Perspect. 116 (2008) 761–768.
[89] D.O. Carpenter, K. Arcaro, D.C. Spink, Environ. Health Perspect. 110 (2002) 25–42.
[90] United States Environmental Protection Agency, Guidance for Assessing Chemical Contaminant Data for Use in Fish Advisories, vol. 2, 2000. EPA 823-B-00-008.
[91] P. Gustavsson, F. Nyberg, G. Pershagen, P. Schéele, R. Jackson, N. Plato, Am. J. Epidemiol. 155 (2002) 1016–1022.
[92] J.W. Chang, H.L. Chen, H.J. Su, P.C. Liao, H.R. Guo, C.C. Lee, J. Hazard. Mater. 185 (2011) 749–755.

[93] M. Haag-Grönlund, N. Johansson, R. Fransson-Steen, H. Håkansson, G. Scheu, L. Wärngård, Toxicol. Appl. Pharmacol. 152 (1998) 153–165.

[94] C.D. Johnson, Y. Balagurunathan, M.G. Tadesse, M.H. Falahatpisheh, M. Brun, M.K. Walker, E.R. Dougherty, K.S. Ramos, Environ. Health Perspect. 112 (2004) 403–412.

[95] M.S.V. Maier, M.E. Legare, W.H. Hanneman, NeuroToxicol. 28 (2007) 594–612.

[96] C.M. Vezina, N.J. Walker, J.R. Olson, Environ. Health Perspect. 112 (2004) 1636–1644.

[97] S. Ghosh, S. Zang, P.S. Mitra, S. Ghimbovschi, E.P. Hoffman, S.K. Dutta, Environ. Int. 37 (2011) 838–857.

[98] A. Baccarelli, V. Bollati, Curr. Opin. Pediatr. 21 (2009) 243–251.

[99] K.Y. Kim, D.S. Kim, S.K. Lee, I.K. Lee, J.H. Kang, Y.S. Chang, D.R. Jacobs Jr., M. Steffes, D.H. Lee, Environ. Health Perspect. 118 (2010) 370–374.

[100] J.A. Rusiecki, A. Baccarelli, V. Bollati, L. Tarantini, L.E. Moore, E.C. Bonefeld-Jorgensen, Environ. Health Perspect. 116 (2008) 1547–1552.

Chapter 11

Long-Range and Regional Atmospheric Transport of POPs and Implications for Global Cycling

Kimberly J. Hageman,[1,*] Christian Bogdal[2] and Martin Scheringer[2,3]

[1]*Department of Chemistry, University of Otago, Dunedin, New Zealand;* [2]*Institute for Chemical and Bioengineering, ETH Zürich, Switzerland;* [3]*Institute for Sustainable and Environmental Chemistry, Leuphana University, Lüneburg, Germany*

**Corresponding author: E-mail: khageman@chemistry.otago.ac.nz*

Chapter Outline

1. INTRODUCTION

The Stockholm Convention on persistent organic pollutants (POPs) [1], governed by the United Nations Environmental Programme (UNEP), defines POPs as chemicals that persist for long periods of time in the environment, bioaccumulate and biomagnify in living organisms, are toxic to humans and wildlife, and become widely distributed in the environment as a result of natural processes. Three types of natural processes are responsible for the transport of chemicals to locations far from their primary emission sources. These processes are atmospheric transport, transport in ocean currents and rivers, and transport as bioaccumulated chemicals in migratory animals (biovectors) [2]. This chapter focuses on long-range and regional atmospheric transport as mechanisms leading to the widespread distribution of POPs in the environment.

Comprehensive Analytical Chemistry, Vol. 67. http://dx.doi.org/10.1016/B978-0-444-63299-9.00011-9

"Long-range" atmospheric transport occurs when chemicals travel distances of several *thousands* of kilometers, such as between continents or to polar regions. In contrast, "regional" atmospheric transport implies transport distances of several *hundreds* of kilometers. Regional atmospheric transport may make a significant contribution to the POP burden in alpine ecosystems located upwind of cities or agricultural regions. "Local" or short-range atmospheric transport (<100 km) is not a focus of this chapter, but is also responsible for the dispersion of POPs in the vicinity of their usage or production zones. Local atmospheric transport may contribute to POP accumulation in alpine or polar ecosystems when POP sources, such as roads [3] or research stations [4], are located in such regions. From both a scientific and regulatory standpoint, it is often important to distinguish POP sources as being local, regional, or long-range.

The Stockholm Convention on POPs currently lists 23 individual chemicals and substance classes, including organochlorine pesticides (OCPs), polychlorinated biphenyls (PCBs), polychlorinated dibenzo-*p*-dioxins and dibenzofurans (PCDD/Fs), certain brominated flame retardants, and perfluorooctane sulfonate (PFOS) and related chemicals. Several additional chemical classes are under review for possible inclusion in the convention. A number of different tools are used to screen large lists of industrial chemicals for POP characteristics [5–7]. It is important to note that polycyclic aromatic hydrocarbons (PAHs), which are known to undergo long-range and regional atmospheric transport, are not listed as POPs under the Stockholm Convention but are listed in the Aarhus Protocol of the United Nations Economic Commission for Europe (UNECE) and Convention on Long-Range Transboundary Air Pollution (CLRTAP) [8,9]. In fact, there are a number of chemicals that undergo atmospheric transport but that are not officially listed by the Stockholm Convention [10], either because they do not meet *all* criteria or consensus for their listing has not yet been reached. Some researchers refer to "POP-like chemicals" for inclusivity and others refer to them as persistent/bioaccumulative/toxic (PBT) chemicals. In this chapter, we will use the term "POPs" in its most general sense, inclusive of all official POPs as well as PBTs and other chemicals with POP-like properties.

The intercontinental transport of POPs was first observed in the 1960s and this included the detection of OCPs and PCBs in the Arctic and Antarctic [11,12]. In 1975, Goldberg [13] discussed the potential of OCPs, such as dichlorodiphenyltrichloroethane (DDT), applied in the Tropics to volatilize and undergo airborne transport over long distances. From the 1980s on, more and more evidence demonstrated the truly global extent of environmental and human exposure to POPs caused by long-range transport. This is reflected by field data from all parts of the world, including the deep sea [14–22]. Earlier studies mostly focused on OCPs, PCBs, and PCDD/Fs. However, the short-chain chlorinated paraffins, which are currently under review by the POP Review Committee [1], were reported as global pollutants as early as 1987 [23]. In the early 2000s, the global presence of PFOS and perfluorooctanesulfonic acid was demonstrated [24]. Typical concentrations of POPs in regions without primary emissions are on the order of several picograms per cubic meter in air (e.g., PCBs [22,25], hexachlorocyclohexanes

(HCHs) [25,26] and endosulfan [26]), from several picograms per liter (e.g., hexachlorobenzene (HCB) [27]) to nanograms per liter (e.g., endosulfan [28]) in seawater, and several nanograms per gram (dry weight) in soil (e.g., PCBs [20]). The presence of POPs in seawater presents a particular concern because their uptake by sea life is followed by biomagnification in the marine food chain, leading to chronic exposure of a wide range of species from plankton to top predators, including seafood-consuming humans in many regions of the world.

The Task Force on Hemispheric Transport of Air Pollution [8] has compiled a list of the criteria used in different regulatory frameworks to classify chemicals as POPs. The regulatory frameworks include the UNEP Stockholm Convention, the UNECE CLRTAP, the United States (US) Environmental Protection Agency's Toxic Release Inventory, the Canadian Environmental Protection Act, and The European Union's regulation on Registration, Evaluation, Authorization, and Restriction of Chemicals (REACH). In addition to persistence criteria (i.e., half-lives in water, soil, and sediment), the Stockholm Convention and the CLRTAP use evidence from monitoring data as part of their system for classifying chemicals as POPs. Currently, the monitoring of POPs is performed by several institutions. Long-term monitoring stations using active air samplers are run by the Integrated Atmospheric Deposition Network (IADN) in the Great Lakes region in Canada and the United States, by the European Monitoring and Evaluation Programme (EMEP), and by the Arctic Monitoring and Assessment Programme (AMAP). Furthermore, under the Global Monitoring Plan (GMP) of the Stockholm Convention [1], POPs are measured in air and human milk in many parts of the world. Most of the measurements in air are performed with passive samplers. For example, the Global Atmospheric Passive Sampling (GAPS) Network [29] provides a growing body of POP concentration data in air from a global network of samplers [30,31].

The objectives of this chapter are to review current knowledge regarding atmospheric transport potential, the processes that control the latitudinal and long-term distributions of POPs, the atmospheric transport of POPs to alpine ecosystems, and approaches for determining the sources of POPs in remote ecosystems. An important related topic, which is addressed in detail in Chapter 13, is the long-range atmospheric transport of POPs to polar regions.

2. UNDERSTANDING ATMOSPHERIC TRANSPORT POTENTIAL

To understand which of the multitude of organic contaminants in the environment are prone to widespread distribution via atmospheric transport, it is useful to visualize this distribution process as occurring in a series of steps (Figure 1) [32]. First, the chemical enters the atmosphere, either by volatilizing from the media into which it was emitted (e.g., pesticides volatilizing from soils or plant surfaces) or being directly emitted to the atmosphere (e.g., PCDD/Fs emission during waste incineration). The second step is transport through the atmosphere with air circulation. For this step to occur, the chemical must be persistent

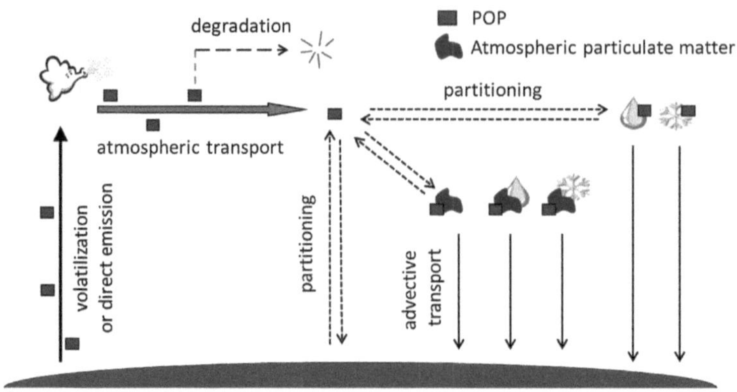

FIGURE 1 Processes involved in the atmospheric transport and deposition of persistent organic pollutants (POPs). Following volatilization, the fate of POPs in the atmosphere depends on the interplay between temperature-driven thermodynamic partitioning, advective transport from the air to surface media, and atmospheric degradation.

enough in air to not undergo significant degradation before reaching the receptor site. For this reason, organizations that classify chemicals as POPs employ an atmospheric half-life criterion of >2 days [8]. The third step, which can occur via several different specific routes, is atmospheric deposition such that the chemical returns to the Earth's surface at certain points during its residence time in air. This step is critical in the distribution process because POPs, by definition, are chemicals that impact organisms via direct contact. The chlorofluorocarbons (CFCs) are an example of a class of chemicals that are volatile enough to enter the atmosphere (step 1) and persistent enough to undergo long-range atmospheric transport (step 2) but that are too volatile to undergo substantial atmospheric deposition (step 3). Such chemicals may influence atmospheric processes, such as the depletion of stratospheric ozone in the case of the CFCs [33]; however, their negative impacts are not due to direct contact with organisms. Thus, we can conclude that POPs have a unique set of physicochemical properties that make them volatile enough to enter the atmosphere but that also allow them to be deposited in substantial amounts to the Earth's surface. For this reason, POPs belong to the larger family of chemicals known as the semivolatile organic compounds.

The classification of chemicals as semivolatile has traditionally been based on vapor pressure. Vapor pressure is the physical property that describes the distribution of a chemical between air and its pure condensed phase (solid or liquid) at thermodynamic equilibrium. However, chemicals are rarely found in their pure forms in the environment and therefore, vapor pressure has limited usefulness in predicting the degree to which chemicals volatilize in the real environment. For example, vapor pressure is a poor predictor of the degree to which recently applied pesticides volatilize from soils [34]. Thus, equilibrium partition coefficients using

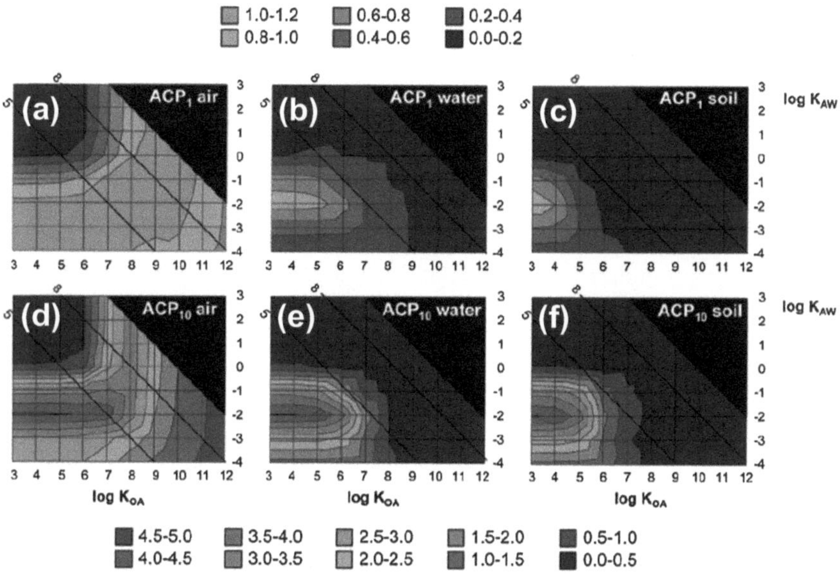

FIGURE 2 Chemical space diagrams are used to show how the predicted Arctic Contaminant Potential (ACP) varies among hypothetical, perfectly persistent chemicals with different combinations of log K_{aw} and log K_{oa} values. The diagonal lines indicate log K_{ow} values of 5 and 8. The matrix of chemical space diagrams shows how emission scenarios and time affect ACP. Emission was assumed to occur into the atmosphere (a,d), freshwater (b,e), or cultivated soils (c,f). Steady emissions occurred for 1 year (a,b,c) or 10 years (d,e,f). The colored scale above the figure refers to panels a-c while that below the figure refers to panels d-f. *Reprinted (adapted) with permission from Ref. [32]. Copyright (2003) American Chemical Society.*

air as one of the media (e.g., air–water, air–soil, or air–vegetation partition coefficients) are typically more useful than vapor pressure alone for determining which chemicals have a sufficient distribution in air to undergo atmospheric transport. However, partition coefficients describe the distribution of a chemical between two media, whereas the environment is composed of *multiple* types of media. For this reason, chemical space diagrams, which enable chemical behavior classification based on two partition coefficients, are often used to illustrate how partitioning affects atmospheric transport potential. For example, Wania [32] used chemical space diagrams to illustrate how the partitioning properties of perfectly persistent chemicals affect their long-range atmospheric transport potential to the Arctic under different emission scenarios (Figure 2). The Arctic Contamination Potential (ACP) was defined as the percentage of chemical emitted into the global environment that accumulated in Arctic surface media, as calculated with the Globo-POP model (Figure 2).

Temperature plays a critical role in the atmospheric transport behavior of chemicals. This is because each of the three main steps involved in the atmospheric transport and deposition of chemicals are affected by temperature. First, a chemical is more likely to enter the atmosphere at higher temperatures because, as the van't Hoff equation indicates, the distribution of a chemical in the gas phase increases substantially relative to that in a condensed phase as temperature increases [35]. Second, degradation rate constants increase with increasing temperature as indicated by the Arrhenius equation, meaning that higher temperatures lead to shorter half-lives and thereby less atmospheric transport. Finally, atmospheric deposition is enhanced at lower temperatures, which is also due to temperature's strong effect on the partitioning of chemicals both directly to terrestrial and aquatic surfaces as well as to atmospheric particulate matter, raindrops, and snowflakes in the atmosphere. Thus, diurnal and seasonal changes in temperature create a cycling effect for POPs in which they volatilize and move through the atmosphere when it is warm and deposit when it is cold, creating what has been referred to as "the grasshopper effect" [36].

At this point, we have established that the atmospheric transport potential of chemicals, and therefore their likelihood of being classified as POPs, depends on several physicochemical properties (including equilibrium partition coefficients and degradation rate constants) as well as temperature, which affects each of these physicochemical properties. However, atmospheric transport potential also depends on the emission scenario and a multitude of environmental factors (e.g., the specific composition and properties of the environmental compartments). An additional complexity in predicting atmospheric transport potential is that chemical equilibrium is not necessarily achieved in the environment. Therefore, kinetic controls on the distribution of chemicals in the environment, such as atmospheric wet and dry deposition processes, must also be considered [37,38]. Thus, the development and use of chemical fate models, which incorporate the many relevant properties and processes that affect chemical behavior in the environment, have become increasingly important in this research area [39]. Readers are referred to Scheringer [40] for a review of the uses of multimedia box models, spatially resolved multimedia box models, and atmospheric transport models to investigate the long-range transport of organic chemicals.

3. PROCESSES CONTROLLING THE LATITUDINAL AND LONG-TERM DISTRIBUTION OF POPs ON THE GLOBAL SCALE

The behavior of POPs in the atmosphere and the likelihood that they will undergo long-range atmospheric transport depends on the interplay of (1) thermodynamic temperature-driven partitioning of POPs between environmental media; (2) kinetic advective phase-transfer processes such as deposition from air to surface media and from surface ocean water to deep water; and (3) transformation and degradation (Figure 1). Our understanding of the roles that these conflicting processes play in controlling the global-scale and long-term distributions of POPs with different physicochemical properties has evolved over time, especially as more advanced global chemical fate modeling concepts have been developed.

Most POPs are multimedia chemicals, i.e., they partition to some degree into all media, including air, water, soil, vegetation, sediment, biota, snow, ice, atmospheric particulate matter, suspended particles in water, etc. [41]. Phase partitioning is a thermodynamically controlled process and is described by a chemical's equilibrium partition coefficients for octanol–water (K_{ow}), air–water (K_{aw} or Henry's law constant), and octanol–air (K_{oa}), with octanol representing soil organic matter and lipid tissue in organisms. On the global scale, the effect of phase partitioning is nicely visible in the south–north concentration profiles of volatile persistent substances such as the CFCs and HCB. These chemicals show a uniform distribution in air on the global scale and concentrations in water or in vegetation that increase toward colder regions [19,42]. This increase is driven by the decrease in vapor pressure and Henry's law constant and the increase in the K_{oa} that is caused by decreasing temperature. However, most POPs are much less volatile than CFCs and even HCB, and for them the effect of phase partitioning is less visible in their global distribution patterns. In 1995, Mackay and Wania [43] hypothesized that also for the less-volatile POPs, their long-term and large-scale distribution in the environment is governed by phase partitioning. They predicted that POPs would be enriched in cold regions because of the temperature dependence of the partition coefficients, i.e., lower vapor pressure, lower Henry's law constant, and higher K_{oa} in colder regions. They called this the *global distillation hypothesis* and proposed that, "It is this thermodynamic or equilibrium effect that drives the condensation phenomenon in cold climates" [43].

In recent years, the role that kinetic processes, which often counteract phase partitioning, play in controlling the global-scale distribution of many POPs has become increasingly recognized [22,37,38]. The relevant kinetic processes are deposition from air to surface media, deposition from surface ocean water to deeper water, and degradation. Dachs et al. [22] and Galbán-Malagón et al. [44] described the effect of the "biological pump" on the global distribution of POPs. This "pump" is driven by plankton-derived marine particulate matter, which continuously sinks to the deeper water and removes POPs from the mixed surface layer of the oceans. The POP concentrations in the mixed layer are reduced by the downward flux and the chemicals are prevented from reaching air–water equilibrium, in particular, in regions with high plankton productivity and for POPs with high K_{ow} (e.g., heavy PCBs). For these substances and in these regions, the air-to-water deposition flux is too slow to compensate for the strong downward flux driven by sinking particulate matter and the POP concentrations in the water are lower than what air–water equilibrium partitioning predicts. Dachs et al. [22] stated that "The influence of biogeochemical processes controlling water column concentrations and vertical fluxes of particle-associated POPs may be as important as temperature as factors driving the global fate, transport, and sinks of POPs."

Deposition from air to surface media has also been found to be more important as a controlling factor of the large-scale distribution of POPs than temperature-dependent phase partitioning. Von Waldow et al. [37] investigated whether the gradient of PCB concentrations in air along two south–north transects in Europe from 50 to 80 °N were correlated with the decrease in temperature along these transects. They did not find a correlation between measured PCB concentrations in air and temperature, but instead found strong correlations between PCB concentrations and distance from the PCB emission sources. This finding illustrates the effect of continuous removal of the PCBs from the moving air by wet and dry deposition along their way northwards. This effect is more pronounced for heavier PCBs than for lighter ones because heavier PCBs sorb more strongly to atmospheric particulate matter and are removed more efficiently from the air. Von Waldow et al. [37] called this effect "differential removal" and concluded that "the interpretation of modeling and empirical data for PCBs as supporting the global distillation hypothesis appears to be erroneous, and stems from overlooking the correlation of temperature with remoteness gradients."

The global distillation hypothesis was further scrutinized by Schenker et al. [38] using a global multimedia environmental fate model able to incorporate long-range transport, temperature-driven partitioning between multiple phases, advective kinetically controlled phase-transfer processes (e.g., wet deposition or settling to the deep ocean), and degradation. The global distribution of a suite of chemicals was modeled under two contrasting hypothetical scenarios. The first scenario was a world where long-range transport and partitioning existed but degradation and the advective phase-transfer processes did not. In this model world, POP distributions as predicted by equilibrium partitioning, i.e., the global distillation hypothesis, were produced. An interesting aspect of such a world is that equilibrium partitioning and long-range transport are counteracting processes. Equilibrium partitioning between, for example, air and water prescribes higher air–water concentration ratios in warmer regions than in colder regions, producing a concentration gradient as is observed for CFCs in seawater [42]. Long-range transport, on the other hand, tends to even out spatial concentration gradients. Therefore, for each chemical, equilibrium partitioning and long-range transport reach a balance that depends on the partition coefficients of the specific chemical. Accordingly, the term, Globally Balanced State (GBS), was used to describe the global distribution of chemicals in this modeled world where no degradation or advective transport could occur [38].

The second modeled world was the Temporal Remote State (TRS) [45], which describes the long-term distribution of a chemical in a given system. In this modeled world, all four processes were active and the global distributions of chemicals were calculated at times sufficiently long after the end of emissions. In this state, the chemical persists only in its most long-lived environmental reservoir. For many POPs, this is a reservoir in Arctic soils because the chemicals are most persistent in such an environment. For a wide range of chemicals, Schenker et al. [38] systematically compared their global distributions in the GBS (i.e., where distribution is controlled by the global distillation hypothesis) and the TRS (i.e., the actual long-term distribution of chemicals for which primary emission has ceased). They found that for most POPs, the TRS is very different from the GBS. In other words, the long-term distribution of POPs in the global environment is mainly driven by degradation processes and the formation of long-lived reservoirs and *not* by equilibrium partitioning.

In conclusion, both field data and modeling studies show that there is a global fractionation of POPs, which means that different POPs have different mobility and different extents of long-range transport. This also implies that there is a compositional shift in a mixture of POPs, such as different PCB homologues, between the point of release of the mixture and points of measurement at remote locations. However, "cold condensation" as suggested by Mackay and Wania [43] occurs only for highly volatile and persistent substances such as HCB and the CFCs; for less-volatile substances, it is not a driver of their long-term and large-scale distribution. The main reason for this is that the transport processes leading to a global distribution according to the partition coefficients of a POP are generally slower than degradation and advective phase-transfer processes, so that these other processes outcompete the distribution process that, in the long term, would lead to equilibrium partitioning.

4. LONG-RANGE AND REGIONAL ATMOSPHERIC TRANSPORT OF POPs TO ALPINE REGIONS

In many scientific studies, the atmospheric transport of POPs along elevational transects has been compared to that along latitudinal gradients [46], mainly because both are associated with strong temperature gradients. However, a key difference is that distances between emission and receptor sites are typically at least an order of magnitude lower for alpine regions than polar regions. Whereas the Arctic, and especially the Antarctic, are located far (thousands of kilometers) from the main industrial and agricultural POP emission sources, many mountain regions are located relatively close (hundreds of kilometers or even less) to these regions. For industrial chemicals, such as PCBs, emission intensity is correlated with population density, degree of industrialization, and economic wealth whereas pesticide POP emission intensity is associated with agricultural activities, including those in populated regions [37,47–51].

Many of the Earth's large mountain systems lie within close proximity to regions with high population and/or agricultural intensity. For example, the Himalaya, with some of the highest peaks on Earth, crosses several of the most densely populated countries, including Bangladesh, China, India, and Pakistan. Similarly, the Alps stretch across the densely populated and highly industrialized regions of Austria, Southern Germany, Northern Italy, and Switzerland. In North America, many parts of the Rocky and Sierra Nevada Mountains are located upwind of POP sources, with California's Sierra Nevada being particularly close to POP source regions [52,53]. Thus, there are many places in the world in which the transport distances from populated and agricultural source areas to remote mountains can be overcome by atmospheric currents within hours to days. Additionally, settlements in mountainous regions can be sources of POPs, especially combustion-related POPs, such as PCDD/Fs and PAHs. Therefore, mountain regions can be affected by local and regional atmospheric transport, in addition to global-scale long-range atmospheric transport, of POPs.

Due to the relative proximity of mountains to POP source regions, the POP concentrations in air entering mountain systems are expected to be much higher than those entering the Arctic [25]. The extent to which POPs are transported into a mountain region is strongly affected by meteorology. The atmosphere is stratified such that it consists of the planetary boundary layer where POP emissions occur, followed by the free troposphere (Figure 3) and above it, the stratosphere. The height of the planetary boundary layer varies from 1000 to 2000 m above sea level, depending on location and season [54]. Sites located in the planetary boundary layer for the entire year are mainly affected by local- and regional-scale atmospheric transport of POPs from sources in the vicinity. In contrast, sites located above the boundary layer, in the free troposphere, are more affected by the long-range transport of POPs from distant source regions (Figure 3(a)). For example, POP concentrations in the Italian Alps were lower in air and surface media at high-elevation sites in the free troposphere compared to those in the planetary boundary layer [55]. In summer, thermally induced convective mixing can result in planetary boundary layer air reaching much higher elevations and, therefore, POPs may also be carried to these sites by local and regional atmospheric transport from nearby sources [56]. The contribution

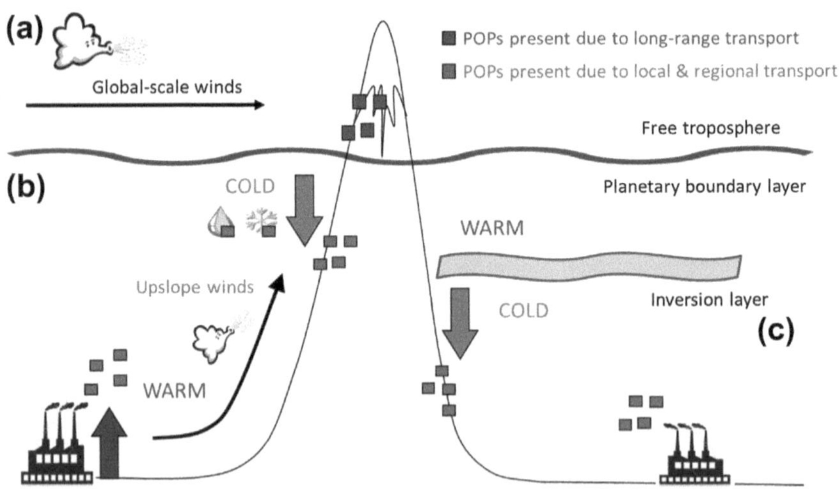

FIGURE 3 Processes affecting the atmospheric transport of persistent organic pollutants (POPs) in mountains where (a) POPs from long-range transport are deposited at high mountain sites in the free troposphere, (b) regional and local winds result in the upslope transport of POPs followed by wet and dry deposition, and (c) temperature inversions limit the vertical transport of POPs to higher elevations.

of local air masses reaching high-elevation sites in summer, due to the vertical transport of polluted boundary-layer air has also been frequently observed for inorganic gaseous air pollutants [55,57].

Mountains are affected by diurnal wind patterns that result from elevational temperature differences. A typical pattern involves air masses flowing up into mountain valleys during the day and down these valleys at night [58]. The daytime upslope transport of air masses causes cooling of this air, enhanced formation of clouds, and an increased likelihood of precipitation, i.e., the conditions become ideal for enhanced wet and dry deposition of POPs (Figure 3(b)). Moreover, snowfall, which often occurs at high elevations when it is raining at nearby lower elevations, scavenges POPs more efficiently from the air than rain [59]. Therefore, when polluted air masses are transported into mountain regions during the day from populated areas in the plains, both wet and dry deposition of POPs are enhanced at the high elevations. During the night, the lower temperatures mean reduced evaporation from surface media and, thus, air masses transported downslope carry less pollutants back into the valley. The difference between the amounts of chemicals transported upslope and those transported downslope results in a net accumulation of POPs at higher elevations [60].

The enhanced accumulation of chemicals at higher, colder elevations is referred to as *mountain cold trapping*. Cold trapping is caused by the temperature dependence of the partitioning of POPs between the atmospheric gaseous and particulate phase [61]. For certain POPs, wet deposition may be significantly enhanced at higher elevations (i.e., where temperatures are lower, precipitation rates are higher, and it may be snowing) compared to those at lower elevations (i.e., where temperatures are higher, precipitation rates are lower, and rain predominates). Model-based simulations have shown that POPs with octanol–air partition coefficients of $8.5 < \log K_{oa} < 11.5$ and air–water partition coefficients of $-6 < \log K_{aw} < -3.5$ are mainly affected by the mountain cold-trapping effect. Higher-chlorinated PCBs, for instance, are POPs with physicochemical properties that make them prone to mountain cold trapping [61]. The cold-trapping effect has also been investigated and illustrated by field studies. Empirically, the relevance of the cold-trapping effect has been indicated by higher concentrations of POPs in air and soils reported for sites where precipitation is higher and temperatures are lower in the Alps [62–65], the Himalaya [66], and in the Andes [67]. Interestingly, cold-trapping trends are not always observed where they might be expected. In particular, several studies have observed cold trapping for certain POPs but not others along the same mountain slopes. For example, chemicals such as HCB, which are ubiquitously and homogeneously distributed in the global atmosphere, often display cold-trapping trends whereas POPs with primary sources that are still significant, such as PCBs or combustion-related PCDD/Fs [62,66,67] do not. These differences can be explained by local or regional emission sources at mountain sites overriding the potential effects of a cold-condensation trend.

Atmospheric stratification can also play a critical role in controlling the extent to which POPs undergo atmospheric transport to high-elevation sites. Whereas the mountain cold-trapping effect can lead to an increase in deposition with increasing elevation for some POPs, a stable atmospheric stratification can significantly reduce the transport of POPs to higher elevations (Figure 3(c)). In mountain regions, temperature inversions are particularly common in winter when they can form at altitudes of typically 1000–3000 m and persist for several days to weeks [56]. Also, in summer, so-called nocturnal boundary layers are regularly formed at low elevation (typically around 200–800 m) and follow a diurnal pattern with formation at night and breakup during the day. Such temperature inversions and the resulting formation of stable boundary layers strongly limit vertical transport of air masses. Thus, even for POPs for which deposition at higher elevations is theoretically favored, the formation of stable boundary layers can limit the mountain cold-trapping effect. Particularly for POPs with sources located at low elevation, atmospheric stratification plays a critical role in controlling the extent of transport into mountains.

From the existing studies, an interest and concern about the *postdepositional* fate of POPs in mountain ecosystems has emerged [68–70]. For example, the exchange of chemicals between snowpack, especially fresh snowpack, and the overlying atmosphere can be significant, particularly because snow covers large areas at high elevation or high latitude [68,71]. The reemission of chemicals from the snowpack to the atmosphere can be driven thermodynamically by a decrease in the storage capacity of the snowpack, which is caused by the snow surface area decreasing during aging [70,72]. Reemissions can also be driven kinetically by wind ventilation through the porous snowpack [70,73,74]. Loss of chemicals from an aging snowpack to the atmosphere has been observed in field studies [75–77] and simulated by models [71,78–80]. Chemicals that are not volatilized from the snowpack may end up stored in glaciers when snow undergoes transformation to ice (firnification) [80,81] or released from the snowpack when it melts in spring [81a]. Upon melting, snow-entrapped chemicals may undergo volatilization, sorption to particles, or dissolution in meltwater; the relative importance of these fate processes depends on both the properties of the chemical and the snowpack [78].

Mountain regions also have unique environmental properties that can affect the fate of POPs transported from source regions. For example, the retention capacity of POPs in soils can vary significantly for lowland and mountain soils at similar latitudes. On the one hand, the slower turnover of carbon in colder high-elevation sites for the same soil type can lead to an organic carbon content pool that increases with increasing elevation [66], causing the retention capacity for POPs to increase with elevation [66,82–85]. On the other hand, a decreasing amount of vegetation (e.g., crossing of the tree line) or change in vegetation type (e.g., a change from deciduous to coniferous forest) can result in decreasing organic carbon content in soils with increasing elevation [86]. The retention capacity of soils is an important factor during the snowmelt season when POPs released with meltwater from the seasonal snowpack can either be retained on mountain soils or transported to lowlands with meltwater. For example, higher POP concentrations were observed in mountain soils after the snowmelt period [87] and the PAH concentration pulse was markedly different in a stream draining a forested mountain slope compared to one draining a nonforested slope [88]. The amplification of POP concentrations in snowmelt, which has been observed in several studies and investigated with modeling, has been discussed in detail in a recent review article [70].

The extent of environmental degradation of POPs can also be different in mountains compared to lowland regions; however, this topic needs much more investigation. The increasing ultra-violet irradiation with elevation results in an increase in OH radicals in air. Therefore, OH radical reactions with POPs (i.e., indirect photochemical transformation), which represent the main atmospheric degradation pathway for some POPs [89], can be expected to gain importance with elevation. Likewise, direct photolysis in air or on surfaces, such as snow, might be more relevant at higher elevation, particularly for brominated POPs [90,91]. Photolytic degradation has been shown to occur in artificial snow under well-controlled experimental conditions [92,93]. However, the environmental relevance of this process is still poorly understood [92,94,95], especially because chemicals deposited in fresh snow are only exposed to sunlight for a relatively short period of time before they are buried in deeper snow layers.

The discussion above indicates that the effect of mountains on the local and regional cycling of POPs is determined by complex processes. Several studies have shown that mountains can act as barriers to the atmospheric transport of POPs [96,97] and there is both empirical evidence and mechanistic descriptions of mountain slopes acting as a kind of filter for POPs due to POPs being washed out from the atmosphere. Monitoring of POPs in mountain regions shows that inside mountain chains, concentrations of POPs in environmental media are often lower than on the border of the mountains. This was shown, for instance, in Europe where concentrations of several POPs in the humus layers of soil and in spruce needles were higher in the peripheral zones of the Alps, while the central or middle part of the Alps was less polluted [64]. The effect of mountains on the global cycling of POPs is more difficult to assess [98]. Although atmospheric scavenging of POPs by snow is important in mountain regions, the seasonal snow and permanent ice cover result in less retention capacity compared to when these surfaces are covered by soils.

5. APPROACHES FOR DETERMINING POP SOURCES IN REMOTE ECOSYSTEMS

The determination of geographic and emission sources of POPs in remote ecosystems has been a component of many studies focused on the atmospheric transport of POPs. Although sometimes linked or investigated together, these two source classifications are distinct. *Geographic* source determination is of interest when the *emission* source is known (e.g., for pesticides that volatilize from sprayed crops) but when the specific region from which the POPs originated is unknown. For example, Lavin et al. [99] investigated the sources of POPs in air at a remote alpine site in New Zealand and found that the pesticide endosulfan was associated with long-range atmospheric transport from Australia (~2000 km away) whereas the pesticide chlorpyrifos was present at the same site due to upslope winds from an agricultural regional located ~70 km away. In contrast, the determination of a specific *emission* source is often of interest for chemicals that are formed and emitted in multiple ways. For example, PAHs are emitted to the environment by multiple sources, including residential heating (e.g., coal and wood burning), industrial processes (e.g., aluminum plants, asphalt production), municipal and commercial incineration, open burning (e.g., agricultural fires, forest fires), power generation, and mobile sources (gasoline and diesel engines) [100,101]. An example of a study in which PAH emission sources were identified at a variety of sites, including remote alpine sites, showed that emission sources shifted from being primarily associated with biomass burning (wood and grass) in the more urban areas to vehicle emissions in the more remote areas of the Fraser Valley, Canada [102].

Researchers, regulators, and the public have a number of motivations for seeking information about the sources of POPs in specific remote ecosystems. The Western Airborne Contaminants Assessment Project (WACAP) [103] is an example of a project that included geographic source determination of POPs as one of its goals. In this case, the general aim was to identify the sources of air masses most likely to have transported POPs to remote lake catchments in western US national parks. A specific objective was to determine if those sources were regional or long range. A key impetus for this aim revolved around emerging evidence at the time that pollutants could travel through the atmosphere from eastern Asia to the western US in as little as ~6 days [104]. Although it was eventually shown that the large majority of the POP burden in the mountains of the continental US came from US sources [51,103], stakeholders and funders were originally keen to determine if POPs *of Asian origin* were accumulating in *US mountains* [103].

Source determination of POPs in remote ecosystems is often of practical importance in the development of local, regional, and global POP reduction strategies. For example, the WACAP study showed that PAHs in an alpine lake catchment on the western side of Glacier National Park, Montana, mainly originated from an aluminum smelter located 45 km away [105]. This study showed that in 2003, 92% of the PAHs in the Snyder Lake Catchment originated from the smelter and only 8% from wood combustion. This type of result indicates a need for local action. In contrast, endosulfan found in WACAP's Arctic Alaskan parks clearly suggested long-range atmospheric transport as the source [51,103], a result that suggested need for action by a global convention. The results from this study and others [106,107] played a role in the final decision by the Stockholm Convention to list endosulfan as a POP in 2011 [1].

A number of different approaches have been developed for geographic and/or emission source apportionment of POPs in remote ecosystems. These approaches have varied much in logistical, analytical, mathematical, and conceptual complexity. A simple conceptual approach for determining the relative contributions of regional versus long-range atmospheric transport on POP quantities in remote alpine ecosystems was utilized in an early study by WACAP [51]. This approach involved comparing pesticide concentrations at sites influenced by a *combination* of regional and long-range atmospheric transport (i.e., those in the continental US) to those where the POP burden was due to long-range atmospheric transport *only* (i.e., those in Arctic Alaska for which there was no regional agriculture). The pesticide concentrations at the sites in Arctic Alaska were assumed to represent a hemispheric average for pesticide concentrations due to long-range transport. Concentrations of POPs attributed to regional transport at the continental sites were calculated by subtracting the mean concentrations at the Arctic Alaskan sites. While simple in concept, this approach required a complex, large-scale sampling campaign and sufficiently large differences in concentrations between sites for them to be meaningful.

A more complex conceptual approach for assessing the contributions of long-range atmospheric transport of POPs in remote ecosystems involves using diagnostic ratios of chemical species. The most commonly used diagnostic ratios for this purpose have been those of (a) isomers found in technical products, (b) enantiomeric pairs of chiral compounds, and (c) parent/metabolite pairs. The ways in which these diagnostic ratios can be used, in general, as chemical tracers for sources and transport pathways of POPs was reviewed in the Task Force on Hemispheric Transport of Air Pollution report [8]. The ratios of stable carbon isotopes have also been used to identify POP sources in remote ecosystems [108]. Although this approach is currently limited by challenges associated with accurately measuring ratios of POP isotopes in complex environmental matrices, these challenges are increasingly being overcome [109].

Isomeric ratios have been used in a wide variety of studies to investigate the sources and transport pathways of legacy OCPs, specifically the HCHs, DDTs, and chlordanes, whose technical products included a mix of isomers. The HCHs provide an interesting example. The original HCH technical mixture, whose commercial production began in 1943 [110], was composed of approximately 60–70% α-HCH, 7–10% β-HCH, and 14–15% γ-HCH [111]. In contrast, the product lindane contained only the γ-HCH isomer. Thus, relatively high α-/γ-HCH ratios suggest that the technical mixture is the primary source whereas low values implicate lindane usage. Moreover, because lindane was adopted earlier or to a greater extent in some parts of the world, the α-/γ-HCH ratio has also been used to identify geographic sources of POPs. For example, Shen et al. [112] observed a marked difference in the α-/γ-HCH ratio measured in passive air samplers located at mountain sites on the western versus eastern slopes of the Canadian Rocky Mountains. This difference was attributed to the long-range atmospheric transport of technical HCH from Asia influencing the western slopes versus regional use of lindane in the Canadian Prairies influencing the eastern slopes. However, it is important to note that non-enantiomeric isomers, such as those in the HCH technical mixture [110], do not have identical physicochemical properties or degradation half-lives and this must be taken into account when using such ratios for POP source identification in remote areas [8]. For example, Sheng et al. [113] reported high α-/γ-HCH ratios in Tibet, but rather than attributing this to residues of technical HCH in downwind regions, they suggested that the observed ratio was due to preferential OH radical degradation and wet deposition of γ-HCH compared to α-HCH.

Enantiomeric ratios of chiral POPs have also been used in a number of studies to investigate the influence of long-range atmospheric transport on POP quantities at remote sites. The most commonly investigated chiral POPs and POP degradation products are α-HCH, *o,p′*-DDT, *o,p′*-DDD, the components of technical chlordane, and several of the brominated flame retardants. General reviews on the use of enantiomeric ratios to assess source and transport pathways are available [8,114,115] and describe the use of enantiomeric ratios to discern agricultural versus nonagricultural sources, recent versus old sources, and primary versus secondary sources of POPs [114]. A key advantage of using enantiomeric ratios for source determination is that enantiomers have identical physicochemical properties, which is not the case for other types of isomers as discussed above for HCHs. Therefore, enantiomeric pairs of compounds are affected identically by *abiotic* transport and transformation processes; however, they interact differently with other chiral molecules, such as enzymes and biological receptors. Thus, enantiomers undergo differential *biotic* degradation, which leads to changes in enantiomer ratios. For example, soil microorganisms degrade differently certain POP enantiomers [116,117] and this can lead to changing enantiomeric fractions with POP aging in soils [118,119]. For the source identification of POPs in remote ecosystems, it is usually assumed that nonracemic mixtures indicate revolatilization from regional soils as the source and racemic mixtures indicate a fresh source or one in which only abiotic transformation has taken place. In an example of a project focused on source identification in remote ecosystems, Genualdi et al. used enantiomeric ratios of several OCPs to confirm that their sources in mountains in the continental US were of regional origin but that those in the Arctic Alaska were from long-range transport [120].

Chemical profile analysis, also called compositional analysis, has also been used to investigate the contributions of long-range atmospheric transport on POP quantities in remote ecosystems. Profiling approaches are similar to those that use diagnostic ratios; however, instead of focusing on the ratio between two diagnostic chemicals, "chemical fingerprints" are compared between those of potential sources and those measured at receptor sites. Chemical profile analysis has been commonly used for chemicals released to the environment as mixtures of congeners or homologues. Although multivariate analysis methods for pollutant profiles have been described [121,122], straightforward comparisons

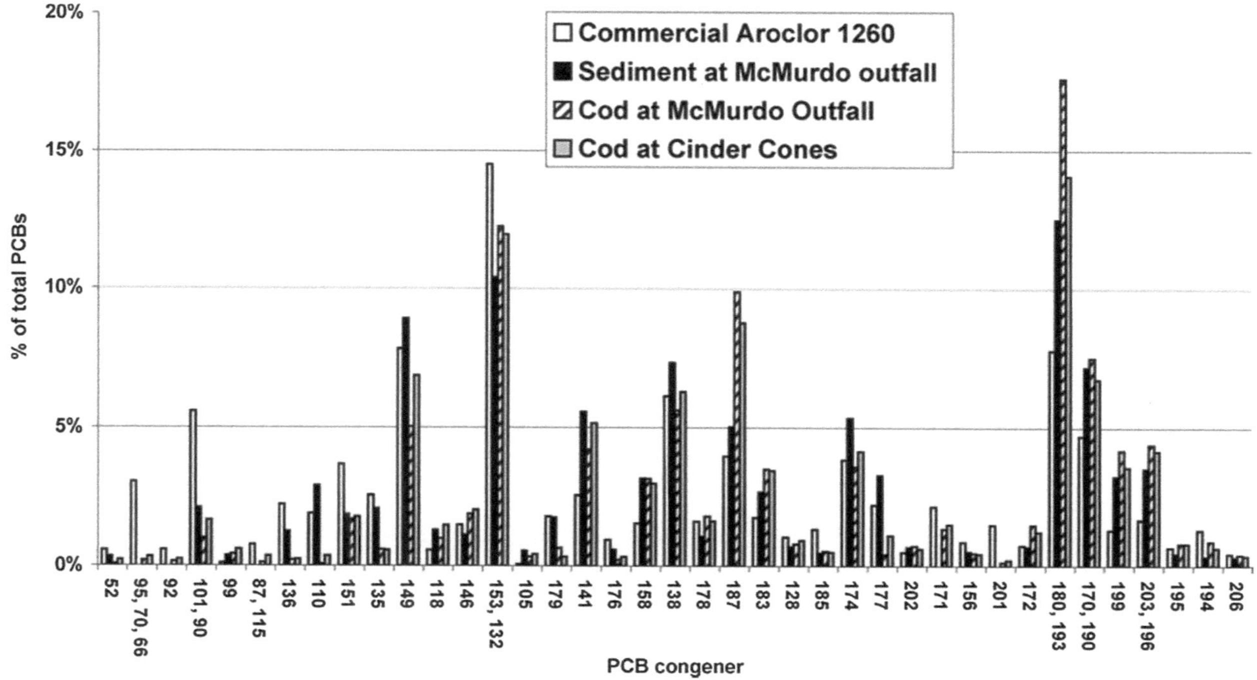

FIGURE 4 Comparison of the polychlorinated biphenyl (PCB) conger "fingerprints" of the commercial Aroclor 1260 product, sediment, and cod from the vicinity of the McMurdo wastewater outfall, and from cod at Cinder Cones, located 12 km away. PCB profiles in fish and sediment resembled Aroclor 1260, released at McMurdo decades ago. The skewing toward the more chlorinated congeners in sediments and fish from the Aroclor signature was attributed to a combination of weathering, selective uptake, and biotransformation processes. *Reprinted (adapted) with permission from Ref. [4]. Copyright (2008) American Chemical Society.*

between chemical profiles are common. For example, PCB and PBDE congener profiles in various aquatic matrices and in potential technical mixtures, in the case of PCBs, were compared in McMurdo Sound, Antarctica, to determine if these chemicals mainly originated from activities at the bases or from long-range atmospheric transport (Figure 4) [4]. As with isomer ratios, one drawback associated with using chemical profiles is that each chemical species in the profile has a different set of physicochemical properties, meaning that differences between profiles may be due to physiochemical processes affecting chemicals differently or may be due to differences in the sources of the chemicals. For example, in the Antarctic study mentioned above, the skewing of the profile away from the technical PCB Aroclor mixture's pattern and toward a more chlorinated profile was attributed to a combination of weathering, selective uptake, and biotransformation processes. An example of a more quantitative approach for dealing with preferential uptake of some chemicals being used in profile analysis was presented by Lavin et al. [97], who modified the profiles in their analysis to account for the temperature-dependent uptake of different chemicals in pine needles, which were used as natural passive samplers, along an elevational gradient.

Yet another common approach for identifying the geographic sources of POPs in remote ecosystems involves air mass transport modeling. A number of specific approaches have been developed and used for this purpose. Westgate and Wania [123] provided an overview of many of the different approaches for calculating contaminant "airsheds" from air mass back trajectories data. Large-grid size modeling programs, such as the Hybrid Single Particle Lagrangian Integrated Trajectory Model (HYSPLIT), have often been used in studies to investigate the sources of POPs in remote locations (e.g., by WACAP [50]). However, higher-resolution Lagrangian particle dispersion models, such as the Flexible particle dispersion model (FLEXPART), offer much better topographic representation and are increasingly being used to address questions about the sources of POPs in remote ecosystems [25,124,125].

6. CONCLUSIONS AND PERSPECTIVES

POPs undergo widespread distribution in the environment when they volatilize from source regions, undergo transport through the atmosphere to distant locations, and are then deposited to surface media by wet or dry deposition. This series of processes is only possible for chemicals with a specific combination of physicochemical properties, which especially concern partitioning behavior and persistence in air. However, atmospheric transport potential is also affected by the emission scenario and a number of environmental parameters, with temperature playing a significant role. Thus, chemical fate modeling is useful for understanding and predicting atmospheric transport potential under different scenarios by combining the many parameters and processes of relevance. The effect of temperature on thermodynamically controlled partitioning behavior was originally used to explain latitudinal trends in POP concentrations; however, the importance of kinetically controlled advective transport processes in controlling these trends is increasingly being recognized.

A combination of local, regional, and long-range atmospheric transport is responsible for the accumulation of POPs in alpine ecosystems. While the concept of mountain cold trapping is often useful for explaining POP concentration trends along elevation gradients, a number of meteorological processes can also have a profound effect on these trends. The postdepositional fate of POPs in alpine and Arctic ecosystems is of particular interest and is likely to receive further research attention due to the potential ecological impacts of POPs on sensitive alpine species. Finally, a number of approaches have been developed to determine the geographic and emission sources of POPs in remote ecosystems. These approaches have used comparisons of POP concentrations at different sites, diagnostic ratios between isomers in technical mixtures, enantiomeric ratios, chemical profile analysis, and air mass modeling. Advancements in source apportionment methodology continue to be made, in particular by using POP isotopes, multivariate analyses, and high-resolution air mass modeling.

REFERENCES

[1] Stockholm Convention (United Nations Environmental Programme), http://chm.pops.int/ (accessed 20.07.14).

[2] J.M. Blais, R.W. Macdonald, D. Mackay, et al., Biologically mediated transport of contaminants to aquatic systems, Environ. Sci. Technol. 41 (2007) 1075–1084.

[3] S.D. Choi, C. Shunthirasingham, G.L. Daly, et al., Levels of polycyclic aromatic hydrocarbons in Canadian mountain air and soil are controlled by proximity to roads, Environ. Pollut. 157 (2009) 3199–3206.

[4] R.C. Hale, S.L. Kim, E. Harvey, et al., Antarctic research bases: local sources of polybrominated diphenyl ether (PBDE) flame retardants, Environ. Sci. Technol. 42 (2008) 1452–1457.

[5] M. Nendza, S. Gabbert, R. Kuhne, et al., A comparative survey of chemistry-driven in silico methods to identify hazardous substances under REACH, Regul. Toxicol. Pharm. 66 (2013) 301–314.

[6] S. Strempel, M. Scheringer, C.A. Ng, et al., Screening for PBT chemicals among the "existing" and "New" chemicals of the EU, Environ. Sci. Technol. 46 (2012) 5680–5687.

[7] D.C.G. Muir, Are there other persistent organic pollutants? A challenge for environmental chemists, Environ. Sci. Technol. 40 (2006) 7157–7166.

[8] Task Force on Hemispheric Transport of Air Pollution, Hemispheric Transport of Air Pollution 2010 Part C: Persistent Organic Pollutants, United Nations, New York and Geneva, 2010.

[9] Aarhus Protocol (United Nations Economic Commission for Europe), http://www.unece.org/env/lrtap/pops_h1.html (accessed 07.09.14).

[10] M. Scheringer, S. Strempel, S. Hukari, et al., How many persistent organic pollutants should we expect? Atmos. Pollut. Res. 3 (2012) 383–391.

[11] R.W. Risebrough, P. Rieche, D.B. Peakall, et al., Polychlorinated biphenyls in the global ecosystem, Nature 220 (1968) 1098–1102.

[12] J.O.G. Tatton, J.H. Ruzicka, Organochlorine pesticides in Antarctica, Nature 215 (1967) 346–348.
[13] E.D. Goldberg, Synthetic organohalides in the sea, Proc. R. Soc. London B 189 (1975) 277–289.
[14] K. Ballschmiter, R. Wittlinger, Interhemisphere exchange of hexachlorocyclohexanes, hexachlorobenzene, polychlorobiphenyls, and 1,1,1-trichloro-2,2-bis(para-chlorophenyl) ethane in the lower troposphere, Environ. Sci. Technol. 25 (1991) 1103–1111.
[15] K. Ballschmiter, O. Froescheis, W.M. Jarman, et al., Contamination of the deep sea, Mar. Pollut. Bull. 34 (1997) 288–289.
[16] T. Bidleman, E.L. Atlas, R. Atkinson, et al., in: A.J. Knap (Ed.), The Long-range Atmospheric Transport of Natural and Contaminated Substances, Kluwer, Dordrecht, 1990, pp. 259–301.
[17] H. Iwata, S. Tanabe, N. Sakai, et al., Distribution of persistent organochlorines in the oceanic air and surface seawater and the role of ocean on their global transport and fate, Environ. Sci. Technol. 27 (1993) 1080–1098.
[18] S.L. Simonich, R.A. Hites, Global distribution of persistent organochlorine compounds, Science 269 (1995) 1851–1854.
[19] D. Calamari, E. Bacci, S. Focardi, et al., Role of plant biomass in the global environmental partitioning of chlorinated hydrocarbons, Environ. Sci. Technol. 25 (1991) 1489–1495.
[20] S.N. Meijer, W.A. Ockenden, A. Sweetman, et al., Global distribution and budget of PCBs and HCB in background surface soils: implications for sources and environmental processes, Environ. Sci. Technol. 2003 (2003) 667–672.
[21] R. Lohmann, W.A. Ockenden, J. Shears, et al., Atmospheric distribution of polychlorinated dibenzo-p-dioxins, dibenzofurans (PCDD/Fs), and non-ortho biphenyls (PCBs) along a North-South Atlantic transect, Environ. Sci. Technol. 35 (2001) 4046–4053.
[22] J. Dachs, R. Lohmann, W.A. Ockenden, et al., Oceanic biogeochemical controls on global dynamics of persistent organic pollutants, Environ. Sci. Technol. 36 (2002) 4229–4237.
[23] W. Kraemer, K. Ballschmiter, Detection of a new class of organochlorine compounds in the marine environment: the chlorinated paraffins, Fresen. Z. Anal. Chem. 327 (1987) 47–48.
[24] J.P. Giesy, K. Kannan, Perfluorochemical surfactants in the environment, Environ. Sci. Technol. 36 (2002) 146A–152A.
[25] A.K. Halse, M. Schlabach, S. Eckhardt, et al., Spatial variability of POPs in European background air, Atmos. Chem. Phys. 11 (2011) 1549–1564.
[26] H. Hung, R. Kallenborn, K. Breivik, et al., Atmospheric monitoring of organic pollutants in the Arctic under the Arctic monitoring and assessment programme (AMAP): 1993–2006, Sci. Total Environ. 408 (2010) 2854–2873.
[27] C.J. Galbán-Malagón, N. Berrojalbiz, R. Gioia, et al., The "degradative" and "biological" pumps controls on the atmospheric deposition and sequestration of hexachlorocyclohexanes and hexachlorobenzene in the North Atlantic and Arctic Oceans, Environ. Sci. Technol. 47 (2013) 7195–7203.
[28] J. Weber, C.J. Halsall, D.C.G. Muir, et al., Endosulfan and gamma-HCH in the arctic: an assessment of surface seawater concentrations and air-sea exchange, Environ. Sci. Technol. 40 (2006) 7570–7576.
[29] Global Atmospheric Passive Sampling Network, https://www.ec.gc.ca/rs-mn/default.asp?lang=En&n=22D58893-1 (accessed 26.09.14).
[30] K. Pozo, T. Harner, S.C. Lee, et al., Seasonally resolved concentrations of persistent organic pollutants in the global atmosphere from the first year of the GAPS study, Environ. Sci. Technol. 43 (2009) 796–803.
[31] K. Pozo, T. Harner, F. Wania, et al., Toward a global network for persistent organic pollutants in air: results from the GAPS study, Environ. Sci. Technol. 40 (2006) 4867–4873.

[32] F. Wania, Assessing the potential of persistent organic chemicals for long range transport and accumulation in polar regions, Environ. Sci. Technol. 37 (2003) 1344–1351.

[33] G.W. VanLoon, S.J. Duffy, Environmental Chemistry : A Global Perspective, third ed., Oxford University Press, Oxford, New York, 2011.

[34] C.L. Davie-Martin, K. Hageman, Y.-P. Chin, An improved screening tool for predicting volatilization of pesticides applied to soils, Environ. Sci. Technol. 47 (2013) 868–876.

[35] R.P. Schwarzenbach, P.M. Gschwend, D.M. Imboden, Environmental Organic Chemistry, second ed., John Wiley & Sons, Inc, Hoboken, New Jersey, 2003.

[36] T. Gouin, D. Mackay, K.C. Jones, et al., Evidence for the "grasshopper" effect and fractionation during long-range atmospheric transport of organic contaminants, Environ. Pollut. 128 (2004) 139–148.

[37] H. von Waldow, M. MacLeod, K. Jones, et al., Remoteness from emission sources explains the fractionation pattern of polychlorinated biphenyls in the northern hemisphere, Environ. Sci. Technol. 44 (2010) 6183–6188.

[38] S. Schenker, M. Scheringer, K. Hungerbühler, Do persistent organic pollutants reach a thermodynamic equilibrium in the global environment? Environ. Sci. Technol. 48 (2014) 5017–5024.

[39] M. MacLeod, M. Scheringer, T.E. McKone, et al., The state of multimedia mass-balance modeling in environmental science and decision-making, Environ. Sci. Technol. 44 (2010) 8360–8364.

[40] M. Scheringer, Long-range transport of organic chemicals in the environment, Environ. Toxicol. Chem. 28 (2009) 677–690.

[41] F. Wania, D. Mackay, The evolution of mass balance models of persistent organic pollutant fate in the environment, Environ. Pollut. 100 (1999) 223–240.

[42] R.J. Hunter-Smith, P.W. Balls, P.S. Liss, Henry's law constants and the air-sea exchange of various low molecular weight halocarbon gases, Tellus B 35B (1983) 170–176.

[43] D. Mackay, F. Wania, Transport of contaminants to the Arctic: Partitioning, processes and models, Sci. Total Environ. 160/161 (1995) 25–38.

[44] C. Galbán-Malagón, N. Berrojalbiz, M.J. Ojeda, et al., The oceanic biological pump modulates the atmospheric transport of persistent organic pollutants to the Arctic, Nat. Commun. 3 (2012) 862.

[45] M. Stroebe, M. Scheringer, K. Hungerbühler, Measures of overall persistence and the temporal remote state, Environ. Sci. Technol. 38 (2004) 5665–5673.

[46] R. Kallenborn, Persistent organic pollutants (POPs) as environmental risk factors in remote high-altitude ecosystems, Ecotoxicol. Environ. Saf. 63 (2006) 100–107.

[47] J.M. Pacyna, K. Breivik, J. Munch, et al., European atmospheric emissions of selected persistent organic pollutants, 1970–1995, Atmos. Environ. 37 (2003) S119–S131.

[48] Y.F. Li, R.W. Macdonald, Sources and pathways of selected organochlorine pesticides to the Arctic and the effect of pathway divergence on HCH trends in biota: a review, Sci. Total Environ. 342 (2005) 87–106.

[49] K. Breivik, A. Sweetman, J.M. Pacyna, et al., Towards a global historical emission inventory for selected PCB congeners – a mass balance approach. 3. An update, Sci. Total Environ. 377 (2007) 296–307.

[50] K.J. Hageman, W. Hafner, D.H. Campbell, et al., Variability in pesticide deposition and source contributions to snowpack in western U.S. national parks, Environ. Sci. Technol. 44 (2010) 4452–4458.

[51] K.J. Hageman, S.L. Simonich, D.H. Campbell, et al., Atmospheric deposition of current-use and historic-use pesticides in snow at national parks in the Western United States, Environ. Sci. Technol. 40 (2006) 3174–3180.

[52] A.M. Mast, D.A. Alvarez, S.D. Zaugg, Deposition and accumulation of airborne organic contaminants in Yosemite National Park, California, Environ. Toxicol. Chem. 31 (2012) 524–533.
[53] C. Davidson, Declining downwind: Amphibian population declines in California and historical pesticide use, Ecol. Appl. 14 (2004) 1892–1902.
[54] J.M. Wallace, P.V. Hobbs, Atmospheric Science: An Introductory Survey, second ed., Academic Press, Amsterdam, 2006.
[55] P. Seibert, H. Kromp-Kolb, A. Kasper, et al., Transport of polluted boundary layer air from the Po Valley to high-alpine sites, Atmos. Environ. 32 (1998) 3953–3965.
[56] R.B. Stull, An Introduction to Boundary Layer Meteorology, Kluwar Academic Publishers, Boston, MA, 1988.
[57] J. Forrer, R. Ruttiman, D. Schneiter, et al., Variability of trace gases at the high-Alpine site Jungfraujoch caused by meteorological transport processes, J. Geophys. Res. 105 (2000) 12241–12251.
[58] A. Kaiser, Origin of polluted air masses in the Alps. An overview and first results for MONARPOP, Environ. Pollut. 157 (2009) 3232–3237.
[59] Y.D. Lei, F. Wania, Is rain or snow a more efficient scavenger of organic chemicals? Atmos. Environ. 38 (2004) 3557–3571.
[60] G.L. Daly, F. Wania, Organic contaminants in mountains, Environ. Sci. Technol. 39 (2005) 385–398.
[61] F. Wania, J.N. Westgate, On the mechanism of mountain cold-trapping of organic chemicals, Environ. Sci. Technol. 42 (2008) 9092–9098.
[62] F.M. Jaward, A. Di Guardo, L. Nizzetto, et al., PCBs and selected organochlorine compounds in Italian mountain air: the influence of altitude and forest ecosystem type, Environ. Sci. Technol. 39 (2005) 3455–3463.
[63] P. Tremolada, S. Villa, P. Bazzarin, et al., POPs in mountain soils from the Alps and Andes: suggestions for a 'precipitation effect' on altitudinal gradients, Water Air Soil Pollut. 188 (2008) 93–109.
[64] I. Offenthaler, R. Bassan, C. Belis, et al., PCDD/F and PCB in spruce forests of the Alps, Environ. Pollut. 157 (2009) 3280–3289.
[65] P. Tremolada, M. Parolini, A. Binelli, et al., Preferential retention of POPs on the northern aspect of mountains, Environ. Pollut. 157 (2009) 3298–3307.
[66] R. Yang, S. Zhang, A. Li, et al., Altitudinal and spatial signature of persistent organic pollutants in soil, lichen, conifer needles, and bark of the southeast Tibetan Plateau: implications for sources and environmental cycling, Environ. Sci. Technol. 47 (2013) 12736–12743.
[67] V.H. Estellano, K. Pozo, T. Harner, et al., Altitudinal and seasonal variations of persistent organic pollutants in the Bolivian Andes Mountains, Environ. Sci. Technol. 42 (2008) 2528–2534.
[68] C.J. Halsall, Investigating the occurrence of persistent organic pollutants (POPs) in the Arctic: their atmospheric behaviour and interaction with the seasonal snow pack, Environ. Pollut. 128 (2004) 163–175.
[69] B.M.J. Herbert, S. Villa, C.J. Halsall, Chemical interactions with snow: understanding the behavior and fate of semi-volatile organic compounds in snow, Ecotoxicol. Environ. Saf. 63 (2006) 3–16.
[70] A.M. Grannas, C. Bogdal, K.J. Hageman, et al., The role of the global cryosphere in the fate of organic contaminants, Atmos. Chem. Phys. 13 (2013) 3271–3305.
[71] J. Stocker, M. Scheringer, F. Wegmann, et al., Modeling the effect of snow and ice on the global environmental fate and long-range transport potential of semivolatile organic compounds, Environ. Sci. Technol. 41 (2007) 6192–6198.

[72] A. Cabanes, L. Legagneux, F. Domine, Rate of evolution of the specific surface area of surface snow layers, Environ. Sci. Technol. 37 (2003) 661–666.

[73] S.C. Colbeck, Air movement in snow due to windpumping, J. Glaciol. 35 (1989) 209–213.

[74] G.K.C. Clarke, E.D. Waddington, A 3-dimensional theory of wind pumping, J. Glaciol 35 (1991) 209–213.

[75] B.M.J. Herbert, C.J. Halsall, S. Villa, et al., Rapid changes in PCB and OC pesticide concentrations in Arctic snow, Environ. Sci. Technol. 39 (2005) 2998–3005.

[76] A. Finizio, S. Villa, F. Raffaele, et al., Variation of POP concentration in fresh-fallen snow and air on an Alpine glacier (Monte Rosa), Ecotoxicol. Environ. Saf. 63 (2006) 25–32.

[77] D.A. Burniston, W.J.M. Strachan, J.T. Hoff, et al., Changes in surface area and concentrations of semivolatile organic contaminants in aging snow, Environ. Sci. Technol. 41 (2007).

[78] G.L. Daly, F. Wania, Simulating the influence of snow on the fate of organic compounds, Environ. Sci. Technol. 38 (2004) 4176–4186.

[79] K.M. Hansen, C.J. Halsall, J.H. Christensen, A dynamic model to study the exchange of gas-phase persistent organic pollutants between air and a seasonal snowpack, Environ. Sci. Technol. 40 (2006) 2644–2652.

[80] C. Steinlin, C. Bogdal, M. Scheringer, et al., Polychlorinated biphenyls in glaciers. 2. Model results of deposition and incorporation processes, Environ. Sci. Technol. 48 (2014) 7849–7857.

[81] P.A. Pavlova, P. Schmid, C. Bogdal, et al., Polychlorinated biphenyls in glaciers. 1. Depositional history from an alpine ice core, Environ. Sci. Technol. 48 (2014) 7842–7848.

[81a] T. Meyer, Y.D. Lei, F. Wania, Transport of polycyclic aromatic hydrocarbons and pesticides during snowmelt within an urban watershed, Water Res. 45 (2011) 1147–1156.

[82] D.C. Girvin, A.J. Scott, Polychlorinated biphenyl sorption by soils: measurement of soil-water partition coefficients at equilibrium, Chemosphere 35 (1997) 2007–2025.

[83] R. Seth, D. Mackay, J. Muncke, Estimating the organic carbon partition coefficient and its variability for hydrophobic chemicals, Environ. Sci. Technol. 33 (1999) 2390–2394.

[84] C. Moeckel, L. Nizzetto, A. Di Guardo, et al., Persistent organic pollutants in boreal and montane soil profiles: distribution, evidence of processes and implications for global cycling, Environ. Sci. Technol. 42 (2008) 8374–8380.

[85] L. Nizzetto, X. Liu, G. Zhang, et al., Accumulation kinetics and equilibrium partitioning coefficients for semivolatile organic pollutants in forest litter, Environ. Sci. Technol. 48 (2014) 420–428.

[86] L. Nizzetto, C. Cassani, A. Di Guardo, Deposition of PCBs in mountains: the forest filter effect of different forest ecosystem types, Ecotoxicol. Environ. Saf. 63 (2006) 75–83.

[87] N. Guazzoni, R. Comolli, L. Mariani, et al., Meteorological and pedological influence on the PCBs distribution in mountain soils, Chemosphere 83 (2011) 186–192.

[88] P. Shahpoury, K.J. Hageman, C.D. Matthaei, et al., Increased concentrations of polycyclic aromatic hydrocarbons in alpine streams during annual snowmelt: investigating the effects of sampling method, site characteristics and meteorology, Environ. Sci. Technol. 48 (2014) 11294–11301.

[89] P.N. Anderson, R.A. Hites, OH radical reactions: the major removal pathway for polychlorinated biphenyls from the atmosphere, Environ. Sci. Technol. 30 (1996) 1756–1763.

[90] P.L. Andersson, K. Oberg, U. Orn, Chemical characterization of brominated flame retardants and identification of structurally representative compounds, Environ. Toxicol. Chem. 25 (2006) 1275–1282.

[91] J.D. Raff, R.A. Hites, Deposition versus photochemical removal of PBDEs from Lake Superior air, Environ. Sci. Technol. 41 (2007) 6725–6731.

[92] P. Klán, I. Holoubek, Ice (photo)chemistry. Ice as a medium for long-term (photo)chemical transformations–environmental implications, Chemosphere 46 (2002) 1201–1210.
[93] N. Matykiewiczová, J. Klánová, P. Klán, Photochemical degradation of PCBs in snow, Environ. Sci. Technol. 41 (2007) 8308–8314.
[94] A.M. Grannas, A.E. Jones, J. Dibb, et al., An overview of snow photochemistry: evidence, mechanisms and impacts, Atmos. Chem. Phys. 7 (2007) 4329–4373.
[95] V.F. McNeill, A.M. Grannas, J.P.D. Abbatt, et al., Organics in environmental ices: sources, chemistry, and impacts, Atmos. Chem. Phys. 12 (2012) 9653–9678.
[96] S. Usenko, D.H. Landers, P.G. Appleby, et al., Current and historical deposition of PBDEs, pesticides, PCBs, and PAHs to Rocky Mountain National Park, Environ. Sci. Technol. 41 (2007) 7235–7241.
[97] K.S. Lavin, K.J. Hageman, Contributions of long-range and regional atmospheric transport on pesticide concentrations along a transect crossing a mountain divide, Environ. Sci. Technol. 47 (2013) 1390–1398.
[98] M. Vighi, Special issue: the role of high mountains in the global transport of persistent organic pollutants, Ecotoxicol. Environ. Saf. 63 (2006) 1–2.
[99] K.S. Lavin, K.J. Hageman, S.K. Marx, et al., Using trace elements in particulate matter to identify the sources of semivolatile organic contaminants in air at an alpine site, Environ. Sci. Technol. (2012) 268–276.
[100] S.O. Baek, R.A. Field, M.E. Goldstone, et al., A review of atmospheric polycyclic aromatic-hydrocarbons – sources, fate and behavior, Water Air Soil Pollut. 60 (1991) 279–300.
[101] K. Ravindra, R. Sokhi, R. Van Grieken, Atmospheric polycyclic aromatic hydrocarbons: source attribution, emission factors and regulation, Atmos. Environ. 42 (2008) 2895–2921.
[102] M.B. Yunker, R.W. Macdonald, R. Vingarzan, et al., PAHs in the Fraser River basin: a critical appraisal of PAH ratios as indicators of PAH source and composition, Org. Geochem. 33 (2002) 489–515.
[103] D.H. Landers, S.M. Simonich, D. Jaffe, et al., The Western Airborne Contaminant Assessment Project (WACAP): an interdisciplinary evaluation of the impacts of airborne contaminants in western US national parks, Environ. Sci. Technol. 44 (2010) 855–859.
[104] D. Jaffe, T. Anderson, D. Covert, et al., Transport of Asian air pollution to North America, Geophys. Res. Lett. 26 (1999) 711–714.
[105] S. Usenko, S.L.M. Smonich, K.J. Hageman, et al., Sources and deposition of polycyclic aromatic hydrocarbons to western US national parks, Environ. Sci. Technol. 44 (2010) 4512–4518.
[106] J. Weber, C.J. Halsall, D.C.G. Muir, et al., Endosulfan, a global pesticide: a review of its fate in the environment and occurrence in the Arctic, Sci. Total Environ. 408 (2010) 2966–2984.
[107] L. Becker, M. Scheringer, U. Schenker, et al., Assessment of the environmental persistence and long-range transport of endosulfan, Environ. Pollut. 159 (2011) 1737–1743.
[108] W. Vetter, S. Gaul, W. Armbruster, Stable carbon isotope ratios of POPS – a tracer that can lead to the origins of pollution, Environ. Int. 34 (2008) 357–362.
[109] Y.H. Zeng, X.J. Luo, H.S. Chen, et al., Method for the purification of polybrominated diphenyl ethers in sediment for compound-specific isotope analysis, Talanta 111 (2013) 93–97.
[110] K. Walker, D.A. Vallero, R.G. Lewis, Factors influencing the distribution of lindane and other hexachlorocyclohexanes in the environment, Environ. Sci. Technol. 33 (1999) 4373–4378.
[111] World Health Organization. Lindane, Environmental Health Criteria No. 124, www.who.int/ipcs/publications/ehc (accessed 20.07.14).

[112] L. Shen, F. Wania, Y.D. Lei, et al., Hexachlorohexanes in the North American atmosphere, Environ. Sci. Technol. 38 (2004) 965–975.

[113] J. Sheng, X. Wang, P. Gong, et al., Monsoon-driven transport of organochlorine pesticides and polychlorinated biphenyls to the Tibetan Plateau: three year atmospheric monitoring study, Environ. Sci. Technol. 47 (2013) 3199–3208.

[114] T. Bidleman, L.M. Jantunen, P.B. Kurt-Karakus, et al., Chiral persistent organic pollutants as tracers of atmospheric sources and fate: review and prospects for investigating climate change influences, Atmos. Pollut. Res. 3 (2012) 371–382.

[115] S. Harrad, Persistent Organic Pollutants, Wiley, Chichester, West Sussex, United Kingdom, 2010.

[116] T.A. Müller, H.P. Kohler, Chirality of pollutants – effects on metabolism and fate, Appl. Microbiol. Biot 64 (2004) 300–316.

[117] M. Suar, A. Hauser, T. Poiger, et al., Enantioselective transformation of alpha-hexachlorocyclohexane by the dehydrochlorinases LinA1 and LinA2 from the soil bacterium Sphingomonas paucimobilis B90A, Appl. Environ. Microbiol. 71 (2005) 8514–8518.

[118] P.B. Kurt-Karakus, J.L. Stroud, T. Bidleman, et al., Enantioselective degradation of organochlorine pesticides in background soils: variability in field and laboratory studies, Environ. Sci. Technol. 41 (2007) 4965–4971.

[119] M. Robson, S. Harrad, Chiral PCB signatures in air and soil: implications for atmospheric source apportionment, Environ. Sci. Technol. 38 (2004) 1662–1666.

[120] S.A. Genualdi, K.J. Hageman, L.K. Ackerman, et al., Sources and fate of chiral organochlorine pesticides in western U.S. National Park ecosystems, Environ. Toxicol. Chem. 30 (2011) 1533–1538.

[121] D. Howel, Multivariate data analysis of pollutant profiles: PCB levels across Europe, Chemosphere 67 (2007) 1300–1307.

[122] C. Real, J.A. Fernandez, J.R. Aboal, et al., The analysis of pollutant profiles by means of compositional analysis. An application to PCDD/F data, Chemosphere 77 (2009) 1177–1183.

[123] J.N. Westgate, F. Wania, On the construction, comparison, and variability of airsheds for interpreting semivolatile organic compounds in passively sampled air, Environ. Sci. Technol. 45 (2011) 8850–8857.

[124] S. Ubl, M. Scheringer, A. Stohl, et al., Primary source regions of polychlorinated biphenyls (PCBs) measured in the Arctic, Atmos. Environ. 62 (2012) 391–399.

[125] A.K. Halse, S. Eckhardt, M. Schlabach, et al., Forecasting long-range atmospheric transport episodes of polychlorinate biphenyls using FLEXPART, Atmos. Environ. 71 (2013) 335–339.

Chapter 12

Occurrence and Ecological Risk of Halogenated Flame Retardants (HFRs) in Coastal Zones

James C.W. Lam[1,2,*] and Paul K.S. Lam[1,2,3,*]

[1]*State Key Laboratory in Marine Pollution, City University of Hong Kong, Hong Kong SAR, China;* [2]*Research Centre for the Oceans and Human Health, Shenzhen Key Laboratory for Sustainable Use of Marine Biodiversity, City University of Hong Kong Shenzhen Research Institute Building, Shenzhen, PR China;* [3]*Department of Biology and Chemistry, City University of Hong Kong, Hong Kong SAR, China*

**Corresponding authors: E-mail: james.lam@cityu.edu.hk or mail.jameslam@gmail.com; bhpksl@cityu.edu.hk*

Chapter Outline

Comprehensive Analytical Chemistry, Vol. 67. http://dx.doi.org/10.1016/B978-0-444-63299-9.00012-0

1. OVERVIEW OF HALOGENATED FLAME RETARDANTS

1.1 Polybrominated Diphenyl Ethers

Flame retardants are considered a crucial component in manufacturing a wide variety of consumer products because they can reduce fire hazards by lowering the rate of burning and often prevent fire from spreading. Of the flame retardants on the market, brominated derivatives or brominated flame retardants (BFRs) have been extensively used for several decades. These chemicals have been marketed as technical mixtures under different commercial brands consisting of different chemical compounds within each class. In view of their high production and broad use, several groups of BFRs including polybrominated diphenyl ethers (PBDEs), polybrominated biphenyls, hexabromocyclododecanes (HBCDs), and tetrabromobisphenol A have been extensively investigated and reviewed [1–4], indicating their ubiquity in the indoor and outdoor environment. Of the BFRs, increasing concentrations of PBDEs have been reported on a global scale in biota, sediment, and even in humans [1]. A large number of studies have reported on the environmental ubiquity of PBDEs and they are present in almost every environmental compartment. PBDEs have been detected in sediment, sludge, soil, indoor and outdoor air, house dust, several food commodities, marine and terrestrial wildlife, and human samples [4–6].

Over the past several decades, levels of PBDEs have significantly increased and five congeners namely tetra-, penta-, and hexa-BDE congeners (BDE-47, -99, -100, -153, -154) were predominate in the environment and in human tissues [1,6,7]. These chemicals have been characterized by their persistence, toxicity, bioaccumulation, and long-range transport potential, and in fact, two important technical mixtures, penta- and octa-BDE, were listed as global Persistent Organic Pollutants (POPs) under the Stockholm Convention (http://chm.pops.int/Convention/ThePOPs/TheNewPOPs/tabid/2511/Default.aspx) in 2009. In addition, the use of fully brominated BDE-209 in the Deca-BDE technical mixture in electronic products was banned in the European Union in 2008 and was also phased out by important US producers and importers since the end of 2013 [8,9].

1.2 Hexabromocyclododecanes

Another important group of BFRs are HBCDs, which are nonaromatic brominated cyclic alkanes and which are not chemically bound to products. This additive BFR has been used for about 20 years and is primarily applied in polystyrene insulation products and textile coatings [2,4,10]. It has been applied as a flame retardant in expandable polystyrene (EPS) and extruded polystyrene (XPS) in the construction industry, as well in automotive interior textiles and furniture. It is also used in high impact polystyrene in electrical and electronic parts [4].

HBCD theoretically consists of 16 stereoisomers, while commercial HBCD product is mainly composed of three diastereoisomers: α-(10–13%), β-(1–12%), and γ-HBCD (75–89%) [2]. It has been reported that a shift in the relative diastereomer contribution was observed with increasing trophic species in aquatic food webs [2,4]. In general, sediment and organisms in lower trophic levels contain mostly γ-HBCD similar to the commercial mixture, whereas α-HBCD dominates the diastereomer pattern in top predators in the food web. It has also been reported that HBCDs undergo thermal rearrangement at temperatures above 160°C; this could occur during the processing of materials containing HBCDs [2]. Consequently, the relative diastereoisomer pattern in the environment may be different from the technical products.

Global concern for HBCDs has risen as it has been detected not only in animals [2,10], but relatively high levels were also identified in human samples [11–13]. The presence of HBCD in polar ecosystems provides unambiguous evidence of long-range transport of this compound [14]. Temporal studies indicate that levels of HBCD in the environment, humans, and wildlife are increasing [15–17]. Field studies in marine mammals and aquatic wildlife have suggested that HBCD is also highly bioaccumulative [4,17].

The occurrence of high levels of HBCDs has resulted in adverse health effects as demonstrated in laboratory studies. Neurodevelopmental toxicity has been linked to HBCD exposure in mice [18]. Increased thyroid, liver, and pituitary were observed in rats [19]. In a one-generation exposure study in rat, decreased concentration of apolar retinoids in the liver, increased immune response, and mineral bone density in females were reported [20]. In view of all its POPs characteristics, banning the use of HBCD was recently proposed by the Stockholm Convention in October 2012 and was subsequently listed as a POP for elimination (in Annex A3 of the Convention) in May 2013 (http://chm.pops.int/Convention/ThePOPs/TheNewPOPs/tabid/2511/Default.aspx), with time-limited exemptions for production and use in EPS and XPS in buildings.

1.3 Halogenated PBDE Alternatives

Since the worldwide restriction on the production and use of PBDEs [3], the market demand for the controlled flame retardants is expected to decline, for instance, the total market demand of penta-BDE has decreased from 8500 tons in 1999 to 7500 tons in 2001 (www.bsef.com), whereas that for its alternatives is projected to increase. The usage of alternatives for PBDE formulations is projected to increase in regions with increasing population and rapid economic development and industrial activities. Recent studies have reported several representative halogenated flame retardants (HFRs) used as alternatives for the banned formulations of BFRs, e.g., 1,2-bis(2,4,6-tribromophenoxy)ethane (BTBPE), decabromodiphenyl ethane (DBDPE), Dechlorane Plus (DP), hexabromobenzene, hexachlorocyclopentadienyl-dibromocyclooctane, pentabromoethylbenzene, pentabromotoluene, 2-ethylhexyl 2,3,4,5-tetrabromobenzoate, 1,2-dibromo-4-(1,2-dibromoethyl)cyclohexane (TBECH), and bis-(2-ethylhexyl)-tetrabromophthalate. The chemical structures of these HFRs are shown in Figure 1. The background information such as production and use of these chemicals has been comprehensively reviewed by Covaci et al. [3], de Wit et al. [14], and Sverko et al. [21]. With the large global demand for HFRs, discharges from manufacturing processes, as well as leaching from existing products that are in service or have been disposed of in landfills or electronic-waste (e-waste) sites could elevate the levels of both conventional and alternative HFRs in the aquatic environment [22–25]. Up till now, little is known about the presence of halogenated PBDE alternatives in the riverine and marine environments, although some of these compounds have been reported in the aquatic systems such as estuaries [25–27] and lakes [28–30]. For most of the regions, however, there is still a lack of information on the recent status of these emerging compounds. More importantly, there is currently a paucity of information regarding the assessment of ecological risks of these replacements in the environment.

This chapter discusses current knowledge on environmental contamination by the halogenated PBDE alternatives in the sediment and marine mammals and describes the preliminary assessment of the risk of adverse biological effects in marine mammals due to the HBCDs, one of the halogenated PBDE alternatives,

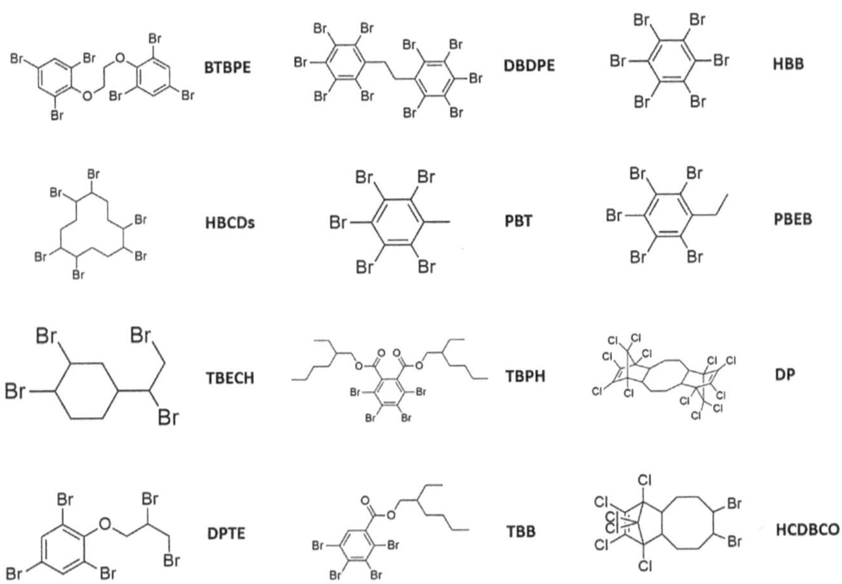

FIGURE 1 Chemical structure of potential halogenated polybrominated diphenyl ether alternatives. HBCD, Hexabromocyclododecane; BTBPE, 1,2-bis(2,4,6-tribromophenoxy)ethane; TBPH/BEHTBP, Bis-(2-ethylhexyl)-tetrabromophthalate; TBECH, 1,2-Dibromo-4-(1,2-dibromoethyl) cyclohexane; DPTE, 2,3-Dibromopropyl-2,4,6-tribromophenyl ether; DBDPE, Decabromodiphenyl ethane; TBB, 2-Ethylhexyl 2,3,4,5-tetrabromobenzoate; HBB, Hexabromobenzene; PBEB, Pentabromoethylbenzene; PBT, Pentabromotoluene; DP, Dechlorane Plus; HCDBCO, Hexachlorocyclopentadienyldibromocyclooctane.

in the blubber. This evaluation involves the comparison of the measured concentrations in the tissues to the predicted no-effect or threshold concentrations derived from the available toxicological information reported.

2. HALOGENATED PBDE ALTERNATIVES IN THE ENVIRONMENT

2.1 Sediment

A wide variety of contaminants are discharged into the aquatic environment from industrial, agricultural, and municipal sources. The contaminants that are adsorbed onto suspended particles and then accumulated in sediments can cause adverse effects on the aquatic ecosystem especially the benthic community that lives in the sediments and subsequently can affect human health indirectly through the food chain. Accordingly, it is important to study the occurrence of toxic and hydrophobic chemicals in the sediment, which also can act as a good HFR contamination indicator for worldwide comparison.

Hydrophobic emerging HFRs are expected to be prevalent in riverine and coastal sediment. Since PBDEs have been extensively reviewed [1,4], this chapter mainly concentrates on the spatial distribution of potential halogenated PBDE alternatives in the coastal regions. Concentrations of potential halogenated PBDE alternatives in the marine and river sediment have been determined by researchers in different countries [23,25–28,30–41]; however, there has been comparatively little information on non-PBDE HFRs in sediment. Among halogenated PBDE alternatives measured in sediment in different regions, geographical variations can be distinguished, as shown in Figure 2.

2.1.1 HBCDs in Sediment

HBCDs have been extensively reported in sediment samples covering the regions of North America, Asia, Europe, and Africa [23,25,30,32–34,36,37,39–41]. These additive HFRs, technically consisting of α-, β-, and γ-isomers, have been recently proposed for inclusion in the Stockholm Convention on POPs because of its similar characteristics to POPs. Among the three diastereoisomers, γ-HBCD was found to be dominant the isomeric profile in sediment, whereas β-HBCD was not detected in all of the samples [2]. Highest levels of these compounds were found in samples from South Africa followed by Japan, Spain, and China [32,33,37,41].

2.1.2 DBDPE in Sediment

Levels of DBDPE were found to be the highest in samples from a natural pond at an e-waste recycling site and the Dongjiang River from South China [23,26] followed by sediment samples collected from the coastal region of South Africa [33]. In the European region, DBDPE concentrations were most prevalent in samples from the Netherlands followed by Spain and Sweden. It

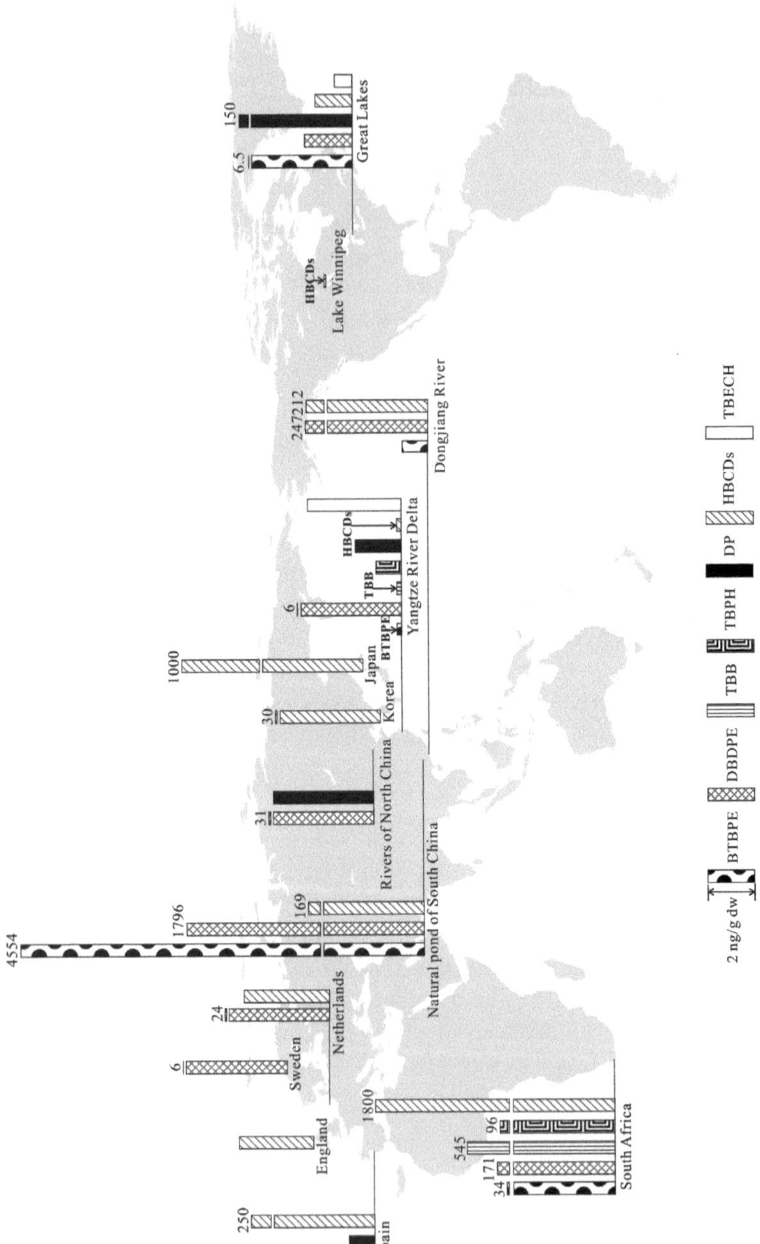

FIGURE 2 Concentrations (ng/g dw) of several emerging halogenated polybrominated diphenyl ether alternatives in aquatic sediment.

should be noted that Deca-BDE (BDE209) was typically found to be the most abundant congener in the sediment samples [25]. DBDPE, which is structurally similar to BDE209, is produced as an alternative to technical Deca-BDE [14]. In general, DBDPE is the most prevalent HFRs followed by HBCDs (Figure 2). In addition, significant positive correlations between BDE209 and DBDPE were consistently found in the sediment samples [25,27]. Therefore, it is expected that a high occurrence pertaining to the level and frequency of DBDPE will be observed in sediment after the phasing out of Deca-BDE technical mixture.

2.1.3 TBPH and TBB

TBPH and TBB are the major components of technical mixtures Firemaster-550 (FM-550), Firemaster BZ-54 (BZ-54), and DP-45 and are produced as replacements for penta-BDE in polyurethane foam application [42,43]. Both of these HFRs have been recently identified in sewage sludge [44], house dust in the US [43], and marine mammals from South China [17,45]. To date, information on these two HFRs is still very limited. These PBDE replacements were detected in sediment from South Africa and eastern part of China where a lot of textile and electronic industries are found [25,33]. In addition, levels of TBB were the second most prevalent PBDE alternative followed by HBCDs in the sediment collected from South Africa [33]. It should be noted that TBPH and TBB, recently detected in both bivalve and gastropod samples collected from a river near a textile mill, had ratios of TBPH:TBB that varied between the sampling locations [42]; however, there is a lack of knowledge of the exposure pathways for biota to these chemicals.

2.1.4 DP in Sediment

DP, consisting of syn- and anti-isomers, is an additive HFR [21] and was the only chlorinated PBDE alternative detected in sediment (Figure 2). Many studies have demonstrated the stereoselective enrichment of anti-DP in the environment when compared to the technical mixture and this may indicate that syn-DP is more susceptible to environmental degradation [21,46]. Therefore, concentrations of anti-DP found in the sediment were more predominant than the syn-isomer in the coastal sediment and suggest that the anti-isomer is more environmentally persistent in comparison to the syn-isomer [25,46]. Highest levels of DP were observed in sediment samples collected from the Great Lakes in North America followed by China and Spain. A high concentration of DP found in Lake Ontario may be attributed to the higher contamination of DP-located downstream of a DP manufacturing plant [47].

2.1.5 TBECH in Sediment

TBECH is an additive flame retardant and is extensively used in electrical cable coatings, extruded polystyrene, and high-impact plastics [14,48].

The technical mixture of TBECH is comprised of two equal amounts of diastereomers, α- and β-TBECH, which can be thermally rearranged to γ- and δ-TBECH at temperatures above 120 °C [48]. There is very little information on the environmental occurrence of TBECH, which have only been investigated in sediment samples from China and the US. Studies showed comparatively higher levels of TBECH in samples from the estuarine area of the eastern part of China [25] than in the US. Furthermore, high levels of TBECH were observed in offshore marine sediment located far away from the territory of eastern part of China and no potential contamination sources were found in the vicinity of these sampling locations [25]. Elevated levels in offshore samples could be likely due to their long-range atmospheric transport and these results were also in agreement with the low degradation rate and long-range atmospheric transport potential predicted by another study [49]. It is also important to note that these compounds were even detected in biota in the Arctic region [48].

2.1.6 BTBPE in Sediment

BTBPE were detected in Great Lakes, China, and South Africa. Highest levels were found in the samples collected from a natural pond in e-waste recycling site, South China [23]. E-waste sites contain high burden of PBDEs and various HFRs, and the recycling process would cause high amount of these compounds into the environment.

2.2 Marine Mammals

Sediment is well known as excellent sinks for many classes of anthropogenic contaminants and it may allow for the mapping of possible sources of pollution; however, it is not able to show whether contaminants are bioavailable. In contrast, the study of biota in the coastal environment can give an indication of the bioavailable fraction of chemicals and also the direct exposure to environmental contamination. In the surveillance of marine pollution, higher trophic animals such as marine mammals and birds are the target species for the global monitoring survey of POPs [50]. Marine mammals are considered to be particularly sensitive to the effects of contaminants because of their relatively long life span and the layer of blubber that allows for the bioaccumulation and integration of lipophilic xenobiotics, which reflect environmental changes across large spatial and over long temporal scales [51–54]. Hence, the investigation of levels of POPs and related contaminants in the marine mammals is of great importance for revealing the current status and bioavailability of the contaminants, which can provide additional information to safeguard human health and ecological systems.

As the majority of the total body burden of lipophilic contaminants such as PBDEs is accumulated in the blubber of marine mammals [55], blubber is used for elucidation of the contamination status of halogenated PBDE alternatives in this chapter Information on halogenated PBDE alternatives in marine mammals is very scarce. Figure 3 shows a global comparison of the baseline concentrations of several important halogenated PBDE alternatives in marine mammals, which include the harbor seal from northwest Atlantic [56], white whale from Svalbard [57], ringed seal from Greenland [58], harbor porpoise from UK [59,60], Indo-Pacific humpback dolphin and finless porpoise from the South China Sea [17,45], harbor porpoise and seal from the North Sea Basin [61], striped dolphins from Japan [55], Franciscana dolphin from Brazil [62], and Canadian Arctic beluga and ringed seal from the Canadian Arctic [63,64].

2.2.1 HBCDs in Marine Mammals

Of the halogenated PBDE alternatives, HBCDs were the most intensively investigated in the marine mammals. They were found in the body burden of the marine mammals from North America, Asia, and European areas, with the highest levels in harbor porpoise from UK, whereas the lowest levels were found in marine mammals collected from the Arctic regions such as Svalbard and Greenland (Figure 3). It has been reported that α-isomer is the most bioaccumulative among the three isomers, such that α-HBCD is the most prevalent in the biological samples especially in the top predators of the food web [2]. This was also the case with regard to species of marine mammals, although the composition of the isomers varied depending on the sampling location [17,55,59]. In general, the concentrations of HBCDs were one to two orders of magnitude lower than the concentrations of PBDEs reported for marine mammal species collected in American and Asia regions [17,55,56], while the levels of PBDEs were similar to that of HBCDs in harbor porpoise from the UK [59].

2.2.2 DP in Marine Mammals

The second most studied halogenated PBDE alternatives in marine mammals is DP, which were detected in harbor seal, Franciscana dolphin, harbor porpoise, Indo-Pacific humpback dolphin, and finless porpoise collected from San Francisco Bay [65], Southeastern and Southern Coast of Brazil [62], UK [60], and South China Sea [45], respectively. The highest levels were found in the samples collected from South China Sea, up to 60 ng/g lipid wt [45], which was two orders of magnitude higher than the levels detected in the porpoise samples collected from the UK [60]. These patterns were similar to the distribution of DP observed in the sediment samples (Figure 2), in that DP was relatively higher in the samples from China compared with other regions, with the only exception of samples collected from the Great Lakes. It is important to note that an extremely high concentration (7590 ng/g dw) of DP was found in surficial sediment sampled from a large-scale e-waste site in southern China [66]. China is now one of the most fastest-growing countries in terms of its population, economy, and industrial activities, particularly the coastal region; rapid industrialization, urbanization, and unregulated disposal of e-waste have been recognized as a cause of

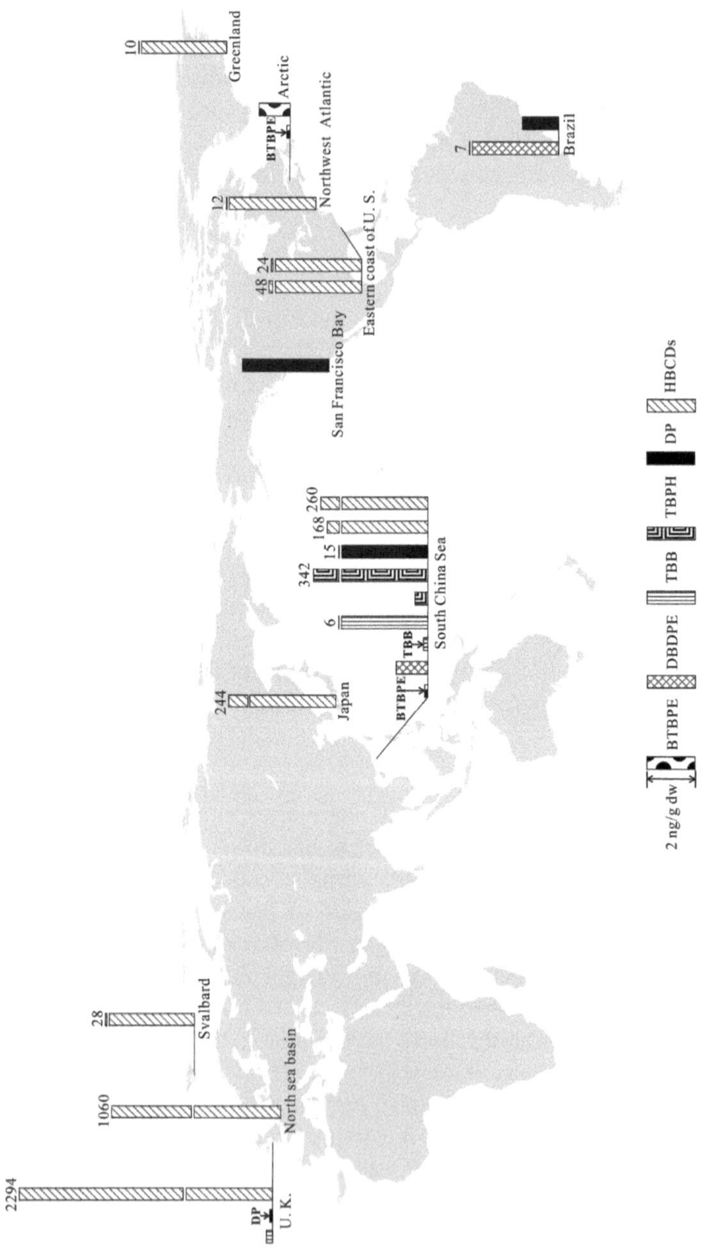

FIGURE 3 Concentrations (ng/g lipid wt) of several emerging halogenated polybrominated diphenyl ether alternatives in marine mammals.

increased release of these environmental contaminants. The large-scale industrial development in terms of electronic, plastic, and manufacturing operations in these regions potentially constitutes important sources of HFRs.

2.2.3 Other Halogenated PBDE Alternatives

Besides HBCDs and DP, other halogenated PBDE alternatives are seldom investigated in marine mammals. DBDPE were recently analyzed in the Franciscana dolphin from Southeastern and Southern Coast of Brazil [62] and Indo-Pacific humpback dolphin and finless porpoise from the South China Sea [45]. The levels were in the same other of magnitude ranging from 1.3 to 7.3 in both regions. Due to the larger molecular weight of DBDPE (971.23), relatively lower bioavailability and bioaccumulation potential of DBDPE as compared with Deca-BDE was expected; however, DBDPE was still extensively detected in the high trophic marine organisms. Bioaccumulation of DBDPE was observed in freshwater fish samples from an industrialized area of South China and the calculated bioaccumulation factor of DBDPE was similar to that of PBDEs and even higher than Deca-BDE in the freshwater system [67]. This may explain the presence of this compound in top predators of the food web. However, there is a lack of information on the bioaccumulation/biomagnification of this compound in the marine ecosystem and further investigation is needed.

BTBPE has been widely identified in biological samples such as fish, bivalve, and birds [42,68], but the levels in the blubber of marine mammals were comparatively lower than PBDEs, DP, and DBDPE in the samples collected from the South China Sea [45]. This may indicate that the use of this flame retardant is lower than other PBDE alternatives in the region. The concentrations of BTBPE in top predators from other locations such as in ringed seals from the Canadian Arctic [63] and glaucous gulls from the Norwegian Arctic [69] were similar to the marine mammals from the South China Sea and ranged from not detected to ~1 ng/g, lipid weight.

Both TBB and TBPH were detected in the Indo-Pacific humpback dolphin and finless porpoise from the South China Sea [17,45], while only TBB was determined in the harbor porpoise from the UK, ranging from not detected to 3.4 ng/g lipid wt [60]. TBPH and TBB were recently detected in both bivalve and gastropod samples collected from a river near a textile mill and the ratios of TBPH:TBB varied between the sampling locations [42]. These ratios also varied between the blubber samples of the two cetacean species from the South China Sea [45]. Data on the environmental fate and impact of these two compounds are still very scarce.

3. POTENTIAL RISK OF HALOGENATED PBDE ALTERNATIVES TO MARINE MAMMALS

Marine mammals are often exposed to high levels of lipophilic environmental contaminants. As such, these species can serve as useful indicators of ecosystem contamination. Risk managers often use the indicator or sentinel species to detect risks to human beings by providing advance warning of a danger of a toxic chemical. Ecological risk assessment was carried out to evaluate the likelihood that adverse effects may occur to the selected target species as a result of exposure to chemical stressors in the environment. Information from the chemical analyses, combined with that from previous studies, will provide a more complete dataset on the exposure levels of the sentinel species to the PBDEs alternatives and will form the basis of an assessment of risks to the target due to these contaminants.

The internal concentrations of toxicants in biota have some advantages in their usage for risk assessment. Firstly, the biotic concentrations are more representative of the levels at the target active site than environmental concentrations because bioaccumulation, losses, and so on are all taken into account and thus can reduce the confounding effects of these factors when interpreting the results. However, limited information is available on the effects of internal concentrations on biota through observations in the field or the laboratory especially for the contaminants of emerging concern. In addition, there are difficulties in the collection of samples of whole protected wildlife for the evaluation of toxicants in body tissues. Such collections, even if possible, may have an adverse effect on the existing population and may be prevented by legislation. Collections are thus often made of carcasses, eggs, feathers, and nestlings to provide appropriate samples for evaluating exposure.

Apart from analyzing environmental contaminants in the blubber samples of carcasses of (stranded) marine mammals, a novel technique has been developed and applied in recent years to collect subcutaneous samples (skin and blubber) from free-ranging marine mammals. This nonlethal sampling tool uses a minimally invasive biopsy darting technique to collect samples. The application of the blubber biopsy technique has been validated for the monitoring of environmental contaminants in cetaceans. The absence of significant differences in the lipid-normalized concentrations of contaminants in the blubber suggests that concentrations of contaminants such as polychlorinated biphenyls and organochlorines in the blubber mantle are principally influenced and determined by lipid content. A biopsy that includes only the outer region of the blubber mantle is representative of the contaminant concentrations in the body of the cetacean as a whole [70]. In addition, the skin from the biopsy samples can potentially be used for assessing the ecotoxicological risk of contaminants to cetaceans [71]. Indeed, the biopsy technique has been employed in a number of studies to investigate the contamination status in cetaceans from various regions around the world [72–76].

For conducting an ecological risk assessment, two approaches are commonly applied, namely probabilistic risk assessment (PRA) and the hazard quotient approach (HQ). In this chapter, these two approaches are briefly discussed. To carry out an ecological risk assessment on marine mammal due to the exposure to HBCDs, only the HQ approach was adopted in this study because of the limited toxicological information and the lack of raw data on the measured concentrations of HBCDs in marine mammal studies conducted elsewhere.

The most sensitive toxicity endpoint for these chemicals is usually taken as the "no observed adverse effect level" (NOAEL) for reproductive and developmental effects in wildlife. The effects of conventional pollutants such as DDE to some species are well known [77]; however, the effects of emerging pollutants, such as HBCDs, to wildlife are not known, so derivation and extrapolation to the thresholds for the ecological risk assessment are needed.

3.1 Hazard Quotient Risk Assessment

The HQ approach is a method that is commonly used by researchers. The principle of estimating the potential risks using the HQ approach generally followed the methods described by Fung et al. [78], with some modifications. Briefly, the HQ was determined as follows:

$$HQ = \frac{\text{Measured Environmental Concentrations (MEC) in tissue samples}}{\text{Critical Effect Level of concern}}$$

The critical effect levels might include toxicity data such as NOAEL, LOAEL (Lowest-Observable-Adverse-Effect-Level), and LD_{50} (Lethal Dose that cause 50% mortality of the population) in which these data can be obtained from literature.

The estimation of the HQ can be further elaborated by two approaches that provide a better estimation of the risks posed by the pollutants, as follows:

$$HQ_{bcs} = \frac{MEC_{min} \text{ in tissue samples}}{\text{Critical Effect Level of concern}}$$

$$HQ_{wcs} = \frac{MEC_{max} \text{ in tissue samples}}{\text{Critical Effect Level of concern}}$$

where HQ_{bsc} represents the best-case HQ;

HQ_{wcs} represents the worst-case HQ;

MEC_{min} represents the minimum measured environmental concentrations and;

MEC_{max} represents the maximum measured environmental concentrations.

This approach can be used to screen contaminants that may require a more refined risk assessment. When the HQ_{bsc} for a certain pollutant is greater than unity, this indicates that the chemical of concern poses a potential risk and therefore appropriate control or management might be necessary. In contrast, when the HQ_{wcs} for a certain pollutant is less than unity, there is little or negligible risk to the target organisms due to exposure to that pollutant.

3.2 Probabilistic Risk Assessment

PRA uses the probability of the occurrence of a particular contaminant concentration to determine whether the contaminant poses a risk. In contrast to the HQ approach, probabilistic approaches can be used to evaluate both the exposure and effects in an ecotoxicological risk assessment, leading to a final result in probabilistic terms. Examples of the use of PRA for evaluating risk on wildlife due to trace elements and organochlorines can be found in Hung et al. [79] and Lam et al. [77], respectively.

The average concentration of a chemical in an environmental compartment can often be estimated from knowledge of partition mechanisms and other environmental processes. However, other processes often influence the distribution of contaminant concentrations around the average value. A large number of different processes can be involved. For example, random processes can govern the entry of a contaminant into the environment. After entry, random movements of water, sediment, air, and other transport mechanisms govern the distribution of contaminant concentrations. Thus, the distribution of the concentrations of a contaminant is likely to follow a normal distribution in any particular situation. From these observed distributions it is possible to obtain relationships of probability of occurrence of a particular concentration against concentrations in the environment. Thus, linear relationships would be expected from plots of cumulative probability of the occurrence of an environmental concentration against a log of concentration. Such plots are extremely useful since they can be used to give the cumulative probability of occurrence of a particular concentration of interest. This approach has been taken in the risk assessment of waterbirds and marine mammals [77,79].

The probabilistic approach of risk assessment involves the plot of cumulative frequency of a ranked set of concentrations against a function of log concentrations. The procedures for plotting such a graph are simplified and summarized as follows:

The cumulative frequency of occurrence of each data point is calculated by the following equation:

$$\text{Cumulative frequency} = \frac{100 \times i}{n + 1}$$

where i = the rank of a particular datum in the sequence and
n = the total number of data used for the plot.

The cumulative frequency of occurrence of each data point can be then plotted against the corresponding concentration. The plot can then be used to estimate the probability of exceedance for a fixed effect assessment criterion.

3.3 Risk Due to HBCDs

HBCDs have been detected in many environmental compartments and in a variety of species worldwide. The potential risks from HBCDs causing harmful effects to the environment in the short term or long term cannot be neglected as the environmental concentration of HBCDs continue to increase.

Although HBCDs are globally widespread, studies on the toxicity of HBCDs to marine mammals are very limited. In the present study, the threshold level of HBCDs for marine mammals used in the preliminary hazard evaluation was taken from studies on the reproduction and development of mice or rats [19,80,81] after exposure to HBCDs as shown in Table 1.

In this evaluation, the lowest levels of estimated "body burden" at which critical harmful effects were observed in mammalian toxicological studies (hereafter referred to as tentative critical concentration, TCC) were compared with the levels of HBCDs measured in the blubber. The rationale for using rat as a toxicological surrogate was that contaminants measured represent the body burden of target species and that the exposure scenario of marine mammals may be considered, to some degree, similar to that of the rat. Since the extrapolation between species and tissue types involve many uncertainties, the results of the assessment will require further refinement when more data becomes available.

Oral absorption percentages of HBCDs (approximately 90%) were based on a study on rodent [82]. To extrapolate the TCC in rodents to marine mammals, a body weight scaling factor was used to account for differences between mammalian surrogates following the method of Sample et al. [83]:

$$TCC_{\text{marine mammal}} = TCC_{\text{rodent}} \left(BW_{\text{rodent}}/BW_{\text{marine mammal}}\right)^{1/4}$$

where BW stands for body weight (kg wet wt). The body weight values used for most of the marine mammals with data of body burden in this study were small cetacean and seal ranging from 60 to 180 kg. For the evaluation of the potential

TABLE 1 Summary of NOAELs for Hexabromocyclododecanes in Rodents and the Endpoint Affected

Dose (=NOAEL) mg/kg body wt	End-point	References
10.0	Increase liver, thyroid and pituitary weights, induction of hepatic CYP 2B	[19]
10.2	Size of thyroid follicles in both sexes	[80]
8.1–21.3	Reduced habituation, decreased locomotion, and decreased rearing in neonatal male mice	[81]

NOAEL, no observed adverse effect level.

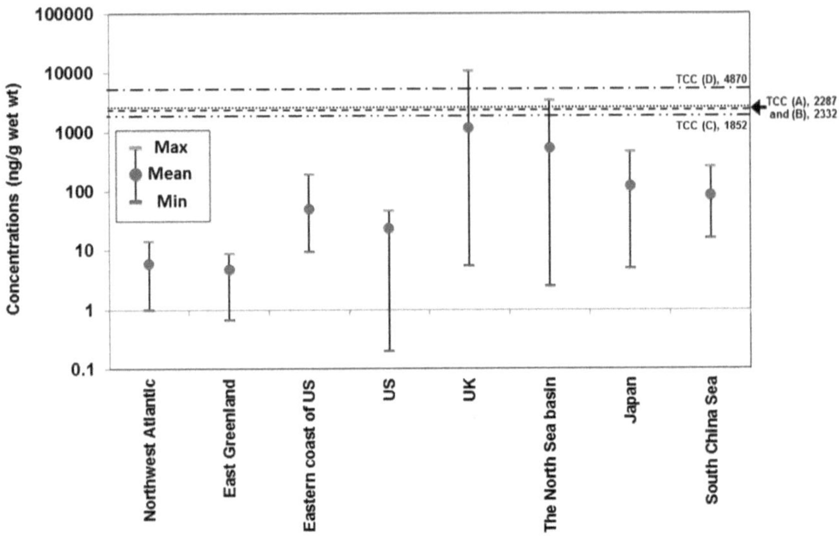

FIGURE 4 Plot of mean, minimum, and maximum concentrations of hexabromocyclododecanes in marine mammals from different regions in comparison to the four calculated tentative critical concentrations.

risk of exposure to HBCDs, 120 kg was used for small marine mammals, while the body weight of male rats (0.5 kg) was used as the value for rodent body weight.

In this preliminary risk evaluation, a lipid percentage of 50 for blubber was assumed for the small marine mammals. Figure 4 shows the mean, minimum, and maximum concentrations of HBCDs in the marine mammals from different regions and the four calculated TCCs from Table 2. The maximum concentrations of HBCDs in the marine mammals from northwest Atlantic, east Greenland, US, Japan, and the South China Sea were below all the TCC values, while some marine mammals from the UK and the North Sea Basin were higher than the four TCCs shown in Figure 4.

In characterizing the risks of HBCDs, only the HQ approach was applied since the data on the concentrations of HBCDs for each individual was not available. In the case that more than one TCC was available, the most sensitive endpoint was chosen for evaluating the risk (i.e., TCC of 1852 μg/g wet wt). In addition, an uncertainty factor of 10 was applied to the calculation to account for the cross-species extrapolation involved in the assessment process. The HQ_{bcs} and HQ_{wcs} were calculated and are summarized in Table 3.

The results of the assessment revealed that all HQ_{bsc} values were below unity, while HQ_{wcs} for white-sided dolphin from the eastern coast of the US, harbor porpoise from the UK and the North Sea Basin, striped dolphins from Japan, and the Indo-Pacific humpback dolphin from the South China Sea exceeded unity, suggesting that blubber concentrations of HBCD pose potential risks to these species.

TABLE 2 Calculated Internal Concentrations and Estimated Tentative Critical Concentrations (TCCs) of Hexabromocyclododecanes for Small Cetacean

	TCC (A)	TCC (B)	TCC (C)	TCC (D)
NOAEL (mg/kg BW)	10.0	10.2	8.1	21.3
Fraction absorbed	0.90	0.90	0.90	0.90
Internal concentration (ng/g wet wt)	9000	9180	7290	19,170
Body weight of test animal (kg)	0.5	0.5	0.5	0.5
Body weight of small marine mammal (kg)	120	120	120	120
TCCs (ng/g wet wt)	2287	2332	1852	4870

TABLE 3 Risk Assessment of Hexabromocyclododecanes to the Marine Mammals Using the Hazard Quotient (HQ) Approach

Species	Location	MEC_{max}	MEC_{min}	HQ_{wcs}	HQ_{bcs}
Harbor seal	Northwest Atlantic	14.5	1	<1	<1
Ringed seal	East Greenland	9.05	0.675	<1	<1
White-sided dolphin	Eastern coast of US	190	9.5	**1.03**	<1
California sea lions	US	48	0.2	<1	<1
Harbor porpoise	U.K.	10,650	5.4	**57.5**	<1
Harbor porpoise	The North sea basin	3400	2.5	**18.4**	<1
Striped dolphins	Japan	470	5	**2.54**	<1
Indo-Pacific humpback dolphin	South China sea	259.5	16	**1.40**	<1

Text in bold means that the HQ exceeds unity.
MEC, minimum measured environmental concentration.

The greatest HQ_{wcs} was found for harbor porpoise from the UK (HQ_{wcs}=57.5). These findings suggest that a more refined risk assessment may be required.

There are possible uncertainties in the risk assessment process. These include the following: (1) the endpoints employed were obtained from different species and thus might overestimate or underestimate the actual risks; (2) different contaminants may interact additively, synergistically, or antagonistically; and (3) the physiology of individual mammals was not taken into account during the risk assessment process.

4. CONCLUSIONS

In general, halogenated PBDE alternatives including HBCDs, DBDPE, DP, TBECH, BTBPE, TBB, and TBPH were found extensively in coastal areas worldwide, indicating their ubiquity in the aquatic environment, and some of these emerging contaminants were even identified in the sediment and marine mammal samples from the Arctic region, suggesting their potential for long-range transport. The results of the present preliminary ecological risk assessment suggests the presence of potential risks due to HBCDs to the marine mammals from the US the UK, the North Sea Basin, Japan, and the South China Sea. Notwithstanding, further studies should be carried out to refine these assessments, and such efforts should focus on the derivation of more relevant threshold values for the assessment when more data becomes available. It is further acknowledged that a number of emerging HFRs such as DP and DBDPE were not included in this assessment. To date, little is known about their ecotoxicity, exposure pathways to organisms, and information on temporal trends in concentrations is also not well known. Recommendations for future work include research on food web studies, and more toxicological information on these PBDE alternatives is urgently needed.

ACKNOWLEDGMENTS

The work described in this chapter was funded by the National Natural Science Foundation of China (China) (41206080) and the Environment and Conservation Fund (Hong Kong SAR) (ECF Project 15/2012). This project was also supported by a Hong Kong Research Grants Council (CityU 160613; CityU 11100614) and Seed Collaborative Research Fund (SCRF) by the State Key Laboratory in Marine Pollution (City University of Hong Kong).

REFERENCES

[1] C.A. de Wit, Chemosphere 46 (2002) 583–624.
[2] A. Covaci, A.C. Gerecke, R.J. Law, S. Voorspoels, M. Kohler, N.V. Heeb, H. Leslie, C.R. Allchin, J. de Boer, Environ. Sci. Technol. 40 (2006) 3679–3688.
[3] A. Covaci, S. Harrad, M.A.E. Abdallah, N. Ali, R.J. Law, D. Herzke, C.A. de Wit, Environ. Int. 37 (2011) 532–566.
[4] R.J. Law, C.R. Allchin, J. de Boer, A. Covaci, D. Herzke, P. Lepom, S. Morris, J. Tronczynski, C.A. de Wit, Chemosphere 64 (2006) 187–208.

[5] P.O. Darnerud, G.S. Eriksen, T. Johannesson, P.B. Larsen, M. Viluksela, Environ. Health Perspect. 109 (Suppl. 1) (2001) 49–68.
[6] M. Petreas, J. She, F.R. Brown, J. Winkler, G. Windham, E. Rogers, G. Zhao, R. Bhatia, M.J. Charles, Environ. Health Perspect. 111 (2003) 1175–1179.
[7] A. Sjödin, R.S. Jones, J.F. Focant, C. Lapeza, R.Y. Wang, E.E. McGahee, Y. Zhang, W.E. Turner, B. Slazyk, L.L. Needham, D.G. Patterson Jr., Environ. Health Perspect. 112 (2004) 654–658.
[8] European Court of Justice, European Joined Cases C-14/06 and C-295/06, Judgement of the Court, April 1, 2008. Directive 2002/95/EC and Commission Decision 2005/717/EC.
[9] P.A. Wäger, M. Schluep, E. Müller, R. Gloor, Environ. Sci. Technol. 46 (2012) 628–635.
[10] G.T. Tomy, W. Budakowski, T. Halldorson, D.M. Whittle, M.J. Keir, C. Marvin, G. MacInnis, M. Alaee, Environ. Sci. Technol. 38 (2004) 2298–2303.
[11] L.S. Birnbaum, D.F. Staskal, Environ. Health Perspect. 112 (2004) 9–17.
[12] L. Meijer, L. Weiss, M. Van Velzen, A. Brouwer, A. Bergman, P.J. Sauer, Environ. Sci. Technol. 42 (2008) 3428–3433.
[13] C. Thomsen, P. Molander, H.L. Daae, K. Jana´k, M. Froshaug, V.H. Liane, S. Thorud, G. Becher, E. Dybing, Environ. Sci. Technol. 41 (2007) 5210–5216.
[14] C.A. de Wit, D. Herzke, K. Vorkamp, Environ. Sci. Technol. 408 (2010) 2885–2918.
[15] H.M. Hermanson, E. Isaksson, S. Forsström, C. Teixeira, D.C. Muir, V.A. Pohjola, R.S. van de Wal, Environ. Sci. Technol. 44 (2010) 7405–7410.
[16] K. Kakimoto, K. Akutsu, Y. Konishi, Y. Tanaka, Chemosphere 71 (2008) 1110–1114.
[17] J.C.W. Lam, R.K.F. Lau, M.B. Murphy, P.K.S. Lam, Environ. Sci. Technol. 43 (2009) 6944–6949.
[18] P. Eriksson, C. Fischer, M. Wallin, E. Jakobsson, A. Fredriksson, Environ. Toxicol. Pharmacol. 21 (2006) 317–322.
[19] L.T.M. Van der Ven, A. van de Kuil, A. Verhoef, R. Fernandez-Canton, S. Germer, H. Lilienthal, D. Schrenk, V.M. an den Berg, A.H. Piersma, P.E.G. Leonards, J.G. Vos, Toxicol. Sci. 94 (2006) 281–292.
[20] H. Lilienthal, L.T. van der Ven, A.H. Piersma, J.G. Vos, Toxicol. Lett. 185 (2009) 63–72.
[21] E.D. Sverko, G.T. Tomy, E.J. Reiner, Y.F. Li, B.E. McCarry, J.A. Arnot, R.J. Law, R.A. Hites, Environ. Sci. Technol. 45 (2011) 5088–5098.
[22] W.L. Han, J.L. Feng, Z.P. Gu, D.H. Chen, M.H. Wu, J.M. Fu, Bull. Environ. Contam. Toxicol. 83 (2009) 783–788.
[23] J.P. Wu, Y.T. Guan, Y. Zhang, X.J. Luo, H. Zhi, S.J. Chen, B.X. Mai, Environ. Sci. Technol. 44 (2010) 5490–5495.
[24] G.F. Zhao, H.D. Zhou, D.H. Wang, J.M. Zha, Y.P. Xu, K.F. Rao, M. Ma, S.B. Huang, Z.J. Wang, Sci. Total Environ. 407 (2009) 2565–2575.
[25] B.Q. Zhu, J.C.W. Lam, S.Y. Yang, P.K.S. Lam, Chemosphere 93 (2013) 555–560.
[26] T. Shi, S.J. Chen, X.J. Luo, X.L. Zhang, C.M. Tang, Y. Luo, Y.J. Ma, J.P. Wu, X.Z. Peng, B.X. Mai, Chemosphere 74 (2009) 910–916.
[27] X.L. Zhang, X.J. Luo, S.J. Chen, J.P. Wu, B.X. Mai, Environ. Pollut. 157 (2009) 1917–1923.
[28] N. Ricklund, A. Kierkegaard, M.S. McLachlan, Environ. Sci. Technol. 44 (2010) 1987–1991.
[29] F.C. Wu, J.Y. Guo, H. Chang, H.Q. Liao, X.L. Zhao, B.X. Mai, B.S. Xing, Environ. Pollut. 162 (2012) 262–268.
[30] R.Q. Yang, H. Wei, J.H. Guo, A. Li, Environ. Sci. Technol. 46 (2012) 3119–3126.
[31] E. Barón, G. Santín, E. Eljarrat, D. Barceló, J. Hazard. Mater. 265 (2014) 288–295.
[32] E. Eljarrat, A. de la Cal, D. Raldua, C. Duran, D. Barceló, Environ. Sci. Technol. 38 (2004) 2603–2608.
[33] M.J. La Guardia, R.C. Hale, B. Newman, Environ. Sci. Technol. 47 (2013) 9643–9650.
[34] T. Harrad, M.A.E. Abdallah, N.L. Rose, S.D. Turner, T.A. Davidson, Environ. Sci. Technol. 43 (2009) 9077–9083.

[35] G. Hu, Z. Xu, J. Dai, H. Cao, J. Wang, Z. Shi, M. Xu, J. Environ. Sci. 22 (2010) 1833–1839.
[36] K. Law, T. Halldorson, R. Danell, G. Stern, S. Gewurtz, M. Alaee, C. Marvin, M. Whittle, G. Tomy, Environ. Toxicol. Chem. 25 (2006) 2177–2186.
[37] S. Managaki, I. Enomoto, S. Masunaga, J. Environ. Pollut. 14 (2012) 901–907.
[38] A. Kierkegaard, J. Björklund, U. Fridén, Environ. Sci. Technol. 38 (2004) 3247–3253.
[39] H.J. Klamer, P.E. Leonards, M.H. Lamoree, L.A. Villerius, J.E. Kerman, J.F. Bakker, Chemosphere 58 (2005) 1579–1587.
[40] K. Ramu, T. Isobe, S. Takahashi, E.Y. Kin, B.Y. Min, S.U. We, S. Tanabe, Chemosphere 79 (2010) 713–719.
[41] X. Zhang, D.D. Zhang, Z.X. Luo, L.F. Lin, C.Z. Yan, Environ. Chem. 8 (2010) 561–568.
[42] M.J. La Guardia, R.C. Hale, E. Harvey, T.M. Mainor, S. Ciparis, Environ. Sci. Technol. 46 (2012) 5798–5805.
[43] H.M. Stapleton, J.G. Allen, S.M. Kelly, A. Konstantinov, S. Klosterhaus, D. Watkins, M.D. McClean, T.F. Webster, Environ. Sci. Technol. 42 (2008) 6910–6916.
[44] M.J. La Guardia, R.C. Hale, E. Harvey, D. Chen, Environ. Sci. Technol. 44 (2010) 4658–4664.
[45] B. Zhu, N.L.S. Lai, T.C. Wai, L.L. Chan, J.C.W. Lam, P.K.S. Lam, Environ. Int. 66 (2014) 65–70.
[46] E.D. Sverko, G.T. Tomy, C.H. Marvin, D. Zaruk, E. Reiner, P.A. Helm, B. Hill, B.E. McCarry, Environ. Sci. Technol. 42 (2008) 361–366.
[47] X.H. Qiu, C.H. Marvin, R.A. Hites, Environ. Sci. Technol. 41 (2007) 6014–6019.
[48] G.T. Tomy, K. Pleskach, G. Arsenault, D. Potter, R. McCrindle, C.H. Marvin, E.D. Sverko, S. Tittlemier, Environ. Sci. Technol. 42 (2008) 543–549.
[49] D.C.G. Muir, P.H. Howard, Environ. Sci. Technol. 40 (2006) 7157–7166.
[50] S. Tanabe, Mar. Pollut. Bull. 45 (2002) 69–77.
[51] P.A. Fair, J. Adams, G. Mitchum, T.C. Hulsey, J.S. Reif, M. Houde, D. Muir, E. Wirth, D. Wetzel, E. Zolman, W. McFee, G.D. Bossart, Sci. Total. Environ. 408 (2010) 1577–1597.
[52] M.C. Fossi, L. Marsili, N. Giovanni, A. Natoli, E. Politi, S. Panigada, Mar. Pollut. Bull. 46 (2003) 972–982.
[53] S.E. Moore, J. Mammal. 89 (2008) 534–540.
[54] R.S. Wells, H.L. Rhinehart, L.J. Hansen, J.C. Sweeney, F.I. Townsend, R. Stone, D.R. Casper, M.D. Scott, A.A. Hohn, T.K. Rowles, EcoHealth 1 (2004) 246–254.
[55] T. Isobe, Y. Ochi, K. Ramu, T. Yamamoto, Y. Tajima, T.K. Yamada, M. Amano, N. Miyazaki, S. Takahashi, S. Tanabe, Mar. Pollut. Bull. 58 (2009) 396–401.
[56] S.D. Shaw, M.L. Berger, L. Weijs, A. Covaci, Environ. Int. 44 (2012) 1–6.
[57] G.D. Villanger, C. Lydersen, K.M. Kovacs, E. Lie, J.U. Skaare, B.M. Jenssen, Sci. Tot. Environ. 409 (2011) 2511–2524.
[58] K. Vorkamp, F.F. Rigét, R. Bossi, R. Dietz, Environ. Sci. Technol. 45 (2011) 1243–1249.
[59] R.J. Law, P. Bersuder, C.R. Allchin, J. Barry, Environ. Sci. Technol. 40 (2006) 2177–2183.
[60] R.J. Law, S. Losada, J.L. Barber, P. Bersuder, R. Deaville, A. Brownlow, R. Penrose, P.D. Jepson, Environ. Int. 60 (2013) 81–88.
[61] S. Morris, C.R. Allchin, B.N. Zegers, J.J.H. Hafrka, J.P. Boon, C. Belpaire, P.E.G. Leonards, S.P.J. van Leeuwen, J. de Boer, Environ. Sci. Technol. 38 (2004) 5497–5504.
[62] A. De la Torre, M.B. Alonso, M.A. Martínez, P. Sanz, L. Shen, E.J. Reiner, J. Lailson-Brito, L. Barbosa, M. Cremer, E. Secchi, O. Malm, E. Eljarrat, D. Barceló, Environ. Sci. Technol. 46 (2012) 12364–12372.
[63] D. Muir, M. Kwan, M. Evans, in: S. Smith, J. Stoe (Eds.), Synopsis of Research Conducted under the 2006–2007 Northern Contaminants Program, Indian Affairs and Northern Development, Ottawa, ON, Canada, 2007, pp. 162–169.

[64] G.T. Tomy, N. Ismail, K. Pleskach, R. Danell, G. Stern, Extended Abstract, 4th International Workshop on Brominated Flame Retardants (BFR 2007), 24–27 April, Amsterdam, the Netherlands, 2007.

[65] S.L. Klosterhaus, H.M. Stapleton, M.J. La Guardia, D. Greig, Environ. Int. 47 (2012) 56–65.

[66] J.P. Wu, Y. Zhang, X.J. Luo, J. Wang, S.J. Chen, Y.T. Guan, B.X. Mai, Environ. Sci. Technol. 44 (2010) 606–611.

[67] M.J. He, X.J. Luo, M.Y. Chen, Y.X. Sun, S.J. Chen, B.X. Mai, Sci. Tot. Environ. 419 (2012) 109–115.

[68] E. Hoh, L. Zhu, R.A. Hites, Environ. Sci. Technol. 39 (2005) 2472–2477.

[69] J. Verreault, W.A. Gebbink, L.T. Gauthier, G.W. Garbrielsen, R.J. Letcher, Environ. Sci. Technol. 41 (2007) 4925–4931.

[70] J.M. Gauthier, C.D. Metcalfe, R. Sears, Mar. Environ. Res. 43 (1997) 157–179.

[71] M.C. Fossi, L. Marsili, G. Neri, S. Casini, G. Bearzi, E. Politi, M. Zanardelli, S. Panigada, Mar. Environ. Res. 50 (2000) 517–521.

[72] J.M. Gauthier, C.D. Metcalfe, R. Sears, Mar. Environ. Res. 44 (1997) 201–223.

[73] K.E. Hobbs, D.C.G. Muir, R. Michaud, P. Béland, R.J. Letcher, R.J. Norstrom, Environ. Pollut. 122 (2003) 291–302.

[74] C.C.M. Leung, T.A. Jefferson, S.K. Hung, G.J. Zheng, L.W.Y. Yeung, B.J. Richardson, P.K.S. Lam, Mar. Pollut. Bull. 50 (2005) 1713–1744.

[75] M. Reddy, S. Echols, B. Finklea, D. Busbee, J. Reif, S. Ridgway, Mar. Pollut. Bull. 36 (1998) 892–903.

[76] G.M. Ylitalo, C.O. Matkin, J. Buzitis, M.M. Krahn, L.L. Jones, T. Rowles, J.E. Stein, Sci. Tot. Environ. 281 (2001) 183–203.

[77] J.C.W. Lam, M.B. Murphy, Y. Wang, S. Tanabe, J.P. Giesy, P.K.S. Lam, Environ. Sci. Technol. 42 (2008) 6296–6302.

[78] C.N. Fung, J.C.W. Lam, G.J. Zheng, D.W. Connell, I. Monirith, S. Tanabe, B.J. Richardson, P.K.S. Lam, Environ. Pollut. 127 (2004) 206–216.

[79] C.L.H. Hung, R.C.F. Lau, J.C.W. Lam, T.A. Jefferson, S.K. Hung, M.H.W. Lam, P.K.S. Lam, Chemosphere 66 (2007) 1175–1182.

[80] M. Ema, S. Fujii, M. Hirata-Koizumi, M. Matsumoto, Reprod. Toxicol. 25 (2008) 335–351.

[81] Y. Saegusa, H. Fujimoto, G.H. Woo, K. Inoue, M. Takahashi, K. Mitsumori, M. Hirose, A. Nishikawa, M. Shibutani, Reprod. Toxicol. 28 (2009) 456–467.

[82] D.T. Szabo, J.J. Diliberto, H. Hakk, J.K. Huwe, L.S. Birnbaum, Toxicol. Sci. 121 (2011) 234–244.

[83] B.E. Sample, D.M. Opresko, G.W. Suter, Toxicological Benchmarks for Wildlife: 1996 Review, Report No. ES/ER/TM-86/RE. Prepared by the Risk Assessment Program, Health Science Research Division for the US Department of Energy, Office of Environmental Management under budget and reporting code EW-20, Oakridge National Laboratory, Oakridge, TN, 1996.

Chapter 13

Atmospheric Long-Range Transport of Persistent Organic Pollutants (POPs) into Polar Regions

Roland Kallenborn[1],* Hayley Hung[2] and Eva Brorström-Lundén[3]

[1]Institute for Chemistry Biotechnology and Food Sciences (IKBM), Faculty of Veterinary Medicine and Biosciences (VetBio), Norwegian University of Biosciences, Ås, Norway; [2]Air Quality Research Division, Science and Technology Branch, Environment Canada, Toronto, Ontario, Canada; [3]IVL-Swedish Environmental Research Institute, Aschebergsgatan, Göteborg, Sweden

**Corresponding author: E-mail: roland.kallenborn@nmbu.no*

Chapter Outline

Comprehensive Analytical Chemistry, Vol. 67. http://dx.doi.org/10.1016/B978-0-444-63299-9.00013-2

1. BACKGROUND

Persistent organic pollutants (POPs) have been a topic of interest in environmental sciences for more than 50 years to date. In 1964, Rachel Carson already pinpointed the considerable adverse effects of this pollutant group in a comprehensive way [1]. POPs in the Arctic atmosphere have been investigated since the 1970s when the first atmospheric measurements revealed the presence of POPs in this pristine polar environment [2,3]. Comprehensive research under the auspice of the Arctic Monitoring and Assessment Programme (AMAP) and culminating in a comprehensive review [4] showed that winds and ocean currents provide effective means to transport contaminants to the Arctic. Within the Arctic, a number of concentrating processes lead to high exposure of classical POPs especially in top predators of the Arctic's marine food webs [5]. These exposures have also been shown to have consequences for the health and well being of the indigenous people of the North [6]. Therefore, the atmospheric POP distribution and fate is continuously surveyed in a comprehensive circum-Arctic monitoring program of AMAP since the early 1990s [7]. The AMAP atmospheric monitoring network is illustrated in Figure 1.

1.1 Atmospheric Monitoring

Today's levels of anthropogenic contaminants found in the polar environments cannot be explained by known use and/or release from sources within the respective region. Based on continuous monitoring and surveillance, atmospheric long-range transport from lower latitudes is known to be one of the most important sources for the presence of many persistent organic and inorganic pollutants in the Arctic environment today [8]. During the past decades, international long-term atmospheric monitoring programs have proven to be versatile tools for the demanding task of identification and evaluation of hazardous chemicals in the environment. Modern tools for atmospheric monitoring especially in background regions like the polar regions are designed to provide scientific information to guide and assess international regulatory measures, determine baseline levels as well as the environmental distribution and spatial and temporal trends of pollutants, and to serve as an "early warning tool" for emerging chemicals of concern. Even minimal changes in patterns and distributions may be a first signal for potential severe environmental changes [4,7–13].

This concept of long-term atmospheric monitoring has proven highly effective and useful. Many examples illustrate the importance of continuous atmospheric monitoring of anthropogenic contamination in the cold Polar Regions. The principles and pathways of global atmospheric long-range transport of persistent semivolatile contaminants across several climate zones into the Polar Regions have been revealed and confirmed by long-term monitoring of POPs from the temperate zones of the globe into the polar atmosphere [14,15].

In order to illustrate the significant contribution of Arctic/Polar long-term monitoring programs to the elucidation of hemispherical and even global POP

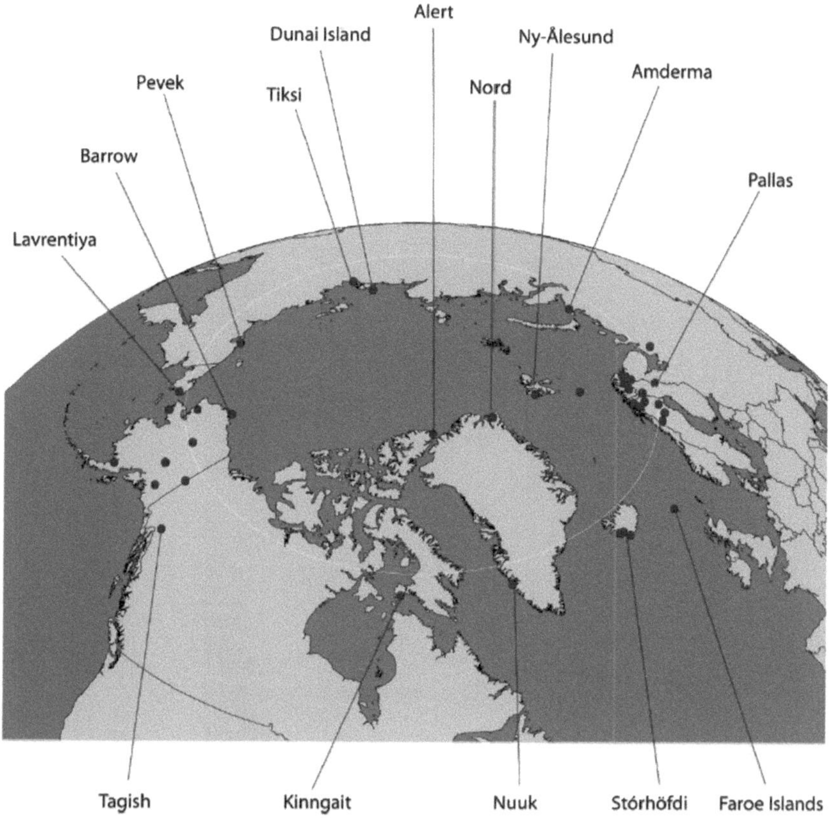

FIGURE 1 The stations of the Arctic atmospheric monitoring network reporting to the Arctic Monitoring and Assessment Programme (AMAP). *The figure is reproduced with courtesy of the journal* Atmospheric Pollution Research *[20].*

distribution process, polychlorinated biphenyls (PCBs) may serve as an illustrative example. PCBs have been used and applied as industrial chemicals, as oils in transformers and in capacitors. About 1,000,000 tons have been produced and applied since the 1940s [16]. In the 1960s, there was evidence of the presence of PCBs in the biological environment [17]. The highly toxic effects of PCBs in the Arctic environment lead consequently to the international ban of these chemicals as industrial products in the 1980s [6,18]. Modern monitoring programs are tracking this type of atmospheric contaminants and contribute to the evaluation of suitable regulatory measures and have shown that even more than a decade of continuous monitoring reveals no clear decreasing tendency for these persistent contaminants in the Arctic and Antarctic air [8]. The atmospheric distribution patterns for PCBs (sum 10AMAP PCB = PCB congeners 28, 52, 99, 101, 118, 138, 153, 170, 180, 194) are compared in Figure 2.

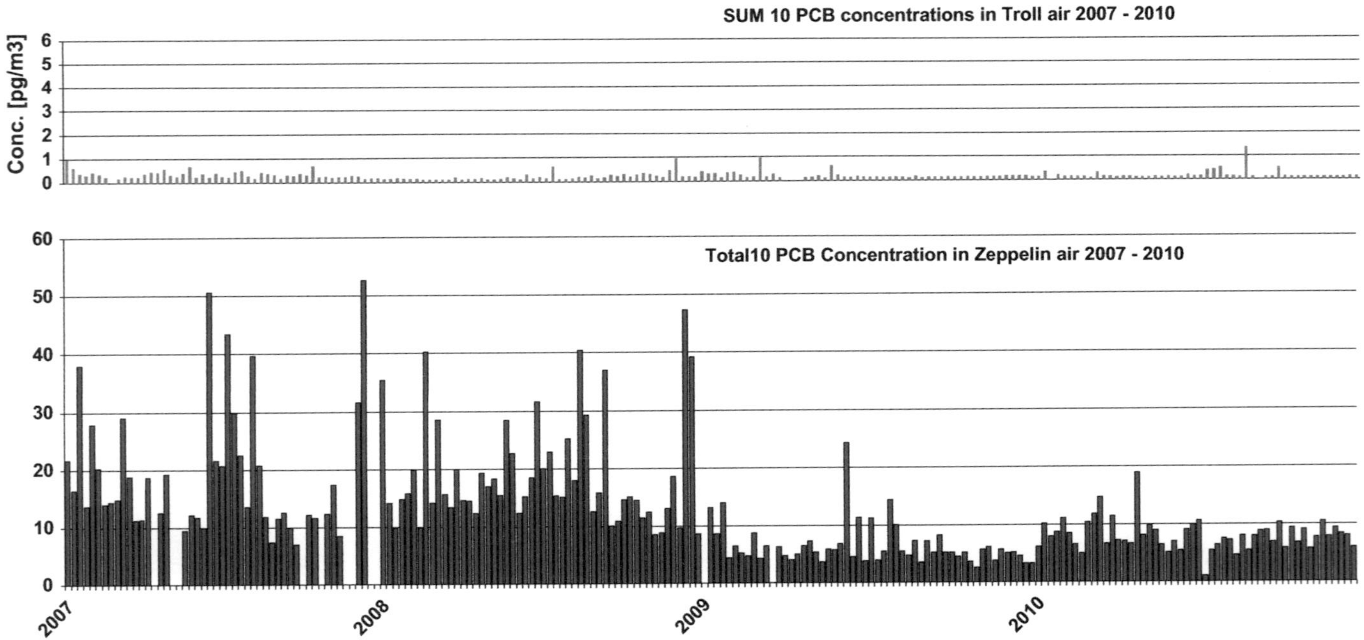

FIGURE 2 Sum polychlorinated biphenyls (10 AMAP PCB) monitored in the period 2007–2010. Comparison between atmospheric polychlorinated biphenyl (PCB) levels from the Arctic research station "Zeppelin mountain" (Svalbard, Norway) and the Norwegian Antarctic Troll station (Queen Maud Land). For comparison, the Troll data are presented a considerable lower concentration range compared to the Zeppelin PCB concentrations (green (light gray in print versions) colored area). Please note the different dimension on the Y-Axis, the values for the Zeppelin station are 10 times higher than reported for the Troll atmospheric monitoring.

2. IMPORTANCE OF POP MONITORING IN POLAR REGIONS

The described concept for atmospheric pollutant monitoring has been proven to be highly effective and useful for control and regulation of anthropogenic organic pollutants distributed throughout the globe via transport in the atmosphere. Many examples illustrate the importance and the relevance of continuous atmospheric monitoring for research and regulatory purposes of anthropogenic contamination in the cold Polar Regions ([7,8,19,20]). Only through this type of long-term international atmospheric POPs monitoring in the North, the global distribution patterns of long-range transported POPs could be revealed and appropriate policies and regulatory measures could be put in place. Thus, the principles and pathways of global atmospheric long-range transport of persistent semivolatile contaminants across several climate zones into the Polar Regions have been identified and investigated comprehensively using long-term atmospheric monitoring data for POPs. Thus, the mere confirmation of existence for a POP-like substance in Arctic monitoring data is considered by the United Nations Environment Programme (UNEP) Stockholm Convention on POPs as a crucial criterion during evaluation processes for possible international regulation. As a direct result of the above-mentioned international efforts on atmospheric contaminant monitoring in Polar Regions (especially in the Arctic), large monitoring programs such as the Canadian Northern Contaminants Program (NCP), the Arctic Monitoring and Assessment Programme (AMAP), the European Monitoring and Evaluation Program (EMEP), and many others have been established covering exclusively or including Polar environments as target regions.

Several recently published reviews and research papers on atmospheric POP monitoring in the Arctic reports the status of POPs, their distribution pathways, and levels on a regional basis [7,8,21–24]. However, these papers usually refer to regional aspects and specific monitoring programs. Our chapter will encompass monitoring and screening activities in Polar region (Arctic and Antarctic). In addition, principle processes (distribution, deposition, storage, transformation) will be covered in the following sections.

3. THE PRINCIPLES OF ATMOSPHERIC POP TRANSPORT TO AND WITHIN POLAR REGIONS

Atmospheric long-range transport of pollutants is not unique for POPs. Forty years ago sulfur dioxide (SO_2), nitrite, and related compounds (NO_x) originated from the industrial zones of central and Eastern Europe made their way into Scandinavia via the atmosphere. This phenomenon resulted in "acid rain" that caused acidification of a large number of Scandinavian lakes and ponds. Another well-known example for atmospheric transport of pollutants is the vertical transport of chlorofluorocarbons into the polar atmosphere, which caused the depletion of the ozone layer. Due to favorable weather conditions over the Antarctic continent, compared to the Artic, depletion of the ozone layer mainly occurred over Antarctica. A much

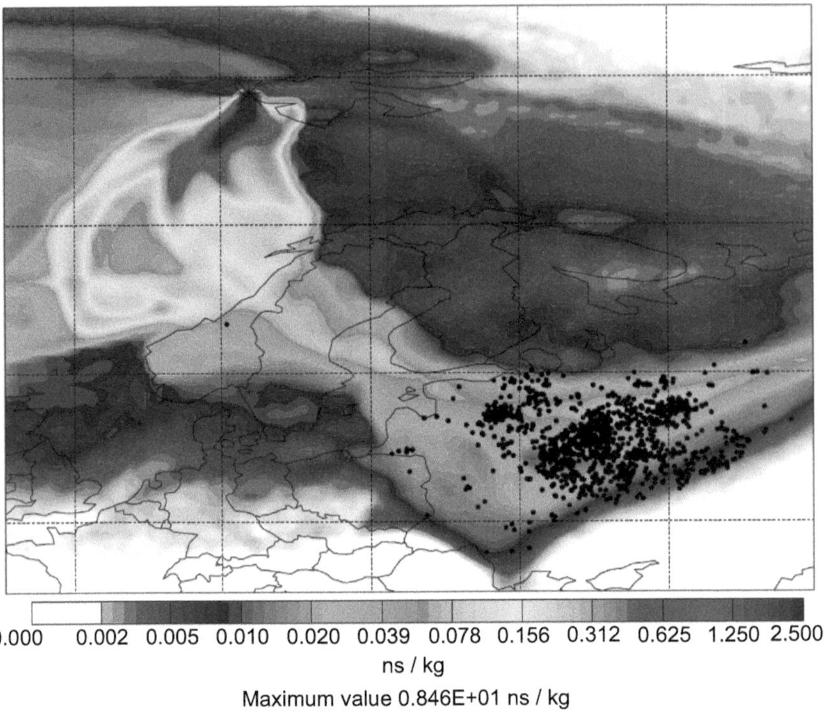

FIGURE 3 Potential emission sensitivity (PES) footprint map (0–100 m) for air masses arriving at Zeppelin station in the period 01.05.2006–03.05.2006. The black marks illustrate satellite confirmed fire detections (MODIS) for the same time period identified as potential polychlorinated biphenyl sources [25]. *The figure is reproduced with courtesy of the journal* Atmospheric Chemistry and Physics.

earlier indication for atmospheric long-range transport is dated back to the 1950s when airplane pilots reported smoglike phenomena over the surface of the Arctic region during their flights across the North Pole. This Arctic smog is still to be seen during intercontinental flights across the Arctic. The smog phenomenon is mainly caused by sulfate-containing dust particles originated from industrial regions in Europe, USA, and Asia, transported via the atmosphere for several 1000 kilometers and deposited in the Arctic. The phenomenon is also known as "Arctic haze." In a recent study, the occurrence of Arctic haze has been associated with elevated PCB levels in the Arctic atmosphere [25] as demonstrated in Figure 3.

One of the most well-known hypotheses in global atmospheric transport of POPs is the process of global distillation, forcing contaminants out of warm reservoirs in temperate industrial areas and into cold reservoirs at the poles, resulting in a complex POP distribution cycle. Together with marine ocean currents and rivers, the atmosphere is considered as one of the most important transport pathways for POPs into the pristine Polar Regions. Under ambient temperature conditions, many POP-like compounds have partitioning properties that would allow them to move from one environmental medium to another [26]. That means, small temperature changes will determine whether the substance will be associated with the gas phase or the particulate/colloidal phase during atmospheric long-range transport (LRAT) events [27]. Therefore, these chemicals tend also to vaporize at elevated ambient temperature and condense at lower air temperatures at higher latitudes. This specific feature has been named earlier as global distillation [28]. Compounds that volatilize faster are expected to have larger atmospheric mobility than those chemicals requiring higher temperatures to volatilize in order to stay in the gas phase. This phenomenon driven by the chemical's physicochemical properties resulted in organic compounds being fractionated latitudinally, with condensation occurring at different temperatures towards the poles. This process is known as global fractionation and cold condensation [27]. The principle is compared with a standard distillation process and is used to explain different POP composition in the atmosphere in the source regions compared to the Polar recipients. The term "the grasshopper effect" again is describing the selective process of POP migration in jumps (repeated volatilization and deposition dependent on ambient conditions), which in turn is highly dependent on the inherent physicochemical properties (i.e., vapor pressure, air-water partition coefficients, etc.) of the respective chemicals [27].

POPs may be transported in the atmosphere by association with particles [29–31]. While transport events of large amounts of fine sandy particles from the Sahara desert up to middle Europe visually demonstrate clear potential for atmospheric transport of pollutants over vast distances [32–36], these events are usually not directly associated with POP transport. For the European Arctic, emissions from anthropogenic sources in Eurasia have been identified as main contributors to air pollution, at altitudes up to 2 km. It is assumed that POP-associated atmospheric transport is taking place at altitudes around 1–2 km in the troposphere. The atmospheric long-range transport of polycyclic aromatic hydrocarbons (PAH), produced and released into the environment by various combustion processes, is known to be closely associated with the particulate phase in the moving air masses [29,37–40]. Also for the majority of POPs including PCBs and selected chlorinated pesticides, an association of the atmospheric levels and particle content is assumed. Gas-particle partitioning of polychlorinated pollutants (e.g., chlorinated benzenes, pentachloroanisole, hexachlorocyclohexanes, PCB, dichlorodiphenyltrichloroethane-derivatives like *p,p'*-DDE, *p,p'*-DDD and *p,p'*-DDT) were investigated in various studies [41–47]. Relatively high contribution of the pollutant content in the particulate phase compared to the gaseous phase was found in real atmospheric samples [7,48,49].

All substances that are known to be transported over large distances into the pristine Polar environments express high stability in the environment and therefore are considered as persistent. In addition to the property of the respective compound, the special geographic and meteorological conditions must be taken into account when it comes to determining the dominant long-range transport pathway.

For less water-soluble POPs like α-, γ-hexachlorocyclohexane (HCH), PCBs, as well as DDT and metabolites (DDE, DDD), the atmosphere is the preferred transport pathway into the remote Arctic. During favorable weather conditions, air masses can transport contaminants into the Arctic within a few days or weeks. The above-mentioned compounds may also sorb on fine particles. The particles are then transported into the northern regions. Volatile organic compounds like the lighter PCBs, chlorinated benzenes, etc., are directly transported in the gaseous phase.

The "global fractionation" theory describes long-range transport as a complex phenomenon depending on various meteorological, climate, geographical, and physicochemical parameters. As already mentioned earlier, atmospheric circulation as well as ocean currents and river-related transport are single elements in the complicated process describing the long-range transport of certain persistent pollutants into the remote and pristine Polar Regions.

The role of characteristic meteorological conditions in the respective climate zones must not be underestimated. Strong seasonal differences in temperature and precipitation regimes rule the global climate. The barrier between the cold northern air masses and the temperate southern air is called the polar front (for both the Arctic and Antarctic atmosphere). The polar front plays a central role in explaining seasonal dependencies in atmospheric long-range transport processes. During winter, the polar front in the Arctic is expanding into the South covering Central Europe, the North Americas, Central Asia, and Eastern Europe. In these regions, the majority of the source areas for persistent pollutants transported into the North are located in these regions. Therefore, during the winter season, the Arctic polar front is not hampering the atmospheric transport from the sources into the Arctic region. In addition, often southerly winds are dominating during winter at the Northern Hemisphere. During summer, on the other hand, the Arctic polar front is situated far North and creates a meteorological barrier often difficult to penetrate for air masses transported from southern regions. Thus, the polar front represents an obvious transport limitation for airborne POPs into the Arctic. An additional factor supporting atmospheric long-range transport during winter is the fact that significantly lower numbers of individual deposition events are reported for European, American, and Asian source regions during wintertime. Thus, the probability that POPs are washed out from the atmosphere during precipitation is lower in winter than in summer. As a general conclusion, it can be stated that the likelihood of a long-range transport episode arriving in the receiving Arctic environment during winter and early spring is higher than during the summer.

As described in the "global fractionation" theory, depending on the physicochemical properties of the contaminant, a water-soluble compound can be transported particle bound or in colloidal form in the atmosphere before deposited and further transported with ocean currents or finally deposited in the marine sediments or in soil. Already minor differences in volatility and solubility of the contaminants can make a significant difference for the preferred transport pathway and thus, whether the substance is transported dissolved in the water column or particle bound in the atmosphere.

4. DETERMINATION OF SOURCE REGIONS BY BACK TRAJECTORY MODELING

A rapid atmospheric transport event leading to an effective transfer of contaminated air masses into the receiving polar atmosphere is called a LRAT event. Such LRAT events are usually characterized with specific elevated POP levels and are often subject for interdisciplinary studies involving chemical, meteorological, as well as modeling expertise. The congener-/isomer-specific distribution in air masses are reported in concentration data from atmospheric monitoring locations, and characteristic air mass back trajectory calculations are performed in order to identify potential source regions. Air mass back trajectories are theoretical calculations mainly based on empirical data about wind velocity, wind direction, temperature etc., which back calculate the way the air parcel was traveling in the past 5–10 days period prior to sampling.

However, one must also bear in mind that not all POPs present in the Arctic originate from outside the region. There are also potential local contamination sources in Polar Regions. For instance, PCBs may have originated from Distant Early Warning line sites [50,51]. In addition, nowadays, many POPs are associated with consumer products, e.g., flame retardants polybrominated diphenyl ethers (PBDEs) and water- and oil-spill repellents per- and polyfluoroalkyl substances (PFASs). The sources of these substances can be local communities and dump sites. A thorough investigation of contaminant sources should account for both potential local sources and LRAT [52].

5. POPs IN THE ARCTIC ATMOSPHERE

Major pollutant transport routes to the Arctic include atmospheric and oceanic transport, as well as inflow from rivers and transport via sea ice. Local sources as mentioned above as well as heating of buildings and other local combustion processes (power generators, industry, vehicle and ship exhausts etc.) contribute significantly to the input of certain pollutants like PAH in the High Arctic. The transport efficiency differs among transport pathways, while atmospheric transport can occur in days or weeks, from the source regions into the Arctic. Transport via sea currents, however, can take several years.

For the past 30 years, the distribution and transport patterns of POPs to the Arctic regions have been monitored and investigated in detail. The confirmation of the atmospheric long-range transport potential of a chemical is considered as an important criterion for classifying a substance as a POP and is subjected to international control and priority monitoring in international conventions [53]. Persistent organochlorine pesticides (e.g., DDT, HCH), previously highly important as crop protection agents worldwide, are transported through the atmosphere over several climate zones into the Arctic regions with significant implications for the health and well being of the indigenous people of the North [6,54].

The Arctic is no longer considered as a highly pristine environment, although the air, water, soil, and sediment concentrations exhibit considerably lower levels of contaminants compared to those found in near-source regions [8]. However, certain characteristic features of the Arctic (low temperatures, ice coverage, snow precipitation, extended periods of darkness during the Arctic winter) mean that it has the potential to accumulate certain globally transported contaminants including POPs [7].

5.1 The Arctic as Deposition Area for POPs

In the Northern and Southern Hemispheres, the air masses in the temperate zones try to reach a temperature balance with the cold polar air masses. Due to low average annual temperature and the special seasonal daylight conditions in Polar Regions, suitable conditions for the deposition of persistent pollutants are present. The following parameters demonstrate why the Arctic is especially exposed to anthropogenic POPs [8].

5.1.1 Daylight Conditions

The polar night at 80° northern latitude spans from October until March and changes within a few weeks into the midnight sun period. This unique phenomenon has a tremendous effect on the photochemical degradation of some persistent pollutants. Due to the lack of photochemical degradation processes, long-range transport of certain volatile organic compounds is only observed during the winter. Chemicals like short-chain alkanes and alkenes (ethane, propane, ethene, etc.) are usually quickly degraded by photochemical processes during the summer time [55,56]. Therefore, only low concentration levels are usually observed in Arctic air during the warm period. In winter, on the other hand, concentrations of this type of compounds are usually elevated by one or two order of magnitudes [7]. Indications for a seasonal dependent photochemical degradation were also found for the persistent pesticide trans-chlordane. A specific annual distribution profile in the chlordane patterns was found during a year round ambient air sampling campaign [10] at Zeppelin Mountain (Ny-Ålesund, Svalbard).

5.1.2 Temperature Conditions

The low annual average temperature of the Polar Regions is the main reason why microbiological degradation processes of organic compounds slow down to a minimum. Therefore, microbiological metabolism of persistent pollutants in the Arctic is considered as minimal compared with middle latitude environments [57]. This extends the lifetime of these compounds exponentially in the Arctic. Compounds, which are known to degrade in temperate regions within weeks or months, express much higher persistence in the Polar Regions, with potential half-lives in the scale of years [14].

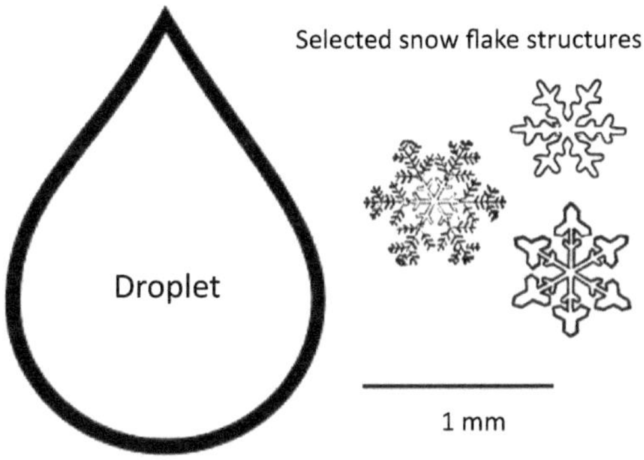

FIGURE 4 Comparison of structural shapes for a droplet and snow flakes illustrating the large surface adsorption potential of snow flakes as deposition vehicles at low ambient temperatures.

5.1.3 Deposition

Due to the low annual average temperature, snow is the predominating form of precipitation in Polar Regions most of the year. The snow crystal possesses a large surface and adsorbs/absorbs particles and contaminants to its crystalline surface more effectively than a water droplet (Figure 4). Therefore, snow precipitation can transport contaminants better to the surface (soil, water surface) than rain [58]. In contrast to rain, the surface properties of snow crystals change during aging on the surface. The surface of a snow crystal diminishes dramatically after a short time on the ground due to the overall pressure of the surrounding snow as well as caused by weather and climate conditions during the aging process. Thus, due to surface reduction the adsorbed contaminant is either released into the soil/ground, stays absorbed/adsorbed to the snow crystal, or revolatilize into the atmosphere [59].

5.2 PCBs as an Indicator POP in the Arctic Atmosphere

PCB is a mixture of 209 compounds (Figure 5). The different PCB congeners contain the same biphenyl structure, but differ in number and substitution pattern of chlorines. PCB mixtures were used as flame retardant in paints, as additive for insulation purposes, and as dielectric fluid in capacitors and transformers. In general, between 50 and 100 congeners are released into the environment usually during destruction and decommissioning of electrical equipment and buildings/ships. More than 1,000,000 tons of various PCB mixtures are reported produced on a global basis. PCBs were used extensively during the 1950s, 1960s, and 1970s, mostly in industrial applications and as additives and sealants in building materials. After the 1970s, the production and use of PCBs were restricted in many countries, and global emissions started to decrease. However, the decline in emissions lags strongly behind the rate of phase-out since in-use products and materials continue to release PCBs to the atmosphere throughout

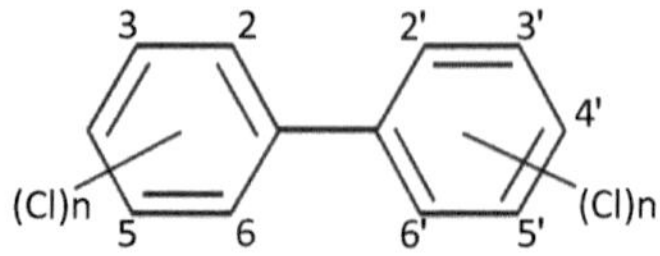

FIGURE 5 The basic structure of polychlorinated biphenyls consists of a biphenyl structure with different chlorination patterns (209 different congeners).

their lifetime and are continuously emitted from sources such as waste dumps, industrial areas, and decommissioned military and utility sites. PCBs have been monitored in the Arctic for several decades already. Such monitoring includes all typical environmental compartments such as atmosphere, biota, water (freshwater and seawater), sediment, and soil. Moreover, the PCB congeners cover a very wide range of physicochemical properties, which dictate their transport pathways and environmental fate. The PCBs have sufficiently reliable physicochemical properties that can be reliably modeled [60,61]. Therefore, an abundance of information on trends, fate and distribution pathway is available in the scientific literature. Therefore, PCBs are considered today as Arctic indicator contaminants for trend and risk assessments.

Furthermore, secondary emissions of PCBs that have been accumulated in environmental reservoirs (sediment, water, soil, snow, and ice) can be remobilized and thus contribute further to slow down the rate of decline of PCBs in the global atmosphere. In a global, long-term perspective, re-emissions from secondary sources are becoming more important than primary sources [62].

A global emission inventory of 22 different PCB congeners was developed by [16] for the period from 1930 to 2100. Primary emissions were obtained by estimates of past and future production of PCB containing goods, combined with assumptions on their life cycle until final disposal [16].

The average PCBs concentrations (for 10AMAP PCBs) at the four major Arctic atmospheric monitoring sites Alert (Canada), Pallas (Finland), Storhofdi (Iceland), and Zeppelin (Svalbard/Norway) are compared in a recent overview. The congener profile (concentration and patterns) may vary due to their geographical location (latitude and longitude) but also on factors such as locality (marine influence or continental sites) and the predominant meteorology for the respective stations [7]. During 2009, the concentrations of PCBs at the Arctic stations, Zeppelin (east) and Alert (west) were at the same level in contrast to the period 1993–2006 when PCB levels were higher at Zeppelin in comparison to Alert.

5.2.1 Time Trends

The atmospheric PCB concentrations in the Arctic have shown a continuous decreasing trend over the past decades, after the international regulation

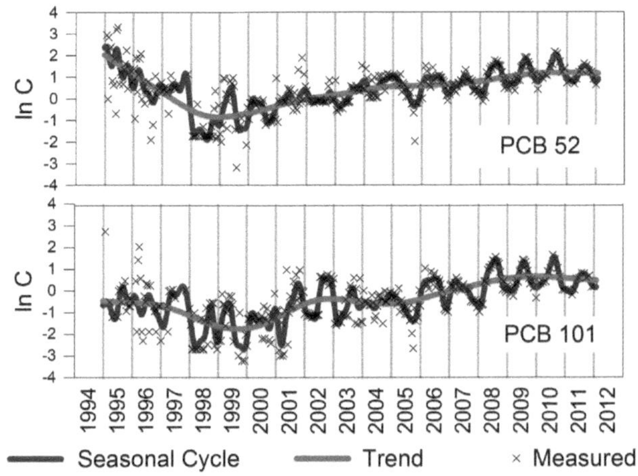

FIGURE 6 Trends for PCB 52 and 101 measured in air at Storhofdi, Iceland in the period 1994–2012.

of PCB usage was enforced as a part of the UNEP Stockholm Convention. Hung et al. [7] showed a general decline in the concentrations at Pallas, Alert, and Zeppelin during the period 1998–2011. Declining PCBs concentrations have also been identified at Pallas for the period 1996–2008 (about 3% per year for Σ_7PCB), which was similar to decline at the Swedish west coast. The concentrations of both PCB28 (tri-CB) and PCB153 (hexa-CB) were generally higher at Zeppelin in comparison to Pallas and Alert for several of the measurement years while the concentrations were similar during recent years. For the Zeppelin station, increasing trends for medium-chlorinated (penta- to hexa-chlorinated) CBs were reported in the period 2005/2006 (decreasing PCB levels after 2009). The increasing trends during this period were associated with the occurrence of boreal forest fires from Siberia and the North American Arctic and the direct transport of contaminated air into the Svalbard region [25]. Like Zeppelin the PCB concentrations at Pallas increased during 2006 but with a decreasing trend after that. Increasing trends were observed for PCB 52 and 101 at Storhofdi (Figure 6) especially after 2000, which may be related to PCB revolatilization due to sea ice retreat around Iceland and deglaciation of ice caps in proximity to the station (AMAP, 2014).

6. POPs IN THE ANTARCTIC ATMOSPHERE

Compared to studies on POP distribution in the Arctic atmosphere, investigations on atmospheric POPs in Antarctica have been surprisingly few due to reasons like remoteness and cost-intensive logistics related to sampling campaigns and logistical requirements. However, during the past two decades, several campaign-based studies on short-term (weeks–month) atmospheric monitoring of POPs in Antarctica have been reported in the literature [63–65]. The above-mentioned studies indicated that potential primary sources (agriculture, industrial releases, etc.) in the southern hemisphere contribute to the POP levels in the Antarctic atmosphere. Nevertheless, due to the remoteness of Antarctica and considerably lower number of significant industrial source regions in the southern hemisphere, levels of POPs have been previously considered as extremely low–much lower than in the Arctic.

During a long-term monitoring exercise at the Norwegian Troll station, a 4-year long data set for standard POPs (similar as measured for the Arctic Zeppelin monitoring station, Ny-Ålesund, Svalbard) was elucidated [23]. The most abundant POP quantified in the Troll atmosphere was α-HCH with concentrations varying between 0.02 and 0.46 pg/m^3. The α-HCH isomer was identified in all 196 weekly samples analyzed during the 4-year monitoring campaign. Similar to the Arctic station of Zeppelin [8,12,20], α-HCH had considerably higher concentrations in all Troll air samples than the related γ-HCH isomer (α-/γ-HCH ratio: 2–24). In an earlier study [65], a slight predominance of γ-HCH was identified in atmospheric samples from Signy Island (Antarctic Peninsula). However, in the recent data set presented by [23], α-HCH was the most abundant HCH isomer. The most obvious reason for this pattern shift along with significant overall concentration reductions of about a factor 100 compared to the earlier campaigns is obviously the national ban of lindane (>99% γ-HCH) in Asian countries (i.e., China and India) around the year 2000. Even after a considerable background level reduction for HCHs in Antarctic air is observed, atmospheric LRAT events are reported for the Norwegian Troll station by [23]. Therefore, potential remaining sources (secondary and primary) are assumed still contributing to these emissions. Potential reasons such as increased emissions from γ-HCH deposited in South Africa may be considered as future research priority in this context (see below).

A first comparison of currently available monitoring data on atmospheric POP levels in the Antarctic atmosphere [23] revealed considerable concentration reduction during the past decade for almost all POP-related compounds reported (including DDT-derivatives, PCB congeners, HCH isomers, and cyclodiene pesticides). However, for HCB similar concentration levels were found as registered in the 1993 sampling campaign at Terra-Nova-Bay [23] as well as for the Signy-Island campaign (in collaboration with the British Antarctic Survey) in the following year 1994–1995 [65].

During the monitoring of atmospheric POPs at the Norwegian Troll station, indications for rapid atmospheric transport of organochlorine pesticides (i.e., γ-HCH) from potential South American sources were identified [23]. For this investigation a backward air mass movement simulation modeled with the FLEXPART program was utilized [66] as presented in Figure 7.

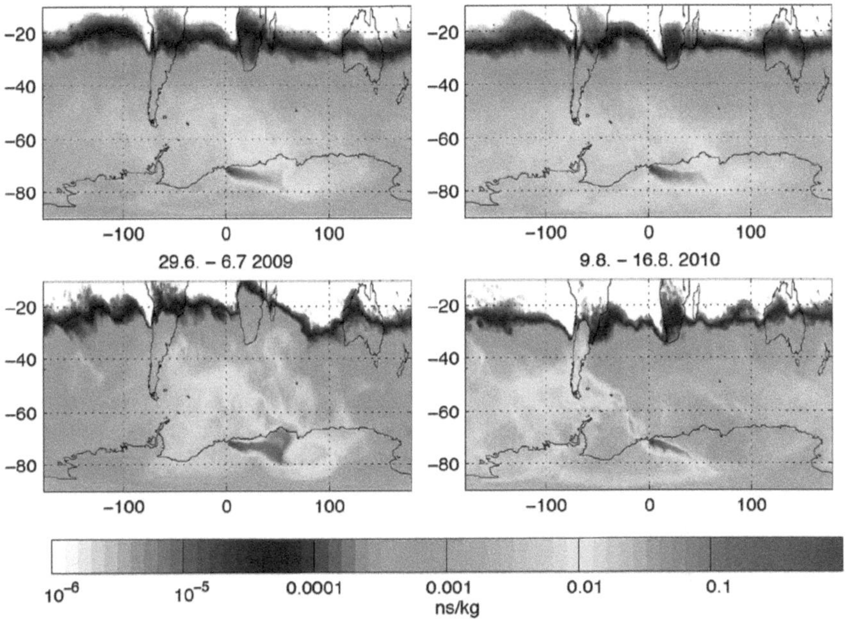

FIGURE 7 Footprint emission sensitivities derived from FLEXPART modeling: Upper line: Average monthly simulations (June and August for the entire sampling period (2007–2010)). Lower line: 20-days backward simulations for two weekly persistent organic pollutant samples from the Troll station (Queen Maud land, Antarctica) during week 27/2009 and 33/2010. *Figure reproduced with courtesy of the journal* Atmospheric Chemistry and Physics.

7. ARCTIC ATMOSPHERIC POPs AND CLIMATE CHANGE

Persistent anthropogenic pollutants are transported via the atmosphere, ocean currents, and rivers into the Arctic. After entering the Polar environment, the chemicals are immediately redistributed within the region by the same transport pathways, in addition to transpolar ice transport as well as incorporation into biological systems through accumulation in the food web. Every step along these transport and redistribution pathways to and within the Arctic is influenced by the current observed climate change because reactivity, adsorption processes, as well as accumulation are temperature-dependent processes. At the same time, the composition and the geophysical properties of transport media such as atmospheric and ocean currents are expected to change, sea and land ice are expected to change (and, in fact, are already changing), and so is the extent and composition of potential intermediate storage compartments for pollutants, such as forests and soils in sub-Arctic areas. Summing up all these processes, changes in global climate and the associated environmental change in the Arctic are expected to have significant consequences for contaminant pathways [67].

During the past decade, circum-Arctic national and international scientific and monitoring programs including AMAP have delivered a wealth of evidence for the presence and distribution of xenobiotics (e.g., contaminants originating from anthropogenic sources) in the Arctic environment. Extensive knowledge and on-going research activities on POPs fate and distribution pathways in the Arctic has been gathered during many circum-Arctic coordinated research activities over recent decades (including the International Polar Year (IPY) 2007–2009).

The cryosphere of the Polar Regions (Arctic and Antarctic) are currently undergoing comprehensive climate change-related changes. They are therefore today considered as early warning regions for the scientific elucidation of environmental consequences related to climate change. This includes also the elucidation of the fate of atmospheric long-range transported POPs. It is, thus not surprising that many of the research-based findings on POPs and climate change reported in the current scientific literature are based on scientific results obtained during studies conducted in Polar Regions [20]. A list of potential effects of climate change on pollutant distribution and fate is given in Table 1.

Within the coming three decades, seasonal sea ice coverage is expected to disappear during the Arctic summer as a consequence of the overall observed ambient temperature increase. Therefore, remobilization of previously stored contaminants (secondary sources) is expected to contribute more significantly in the coming years (Figure 1), leading to increased exposure of local indigenous human populations and wildlife in the future over certain time periods [19,20]. It is therefore expected that climate change will have significant impacts on the complex global contaminant pathways, storage and distribution processes into and within the Polar environments but specifically with respect to Arctic climate change.

As previously reported, for many of the air monitoring stations, for example weekly/monthly concentration data are now available for several decades already. For the Arctic, the longest data series exist for the monitoring stations of Alert (Ellesmere Island, Canada) and the Zeppelin Atmospheric Research Station in Ny-Ålesund (Svalbard, Norway) dating back to the early 1990s [7]. In 2007, also in Antarctica a similar long-term monitoring program was established at the Norwegian Troll station closely coordinated with the sampling and measuring program at the Norwegian Zeppelin station [23].

From this comprehensive information, it is possible to identify examples where perturbations to the longer-term temporal trend in concentrations as a possible consequence of a changing climate seem obvious. For HCB (a formerly used fungicide, now emitted as by-product of numerous chemical production lines) downward trends were reported in the first decade of sampling for all programs and stations [10]. However, during the past decade the HCB levels for the Zeppelin station (Ny-Ålesund, Spitsbergen) and Storhofdi (Iceland) have been increasing continuously again (AMAP, 2014). One plausible explanation is that the western coast of Spitsbergen was ice-free during the winters 2004–2010, which could allow for the volatilization of HCB present in surface marine waters accounting for the observed trend of HCB in the air [8,20]. Sea ice and glacier retreat has also been experienced in Iceland.

TABLE 1 Expected Influences of Climate Change on Environmental Processes Related to Transport and Distribution of Persistent Organic Pollutants

Parameters	Effects	Consequences (Examples)
Transport and distribution	Compound specific changing of distribution profile (also influencing exchange between compartments)	Distribution pathways will change
Sources	Changing emissions from secondary sources	Stronger contribution from secondary sources previously considered as sinks
Bioavailability	Compound- and species-dependent adaptations to bioavailability	Reduced bioavailability due to reduced half-life
Environmental stability (persistence)	Changes in compound composition stability properties	Reduced persistence due to reduced half-life
Transformation and degradation	Changing transformation/ degradation processes and resulting transformation products	Creation of stable transformation products
Toxicity/ environmental toxicity	Changes in toxicity potential for pollutants	Reduced toxicity due to reduced half-life

8. CONCLUSIONS AND PERSPECTIVES

POPs are still considered as high-priority pollutants in the polar environment, specifically in the Arctic, although most of the long-monitored compounds are under considerable restrictions and even banned for decades. The role of these long-term atmospheric monitoring activities in Polar Regions as a versatile tool for the evaluation and assessment of compound-specific distribution and transport processes is today fully recognized by the international scientific community as well as by regulatory authorities.

The current presence of POPs in polar air samples is explained by their extreme environmental stability, their resistance against natural transformation processes, and reemission and remobilization from secondary sources (sediment, soil, water masses).

However, along with the continuous development of new technologies and hyphenated techniques for ultra-trace quantitative analysis of persistent organic pollutants, many new chemicals are classified as chemicals of emerging concern and are introduced into the list of priority substances to be evaluated as potential persistent, bioaccumulative, and toxic pollutants. Thus, the research focus is gradually moving today from the so-called legacy POPs towards new POPs of emerging concern in the polar atmosphere. The classical list of chlorinated organic pollutants identified in arctic atmospheric samples such as pesticides and industrial (by-) products are today expanded with brominated flame retardants (such as PBDEs), PFASs, polysiloxanes, and chemicals in personal care products. The list of priority substances to be analyzed in atmospheric samples from the polar environments will continue to grow in the future.

In addition, the influence of climate change on the distribution and transformation patterns of POPs in the polar atmosphere is developing into a priority research area for polar environmental scientists. Local investigations on atmospheric POP distribution patterns reveal expected changes in source strength, distribution, and deposition pathways in Northern environments. It is expected that Arctic studies on climate change influence on atmospheric POP distribution pathways will provide scientific insight, which can be extended to understanding climate change impacts on contaminants in lower-latitude regions.

Even after more than 50 years of research, the monitoring of POPs and compounds of emerging concern is still an important research topic for environmental scientists and national and international regulators. Long-term air monitoring data sets from national and international Arctic monitoring programs will continue to allow for in-depth interdisciplinary research on contaminant pattern profiles, spatial and temporal trend studies, environmental fate and distribution modeling, and understanding of regional transport pathways in the future.

ACKNOWLEDGMENTS

The help of many colleagues during the realization of the presented book chapter is highly appreciated. We thank Robie Macdonald (Fisheries and Oceans Canada), Jianmin Ma (Environment Canada), Lars-Otto Reiersen, Janet Pawlak, Simon Wilson, and Pernilla Carlsson (AMAP Secretariat, Oslo, Norway) for inspiring discussions and advise during the development of the manuscript.

This review was financially supported by the combined Nordic Council of Ministers (NCM) and Arctic Monitoring and Assessment Programme (AMAP) project "Combined effects of pollutants and climate change in the Arctic," The Norwegian NorACIA programme, and the FP7 EU project ArcRisk "Arctic Health Risks: Impacts on health in the Arctic and Europe owing to climate-induced changes in contaminant cycling" (www.arcrisk.eu).

REFERENCES

[1] R. Carson, Silent Spring, Fawcett Crest, New York, 1964.

[2] T.F. Bidleman, E.J. Christensen, W.N. Billings, R. Leonard, Atmospheric transport of organochlorines in the North-Atlantic Gyre, J. Marine Res. 39 (1981) 443–464.

[3] M. Oehme, B. Ottar, The long-range transport of polychlorinated hydrocarbons to the Arctic, Geophys. Res. Lett. 11 (1984) 1133–1136.

[4] I.C. Burkow, R. Kallenborn, Sources and transport of persistent pollutants to the Arctic, Toxicol. Lett. 112–113 (2000) 87–92.
[5] R.W. Macdonal, L.A. Barrie, T.F. Bidleman, M.L. Diamond, D.J. Gregor, R.G. Semkin, W.M. Strachan, Y.F. Li, F. Wania, M. Alaee, L.B. Alexeeva, S.M. Backus, R. Bailey, J.M. Bewers, C. Gobeil, C.J. Halsall, T. Harner, J.T. Hoff, L.M. Jantunen, W.L. Lockhart, D. Mackay, D.C. Muir, J. Pudykiewicz, K.J. Reimer, J.N. Smith, G.A. Stern, Contaminants in the Canadian Arctic: five years of progress in understanding sources, occurrence and pathways, Sci. Total Environ. 254 (2000) 93–234.
[6] J.C. Van Oostdam, E. Dewailly, A. Gilman, J.C. Hansen, J.O. Odland, V. Chashchin, J. Berner, J. Butler-Walker, B.J. Lagerkvist, K. Olafsdottir, L. Soininen, P. Bjerregard, V. Klopov, J.P. Weber, Circumpolar maternal blood contaminant survey, 1994–1997 organochlorine compounds, Sci. Total Environ. 330 (2004) 55–70.
[7] H. Hung, R. Kallenborn, K. Breivik, Y. Su, E. Brorstrom-Lunden, K. Olafsdottir, J.M. Thorlacius, S. Leppanen, R. Bossi, H. Skov, S. Mano, G.W. Patton, G. Stern, E. Sverko, P. Fellin, Atmospheric monitoring of organic pollutants in the Arctic under the Arctic Monitoring and Assessment Programme (AMAP): 1993–2006, Sci. Total Environ. 408 (2010) 2854–2873.
[8] R. Kallenborn, L.O. Reiersen, C.D. Olseng, Long-term atmospheric monitoring of persistent organic pollutants (POPs) in the Arctic: a versatile tool for regulators and environmental science studies, Atmos. Pollut. Res. 3 (2012) 485–493.
[9] S. Becker, C.J. Halsall, W. Tych, R. Kallenborn, M. Schlabach, S. Mano, Changing sources and environmental factors reduce the rates of decline of organochlorine pesticides in the Arctic atmosphere, Atmos. Chem. Phys. 12 (2012) 4033–4044.
[10] T. Berg, R. Kallenborn, S. Mano, Temporal trends in atmospheric heavy metal and organochlorine concentrations at Zeppelin, Svalbard, Arctic Antarctic Alpine Res. 36 (2004) 284–291.
[11] G. Hansen, K. Aspmo, T. Berg, K. Edvardsen, M. Fiebig, R. Kallenborn, T. Krognes, C. Lunder, K. Stebel, N. Schmidbauer, S. Solberg, K.E. Yttri, Atmospheric monitoring at the Norwegian Antarctic station troll: measurement programme and first results, Polar Res. 28 (2009) 353–363.
[12] R. Kallenborn, Persistent organic pollutants (POPS) as environmental risk factors in remote high-altitude ecosystems, Ecotoxicol. Environ. Saf. 63 (2006) 100–107.
[13] R. Kallenborn, G. Christensen, A. Evenset, M. Schlabach, A. Stohl, Atmospheric transport of persistent organic pollutants (POPs) to Bjornoya (Bear island), J. Environ. Monit. 9 (2007) 1082–1091.
[14] G. Czub, F. Wania, M.S. Mclachlan, Combining long-range transport and bioaccumulation considerations to identify potential Arctic contaminants, Environ. Sci. Technol. 42 (2008) 3704–3709.
[15] F. Wania, D. Mackay, Global fractionation and cold condensation of low volatility organochlorine compounds in polar-regions, Ambio 22 (1993) 10–18.
[16] K. Breivik, A. Sweetman, J.M. Pacyna, K.C. Jones, Towards a global historical emission inventory for selected PCB congeners – a mass balance approach 3. An update, Sci. Total Environ. 377 (2007) 296–307.
[17] S. Jensen, A.G. Johnels, M. Olsson, G. Otterlind, DDT and PCB in marine animals from Swedish waters, Nature 224 (1969) 247–250.
[18] J.O. Odland, E. Nieboer, Human biomonitoring in the Arctic. Special challenges in a sparsely populated area, Int. J. Hyg. Environ. Health 215 (2012) 159–167.
[19] J.M. Ma, H.L. Hung, C. Tian, R. Kallenborn, Revolatilization of persistent organic pollutants in the Arctic induced by climate change, Nat. Clim. Change 1 (2011) 255–260.

[20] R. Kallenborn, C. Halsall, M. Dellong, P. Carlsson, The influence of climate change on the global distribution and fate processes of anthropogenic persistent organic pollutants, J. Environ. Monit. 14 (2012) 2854–2869.
[21] S. Becker, C.J. Halsall, M. Macleod, M. Scheringer, K.C. Jones, K. Hungerbuhler, Empirical investigation of the Junge variability-lifetime relationship using long-term monitoring data on polychlorinated biphenyl concentrations in air, Environ. Sci. Technol. 43 (2009) 2746–2752.
[22] H. Hung, P. Blanchard, C.J. Halsall, T.F. Bidleman, G.A. Stern, P. Fellin, D.C. Muir, L.A. Barrie, L.M. Jantunen, P.A. Helm, J. Ma, A. Konoplev, Temporal and spatial variabilities of atmospheric polychlorinated biphenyls (PCBs), organochlorine (OC) pesticides and polycyclic aromatic hydrocarbons (PAHs) in the Canadian Arctic: results from a decade of monitoring, Sci. Total Environ. 342 (2005) 119–144.
[23] R. Kallenborn, K. Breivik, S. Eckhardt, C.R. Lunder, S. Mano, M. Schlabach, A. Stohl, Long-term monitoring of persistent organic pollutants (POPs) at the Norwegian troll station in Dronning Maud Land, Antarctica, Atmos. Chem. Phys. 13 (2013) 6983–6992.
[24] NCP, in: D. Muir, P. Kurt-Karakus, J. Stow (Eds.), Canadian Arctic Contaminants Assessment Report III: Persistent Organic Pollutants in Canada's North, Aboriginal Affairs and Northern Development Canada, Ottawa, 2013.
[25] S. Eckhardt, K. Breivik, S. Mano, A. Stohl, Record high peaks in PCB concentrations in the Arctic atmosphere due to long-range transport of biomass burning emissions, Atmos. Chem. Phys. 7 (2007) 4527–4536.
[26] T. Gouin, F. Wania, Time trends of Arctic contamination in relation to emission history and chemical persistence and partitioning properties, Environ. Sci. Technol. 41 (2007) 5986–5992.
[27] F. Wania, D. Mackay, Peer reviewed: tracking the distribution of persistent organic pollutants, Environ. Sci. Technol. 30 (1996) 390A–396A.
[28] S.L. Simonich, R.A. Hites, Global distribution of persistent organochlorine compounds, Science 269 (1995) 1851–1854.
[29] G. Lammel, A.M. Sehili, T.C. Bond, J. Feichter, H. Grassl, Gas/particle partitioning and global distribution of polycyclic aromatic hydrocarbons—a modelling approach, Chemosphere 76 (2009) 98–106.
[30] W. Perrie, Z.X. Long, H. Hung, A. Cole, A. Steffen, A. Dastoor, D. Durnford, J.M. Ma, J.W. Bottenheim, S. Netcheva, R. Staebler, J.R. Drummond, N.T. O'neill, Selected topics in Arctic atmosphere and climate, Clim. Change 115 (2012) 35–58.
[31] Y.S. Su, H.L. Hung, P. Blanchard, G.W. Patton, R. Kallenborn, A. Konoplev, P. Fellin, H. Li, C. Geen, G. Stern, B. Rosenberg, L.A. Barrie, A circumpolar perspective of atmospheric organochlorine pesticides (OCPs): results from six Arctic monitoring stations in 2000–2003, Atmos. Environ. 42 (2008) 4682–4698.
[32] V.H. Garrison, M.S. Majewski, W.T. Foreman, S.A. Genualdi, A. Mohammed, S.L. Massey Simonich, Persistent organic contaminants in Saharan dust air masses in West Africa, Cape Verde and the eastern Caribbean, Sci. Total Environ. 468–469 (2014) 530–543.
[33] A. Karanasiou, N. Moreno, T. Moreno, M. Viana, F. De Leeuw, X. Querol, Health effects from Sahara dust episodes in Europe: literature review and research gaps, Environ. Int. 47 (2012) 107–114.
[34] L. Perez, A. Tobias, X. Querol, J. Pey, A. Alastuey, J. Diaz, J. Sunyer, Saharan dust, particulate matter and cause-specific mortality: a case-crossover study in Barcelona (Spain), Environ. Int. 48 (2012) 150–155.
[35] E. Remoundaki, A. Bourliva, P. Kokkalis, R.E. Mamouri, A. Papayannis, T. Grigoratos, C. Samara, M. Tsezos, PM10 composition during an intense Saharan dust transport event over Athens (Greece), Sci. Total Environ. 409 (2011) 4361–4372.

[36] T. Sandstrom, B. Forsberg, Desert dust: an unrecognized source of dangerous air pollution? Epidemiology 19 (2008) 808–809.

[37] E. Aamot, E. Steinnes, R. Schmid, Polycyclic aromatic hydrocarbons in Norwegian forest soils: impact of long range atmospheric transport, Environ. Pollut. 92 (1996) 275–280.

[38] A. Arruti, I. Fernandez-Olmo, A. Irabien, Evaluation of the urban/rural particle-bound PAH and PCB levels in the northern Spain (Cantabria region), Environ. Monit. Assess. 184 (2012) 6513–6526.

[39] C.L. Friedman, J.R. Pierce, N.E. Selin, Assessing the influence of secondary organic versus primary Carbonaceous aerosols on long-range atmospheric polycyclic aromatic hydrocarbon transport, Environ. Sci. Technol. 48 (2014) 3293–3302.

[40] G. Lammel, A. Heil, I. Stemmler, A. Dvorska, J. Klanova, On the contribution of biomass burning to POPs (PAHs and PCDDs) in air in Africa, Environ. Sci. Technol. 4 (2013) 11616–11624.

[41] S.S. Cindoruk, Y. Tasdemir, Dynamics of atmospheric polychlorinated biphenyls (PCBs): concentrations, patterns, partitioning, and dry deposition level estimations in a residential site of Turkey, Environ. Monit. Assess. 162 (2010) 67–80.

[42] C.W. Gotz, M. Scheringer, M. Macleod, C.M. Roth, K. Hungerbuhler, Alternative approaches for modeling gas-particle partitioning of semivolatile organic chemicals: model development and comparison, Environ. Sci. Technol. 41 (2007) 1272–1278.

[43] E. Jurado, F.M. Jaward, R. Lohmann, K.C. Jones, R. Simo, J. Dachs, Atmospheric dry deposition of persistent organic pollutants to the Atlantic and inferences for the global oceans, Environ. Sci. Technol. 38 (2004) 5505–5513.

[44] J. Li, X. Liu, G. Zhang, X.D. Li, Particle deposition fluxes of BDE-209, PAHs, DDTs and chlordane in the Pearl River Delta, South China, Sci. Total Environ. 408 (2010) 3664–3670.

[45] J. Radonic, M.T. Sekulic, M.V. Miloradov, P. Cupr, J. Klanova, Gas-particle partitioning of persistent organic pollutants in the Western Balkan countries affected by war conflicts, Environ. Sci. Pollut. Res. Int. 16 (2009) 65–72.

[46] S. Siddik Cindoruk, Y. Tasdemir, Characterization of gas/particle concentrations and partitioning of polychlorinated biphenyls (PCBs) measured in an urban site of Turkey, Environ. Pollut. 148 (2007) 325–333.

[47] P. Sun, P. Blatnchard, K. Brice, R.A. Hites, Atmospheric organochlorine pesticide concentrations near the Great Lakes: temporal and spatial trends, Environ. Sci. Technol. 40 (2006) 6587–6593.

[48] A. Gusev, M. Macleod, P. Bartlett, Intercontinental transport of persistent organic pollutants: a review of key findings and recommendations of the task force on hemispheric transport of air pollutants and directions for future research, Atmos. Pollut. Res. 3 (2012) 463–465.

[49] L. Lamon, M.D. Valle, A. Critto, A. Marcomini, Introducing an integrated climate change perspective in POPs modelling, monitoring and regulation, Environ. Pollut. 157 (2009) 1971–1980.

[50] AMAP, AMAP Assessment Report: Arctic Pollution Issues, first ed., 1998. Arctic Monitoring and Assessment Programme. Report.

[51] C.J. Halsall, Investigating the occurrence of persistent organic pollutants (POPs) in the Arctic: their atmospheric behaviour and interaction with the seasonal snow pack, Environ. Pollut. 128 (2004) 163–175.

[52] R.C. Hale, S.L. Kim, E. Harvey, M.J. La Guardia, T.M. Mainor, E.O. Bush, E.M. Jacobs, Antarctic research bases: local sources of polybrominated diphenyl ether (PBDE) flame retardants, Environ. Sci. Technol. 42 (2008) 1452–1457.

[53] J.A. Arnot, J.M. Armitage, L.S. Mccarty, F. Wania, I.T. Cousins, L. Toose-Reid, Toward a consistent evaluative framework for POP risk characterization, Environ. Sci. Technol. 45 (2011) 97–103.

[54] S.G. Donaldson, J. Van Oostdam, C. Tikhonov, M. Feeley, B. Armstrong, P. Ayotte, O. Boucher, W. Bowers, L. Chan, F. Dallaire, R. Dallaire, E. Dewailly, J. Edwards, G.M. Egeland, J. Fontaine, C. Furgal, T. Leech, E. Loring, G. Muckle, T. Nancarrow, D. Pereg, P. Plusquellec, M. Potyrala, O. Receveur, R.G. Shearer, Environmental contaminants and human health in the Canadian Arctic, Sci. Total Environ. 408 (2010) 5165–5234.

[55] R. Mabilia, V. Di Palo, C. Cassardo, C. Ciuchini, A. Pasini, M. Possanzini, Measurements of lower carbonyls and hydrocarbons at Ny-Ålesund, Svalbard, Ann. Chim. 97 (2007) 1027–1037.

[56] J. Pollmann, D. Helmig, J. Hueber, C. Plass-Dulmer, P. Tans, Sampling, storage, and analysis of C_2–C_7 non-methane hydrocarbons from the US National Oceanic and Atmospheric Administration Cooperative Air Sampling Network glass flasks, J. Chromatogr. A 1188 (2008) 75–87.

[57] C.J. Galban-Malagon, N. Berrojalbiz, R. Gioia, J. Dachs, The "degradative" and "biological" pumps controls on the atmospheric deposition and sequestration of hexachlorocyclohexanes and hexachlorobenzene in the North Atlantic and Arctic Oceans, Environ. Sci. Technol. 47 (2013) 7195–7203.

[58] A.M. Grannas, C. Bogdal, K.J. Hageman, C. Halsall, T. Harner, H. Hung, R. Kallenborn, P. Klan, J. Klanova, R.W. Macdonald, T. Meyer, F. Wania, The role of the global cryosphere in the fate of organic contaminants, Atmos. Chem. Phys. 13 (2013) 3271–3305.

[59] B.M.J. Herbert, C.J. Halsall, K.C. Jones, R. Kallenborn, Field investigation into the diffusion of semi-volatile organic compounds into fresh and aged snow, Atmos. Environ. 40 (2006) 1385–1393.

[60] N. Li, F. Wania, Y.D. Lei, G.L. Daly, A comprehensive and critical compilation, evaluation, and selection of physical–chemical property data for selected polychlorinated biphenyls, J. Phys. Chem. Ref. Data 32 (2003).

[61] U. Schenker, M. Macleod, M. Scheringer, K. Hungerbuhler, Improving data quality for environmental fate models: a least-squares adjustment procedure for harmonizing physicochemical properties of organic compounds, Environ. Sci. Technol. 39 (2005) 8434–8441.

[62] I. Stemmler, G. Lammel, Long-term trends of continental-scale PCB patterns studied using a global atmosphere-ocean general circulation model, Environ. Sci. Pollut. Res. Int. 19 (2012) 1971–1980.

[63] R.M. Dickhut, A. Cincinelli, M. Cochran, H.W. Ducklow, Atmospheric concentrations and air-water flux of organochlorine pesticides along the Western Antarctic Peninsula, Environ. Sci. Technol. 39 (2005) 465–470.

[64] A. Gambaro, L. Manodori, R. Zangrando, A. Cincinelli, G. Capodaglio, P. Cescon, Atmospheric PCB concentrations at terra Nova Bay, Antarctica, Environ. Sci. Technol. 39 (2005) 9406–9411.

[65] R. Kallenborn, M. Oehme, D.D. Wynn-Williams, M. Schlabach, J. Harris, Ambient air levels and atmospheric long-range transport of persistent organochlorines to Signy Island, Antarctica, Sci. Total Environ. 220 (1998) 167–180.

[66] X.L. Pan, Y. Kanaya, Z.F. Wang, X. Tang, M. Takigawa, P. Pakpong, F. Taketani, H. Akimoto, Using Bayesian optimization method and FLEXPART tracer model to evaluate CO emission in East China in springtime, Environ. Sci. Pollut. Res. Int. 21 (2014) 3873–3879.

[67] R.W. Macdonald, T. Harner, J. Fyfe, Recent climate change in the Arctic and its impact on contaminant pathways and interpretation of temporal trend data, Sci. Total. Environ. 342 (1–3) (2005) 5–86.

Chapter 14

Bioaccumulation and Biotransformation of Brominated Flame Retardants

Liesbeth Weijs,[1,2,3] Alin C. Dirtu,[1,4] Govindan Malarvannan[1] and Adrian Covaci[1,*]

[1]*Toxicological Centre, University of Antwerp, Universiteitsplein, Wilrijk, Belgium;* [2]*Systemic Physiological and Ecotoxicological Research (SPHERE), Department of Biology, University of Antwerp, Groenenborgerlaan, Antwerp, Belgium;* [3]*National Research Centre for Environmental Toxicology (Entox), The University of Queensland, Coopers Plains, QLD, Australia;* [4]*Department of Chemistry, "Al. I. Cuza" University of Iasi, Iasi, Romania*

**Corresponding author: E-mail: adrian.covaci@uantwerpen.be*

Chapter Outline

1. INTRODUCTION

Due to their presence in the environment and to their persistence and toxicity in wildlife and humans, brominated flame retardants (BFRs) have been under scientific scrutiny during the past decade. BFRs prevent the spread of fire by quenching the radical oxidation reactions that take place during combustion by reacting with the H· and HO· radicals. These compounds can enter in the environment by

Comprehensive Analytical Chemistry, Vol. 67. http://dx.doi.org/10.1016/B978-0-444-63299-9.00014-4

a number of different pathways, such as emission during their production or the manufacture of flame-retarded products, by leaching from consumer products, or following disposal. There is already a lot of knowledge on the physicochemical parameters characterizing both "classical" BFRs, such as polybrominated diphenyl ethers (PBDEs) and hexabromocyclododecanes (HBCDs), and "novel" BFRs (NBFRs) [31]. The similarity in characteristics between these classes of contaminants suggests similar environmental behavior. Therefore, positive relationships obtained for low trophic level (TL) species between log-transformed bioaccumulation factors (log BAF) and log-transformed octanol–water partition coefficients (log K_{OW}) or molecular weight for NBFRs, indicate that physicochemical properties play a key role in the bioaccumulation processes [177]. A summary of physicochemical properties of targeted BFRs are given in Table 1. Several BFRs are found in quantifiable levels in wildlife and in humans and have been extensively investigated in the past decade; however, we still lack extensive information on the health effects caused by these compounds. Generally, most of the BFRs are absorbed from the gastrointestinal (GI) tract and accumulate in fatty tissues. Even if they do not necessarily cause immediate symptoms from acute toxicity at average doses, their health effects from chronic exposure are of more concern, especially when they are related to the exposure of developing infants and wildlife [21]. The BFRs included in this chapter are PBDEs, HBCDs, tetrabromobisphenol A (TBBPA), and some NBFRs, such as decabromodiphenyl ethane (DBDPE), 1,2-bis(2,4,6-tribromophenoxy)ethane (BTBPE), 2-ethylhexyl 2,3,4,5-tetrabromobenzoate (TBB) and di(2-ethylhexyl) tetrabromophthalate (TBPH).

PBDEs are additive flame retardants (FRs) that are/were used in a wide array of household products in concentrations up to 30% by weight, typically between 2% and 6%. They are structurally related to polychlorinated biphenyls (PCBs) and are produced commercially as mixtures. The three commercially produced mixtures of PBDEs are penta-BDE, octa-BDE, and deca-BDE according to the number of bromine atoms in the dominating congeners of the mixtures. PBDEs have been classified as contaminants of concern and the penta- and octa-BDE mixtures were added to the Stockholm Convention list of persistent organic pollutants (POPs) [44,150,151].

As an additive BFR, HBCD easily reaches the environment through production, usage as BFR in a wide variety of household consumer products, and recycling of HBCD-containing materials [22]. There are three main stereo isomers in the commercial mixture: α-HBCD (10–13%), β-HBCD (1–12%), and γ-HBCD (75–89%) [66]. Although its usage is not restricted, HBCD has been included in the PBT (persistent, bioaccumulative, and toxic) list of the European Chemical Substance Information System. The γ-HBCD isomer dominates the commercial mixture and in abiotic matrices, whereas biota shows a predominance of α-HBCD [99]. This finding has driven the research in the direction of elucidation of the processes which would explain such behavior.

TBBPA (4,4′-isopropylidenebis(2,6-dibromophenol)) is the most widely used and distributed BFR worldwide [24]. TBBPA can be used as an additive

TABLE 1 Summary of Important Physicochemical Properties of Major NBFRs and PBDE congeners

Name	Acronyms	Molecular Formula	Molecular Weight (g/mol)	Melting Point (°C)	Boiling Point (°C)	Flash Point (°C)	Density (g/cm³)	Vapor Pressure (Pa, 25 °C)	Water Solubility (μg/L (25 °C))	Octanol–Water Coefficient (log K_{OW})
Decabromodiphenylethane	DBDPE	$C_{14}H_4Br_{10}$	971.2	351–355	676±50	286.8	3.06	6.0E-15	0.21	11.64
1,2-Bis(2,4,6-tribromophenoxy)ethane	BTBPE	$C_{14}H_8Br_6O_2$	687.6	NA	566.4±50	238.8	2.33	3.88E-10	19.0	9.10
Bis(2-ethylhexyl) tetrabromophthalate	TBPH	$C_{24}H_{34}Br_4O_4$	706.2	NA	584.8±45	307.5	1.53	1.55E-11	1.60	10.93
2-Ethylhexyl-2,3,4,5-tetrabromobenzoate	TBB	$C_{15}H_{18}Br_4O_2$	549.9	NA	497.3	254.5	NA	NA	NA	NA
2,4,6-Tribromophenol	TBP	$C_6H_3Br_3O$	330.8	94–95	286.8	NA	NA	0.41	0.968E+6	4.0
Tetrabromobisphenol A bis(2,3-dibromopropyl ether)	TBBPA-DBPE	$C_{21}H_{20}Br_8O_2$	943.6	90–105	676.5±55.0	287.7	2.17	1.60 E–07	0.17	12.12
2,2′,4,4′-Tetrabromodiphenyl ether	BDE 47	$C_{12}H_6Br_4O$	485.8	–5	>300	>200	2.3	1.9 E–04	15	6.01–6.77
2,2′,4,4′,5-Pentabromodiphenyl ether	BDE 99	$C_{12}H_5Br_5O$	564.7	–5	>300	<200	2.3	1.8 E–05	9	6.53–7.66
2,2′,4,4′,5,5′-Hexabromodiphenyl ether	BDE 153	$C_{12}H_4Br_6O$	643.6	87		>330	2.8	2.1 E–06	1	7.62–7.90
2,2′,3,3′,4,4′,5,5′,6,6′-Decabromodiphenyl ether	BDE 209	$C_{12}Br_{10}O$	959.2	300		>320	3.2	1.2 E–10	<0.1	12.6

BFR as an alternative to octa-BDE in television casings, printer components, photocopiers, coffee makers and plugs/sockets [24]. TBBPA is nonvolatile [60] and degrades very slowly, photochemically [38]. TBBPA has a log K_{OW} of 4.75 at pH 7.6 [93]. The widespread use of TBBPA and its detection in dust, sediments, and biota has led to increasing concerns regarding its effects on wildlife and humans [24]. Since it was measured in humans, TBBPA may be a matter of concern to human health [16].

NBFRs have been monitored only recently in the environment [23]. These chemicals have been produced in lower volumes than the major traditional BFRs, usually either as alternatives for these traditional BFRs or with a very specific intended use and, not so frequently, as replacements for the FRs that have been banned or phased out of use. DBDPE is a high production volume chemical produced in the USA and China that has been proposed as replacement for BDE 209 [176]. As BDE 209 is gradually removed from the market, the use of DBDPE has consequently increased [168]. BTBPE has been produced since the mid-1970s, in a high production volume in the USA and a low production volume in the EU [42]. In the USA, it was marketed by the Great Lakes Chemical Corporation as FF-680 with the intended use as additive FR for thermoplastic and thermoset resin systems (e.g., ABS, HIPS, and polycarbonate). As of 2005, Chemtura used the FF-680 as a replacement for the octa-BDE formulation [72]. TBB and TBPH are produced by the Chemtura Chemical Corporation and used together in two FR formulations, namely Firemaster 550 (TBB:TBPH in a ratio of 4:1, by mass) [138] and Firemaster BZ-54 (TBB:TBPH in a ratio of 2.5:1, by mass) [9]. Firemaster 550 was first introduced as a replacement for the penta-BDE mixture [138]. TBPH is also used in the DP-45 formulation, also marketed by Chemtura. In the USA, FM550 is classified as a high production volume chemical [65].

In this chapter, we focused on the environmental behavior of BFRs and provided evidence that they behave similarly to POPs: they bioaccumulate in either aquatic or terrestrial ecosystems, while biomagnification occurs among different trophic chains. Therefore, we systematically reviewed the literature presenting data on bioaccumulation/biomagnification of BFRs and also we have reviewed the information on their degradation pathways and transformation products (TPs).

2. BIOACCUMULATION OF BFRs IN AQUATIC ORGANISMS AND FOOD CHAINS

2.1 Bioaccumulation

The occurrence of BFRs in various species has been documented in reviews by [29] for both PBDEs as HBCDs, by [96] for PBDEs and by [22] for HBCDs. To avoid repetition, we build on these reviews and include most important findings of studies published afterward. BDEs 28, 47, 99, 100, 153, 154, 183, and 209 are recommended by the European Food Safety Authority (EFSA) for monitoring of PBDE contamination in food.

Invertebrates. Although at the bottom of the food chain, BFRs accumulate in aquatic invertebrates (Dinn et al., 2012). The results of PBDEs in invertebrates are difficult to compare between species. Apart from the different (number of) congeners included in the totals or sums, results can be expressed in wet weight (ww) [20,152,158], lipid weight (lw) [144], or even dry weight (dw) [12,103]. It is also impossible to compare a single/major congener, as this can differ among studies.

Van Leeuwen et al. [158] investigated PBDEs and HBCDs in three shrimp species from Asia and the Netherlands. With a low mean lipid percentage of 1.2%, these organisms are theoretically not the preferred sinks for lipophilic compounds, such as PBDEs and HBCDs. According to the same study, the PBDE levels in shrimps were similar to the levels in tilapia and pangasius, but one and two orders of magnitude lower than the levels in trout and salmon, respectively. PBDE concentrations ranged from 0.15 to 1.8 ng/g ww in zebra mussels from Belgium [20]; were up to 6.9 ng/g dw in Mediterranean mussels [12]; and ranged from 27.0 to 83.7 ng/g dw in green-lipped mussels from Hong Kong [103]. Van Ael et al. [152] investigated a range of aquatic invertebrates such as crab, shrimp, mussel, and worms in the Scheldt Estuary (Belgium) and found the lowest body burdens of PBDEs in brown shrimp (0.05–0.47 ng g ww).

Tomy et al. [148] did not find β-HBCD at detectable levels in any sample, but found consistently higher levels of α-HBCD compared to γ-HBCD in fish and invertebrates from the Lake Ontario food web. Sums of α-HBCD and γ-HBCD were on average 0.06, 0.14, and 0.16 ng/g ww in three species of zooplankton [148], whereas HBCD levels ranging from 6.0 to 500 ng/g lw were found in blue mussels in Korea [122]. Nakagawa et al. [113] quantified the three HBCD stereoisomers in several invertebrate species and found the highest levels in octopus (0.93 ng/g ww), followed by farmed oysters (0.19 ng/g ww) and wild Pacific flying squid (0.16 ng/g ww). α-HBCD was dominant in farmed oysters and γ-HBCD in octopus and flying squid. The same study was not able to detect HBCDs in shrimp.

Profiles in sediments can sometimes differ from profiles in benthic invertebrates. BDE 209 was dominant in sediment samples of the Scheldt Estuary, while BDE 47 was dominant in all biota samples from the same locations [152]. BDE 209 is the precursor for lower brominated, but highly persistent, compounds in a range of fish species, for example, rainbow trout, common carp in [139], or fathead minnows in [115]. Timing is therefore of importance for measuring BDE 209 and/or its degradation products in biota. BDE 209 was the predominant congener detected in shrimp originating from Asia [158] and in green and freshwater mussels from Jakarta Bay [144]. Compared to the findings from [152], these results probably indicate a more recent release of BDE 209 (and PBDEs) in Asian waters than in the Scheldt Estuary. Liu et al. [103] have found different predominant PBDE congeners depending on sampling sites in green-lipped mussels from several sites in Hong Kong. These examples show that location, in terms of PBDE usages by different countries as well as distance from sources within a specific country, has an influence on the PBDE profiles.

Concerning HBCDs, ragworms were found to accumulate HBCDs only from their feed (e.g., mussel) and not from the sediment [63]. α-HBCD is the most dominant stereoisomer in invertebrates (Table 2), although γ-HBCD has been the most prevalent in octopus and squid from Japan. Since this is generally different from the dominance of γ-HBCD in the technical HBCD mixtures and the HBCD profiles in sediment, it suggests that the conversion of γ-HBCD to α-HBCD: (1) already occurs by abiotic processes before HBCDs enter the food chains and/or (2) occurs already in aquatic invertebrates.

Teleost fish. Although some species have a top predator status, particularly in freshwater systems, most teleost fish occupy the middle portion of the aquatic food webs. As such, they are a vital part of the biomagnification process and are capable of changing the pollutant's profiles in food webs through species-specific metabolic capacities. Several teleost fish species are preferred model organisms in exposure experiments, both in laboratory settings and in the wild (e.g., through fish farms or cage experiments). Teleost fish are also consumed by humans which makes that they are often included in studies about dietary exposure of humans to pollutants. Consequently, there are numerous reports in the literature about PBDEs and HBCDs in fish.

Overall, PBDE levels are low (<1–2 ng/g ww) in teleost species from the Congo River Basin [159]. Although results in Table 2 are not given in the same unit (lw or ww), PBDE levels varied greatly among fish species from Europe, South America, Australia, and Asia: 1–4 ng/g lw in [137]; <1–446 ng/g ww in [37]; 24–115 ng/g lw in [104]; 3–1240 ng/g lw in [120]; and 12–24 ng/g lw in [95]. In these studies, the highest levels could often be attributed to known nearby sources of industry. In [120]; the lowest levels were found in an agricultural site. For the USA, very low levels have been reported in teleost species from wild or farmed situations in more recent studies (<1–38 ng/g ww in [64]), as well as elevated levels in earlier studies (369–1395 ng/g lw measured in 2000 in [184]). This latter study reported an exponential increase in PBDE levels in the Great Lakes from 1980 to 2000 [184]. Nevertheless, as mentioned earlier, conclusions about potential temporal trends in fish species from the USA cannot be drawn from all studies mentioned here as they all include different species from different places around the US marine and freshwater systems. For HBCDs, levels range from <LOQ to 77 ng/g ww in fish species from Japan [113], from 1 to 3 ng/g lw in fish species from Sydney harbor, Australia and in polar cod from Svalbard [104,137] and from not detected to 1172 ng/g ww in barbel from Spain [36].

Barbels from Spain [36] and several fish species from the Congo River Basin [159] have a different profile compared to other studies (Table 2). Both studies investigated fish species from river systems, but also have different predominant PBDE congeners in the muscle samples of the teleost species, namely BDE 183 and 153 in [36] and BDE 99 in [159]. BDE 47 is consistently predominant in US waters [64,132], in South America (Chile, Montory and Barra, 2006) and Australia [104]. Zhu and Hites [184] reported a shift in PBDE profile over

TABLE 2 Literature Overview Regarding the Bioaccumulation of PBDEs and HBCDs in the Aquatic Environment. Levels of PBDEs and HBCDs in This Table are Minimum–Maximum Ranges Unless Only Mean ± SD Was Provided

Species	Location	n	Tissue	PBDEs (ng/g)	Major PBDE	HBCDs (ng/g)	Major HBCD	References
Invertebrates								
Bristle worms	Belgium/The Netherlands	7 pools	Whole soft body	<1–2 (ww)	BDE 47			[152]
Blue mussel	Belgium/The Netherlands	6 pools	Whole soft body	<1–1 (ww)	BDE 47, BDE 100			
Brown shrimp	Belgium/The Netherlands	18 pools	Muscle	<1	BDE 47, BDE 99			
Shore crab	Belgium/The Netherlands	12	Muscle	<1–1 (ww)	BDE 47			
Chinese mitten crab	Belgium/The Netherlands	13	Muscle	1–3 (ww)	BDE 47			
Green-lipped mussels	Hong Kong	30	Whole soft body	27–84 (dw)	BDE 197, BDE 99, BDE 153, BDE 207, BDE 183, BDE 28 (depending on site)			[103]
Zebra mussels	Belgium	950 indiv. in 38 pools	Whole soft body	<1–2 (ww)	BDE 47, BDE 99			[20]

Continued

TABLE 2 Literature Overview Regarding the Bioaccumulation of PBDEs and HBCDs in the Aquatic Environment. Levels of PBDEs and HBCDs in This Table are Minimum–Maximum Ranges Unless Only Mean ± SD Was Provided—cont'd

Species	Location	n	Tissue	PBDEs (ng/g)	Major PBDE	HBCDs (ng/g)	Major HBCD	References
Mediterranean mussel	Italy	40 pools	Whole soft body	<LOQ–7 (dw)	BDE 47, BDE 99			[12]
Shrimp	Congo	4	Whole soft body	<LOQ–8 (lw)	BDE 99			[159]
Invertebrates	Svalbard	4		<1–15 (lw)	BDE 47 and 99			[137]
Blue swimmer crab	Sydney, Australia	5 pools	Muscle	16 ± 4 (lw)	BDE 47	1 ± 0 (lw)		[104]
Squid	Sydney, Australia	4 pools	Muscle	6 ± 2 (lw)	BDE 47	1 ± 0 (lw)		
Calanoid copepods	Svalbard	4	Whole body	<1–15 (lw)	BDE 209			[137]
Mysids	Lake Ontario	2 pools	Whole body			<1–1 (ww)	α	[148]
Amphipods	Lake Ontario	2 pools	Whole body			<1–1 (ww)	α	
Zooplankton fauna	Lake Ontario	2 pools	Whole body			<1–1 (ww)	α	

Octopus and squid	Japan	9	Edible parts			< LOQ–1 (ww)	γ	[113]
Shrimp	Japan	108	Edible parts			<LOQ		
Oysters	Japan	46	Edible parts			<1 (ww)	α	
Teleost Fish								
Polar cod	Svalbard	7	Whole body	1–4 (lw)	BDE 47	1–3 (lw)		[137]
Barbel	Spain	23	Muscle	<1–436 (ww)	BDE 183, BDE 153	ND–1127 (ww)		[36]
Barbel	Spain	23	Liver	<1–446 (ww)	BDE 153, BDE 47	ND–1172 (ww)		
Flounder	Sydney, Australia	5 pools	Muscle	115 ± 64 (lw)	BDE 47	1 ± 0 (lw)		[104]
Tailor	Sydney, Australia	4 pools	Muscle	107 ± 41 (lw)	BDE 47	2 ± 1 (lw)		
Yellowfin bream	Sydney, Australia	5 pools	Muscle	88 ± 45 (lw)	BDE 47	1 ± 1 (lw)		
Fanbelly leatherjacket	Sydney, Australia	4 pools	Muscle	24 ± 5 (lw)	BDE 47	3 ± 3 (lw)		
Sea mullet	Sydney, Australia	5 pools	Muscle	28 ± 14 (lw)	BDE 47	3 ± 1 (lw)		
Luderick	Sydney, Australia	5 pools	Muscle	24 ± 14 (lw)	BDE 47	1 ± 0		

Continued

TABLE 2 Literature Overview Regarding the Bioaccumulation of PBDEs and HBCDs in the Aquatic Environment. Levels of PBDEs and HBCDs in This Table are Minimum–Maximum Ranges Unless Only Mean ± SD Was Provided—cont'd

Species	Location	n	Tissue	PBDEs (ng/g)	Major PBDE	HBCDs (ng/g)	Major HBCD	References
Unspecified fish	Taiwan	60	Muscle and tissue	3–1240 (lw)	BDE 47, BDE 154			[120]
Alaskan King salmon (W)	Maryland and Washington, USA	2	Muscle	1 (ww)	BDE 47			[64]
Alaskan coho (W)	Maryland, USA	2	Muscle	<1 (ww)	BDE 47			
Alaskan sockeye (W)	Maryland, USA	2	Muscle	<1 (ww)	BDE 47			
Atlantic salmon (F)	Maryland and Washington, USA	5	Muscle	<1–2 (ww)	BDE 47			
Icelandic salmon (F)	Maryland, USA	1	Muscle	1 (ww)	BDE 47			
Rockfish (W)	Maryland and Washington, USA	5	Muscle	2–12 (ww)	BDE 47			
Bluefish (W)	Maryland, North Carolina, and Washington, USA	5	Muscle	1–38 (ww)	BDE 47			

Atlantic salmon (F)	Maine (USA), Canada, and Norway	54	Fillet	<1–1 (ww)	BDE 47			[132]
Alaskan Chinook salmon (W)	Alaska	9	Fillet	<1–1 (ww)	BDE 47			
Several fish species[a]	Great Lakes, USA	14 pools	Unspecified fish tissue	369–1395 (lw)	BDE 47+BDE 99+BDE 100			[184]
Atlantic salmon (F)	Chile	14	Muscle	12–24 (lw)	BDE 47			[95]
Carp	Yangtze River, China	10	Muscle	17–1100 (lw)	BDE 47	11–330 (lw	α	[179]
Bream		2	Muscle	9–14 (lw)	BDE 47	–	–	
Marcusenius sp	Congo	18	Muscle	<1–2 (ww)	BDE 99			[159]
Shoulderspot catfish	Congo	17	Muscle	<1 (ww)	BDE 99			
Blackline glass catfish	Congo	7	Muscle	<1 (ww)	BDE 99			
Bigeye squeaker	Congo	4	Muscle	<1–2 (ww)	BDE 99			
Spottail robber	Congo	8	Muscle	<1–1 (ww)	BDE 99			
Sharktail distichodus	Congo	16	Muscle	<1 (ww)	BDE 99			

Continued

TABLE 2 Literature Overview Regarding the Bioaccumulation of PBDEs and HBCDs in the Aquatic Environment. Levels of PBDEs and HBCDs in This Table are Minimum–Maximum Ranges Unless Only Mean ± SD Was Provided—cont'd

Species	Location	n	Tissue	PBDEs (ng/g)	Major PBDE	HBCDs (ng/g)	Major HBCD	References
Silver perch	Florida	6	Muscle	4–14 (lw)	BDE 47			[81]
Striped mullet	Florida	6	Muscle	2–63 (lw)	BDE 47			
Spotted seatrout	Florida	7	Muscle	12–78 (lw)	BDE 47			
Red drum	Florida	11	Muscle	3–154 (lw)	BDE 47			
Hardhead catfish	Florida	8	Muscle	2–306 (lw)	BDE 47			
Barracuda	Japan	10	Edible parts			1–5 (ww)	α	[113]
Butter fish	Japan	7	Edible parts			<LOQ		
Horse mackerel	Japan	23	Edible parts			<1–5(ww)	α	
Japanese seabass	Japan	28	Edible parts			2–26 (ww)	α–γ	
Japanese Spanish mackerel	Japan	4	Edible parts			<1–26 (ww)	γ	
Pacific mackerel	Japan	16	Edible parts			<1–23 (ww)	α	

Sea bream	Japan	10	Edible parts			<LOQ–8 (ww)	α	
Sillago	Japan	25	Edible parts			<1 (ww)	α	
Whitespotted conger	Japan	27	Edible parts			<LOQ–<1 (ww)	α	
Marbled rockfish	Japan	8	Edible parts			<LOQ		
Japanese red sea perch	Japan	7	Edible parts			<1 (ww)	γ	
Flatfish	Japan	17	Edible parts			<LOQ		
Silver-stripe round herring	Japan	191	Edible parts			<1 (ww)	α	
Sardines	Japan	83	Edible parts			<1–77 (ww)	α–γ	
Japanese halfbeak	Japan	14	Edible parts			<1 (ww)	α	
Flathead mullet	Japan	2	Edible parts			1 (ww)	α	
Salmon	Japan	21	Edible parts			1 (ww)	α	
Sea bream	Japan	6	Edible parts			<1–1 (ww)	α	
Yellow tail	Japan	1	Edible parts			<1 (ww)	α	

Continued

TABLE 2 Literature Overview Regarding the Bioaccumulation of PBDEs and HBCDs in the Aquatic Environment. Levels of PBDEs and HBCDs in This Table are Minimum–Maximum Ranges Unless Only Mean ± SD Was Provided—cont'd

Species	Location	n	Tissue	PBDEs (ng/g)	Major PBDE	HBCDs (ng/g)	Major HBCD	References
Elasmobranchs								
Bull shark	Florida, USA	6	Muscle	16–287 (lw)	BDE 209	9–413 (lw)	γ	[80,81]
Bull shark	Florida, USA	7	Muscle	12–4190 (lw)	BDE 209	17–310 (lw)	γ	
Atlantic sharpnose shark	Florida, USA	5	Muscle	45–1930 (lw)	BDE 209	2–156 (lw)	γ	
Atlantic stingray	Florida, USA	7	Muscle	2–83 (lw)	BDE 47			
Spiny dogfish	Florida, USA	5	Muscle	16–60 (lw)	BDE 209			
Scalloped hammerhead shark	Japan	3	Blood	8 ± 1 (pg/g ww)	BDE 47			[114]
Scalloped hammerhead shark	Japan	4	Blood	12 ± 9 (pg/g ww)	BDE 47			

Atlantic stingray	Florida, USA	16	Liver	<LOQ–46 (lw)	BDE 47[b]			[170]
Bonnethead sharks	Florida, USA	19	Liver	4–100 (lw)	BDE 47			
Bull sharks	Florida, USA	8	Liver	60–835 (lw)	BDE 47			
Lemon sharks	Florida, USA	12	Liver	12–103 (lw)	BDE 47			
Marine Mammals								
Polar bear	Svalbard	4	Fat	21–45 (lw)	BDE 47	5–17 (lw)		[137]
Polar bear	Beaufort Sea, Alaska	8	Liver	<1–2 (ww)	BDE 47[b]			[84]
Polar bear	Chukchi Sea, Alaska	27	Liver	<1–3 (ww)	BDE 47[b]			
Polar bear	Svalbard, Norway	15	Blood	3–10 (ww)	BDE 47	<1–1 (ww)		[160]
Polar bear	Alaska	15	Fat	3–17 (lw)	BDE 47	<1–35 (lw)		[112]
Polar bear	Canada	65	Fat	4–56 (lw)	BDE 47			
Polar bear	Greenland	44	Fat	35–132 (lw)	BDE 47	32–59 (lw)		
Polar bear	Svalbard	15	Fat	27–114 (lw)	BDE 47	18–109 (lw)		

Continued

TABLE 2 Literature Overview Regarding the Bioaccumulation of PBDEs and HBCDs in the Aquatic Environment. Levels of PBDEs and HBCDs in This Table are Minimum–Maximum Ranges Unless Only Mean ± SD Was Provided—cont'd

Species	Location	n	Tissue	PBDEs (ng/g)	Major PBDE	HBCDs (ng/g)	Major HBCD	References
Polar bear	East Greenland	92	Fat	22–192 (lw)	NM			[136]
Polar bear	East Greenland	20	Liver	127–936 (lw)	NM			
Polar bear	East Greenland	20	Brain	<1–36 (lw)	NM			
Polar bear	East Greenland	20	Blood	38–146 (lw)	NM			
Killer whale (northern resident)	NE Pacific Ocean	21	Skin + blubber	11–1357 (ww)	BDE 47			[123]
Killer whale (southern resident)	NE Pacific Ocean	5	Skin + blubber	116–870 (ww)	BDE 47			
Killer whale (transient)	NE Pacific Ocean	13	Skin + blubber	25–1106 (ww)	BDE 47			
Killer whale	Norway	9	Fat	348–647 (lw)[c]	BDE 47			[174]

Beluga whales	Canada	54	Fat	17–935 (ww)	BDE 47			[100]
Beluga	Cook Inlet, Alaska	27	Fat	7–46 (lw)				[71]
Beluga	Chukchi Sea, Alaska	40	Fat	2–32 (lw)				
Long-finned pilot whales	Tasmania, Australia	108	Fat	2–54 (lw)	BDE 47			[173]
Long-beaked common dolphins	Korea	22	Fat	140–3100 (lw)	BDE 47			[111]
Long-beaked common dolphins	Korea	22	Liver	20–320 (lw)	BDE 47			
Indo-Pacific humpback dolphins	China	17	Fat	280–51100 (lw)	NM	32–519 (lw)	α	[94]
Bottlenose dolphins	Florida, USA	58	Fat	196–3790 (lw)	BDE 47			[45]
Bottlenose dolphins	South Carolina, USA	53	Fat	429–22783 (lw)	BDE 47			

Continued

TABLE 2 Literature Overview Regarding the Bioaccumulation of PBDEs and HBCDs in the Aquatic Environment. Levels of PBDEs and HBCDs in This Table are Minimum–Maximum Ranges Unless Only Mean ± SD Was Provided—cont'd

Species	Location	n	Tissue	PBDEs (ng/g)	Major PBDE	HBCDs (ng/g)	Major HBCD	References
Guiana dolphins	Brazil	20	Liver	13–1620 (lw)	BDE 47			[34]
Atlantic spotted dolphin	Brazil	7	Liver	96–2440 (lw)	BDE 47			
False killer whale	Brazil	2	Liver	1210–5960 (lw)	BDE 47			
Bottlenose dolphins	Brazil	3	Liver	270–1350 (lw)	BDE 47			
Rough-toothed dolphin	Brazil	3	Liver	360–1600 (lw)	BDE 47			
Short-beaked common dolphin	Brazil	2	Liver	125–240 (lw)	BDE 47			
Pantropical spotted dolphin	Brazil	1	Liver	1215 (lw)	BDE 47			
Spinner dolphin	Brazil	1	Liver	150 (lw)	BDE 47			
Striped dolphin	Brazil	1	Liver	210 (lw)	BDE 47			
Fraser's dolphin	Brazil	9	Liver	1–28 (lw)	BDE 47			

Bottlenose dolphin	Florida	20	Fat	30–4500 (lw)	BDE 47			[81]
Striped dolphin	Florida	1	Fat	660 (lw)	BDE 47			
Finless porpoises	China	33	Fat	103–6789 (lw)	NM	4–501 (lw)	α	[94]
Harbor porpoises	North Sea	34	Fat	220–5930 (lw)	BDE 47			[172]
Harbor porpoises	Black Sea	20	Fat	44–85 (lw)	BDE 47			[171]
Harbor porpoises	UK	61	Fat	<LOQ–6900 (ww)	BDE 47			[97]
Grey seals	Canada	5	Fat	112 ± 55 (lw)	BDE 47			[74]
Harbor seals	Canada	20	Fat	28–319 (lw)	BDE 47			
Harbor seals	North Sea	28	Fat	90–1150 (lw)	BDE 47			[172]
Ringed seal	Svalbard	6	Fat	42–94 (lw)	BDE 47	15–35 (lw)		[137]
California sea lions	California, USA	25	Fat	451–4739 (ww)	BDE 47	<1–12 (ww)	α	[143]

Continued

TABLE 2 Literature Overview Regarding the Bioaccumulation of PBDEs and HBCDs in the Aquatic Environment. Levels of PBDEs and HBCDs in This Table are Minimum–Maximum Ranges Unless Only Mean ± SD Was Provided—cont'd

Species	Location	n	Tissue	PBDEs (ng/g)	Major PBDE	HBCDs (ng/g)	Major HBCD	References
California sea lions Harbor seals Northern elephant seals	California, USA	63 9 16	Fat Fat Fat	40–33700 (ww) 30–2410 (ww) 10–450 (ww)	BDE 47 BDE 47 BDE 47			[110]
Steller sea lions	Canada	22	Fat	50–3776 (lw)	BDE 47			[4]
Minke whales	Korea	27	Fat	30–430 (lw)	BDE 47			[111]
Minke whales	Korea	27	Liver	3–90 (lw)	BDE 47			
Humpback whale	North Pacific	67	Biopsy	<LOQ–330 (lw)	NM			[35]
Humpback whale	North Atlantic	20	Biopsy	520–1500 (lw)	NM			
Dugong	Queensland, Australia	1	Fat	1 (lw)	BDE 47			[69]

Note: PBDEs does not necessarily include the same number, nor the same congeners, HBCDs does not necessarily include all three stereoisomers.
NM = not mentioned, ND = not detected.
[a] *Only fish species from year 2000 were included.*
[b] *BDE 47 was the only congener detected in this species.*
[c] *Not a min–max range but 95% low and high values.*

time in fish species from the Great Lakes, USA. More precisely, they reported an increase in proportion of BDE 47 from 1984 to 1996 and the opposite trend for BDE 153 which is probably due to a change in the use of different PBDE mixture over time, combined with a higher bioavailability for less brominated congeners compared to higher brominated congeners. For HBCDs, the α-isomer was dominant in most fish species from Japan, although there were also species with a higher proportion of γ-HBCD [113]. Unfortunately, several studies do not mention individual HBCD isomers, and present only the sum HBCDs (Table 2).

Different distribution of PBDE congeners together with differences in the total PBDE content were observed between various fish tissues from three different farming types, such as freshwater farmed, seawater farmed, and seawater wild fish (collected from the Pearl River Delta of South China) [55]. The dry weight based concentrations of sum 10 PBDE (sum of BDE-28, -47, -66, -85, -99, -100, -138, -153, -154, and -183) in fish tissues followed the sequence of liver > gill > skin > GI tract > muscle. In another study in carp [179], no differences in the PBDE concentrations could be observed among the tissues of the same species when lipid normalized. Nevertheless, when concentrations were expressed in ng/g ww, PBDE levels in liver and eggs were significantly higher than that in the muscle.

Elasmobranchs. Compared to other species, sharks and rays have been the focus of only few BFR-related studies. Sharks and rays are a very interesting group in terms of bioaccumulation and biomagnification as they can occupy several levels in aquatic food chains. Rays and some shark species are typical benthic organisms, whereas some larger shark species are marine mammal-eating apex predators in aquatic ecosystems. Obviously, such different life histories have an impact on the overall bioaccumulation in a specific species.

The highest PBDE and HBCD levels were measured in liver samples of two different bull sharks from the coastal waters in Florida and were 4190 and 416 ng/g lw, respectively (Table 2; [81]). The lowest PBDE levels were measured in Atlantic stingrays (Table 2; [170]) and the lowest HBCD levels were found in Atlantic sharpnose sharks [80]. An increase in PBDE levels in muscle of bull sharks from Florida was observed between 1993 and 2004 [81]. Yet, levels in the liver of the same species were lower in samples from 2009 to 2012 [170]. Nomiyama et al. [114] found relatively low levels of PBDEs (pg/g ww) in blood of scalloped hammerhead sharks from Japanese waters.

BDE 209 was dominant in all shark species, except for the muscle of Atlantic stingrays where BDE 47 was predominant [81]. In the study of [170], BDE 47 was the dominant PBDE in all investigated species from Florida. Johnson-Restrepo et al. [80] have investigated HBCDs in elasmobranch species, namely bull sharks and Atlantic sharpnose sharks. In both species, γ-HBCD was the most dominant stereoisomers among all three HBCDs. Since these shark species are top predators, and γ-HBCD is supposed to be rapidly metabolized and eliminated in the lower levels of the food chain [146], this finding raises the question about where these sharks accumulated γ-HBCD. This finding also shows that, while mammals (mice) are capable of converting γ-HBCD into α-HBCD [146], top predator shark species might be unable to perform the same conversion.

Marine mammals. Marine mammals are long lived which obviously provides the opportunity to bioaccumulate lipophilic BFRs over several decades. Additionally, the blubber layers that are typically present in marine mammals are preferred sinks for lipophilic compounds. Furthermore, these animals pass on high levels of lipophilic BFRs to their offspring through lipid-rich milk, thereby effectively accelerating the bioaccumulation process in newborns.

BFRs contaminate marine mammals primarily through dietary intake and this pathway can be used to divide marine mammals roughly into four groups: the ones that are herbivorous (Sirenia), those that eat planktonic organisms (Mysticetes), those that eat fish (Odontocetes, pinnipeds), and those that prey upon fish and, to some extent, marine mammals (killer whales and polar bears). Although levels should theoretically be higher in the last group, the highest concentration of PBDEs was found in an Indo-Pacific humpback dolphin from China (Table 2; 51 μg/g lw) [94], followed by a bottlenose dolphin from South Carolina, USA (23 μg/g lw) [45]. Compared to these levels, concentrations of total PBDEs in blubber of killer whales and polar bears were about 50–100 times and 250–3000 times lower, respectively. Muir et al. [112] found higher levels of PBDEs in fat of polar bears than [137], but these higher levels were still approximately 100 times lower than the levels of PCBs. Nevertheless, levels of PBDEs in Arctic organisms have been reported to increase over time which puts these top predators at risk.

In the group of fish-eating marine marine mammals, species from the North Sea in Europe, US coastal waters, Brazilian waters, and Asian coastal waters have high levels of PBDEs exceeding 1000 ng/g lw. Fish-eating marine mammals from Australia, the Black Sea, and Artic regions (Norway, Canada, and Alaska) accumulate lower amounts of PBDEs in their fatty tissues. Although this probably reflects the distribution of PBDEs globally, there is also the issue of PBDE bioaccumulation in marine mammals from estuarine/coastal areas versus marine mammals from open waters. Dorneles et al. [34] investigated 10 marine mammal species from different areas (estuarine–continental shelf–oceanic) and found much lower levels in oceanic species than in species closer to shore. In addition to these spatial differences, there are also temporal differences that were not taken into account here. Although lower PBDE levels were expected in plankton-eating marine mammals, levels of PBDEs can still be as high as 1500 ng/g lw as found in a humpback whale from the North Atlantic [35]. Such a PBDE level fits within the levels found for fish-eating marine mammals, indicating that Mysticetes should not be ignored in biomonitoring studies. Low levels (1 ng/g lw) was found in fat of dugong, a herbivorous species from Queensland, Australia [69]. However, PBDE levels were also relatively low in Australian long-finned pilot whales [173] indicating that PBDEs are probably not ubiquitous in the Australian environment. Few studies have analyzed HBCDs in marine mammals. HBCD results show that levels are lower in polar bears [112] compared to porpoises and Indo-Pacific humpback dolphins [94], but higher than in sea lions [110,143] and ringed seals [137].

In most studies, BDE 47 was predominant, without any exception (Table 2). This may indicate that higher brominated PBDEs are already broken down to lower brominated PBDEs before reaching the top of the aquatic food chains. On the other hand, it might also be possible that some biotransformation occurs in marine mammals. Wolkers et al. [174] have suggested that killer whales are capable of metabolizing PBDEs and toxaphenes. In addition, BDE 209 has been found in several shark species [81] and [80]. As these species are top predators too, this shows that also higher brominated PBDEs are able to bioaccumulate in the higher regions of the food webs. However, metabolic breakdown has been shown so far only in teleost fish species (e.g., rainbow trout, common carp in [143]; fathead minnows in [115]). For HBCDs, all studies found a predominance of α-HBCD, in accordance with the short half-life of γ-HBCD and the time necessary for a compound to move up through the food chain.

2.2 Biomagnification

Biomagnification factors. There are several ways to calculate biomagnification factors (BMFs). A method where the TLs are directly taken into account in the calculations is the method used by [86]. Kelly et al. [86] calculated the trophic magnification factors (TMFs) as the antilog of the empirical slope m from the log-linear regression between the concentration in biota (C_B; expressed on a lipid basis) and the TLs for which $\delta^{15}N$ measurements were used.

$$\log C_B = (m \times TL) + b \ \textbf{with TMF} = \mathbf{10^m}$$

The easiest method is probably the one where the BMF is calculated as the ratio between the concentration of a certain pollutant in the predator and the concentration in the prey.

$$BMF = \frac{C_{predator}}{C_{prey}}$$

As the prey is often from a lower TL, this approach takes the TL into account only indirectly. Both approaches are often used, although they do not represent the same thing. TMFs are more general and represent an average biomagnification potential in a food web, whereas BMFs are more specific and represent the biomagnification potential on a predator–prey basis.

Freshwater. Although river dolphins exist, studies investigating biomagnification of PBDEs or HBCDs from invertebrates to marine mammals in freshwater ecosystems are lacking. Some of the larger shark top predators like bull sharks are known to forage in freshwater as well as marine water. However, such a complicated and mixed lifestyle and diet is hard to explain in terms of BMFs. Consequently, biomagnification studies in freshwater ecosystems have been done mainly between teleost predators and their invertebrate prey. Van Ael et al. [152] calculated biota-sediment accumulation factors (BSAF) for PBDEs, but no BMFs. However, for two realistic prey–predator combinations, namely brown shrimp–common sole and brown shrimp–European flounder, BMFs for PBDEs would be on average 3 and 7, respectively. In the study of [159], all TMFs calculated for PBDEs in rivers from the Congo River Basin ranged between 1.6 and 3.6 indicating that biomagnification of PBDEs occurs in these rivers. Tomy et al. [148] reported a TMF value of 6.3 for HBCDs in the Lake Ontario pelagic food web, which was higher than the TMFs for p,p'-DDE and PCBs.

Marine ecosystems. Johnson-Restrepo et al. [81] calculated mean BMFs for PBDEs between primary consumers (silver perch, spotted seatrout, striped mullet, and Atlantic stingray) and their predators from the marine food web in Florida. Mean BMFs for PBDEs between primary consumers and top-predators (Atlantic sharpnose shark, bull shark, and bottlenose dolphin) were 31 (Atlantic sharpnose shark), 85 (bull shark), and 63 (bottlenose dolphin) [81]. Sørmo et al. [137] calculated BMFs for PBDEs and HBCDs between polar cod (prey) and ringed seals (predator) which resulted in values of 11 and 37 for HBCDs and PBDEs, respectively. The same study also calculated BMFs for PBDEs and HBCDs between ringed seals (prey) and polar bears (predator) and found BMFs of 0.5 and 0.7, respectively, indicating that both contaminants biomagnify more between fish and marine mammals as between two marine mammal species. Losada et al. [104] investigated biomagnification of PBDEs in a marine food web of Sydney harbor, Australia and reported that the highest TMF value of all studied compound classes was found for PBDEs (TMF = 3.9). Although food webs do not have similar composition, and TMFs and BMFs are, in essence, uncomparable, it is clear that both PBDEs as HBCDs biomagnify in marine as well as freshwater food webs.

3. BIOACCUMULATION OF BFRs IN TERRESTRIAL ORGANISMS

Since BFRs leak from consumer products, the bulk of released BFRs remain in various compartments of the terrestrial environment [99]. It was previously evidenced that the soil–air partition coefficients for some BFRs increased over time after they were added to the soil suggesting that their volatility decreases as the chemicals become more strongly bound to the matrix [175]. Contaminated matrices, such as indoor dust, waste disposal sites of electric and electronic equipment (e-waste sites) or soil, could become secondary emission sources. As such, these matrices may become a risk for contamination of terrestrial organisms [106,175].

Since their physicochemical properties are comparable to those of the legacy POPs [31], BFRs are likely to undergo low biodegradation processes in the environment and organisms. Independently of the trophic position of an organism, the BFRs intake is higher when compared to their elimination due to slow biotransformation, thus generating their bioaccumulation [52]. However, not all BFRs show similar behavior. Their structure and properties, and the metabolic processes taking place within the exposed organisms, are of great importance in determining their bioaccumulation profile [32]. The main exposure pathway to BFRs in terrestrial systems seems to be the indoor environment, mainly indoor dust [62], and therefore their bioaccumulation potential should be carefully evaluated. However, they are also known to biomagnify in terrestrial food chains as there is evidence of increasing levels from prey to predator [32].

In order to evaluate the bioaccumulation/biomagnification potential of BFRs, standardized test systems were developed. However, these tests are available mostly for aquatic organisms (OECD Guideline 305: Bioconcentration: Flow-through Fish Test 1996 [117]), and their extrapolation to terrestrial species is not necessarily realistic. For a comprehensive assessment of the risks posed by the presence of bioaccumulative BFRs into different ecosystems, specific food-chain models should also be developed for the terrestrial compartment [10]

3.1 Bioaccumulation

Bioaccumulation studies have become increasingly important in environmental research and risk analysis [10]. To document the bioaccumulation potential of BFRs, their levels should be monitored in various organisms over a certain period of time. Time trend studies would provide valuable information regarding the increasing levels of BFRs in various organisms.

In order to evaluate the bioaccumulation potential of BFRs, experiments were conducted directly under simulated indoor conditions where these chemicals are primarily released. Insects exposed to PBDE-containing polyurethane foam might bioaccumulate these chemicals [52]. Insects are the most abundant invertebrate organisms that facilitate the trophic transfer of contaminants by breaking down complex organic matter (including discarded polymers) and which serve as prey for other organisms. Experiments suggested that substantial PBDE burdens may be incurred by insects in contact with current-use and derelict-treated polymers within human spaces and solid waste disposal sites (e.g., landfills and automotive dumps). Even if the estimated PBDE bioaccumulation factors were low, ingested, but not absorbed, PBDE burdens may be dispersed within proximate terrestrial food webs via the insect's movements and/or predation [52].

3.1.1 Bird's Tissues and Eggs

Studies on the temporal trends recorded for BFRs in different terrestrial organisms could be a good indicator for their bioaccumulation potential [51]. Several studies focused on PBDE levels recorded in peregrine falcon eggs sampled in the USA starting from 1980s and up to present [17,119,121]. These studies

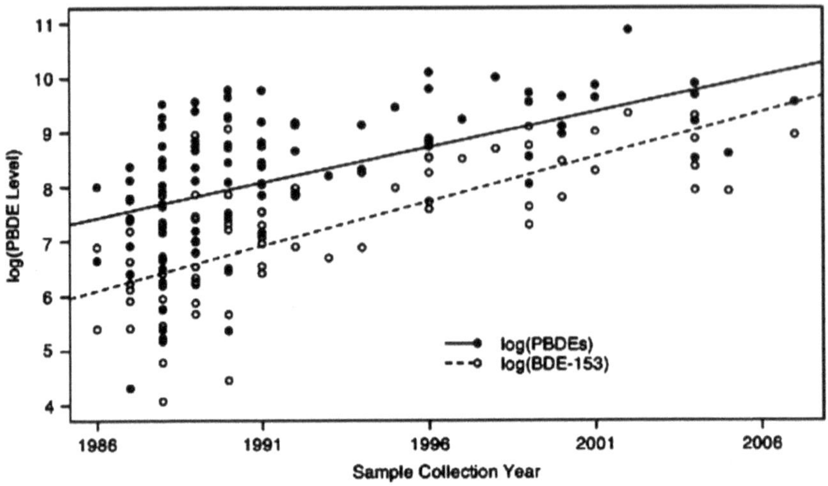

FIGURE 1 Temporal changes in the ln (natural logarithm) of levels (ng/g lipid weight) of ΣBDEs and BDE153 in peregrine falcon eggs from California. *Reprinted and adapted from [119] and [32].*

reported a significant increase over each decade for both lower and higher brominated compounds (Figure 1). Besides increasing levels over time, a change in the PBDE profile was also noticed as function of the contamination source: BDE 209 together with hexa- to nona-brominated congeners were the dominant compounds in eggs from cities, while BDE47 and BDE99 were dominant in coastal eggs [17,119]. Reported PBDE levels in common magpie feathers sampled in Belgium have also shown regional differences: BDE47 had a higher contribution in feathers from the urban area, while BDE99 was the most prominent congener in feathers from the rural areas [76]. Similarly, regional variations in the PBDE levels were also found for great tits and starlings eggs sampled in Belgium, where PBDE concentrations were shown to be significantly higher in the urban locations than in the rural and remote locations (urban > rural > remote) [155–157].

Fangstrom et al. [47] indicated only limited contamination with PBDE congeners, including BDE 209 in eggs, fat, and muscle of fulmars (Fulmarus glacialis) collected in 2000–2001 from the Faroe Islands. PBDE levels in fulmar eggs collected at Vidareidi on the Faroe Islands in 2003 were similar to those reported earlier by [47] and ranged from 16 to 43 ng/g lw [85]. BTBPE was detected in eight of nine samples up to 0.17 ng/g lw and BDE 209 was detected in only one of nine samples at 7.2 ng/g lw. Knudsen et al. [91] determined PBDEs, PBBs and HBCD in liver of fulmars breeding on Bjornoya in the Norwegian Arctic. Determinations were also made in blood plasma, but the authors concluded that this was a poor matrix for monitoring POPs burdens, at least in the breeding season when these samples were taken. In [162] maternal transfer of six PBDEs to the eggs of glaucous gulls collected in 2002 from Bear

Island in the Norwegian Arctic was studied. PBDE concentrations were consistently higher (on average by a factor of 2; actual concentrations not given) in the plasma of laying females than in their eggs. Intra-clutch profiles (on a lipid basis) showed no difference in concentrations between the three eggs. Later, [161] also determined 10 PBDEs and a range of MeO- and HO-PBDE congeners in blood samples taken from glaucous gulls at Bear Island during 2004. PBDE concentrations ranged from 8.7 to 68 and 8.2 to 35 ng/g ww in males and females, respectively. Concentrations of MeO–PBDE and HO–PBDE were lower, ranging from 0.31 to 4.3 and 0.06 to 1.1 ng/g ww, respectively.

Jimenez et al. [78] determined PBDEs in unhatched eggs of peregrine falcons (Falco peregrinus) from Madrid and Guadalajara, Spain, collected in 2000–2001. Dominant congeners in the eggs were BDE153 and BDE99, as compared to BDE47 which is dominant in piscivorous birds. Perhaps differences in metabolic capacity could help to explain these differences in congener patterns. Dauwer et al. [26] and Van den Steen et al. [154] studied PBDEs in eggs and nestlings of the great tit (Parus major) from Belgium collected in 2005 and 2000. In the study by [154], PBDE concentrations in pooled egg samples ranged from 57 to 609 ng/g lw. In the study by [26], PBDE concentrations in great tit eggs were 4–6 times higher than those in nestlings, corresponding to concentrations predicted by a bioenergetics-based model. Most of the POPs in 15-day-old nestlings were still of maternal origin.

Van den Steen et al. [153] also studied BDE 209 biotransformation in starlings (Sturnus vulgaris) following exposure using implants. No lower brominated congeners could be detected in blood; however, octa- and nona-BDEs were detected in liver and, to a lesser degree, muscle. For the nona-BDEs, the abundance was BDE207 > BDE208 > BDE206. Watanabe et al. [169] determined PBDEs and PBBs in the livers and eggs of common cormorants ($n=10$ of each) from Japan sampled in 2000. BDE47 comprised 40–50% of the total PBDE content in both liver and eggs. PBDE concentrations were lower than those reported earlier in cormorants from the UK and the Netherlands. Chen et al. [18] determined PBDEs in birds of prey from northern China. BDE 209 was the dominant congener in tissues from some buzzards, scops owls, and little owls. Most of the studied raptors were from urbanized areas and may have been exposed via exposure to flame-retarded products and/or via their diet. These results support the growing view that animals at the top of terrestrial food chains have greater exposure to BDE 209.

Measured concentrations of BDE 209 in peregrine falcon eggs from the USA were significantly correlated with the human population density of the area surrounding the nest, but such correlations were not significant for ΣPBDE [121]. A change in the PBDE patterns was also observed in many birds that yielded multiple eggs over time: the high proportions of BDE 209 and higher brominated BDEs (with short half-lives) in young birds contrasted with increasingly higher proportions of BDE153 (long half-life) and other lower brominated PBDEs in adult birds (Figure 1). These results suggested that BDE 209 is taken up by wildlife (particularly in urban locations) and further undergoes metabolic debromination to the lower brominated BDE congeners [32,119].

The bioaccumulation potential of BFRs in eggs might have implications for terrestrial birds since exposure studies reported several effects including delayed egg laying, smaller size of laid eggs, thinner eggshells, differential weight loss during embryonic development, and reduced fertility and reproductive success [48,49,79]. Therefore, it was noticed that the thickness of the eggshell declined with increasing concentrations of most PBDEs and α-HBCD tested in exposure experiments of adult American kestrels. Delayed egg laying, reduced eggshell mass and reduced fledging success was also observed to take place with the increasing concentrations of BDE153, 154, 28, and 17, or ΣPBDEs [48,49].

3.1.2 Terrestrial Wildlife

Apart from birds, bioaccumulation of BFRs was also shown to occur in other terrestrial wildlife. Concentrations of higher brominated BDE congeners including BDE 209 in terrestrial animals have been found to be relatively high, which indicates their bioaccumulation capacity in terrestrial animals, such as cows, red fox, and grizzly bears [19,87,165]. D'Have et al. [25] studied 7 PBDE congeners and BB153 in tissues of the European hedgehog. Road kills and carcasses from wildlife rescue centers in Belgium were used to investigate relationships between concentrations of BFRs in hair and internal tissues (liver, kidney, muscle, and adipose tissue). The relatively high PBDE levels in hedgehogs compared to other terrestrial wildlife probably result from their higher trophic position (insectivorous rather than grazers) in the food chain [165] determined seven PBDEs in adipose tissue, liver, and muscle of red fox collected in Belgium. Red foxes are opportunistic feeders, their diet is diverse and depends on the environment and the time of year. Generally, these concentrations can be considered low in relation to those in their prey. In some cases, BDE 209 was detected (40% of samples) and, for these cases, it dominated the PBDE profiles at concentrations up to 200, 760, and 290 ng/g lw in adipose tissue, liver, and muscle, respectively. BDE 209 in these samples contributed, on average, 70% of the total BDE content. This confirms that BDE 209 can biomagnify in a number of terrestrial top predators, including the red fox.

PBDE levels in wildlife from China were generally higher than those from other parts in Asia, and were comparable to those from Europe, but were lower than those from North America. However, wildlife from surroundings of e-waste recycling sites in South China and East China contained much higher PBDEs compared to other reports around the world, suggesting the heavy contamination of PBDEs in these regions. NBFRs were also detected in wildlife, revealing that the animals are exposed to these chemicals, in addition to PBDEs [178].

The bioaccumulative potential of BFRs applies not only for birds and wildlife, but also for humans. Although it is more difficult to set up human exposure studies, prospective studies have concluded that the PBDE concentrations have increased in humans by a factor of ~100 during the past 30 years [70]. The regression analysis of these data shows a positive linearity with time, despite the disparate sample types, the different continents of origin, and the various congeners measured. This approach shows that samples originating from North America were always above the regression line (in recent years by a factor of >10), while the Japanese samples are usually below the regression line (by a factor of ~5) indicating that the results are related to the exposure level of the general population from studied areas [21,70].

Given the rising trend observed in HBCD concentrations in porpoises from the UK in recent years, which was not confounded by small sample size or other factors, additional studies on marine mammal diet is indicated. Covaci et al. [22] also recommended that more emphasis be placed upon the terrestrial environment, for which limited HBCD data are available. Also, more detailed information concerning the kinetics, toxicology, pathways of exposure, and bioavailability of HBCD is needed, as well as the study of possible metabolites.

3.2 Biomagnification

BFRs are chemically and biologically persistent and furthermore lipophilic, which results in their bioaccumulation in fatty tissues of organisms and enrichment throughout food chains [96]. As a consequence of biomagnification, increasing concentrations of PBDEs can be found with increasing TL, leading to the highest concentrations in top predators [98]. Until this date, only few studies have discussed biomagnification of BFRs in terrestrial environments.

Sellstrom et al. [131] determined PBDEs, including octa-BDE congeners, in earthworm samples collected in 2000 from three research stations and two farms in Sweden. BSAFs declined in the following order, and with molecular size: tetra-BDE > penta-BDE > hexa-BDE > octa-BDE > nona-BDE > BDE 209. This shows that higher brominated BDEs, including BDE 209 are bioavailable from soils and accumulate in earthworms, presenting an exposure pathway into the terrestrial foodweb, as suggested by [98]. Gaylor et al. [53] demonstrated bioaccumulation of PBDEs in earthworms following ingestion of soil amended with biosolids, polyurethane foam microparticles or soils spiked with penta-BDE mixture. Worms accumulated up to 3740 μg/g lw PBDE burdens from all substrates. BSAFs generally decreased with increasing K_{OW} and substrate dosage. These results demonstrated that PBDEs may accumulate in organisms ingesting soils containing biosolids or waste plastics. Earthworms may further transfer their burdens to predators or translocate them from the site of application/disposal.

The biomagnification potential of the most commonly reported PBDE congeners was assessed in three small terrestrial food chains [164], by combining PBDE data on passerines and rodents with previously published data on terrestrial predators from the same area; that is, birds of prey [77,165] and red fox [166]. Because birds of prey occupy top positions in the food chain, it is most likely that biomagnification of persistent and lipophilic PBDEs through their prey would be substantial. BMFs were calculated using the median concentrations of PBDEs in prey and predator species. The BMF values for PBDEs in buzzard and sparrowhawk ranged from 2 to 34, depending on the congener, thus evidencing biomagnification. All PBDE congeners that could be determined in both prey and predator were biomagnified (BMF > 1). Surprisingly, no biomagnification could be observed from rodents to foxes. The median levels of sum PBDEs measured in fox were even lower than those in rodents.

The correct characterization of contaminants in different food webs makes their behavior in the environment more understandable, and trophic magnification can lead the chronic exposure and toxicological effects in organisms. The potential risk to human exposure considering our omnivorous diet is of special concern.

4. TRANSFORMATION PRODUCTS OF BFRs

4.1 Transformation Products of BFRs in Abiotic Matrices

Several studies have shown that BDE 209 and other higher brominated PBDEs may decompose in photolytic conditions due to natural sunlight or artificial UV light exposure (Table 3). The degradation of BDE 209 was observed independently of the exposure conditions or tested media, such as solvent–water mixtures, silica-gel, sand, sediment, or soil samples [3,38,46,134,135]. Even if the profile of TPs consistently indicated formation of tri- to nona-BDEs [3,135], the photodegradation kinetics were highly variable among different matrices. Also, nona-BDEs may undergo rapid photodegradation in organic solvents [27]. Photodegradation of BDE 209 in dust led to the formation of all nona- and several octa-BDE congeners [140]. Since the octa-BDE mixture may contain BDE201 only in a very small amount (<0.8%), and BDE202 is usually not present in the mixture, it was therefore suggested that the presence of BDE201 and 202 in house dust is a marker of environmental debromination of BDE 209 [140].

Controlled dehalogenation has been described for PBDEs using nanoscale zero-valent iron particles (nZVI), composite particles, and variants [185,186]. Microbial degradation using mixed microbes or pure bacteria strains, like Sulfurospirillum multivorans or Dehalococcoides species [89,126,133] has been reported as a cheaper alternative for PBDE degradation than physical or chemical treatments.

For HBCD, two degradation mechanisms were reported: debromination with elimination of either HBr or Br_2 and further generation of lower brominated dodecadiene species and isomerization processes in which stereoisomer interconversion takes place without the loss of any other secondary products (Table 4).

Thermal degradation showed an auto-accelerating behavior, possibly catalyzed by organic intermediates formed in the decomposition process [7,8]. Due to thermal stress, around 35% of the total HBCD was degraded, from which about 75% of bromine (by weight) was released as HBr, while 25% was

TABLE 3 Transformation Pathways of Selected BFRs in Various Abiotic Media

Compound	Transformation Pathways/Transformation Products		Conditions	Comments	Reference
	Debromination	Other Transformation Pathway			
BDE 28	BDE15	–	UV	Experiments in n-hexane	[46]
BDE 47	BDE 28, 15	–	UV	Experiments in n-hexane	[46]
BDE 99	BDE 28, 49, 47, 66	–	UV	Experiments in n-hexane	[46]
BDE 100	BDE 28, 75, 47	–	UV	Experiments in n-hexane	[46]
BDE 153	BDE 28, 47, 118, 49, 66, 99	–	UV	Experiments d in n-hexane	[46]
BDE 183	BDE 28, 47, 99, 154, 138, 49, 66, 118, 153	–	UV	Experiments in n-hexane	[46]
BDE 206, 207, 208	Octa- and hepta-BDEs	–	Exposure to natural sunlight, while compounds kept in different solvents	BDE207 rapid degradation BDE206 slow degradation	[27]

Continued

TABLE 3 Transformation Pathways of Selected BFRs in Various Abiotic Media—cont'd

Compound	Transformation Pathways/Transformation Products		Conditions	Comments	Reference
	Debromination	Other Transformation Pathway			
BDE 209	Lower brominated BDEs	–	Bacterial aerobic biotransformation by Lysinibacillus fusiformis strain DB-1	Carbon sources (lactate, pyruvate, acetate)	[30]
	Nona-BDEs: 206, 207, 208 Octa-BDEs: 196, 197, 201, 202, 203/200	–	Exposure to direct sunlight for over 200 h	Degradation study conducted in house dust	[140]
	Nona- and octa-BDEs	Tetra- to octa-brominated dibenzofurans	Exposure to natural sunlight	BDE 209 incorporated into high-impact polystyrene and TV casings	[82]
HBCDs	–	75% of bromine is released as HBr	Thermal degradation in N_2 and in air at moderate heating rates (10 °C/min)	No influence of O_2 on the HBCD decomposition	[7,8]
	–	γ-HBCD isomerization to α-HBCD	Process influenced by temperature increase to above 110 °C	Process observed in toluene	[83]

			experiments in a quartz tube (600 °C)	polystyrene samples containing DBDPE	
	Hepta- to nona-BDPEs	–	Irradiation with UV light	Experiments conducted in acetone/ tetrahydrofuran/ toluene (20/30/50%) mixture	[167]
TBPH	Debrominated products	–	Exposure to natural sunlight	Test compound kept in different solvents	[27]
TBB	Debrominated products	–	Exposure to natural sunlight	Test compound kept in different solvents	[27]
BTBPE	–	2,4,6-Tribromophenol; vinyl 2,4,6-tribromophenyl ether	Isothermal conditions at 240 and 340 °C	Inert atmosphere	[6]
	–	4-Isopropyl-2,6-dibromophenol, 4-isopropylene-2,6-dibromophenol,4-(2-HO-isopropyl)-2,6-dibromophenol	Irradiation with UV light	Experiments conducted in aqueous solutions at different pH values	[38]
TBBPA	5,5′-propane-2,2-diyldibenzene-1,2,3-triol,5,5′-(1,3-di-HO-propane-2,2-diyl)dibenzene-1,2,3-triol	2,6-Dibromo-4-isopropenylphenol,2,6-dibromo-4-isopropylidenecyclohexa-2,5-dien-1-one, 2,6-dibromo-4-(1-HO-1-methylethyl)phenol	Photocatalytic degradation by irradiation with UV–VIS light	In the presence of bismuth oxybromide (BiOBr)	[180]

TABLE 4 Transformation Pathways of Selected BFRs in Various Biotic Matrices

Compound	Transformation Pathways/Transformation Products			Conditions	Comments	Reference
	Debromination (Loss of at Least One Br Atom)	Oxidative Metabolism	Other Transformation Pathway			
Penta-BDE	BDE 47, 100	11 HO-PBDEs	–	Dietary exposure studies in common carp (*Cyprinus carpio*)	Low ratio HO-PBDEs to precursor PBDEs	[142,182]
Octa-BDE	BDE 154, 155, 149, 153	–	–	Dietary exposure studies in common carp (*C. carpio*)		[142,182]
Deca-BDE	BDE 154, 155, 149, 188, 179, 202	–	–	Dietary exposure studies in common carp (*C. carpio*)		[142,182]
BDE 47	BDE 4	5 isomers of HO-BDE47	2 isomers Of HO-tri-BDE	Orallt dised Jumbo Cornish × Rock cross male chickens at 7 weeks of age		[57]
	–	5 HO-BDEs, majors: 4'-OH-BDE-49 and 3-OH-BDE-47	–	In vitro testing with rat hepatic microsomes	Oxidative metabolism of BDE 99 was greater than that of BDE-47	[39]

BDE 99	BDE 47	–	–	Experiments on intestine microflora and intestine and liver microsomes of common carp (*C. carpio*)	Intestinal microflora not responsible for BDE99 debromination	[11]
	BDE 49	–	–	Experiments on Chinook salmon (*Onchorhynchus tshawytscha*) liver fractions	Process not NADPH-dependent, indicating a lack of cytochrome P450 involvement	[14]
	–	7 HO-BDEs, among of which more important were 4-OH-BDE-90 and 6′-OH-BDE-99	–	In vitro testing with rat hepatic microsomes	Oxidative metabolism of BDE 99 was greater than that of BDE 47	[39]
	–	Several HO-BDEs and diHO-BDEs	2,4,5-Tribromo-phenol	In vitro testing with human hepatic microsomes		[40,107]

Continued

TABLE 4 Transformation Pathways of Selected BFRs in Various Biotic Matrices—cont'd

Compound	Transformation Pathways/Transformation Products			Conditions	Comments	Reference
	Debromination (Loss of at Least One Br Atom)	Oxidative Metabolism	Other Transformation Pathway			
BDE 209	BDE 202	–	–	In vivo exposure on common carp (*C. carpio*) and rainbow trout (*Onchorhynchus mykiss*)		[139]
	Octa- and nona-BDEs	–	–	In vitro tests on liver microsomal fractions from rainbow trout (*Onchorhynchus mykiss*)	Up to 22% by mass of the BDE 209 was biotransformed	[139]
	Hexa- to nona-BDEs	–	–	In vitro tests on liver microsomal fractions from common carp (*C. carpio*)	Up to 65% by mass of the BDE 209 was biotransformed	[139]

HBCD	PBCDs, TBCDs (unclear if they are metabolic products in humans)	–	Enantioselective enrichment of (−)-α-HBCD	Observation on human milk samples		[1]
	–	–	Interconversion from γ- to α-HBCD	Repeated exposure in adult female C57BL/6 mice	Shorter half-life recorded for from γ- compared to α-HBCD	[145,146]
	Dehydrobromination products through the loss of [-HBr] and [-2HBr],[-3HBr]	–	–	In vitro hepatic assay using microsomes extracted from: polar bear, beluga whale, ringed seal, rat	Regardless of species, the rate of γ-HBCD metabolism is more rapid than α-HBCD	[108]
	–	Mono HO-HBCDs; diHO-HBCDs only in case of α- and γ-HBCD	–	In vitro, rat liver microsomes from male Spargue-Dawley rats preinduced with phenobarbital/β-naphtoflavone	Preferential metabolism of γ-HBCD	[41]
	PBCD as minor metabolites	HO-HBCDs as major metabolites	HO-PBCDs as minor metabolites	In vitro, phenobarbital induced rat microsomes		[128,129]
	PBCDs and TBCDs	Mono HO-HBCDs and diHO-HBCDs	–	Observations on wildlife species (tern egg, seal, and flounder) and in rats exposed to 30 and 100 mg HBCD/kg bw/day for 28 days	Differences in metabolism observed among species	[13]

Continued

TABLE 4 Transformation Pathways of Selected BFRs in Various Biotic Matrices—cont'd

Compound	Transformation Pathways/Transformation Products			Conditions	Comments	Reference
	Debromination (Loss of at Least One Br Atom)	Oxidative Metabolism	Other Transformation Pathway			
DBDPE	Lower BDPEs followed by phase I and phase II metabolism	–	$MeSO_2$-nona-BDPE, $EtSO_2$-nona-BDPE	Oral exposure of male Sprague–Dawley rats (21 days old)	Oral administration of 100 mg/kg bw/day for 90 days	[167]
TBPH	–	–	Mono(2-ethylhexyl) tetrabromophthalate	In vitro, rat and human liver microsomes	Enzymes involved in metabolism: Carboxylesterases	[124]
TBB	–	–	2,3,4,5-Tetrabromo-benzoic acid	In vitro, rat and human liver microsomes	Enzymes involved in metabolism: carboxylesterases	[124]
	–	–	2,3,4,5-Tetrabromo-methylbenzoate	In hepatic subcellular fractions (i.e., S9, microsomes and cytosol) in the fathead minnow, common carp, mouse and snapping turtle	Differences in metabolism observed among species	[9]

BTBPE	Multiple metabolites: monoHO-, monoHO- with debromination, diHO-/debrominated on a single aromatic ring, monoHO- on each aromatic ring with accompanying debromination, and cleavage on either side of the ether linkage to yield tribromophenol and tribromophenoxyethanol			Single dose administrated to male Sprague–Dawley rats	Limited absorption and metabolism occur by ingestion in mammals	[58]
TBBPA	–	–	TBBPA-glucuronide, TBBPA-glucuronide-sulfate, TBBPA-sulfate, TBBPA-disulfate	In vivo, exposure of *Xenopus laevis* tadpoles	Metabolism of >94% of ^{14}C-TBBPA within 8 h	[50]
	–	–	TBBPA-glucuronide, TBBPA-sulfate	Studies of the TBBPA metabolism in mammals, including humans	Predominance of phase II biotransformation pathway products	[90,92,130]
	TribromobisphenolA	–	–	^{14}C-TBBPA intraperitoneally exposure of female Wistar rats	High exposure doses of 250 and 1000 mg/kg	[147]

involved in the formation of high-molecular-weight brominated decomposition compounds. Photodegradation of HBCD through a debromination mechanism was reported in aqueous solutions containing methanol [183]. In indoor dust, photodegradation and/or isomerization of HBCDs have been reported to occur possibly via elimination of HBr. The identified TPs were pentabromocyclododecenes (four isomers) and tetrabromo-cyclododecadienes (two isomers) [2]. Substantial debromination of α-, β-, and γ-HBCDs via dihaloelimination was observed in wastewater sludge and in freshwater aquatic sediment microcosms, each prepared under aerobic and anaerobic conditions [28]. No significant difference in the degradation of the HBCD isomers was seen. However, almost a double half-life was reported for α-HBCD degradation under anaerobic conditions in digester sludge compared to β- and γ-HBCDs [54].

Temperature is an important factor for the isomerization of HBCDs from various products; for example, expanded (EPS) or extruded polystyrenes (XPS). The increase in temperature from 140 to 160 °C led to decreased proportions of γ-HBCD over time from 80% to 21% in exposed EPS, whereas those of α- and β-HBCDs increased from 6% to 46% and from 13% to 30%, respectively, with δ-HBCD remaining at 1–3% [68]. Such results are consistent with previous data which showed that under kinetic control (110–140 °C) enantiomerically pure HBCDs isomerize regio- and stereoselectively with (+)γ-HBCD converting to (+)α-HBCD and (−)γ-HBCD converting to (−)α-HBCD [67]. Such isomerization processes were observed only in EPS, but not in XPS. Probably, the XPS sample was already rich in α-HBCD before thermal exposure and proportions increased only slightly from 43% to 54%, whereas those of β- and γ-HBCDs decreased from 36% to 25% and from 21% to 19%, respectively, with δ-HBCD remaining constant at ≈1%. Temperatures above 110 °C during fabrication processes were already reported to influence the isomerization of HBCDs leading to higher percentages of α-HBCD [83].

During direct photolysis of TBBPA with UV irradiation and oxidation by MnO, intermediate phenols, and ketones may be formed [38,61,73,102]. The reductive degradation of TBBPA over iron–silver bimetallic nanoparticles was assumed to be governed by debromination [105]. Lin et al. [101] indicated that more than 86% of TBBPA was debrominated by nZVI in methanol–water to tri-BBPA, di-BBPA, BBPA, and BPA. The transformation mechanisms involve debromination, hydroxylation, or demethylation processes which lead the mineralization of TBBPA [180]. Brominated light hydrocarbons, as well as phenolic derivatives, were observed during thermal decomposition (combustion and pyrolysis) of TBBPA [118].

A reduction of TBBPA was observed over 45 days of anaerobic degradation of a highly contaminated soil [127]. TBBPA converted into bisphenol A (BPA) by reductive debromination under anaerobic conditions, such as anoxic estuarine sediments [163]. Complete debromination of TBBPA to BPA was also observed under both methanogenic and sulfate-reducing conditions. Various bacteria and fungi of different origin were also capable of simultaneous debromination and aerobic mineralization of TBBPA [5,149]. High degradation and debromination efficiencies were achieved within 72 h.

DBDPE undergoes thermal degradation at a comparable temperature to BDE 209 but besides debromination or intermolecular ring closure, it also generates bromotoluenes [75]. Two nona-brominated degradation products of DBDPE were observed in the technical mixture [88]. When exposed to UV (125 W), DBDPE degraded to a number of hepta- to nona-BDPEs [167]. Under natural sunlight condition, this degradation was not observed in plastic matrices [82]. The main pyrolysis decomposition products of BTBPE were TBP and vinyl-2,4,6-tribromophenyl ether [6]. However, prolonged contact at high temperature with a heat source can also produce many brominated species, including polybrominated dibenzodioxins, which are even more toxic than BTBPE. TBB and TBPH are prone to photodegradation in organic solvents with formation debrominated analogs having two or three bromine atoms or compounds without the alkyl side chain [27].

4.2 Biotransformation Pathways of BFRs

The main biotransformation pathways can be initially assessed by in vitro research using microsomes together with a wide variety of enzymes, while in vivo exposure studies are valuable in identifying major pathways of metabolism. Further, results obtained through either in vitro or in vivo research might be extrapolated when BFRs or their TPs are measured in biota or human samples.

The main transformation pathways reported for the PBDEs in biota are: (a) *reductive debromination* in which hepta- to deca-BDE congeners lead to tetra- to hexa-PBDEs [115]; (b) *oxidative metabolism* leading to the formation of hydroxylated PBDEs (HO-PBDEs) or other phenolic compounds (Figure 2). Other metabolic pathways may include phase II conjugation (glucuronidation or sulfation).

In common carp (Cyprinus carpio), BDE 47 and 100 were the major congeners resulting after exposure to penta-BDE; BDE154, 155, 149, and 153 from the octa-BDE exposure; and BDE154, 155, 149, 188, 179, and 202 from the deca-BDE exposure [142,182]. Congeners with at least one meta- or para-double flanked bromine atoms easily undergo metabolic debromination in common carp [11,182], while PBDE congeners without one meta-substituted bromine (BDEs 28, 47, and 100) usually do not undergo debromination in fish [125].

Using in vitro aerobic and anaerobic experiments on intestinal microflora of carp, Benedict et al. [11] have shown that debromination of BDE99 to 47 did not take place. In vivo exposure experiments with BDE 209 in common carp and rainbow trout (*Onchorhynchus mykiss*) showed that BDE202 was a common debromination product in both organisms. Up to 22% by mass of BDE 209 was biotransformed in rainbow trout, primarily to octa- and nona-BDEs, while carp liver microsomes biotransformed up to 65% of BDE 209 primarily down to hexa-BDEs [139].

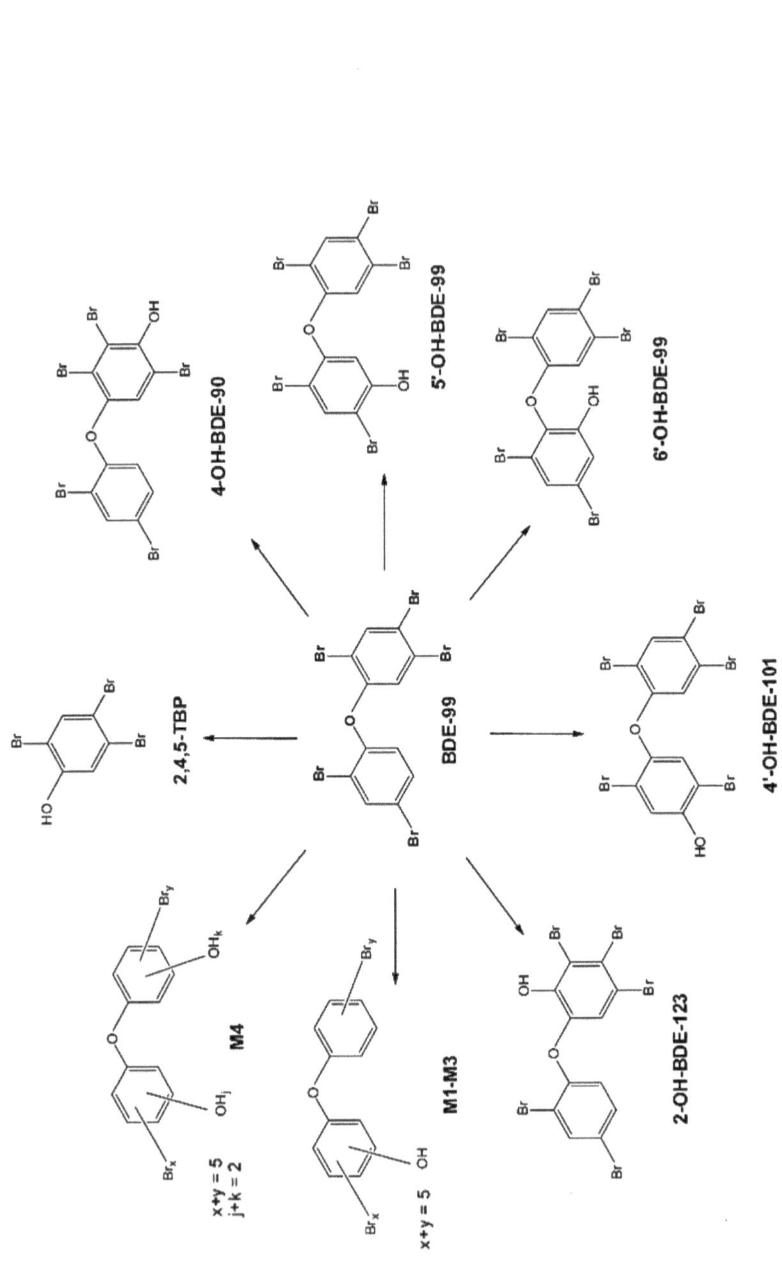

FIGURE 2 Scheme showing the chemical structures of metabolites formed following the incubation of human liver microsomes with 2,2′,4,4′,5-pentabromodiphenyl ether (tBDE-99). General structures for the metabolites M1–M4 are also shown. *Reproduced with permission from the publisher of [40].*

In vitro experiments on intestinal microflora and liver fractions of wildlife species and in vivo exposure studies have indicated that debromination pathways were species specific [11,116,125], and therefore findings are difficult to extrapolate between species.

Another important pathway is the **oxidative metabolism** with the formation of slightly more polar compounds, such as HO-PBDEs and other phenolic compounds. Oxidative metabolism is of toxicological interest because HO-PBDE metabolites may have greater biological and toxicological activity than the parent compounds [15,109]. In vitro testing with rat hepatic microsomes showed that the rate of oxidative metabolism of BDE99 was greater than that of BDE47, while the major metabolic pathway for both PBDE congeners was a para-hydroxylation involving an NIH-shift mechanism [39]. These results were in agreement with the PBDEs profiles in biota, where BDE99 has a lower contribution to the total PBDEs compared to BDE47 [62]. For BDE47, five HO-PBDEs were identified, while for BDE99, seven HO-PBDEs were formed [39]. Using human liver microsomes (HLMs), a higher number of hydroxylated metabolites of BDE99 were identified than for rat liver microsomes [40]. Apart from HO-PBDEs, 2,4,5-tribromophenol (TBP) was also formed from BDE 99 [33,40,107,141]. The oxidative metabolism of BDE47 and 99 is mediated by distinct, but overlapping, sets of CYP enzymes (e.g., CYP2A2 and CYP3A1) and represents a key process that dictates the bioaccumulation of BDE47 and 99 in mammals [33,39]. In HLMs, the oxidative metabolism of BDE99 was catalyzed solely by CYP2B6 and is an important determinant of the toxicity and bioaccumulation of BDE99 in humans [40]. Consequently, due to the variable composition of CYP enzymes in liver of different species, chemicals might be metabolized in different ways, and therefore interspecies differences in profiles of TPs are expected.

No hydroxylated or debrominated metabolites were observed in HLMs exposed in vitro to BDE 209 [141]. This suggests that BDE 209 was not metabolized, that non-extractable, covalently protein-bound metabolites were formed, or that the exposure time (24–72h) was not long enough for BDE 209 to diffuse into the cells [141]. Oxidative metabolism of BDE 209 would imply the formation of the arene-oxide intermediate, but its formation is unlikely on the fully brominated rings, and therefore debromination probably precedes oxidation [60]. The lower levels of BDE 209 in biota are not necessarily explained by its metabolism in organisms, but may be due to its relatively lower absorption rate and bioavailability compared to the other PBDEs.

The biotransformation of HBCD leads to the formation of various metabolites through *oxidative metabolism, dehydrobromination* or *debromination*, and *stereospecific metabolism (bioizomerization)* with isomeric interconversion (Table 4).

An in vitro hepatic assay using microsomes extracted from different animal species (polar bear, beluga whale, ringed seal, and rat) tested for α-HBCD and γ-HBCD in vitro depletion/metabolite formation [108]. In all species, debromination of HBCD isomers appeared to be the predominant metabolic pathway, but no other metabolites, for example, HO-HBCDs, were investigated. Dehydrobromination was reported to take place through the loss of the [-HBr] and [-2HBr], but [-3HBr] products were also detected in some of the incubation extracts. As expected, α-HBCD was by far the most predominant parent isomer relative to β-HBCD or γ-HBCD. However, other studies showed that, besides debromination, oxidative metabolism is one of the most important biotransformation pathways for HBCDs [41,129].

By using subcellular rat and HLMs, the major phase I metabolites of α-HBCD and γ-HBCDs were reported to be OH-HBCDs, followed by intermediates like HO-PBCDs and PBCDs as minor metabolites [128,129]. The formation of similar metabolites was also confirmed by [41] after incubation of rat liver microsomes with enantiomers of α-, β-, and γ-HBCDs. Similar patterns of mono-HO-HBCDs were found in environmental samples (e.g., pollack, mackerel muscle, and herring gull eggs) [41]. The formation of the mono- and di-HO metabolites of HBCDs, together with the presence of PBCDs and TBCDs, was tested in three wildlife species (tern egg, seal, and flounder), and also in Wistar rats exposed to technical HBCD for 28 days [13]. While mono-HO-HBCDs and PBCDs were detected in tern eggs from the Western Scheldt and in the blubber of harbor seals (Wadden Sea), no HO-HBCDs could be detected in the tissue of flounder (Wadden Sea) [13].

The interconversion from γ-HBCD to α-HBCD through bioisomerization was investigated through exposure of adult mice female to γ-HBCD [145]. γ-HBCD was rapidly metabolized and eliminated in urine and feces. In vivo stereoisomerization of γ-HBCD to β-HBCD was observed in liver and brain tissues and to α-HBCD and β-HBCD in fat and feces. However, in a similar exposure study, α-HBCD was excreted in the feces as parent compound and metabolites, whereas urine contained only metabolites [146]. In mice exposed to α-HBCD, four hydroxylated metabolites were detected in fecal extracts, and one of these metabolite isomers was consistently characterized in liver, brain, and adipose tissue [56]. In contrast, fecal extracts from mice exposed to γ-HBCD contained multiple isomers of hydroxylated debromination products, while in liver and adipose tissue only a single monohydroxy-pentabrominated metabolite was observed.

Fini et al. [50] observed the extensive and rapid uptake of ^{14}C-TBBPA in tadpoles, with >94% of ^{14}C-TBBPA being metabolized within 8 h to TBBPA glucuronide, TBBPA glucuronide sulfate, TBBPA sulfate, and TBBPA disulfate. These metabolites are identical to the TBBPA conjugates characterized in mammals, including humans. Most radioactivity (>75%) was associated with the sulfated conjugates.

An intraperitoneal dose of ^{14}C-TBBPA administered to female rats [147] led to urine and fecal excretion for 72 h. Cumulative urine excretion was very low (<1%), and fecal excretion was 50–65% of the administered dose. The majority of the fecal radioactivity was identified as unchanged TBBPA (>90%) and approximately 10% of the fecal ^{14}C-TBBPA was the debrominated tribromo-bisphenolA (TriBBPA). In male Sprague–Dawley rats, ^{14}C-TBBPA was readily absorbed from the GI tract [59], with 90% of a single oral dose of TBBPA being excreted in feces as unchanged TBBPA. No metabolites of TBBPA could be detected in the fecal extracts, but three major metabolites (glucuronide ether, the diglucuronide ether, and the glucuronide ether–sulfate ester diconjugate) have been identified as polar conjugates in rat bile [59]. Schauer et al. [130] also reported a predominance of phase II biotransformation pathway products, concluding that TBBPA was completely metabolized in rats and humans into sulfates and glucuronides and then excreted in urine. Absorption of TBBPA from the GI tract and rapid metabolism of the absorbed TBBPA by conjugation resulted in a low systemic bioavailability of TBBPA. Biotransformation studies of TBBPA by human and rat subcellular liver fractions have shown that TBBPA undergoes oxidative cleavage at the quaternary carbon atom to give brominated phenols, including 2,6-dibromohydroquinone [181].

Wang et al. [167] compared peaks corresponding to the main photodegradation products of DBDPE with peaks from an extract of DBDPE-exposed rat liver tissue and concluded that debromination alone is not the main biotransformation pathway (Table 4). Two metabolites (methyl sulfone substituted nona-BDPE and the ethyl sulfone analog) were tentatively identified. Apparently, reductive debromination was followed by phase I and II (glutationation + transformation to methylsulfone) metabolism.

Most of the ^{14}C-BTBPE (~94%) administered through was found in feces within 72 h. Tissue retention was minimal, in the most lipophilic tissues, and the metabolites were excreted in urine, bile, and feces, at very low levels [58]. The most significant TPs were the fecal ones (Table 2). The predominant compounds were those resulted after hydroxylation, debromination, and cleavage on either side of the ether linkage. By correlating the observed metabolites with the information available on the main biochemical pathways in rats, the authors have proposed a pathway for the metabolism of BTBPE in rats (Figure 3).

TBB and TBPH have shown a slightly different metabolism during in vitro studies (Table 4); the first compound was rapidly metabolized in rat microsomes and HLMs to 2,3,4,5-tetrabromobenzoic acid (TBBA) [124], probably by carboxylesterases. The same enzymes also cleaved one of the ester bonds in TBPH, forming mono(2-ethylhexyl) tetrabromophthalate. However, when [9] exposed fathead minnows to Firemaster BZ-54, 10 potential metabolites were observed. Only two of them were identified: 2-ethylhexyl dibromobenzoate and 2,3,4,5-tetrabromo methylbenzoate. No TBBA was detected. The authors hypothesized that TBBA undergoes methylation by methyltransferase enzymes, and the process occurs rapidly, preventing thus the accumulation of TBBA.

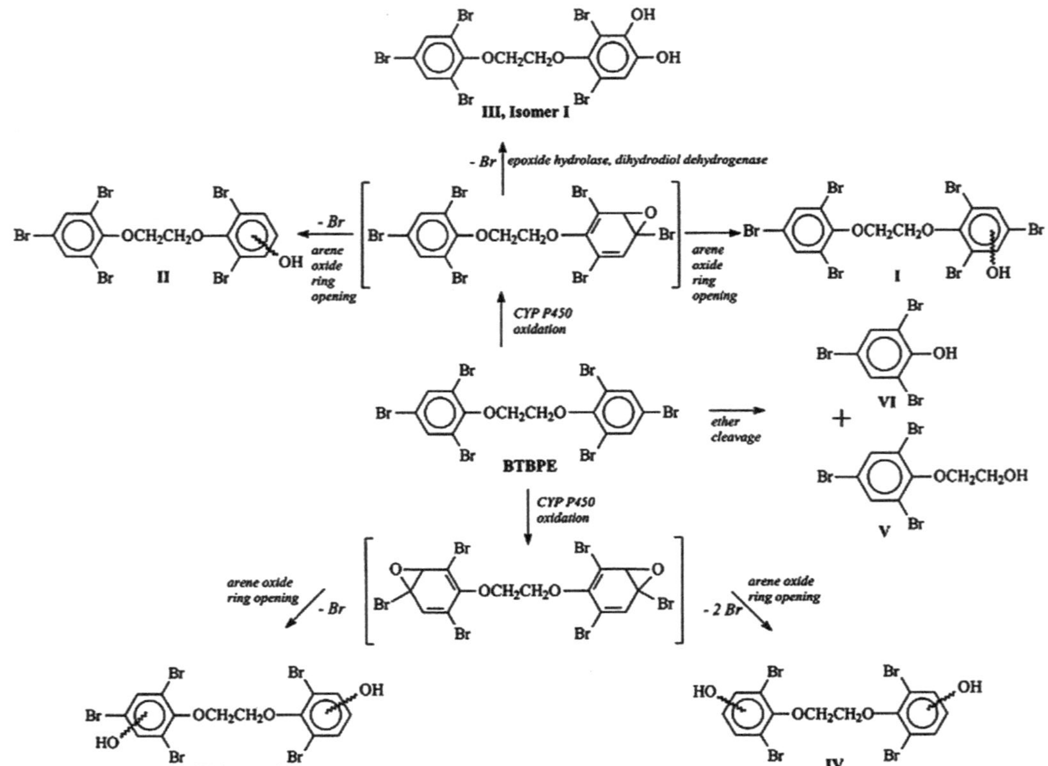

FIGURE 3 Proposed pathway for the metabolism of 1,2-bis(2,4,6-tribromophenoxy)ethane (BTBPE) in male rats based on the metabolism identified by mass spectrometry in feces. Oxidation, oxidative debromination, and ether cleavage are the favored biotransformations. *Reproduced with permission from the publisher of [58].*

5. CONCLUDING REMARKS AND FUTURE PERSPECTIVES

To better explain the accumulation profiles of BFRs in wildlife and humans, the lower absorption potential of the higher brominated compounds together with species differences in their metabolism are important factors explaining the common findings in biota. Additionally, debromination of the higher brominated compounds, for example, BDE 209 leading to more bioaccumulative and toxic compounds should be considered as an important aspect when addressing the regulatory policies for such chemicals [43]. Clearly, there are differences in the degree of bioaccumulation and/or biomagnification of BFRs between the aquatic and terrestrial food chains, and various aspects should be considered, including the physicochemical properties of the investigated BFRs, their structure and persistence in various media, and the media. The findings of the above-mentioned studies have suggested that in many cases, TPs may represent a higher risk in terms of toxicity than the parent products. Elucidating the transformation or metabolic pathways of BFRs would help to better explain the common findings for the congener-specific profiles in environment/biota. Since unregulated chemicals may degrade in the environment and biota to restricted or banned substances, regulation policies should better be extended also over the first group. Additional restriction criteria of their use in specific consumer products should also consider conditions that favor the degradation to regulated compounds.

While there is a clear shift in the production and use of BFRs, with PBDEs gradually disappearing from the markets, there are a large number of NBFRs and non-halogenated FRs which have recently seen an increase in their use in the past decade. Such compounds need to be closely investigated in the furutre studies, since much less information is available regarding their persistency, bioaccumulation potential, and toxicity.

ACKNOWLEDGMENTS

L.W. and A.C.D. were financially supported through a PhD and postdoctoral fellowships, respectively, from the Research Scientific Foundation-Flanders (FWO). G.M. thanks the University of Antwerp (UA) for a post-doctoral fellowship (GOA project). A.C. acknowledges financial support from the FWO and UA.

REFERENCES

[1] M.A.E. Abdallah, S. Harrad, Tetrabromobisphenol-A, hexabromocyclododecane and its degradation products in UK human milk: relationship to external exposure, Environ. Int. 37 (2011) 443–448.

[2] M.A.E. Abdallah, C. Ibarra, H. Neels, S. Harrad, A. Covaci, Comparative evaluation of liquid chromatography–mass spectrometry versus gas chromatography–mass spectrometry for the determination of hexabromocyclododecanes and their degradation products in indoor dust, J. Chromatogr. A 1190 (2008) 333–341.

[3] M.Y. Ahn, T.R. Filley, C.T. Jafvert, L. Nies, I. Hua, J. Bezares-Cruz, Photodegradation of decabromodiphenyl ether adsorbed onto clay minerals, metal oxides, and sediment, Environ. Sci. Technol. 40 (2006) 215–220.

[4] J.J. Alava, D. Lambourn, P. Olesiuk, M. Lance, S.J. Jeffries, F.A.P.C. Gobas, P.S. Ross, PBDE flame retardants and PCBs in migrating Steller seal lions (Eumatopias jubatus) in the Strait of Georgia, British Columbia, Canada, Chemosphere 88 (2012) 855–864.

[5] T. An, L. Zu, G. Li, S. Wan, B. Mai, P.K. Wong, One-step process for debromination and aerobic mineralization of tetrabromobisphenol-A by a novel Ochrobactrum sp. T isolated from an e-waste recycling site, Bioresour. Technol. 102 (2011) 9148–9154.

[6] A.I. Balabanovich, M.P. Luda, G. Camino, A. Hornung, Thermal decomposition behaviour of 1,2-bis-(2,4,6-tribromophenoy)ethane, J. Anal. Appl. Pyrol. 67 (2003) 95–107.

[7] F. Barontini, V. Cozzani, A. Cuzzola, L. Petarca, Investigation of hexabromocyclododecane thermal degradation pathways by gas chromatography/mass spectrometry, Rapid Commun. Mass Spectrom. 15 (2001) 690–698.

[8] F. Barontini, V. Cozzani, L. Petarca, Thermal stability and decomposition products of hexabromocyclododecane, Ind. Eng. Chem. Res. 40 (2001) 3270–3280.

[9] J.S. Bearr, C.L. Mitchelmore, S.C. Roberts, H.M. Stapleton, Species specific differences in the in vitro metabolism of the flame retardant mixture Firemaster BZ-54, Aquat. Toxicol. 124–125 (2012) 41–47.

[10] B. Beek, Bioaccumulation. New Aspects and Developments, The Handbook of Environmental ChemistrySpringer-Verlag, Berlin Heidelberg, 2000, ISBN: 3-540-62575-5.

[11] R.T. Benedict, H.M. Stapleton, R.J. Letcher, C.L. Mitchelmore, Debromination of polybrominated diphenyl ether-99 (BDE-99) in carp (Cyprinus carpio) microflora and microsomes, Chemosphere 69 (2007) 987–993.

[12] G. Bianco, G. Novario, G. Anzilotta, A. Palma, A. Mangone, T.R. Cataldi, Polybrominated diphenyl ethers (PBDEs) in Mediterranean mussels (Mytilus galloprovincialis) from selected Apulia coastal sites evaluated by GC-HRMS, J. Mass Spectrom. 45 (2010) 1046–1055.

[13] S.H. Brandsma, L.T.M. van der Ven, J. de Boer, P.E.G. Leonards, Identification of hydroxylated metabolites of hexabromocyclododecane in wildlife and 28-days exposed Wistar rats, Environ. Sci. Technol. 43 (2009) 6058–6063.

[14] E.P. Browne, H.M. Stapleton, S.M. Kelly, S.C. Tilton, E.P. Gallagher, In vitro hepatic metabolism of 2,2′,4,4′,5-pentabromodiphenyl ether (BDE 99) in Chinook salmon (Oncorhynchus tshawytscha), Aquat. Toxicol. 92 (2009) 281–287.

[15] C.M. Butt, D. Wang, H.M. Stapleton, Halogenated phenolic contaminants inhibit the in vitro activity of the thyroid-regulating deiodinases in human liver, Toxicol. Sci. 124 (2011) 339–347.

[16] R. Cariou, J.P. Antignac, D. Zalko, A. Berrebi, J.P. Cravedi, D. Mamume, P. Marchand, F. Monteau, A. Riu, F. Andre, B. Lebizec, Exposure assessment of French women and their newborns to tetrabromobisphenol-A: occurrence measurements in maternal adipose tissue, serum, breast milk and cord serum, Chemosphere 73 (2008) 1036–1041.

[17] D. Chen, M.J. La Guardia, E. Harvey, M. Amaral, K. Wohlfort, R.C. Hale, Polybrominated diphenyl ethers in peregrine falcon (Falco peregrinus) eggs from the Northeastern US, Environ. Sci. Technol. 42 (2008) 7594–7600.

[18] D. Chen, B. Mai, J. Song, Q. Sun, Y. Luo, X. Luo, E.Y. Zeng, R.C. Hale, Polybrominated diphenyl ethers in birds of prey from northern China, Environ. Sci. Technol. 41 (2007) 1828–1833.

[19] J.R. Christensen, M. MacDuffee, R.W. MacDoanld, M. Whiticar, P. Ross, Persistent organic pollutants in British Columbia grizzly bears: consequence of divergent diets, Environ. Sci. Technol. 39 (2005) 6952–6960.

[20] A. Covaci, L. Bervoets, P. Hoff, S. Voorspoels, J. Voets, K. Van Campenhout, R. Blust, P. Schepens, Polybrominated diphenyl ethers (PBDEs) in freshwater mussels and fish from Flanders, Belgium. J. Environ. Monitor. 7 (2005) 132–136.

[21] A. Covaci, A.C. Dirtu, Brominated flame retardants: analytical, toxicological and environmental aspects, in: Applications of Mass Spectrometry in Life Safety, Springer Science+Business Media B.V., 2008, pp. 153–184.
[22] A. Covaci, A.C. Gerecke, R.J. Law, S. Voorspoels, M. Kohler, N.V. Heeb, H. Leslie, C.R. Allchin, J. de Boer, Hexabromocyclododecanes (HBCDs) in the environment and humans: a review, Environ. Sci. Technol. 40 (2006) 3679–3688.
[23] A. Covaci, S. Harrad, M.A.E. Abdallah, N. Ali, R.J. Law, D. Herzke, C.A. de Wit, Novel brominated flame retardants: a review of their analysis, environmental fate and behaviour, Environ. Int. 37 (2011) 532–556.
[24] A. Covaci, S. Voorspoels, M.A.E. Abdallah, T. Geens, S. Harrad, R.J. Law, Analytical and environmental aspects of the flame retardant tetrabromobisphenol- A and its derivatives, J. Chromatogr. A 1216 (2009) 346–363.
[25] H. D'Have, A. Covaci, J. Scheirs, P. Schepens, R. Verhagen, W. de Coen, Hair as an indicator of endogenous tissue levels of brominated flame retardants in mammals, Environ. Sci. Technol. 39 (2005) 6016–6020.
[26] T. Dauwe, V.L.B. Jaspers, A. Covaci, M. Eens, Accumulation of organochlorines and brominated flame retardants in the eggs and nestlings of the great tit, Parus major, Environ. Sci. Technol. 40 (2006) 5297–5303.
[27] E.F. Davis, H.M. Stapleton, Photodegradation pathways of nonabrominated diphenyl ethers, 2-ethylhexyltetrabromobenzoate and di(2-ethylhexyl)tetrabromophthalate: identifying potential markers of photodegradation, Environ. Sci. Technol. 43 (2009) 5739–5746.
[28] J.W. Davis, S.J. Gonsior, D.A. Markham, U. Friederich, R.W. Hunziker, J.M. Ariano, Biodegradation and product identification of [^{14}C]Hexabromocyclododecane in wastewater sludge and freshwater aquatic sediment, Environ. Sci. Technol. 40 (2006) 5395–5401.
[29] C.A. De Wit, An overview of brominated flame retardants in the environment, Chemosphere 46 (2002) 583–624.
[30] D.Y. Deng, J. Guo, G.P. Sun, X.J. Chen, M.D. Qiu, M.Y. Xu, Aerobic debromination of deca-BDE: isolation and characterization of an indigenous isolate from a PBDE contaminated sediment, Int. Biodeter. Biodegr. 65 (2011) 465–469.
[31] A.C. Dirtu, M. Abdallah, A. Covaci, Advances in the sample preparation of brominated flame retardants and other brominated compounds, Trends Anal. Chem. 43 (2013) 189–203.
[32] A.A. Dominguez, R.J. Law, D. Herzke, J. de Boer, Bioaccumulation of Brominated Flame Retardants. Brominated Flame Retardants, Handbook of Environmental Chemistry Series, Springer-Verlag Berlin, vol. 16, 2011, pp. 141–185.
[33] H. Dong, Z. Li, X. Man, J. Zhou, H. Lu, S. Wang, Identification of the metabolites of BDE-99 and its related P450s, J. Biomed. Res. 24 (2010) 223–232.
[34] P.R. Dorneles, J. Lailson-Brito, A.C. Dirtu, L. Weijs, A.F. Azevedo, J.P.M. Torres, O. Malm, H. Neels, R. Blust, K. Das, A. Covaci, Anthropogenic and naturally-produced organobrominated compounds in marine mammals from Brazil, Environ. Int. 36 (2010) 60–67.
[35] C.T. Elfes, G.R. VanBlaricom, D. Boyd, J. Calambokidis, P.J. Clapham, R.W. Pearce, J. Robbins, J.C. Salinas, J.M. Straley, P.R. Wade, M.M. Krahn, Geographic variation of persistent organic pollutant levels in humpback whale (Megaptera novaeangliae) feeding areas of the North Pacific and North Atlantic, Environ. Toxicol. Chem. 29 (2010) 824–834.
[36] E. Eljarrat, A. De La Cal, D. Raldua, C. Duran, D. Barcelo, Occurrence and bioavailability of polybrominated diphenyl ethers and hexabromocyclododecane in sediment and fish from the Cinca river, a tributary of the Ebro river (Spain), Environ. Sci. Technol. 38 (2004) 2603–2608.

[37] E. Eljarrat, P. Guerra, E. Martinez, M. Farre, J. Alvarez, M. Lopez, D. Barcelo, Hexabromocyclododecane in human breast milk: levels and enantiomeric patterns, Environ. Sci. Technol. 43 (2009) 1940–1946.

[38] J. Eriksson, N. Green, G. Marsh, Å. Bergman, Photochemical decomposition of 15 polybrominated diphenyl ether congeners in methanol/water, Environ. Sci. Technol. 38 (2004) 3119–3125.

[39] C.A. Erratico, S.C. Moffatt, S.M. Bandiera, Comparative oxidative metabolism of BDE-47 and BDE-99 by rat hepatic microsomes, Toxicol. Sci. 123 (2011) 37–47.

[40] C.A. Erratico, A. Szeitz, S.M. Bandiera, Oxidative metabolism of BDE-99 by human liver microsomes: predominant role of CYP2B6, Toxicol. Sci. 129 (2012) 280–292.

[41] S. Esslinger, R. Becker, R. Maul, I. Nehls, Hexabromocyclododecane enantiomers: microsomal degradation and patterns of hydroxylated metabolites, Environ. Sci. Technol. 45 (2011) 3938–3944.

[42] European Chemical Substance Information System (ECSIS), European Commission, Joint Research Centre, European Chemicals Bureau. Available at: http://esis.jrc.ec.europa.eu/ (accessed in 2012).

[43] European Chemicals Agency (ECHA), Proposals to Identify Substances of Very High Concern Current Consultations: Bis(pentabromophenyl) Ether (DecaBDE), 2012. http://echa.europa.eu/proposals-to-identify-substances-of-very-high-concern.

[44] European Union (EU), The Ban of Marketing the Penta- and Octa-BDE Mixtures, Directive 2003/11/EC of the European parliament and of the council of 6 February 2003 amending for the 24th time Council Directive 76/769/EEC relating to restrictions on the marketing and use of certain dangerous substances and preparations, 2003.

[45] P.A. Fair, G. Mitchum, T.C. Hulsey, J. Adams, E. Zolman, W. McFee, E. Wirth, G.D. Bossart, Polybrominated diphenyl ethers (PBDEs) in blubber of free-ranging bottlenose dolphins (Tursiops truncatus) from two southeast Atlantic estuarine areas, Arch. Environ. Contam. Toxicol. 53 (2007) 483–494.

[46] L. Fang, J. Huang, G. Yu, L. Wang, Photochemical degradation of six polybrominated diphenyl ether congeners under ultraviolet irradiation in hexane, Chemosphere 71 (2008) 258–267.

[47] B. Fangstrom, M. Athanasiadou, I. Athanassiadis, A. Bignert, P. Grandjean, P. Weihe, A. Bergman, Polybrominated diphenyl ethers and traditional organochlorine pollutants in fulmars (Fulmarus glacialis) from the Faroe Islands, Chemosphere 60 (2005) 836–843.

[48] K.J. Fernie, J.L. Shutt, R.J. Letcher, I.J. Ritchie, D.M. Bird, Environmentally relevant concentrations of DE-71 and HBCD alter eggshell thickness and reproductive success of American kestrels, Environ. Sci. Technol. 43 (2009) 2124–2130.

[49] K.J. Fernie, J.L. Shutt, R.J. Letcher, J.I. Ritchie, K. Sullivan, D.M. Bird, Changes in reproductive courtship behaviors of adult American kestrels (Falco sparverius) exposed to environmentally relevant levels of the polybrominated diphenyl ether mixture, DE-71, Toxicol. Sci. 102 (2008) 171–178.

[50] J.B. Fini, A. Riu, L. Debrauwer, A. Hillenweck, S.L. Mével, S. Chevolleau, A. Boulahtouf, K. Palmier, P. Balaguer, J.P. Cravedi, B.A. Demeneix, D. Zalko, Parallel biotransformation of tetrabromobisphenol A in Xenopus laevis and mammals: Xenopus as a model for endocrine perturbation studies, Toxicol. Sci. 125 (2012) 359–367.

[51] L.T. Gauthier, D. Potter, C.E. Hebert, R.J. Letcher, Temporal trends and spatial distribution of non-polybrominated diphenyl ether flame retardants in the eggs of colonial populations of Great Lakes herring gulls, Environ. Sci. Technol. 43 (2009) 312–317.

[52] M.O. Gaylor, E. Harvey, R.C. Hale, House crickets can accumulate polybrominated diphenyl ethers (PBDEs) directly from polyurethane foam common in consumer products, Chemosphere 86 (2012) 500–505.

[53] M.O. Gaylor, E. Harve, R.C. Hale, Polybrominated diphenyl ether (PBDE) accumulation by earthworms (Eisenia fetida) exposed to biosolids-, polyurethane foam microparticle-, and penta-BDE-amended soils, Environ. Sci. Technol. 47 (2013) 13831–13839.

[54] A.C. Gerecke, W. Giger, P.C. Hartmann, N.V. Heeb, H.P.E. Kohler, P. Schmid, M. Zennegg, M. Kohler, Anaerobic degradation of brominated flame retardants in sewage sludge, Chemosphere 64 (2006) 311–317.

[55] Y. Guo, X.Z. Meng, H.L. Tang, B.X. Mai, E.Y. Zeng, Distribution of polybrominated diphenyl ethers in fish tissues from the Pearl River Delta, China: levels, compositions, and potential sources, Environ. Toxicol. Chem. 27 (2008) 576–582.

[56] H. Hakk, D.T. Szabo, J. Huwe, J. Diliberto, L.S. Birnbaum, Novel and distinct metabolites identified following a single oral dose of α- or γ-hexabromocyclododecane in mice, Environ. Sci.Technol 46 (2012) 13494–13503.

[57] H. Hakk, J.K. Huwe, K. Murphy, D. Rutherford, Metabolism of 2,2′,4,4′-tetrabromodiphenyl ether (BDE-47) in chickens, J. Agric. Food Chem. 58 (2010) 8757–8762.

[58] H. Hakk, G. Larsen, J. Bowers, Metabolism, tissue disposition and excretion of 1,2-bis(2,4,6-tribromophenoxy)-ethane (BTBPE) in male Sprague-Dawley rats, Chemosphere 54 (2004) 1367–1374.

[59] H. Hakk, G.L. Larsen, Å. Bergman, U. Orn, Metabolism, excretion, and distribution of the flame retardant tetrabromobisphenol-A in conventional and bile-duct Cannulated rats, Xenobiotica 30 (2000) 881–890.

[60] H. Hakk, R.J. Letcher, Metabolism in the toxicokinetics and fate of brominated flame retardants—a review, Environ. Int. 29 (2003) 801–828.

[61] S.K. Han, P. Bilski, B. Karriker, R.H. Sik, C.F. Chignell, Oxidation of flame retardant tetrabromobisphenol A by singlet oxygen, Environ. Sci. Technol. 42 (2009) 166–172.

[62] S. Harrad, C.A. de Wit, M.A.E. Abdallah, C. Bergh, J.A. Bjorklund, A. Covaci, P.O. Darnerud, J. de Boer, M. Diamond, S. Huber, P. Leonards, M. Mandalakis, C. Oestman, L.S. Haug, C. Thomsen, T.F. Webster, Indoor contamination with hexabromocyclododecanes, polybrominated diphenyl ethers, and perfluoroalkyl compounds: an important exposure pathway for people? Environ. Sci. Technol. 44 (2010) 3221–3231.

[63] M. Haukas, A. Ruus, K. Hylland, J.A. Berge, E. Mariussen, Bioavailability of hexabromocyclododecane to the polychaete Hediste diversicolor: exposure through sediment and food from a contaminated fjord, Environ. Toxicol. Chem. 29 (2010) 1709–1715.

[64] D. Hayward, J. Wong, A.J. Krynitsky, Polybrominated diphenyl ethers and polychlorinated biphenyls in commercially wild caught and farm-raised fish fillets in the United States, Environ. Res. 103 (2007) 46–54.

[65] Health and Environmental Horizons (HEH), High Production Volume (HPV) Challenge Program Test Plan for phthalic acid tetrabromo bis 2-ethylhelxyl ester (CAS#26040-51-7), Prepared for the Brominated Phthalate Ester Panel, American Chemical Council, 2004.

[66] N.V. Heeb, W.B. Schweizer, M. Kohler, A.C. Gerecke, Structure elucidation of hexabromocyclododecanes-A class of compounds with a complex stereochemistry, Chemosphere 61 (2005) 65–73.

[67] N.V. Heeb, W.B. Schweizer, P. Mattrel, R. Haag, A.C. Gerecke, P. Schmid, M. Zennegg, H. Vonmont, Regio- and stereoselective isomerization of hexabromocyclododecanes (HBCDs): kinetics and mechanism of γ-to α-HBCD isomerization, Chemosphere 73 (2008) 1201–1210.

[68] N.V. Heeb, H. Graf, W.B. Schweizer, P. Lienemann, Thermally-induced transformation of hexabromocyclo dodecanesand isobutoxypenta bromocyclododecanes in flame-proofed polystyrene materials, Chemosphere 80 (2010) 701–708.

[69] S. Hermanussen, V. Matthews, O. Päpke, C.J. Limpus, C. Gaus, Flame retardants (PBDEs) in marine turtles, dugongs and seafood from Queensland, Australia, Mar. Pollut. Bull 57 (2008) 409–418.

[70] R.A. Hites, Polybrominated diphenyl ethers in the environment and in people: a meta-analysis of concentrations, Environ. Sci. Technol. 38 (2004) 945–956.

[71] J. Hoguet, J.M. Keller, J.L. Reiner, J.R. Kucklick, C.E. Bryan, A.J. Moors, R.S. Pugh, P.R. Becker, Spatial and temporal trends of persistent organic pollutants and mercury in beluga whales (Delphinapterus leucas) from Alaska, Sci. Total Environ. 449 (2013) 285–294.

[72] E. Hoh, L. Zhu, R.A. Hites, Novel flame retardants, 1,2-Bis(2,4,6-tribromophenoxy)ethane and 2,3,4,5,6-pentabromoethylbenzene, in United States' environmental samples, Environ. Sci. Technol. 39 (2005) 2472–2477.

[73] S. Horikoshi, T. Miura, M. Kajitani, N. Horikoshi, N. Serpone, Photodegradation of tetrahalobisphenol-A (X=Cl, Br) flame retardants and delineation of factors affecting the process, Appl. Catal. B-Environ. 84 (2008) 797–802.

[74] M.G. Ikonomou, R.F. Addison, Polybrominated diphenyl ethers (PBDEs) in seal populations from eastern and western Canada: an assessment of the processes and factors controlling PBDE distribution in seals, Mar. Environ. Res. 66 (2008) 225–230.

[75] E. Jakab, M.A. Uddin, T. Bhaskar, Y. Sakata, Thermal decomposition of flame-retarded high-impact polystyrene, J. Anal. Appl. Pyrol. 68-9 (2003) 83–99.

[76] V.L.B. Jaspers, A. Covaci, P. Deleu, M. Eens, Preen oil as the main source of external contamination with organic pollutants onto feathers of the common magpie (Pica pica), Environ. Int. 34 (2008) 741–748.

[77] V.L.B. Jaspers, A. Covaci, S. Voorspoels, T. Dauwe, M. Eens, P. Schepens, Brominated flame retardants and organochlorine pollutants in aquatic and terrestrial predatory birds of Belgium: levels, patterns, tissue distribution and condition factors, Environ. Pollut. 139 (2006) 340–352.

[78] B. Jimenez, M. Alaee, R. Merino, G. Pacepavicius, Higher brominated diphenyl ethers and organochlorines (PCDDs, PCDFs, PCBs, DDTs) in peregrine falcon (Falco peregrinus) breeding in Spain, Organohalogen Compd. 67 (2005) 518–521.

[79] A.K. Johansson, U. Sellstr₡om, P. Lindberg, A. Bignert, C.A. de Wit, Polybrominated diphenyl ether congener patterns, hexabromocyclododecane, and brominated biphenyl 153 in eggs of peregrine falcons (Falco peregrinus) breeding in Sweden, Environ. Toxicol. Chem. 28 (2009) 9–17.

[80] B. Johnson-Restrepo, D.H. Adams, K. Kannan, A. Tetrabromobisphenol, (TBBPA) and hexabromocyclododecanes (HBCDs) in tissues of humans, dolphins and sharks from the United States, Chemosphere 70 (2008) 1935–1944.

[81] B. Johnson-Restrepo, K. Kannan, R. Addink, D.H. Adams, Polybrominated diphenyl ethers and polychlorinated biphenyls in a marine foodweb of coastal Florida, Environ. Sci. Technol. 39 (2005) 8243–8250.

[82] N. Kajiwara, Y. Noma, H. Takigami, Photolysis studies of technical decabromodiphenyl ether (DecaBDE) and ethane (DeBDethane) in plastics under natural sunlight, Environ. Sci. Technol. 42 (2008) 4404–4409.

[83] N. Kajiwara, M. Sueoka, T. Ohiwa, H. Takigami, Determination of flame-retardant hexabromocyclododecane diastereomers in textiles, Chemosphere 74 (2009) 1485–1489.

[84] K. Kannan, S.H. Yun, T.J. Evans, Chlorinated, brominated, and perfluorinated contaminants in livers of polar bears from Alaska, Environ. Sci. Technol. 39 (2005) 9057–9063.

[85] M. Karlsson, I. Ericson, B. van Bavel, J.-K. Jensen, M. Dam, Levels of brominated flame retardants in Northern Fulmar (Fulmarus glacialis) eggs from the Faroe Islands, Sci. Total Environ. 367 (2006) 840–846.

[86] B.C. Kelly, M.G. Ikonomou, J.D. Blair, F.A.P.C. Gobas, Bioaccumulation behaviour of polybrominated diphenyl ethers (PBDEs) in a Canadian Arctic marine food web, Sci. Total Environ. 401 (2008) 60–72.

[87] A. Kierkegaard, L. Asplund, C.A. de Wit, M.S. McLachlan, Fate of higher brominated PBDEs in lactating cows, Environ. Sci. Technol. 41 (2007) 417–423.

[88] A. Kierkegaard, PBDEs in the Environment. Time Trends, Bioaccumulation and the Identification of Their Successor, Decabromodiphenyl Ethane (Ph.D. thesis), Stockholm University, 2007.

[89] Y.M. Kim, I.H. Nam, K. Murugesan, S. Schmidt, D.E. Crowley, Y.S. Chang, Biodegradation of diphenyl ether and transformation of selected brominated congeners by Sphingomonas sp. PH-07, Appl. Microbiol. Biotechnol. 77 (2007) 187–194.

[90] G.A. Knudsen, L.M. Jacobs, R.K. Kuester, I.G. Sipes, Absorption, distribution, metabolism and excretion of intravenously and orally administered tetrabromobisphenol A [2,3-dibromopropyl ether] in male Fischer-344 rats, Toxicology 237 (2007) 158–167.

[91] L.B. Knudsen, K. Borga, E.H. Jorgensen, B. van Bavel, M. Schlabach, J. Verreault, G.W. Gabrielsen, Halogenated organic contaminants and mercury in northern fulmars (Fulmarus glacialis): levels, relationships to dietary descriptors and blood to liver comparison, Environ. Pollut. 146 (2007) 25–33.

[92] R.K. Kuester, A.M. Solyom, V.P. Rodriguez, I.G. Sipes, The effects of dose, route, and repeated dosing on the disposition and kinetics of tetrabromobisphenol A in male F-344 rats, Toxicol. Sci. 96 (2007) 237–245.

[93] H. Kuramochi, K. Kawamoto, K. Miyasaki, K. Nagahama, K. Maeda, X.W. Li, E. Shibata, T. Nakamura, S.I. Sakai, Determination of physicochemical properties of tetrabromobisphenol A, Environ. Toxicol. Chem. 27 (2008) 2413–2418.

[94] J.C.W. Lam, R.K.F. Lau, M.B. Murphy, P.K.S. Lam, Temporal trends of hexabromocyclododecanes (HBCDs) and polybrominated diphenyl ethers (PBDEs) and detection of two novel flame retardants in marine mammals from Hong Kong, South China, Environ. Sci. Technol. 43 (2009) 6944–6949.

[95] M. Montory, R. Barra, Preliminary data on polybrominated diphenyl ethers (PBDEs) in farmed fish tissues (Salmo salar) and fish feed in Southern Chile, Chemosphere 63 (2006) 1252–1260.

[96] R.J. Law, M. Alaee, C. Allchin, J.P. Boon, M. Lebeuf, P. Lepom, G.A. Stern, Levels and trends of Polybrominated diphenylethers and other brominated flame retardants in wildlife, Environ. Int. 29 (2003) 757–770.

[97] R.J. Law, C.R. Allchin, M.E. Bennett, S. Morris, E. Rogan, Polybrominated diphenyl ethers in two species of marine top predators from England and Wales, Chemosphere 46 (2002) 673–681.

[98] R.J. Law, C.R. Allchin, J. de Boer, A. Covaci, D. Herzke, P. Lepom, S. Morris, J. Tronczynski, C.A. de Wit, Levels and trends of brominated flame retardants in the European environment, Chemosphere 64 (2006) 187–208.

[99] R.J. Law, D. Herzke, S. Harrad, S. Morris, P. Bersuder, C.R. Allchin, Levels and trends of HBCD and BDEs in the European and Asian environments, with some information for other BFRs, Chemosphere 73 (2008) 223–241.

[100] M. Lebeuf, B. Gouteux, L. Measures, S. Trottier, Levels and temporal trends (1988-1999) of polybrominated diphenyl ethers in beluga whales (Delphinapterus leucas) from the St. Lawrence Estuary, Canada, Environ. Sci. Technol. 38 (2004) 2971–2977.

[101] K. Lin, J. Ding, X. Huang, Debromination of tetrabromobisphenol A by nanoscale zerovalent iron: kinetics, influencing factors, and pathways, Ind. Eng. Chem. Res. 51 (2012) 8378–8385.

[102] K. Lin, W. Liu, J. Gan, Reaction of tetrabromobisphenol A (TBBPA) with manganese dioxide: kinetics, products, and pathways, Environ. Sci. Technol. 43 (2009) 4480–4486.

[103] Y. Liu, G.J. Zheng, H. Yu, M. Martin, B.J. Richardson, M.H.W. Lam, P.K.S. Lam, Polybrominated diphenyl ethers (PBDEs) in sediments and mussel tissues from Hong Kong marine waters, Mar. Pollut. Bull 50 (2005) 1173–1184.

[104] S. Losada, A. Roach, L. Roosens, F.J. Santos, M.T. Galceran, W. Vetter, H. Neels, A. Covaci, Biomagnification of anthropogenic and naturally-produced organobrominated compounds in a marine food web from Sydney harbour, Australia, Environ. Int. 35 (2009) 1142–1149.

[105] S. Luo, S.G. Yang, X.D. Wang, C. Sun, Reductive degradation of tetrabromobisphenol A over iron–silver bimetallic nanoparticles under ultrasound radiation, Chemosphere 79 (2010) 672–678.

[106] X.-J. Luo, J. Liu, Y. Luo, X.-L. Zhang, J.-P. Wu, Z. Lin, S.-J. Chen, B.-X. Mai, Z.-Y. Yang, Polybrominated diphenyl ethers (PBDEs) in free-range domestic fowl from an e-waste recycling site in South China: levels, profile and human dietary exposure, Environ. Int. 35 (2009) 253–258.

[107] S.J. Lupton, B.P. McGarrigle, J.R. Olson, T.D. Wood, D.S. Aga, Human liver microsome-mediated metabolism of brominated diphenyl ethers 47, 99, and 153 and identification of their major metabolites, Chem. Res. Toxicol. 22 (2009) 1802–1809.

[108] G.A. MacInnis, R.J. Letcher, M.A. McKinney, G.T. Tomy, M. Lebeuf, J. Fång, Å. Bergman, C.H. Marvin, Investigation of HBCD metabolism in marine mammals from Canada using hepatic microsomal in vitro bioassay approach and comparison with field samples, Organohalog. Compd. 72 (2010) 1097–1100.

[109] I. Meerts, R.J. Letcher, S. Hoving, G. Marsh, Å. Bergman, J.G. Lemmen, B. van der Burg, A. Brouwer, In vitro estrogenicity of polybrominated diphenyl ethers, hydroxylated PBDEs, and polybrominated bisphenol A compounds, Environ. Health Perspect. 109 (2001) 399–407.

[110] X.Z. Meng, M.E. Blasius, R.W. Gossett, K.A. Maruya, Polybrominated diphenyl ethers in pinnipeds stranded along the southern California coast, Environ. Pollut. 157 (2009) 2731–2736.

[111] H.B. Moon, K. Kannan, M. Choi, J. Yu, H.G. Choi, Y.R. An, S.G. Choi, J.Y. Park, Z.G. Kim, Chlorinated and brominated contaminants including PCBs and PBDEs in minke whales and common dolphins from Korean coastal waters, J. Hazard. Mater. 179 (2010) 735–741.

[112] D.C. Muir, S. Backus, A.E. Derocher, R. Dietz, T.J. Evans, G.W. Gabrielsen, J. Nagy, R.J. Norstrom, C. Sonne, I. Stirling, M.K. Taylor, R.J. Letcher, Brominated flame retardants in polar bears (Ursus maritimus) from Alaska, the Canadian Arctic, East Greenland, and Svalbard, Environ. Sci. Technol. 15 (2006) 449–455.

[113] R. Nakagawa, S. Murata, Y. Ashizuka, Y. Shintani, T. Hori, T. Tsutsumi, Hexabromocyclododecane determination in seafood samples collected from Japanese coastal areas, Chemosphere 82 (2010) 445–452.

[114] K. Nomiyama, Y. Uchiyama, S. Horiuchi, A. Eguchi, H. Mizukawa, S.H. Hirata, R. Shinohara, S. Tanabe, Organohalogen compounds and their metabolites in the blood of Japanese amberjack (Seriola quinqueradiata) and scalloped hammerhead shark (Sphyrna lewini) from Japanese coastal waters, Chemosphere 85 (2011) 315–321.

[115] P.D. Noyes, D.E. Hinton, H.M. Stapleton, Accumulation and debromination of decabromodiphenyl ether (BDE-209) in juvenile fathead minnows (Pimephales promelas) induces thyroid disruption and liver alterations, Toxicol. Sci. 122 (2011) 265–274.

[116] P.D. Noyes, S.M. Kelly, C.L. Mitchelmore, H.M. Stapleton, Characterizing the in vitro hepatic biotransformation of the flame retardant BDE 99 by common carp, Aquat. Toxicol. 97 (2010) 142–150.

[117] Organisation for Economic Co-operation and Development (OECD), OECD Guideline for Testing of Chemicals No. 305: Bioconcentration: Flow-through Fist Test, June 14, 1996 (Paris).

[118] N. Ortuño, R. Font, J. Moltó, J.A. Conesa, Thermal degradation of tetrabromobisphenol A: emission of polybrominated dibenzo-p-dioxins and dibenzofurans and other organic compounds, Organohalog. Compd. 73 (2011) 511–514.

[119] J.S. Park, A. Holden, V. Chu, M. Kim, A. Rhee, P. Patel, Y.T. Shi, J. Linthicum, B.J. Walton, K. McKeown, N.P. Jewell, K. Hooper, Time-trends and congener profiles of PBDEs and PCBs in California peregrine falcons (Falco peregrinus), Environ. Sci. Technol. 43 (2009) 8744–8751.

[120] J.H. Peng, C.W. Huang, Y.M. Weng, H.K. Yak, Determination of polybrominated diphenyl ethers (PBDEs) in fish samples from rivers and estuaries in Taiwan, Chemosphere 66 (2007) 1990–1997.

[121] K.E. Potter, B.D. Watts, M.J. La Guardia, E.P. Harvey, R.C. Hale, Polybrominated diphenyl ether flame retardants in Chesapeake Bay Region, USA, peregrine falcon (Falco Peregrinus) eggs: urban/rural trends, Environ. Toxicol. Chem. 28 (2009) 973–981.

[122] K. Ramu, N. Kajiwara, T. Isobe, S. Takahashi, E.Y. Kim, B.Y. Min, S.U. We, S. Tanabe, Spatial distribution and accumulation of brominated flame retardants, polychlorinated biphenyls and organochlorine pesticides in blue mussels (Mytilus edulis) from coastal waters of Korea, Environ. Pollut. 148 (2007) 562–569.

[123] S. Rayne, M.G. Ikonomou, P.S. Ross, G.M. Ellis, L.G. Barrett-Lennard, PBDEs, PBBs, and PCNs in three communities of free-ranging killer whales (Orcinus orca) from the northeastern Pacific Ocean, Environ. Sci. Technol. 38 (2004) 4293–4299.

[124] S.C. Roberts, L.J. Macaulay, H.M. Stapleton, In vitro metabolism of the brominated flame retardants 2-ethylhexyl-2,3,4,5-tetrabromobenzoate (TBB) and bis(2-ethylhexyl) 2,3,4,5-tetrabromophthalate (TBPH) in human and rat tissues, Chem. Res. Toxicol. 25 (2012) 1435–1441.

[125] S.C. Roberts, P.D. Noyes, E.P. Gallagher, H.M. Stapleton, Species-specific differences and structure-activity relationships in the debromination of PBDE congeners in three fish species, Environ. Sci. Technol. 45 (2011) 1999–2005.

[126] K.R. Robrock, P. Korytar, L. Alvarez-Cohen, Pathways for the anaerobic microbial debromination of polybrominated diphenyl ethers, Environ. Sci. Technol. 42 (2008) 2845–2852.

[127] A. Ronen, A. Abelovitch, Anaerobic–aerobic process for microbial degradation of tetrabromobisphenol A, Appl. Environ. Microbiol. 66 (2000) 2372–2377.

[128] L. Roosens, C. Erratico, H. Neels, H.M. Stapleton, A. Covaci, New insights in the in vitro metabolism of HBCD by rat liver microsomes, Organohalog. Compd. 73 (2011) 1176–1179.

[129] L. Roosens, H.M. Stapleton, A. Covaci, In vitro metabolism of HBCD isomers by rat liver microsomes, Organohalog. Compd. 72 (2010) 1075–1078.

[130] U.M. Schauer, W. Volkel, W. Dekant, Toxicokinetics of tetrabromobisphenol A in humans and rats after oral administration, Toxicol. Sci. 91 (2006) 49–58.

[131] U. Sellstrom, C.A. de Wit, N. Lundgren, M. Tysklind, Effect of sewage-sludge application on concentrations of higher-brominated diphenyl ethers in soils and earthworms, Environ. Sci. Technol. 39 (2005) 9064–9070.

[132] S.D. Shaw, M.L. Berger, D. Brenner, D.O. Carpenter, L. Tao, C.S. Hong, K. Kannan, Polybrominated diphenyl ethers (PBDEs) in farmed and wild salmon marketed in the Northeastern United States, Chemosphere 71 (2008) 1422–1431.

[133] Y. Shih, H.L. Chou, Y.H. Peng, Microbial degradation of 4-monobrominated diphenyl ether with anaerobic sludge, J. Hazard. Mater. 213-214 (2012) 341–346.

[134] Y.H. Shih, C.K. Wang, Photolytic degradation of polybromodiphenyl ethers under UV-lamp and solar irradiations, J. Hazard. Mater. 165 (2009) 34–38.

[135] G. Söderström, U. Sellström, C.A. De Wit, M. Tysklind, Photolytic debromination of decabromodiphenyl ether (BDE 209), Environ. Sci. Technol. 38 (2004) 127–132.

[136] C. Sonne, R.J. Letcher, T.Ø. Bechshøft, F.F. Rigét, D.C.G. Muir, P.S. Leifsson, E.W. Born, L. Hyldstrup, N. Basu, M. Kirkegaard, R. Dietz, Two decades of biomonitoring polar bear health in Greenland: a review, Acta Veterinaria Scand. 54 (2012) S15.

[137] E.G. Sørmo, M.P. Salmer, B.M. Jenssen, H. Hop, K. Baek, K.M. Kovacs, C. Lydersen, S. Falk-Petersen, G.W. Gabrielsen, E. Lie, J.U. Skaare, Biomagnification of polybrominated diphenyl ether and hexabromocyclododecane flame retardants in the polar bear food chain in Svalbard, Norway, Environ. Toxicol. Chem. 25 (2006) 2502–2511.

[138] H.M. Stapleton, J.G. Allen, S.M. Kelly, A. Konstantinov, S. Klosterhaus, D. Watkins, M.D. McClean, T.F. Webster, Alternate and new brominated flame retardants detected in U.S. house dust, Environ. Sci. Technol. 42 (2008) 6910–6916.

[139] H.M. Stapleton, B. Brazil, R. Holbrook, C. Mitchelmore, R. Benedict, A. Konstantinov, D. Potter, In vivo and in vitro debromination of decabromodiphenyl ether (BDE 209) by juvenile rainbow trout and common carp, Environ. Sci. Technol. 40 (2006) 4653–4658.

[140] H.M. Stapleton, N.G. Dodder, Photodegradation of decabromodiphenyl ether in house dust by natural sunlight, Environ. Toxicol. Chem. 27 (2008) 306–312.

[141] H.M. Stapleton, S.M. Kelly, R. Pei, R.J. Letcher, C. Gunsch, Metabolism of polybrominated diphenyl ethers (PBDEs) by human hepatocytes in vitro, Environ. Health Perspect. 117 (2009) 197–202.

[142] H.M. Stapleton, R.J. Letcher, J. Li, J.E. Baker, Dietary accumulation and metabolism of polybrominated diphenyl ethers by juvenile carp (Cyprinus carpio), Environ. Toxicol. Chem. 23 (2004) 1939–1946.

[143] H.M. Stapleton, N.G. Dodder, J.R. Kucklick, C.M. Reddy, M.M. Schantz, P.R. Becker, F. Gulland, B.J. Porter, S.A. Wise, Determination of HBCD, PBDEs and MeO-BDEs in California sea lions (Zalophus californianus) stranded between 1993–2003, Mar. Pollut. Bull. 52 (2006) 522–531.

[144] A. Sudaryanto, I.E. Setiawan, M. Ilyas, E. Soeyanto, A.S. Riadi, T. Isobe, S. Takahashi, S. Tanabe, Levels of brominated flame retardants in sediments and their bioaccumulation potential in biota from Jakarta Bay and its surroundings, Indonesia, in: Y. Obayashi, T. Isobe, A. Subramanian, S. Suzuki, S. Tanabe (Eds.), Interdisciplinary Studies on Environmental Chemistry—Environmental Research in Asia, TERRAPUB, 2009, pp. 125–131.

[145] D.T. Szabo, J.J. Diliberto, H. Hakk, J.K. Huwe, L.S. Birnbaum, Toxicokinetics of the flame retardant hexabromocyclododecane gamma: effect of dose, timing, route, repeated exposure, and metabolism, Toxicol. Sci. 117 (2010) 282–293.

[146] D.T. Szabo, J.J. Diliberto, H. Hakk, J.K. Huwe, L.S. Birnbaum, Toxicokinetics of the flame retardant hexabromocyclododecane alpha: effect of dose, timing, route, repeated exposure, and metabolism, Toxicol. Sci. 121 (2011) 234–244.

[147] J.A. Szymanska, A. Sapota, B. Frydrych, The disposition and metabolism of tetrabromobisphenol A after a single i.p. dose in the rat, Chemosphere 45 (2001) 693–700.

[148] G.T. Tomy, W. Budakowski, T. Halldorson, M. Whittle, M.J. Keir, C. Marvin, G. MacInnis, M. Alaee, Biomagnification of α- and γ-hexabromocyclododecane isomers in a Lake Ontario food web, Environ. Sci. Technol. 38 (2004) 2298–2303.

[149] B. Uhnakova, R. Ludwig, J. Peknicova, L. Homolka, L. Lisa, L. Šulc, A. Petrickova, F. Elzeinova, H. Pelantova, M. Daniela, V. Kroen, H. Dietmar, M. Ludmila, Biodegradation of tetrabromobisphenol A by oxidases in basidiomycetous fungi and estrogenic activity of the biotransformation products, Bioresour. Technol. 102 (2011) 9409–9415.

[150] UNEP/POPS/COP, Report of the Conference of the Parties of the Stockholm Convention on Persistent Organic Pollutants on the Work of its Fourth Meeting, Geneva, 2009.

[151] US EPA, DecaBDE Phase-out Initiative, 2010. http://www.epa.gov/oppt/existingchemicals/pubs/actionplans/deccadbe.html.

[152] E. Van Ael, A. Covaci, R. Blust, L. Bervoets, Corrigendum to "Persistent organic pollutants in the Scheldt estuary: environmental distribution and bioaccumulation", Environ. Int. 63 (2014) 246–251.

[153] E. Van den Steen, A. Covaci, V.L.B. Jaspers, T. Dauwe, S. Voorspoels, M. Eens, R. Pinxten, Accumulation, tissue-specific distribution and biotransformation of BDE 209 in European starlings (Sturnus vulgaris) after exposure with silastic implants, Organohalogen Compd. 68 (2006) 647–650.

[154] E. Van den Steen, T. Dauwe, A. Covaci, V.L.B. Jaspers, R. Pinxten, M. Eens, Within- and among-clutch variation of organohalogenated contaminants in eggs of great tits (Parus major), Environ. Pollut. 144 (2006) 355–359.

[155] E. Van den Steen, M. Eens, A. Covaci, A.C. Dirtu, V.L.B. Jaspers, H. Neels, R. Pinxten, An exposure study with polybrominated diphenyl ethers (PBDEs) in female European starlings (Sturnus vulgaris): toxicokinetics and reproductive effects, Environ. Pollut. 157 (2009) 430–436.

[156] E. Van den Steen, M. Eens, V.L.B. Jaspers, A. Covaci, R. Pinxten, Effects of laying order and experimentally increased egg production on organic pollutants in eggs of a terrestrial songbird species, the great tit (Parus major), Sci. Total Environ. 407 (2009) 4764–4770.

[157] E. Van den Steen, R. Pinxten, V.L.B. Jaspers, A. Covaci, E. Barba, et al., Brominated flame retardants and organochlorines in the European environment using great tit eggs as a biomonitoring tool, Environ. Int. 35 (2009) 310–317.

[158] S.P.J. Van Leeuwen, M.J.M. Van Velzen, C.P. Swart, I. Van Der Veen, W.A. Traag, J. de Boer, Halogenated contaminants in farmed salmon, trout, tilapia, pangasius and shrimp, Environ. Sci. Technol. 43 (2009) 4009–4015.

[159] V. Verhaert, A. Covaci, S. Bouillon, K. Abrantes, D. Musibono, L. Bervoets, E. Verheyen, R. Blust, Baseline levels and trophic transfer of persistent organic pollutants in sediments and biota from the Congo River Basin (DR Congo), Environ. Int. 59 (2013) 290–302.

[160] J. Verreault, G.W. Gabrielsen, S. Chu, D.C.G. Muir, M. Andersen, A. Hamaed, R.J. Letcher, Flame retardants and methoxylated and hydroxylated polybrominated diphenyl ethers in two Norwegian Arctic top predators: glaucous gulls and polar bears, Environ. Sci. Technol. 39 (2005) 6021–6028.

[161] J. Verreault, C. Bech, R.J. Letcher, E. Ropstad, E. Dahl, G.W. Gabrielsen, Organohalogen contamination in breeding glaucous gulls from the Norwegian Arctic: associations with basal metabolism and circulating thyroid hormones, Environ. Pollut. 145 (2007) 138–145.

[162] J. Verreault, R.A. Villa, G.W. Gabrielsen, J.U. Skaare, R.J. Letcher, Maternal transfer of organohalogen contaminants and metabolites to eggs of Arctic breeding glaucous gulls, Environ. Pollut. 144 (2006) 1053–1060.

[163] L.W. Voordeckers, D.E. Fennell, K. Jones, M.M. Haggblom, Anaerobic biotransformation of tetrabromobisphenol A, tetrachlorobisphenol A, and bisphenol A in estuarine sediments, Environ. Sci. Technol. 36 (2002) 696–701.

[164] S. Voorspoels, A. Covaci, V.L.B. Jaspers, H. Neels, P. Schepens, Biomagnification of PBDEs in three small terrestrial food chains, Environ. Sci. Technol. 41 (2007) 411–416.

[165] S. Voorspoels, A. Covaci, P. Lepom, S. Escutenaire, P. Schepens, Remarkable findings concerning PBDEs in the terrestrial toppredator red fox (Vulpes vulpes), Environ. Sci. Technol. 40 (2006) 2937–2943.

[166] S. Voorspoels, A. Covaci, P. Lepom, V.L.B. Jaspers, P. Schepens, Levels and distribution of polybrominated diphenyl ethers in various tissues of birds of prey, Environ. Pollut. 144 (2006) 218–227.

[167] F. Wang, J. Wang, J. Dai, G. Hu, J. Wang, X. Luo, B. Mai, Comparative tissue distribution and associated biological effects of decabromodiphenyl ethane and decabrominated diphenyl ether in male rats after a 90-day exposure study, Environ. Sci. Technol. 44 (2010) 5655–5660.

[168] I. Watanabe, S. Sakai, Environmental release and behavior of brominated flame retardants, Environ. Int. 29 (2003) 665–682.

[169] K. Watanabe, K. Senthilkumar, S. Masunaga, T. Takasuga, N. Iseki, M. Morita, Brominated organic contaminants in the liver and egg of the common cormorants (Phalacrocorax carbo) from Japan, Environ. Sci. Technol. 38 (2004) 4071–4077.

[170] L. Weijs, N. Briels, D.H. Adams, G. Lepoint, K. Das, R. Blust, A. Covaci, Bioaccumulation of organohalogenated compounds in sharks and rays from the southeastern USA, Environ Res. 137C (2015) 199–207.

[171] L. Weijs, K. Das, H. Neels, R. Blust, A. Covaci, Occurrence of anthropogenic and naturally-produced organohalogenated compounds in tissues of Black Sea harbour porpoises, Mar Pollut Bull. 60 (2010) 725–731.

[172] L. Weijs, A.C. Dirtu, K. Das, A. Gheorghe, P.J. Reijnders, H. Neels, R. Blust, A. Covaci, Inter-species differences for polychlorinated biphenyls and polybrominated diphenyl ethers in marine top predators from the Southern North Sea: Part 1, Environ Pollut. 157 (2009) 437–444.

[173] L. Weijs, D. Tibax, A.C. Roach, T.M. Manning, J.C. Chapman, K. Edge, R. Blust, A. Covaci, Assessing levels of halogenated organic compounds in mass-stranded long-finned pilot whales (Globicephala melas) from Australia, Sci. Total Environ. 461–462 (2013) 117–125.

[174] H. Wolkers, P.J. Corkeron, S.M. Van Parijs, T. Similä, B. Van Bavel, Accumulation and transfer of contaminants in killer whales (Orcinus orca) from Norway: indications for contaminant metabolism, Environ. Toxicol. Chem. 26 (2007) 1582–1590.

[175] F. Wong, P. Kurt-Karakus, T.F. Bidleman, Fate of brominated flame retardants and organochlorine pesticides in urban soil: volatility and degradation, Environ. Sci. Technol. 46 (2012) 2668–2674.

[176] World Health Organization (WHO), Flame Retardants: A General Introduction, Environmental Health Criteria, 192, 1997.

[177] J.P. Wu, Y.T. Guan, Y. Zhang, X.J. Luo, H. Zhi, S.J. Chen, B.X. Mai, Several current-use, non-PBDE brominated flame retardants are highly bioaccumulative: evidence from field determined bioaccumulation factors, Environ. Int. 37 (2011) 210–215.

[178] J.P. Wu, Y. Zhang, X.J. Luo, Y.Z. She, L.H. Yu, S.J. Chen, B.X. Mai, A review of polybrominated diphenyl ethers and alternative brominated flame retardants in wildlife from China: levels, trends, and bioaccumulation characteristics, J. Environ. Sci.-China 24 (2012) 183–194.

[179] Q.M. Xian, K. Ramu, T. Isobe, A. Sudaryanto, X.H. Liu, Z.S. Gao, S. Takahashi, H.X. Yu, S. Tanabe, Levels and body distribution of polybrominated diphenyl ethers (PBDEs) and hexabromocyclododecanes (HBCDs) in freshwater fishes from the Yangtze River, China, Chemosphere 71 (2008) 268–276.

[180] J. Xu, W. Meng, Y. Zhang, L. Li, C. Guo, Photocatalytic degradation of tetrabromobisphenol A by mesoporous BiOBr: efficacy, products and pathway, Appl. Catal. B-Environ. 107 (2011) 355–362.

[181] D. Zalko, C. Prouillac, A. Riu, E. Perdu, L. Dolo, I. Jouanin, C. Canlet, L. Debrauwer, J.P. Cravedi, Biotransformation of the flame retardant tetrabromo-bisphenol A by human and rat sub-cellular liver fractions, Chemosphere 64 (2006) 318–327.
[182] Y.H. Zeng, X.J. Luo, H.S. Chen, L.H. Yu, S.J. Chen, B.X. Mai, Gastrointestinal absorption, metabolic debromination, and hydroxylation of three commercial polybrominated diphenyl ether mixtures by common carp, Environ. Toxicol. Chem. 31 (2012) 731–738.
[183] D.N. Zhou, L. Chen, F. Wu, J. Wang, F. Yang, Debromination of hexabromocyclododecanes in aqueous solutions by UV-C irradiation, Fresen. Environ. Bull. 21 (2012) 107–111.
[184] L.Y. Zhu, R.A. Hites, Temporal trends and spatial distributions of brominated flame retardants in archived fishes from the Great Lakes, Environ. Sci. Technol. 38 (2004) 2779–2784.
[185] Y. Zhuang, S. Ahn, R.G. Luthy, Debromination of polybrominated diphenyl ethers by nanoscale zerovalent iron: pathways, kinetics, and reactivity, Environ. Sci. Technol. 44 (2010) 8236–8242.
[186] Y. Zhuang, S. Ahn, A.L. Seyfferth, Y. Masue-Slowey, S. Fendorf, R.G. Luthy, Dehalogenation of polybrominated diphenyl ethers and polychlorinated biphenyl by bimetallic, impregnated, and nanoscale zerovalent iron, Environ. Sci. Technol. 45 (2011) 4896–4903.

Chapter 15

Bioavailability of Persistent Organic Pollutants in Soils: Concept, Analytical Tools, and Application in the Risk Assessment

Bilqees M. Adedigba and Kirk T. Semple*

Lancaster Environment Centre, Lancaster University, Lancaster, UK

**Corresponding author: E-mail: k.semple@lancaster.ac.uk*

Chapter Outline

1. INTRODUCTION

Historically, the assessment of contaminated land has been based on total contaminant concentrations obtained following an exhaustive solvent extraction, with the assumption that extracted contaminants are 100% bioavailable

Comprehensive Analytical Chemistry, Vol. 67. http://dx.doi.org/10.1016/B978-0-444-63299-9.00015-6

[3,44,124]. Recent advances in the understanding of fate and behavior of hydrophobic organic contaminants (HOCs) have led to the appreciation of the concepts of bioavailability and bioaccessibility in the consideration of potential hazard and risk that may be associated with a given contaminant in soil or sediment. Increasing knowledge about the sorption, sequestration, and ageing processes have shed more light in this area and it is now widely accepted that total contaminant does not give accurate indication of risk posed by organic contaminants to biota, it almost certainly led to overestimation of risk [3,104,114].

Risk assessment is being incorporated into legislation across different parts of the world as a mechanism to establish corporate responsibility and impose liability on those activities creating the risk as well as to promote formulation and implementation of appropriate measures needed to eliminate or control such risk [50]. For instance, risk based approaches to contaminated land regulation in the United Kingdom's contaminated land regime (Part IIa of the Environmental Protection Act, 1990, published in 2000) identifies contaminated land based on the existence of one or more "significant pollutant linkages." In this context, there must be a receptor and a linking pathway to the contaminant source (source–pathway–receptor paradigm), whilst the contaminant must also be present at ecotoxicologically relevant concentrations.

Similarly in the Netherlands, consideration of bioavailability in hazard assessment has employed maximum permissible concentrations (MPCs) to set the concentrations above which the risk of adverse effects can be considered unacceptable [30]. Each environmental compartment has been assigned an MPC value which forms the basis for the Dutch government environmental quality standards [63]. The United States policy also considered bioavailability, through the introduction of bioavailability "adjustment factors" for oral and dermal uptake for the BTEX compounds, certain polycyclic aromatic hydrocarbons (PAHs) and pesticides, as well as metals [97].

The backbone of all the regulatory instruments described above consider the likelihood to cause harm. However, their full implementation has been limited and the extent to which bioavailability concept is implemented vary widely across nations [79]. For instance, in the United States, the default bioavailability adjustment value of one has been adopted assuming that contaminant is 100% bioavailable [97]. The reluctance for the proper incorporation of a bioavailability factor into the contaminated land risk assessment in the United States and elsewhere rests on the concerns over analytical cost, public acceptance, and the absence of formal guidelines at the implementation level [44,81]. Regardless, there is increasing global demand for legislation to address the concept of "actual risk" using bioavailability and bioaccessibility as measures of true exposure of xenobiotic compounds to biota.

The distinction between bioavailability and bioaccessibility is contained in the content of what is being measured, while bioavailability quantifies the biotic interaction with a chemical at a given time (i.e., the rate), bioaccessibility incorporates the bioavailable and the potentially bioavailable contaminants over time (i.e., finite concentration) [122,123]. However, the lack of widely accepted procedures to quantify bioavailable or bioaccessible contaminant fractions creates a stumbling block for researchers to make a case for the proper implementation of bioavailability and/or bioaccessibility for inclusion in the risk assessment and remediation of contaminated land [96].

The aims of this chapter are (1) to introduce the concepts of bioavailability and bioaccessibility; (2) to present a brief overview of the different approaches used to predict bioaccessibility of organic contaminants in soil and sediment; and (3) to consider the role of bioavailability and bioaccessibility in risk assessment and bioremediation of contaminated land.

2. DEFINITIONS: BIOAVAILABILITY AND BIOACCESSIBILITY OF ORGANIC CONTAMINANTS IN SOIL

Bioavailability is relevant to many aspects of risk assessment and bioremediation strategies including exposure and toxicity assessment, and has been used in many scientific fields to describe the relationship between a living cell and a chemical(s), coexisting in a habitat. A number of different definitions have been presented in the peer-reviewed literature; for instance, in pharmacological context, bioavailability has been used to describe the systemic availability of a xenobiotic in exposed organisms following an intravenous or oral dosing [70]. In human health assessment, two operational definitions have been proposed: "absolute" and "relative" bioavailability [44]. Absolute bioavailability defines the fraction of the applied dose that can be absorbed and reaches systemic circulation of organisms, while relative bioavailability basically compares the absorption capacity under two sets of conditions i.e., comparing systemic availability of a chemical following consumption of a soil sample with availability following food ingestion [44]. In toxicological context, bioavailability refers to the portion of total concentration of a chemical in the environment that is potentially available for biological action [133], or the ability of a substrate to interact with the biological system of an organism [147]. Bioavailability from environmental perspective describes the amount of a compound removed from soil under physiological conditions that can be transferred to the circulatory system [73,120].

A critical analysis of the different definitions across several disciplines has been provided by Semple et al. [122]; from this two definitions were offered: bioavailability and bioaccessibility. In their paper, Semple et al. [122] suggested that bioavailability referred to the fraction of contaminant "that which is freely available to cross an organism's (cellular) membrane from the medium the organisms inhabits at a given time"; while bioaccessibility was defined as "that which is available to cross an organisms' (cellular) membrane from the environment it inhabits, if the organism had access to it; however, it may be either physically removed from the organism or only bioavailable after a period of time." Subsequent to this, Reichenberg and Mayer [150] went on to consider that bioavailability "is composed of what is accessible (as defined by mild extraction) and the contaminant's chemical activity (as defined by spontaneous physico-chemical processes, such as diffusion, sorption and partitioning)."

In relation to contaminated land risk assessment and remediation, Semple et al. [123] considered the above definitions, proposed by Semple et al. [122] and Reichenberg and Mayer [150] suggesting that bioavailability (chemically active compound which may be taken up by soil biota) is a good descriptor for potential rates of interaction, while bioaccessible concentration may be a finite and measurable concentration, which could be used in the risk assessment of contaminated land and/or quantifying bioremediation end points.

3. FACTORS CONTROLLING THE BIOAVAILABILITY AND BIOACCESSIBILITY OF ORGANIC CONTAMINANTS IN SOIL

It is widely accepted that as a contaminant ages in soil, it undergoes some physical, biological, and chemical processes which result in a fraction of contaminant becoming less available to biota. Sequestration describes important mechanisms which describe the interaction between the contaminant and the three dimensional structure of the soil, through sorption, diffusion, and partition into the nanopores and micropores within the soil [3,108,111]. This interaction collectively termed "ageing" results in the formation of "bound, or non-extractable residues." Therefore, a contaminated soil may contain three fractions of contaminants: (1) the labile pool which is freely accessible, (2) the partially sorbed, which may become available over time, and (3) the non-extractable residue, which is strongly sequestered and not available for uptake or able to be transported in the soil.

While the readily available pool in the soil-pore phases can be taken as the bioavailable fraction, it has been shown that the slowly desorbing fraction (from the partially sorbed) may become available with time and also contribute to the overall toxicity of organic contaminant in the environment [27]. Hence the concept of bioaccessibility was introduced to describe the totality of contaminant that an organism may be exposed to if there is no physical barrier that may prevent access to the chemical [122]. Furthermore, while bioavailability is a rate value, bioaccessibility is an extent value which includes the aqueous and the rapidly desorbing fractions. A great understanding of the bioaccessibility of aged contaminant is no doubt desirable for effective assessment of the potential impact of HOCs on the environment and also to help the decision making on remediation strategies.

The biodegradability and toxicity of organic contaminants are related to their mobility in the environmental compartments in which the contaminant resides, and is governed by factors controlling the fate and behavior of HOCs in the environment. Important characteristics include soil–contaminant contact time [1,3,55,126], the properties of the contaminants and of the soil [61,95,113,121], the concerned organisms [48,68,123], the micro- and macro-environmental conditions within the soil compartment [112,121,127]. The processes involved in organism–contaminant interactions are complex and intertwined; for instance, contaminant factors affecting bioaccessibility of HOCs in soil include the concentration [29], physico-chemical properties of the contaminant [29,145], presence of cosubstrate(s) [29,60,102,137], presence of non-aqueous phase liquid [41,49,67,72,141]. In such a complex system, it can be difficult to delineate one factor without a mention of other; therefore this section focuses on ageing-related processes and microbial factors that affect bioaccessibility of HOCs in soil.

Ageing describes the variety of physical, chemical, and biological interactions between the organic contaminant and the environmental matrix following the contaminant release into the biosphere. As soil contaminant contact time increases, there is a change in their extractability and toxicity to biota and change in the kinetics of sorption and desorption showed that the compounds are becoming sequestered in the inaccessible microsites within the environmental matrix containing them [2,4]. Increasing scientific evidence had shown that an increasing proportion of contaminant may diffuse into the 3D structure of the soil, and to the remote areas of soil nanopores and micropores, thereby decreasing their bioavailability to microorganisms and hence their biodegradability in the environment [87,114,116,125].

The impact of ageing on bioavailability of organic contaminants have been studied extensively [3,5,21,38,55,89], as such the mechanisms of interaction and importance of soil constituents to the entire ageing process have been elucidated [87,108,109,148,149]. Sequestration of contaminant is controlled by two major processes, (1) intra-organic matter diffusion involving the slow partition of chemicals into the interior region of soil organic material, and (2) intra-particle diffusion involving the entrapment of molecules into the micropores and nanopores. The latter is controlled by pore size and structure [94,134], while the former is regulated by the quality and quantity of soil organic matter [10,12,42,43,55,149].

It is also well established that intra-organic matter diffusion plays the principal role in the sequestration and reduced availability of organic contaminant in soil [108,110,130]. Other soil properties important in the sequestration process include the clay content and repeated cycle of drying and wetting [148], as well as the soil cation exchange capacity [22].

4. APPROACHES FOR MEASURING BIOACCESSIBILITY OF ORGANIC CONTAMINANTS IN SOIL

Chemical approaches for predicting bioaccessibility of organic contaminants in soil and sediments have involved the development of chemical extraction techniques which mimic biological interactions with HOCs in soil [151]. Chemical techniques for assessing bioaccessibility can be broadly categorized into three broad groupings: (1) solvent extractions (2) aqueous-based extractions, and (3) solid-phase extractions. These have been reviewed in detail by a number of authors [35,48,90,115,118,123]. This section provides a brief synopsis of commonly reported chemical extractions for measuring bioaccessibility of a wide range of HOC in complex contaminated systems.

4.1 Organic Solvent Extraction

Organic solvent methodologies employ the use of pressures and/or heat to overcome energy barrier required for desorption of HOCs from the soil sorption domains, thus facilitate the diffusion and partitioning of contaminant into the organic solvent phase, to such an extent that the total contaminant pool may be extracted. Examples of exhaustive procedure is the use of soxhlet apparatus and the accelerated solvent extraction (ASE) procedure [117]. The range of extracting solvent includes dichloromethane, hexane, and hexane and acetone in different proportions. The limitation of this approach is that it extracts the total contaminant fraction. It is widely accepted that this fraction has little relevance to the putative risk posed by organic contaminants in the environment, as it is assumed that all that is extractable is bioavailable, leading to an overestimation of risk [3,117].

Due to the aggressive nature of many organic solvent extraction techniques, other authors have proposed the use of less harsh solvent extraction techniques, either as a sole extractant e.g., *n*-butanol [84,114,140], methanol [69], ethanol [82,143], isopropanol [33,82,142] or as a combination of alcohol and water in different proportions. These methods have been shown to reflect bioaccessibility of organic contaminants to various ecological receptors, but there is a lack of consistency [123,151].

4.2 Supercritical Fluid Extraction

Supercritical fluid extraction (SFE) is a rapid technique for the extraction of organic contaminants from various environmental matrices [59]. The SFE technique offers compound and matrix selectivity simply by varying the extraction conditions, such as temperature, pressure, and type of solvent [9,57]. The extraction procedure appears very harsh, with overall recovery greater than those achieve by sonication, soxhlet, and ASE methodologies [57,59]. It has recently been reported as a confirmatory tool in the Phase III of sediment toxicity identification evaluation where it was demonstrated that SFE treatment applied to sediment may release otherwise non-bioavailable chemicals, thereby reducing toxicity and concentration of non-ionic organic contaminants in some contaminated sediment, while increasing toxicity to exposed biota [15]. Notwithstanding, its relevance for predicting bioaccessibility has been demonstrated, for instance, SFE has predicted bioaccumulation and toxicity of PAHs [8,75,139] and PCBs [52,99] to earthworms or benthic organisms, as well as the biodegradability of PAHs in contaminated soils using composting as bioremediation strategy [17]. Furthermore, Lorenzi et al. [86] and Hawthrone et al. [58] showed that an SFE extraction of progressively increased strength resulted in a stair-step pattern of extractability that reflects two distinct fractions—the loosely bound and the tightly bound—which corresponded well to the biodegradable and residual fractions observed following one year bioremediation process [58]. The suitability of SFE to predict biodegradation of phenanthrene in laboratory spiked soil was demonstrated by Stroud et al. [138]; where carefully selected SFE extraction parameters accurately predicted phenanthrene mineralization by a culture of *Pseudomonas* inoculum. Overall, a high degree of a posteriori fitting of extraction parameters was required for SFE extraction to accurately predict bioaccessibility of contaminant [56,85,138].

4.2.1 Supercritical Water Extraction

Supercritical water extraction use superheated water (up to 200 °C) maintained as liquid under moderate pressure (5–20 psi) to reduce the dielectric constant of water such that the hydrogen bonding is weakened and the polarity of water reduced, thus water behaves more like an organic solvent [131]. The method was adopted by Latawiec et al. [79,80] to measure bioaccessibility of model PAH to catabolic inoculum in two laboratory spiked soils [80] and for a range of PAHs in both laboratory spiked and field contaminated soil and sediment [78]. The authors showed that extraction performed at 160 °C accurately estimated phenanthrene bioaccessibility to catabolic inoculum across a range of soil types. However, their results suggested that the efficiency of this method depend largely on the temperature at which the extraction is performed as well as the physico-chemical properties of the contaminants, mainly the octanol–water partition coefficient (K_{ow}). For example, extractions performed at lower temperatures (40 and 80 °C) underestimated biodegradation, while those performed at 180° and 200° also overestimated biodegradation.

4.3 Solid-Phase Extraction

Solid-phase extraction (SPE) technique using various sorbent has been proposed as another non-exhaustive technique to measure bioaccessibility of nonpolar organic contaminant in soil. The adsorbent used as solid phase is either mixed with contaminated matrix and aqueous solution (water or $CaCl_2$) in a batch system. The contaminant in the aqueous phase is then adsorb onto the nonpolar surface of the adsorbent, leading to further desorption and subsequent entrapment by the "receiver phase" until the all the freely available and loosely sorbed chemical are released from the matrix. Therefore, SPEs mimic the mass transfer limitation that governs the interactions between contaminants and biological receptors [24].

Adsorbents used for SPE include solid-phase microextraction (SPME) coated with different hydrophobic phases [23,37,54,62,65,103,146], octadecyl-modified silica disk—C18 disk [74,143,144], Tenax TA® [6,7,25,34,54,71,85,132],

Amberlite XADs [83,105,129] and, recently, the sorptive extraction with silicone rod [47]. Many of the proposed solid-phase extractants have been demonstrated as suitable for predicting the uptake and bioaccumulation of organic compounds such as PCB, PAHs, total petroleum hydrocarbons (TPHs) and pesticides, in soil and sediment dwelling biota. However, the hydrophobic resins have been particularly well studied across several environmental compartments, major group of contaminants, and across several groups of biological receptors. The more prevalent techniques are described in the following sections.

4.3.1 Tenax-TA Extraction

Tenax TA is a beadlike, porous polymer based on 2,6-diphenyl-*p*-phenylene oxide with characteristic size of 60–80 mesh and pore diameter of c. 177–250 μm. The method was first applied by Pignatello [106,107] to characterize the reversible sorption of aliphatic halocarbons, and later adopted by Cornelissen et al. [26,28] to study the desorption behavior of PCB and PAHs in sediments. Further Kukkonen et al. [76], used Tenax procedure to described desorption kinetics of insecticide Pendimethalin in artificially contaminated sediment. The Tenax extraction provides an infinite sink for desorbable HOCs and conceptually mimics the interaction between organic contaminants and biotic interaction [7,28,107].

Studies by Cornelissen et al. [27,28] showed that a 6-h extraction by Tenax TA corresponded to the rapidly desorbing HOC fraction that can be applied to access bioavailability of HOCs in sediments. Since then the method has gain rapid recognition in the field of ecological risk assessment including the prediction of biota-sediment accumulation factor (BSAF) in soil and sediment oligocheates [7,53,77,85,88,91–93], toxicity to fish and macrophytes [45,92], and biodegradability of petroleum hydrocarbon and chlorinated pesticides [6,7,24,31].

4.3.2 XAD Extraction

Amberlite XAD is the manufacturer's name for a group of polymers generally in the form of white beads. The beads size of 20–60 mesh with slight variations is common within the group but other physical parameters including pore size, surface area, and pore diameter differs greatly and gives individual XAD the unique adsorption characteristics within the group. The distribution of sorption sites around the macrospherical structure enables the XAD to serve as an infinite sink for organic compound across various environmental matrices [19,100]. Chemically, the make-up of XAD also differs and may be polymeric (e.g., XAD-2, XAD-4, XAD-16, XAD-1180, XAD-2000) or acrylic esters (e.g., XAD-7HP, XAD-8). Historically, XADs have been used for the concentration and detection of trace organic compounds from water, and its high retention capacity coupled with the good recoveries of analyte from aqueous solution to the limit of parts per million and up to parts per billion range has been well established [16,18,40,46,64,66,98,119].

Gustafson and Dickhut [51] found that XAD-2 was a better sorbent to determine the freely dissolved concentration in an estuarine system containing higher level of suspended particulate matter [51], and XAD extraction has been proposed as the most cost-effective strategy for the removal of trace organic contaminants from waste water where factors controlling bioavailability may limit the application of biodegradation as a treatment option [128].

Despite the extensive array of scientific literature reporting the ability of amberlite XAD to serve as sink for HOCs in the aqueous phase, its application to predict bioaccessibility of various organic chemicals from soil and sediments has not been particularly well studied. Among the limited available studies, there is contracting evidence on the suitability of XAD extraction as a surrogate for measuring bioaccessibility of organic pollutants in soil and sediment [83,105,129]. For instance in the study of Lei et al., XAD-2 significantly estimated microbial accessibility of 2-, 3-, and 4-ring PAHs in a field contaminated sediment while 5- and 6-ring PAHs were overestimated due to their lack of degradation by the indigenous microflora employed for the study [83]. However, the study by Patterson et al. [105] reported good and poor prediction of naphthalene biodegradation by XAD-4 extraction across five soils representing wide physico-chemical parameters.

4.3.3 Sorptive Bioaccessibility Extraction

Proposed by Gouliarmou et al. [47]; the sorptive bioaccessibility extraction (SBE) is a simple dissolution procedure designed to mimic desorption from the contaminated matrix and uptake by organisms or degradation by microorganisms. HP-β-CD (2-hydroxylpropyl-β-cyclodextrin) was used as the diffusive carrier to enhance desorption from the matrix while a silicone rod was incorporated into the slurry as the dominating sink that concomitantly absorbed the HOC molecules from the cyclodextrin solution [47]. The method has been demonstrated to effectively remove HOC from wood soot, a frequently encountered carbonaceous geosorbent to represent worst environmental case scenario. The SBE method is in way similar to the solid-phase microextraction polydimethylsiloxane (SPME-PDMS), but the major difference is that SPME is nondepletive as the coated fibers have a small capacity and extract negligible fraction, whereas SBE was designed to act as a depleting sorbent. The practicality of employing such seemingly exhaustive methodology for predicting bioaccessibility of organic contaminants in soil and sediment may be questionable, in that the amount of contaminant extracted in the sorptive bioaccessibility assay was 3–24 times higher than that extracted by the aqueous HP-β-CD solution alone [47]. Given the reported performance of HP-β-CD alone, the incorporation of a secondary sink in the SBE methodology may likely lead to overestimation of bioaccessibility. However, the method is still at development stage and its validation has not been reported in soils and sediments.

4.4 Aqueous-Based Cyclodextrin Extraction

It is widely accepted that, under most circumstances, soil and sediment associated contaminants have to enter the soil/sediment pore-water prior to uptake by biota [11,20]. Aqueous-based extraction techniques are approaches that rely on desorption of organic chemicals from the solid phase of the soil to the aqueous phase (labile fraction). Thus, by measuring the contaminant concentration in the pore-water phase and/or what is readily desorbable, the bioaccessibility of organic contaminants may be estimated.

Cyclodextrins are cyclic oligosaccharides produced form enzymatic digestion of starch, with a toroidal shaped hydrophilic exterior and hydrophobic cavity. The basis of the HPCD approach is that mobile organic molecules (desorbed from soil) in to the aqueous phase will be encapsulated into the hydrophobic cavity of cyclodextrin in a guest-host inclusion complex model [14,35,152]. To date most studies have focused on using HP-β-CD to extract organic contaminants from soils; as such, the application of HP-β-CD for predicting bioaccessibility of PAHs is well established [5,7,32,38,39,60,102,117]. Further, the HP-β-CD extraction for predicting TPHs has also been demonstrated in the laboratory spiked soil using a model alkane, hexadecane [136,137], in field contaminated soils [7,31,33] and field contaminated sediments [32].

There are three types of cyclodextrin (α-, β-, and γ-), containing 6, 7, and 8 glucose molecules, respectively. The size of the cavity determines the size of chemicals that can be effectively extracted by a particular cyclodextrin molecule from soil [33]. Studies by Stroud et al. [136,137] revealed that the size specificity of CD used in extractions was important; for example, the extraction of ^{14}C-hexadecane using α-CD correlated well with mineralization of the model ^{14}C-alkane, but not ^{14}C-phenanthrene, while the extraction of ^{14}C-phenanthrene using β-CD correlated with mineralization of ^{14}C-PAH, but not ^{14}C-hexadecane. In line with this observation, other authors have also reported poor correlation between β-CD and TPHs [7,31,33]. Nonetheless, the β-CD extraction has been shown to be robust for predicting PAHs under various contamination scenarios, either in experimental [5,36,38,39,61,116,117] or field contaminated soils [7,38,60,101,135] and sediments [32].

5. CONCLUSIONS

The assessment of contaminated environments has moved away from a generic, total concentration-based approach to consider the risk posed by chemical contamination on a case-by-case basis. There is a greater awareness of the behavior of chemicals in soils and sediments, associated hazards and risk to various receptors; this extends to an understanding of contaminant mobility and bioavailability. However, total extractable concentrations still remain a key component within risk assessment of contaminated soils and sediments, in spite of the multitude of published studies on chemical methods for determining contaminant bioaccessibility [118]. At the moment, the risk assessment community remains reluctant to include bioavailable or bioaccessible measurements overtly within the risk assessment framework for a number of reasons including the protective (sometimes overly protective) measure associated with total extractable concentrations; the number of different chemical approaches, together with the lack of consistent data from the different extraction methodologies in relation to soil and sediment type, target contaminant(s), laboratory and field contamination scenarios and target organisms/receptors. In relation to this, Latawiec et al. [79] investigated the decision-maker's perspective of the use of bioaccessibility as a measure within the risk assessment process used by local authority contaminated land officers in England and Wales. The results showed that 70% thought that bioaccessibility was a useful tool in contaminated land decision-making support and could be important in a more cost-effective management of contaminated land. However, 78% of the contaminated land officers considered that a lack of statutory guidance was a major impediment to the use of such methodologies being used. Key considerations, which may support the inclusion of bioaccessibility measurements, are (1) linking an extraction to a particular biological endpoint/organism (ecological or human health); (2) the target use of extractions for particular contaminants; (3) the development of a suite of standardized extractions each of which are relevant to a particular receptor.

Although, there is very limited uptake of bioavailability/bioaccessibility measurements within the risk assessment process, at the moment, the drive for inclusion of certain measurements is gaining momentum. For example, in the report published by RIVM in the Netherlands, Brand et al., [13] considered the usefulness of a number of extractions, which could be used in the decision-making process to measure the bioaccessible concentration of contaminants in soils. The authors went on to state that a bioaccessibility extraction can only be useful if the measured concentration can be linked to toxicity data for specific biota. In this report, Brand and co-authors suggested that a tiered approach could be adopted, where Tier 1 would link total extractable concentrations with soil quality standards; this would then be followed by a second tier in which bioaccessible concentrations (as determined by Tenax or cyclodextrin extractions) would be measured thereby providing a more realistic level of risk.

It remains unclear if or when such extractions will be included within contaminated land risk assessment framework, but it is clear that additional testing is required to link toxicity data to total and bioaccessible concentrations.

Rigorously testing the robustness and reproducibility of chemical extractions that measure bioaccessibility will enhance confidence in the methodologies for use within any risk assessment framework for contaminated soils and sediments.

ACKNOWLEDGMENTS

Special thanks to the UK Department for International Development (DFID) for funding this project through a doctoral scholarship administered by the Association of Commonwealth Universities (ACU) in the United Kingdom.

REFERENCES

[1] R. Ahmad, R. Kookana, M. Megharaj, A. Alston, Aging reduces the bioavailability of even a weakly sorbed pesticide (carbaryl) in soil, Environ. Toxicol. Chem. 23 (2004) 2084–2089.

[2] M. Alexander, How toxic are toxic-chemicals in soil, Environ. Sci. Technol. 29 (1995) 2713–2717.

[3] M. Alexander, Aging, bioavailability, and overestimation of risk from environmental pollutants, Environ. Sci. Technol. 34 (2000) 4259–4265.

[4] M. Alexander, P. Hatzinger, J. Kelsey, B. Kottler, K. Nam, R. Bajpai, M. Zappi, Sequestration and realistic risk from toxic chemicals remaining after bioremediation, Bioremediation of Surface and Subsurface Contamination 829 (1997) 1–5.

[5] I. Allan, K. Semple, R. Hare, B. Reid, Prediction of mono- and polycyclic aromatic hydrocarbon degradation in spiked soils using cyclodextrin extraction, Environ. Pollut. 144 (2006) 562–571.

[6] T. Baczynski, D. Pleissner, M. Krylow, Bioremediation of chlorinated pesticides in field-contaminated soils and suitability of tenax solid-phase extraction as a predictor of its effectiveness, Clean–Soil Air Water 40 (2012) 864–869.

[7] C. Bernhardt, K. Derz, W. Kördel, et al., Applicability of non-exhaustive extraction procedures with Tenax and HPCD, J. Hazard. Mater. 261 (2013) 711–717.

[8] L. Bielska, K. Smidova, J. Hofman, Supercritical fluid extraction of persistent organic pollutants from natural and artificial soils and comparison with bioaccumulation in earthworms, Environ. Pollut. 176 (2013) 48–54.

[9] E. Bjorklund, T. Nilsson, S. Bowadt, K. Pilorz, L. Mathiasson, S. Hawthorne, Introducing selective supercritical fluid extraction as a new tool for determining sorption/desorption behavior and bioavailability of persistent organic pollutants in sediment, J. Biochem. Biophys. Meth. 43 (2000) 295–311.

[10] B. Bogan, V. Trbovic, Effect of sequestration on PAH degradability with Fenton's reagent: roles of total organic carbon, humin, and soil porosity, J. Hazard. Mater. 100 (2003) 285–300.

[11] T. Bosma, P. Middeldorp, G. Schraa, A. Zehnder, Mass transfer limitation of biotransformation: quantifying bioavailability, Environ. Sci. Technol. 31 (1997) 248–252.

[12] D.C. Bouchard, A.L. Wool, M.L.g Campbell, P. Nkedi-Kizza, P.S.C. Rao, Sorption nonequilibrium during solute transport, J. Contam. Hydrol. 2 (1988) 209–223.

[13] E. Brand, W. Peijnenburg, B. Goenenberg, J. Vink, J. Lijzen, D. ten Hulscher, C. Jonker, P. Romkens, E. Roex, Towards Implementation of Bioavailability Measurements in the Dutch Regulatory Framework, National Institute for Public Health and the Environment Report, 2009. RIVM Report 711701084.

[14] M. Brusseau, X. Wang, W. Wang, Simultaneous elution of heavy metals and organic compounds from soil by cyclodextrin, Environ. Sci. Technol. 31 (1997) 1087–1092.

[15] R. Burgess, S. Hawthorne, M. Perron, M. Cantwell, C. Grabanski, D. Miller, K. Ho, M. Pelletier, Assessment of supercritical fluid extraction use in whole sediment toxicity identification evaluations, Environ. Toxicol. Chem. 30 (2011) 819–827.

[16] A.K. Burnham, G.V. Calder, J.S. Fritz, G.A. Junk, H.J. Svec, R. Willis, Identification and estimation of neutral organic contaminants in potable water, Anal. Chem. 44 (1972) 139–142.

[17] T. Cajthaml, V. Sasek, Application of supercritical fluid extraction (SFE) to predict bioremediation efficacy of long-term composting of PAH-contaminated soil, Environ. Sci. Technol. 39 (2005) 8448–8452.

[18] R. Care, D. Morrison, J.F. Smith, On the limit of detection of traces of volatile organics in water using amberlite XAD-2 resin, Water Res. 16 (1982) 663–665.

[19] K. Carroll, M. Harkness, A. Bracco, R. Balcarcell, Application of permanent polymer diffusional model to the desorption of polychlorinated-biphenyls from Hudson river sediments, Environ. Sci. Technol. 28 (1994) 253–258.

[20] C. Cerniglia, Biodegradation of organic contaminants – overview and examples with polycyclic aromatic hydrocarbons, Abstracts of Papers of the American Chemical Society 199 (1990) 136 (ENVR).

[21] N. Chung, M. Alexander, Differences in sequestration and bioavailability of organic compounds aged in dissimilar soils, Environ. Sci. Technol. 32 (1998) 855–860.

[22] N. Chung, M. Alexander, Effect of soil properties on bioavailability and extractability of phenanthrene and atrazine sequestered in soil, Chemosphere 48 (2002) 109–115.

[23] J. Conder, T. La Point, Solid-phase microextraction for predicting the bioavailability of 2,4,6-trinitrotoluene and its primary transformation products in sediment and water, Environ. Toxicol. Chem. 24 (2005) 1059–1066.

[24] G. Cornelissen, H. Rigterink, M. Ferdinandy, P. Van Noort, Rapidly desorbing fractions of PAHs in contaminated sediments as a predictor of the extent of bioremediation, Environ. Sci. Technol. 32 (1998) 966–970.

[25] G. Cornelissen, H. Rigterink, D. ten Hulscher, B. Vrind, P. van Noort, A simple Tenax (R) extraction method to determine the availability of sediment-sorbed organic compounds, Environ. Toxicol. Chem. 20 (2001) 706–711.

[26] G. Cornelissen, H. Rigterink, B.A. Vrind, D.T.E.M. ten Hulscher, M.M.A. Ferdinandy, P.C.M. van Noort, Two-stage desorption kinetics and in situ partitioning of hexachlorobenzene and dichlorobenzenes in a contaminated sediment, Chemosphere 35 (1997) 2405–2416.

[27] G. Cornelissen, P. Van Noort, H. Govers, Mechanism of slow desorption of organic compounds from sediments: a study using model sorbents, Environ. Sci. Technol. 32 (1998) 3124–3131.

[28] G. Cornelissen, P. van Noort, H. Govers, Desorption kinetics of chlorobenzenes, polycyclic aromatic hydrocarbons, and polychlorinated biphenyls: sediment extraction with Tenax(R) and effects of contact time and solute hydrophobicity, Environ. Toxicol. Chem. 16 (1997) 1351–1357.

[29] N.R. Couling, M.G. Towell, K.T. Semple, Biodegradation of PAHs in soil: Influence of chemical structure, concentration and multiple amendment, Environ. Pollut. 158 (2010) 3411–3420.

[30] T. Crommentuijn, D. Sijm, J. de Bruijn, K. van Leeuwen, E. van de Plassche, Maximum permissible and negligible concentrations for some organic substances and pesticides, J. Environ. Manage. 58 (2000) 297–312.

[31] C. Cuypers, R. Clemens, T. Grotenhuis, W. Rulkens, Prediction of petroleum hydrocarbon bioavailability in contaminated soils and sediments, Soil Sediment Contam. 10 (2001) 459–482.

[32] C. Cuypers, T. Pancras, T. Grotenhuis, W. Rulkens, The estimation of PAH bioavailability in contaminated sediments using hydroxypropyl-beta-cyclodextrin and Triton X-100 extraction techniques, Chemosphere 46 (2002) 1235–1245.

[33] C. Dandie, J. Weber, S. Aleer, E. Adetutu, A. Ball, A. Juhasz, Assessment of five bioaccessibility assays for predicting the efficacy of petroleum hydrocarbon biodegradation in aged contaminated soils, Chemosphere 81 (2010) 1061–1068.

[34] A. de la Cal, E. Eljarrat, T. Grotenhuis, D. Barcelo, Tenax (R) extraction as a tool to evaluate the availability of polybrominated diphenyl ethers, DDT, and DDT metabolites in sediments, Environ. Toxicol. Chem. 27 (2008) 1250–1256.

[35] J. Dean, W. Scott, Recent developments in assessing the bioavailability of persistent organic pollutants in the environment, TrAC 23 (2004) 609–618.

[36] N. Dew, G. Paton, K. Semple, Prediction of [3-C-14]phenyldodecane biodegradation in cable insulating oil-spiked soil using selected extraction techniques, Environ. Pollut. 138 (2005) 316–323.

[37] Y. Ding, P. Landrum, J. You, A. Harwood, M. Lydy, Use of solid phase microextraction to estimate toxicity: relating fiber concentrations to toxicity-part I, Environ. Toxicol. Chem. 31 (2012) 2159–2167.

[38] K. Doick, N. Dew, K. Semple, Linking catabolism to cyclodextrin extractability: determination of the microbial availability of PAHs in soil, Environ. Sci. Technol. 39 (2005) 8858–8864.

[39] K.J. Doick, P.J. Clasper, K. Urmann, K.T. Semple, Further validation of the HPCD-technique for the evaluation of PAH microbial availability in soil, Environ. Pollut. 144 (2006) 345–354.

[40] M. Dressler, Extraction of traces amounts of organic compounds from water with porous organic polymers, J. Chromatogr. 165 (1979) 167–772.

[41] R. Efroymson, M. Alexander, Biodegradation in soil of hydrophobic pollutants in nonaqueous-phase liquids (NAPLS), Environ. Toxicol. Chem. 13 (1994) 405–411.

[42] G. Ehlers, S. Forrester, K. Scherr, A. Loibner, L. Janik, Influence of the nature of soil organic matter on the sorption behaviour of pentadecane as determined by PLS analysis of mid-infrared DRIFT and solid-state C-13 NMR spectra, Environ. Pollut. 158 (2010) 285–291.

[43] G. Ehlers, A. Loibner, Linking organic pollutant (bio)availability with geosorbent properties and biomimetic methodology: a review of geosorbent characterisation and (bio)availability prediction, Environ. Pollut. 141 (2006) 494–512.

[44] L. Ehlers, R. Luthy, Contaminant bioavailability in soil and sediment, Environ. Sci. Technol. 37 (2003) 295A–302A.

[45] M. Feo, C. Corcellas, C. Barata, A. Ginebreda, E. Eljarrat, D. Barcelo, Organic carbon content effects on bioavailability of pyrethroid insecticides and validation of solid phase extraction with poly (2,6-diphenyl-p-phenylene oxide) polymer by Daphnia magna toxicity tests, Sci. Total Environ. 442 (2013) 497–502.

[46] J.S. Fritz, Concentration of solutes from aqueous solution, Acct. Chem. Res. 10 (1977) 67–72.

[47] V. Gouliarmou, P. Mayer, Sorptive bioaccessibility extraction (SBE) of soils: combining a mobilization medium with an absorption sink, Environ. Sci. Technol. 46 (2012) 10682–10689.

[48] W. Guerin, S. Boyd, Differential bioavailability of soil-sorbed naphthalene to 2 bacterial species, Appl. Environ. Microbiol. 58 (1992) 1142–1152.

[49] W.F. Guerin, S.A. Boyd, Bioavailability of naphthalene associated with natural and synthetic sorbents, Water Res. 31 (1997) 1504–1512.

[50] E. Gurzau, A. Gurzau, I. Neamtiu, A. Coman, R. Hull, C. Barbu, N. Goncharova, Integration of metal bioavailability in risk assessment policy decision making, in: Strategies to Enhance Environment Security in Transition Countries, 2007, pp. 349–368.
[51] K. Gustafson, R. Dickhut, Distribution of polycyclic aromatic hydrocarbons in southern Chesapeake Bay surface water: evaluation of three methods for determining freely dissolved water concentrations, Environ. Toxicol. Chem. 16 (1997) 452–461.
[52] P. Hallgren, R. Westbom, T. Nilsson, S. Sporring, E. Bjorklund, Measuring bioavailability of polychlorinated biphenyls in soil to earthworms using selective supercritical fluid extraction, Chemosphere 63 (2006) 1532–1538.
[53] A. Harwood, P. Landrum, M. Lydy, Can SPME fiber and tenax methods predict the bioavailability of biotransformed insecticides? Environ. Sci. Technol. 46 (2012) 2413–2419.
[54] A. Harwood, P. Landrum, D. Weston, M. Lydy, Using SPME fibers and Tenax to predict the bioavailability of pyrethroids and chlorpyrifos in field sediments, Environ. Pollut. 173 (2013) 47–51.
[55] P.B. Hatzinger, M. Alexander, Effect of aging of chemicals in soil on their biodegradability and extractability, Environ. Sci. Technol. 29 (1995) 537–545.
[56] S. Hawthorne, C. Grabanski, Correlating selective supercritical fluid extraction with bioremediation behavior of PAHs in a field treatment plot, Environ. Sci. Technol. 34 (2000) 4103–4110.
[57] S. Hawthorne, C. Grabanski, E. Martin, D. Miller, Comparisons of Soxhlet extraction, pressurized liquid extraction, supercritical fluid extraction and subcritical water extraction for environmental solids: recovery, selectivity and effects on sample matrix, J. Chromatogr. A 892 (2000) 421–433.
[58] S. Hawthorne, D. Poppendieck, C. Grabanski, R. Loehr, PAH release during water desorption, supercritical carbon dioxide extraction, and field bioremediation, Environ. Sci. Technol. 35 (2001) 4577–4583.
[59] S.B. Hawthorne, D.J. Miller, Extraction and recovery of polyaromatic hydrocarbons from environmental solid using supercritical fluids, Anal. Chem. 59 (1987) 4103–4110.
[60] Z. Hickman, A. Swindell, I. Allan, A. Rhodes, R. Hare, K. Semple, B. Reid, Assessing biodegradation potential of PAHs in complex multi-contaminant matrices, Environ. Pollut. 156 (2008) 1041–1045.
[61] J. Hofman, A. Rhodes, K.T. Semple, Fate and behaviour of phenanthrene in the natural and artificial soils, Environ. Pollut. 152 (2008) 468–475.
[62] X. Hu, J. Liu, Q. Zhou, S. Lu, R. Liu, L. Cui, D. Yin, P. Mayer, G. Jiang, Bioavailability of organochlorine compounds in aqueous suspensions of fullerene: evaluated with medaka (*Oryzias latipes*) and negligible depletion solid-phase microextraction, Chemosphere 80 (2010) 693–700.
[63] IWIN, Integrale Normstelling Stoffen. Milieukwaliteitsnormen Boden, Water, Iucht, 1997.
[64] R.L. Jolley, Concentrating organics in water for biological testing, Environ. Sci. Technol. 15 (1981) 874–880.
[65] M. Jonker, S. van der Heijden, J. Kreitinger, S. Hawthorne, Predicting PAH bioaccumulation and toxicity in earthworms exposed to manufactured gas plant soils with solid-phase microextraction, Environ. Sci. Technol. 41 (2007) 7472–7478.
[66] G.A. Junk, J.J. Richard, M.D. Grieser, D. Witiak, J.L. Witiak, M.D. Diguello, R. Vick, H.J. Svec, J.S. Fritz, G.V. Calder, Use of macroreticular resins in the analysis of water for trace organic contaminants, J. Chromatogr. 99 (1974) 745–762.
[67] R. Kanaly, R. Bartha, K. Watanabe, S. Harayama, Enhanced mineralization of benzo[a]pyrene in the presence of nonaqueous phase liquids, Environ. Toxicol. Chem. 20 (2001) 498–501.

[68] A. Katayama, R. Bhula, G. Burns, E. Carazo, A. Felsot, D. Hamilton, C. Harris, Y. Kim, G. Kleter, W. Koedel, J. Linders, J. Peijnenburg, A. Sabljic, R. Stephenson, D. Racke, B. Rubin, K. Tanaka, J. Unsworth, R. Wauchope, D. Whitcare, Bioavailability of xenobiotics in the soil environment, Rev. Environ. Contamin. Toxicol. 203 (2010) 1–86.

[69] M. Khan, S. Cheema, C. Shen, C. Zhang, X. Tang, J. Shi, X. Chen, J. Park, Y. Chen, Assessment of phenanthrene bioavailability in aged and unaged soils by mild extraction, Environ. Monit. Assess. 184 (2012) 549–559.

[70] C.D. Klaassen, Cassarett and Doull's toxicology, in: C.D. Klaassen, M.O. Amdur, J. Doull (Eds.), Cassarett and Doull's Toxicology, Macmillan Publishing Co, New York, 1986.

[71] S. Kohlmeier, M. Mancuso, U. Deepthike, R. Tecon, J. van der Meer, H. Harms, M. Wells, Comparison of naphthalene bioavailability determined by whole-cell biosensing and availability determined by extraction with Tenax, Environ. Pollut. 156 (2008) 803–808.

[72] M. Kraatz, E. Ortiz, R. Luthy, B.M. INST, PAH-bioavailability from NAPLs and significance for soil biotreatment, in: In Situ and On-Site Bioremediation, 1997. vol. 5 4(5), pp. 669–674.

[73] B.K. Kramer, P.B. Ryan, Soxhlet and microwave extraction in determining the bioaccessibility of pesticides from soil and model solids, in: Conference on Hazardous Waste Research, 2000, pp. 196–210.

[74] M. Krauss, W. Wilcke, Biomimetic extraction of PAHs and PCBs from soil with octadecyl-modified silica disks to predict their availability to earthworms, Environ. Sci. Technol. 35 (2001) 3931–3935.

[75] J. Kreitinger, A. Quinones-Rivera, E. Neuhauser, M. Alexander, S. Hawthorne, Supercritical carbon dioxide extraction as a predictor of polycyclic aromatic hydrocarbon bioaccumulation and toxicity by earthworms in manufactured-gas plant site soils, Environ. Toxicol. Chem. 26 (2007) 1809–1817.

[76] J. Kukkonen, P. Landrum, S. Mitra, D. Gossiaux, J. Gunnarsson, D. Weston, Sediment characteristics affecting desorption kinetics of select PAH and PCB congeners for seven laboratory spiked sediments, Environ. Sci. Technol. 37 (2003) 4656–4663.

[77] P. Landrum, S. Robinson, D. Gossiaux, J. You, M. Lydy, S. Mitra, T. Ten Hulscher, Predicting bioavailability of sediment-associated organic contaminants for Diporeia spp. and oligochaetes, Environ. Sci. Technol. 41 (2007) 6442–6447.

[78] A. Latawiec, B. Reid, Sequential extraction of polycyclic aromatic hydrocarbons using subcritical water, Chemosphere 78 (2010) 1042–1048.

[79] A. Latawiec, P. Simmons, B. Reid, Decision-makers' perspectives on the use of bioaccessibility for risk-based regulation of contaminated land, Environ. Int. 36 (2010) 383–389.

[80] A. Latawiec, A. Swindell, B. Reid, Environmentally friendly assessment of organic compound bioaccessibility using sub-critical water, Environ. Pollut. 156 (2008) 467–473.

[81] A. Latawiec, A. Swindell, P. Simmons, B. Reid, Bringing bioavailability into contaminated land decision making: the way forward? Crit. Rev. Environ. Sci. Technol. 41 (2011) 52–77.

[82] L. Lei, R. Bagchi, A. Khodadoust, M. Suidan, H. Tabak, Bioavailability prediction of polycyclic aromatic hydrocarbons in field-contaminated sediment by mild extractions, J. Environ. Eng.-Asce 132 (2006) 384–391.

[83] L. Lei, M.T. Suidan, A.P. Khodadoust, H.H. Tabak, Assessing the bioavailability of PAHs in field-contaminated sediment using XAD-2 assisted desorption, Environ. Sci. Technol. 38 (2004) 1786–1793.

[84] H. Liste, M. Alexander, Butanol extraction to predict bioavailability of PAHs in soil, Chemosphere 46 (2002) 1011–1017.

[85] M. Liu, S. Tian, P. Chen, L. Zhu, Predicting the bioavailability of sediment-associated polybrominated diphenyl ethers using a 45-d sequential Tenax extraction, Chemosphere 85 (2011) 424–431.

[86] A. Loibner, M. Gartner, M. Schlegl, I. Hautzenberger, R. Braun, B.M. Inst, PAHs: rapid estimation of their bioavailability in soil, in: In Situ and On-Site Bioremediation, vol. 5, 1997, pp. 617–622. 4(5).

[87] R. Luthy, G. Aiken, M. Brusseau, S. Cunningham, P. Gschwend, J. Pignatello, M. Reinhard, S. Traina, W. Weber, J. Westall, Sequestration of hydrophobic organic contaminants by geosorbents, Environ. Sci. Technol. 31 (1997) 3341–3347.

[88] E. Mackenbach, Y. Jing, M. Mills, P. Landrum, M. Lydy, Application of a Tenax model to assess bioavailability of PCBs in field sediments, Environ. Toxicol. Chem. 31 (2012) 2210–2216.

[89] C. Macleod, K. Semple, Influence of contact time on extractability and degradation of pyrene in soils, Environ. Sci. Technol. 34 (2000) 4952–4957.

[90] L. McAllister, K.T. Semple, Measurement of bioaccessibility of organic pollutants in soil, in: S.P. Cummings (Ed.), Bioremediation: Methods and Protocols, 2010, pp. 1–14.

[91] W. Mehler, H. Li, J. Pang, B. Sun, M. Lydy, J. You, Bioavailability of hydrophobic organic contaminants in sediment with different particle-size distributions, Arch. Environ. Contamin. Toxicol. 61 (2011) 74–82.

[92] C. Moermond, I. Roessink, M. Jonker, T. Meijer, A. Koelmans, Impact of polychlorinated biphenyl and polycyclic aromatic hydrocarbon sequestration in sediment on bioaccumulation in aquatic food webs, Environ. Toxicol. Chem. 26 (2007) 607–615.

[93] D. Morrison, B. Robertson, M. Alexander, Bioavailability to earthworms of aged DDT, DDE, DDD, and dieldrin in soil, Environ. Sci. Technol. 34 (2000) 709–713.

[94] K. Nam, M. Alexander, Role of nanoporosity and hydrophobicity in sequestration and bioavailability: tests with model solids, Environ. Sci. Technol. 32 (1998) 71–74.

[95] K. Nam, J. Kim, Role of loosely bound humic substances and humin in the bioavailability of phenanthrene aged in soil, Environ. Pollut. 118 (2002) 427–433.

[96] C. Nathanail, R. Smith, Incorporating bioaccessibility in detailed quantitative human health risk assessments, J. Environ. Sci. Health Part A–Tox./hazard. Subst. Environ. Eng. 42 (2007) 1193–1202.

[97] National Research Council, Bioavailability of Contaminants in Soil and Sediment: Processes, Tools, and Applications, National Academic Press, Washington DC, 2003.

[98] U. Niederschulte, K. Ballschmitter, Isolierung und Anreicherung von Chlorphenoxy-essigsauren aus Wassermit makroreticularen Adsorberharzen XAD-2 und XAD-7, Anal. Chem. 269 (1974) 360–363.

[99] T. Nilsson, S. Sporring, E. Bjorklund, Selective supercritical fluid extraction to estimate the fraction of PCB that is bioavailable to a benthic organism in a naturally contaminated sediment, Chemosphere 53 (2003) 1049–1052.

[100] G. Northcott, K. Jones, Experimental approaches and analytical techniques for determining organic compound bound residues in soil and sediment, Environ. Pollut. 108 (2000) 19–43.

[101] A. Papadopoulos, G.I. Paton, B.J. Reid, K.T. Semple, Prediction of PAH biodegradation in field contaminated soils using a cyclodextrin extraction technique, J. Environ. Monit. 9 (2007) 516–522.

[102] A. Papadopoulos, B. Reid, K.T. Semple, Prediction of microbial accessibility of carbon-14-phenanthrene in soil in the presence of pyrene or benzo[a]pyrene using an aqueous cyclodextrin extraction technique, J. Environ. Qual. 36 (2007) 1385–1391.

[103] T. Parkerton, M. Stone, D. Letinski, Assessing the aquatic toxicity of complex hydrocarbon mixtures using solid phase microextraction, Toxicol. Lett. 112 (2000) 273–282.
[104] G. Paton, K. Killham, H. Weitz, K.T. Semple, Biological tools for the assessment of contaminated land: applied soil ecotoxicology, Soil Use Manage. 21 (2005) 487–499.
[105] C. Patterson, K.T. Semple, G. Paton, Non-exhaustive extraction techniques (NEETs) for the prediction of naphthalene mineralisation in soil, Fems Microbiol. Lett. 241 (2004) 215–220.
[106] J. Pignatello, Slowly desorbing fraction of haliphatic halocarbons in sediment 2. Mechanistic aspect, Environ. Toxicol. Chem. 9 (1990) 1117–1126.
[107] J. Pignatello, Slowly reversible sorption of aliphatic halocarbons in soils 1. Formation of residual fraction, Environ. Toxicol. Chem. 9 (1990) 1107–1115.
[108] J. Pignatello, Sorption of organic compounds to soil organic matter: a new concept of mechanism and its possible implications for kinetics, Abstracts of Papers of the American Chemical Society 213 (1997) 86 (GEOC).
[109] J. Pignatello, Soil organic matter as a nanoporous sorbent of organic pollutants, Adv. Colloid Interface Sci. 76 (1998) 445–467.
[110] J. Pignatello, Dynamic interactions of natural organic matter and organic compounds, J. Soils Sediments 12 (2012) 1241–1256.
[111] J. Pignatello, B. Xing, Mechanisms of slow sorption of organic chemicals to natural particles, Environ. Sci. Technol. 30 (1996) 1–11.
[112] T. Pravecek, R. Christman, F. Pfaender, Impact of imposed anaerobic conditions and microbial activity on aqueous-phase solubility of polycyclic aromatic hydrocarbons from soil, Environ. Toxicol. Chem. 24 (2005) 286–293.
[113] T. Pravecek, R. Christman, F. Pfaender, Microbial bioavailability of pyrene in three laboratory-contaminated soils under aerobic and anaerobic conditions, J. Contamin. Hydrol. 86 (2006) 72–86.
[114] B. Reid, K.C. Jones, K.T. Semple, Bioavailability of persistent organic pollutants in soils and sediments – a perspective on mechanisms, consequences and assessment, Environ. Pollut. 108 (2000) 103–112.
[115] B. Reid, K. Jones, K.T. Semple, A. Leeson, B. Alleman, Can bioavailability of PAHs be assessed by a chemical means? Bioremediat. Technol. Polycycl. Aromat. Hydrocarbon Comp. (1999) 253–258.
[116] B. Reid, K.T. Semple, K.C. Jones, R.C. Karlsruhe, Prediction of bioavailability of persistent organic pollutants by a novel extraction technique, Contamin. Soil '98 1–2 (1998) 889–890.
[117] B. Reid, J. Stokes, K.C. Jones, K.T. Semple, Nonexhaustive cyclodextrin-based extraction technique for the evaluation of PAH bioavailability, Environ. Sci. Technol. 34 (2000) 3174–3179.
[118] M.J. Riding, K.J. Doick, F.L. Martin, K.C. Jones, K.T. Semple, Bioavailability/bioaccessibility of organic contaminants in soil: fundamentals to application, J. Hazard. Mater. 261 (2013) 687–700.
[119] J.P. Riley, D. Taylor, Analytical concentration of traces of dissolved organic materials from sea water with Amberlite XAD-1 resin, Anal. Chim. Acta 46 (1969) 307–311.
[120] M. Ruby, A. Davis, R. Schoof, S. Eberle, C. Sellstone, Estimation of lead and arsenic bioavailability using a physiologically based extraction test, Environ. Sci. Technol. 30 (1996) 422–430.
[121] P. Ruggiero, M. Pizzigallo, C. Crecchio, A. Violante, P. Huang, J. Bollag, L. Gianfreda, Effects of Soil Abiotic Processes on the Bioavailability of Anthropogenic Organic Residues, Soil Mineral-Organic Matter-Microorganism Interactions and Ecosystem Health, vol. 28B, 2002, pp. 95–133.

[122] K. Semple, K. Doick, K. Jones, P. Burauel, A. Craven, H. Harms, Defining bioavailability and bioaccessibility of contaminated soil and sediment is complicated, Environ. Sci. Technol. 38 (2004) 228A–231A.

[123] K. Semple, K. Doick, L. Wick, H. Harms, Microbial interactions with organic contaminants in soil: definitions, processes and measurement, Environ. Pollut. 150 (2007) 166–176.

[124] K. Semple, B. Reid, T. Fermor, Impact of composting strategies on the treatment of soils contaminated with organic pollutants, Environ. Pollut. 112 (2001) 269–283.

[125] K. Semple, J. Stokes, B. Reid, A. Leeson, B. Alleman, Temporal chances in the bioavailability of PAHs in soils, Bioremediat. Technol. Polycycl. Aromat. Hydrocarbon Comp. (1999) 247–252.

[126] M. Sharer, J. Park, T. Voice, S. Boyd, Aging effects on the sorption-desorption characteristics of anthropogenic organic compounds in soil, J. Environ. Qual. 32 (2003) 1385–1392.

[127] D. Sijm, R. Kraaij, A. Belfroid, Bioavailability in soil or sediment: exposure of different organisms and approaches to study it, Environ. Pollut. 108 (2000) 113–119.

[128] M. Silva, A. Fernandes, A. Mendes, C. Manaia, O. Nunes, Preliminary feasibility study for the use of an adsorption/bioregeneration system for molinate removal from effluents, Water Res. 38 (2004) 2677–2684.

[129] S. Simpson, V. Burston, D. Jolley, K. Chau, Application of surrogate methods for assessing the bioavailability of PAHs in sediments to a sediment ingesting bivalve, Chemosphere 65 (2006) 2401–2410.

[130] K. Smidova, J. Hofman, A. Ite, K. Semple, Fate and bioavailability of C-14-pyrene and C-14-lindane in sterile natural and artificial soils and the influence of aging, Environ. Pollut. 171 (2012) 93–98.

[131] R. Smith, Extractions with superheated water, J. Chromatogr. A 975 (2002) 31–46.

[132] B. Smreczak, B. Maliszewska-Kordybach, A. Klimkowicz-Pawlas, Assessing the bioavailability of phenanthrene to soil microorganisms using the Tenax extraction method, Environ. GeoChem.Chem. Health 30 (2008) 183–186.

[133] A. Spacie, J.L. Hamelink, Bioaccumulation, in: G. Rand (Ed.), Fundamentals of Aquatic Toxicology, Effects, Environmental Fate and Risk Assessment, Taylor and Francis, Washington DC, USA, 1995.

[134] S. Steinberg, J. Pignatello, B. Sawhney, Persistence of 1,2-Dibromoethane in soils - entrapment in intraparticle micropores, Environ. Sci. Technol. 21 (1987) 1201–1208.

[135] J. Stokes, A. Wilkinson, B. Reid, K. Jones, K. Semple, Prediction of polycyclic aromatic hydrocarbon biodegradation in contaminated soils using an aqueous hydroxypropyl-beta-cyclodextrin extraction technique, Environ. Toxicol. Chem. 24 (2005) 1325–1330.

[136] J. Stroud, G. Paton, K. Semple, Linking chemical extraction to microbial degradation of (14) C-hexadecane in soil, Environ. Pollut. 156 (2008) 474–481.

[137] J. Stroud, G. Paton, K. Semple, Predicting the biodegradation of target hydrocarbons in the presence of mixed contaminants in soil, Chemosphere 74 (2009) 563–567.

[138] J. Stroud, A. Rhodes, K. Semple, Z. Simek, J. Hofman, Using supercritical fluid extraction to measure the desorption and bioaccessibility of phenanthrene in soils, Environ. Pollut. 156 (2008) 664–670.

[139] H. Sun, J. Li, Availability of pyrene in unaged and aged soils to earthworm uptake, butanol extraction and SFE, Water Air Soil Pollut. 166 (2005) 353–365.

[140] A. Swindell, B. Reid, Comparison of selected non-exhaustive extraction techniques to assess PAH availability in dissimilar soils, Chemosphere 62 (2006) 1126–1134.

[141] A. Swindell, B. Reid, The influence of a NAPL on the loss and biodegradation of C-14-phenanthrene residues in two dissimilar soils, Chemosphere 66 (2007) 332–339.

[142] H. Tabak, J. Lazorchak, L. Lei, A. Khodadoust, J. Antia, R. Bagchi, M. Suidan, Studies on bioremediation of polycyclic aromatic hydrocarbon-contaminated sediments: bioavailability, biodegradability, and toxicity issues, Environ. Toxicol. Chem. 22 (2003) 473–482.

[143] J. Tang, H. Liste, M. Alexander, Chemical assays of availability to earthworms of polycyclic aromatic hydrocarbons in soil, Chemosphere 48 (2002) 35–42.

[144] J. Tang, B. Robertson, M. Alexander, Chemical-extraction methods to estimate bioavailability of DDT, DDE, and DDD in soil, Environ. Sci. Technol. 33 (1999) 4346–4351.

[145] M. Towell, G. Paton, K. Semple, The biodegradation of cable oil components: Impact of oil concentration, nutrient addition and bioaugmentation, Environ. Pollut. 159 (2011) 3777–3783.

[146] T. Trimble, J. You, M. Lydy, Bioavailability of PCBs from field-collected sediments: application of Tenax extraction and matrix-SPME techniques, Chemosphere 71 (2008) 337–344.

[147] C.J. van Leeuwen, J.L. Hermens, Terrestial toxicity, in: Risk Assessment of Chemicals: An Introduction, Kluwer Academic, Dordrecht, Netherlands, 1995, pp. 211–216.

[148] J. White, J. Kelsey, P. Hatzinger, M. Alexander, Factors affecting sequestration and bioavailability of phenanthrene in soils, Environ. Toxicol. Chem. 16 (1997) 2040–2045.

[149] B. Xing, J.J. Pignatello, Dual-mode sorption of low-polarity compounds in Glassy Poly(Vinyl Chloride) and soil organic matter, Environ. Sci. Technol. 31 (1997) 792–799.

[150] F. Reichenberg, P. Mayer, Two complementary sides of bioavailability: accessibility and chemical activity of organic contaminants in sediments and soils, Environ. Toxicol. Chem. 25 (2006) 1239–1245.

[151] K.T. Semple, A.W.J. Morriss, G.I. Paton, Bioavailability of hydrophobic organic contaminants in soils: fundamental concepts and techniques for analysis, Euro. J. Soil. Sci. 54 (2003) 809–818.

[152] J.D. Stokes, A. Wilkinson, B.J. Reid, K.C. Jones, K.T. Semple, Validation of the use of an aqueous hydroxypropyl-β-cyclodextrin (HPCD) extraction as an indicator of PAH bioavailability in contaminated soils, Environ. Toxicol. Chem. 24 (2005) 1325–1330.

Chapter 16

Benzotriazoles: History, Environmental Distribution, and Potential Ecological Effects

Mark G. Cantwell,[1,*] Julia C. Sullivan[2] and Robert M. Burgess[1]
[1]*Office of Research and Development, US Environmental Protection Agency, Narragansett, RI, USA;* [2]*Oak Ridge Institute for Science and Education, Narragansett, RI, USA*
**Corresponding author: E-mail: cantwell.mark@epa.gov*

Chapter Outline

1. INTRODUCTION

Benzotriazoles (BZTs) are a class of organic compounds used in a diverse range of household, commercial, and industrial applications and products. To date, there have been several major categories of usage of these compounds. The first is as an additive to provide ultraviolet (UV) stabilization for plastics, paints, coatings, and films. The second has been as an additive to inhibit metal corrosion. These compounds have also been used to a lesser extent in photographic film processes and the preparation and synthesis of specialty chemicals.

Commercial usage and production of BZTs started in the USA in the late 1950s and the initial US patent for BZT focused on the application and utility of BZTs as metal corrosion inhibitors in aqueous (glycol-based) heat exchange

Comprehensive Analytical Chemistry, Vol. 67. http://dx.doi.org/10.1016/B978-0-444-63299-9.00016-8
2015 Published by Elsevier B.V.

fluids [1]. These compounds are used in cooling systems (e.g., automotive radiators and commercial boilers) [2], and in the formulation of petroleum products, such as hydraulic fluids, greases, and lubricants, to prevent oxidation [3,4]. Shortly thereafter, the UV absorbing properties of BZTs were documented, and a US patent describing phenolic BZTs for this use was issued in 1961 [5]. This patent filing centered on use of BZTs as UV light stabilizer additives in plastics and polymers. Since then, many additional patents have been issued in the USA and globally for other BZT compounds for use as UV stabilizers. This has resulted in the evolution and incorporation of the parent BZT structure into many larger and more complex chemical compounds in response to new product applications and evolving formulation requirements. BZT compounds were also used in photographic applications starting in the 1960s, with the primary usage being as an antifogging agent in photographic emulsions [6]. Other associated commercial purposes included use as UV stabilizers in color tinting agents and thermographic photocopying applications [6]. BZTs have also been used to synthesize certain specialty chemicals and pharmaceutical compounds. In the 1990s, BZTs were incorporated as anticorrosion additives in aircraft deicing and anti-icing fluids (ADAFs) [7].

There is a small group of large volume, commercially produced BZT compounds that have been the recent focus of environmental research and monitoring efforts (Figure 1). As discussed below, BZTs are an example of an emerging contaminant with the characteristics of some of the better known industrial legacy persistent organic pollutants (POPs) (e.g., polychlorinated biphenyls (PCBs)). This chapter addresses key environmental issues such as behavior, fate, effects, and persistence associated with the use of high production, commonly used BZTs that are present in the environment at elevated levels.

2. CHEMICAL AND PHYSICAL PROPERTIES OF BZTs

The primary triazole molecule is comprised of a five-member heterocyclic compound with three nitrogen atoms and a vicinal pair of carbon atoms [6]. Two isomers are presented, with the hydrogen present in either the 1-H (Figure 2(a)) or 2-H configuration (Figure 2(b)). When a benzene ring is fused to the triazole, BZT is formed (e.g., Figure 1(a)). Many BZT compounds exist, but those with a high production volume (HPV) and which have been reported as being present in the environment most commonly follow these two configurations.

BZTs used as anticorrosive additives have a hydrogen oriented in the 1-H position (Figure 1(a)). In addition, two BZTs used extensively for this purpose are methylated, with the functional group at either the 4′ or 5′ position (e.g., 4-TT and 5-TT) (Figure 1(b–d)). Conversely, the UV-stabilizing BZTs (UV-BZTs) generally have a phenolic moiety positioned at 2-H (e.g., Figure 2e). Overall, the phenolic BZTs are higher in molecular weight (MW) and structurally more complex. These compounds often include an ester, chlorine atom, or additional functional groups (e.g., phenol and methyl) as part of the molecule

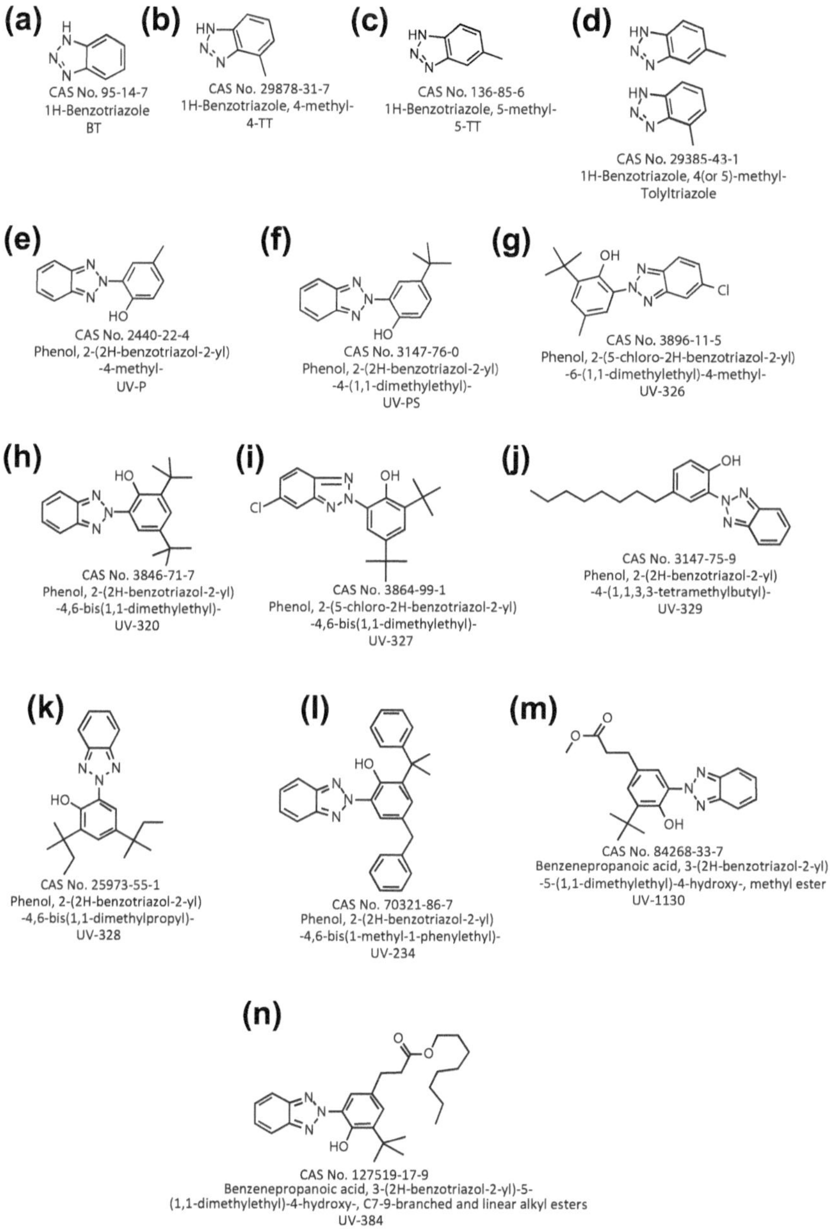

FIGURE 1 Structures, CAS numbers, and names of select BZT compounds (a–n).

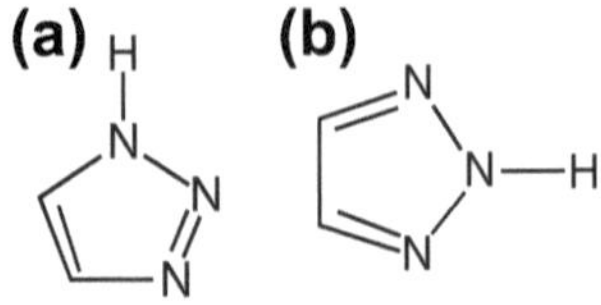

FIGURE 2 Structures of 1-H (a) and 2-H (b) triazole.

(e.g., Figure 1(e–g)). Since the discovery of the original parent BZT molecule, methods of synthesis have evolved with numerous patents issued for preparation, many of which have utilized diazotization. Briefly, BZT can be synthesized by reacting phenylenediamine, sodium nitrite, and acetic acid via diazotization of one of the amine groups [8]. Further details on synthesis are beyond the scope of this chapter, but early information on industrial-scale production of the 1-H BZTs can be found in Levy [9] and Long [10]. Information on 2-H BZTs (e.g., 2-2'-hydroxyphenyl-benzotriazoles), which are characteristic of UV stabilizers, can be found in Heller [5], Boyle and Milionis [11], and more recently Fukuoka et al. [12].

Stability is a key characteristic and desired attribute of most commercially used BZTs. The parent structure will in many instances readily undergo a range of reactions; however, the triazole moiety in most cases remains unchanged [6]. For applications utilizing their anticorrosive properties, BZTs form highly stable metal complexes with numerous metals and alloys, particularly copper, which protect them against corrosive attack. In this application, the BZT–metal complex forms a passive layer on the metal surface, preventing surface reactions [13].

For UV stabilization of plastics, polymers, and coatings, BZTs of the 2-H hydroxyphenyl class are used because they are photochemically stable and capable of absorbing light primarily in the near UV range (300–400 nm) [14]. The primary mechanism of action is for the UV-BZT stabilizer molecules to convert absorbed light energy in the BZT amended product to heat, which is in essence a reversible proton transfer reaction. The absorbance performance varies depending on the BZT used, the composition of the product, and the amount of BZT added to the material.

The physical properties of BZTs vary greatly, with large differences between the UV-BZTs and the ones used in anticorrosive applications. Presented here is a brief compilation of most of the BZT compounds along with their common synonyms that have been researched and reported recently in the peer reviewed environmental sciences literature (Table 1). The anticorrosive compounds (Table 1(a–c)) encompass a narrow range of MW (119–133), while the UV-BZTs have higher MWs overall and span a broader range (225 to over 400) (Table 1). These characteristics translate into their suitability for specific applications, whereas anticorrosive BZTs require high aqueous solubility, as evidenced by the large differences in K_{OW} values between the two.

TABLE 1 Synonyms, CAS Numbers, Molecular Weights (MW), Octanol–Water Partition Coefficients (log K_{OW}), and Acid Dissociation Constants (pKa) of Select BZT Compounds

Systematic Name	CAS Number	Synonyms	MW	log K_{OW}	pKa
Phenol, 2-(2H-benzotriazol-2-yl)-4-methyl	2440-22-4	C1-BZT; drometrizole; UV-P	225.25	4.2	8.15
Phenol, 2-(2H-benzotriazol-2-yl)-4-(1,1-dimethylethyl)	3147-76-0	C4-BZT; UV-PS;	267.33	4.46	8.07
Phenol, 2-(2H-benzotriazol-2-yl)-4-(1,1,3, 3-tetramethylbutyl)-	3147-75-9	Octrizole; UV-329	323.44	6.2	
Phenol, 2-(2H-benzotriazol-2-yl)-4,6-bis (1-methyl-1-phenylethyl)-	70321-86-7	DiMeEtPh-BZT; UV-234	447.58	7.7	
Phenol, 2-(2H-benzotriazol-2-yl)-4,6-bis(1,1-dimethylpropyl)-	25973-55-1	C10-BZT; UV-328	351.49	7.22	8.73
Benzenepropanoic acid, 3-(2H-benzotriazol-2-yl)-5-(1,1-dimethylethyl)-4-hydroxy-, C7-9-branched and linear alkyl esters	127519-17-9	Tbu(C7-9)Est-BZT; UV-384	451.6		
Benzenepropanoic acid, 3-(2H-benzotriazol-2-yl)-5-(1,1-dimethylethyl)-4-hydroxy-, methyl ester	84268-33-7	tBuPrMeEst-BZT; UV-1130	353.42	5.7	
Phenol, 2-(5-chloro-2H-benzotriazol-2-yl)-6-(1,1-dimethylethyl)-4-methyl-	3896-11-5	C5-Cl-BZT; UV-326	315.8	5.52	9.49
Phenol, 2-(5-chloro-2H-benzotriazol-2-yl)-4,6-bis(1,1-dimethylethyl)-	3864-99-1	C8-Cl-BZT; UV-327	357.88	6.75	9.41

Continued

TABLE 1 Synonyms, CAS Numbers, Molecular Weights (MW), Octanol–Water Partition Coefficients (log K_{OW}), and Acid — Cont'd

Systematic Name	CAS Number	Synonyms	MW	log K_{OW}	pKa
Phenol, 2-(2H-benzotriazol-2-yl)-4,6-bis (1,1-dimethylethyl)-	3846-71-7	C8-BZT; UV-320	323.44	7.21	
1H-Benzotriazole	95-14-7	1-H-BZT; 1,2,3-benzotriazole; BT	119.13	1.44	8.37
1H-Benzotriazole, 6(or 7)-methyl-	29385-43-1	Tolyltriazole; TT	133.15		8.8
1b-Benzotriazole, 6-methyl-	136-85-6	5-tolyltriazole; 5-TT	133.15	2.16	
1H-Benzotriazole, 4-methyl-	29878-31-7	4-tolyltriazole; 4-TT	133.15		

3. MAJOR USES AND APPLICATIONS

The primary end use of the phenolic BZTs is as UV stabilizers in plastics, films, paints, inks, and other coatings [14]. These UV-BZTs are utilized in many synthetic polymers, including polyvinyl chloride, acrylics, polyesters, and polycarbonates [15], and are added to materials during the production process. The precise amount of BZT added to a product varies on the product being manufactured and the BZT being specified. Typically, the amount of BZT ranges from 0.1 to 1.0% w/w [14]. A range of UV-BZT compounds exist in solid and liquid formulations, each having different chemical and physical properties to impart the desired performance in the end product [14]. Physical structures of the UV-BZTs vary widely, with different functionalities (e.g., esters and Cl) present and configured in order to impart specific performance characteristics (e.g., Figure 1(e–g)). Structurally, some of the UV-BZTs are singly substituted (e.g., UV-P), while others may be doubly substituted (e.g., UV-328), or have other additional functional groups [16]. UV-BZTs may also be blended with compatible, non-BZT UV light stabilizers, such as hindered amine light stabilizers (HALS), to meet complex product and performance specifications [14]. The UV-BZTs that have been the subject of environmental research and which are examined in this chapter are listed in Figure 1(e–n).

The other major commercial use of BZT-based compounds is as a metal anticorrosion additive in a variety of household and commercial products. Usage in this category utilizes the basic BZT molecule (Figure 1(a)) or a technical mixture with a methyl group at either the 4′ or 5′ position (Figure 1(b–c)). For surface metal protection, the product containing BZT is applied to metal surfaces and is particularly effective for copper and copper bearing alloys. It is also present in many industrial fluids such as hydraulic oils, greases, and lubricants to protect metal machinery components from corrosion and oxidation. To prevent corrosion in heat exchange fluids, such as aqueous antifreezes, BZT is also a key component. BZTs are also added to automatic dishwasher detergent formulations for protecting silverware and metal surfaces in the dishwasher itself [17]. In the past, BZTs have been used as anticorrosive additives in ADAFs, which are used extensively in commercial and military flight operations to remove and prevent icing on flight surfaces of aircraft [18].

4. ESTIMATES OF BZT PRODUCTION

The US EPA collects and releases nonconfidential Chemical Data Reporting (CDR) information on chemical production and use in the USA. Information in this database includes many of the commercially produced BZTs. BZTs produced or imported into the USA in quantities exceeding 10,000 lb annually are reported in this database. The most current CDR data available are from 2012 and reflect recent manufacturing, processing, and usage information. One limitation of the database is that precise production volumes are considered confidential business information (CBI) in many cases. Consequently, much of the publicly accessible production data are binned and presented in broad weight intervals. Despite this limitation, the changes in production over time do provide some insight into usage trends in the USA and likely elsewhere. Presented in Table 2 are records of production for the BZTs listed in the US EPA CDR database from 1986 to 2012. Most, but not all, BZT chemicals (e.g., UV-PS and UV-301) reported in the environmental literature and technical documents were available in this database. Column entries identified as "na" indicate that production values were not available for this period. This could indicate that either production was below 10,000 lb per year or that specific material was not being commercially produced or imported during this period. Overall, the BZTs researched and reported to date as occurring in environmental compartments are limited to a relatively small number of compounds (e.g., UV-P, UV-PS, UV-320, UV-326, UV-327, UV-328, UV-329, BT, and TT). For the UV-BZTs, this may be due to the fact that these compounds have been produced longer and thus present in the environment longer than other BZTs. Most of these compounds have production information available predating the CDR [6], which may explain their inclusion as compounds being researched. Based on the most recent production information available as well as manufacturer marketing information, it is clear that there are newer and more structurally complex BZTs being produced and likely entering the environment.

Production information on the anticorrosive BZTs (e.g., BT and TT) show high-volume production dating back to 1986. One exception is 5-TT, which has not reported production following 1994, possibly due to declining use in photographic applications. More specifically, this may have been due to the increasing availability and declining cost of digital cameras to consumers in the mid-1990s.

5. ANALYTICAL METHODS FOR MEASUREMENT IN ENVIRONMENTAL MATRICES

Currently, the major methods for analysis of BZTs in environmental matrices are by gas chromatography–mass spectrometry (GC-MS) and by liquid chromatography–multiple quadrupole mass spectrometry (LC-MS-MS). The high aqueous solubility of the anticorrosive BZTs, such as 1-H benzotriazole (BT) and tolyltriazole (TT), and their ubiquitous presence in the dissolved phase of natural waters make them well suited for isolation by solid phase extraction (SPE) cartridges and quantitation by LC-MS-MS [19]. Isolation of the phenolic compounds (UV-BZTs) from environmental samples such as sediments, soils, biosolids, and organism tissues has been accomplished by methods generally incorporating manual or automated solvent extraction techniques and subsequent extract purification using silica, florisil, or another suitable substrate [20]. Typically, analysis of the larger MW BZTs (UV-BZTs) has been primarily by GC-MS, although LC-MS-MS has been used recently with the advent of more advanced techniques for separation of these compounds [21].

TABLE 2 Production Volumes of Select BZT Compounds (Pounds per Year)

CAS No.	Compound	1986	1990	1994	1998	2002	2006	2012
70321-86-7	UV-234	1–10M	1–10M	1–10M	1–10M	1–10M	1–10M	1–10M
25973-55-1	UV-328	1–10M	1–10M	1–10M	1–10M	1–10M	1–10M	2,246,476
127519-17-9	UV-384	na	na	500K–1M	10–500K	500K–1M	1–10M	Withheld
3147-75-9	UV-329	500K–1M	1–10M	10–500K	1–10M	1–10M	1–10M	500K–1M
2440-22-4	UV-P	1–10M	1–10M	1–10M	1–10M	500K–1M	500K–1M	605,074
84268-33-7	UV-1130	na	na	10–500K	1–10M	1–10M	1–10M	na
3896-11-5	UV-326	10–500K	10–500K	500K–1M	500K–1M	500K–1M	500K–1M	394,026
3864-99-1	UV-327	10–500K	10–500K	10–500K	10–500K	10–500K	<500K	Withheld
3846-71-7	UV-320	na	10–500K	10–500K	10–500K	10–500K	na	na
29385-43-1	TT	1–10M	1–10M	1–10M	1–10M	10–50M	1–10M	1–10M
95-14-7	BT	500K–1M	1–10M	1–10M	1–10M	1–10M	1–10M	1–10M
136-85-6	5-TT	10–500K	10–500K	10–500K	na	na	na	na

6. ENVIRONMENTAL HISTORY AND DISTRIBUTION OF BZTs

6.1 UV-Stabilizing BZTs

One likely consequence of high-volume chemical production and usage is the eventual entry of these chemicals into the environment. This is true for BZTs, which have been discovered and measured in different environmental compartments. The first reports documenting the presence of phenolic (UV) BZT compounds in the environment were in the late 1970s with peer reviewed journal articles by Jungclaus et al. [22] and Lopez-Avila and Hites [23]. These articles detailed the release of BZTs and numerous other chemicals from a chemical manufacturing facility located in Cranston, RI, USA, which operated until 1985. BZT compounds manufactured at this location were released via industrial wastewater discharged into the Pawtuxet River and subsequently transported through the river and into Narragansett Bay, RI, USA. This resulted in the deposition and accumulation of these compounds in the sediments of both the Pawtuxet River and Narragansett Bay. At this plant, BZTs were produced from approximately 1961 to 1985 and included compounds in Figure 1(e–k) [23].

The highest known levels of BZTs in the environment to date have been reported in the lower Pawtuxet River sediments with concentrations of 4300 mg/kg (dry weight) and 5200 mg/kg for UV-P and UV-327, respectively [24]. Reported sediment concentrations in Narragansett Bay are lower for UV-P, UV-327, and UV-328, ranging from 10 to 25 mg/kg [23,24].

Relative to the bioavailability of these BZTs, Pruell et al. [25] reported the first tissue concentrations of two UV-BZTs, UV-327 and UV-328, in hard shell clams (*Mercenaria mercenaria*) harvested from Narragansett Bay waters. Concentrations ranged from 1.0 to 8.5 μg/kg and 7.0–61 μg/kg wet weight for UV-327 and UV-328, respectively. Subsequent work by Pruell and Quinn [26] reported concentrations of the same two BZTs in sediment cores showing spatial and temporal distributions of both compounds throughout Narragansett Bay. Reddy et al. [24] measured the free versus bound distribution of most of the BZTs (e.g., UV-P, UV-PS, UV-301, UV-326, UV-327, and UV-328) present in Narragansett Bay (Figure 1), focusing on potential sediment-binding mechanisms of these compounds in sediment cores. The amount of bound BZT varied, but overall the percent bound was less than 10% of the total concentration. Hartmann et al. [27] reported on UV-P and UV-328 (Figure 1) levels across Narragansett Bay in sediment cores documenting their widespread spatial and temporal distribution. Finally, research by Cantwell et al. [28] used UV-328 (Figure 1) as a tracer compound for other contaminants discharged from the same chemical manufacturing plant in a dated, high-resolution sediment core correlating their presence with production dates. Combined, these findings provide evidence that long-term preservation and persistence of UV-BZTs is occurring in freshwater and marine sediments.

Until recently, research that has investigated the environmental behavior and fate of UV-BZTs has been limited to Narragansett Bay. Fortunately, recently published studies focus on locations not known to be directly impacted by BZT production activities and provide critically needed information on environmentally relevant concentrations (Table 3). For example, measurements in sediments from the Ariake Sea of Japan report individual concentrations of UV-320, UV-326, UV-327, and UV-328 ranging from less than 1 μg/kg up to 16 μg/kg dry weight [29]. The highest levels in this study were measured in the heavily industrialized Omuta River, with concentrations of UV-328 alone reaching 320 μg/kg. In the Songhua River of China, individual concentrations of UV-326, UV-327, and UV-328 ranged from 0.31 to 7.12 μg/kg, while sites in the Saginaw and Detroit rivers of the USA ranged from 0.22 to 224 μg/kg [20]. Based on this limited field data, we suggest that UV-BZT concentrations are present at levels in sediments comparable to other synthetic compounds also used as plastics additives, such as polybrominated diphenyl ethers (PBDEs). PBDEs are used in plastics as flame retardants and are also recognized as POPs. These compounds are added to a wide range of plastic products, yet are present in the environment due mainly to their release from the end products during and following their use [30]. What remains uncertain, however, is the behavior, concentration, and fate of the BZTs not reported to date in the literature.

More recent information is available on the tissue burdens of UV-BZTs in several species of marine organisms (Table 3). A study by Nakata et al. [31] reported levels of UV-327 and UV-328 in blubber of porpoises from the Ariake Sea of Japan with concentrations ranging from 4.5 to 31 μg/kg and 11 to 64 μg/kg, respectively. Kim et al. [32] measured eight BZTs (UV-P, UV-9, UV-234, UV-320, UV-326, UV-327, UV-328, and UV-329) in a broad range of pelagic and demersal marine organisms in Manila Bay, Philippines. Concentrations of the individual BZTs ranged from below detection to 211 μg/kg lipid weight.

A mussel watch monitoring study was conducted in waters of 11 countries throughout the Asia-Pacific coastline from 2003 to 2007. Four BZTs were measured (UV-320, UV-326, UV-327, and UV-328) in green and blue mussels by Nakata et al. [33]. Reported concentrations of the individual compounds ranged from not detected (ND) to 1500 μg/kg lipid weight. UV-328 was found in the most locations, with concentrations ranging from ND-830 μg/kg lipid weight. UV-326 and UV-327 had slighter lower detection frequencies, but UV-326 was found in the highest concentrations in this study (ND-1500 μg/kg lipid weight). Concentrations of UV-327 were considerably lower, ranging from ND-300 μg/kg lipid weight. UV-320 was found only in Japanese samples at low concentrations (39–86 μg/kg lipid weight).

While early documented releases of BZTs to the environment were associated with production, it is now becoming increasingly apparent that BZT release is occurring post production; that is, during and following the usage of the products amended with both UV and anticorrosive BZTs. For the UV-BZTs,

TABLE 3 Locations, Concentrations, and Specific BZTs Present in Different Environmental Compartments

Sample Type	Location	Compounds	Concentration	References
Sediments	Pawtuxet River, USA	UV-P, UV-327	4300–5200 mg/kg	[24]
	Narragansett Bay, USA	UV-P, UV-327	10–25 mg/kg	[23,24]
	Ariake Sea, Japan	UV-320, UV-326, UV-327, UV-328	1–16 μg/kg	[29]
	Omuta River, Japan	UV-328	320 μg/kg	[29]
	Songhau River, China	UV-326, UV-327, UV-328	0.31–7.12 μg/kg	[29]
	Saginaw and Detroit rivers, USA	UV-326, UV-327, UV-328	0.22–224 μg/kg	[20]
Tissues	Narragansett Bay, USA (hard shell clams)	UV 327, UV-328	1–61 μg/kg	[25]
	Ariake Sea, Japan (porpoise blubber)	UV-327, UV-328	4.5–64 μg/kg	[31]
	Manila Bay, Philippines (various fish)	UV-P, UV-9, UV-234, UV-320, UV-326, UV-327, UV-328, UV-329	nd-211 μg/kg	[32]
	Asia–Pacific coastal waters (blue and green mussels)	UV-320 UV-326 UV-327 UV-328	nd-86 μg/kg nd-1500 μg/kg nd-300 μg/kg nd-830 μg/kg	[33]

WWTP influents	Portugal; Spain	UV-P, UV-320, UV-326, UV-327, UV-328	24–85 ng/L; Nd–65 ng/L	[34]
	Japan	UV-326, UV-327, UV-328	5.6–78 ng/L	[35]
	Australia	UV-326 UV-329	15–35 ng/L 227–414 ng/L	[36]
	Germany	BT TT	17–44 μg/L 1.1–4.9 μg/L	[48]
	Switzerland	BT TT	13–39 μg/L 1.1–5.6 μg/L	[49]
WWTP effluents	Spain	UV-P, UV-326, UV-327, UV-328, UV-329	4–13 ng/L	[21]
	Portugal	UV-327, UV-328	21–31 ng/L	[34]
WWTP biosolids	Japan	UV-320 UV-326 UV-327 UV-328	nd 760–1800 μg/kg 120–200 μg/kg 430–570 μg/kg	[35]
	China	UV-PS UV-326 UV-328	0.7–1.2 μg/kg 23–136 μg/kg 41–5920 μg/kg	[20]
House dust	Spain	UV-P, UV-326, UV-327, UV-328	22–657 μg/kg	[38]
Air	Sweden	UV-P, UV-320, UV-327, UV-328, UV-329	Nd-25 ng/m^3	[40]

Continued

TABLE 3 Locations, Concentrations, and Specific BZTs Present in Different Environmental Compartments—cont'd

Sample Type	Location	Compounds	Concentration	References
Water	Main, Hengtsbach, and Hebach rivers, Germany	BT, TT	290–1200 ng/L	[63]
	Various rivers, Switzerland	BT TT	5440 ng/L 910 ng/L	[64]
	Elbe, Weser, Ems, Rhine, and Scheldt rivers, central Europe	BT TT	55–400 ng/L 77–1100 ng/L	[65]
	German Bight, North Sea	BT TT	1–21 ng/L 1–37 ng/L	[65]

one assumption is that wastewater treatment plants (WWTPs) are a primary source to the environment. Based on end product applications, this appears to be reasonable and is backed with some supporting data. Concentration data of UV-BZTs in WWTP influents are recent and provide a good estimate of current trends. In Portugal, individual influent concentrations of UV-P, UV-320, UV-326, UV-327, and UV-328 ranged from 24 to 85 ng/L, while in Spain they ranged from ND to 65 ng/L, with UV-320 not detected at all [34]. Individual influent concentrations of UV-326, UV-327, and UV-328 ranged from 5.6 to 78 ng/L at five WWTPs in Japan [35]. UV-326 and UV-329 were measured at approximately 25 and 310 ng/L, respectively, in the influent of a WWTP in Australia [36]. These levels overall are low, but consistent with other classes of compounds such as PBDEs [37].

Some of the UV-BZTs (UV-P, UV-326, UV-327, and UV-328) have also been measured at environmentally relevant levels in residential house dusts (22–657 μg/kg) [38]. This provides evidence that UV-BZT-amended products and materials within residential homes are a potentially significant source of BZTs to WWTPs and humans. Again, the UV-BZTs appear to be behaving similarly to PBDEs. Compounds such as PBDEs have been documented to be released over time from plastic products and are present in virtually all commercial and consumer products made with plastic and polymers [39]. PBDEs are present in house dusts at relatively high levels as well [39] and are ubiquitous in domestic WWTP influents.

To our knowledge, there are no peer reviewed studies on BZTs in the atmosphere. However, a government study from the Swedish Environmental Research Institute (IVL) [40] presents results of the major UV stabilizers (e.g., UV-P, UV-320, UV-327, UV-328, and UV-329) in outdoor air. Concentrations ran as high as 25 ng/m^3 in urbanized locations and were lower in less-developed settings. Two compounds, UV-P and UV-327, were found in 100% and 75% of the samples collected, with concentrations ranging from 0.07 to 11 ng/m^3 and <0.01 to 25 ng/m^3, respectively. No direct comparisons were available; however, measured levels were similar to individual PBDE congeners observed and reported by Strandberg et al. [41] at sites across the Great Lakes. Many flame retardants such as PBDEs are not bound to the molecular structure or matrix of the plastics, and this similarity in behavior suggests that the UV-BZTs discussed here are not structurally bound either. Thus, they may have a significant potential to migrate from the plastic over time, as supported by levels reported for house dust, and this raises legitimate concern for their potential exposure and distribution in environmental compartments.

6.2 Anticorrosive BZTs

Compared to the BZTs associated with UV stabilization of plastics, the BZTs used in metal anticorrosive and photographic processing applications comprise a much smaller number of compounds. Additionally, the anticorrosive-based

BZTs have been in production and use longer, but their long-term history in the environment is not as clear. Based on patent dates and available information, it is likely that commercial use of BZT as a metal preservative started around the late 1950s [42]. Usage expanded thereafter to other applications in products including antifreezes, waxes, polishes, and a wide array of lubricants [6].

Primary compounds of interest are BT (Figure 1(a)) and the (4) and (5) methylated BZTs (Figure 1(b–c)). When sold for commercial or industrial use, the methyl BZT often contains equal amounts of both isomer and is recognized and sold generically as tolyltriazole (TT) (Figure 1(d)). Other BT compounds exist and have been sold commercially as salts for special applications (e.g., chloro-BZT and Ag-BZT). These compounds were used in photofinishing operations to improve photographic image quality and were consumed in much lower quantities than the other aforementioned BZTs. These compounds have not been reported in the environmental scientific literature to our knowledge.

In the early 1970s, BZTs were evaluated for use in automatic dishwasher formulations to provide corrosion resistance [6]. Several patents were filed during this period for this application, yet it is unknown precisely when BZTs started to be incorporated into the formulations of these products [6,43,44]. BZTs are added to dishwashing detergents to protect nonferrous metals (e.g., silver) and both BT and its methylated analog, TT, are used in the formulation of dishwasher detergents in differing concentration and ratio, varying by brand [45]. Information (e.g., scientific studies) on the sources and distribution of anticorrosive BZTs in the environment exists, with much of the research being conducted in Europe. For these BZTs, identification of potential sources is better constrained due to their narrowly defined uses. For instance, the use of BT and TT as anticorrosive additives in automatic dishwashing detergents [46] results in their direct and constant discharge to domestic WWTP influents [47]. It appears that this is the most important source and pathway of these types of BZTs to WWTP influents. Janna et al. [17] and Vetter and Lorenz [47] measured BZT concentrations of dishwashing detergents in the UK and Germany, respectively, in order to estimate input loadings of BT and TT to WWTPs. Dishwasher tablets of different brands in Germany varied from 2 to 66 mg of total BZT, while a similar survey of powders and tablets in the UK were somewhat lower, ranging from 1.45 to 28 mg of BZT per tablet. The widespread and constant use of these detergents results in a relatively steady input to the domestic wastewater streams of most industrialized nations. Measured concentrations of BZTs in wastewater influents from several locations in Europe support this assertion (Table 3). As a result of use, influents of three WWTPs in Germany had concentrations of BT and TT ranging from 17 to 44 μg/L and 1.1 to 4.9 μg/L, respectively [48]. In Switzerland, average concentrations were somewhat higher, with BT and TT ranging from 13 to 39 μg/L and 1.1–5.6 μg/L, respectively [49]. In the USA, a survey of surface waters found 45% of waters sampled had measurable concentrations of BT present [50], indicating that these compounds are widespread in natural waters and their sources are likely to be WWTP effluents. A similar survey of 100 rivers in the European Union found BT and TT present in 94% and 81% of the samples, with concentration as high as 8 μg/L and 19 μg/L, respectively [51].

In 1994, a US patent application was filed for an ADAF incorporating BZTs as a corrosion inhibiting additive [7]. Soon after, research articles started to appear in scientific journals detailing the presence of BZTs associated with ADAF usage in the environment [21,52]. These articles documented concern of BZT use and its subsequent release into the environment, with a focus on the potential persistence and toxicity of BT and TT [52–55]. These concerns were partially based on application techniques that involved spraying large volumes of ADAFs containing BZTs onto aircraft and airport infrastructure. All ADAFs are aqueous based and contain glycols as the primary active component for removing and preventing icing [56]. BZTs were added as corrosion inhibitors and present in ADAF products at levels below 1% [57]. For a number of years, ADAFs were deployed over wide areas to aircraft and airport infrastructure without capturing or recovering the overspray or runoff, resulting in BT, TT, and other ADAF components being released to surface waters and, in a number of cases, groundwater [58,59].

Annual US usage of deicing fluids for aircraft is estimated at 9.46×10^7 L per year [56]. This specific use has also resulted in contamination of groundwater supplies in and around application areas [58,59]. In recent years, improved management practices of deicing operations at aviation facilities and storm water handling have led to reductions in the discharge of these compounds. Specifically, most major airports have limited the application of ADAFs to designated areas (e.g., deicing pads) where overspray and runoff are collected and treated [56]. Concerns with regard to the potential toxicity and persistence of BZTs used in ADAFs have resulted in manufacturers gradually phasing out the use of BZT-based ADAF compounds used in the USA and Europe [56].

7. FATE OF BZTs DURING DOMESTIC WASTEWATER TREATMENT

Current information on the fate of UV-BZTs during wastewater treatment is limited. Based on their general physical and chemical properties (e.g., MW and K_{OW}), these compounds are sparingly soluble in the dissolved phase, readily partitioning to and becoming associated with the biosolids component during wastewater treatment (Table 3). The behavior of UV-326 and UV-329 in a secondary level activated sludge WWTP appears to confirm this, with influent concentrations of UV-326 ranging from 15 to 35 ng/L while UV-329 was significantly higher, ranging from 227 to 414 ng/L [36]. Effluent concentrations of both compounds were below detection, indicating high removal efficiency for these compounds [36]. Other studies show measurable levels of multiple UV-BZTs in effluents, with detectable concentrations ranging from 4.0 to 13 ng/L for UV-P, UV-326, UV-327, UV-328, and UV-329 [21], to 21–31 ng/L for UV-327 and UV-328 [34]. Biosolids from five Japanese WWTPs were measured for four of the UV stabilizers. UV-320 was not detected, UV-328 and UV-326 varied from 430 to 570 μg/kg and 760 to 1800 μg/kg, respectively, while UV-327 ranged from 120 to 200 μg/kg [35]. Zhang et al. [20] measured UV-PS, UV-326, and UV-328 at five plants with biosolids concentrations ranging from 0.70 to 1.2, 23 to 136, and 41 to 5920 μg/kg, respectively. Overall, the behavior of the UV-BZTs during domestic wastewater treatment indicates relatively efficient removal from the water column and significant accumulation and enrichment in the biosolids fraction. Effluent concentrations appear to be somewhat variable, possibly due to differences in treatment plant design and operation, with individual levels being discharged to receiving waters in the low ng/L range.

In contrast, more information is available on the presence and removal efficiency of anticorrosive BZTs from domestic WWTPs. As discussed earlier, average concentrations of BT in WWTP influents are in the low to mid-range of μg/L, depending on a number of factors such as the size and population of the sewer shed, source characteristics, and other related variables. A review of the available scientific literature reports that only partial removal of BT and TT occurs, with efficiencies varying on the type and level of treatment process in use. In a survey of three Western European WWTPs, removal of BT and TT averaged 20% and 18%, respectively [60]. In Germany, several secondary activated sludge plants reported removal efficiencies of 29–58% for BT, 19–65% for TT, and negligible removal for 4-TT [48]. Removal rates for this class of BZTs indicate that only partial removal is accomplished at the secondary level of treatment. This is due primarily to their high aqueous solubilities and low K_{OW} values, which allow them to remain predominantly in the dissolved phase, and to pass through WWTPs.

Advanced treatment processes when available may provide increased removal efficiencies of BZTs and other compounds from the wastewater stream. For example, activated carbon treatment has been shown to be somewhat effective in removal of the methylated BTs (e.g., 4-TT and 5-TT) [48]. Ozonation has also shown promise in removing all three anticorrosive and ADAF-associated BZTs as well as other polar contaminants from waters. Weiss et al. [61] has reported removals of up to 99% for BT and TT following ozone treatment. What remains largely unknown, however, is the potential of these BZTs to transform or degrade into more toxic or persistent compounds following advanced treatment [62].

The partial removal of BZTs (e.g., BT and TT) during wastewater treatment results in steady loadings of these compounds to receiving waters. Observed concentrations vary spatially and temporally depending on WWTP performance and characteristics of receiving waters. Several studies of lakes and rivers in Europe provide a good indication of the environmental concentrations present. In a study of the Main, Hengtsbach, and Hegbach rivers in Germany, total median concentrations of BZTs were 290 and 1200 ng/L for the Main and Hengtsbach, respectively. In the Hegbach River, which has no WWTP effluent discharge, most measurements were below detection, evidencing the significance of WWTPs as sources of anticorrosive BZTs [63]. Both BT and TT were detected in 12 rivers in Switzerland at maximum levels of 5440 and 910 ng/L, respectively [64]. A positive relationship between the amount of WWTP effluent and BZT water concentrations was noted in one river, further illustrating the importance of WWTPs as sources of BZT to the environment. In the same study, mass river flow measurements in and out of Lake Greifensee suggest that BZTs are not being eliminated in the lake. Wolschke et al. [65] measured BT and TT in the Elbe, Weser, Ems, Rhine, and Scheldt rivers, all of which discharge to the North Sea. Concentrations of BT and TT were relatively consistent among rivers, ranging from 55 to 400 ng/L and 77 to 1100 ng/L, respectively. Concentrations in the German Bight were up to 21 ng/L and 37 ng/L for BT and TT, respectively, with concentrations for both declining to about 1 ng/L 300 km from the coast [65]. They also determined that decreases in concentration in the North Sea were due to dilution, not degradation, providing evidence of the persistence of these BZTs in natural waters (see discussion below).

8. BIOACCUMULATION AND AQUATIC TOXICITY

As discussed above, BZTs are known to be transported from areas of production and use to the aquatic environment where they are readily detectable in the water column and sediments. Once in the aquatic environment, a relevant concern is the magnitude of BZT's bioavailability to aquatic organisms. As noted earlier, several studies have reported the presence of UV-BZTs in organismal tissues found in the environment, including hard shell clams, green and blue mussels, and porpoises [25,31,33]. The anticorrosive BZTs have not yet been detected in organismal tissues in the environment, but their greater solubility compared to the UV-BZTs (i.e., higher freely dissolved concentrations) would suggest they may be bioavailable. In addition to measuring BZTs in the tissues of organisms found in the environment to demonstrate bioavailability, laboratory studies have been performed to determine the magnitude of BZT bioaccumulation. The magnitude of bioaccumulation is expressed as the bioconcentration factor (BCF) (L/kg):

$$BCF = \frac{C_O}{C_W}$$

where C_O (µg/kg) and C_W (µg/L) are the concentrations of BZTs in the organismal tissues and exposure water, respectively. Generally, the laboratory studies are performed for an extended period of time (weeks to months) so that equilibrium of the contaminant between the organismal and water compartments can be demonstrated or assumed (this is not necessarily the case for organisms collected from the environment). For the UV-BZTs, several BCFs are available from the literature for *Cyprinus carpio*. For the compounds investigated, relatively large BCFs ranging from 370 to 10,000 L/kg are reported (Table 4) [29,66–68]. These observations are supported by the physicochemical

TABLE 4 Bioconcentration Factors of Select BZT Compounds

Specific BZT Compound	Organism	Test Duration	Bioconcentration Factor (BCF) (L/kg)	References
BT	Daphnid ephippia *Daphnia longispina-galeata*	30 days	2.9	[66]
UV-320	Carp *Cyprinus carpio*	10 weeks	2,960–10,000	[29]
		14 weeks	1380–8180	
		98 days	365–2250	[67]
UV-327	Carp *Cyprinus carpio*	63 days	7600	[67]
			6500	
		70 days	900	
			4700	
UV-328	Carp *Cyprinus carpio*	28 days	1120–3960	[68]
		42 days	1680–5580	
		56 days	2110–4760	
		60 days	1000–1700	
		60 days	1800–2400	[67]

characteristics of the chemicals, including low aqueous solubilities and relatively high K_{OW}s. More specifically, organic chemicals that demonstrate limited aqueous solubility and large K_{OW}s will tend to accumulate in the organic carbon of sediments and potentially in the lipids of aquatic organisms rather than remain dissolved in the water. This is especially true for chemicals that do not degrade rapidly (i.e., persistent). For the anticorrosive BZT BT, the data are limited, with BCFs that are relatively small for a freshwater invertebrate (~3 L/kg). As with the UV-BZTs, this finding is also supported by BT's high aqueous solubility and low K_{OW} value. These data would suggest that the UV-BZTs demonstrate the greater potential for concern with regard to bioaccumulation by aquatic organisms in the environment.

Having found that BZTs do bioaccumulate in aquatic organisms, the next step is to determine if these chemicals manifest any adverse effects. One measure of a chemical's impact on the environment is its overall potential to impart toxicity, both acute and sublethal, to living organisms. For most chemicals, including BZTs, it is appropriate to assess their aquatic toxicity under a range of conditions using a variety of organisms. Currently, no aquatic toxicity testing data are available for the UV-BZTs. Conversely, there have been a number of investigations studying the toxicity of ADAFs to fish and other organisms [52–55]. While ADAFs and aqueous-based antifreeze solutions contain a number of different chemicals, including glycols, findings indicate that it is the BZTs that were largely responsible for the observed toxicity in these products [57].

In Table 5, results of toxicity testing are presented for BT, 4-TT, 5-TT, and TT with different organisms (including plants, bacteria, invertebrates, and fish), end points (mortality and sublethal), and test durations ranging from 5 min to 21 days [6,52–55,69–72]. For BT, all results are in the mg/L range and vary from 12 to 231 mg/L. Fish appear to be the most sensitive to BT with median lethality concentrations (LC_{50}s) ranging from 12 to 65 mg/L, followed by the results of the Microtox bacteria test with a sublethal luminescence end point median effects concentration (EC_{50}) of about 41 mg/L. Several invertebrate species had LC_{50}s and EC_{50}s (combining mortality and immobilization end points) ranging from 16 to 170 mg/L and one sublethal toxicity test with the daphnid *Daphnia magna*, using a reproduction end point, resulted in EC_{50}s ranging from 26 to 77 mg/L. Plants were the least sensitive to BT (i.e., EC_{50}s ranging from 100 to 230 mg/L). Data on 4-TT were limited, with one EC_{50} and two LC_{50}s ranging from 21 to 118 mg/L for all three organisms (i.e., Microtox, a daphnid, and a fish). For 5-TT, Microtox results were lowest overall, ranging from 4.2 to 8.7 mg/L, followed by 96 h LC_{50}s of 22 mg/L for the fathead minnow. Invertebrates had the highest LC_{50}s (again combining mortality and immobilization end points), ranging from 8.6 to 152 mg/L. The values for TT, which is the isomer mix of 4-TT and 5-TT and the commercially available methylated mixture, had Microtox EC_{50}s ranging from 6.1 to 7.3 mg/L (similar to the separate isomers), with fish and plant toxicity data (LC_{50}s and EC_{50}s) having similar ranges of 21–65 and 32–62 mg/L, respectively. Finally, the invertebrate LC_{50}s

TABLE 5 Available Ecotoxicity Data for Select BZT Compounds

Specific BZT Compound	Organism	Test Duration	Test Conditions	Organism Life Stage	Test End Point	LC50 or EC50 (mg/L)	References
BT	Plant						
	Green algae *Scenedesmus subspicatus*	72 h	NR	NR	Growth	231	[69]
		72 h	NR	NR	Biomass	102	[69]
	Bacteria						
	Proteobacteria *Vibrio fischeri*	5 min	Microtox	NA	Luminescence	41.13	[52]
		15 min	Microtox	NA	Luminescence	41.65	[52]
	Invertebrate						
	Flatworm *Dugesia japonica*	24 h	Static	NR	Mortality	171.1	[70]
		48 h	Static	NR	Mortality	145.3	[70]
		72 h	Static	NR	Mortality	143.3	[70]
		96 h	Static	NR	Mortality	142.9	[70]
	Daphnid *Ceriodaphnia dubia*	48 h	Static	<24 h	Mortality	102	[53]
	Daphnid *Daphnia magna*	48 h	NR	NR	Mortality	141.6	[69]
		21 days	Static-renewal	<24 h	Reproduction	25.9–76.9	[69]
		48 h	Static	<24 h	Immobilization	91	[69]
		48 h	NR	<24 h	Immobilization	107	[71]

		48 h	NR	<24 h	Immobilization	155.4	[72]
		21 days	Static-renewal		Immobilization	76.9	[69]
	Daphnid *Daphnia galeata*	48 h	NR	<24 h	Immobilization	15.8	[71]
	Fish						
	Zebrafish *Brachydanio rerio*	96 h	Static	NR	Mortality	>100	[69]
	Bluegill *Lepomis macrochirus*	48 h	Static	Juvenile	Mortality	27.5	[6]
		48 h	Flow-through	NR	Mortality	30.2	[6]
		96 h	Static	Juvenile	Mortality	25	[6]
		96 h	Flow-through	NR	Mortality	28	[6]
	Fathead minnow *Pimephales promelas*	48 h	Static	NR	Mortality	27.5	[6]
		48 h	Flow-through	NR	Mortality	30.2	[6]
		96 h	Static	NR	Mortality	25	[6]
		96 h	Flow-through	NR	Mortality	28	[6]
		96 h	Static-renewal	5–7 days	Mortality	65	[53]
	Rainbow trout *Salmo gairdneri*	48 h	Flow-through	Juvenile	Mortality	15	[6]
		96 h	Flow-through	Juvenile	Mortality	12	[6]
		96 h	NR	NR	Mortality	39.4	[69]

Continued

TABLE 5 Available Ecotoxicity Data for Select BZT Compounds—cont'd

Specific BZT Compound	Organism	Test Duration	Test Conditions	Organism Life Stage	Test End Point	LC50 or EC50 (mg/L)	References
4-TT	Bacteria						
	Proteobacteria *Vibrio fischeri*	15 min	Microtox	NA	Luminescence	21	[53]
	Invertebrate						
	Daphnid *Ceriodaphnia dubia*	48 h	Static	<24 h	Mortality	118	[53]
	Fish						
	Fathead minnow *Pimephales promelas*	96 h	Static-renewal	5–7 days	Mortality	63	[53]
5-TT	Bacteria						
	Proteobacteria *Vibrio fischeri*	15 min	Microtox	NA	Luminescence	8.7	[53]
		15 min	Microtox	NA	Luminescence	4.25	[54]
		15 min	Microtox	NA	Luminescence	4.25	[55]
		5 min	Microtox	NA	Luminescence	5.69	[52]
		15 min	Microtox	NA	Luminescence	5.91	[52]

	Invertebrate						
	Flatworm *Dugesia japonica*	24 h	Static	NR	Mortality	152.4	[70]
		48 h	Static	NR	Mortality	90.5	[70]
		72 h	Static	NR	Mortality	80.8	[70]
		96 h	Static	NR	Mortality	74	[70]
	Daphnid *Ceriodaphnia dubia*	48 h	Static	<24 h	Mortality	79	[53]
		48 h	NR	<24 h	Mortality	81.3	[54]
		48 h	Static-renewal	<24 h	Mortality	81.3	[55]
	Daphnid *Daphnia galeata*	48 h	NR	<24 h	Immobilization	8.58	[71]
	Daphnid *Daphnia magna*	48 h	NR	<24 h	Immobilization	51.6	[71]
	Fish						
	Fathead minnow *Pimephales promelas*	96 h	Static-renewal	5–7 days	Mortality	22	[53]
		96 h	NR	Adult	Mortality	22	[54]
		96 h	Static-renewal	4–10 days	Mortality	22	[55]

Continued

TABLE 5 Available Ecotoxicity Data for Select BZT Compounds—cont'd

Specific BZT Compound	Organism	Test Duration	Test Conditions	Organism Life Stage	Test End Point	LC50 or EC50 (mg/L)	References
TT	Plant						
	Green algae *Scenedesmus subspicatus*	72 h	NR	NR	Growth	62	[69]
		72 h	NR	NR	Biomass	32	[69]
	Bacteria						
	Proteobacteria *Vibrio fischeri*	15 min	Microtox	NA	Luminescence	7.3	[53]
		15 min	Microtox	NA	Luminescence	6.08	[55]
	Invertebrate						
	Daphnid *Ceriodaphnia dubia*	48 h	Static	<24 h	Mortality	108	[53]
		48 h	Static-renewal	<24 h	Mortality	80.7	[55]
	Daphnid *Daphnia magna*	48 h	NR	NR	Mortality	73.7	[69]
		21 days	NR	NR	Immobilization	37.6	[69]
		21 days	NR	NR	Reproduction	18.4–23.5	[69]

	Fish						
	Zebrafish *Brachydanio rerio*	96 h	Static	NR	Mortality	65	[69]
	Bluegill *Lepomis macrochirus*	96 h	NR	NR	Mortality	31	[69]
	Fathead minnow *Pimephales promelas*	24 h	Static	NR	Mortality	36.9	[6]
		48 h	Static	NR	Mortality	29.8	[6]
		96 h	Static	NR	Mortality	25.5	[6]
		96 h	NR	NR	Mortality	25.5	[69]
		96 h	Static-renewal	5–7 days	Mortality	38	[53]
		96 h	Static-renewal	4–10 days	Mortality	30.1	[55]
	Rainbow trout *Salmo gairdneri*	48 h	NR	NR	Mortality	38	[69]
		96 h	NR	NR	Mortality	21.4	[69]

ranged from 18 to 108 mg/L with *Daphnia magna* reproduction EC_{50}s ranging from 18 to 24 mg/L, which also encompasses the toxicity range of the individual isomers. These results suggest that TT is the most toxic of the BZTs for which there are data, with an overall range of toxicity values for all species of 6.1–108 mg/L. In relative terms, these data suggest that the BZTs, specifically the anticorrosive BZTs, are not extremely toxic to aquatic organisms. By comparison, some pesticides, designed to be toxic, have LC_{50}s in the 0.10–10 μg/L range [73].

As reported above, the concentrations of the anticorrosive BZTs found in the water column in the environment are well below the toxicity concentrations discussed here: μg/L versus mg/L. This suggests water column toxicity due to this type of BZT alone is unlikely. Further, because the anticorrosive BZTs are relatively water soluble, they are unlikely to accumulate in sediments to concentrations that would cause toxicity to benthic organisms. In contrast, because of the complete lack of toxicity data for the UV-BZTs, which as discussed above, are compounds known to accumulate in sediments, trying to assess risk to water column and benthic organisms or to organisms dependent on the benthos (e.g., fish, birds, and wildlife) is not currently possible.

9. PERSISTENCE OF BZTs UNDER ENVIRONMENTAL CONDITIONS

Currently, the status of both classes of BZTs as POPs is unclear. While they have been in production and measured in the environment for decades, a clear understanding of their behavior and fate is incomplete. Predictions of persistence based on physicochemical data of the major UV-BZTs indicate long-term preservation [74], and there is evidence suggesting that they remain structurally intact for many years in freshwater and marine sediments [23–25,27,29]. The long-term persistence of BT and TT in soils and sediments is unknown, but existing science documents their resistance to biodegradation. Based on available modeled and field data, there does seem to be justifiable concern for the potential of the uptake and accumulation of the UV-BZTs and well-documented aquatic toxicity for the anticorrosive compounds. Research has shown the potential for BZTs to transfer between environmental compartments (e.g., WWTP influent to effluent to sediments to aquatic organisms), yet no data are known to exist for long distance transport to remote regions (e.g., Arctic) that would confirm their global mobility. Currently the list of POPs identified by the United Nations Stockholm Convention on Persistent Organic Pollutants (effective 2004) and recognized by other organizations is very limited (i.e., less than 25 chemicals), yet numerous chemicals, like the BZTs, fit the profile and behavior of a POP.

Recently, the Environmental Ministry of Japan has identified a number of UV-BZTs as POPs (e.g., persistent, potentially toxic, with characteristics to bioaccumulate) in the environment. For example, the Japanese government has listed UV-320 as a class one substance (i.e., a POP). They have banned the production, use, and transport of UV-320 since 2007 based on bioaccumulation, toxicity, and persistence data [29]. They have also identified UV-327 as a type I monitored chemical substance, indicating similarity in behavior and properties of known POPs, with its potential environmental toxicity yet undetermined [29]. Based on BCF values for other UV-BZTs (Table 4), it is possible that more of these compounds will be identified, listed, and treated as POPs in the future.

10. SUMMARY

The state of knowledge to date on BZTs indicates that while basic information exists for some BZTs, more research is needed to better assess their overall environmental risk. Based on their usage, each class of BZT presents it own set of challenges and concerns to the environment. Following initial reporting that was limited to Narragansett Bay in the USA, research has focused on some of the high-volume commercially used UV-BZTs in a number of environmental compartments. Little information on their environmental toxicity exists, especially for the UV-BZTs, and the potential of these same compounds to bioaccumulate is a concern. In addition, other newer, unstudied UV-BZTs are likely present to some extent in the environment, yet little or no information exists on these chemicals or their potential effects. To date, most of the environmental research on BZTs has been focused at the ecosystem or watershed scale. At larger scales, it is currently unknown how susceptible these UV-BZTs are to global transport processes and to what extent global redistribution has or is occurring. Emerging data on their behavior does suggest that they may act like other POPs. Consequently, there is an increasing need to better understand UV-BZTs and their impact on larger-scale ecosystems.

For the BZTs associated with anticorrosion applications, more information exists on their environmental behavior and fate due to the smaller number of compounds in use and their well-defined applications. Concerns do exist with regard to their persistence and their low removal efficiency following most domestic wastewater treatment processes, and emerging evidence indicates that biotransformation may occur in WWTPs employing activated sludge treatment [75]. Their high frequency of detection, elevated concentrations, and likelihood to be bioavailable in natural waters across the USA and Europe generate legitimate concerns for adverse ecological impacts resulting from their demonstrated toxicity to aquatic organisms. This concern also brings into question the safety and integrity of drinking water supplies in industrialized nations that draw waters from suspected sources.

As they stand currently, BZTs meet the criteria as contaminants of emerging concern: chemicals that have been in use and are present in the environment, but whose potential risks to the environment are still largely unknown. Priority research needs for this class of compound include identification of the predominant sources of these chemicals, particularly the UV-BZTs, as well as their spatial and temporal distribution; an improved understanding of their environmental behavior (e.g., transformation and degradation); and more studies investigating their ability to bioaccumulate and cause toxicity to aquatic organisms.

ACKNOWLEDGMENTS

This research was supported in part by an appointment to the Research Participation Program for the US Environmental Protection Agency, Office of Research and Development, administered by the Oak Ridge Institute for Science and Education through an interagency agreement between the US Department of Energy and EPA. This manuscript is contribution number ORD-008062 of the Atlantic Ecology Division of the United States Environmental Protection Agency, Office of Research and Development, National Health Effects Environmental Research Laboratory.

REFERENCES

[1] A. Meighen, Heat Exchange Compositions, US Patent No. 2 803 603 (1957).
[2] Union Carbide Corporation, Free-flowing collapsible capsule corrosion and leakage inhibitors. Netherlands Application No. 6 605 917 (1966), US Application (1965), Chem. Abstr. 1967, 66, 67671a.
[3] E. Donaldson, B.T. Fowler and J.M. Tims, Lubricating Greases, British Patent No. 984 488 (1965).
[4] R.E. Malec, Lubricating Oil Compositions Containing Trialkyl-substituted Phenols and Benzotriazole, US Patent No. 3 900 410 (1975).
[5] H. Heller, Ultra-violet Light-absorbing Composition of Matter, US Patent No. 3 004 896 (1961).
[6] L.N. Davis, J. Santodonato, P. Howard, J. Saxena, Investigation of Selected Potential Environmental Contaminants: Benzotriazoles. Final Report Prepared for Office of Toxic Substances, US Environmental Protection Agency, Washington, DC, 1977.
[7] S.K. Chan, G.L. Hubred and H.E. Todd, Deicer/anti-icer Compositions for Aircraft, US Patent No. 5 435 930 (1995).
[8] R.A. Smiley, sixth ed., Phenylene- and Toluenediamines, Ullmann's Encyclopedia of Industrial Chemistry, vol. 26, Wiley-VCH, Weinheim, 2002, pp. 617–622. http://dx.doi.org/10.1002/14356007.a19_405.
[9] J. Levy, Preparation of Benzotriazole, US Patent No. 3 227 726 (1966).
[10] J.W. Long, Method for Production of Benzotriazole, US Patent No. 3 564 001 (1971).
[11] R.J. Boyle and J.P. Milionis, Benzotriazole Compounds, US Patent No. 3 214 436 (1965).
[12] N. Fukuoka, K. Kubota and K. Iguchi, Method of Preparing 2-phenyl Benzotriazoles, US Patent No. 5 187 289 (1993).
[13] C. Sease, Stud. Conserv 23 (2) (1978) 76–85.
[14] Coatings that stay looking good BASF performance additives. http://www.basf.com/group/corporate/us/en/literature-document:/Brand+Chimassorb-Brochure–Coatings+that+stay+looking+good+BASF+performance+additives-English.pdf, (accessed 01.07.14).
[15] The Phenolic Benzotriazoles Association, Data Summary and Test Plan for Phenolic Benzotriazoles, Submitted to the Environmental Protection Agency under the HPV Challenge Program High Production Volume (HPV) Challenge Program, 2001.

[16] National Toxicology Program, National Institute of environmental health sciences, National Institutes of Health, US Department of Health and Human Services, Chemical Information Review Document for Phenolic Benzotriazoles, Supporting Nomination for Toxicological Evaluation by the National Toxicology Program, 2011.

[17] H. Janna, M.D. Scrimshaw, R.J. Williams, J. Churchley, J.P. Sumpter, Environ. Sci. Technol. 45 (9) (2011) 3858–3864.

[18] D.A. Cancilla, J. Martinez, G.C. Van Aggelen, Environ. Sci. Technol. 32 (23) (1998) 3834–3835.

[19] S. Montesdeoca-Esponda, T. Vega-Morales, Z. Sosa-Ferrera, J. Santana-Rodríguez, TrAC, Trends Anal. Chem. 51 (2013) 23–32.

[20] Z. Zhang, N. Ren, Y.-F. Li, T. Kunisue, D. Gao, K. Kannan, Environ. Sci. Technol. 45 (9) (2011) 3909–3916.

[21] S. Montesdeoca-Esponda, Z. Sosa-Ferrera, J. Santana-Rodríguez, Anal. Bioanal. Chem. 403 (3) (2012) 867–876.

[22] G. Jungclaus, V. Avila, R. Hites, Environ. Sci. Technol. 12 (1) (1978) 88–96.

[23] V. Lopez-Avila, R.A. Hites, Environ. Sci. Technol. 14 (11) (1980) 1382–1390.

[24] C.M. Reddy, J.G. Quinn, J.W. King, Environ. Sci. Technol. 34 (6) (2000) 973–979.

[25] R.J. Pruell, E.J. Hoffman, J.G. Quinn, Mar. Environ. Res. 11 (3) (1984) 163–181.

[26] R.J. Pruell, J.G. Quinn, Estuarine, Coastal Shelf Sci. 21 (3) (1985) 295–312.

[27] P.C. Hartmann, J.G. Quinn, R.W. Cairns, J.W. King, Mar. Pollut. Bull. 50 (4) (2005) 388–395.

[28] M.G. Cantwell, B.A. Wilson, J. Zhu, G.T. Wallace, J.W. King, C.R. Olsen, R.M. Burgess, J.P. Smith, Chemosphere 78 (4) (2010) 347–352.

[29] H. Nakata, S. Murata, J. Filatreau, Environ. Sci. Technol. 43 (18) (2009) 6920–6926.

[30] US Environmental Protection Agency, An Exposure Assessment of Polybrominated Diphenyl Ethers, EPA/600/R-08/086F National Center for Environmental Assessment, Washington, DC, 2010. Available from the National Technical Information Service, Springfield, VA, and online at http://www.epa.gov/ncea.

[31] H. Nakata, R.-I. Shinohara, S. Murata, M. Watanabe, J. Environ. Monit. 12 (11) (2010) 2088–2092.

[32] J.-W. Kim, T. Isobe, B.R. Ramaswamy, K.-H. Chang, A. Amano, T.M. Miller, F.P. Siringan, S. Tanabe, Chemosphere 85 (5) (2011) 751–758.

[33] H. Nakata, R.-I. Shinohara, Y. Nakazawa, T. Isobe, A. Sudaryanto, A. Subramanian, S. Tanabe, M.P. Zakaria, G.J. Zheng, P.K.S. Lam, E.Y. Kim, B.-Y. Min, S.-U. We, P.H. Viet, T.S. Tana, M. Prudente, D. Frank, G. Lauenstein, K. Kannan, Mar. Pollut. Bull. 64 (10) (2012) 2211–2218.

[34] I. Carpinteiro, M. Ramil, I. Rodríguez, J.M.F. Nogueira, J. Sep. Sci. 35 (3) (2012) 459–467.

[35] H. Nakata, R.-I. Shinohara, Interdisciplinary Studies on Environmental Chemistry-environmental Research in Asia, 2010, pp. 239–245.

[36] Y.-S. Liu, G.-G. Ying, A. Shareef, R.S. Kookana, Environ. Pollut. 165 (2012) 225–232.

[37] T.D. Anderson, J.D. MacRae, Chemosphere 62 (2006) 1153–1160.

[38] I. Carpinteiro, B. Abuin, I. Rodriguez, M. Ramil, R. Cela, J. Chromatogr. A 1217 (24) (2010) 3729–3735.

[39] H.A. Jones-Otazo, J.P. Clarke, M.L. Diamond, J.A. Archbold, G. Ferguson, T. Harner, G.M. Richardson, J.J. Ryan, B. Wilford, Environ. Sci. Technol. 39 (14) (2005) 5121–5130.

[40] E. Brorström-Lundén, M. Remberger, L. Kaj, K. Hansson, H. Andersson, P. Haglund, R. Andersson, P. Liljelind, R. Grabic, Screening of Benzothiazoles, Benzenediamines, Dicyclohexylamine and Benzotriazoles 2009, IVL Report B2023, IVL Swedish Environmental Research Institute, Swedish Environmental Protection Agency, 2011.

[41] B. Strandberg, N.G. Dodder, I. Basu, R.A. Hites, Environ. Sci. Technol. 35 (6) (2001) 1078–1083.
[42] R. Liddel, Corrosion Inhibitors and Method of Using Same, US Patent No. 2 877 188 (1959).
[43] J.R. Schaeffer, Detergent Compositions Containing Metal Discoloration Inhibitors, US Patent No. 2 618 606 (1952).
[44] R.E. Zimmerer, Detergent Composition, UK Patent No. GB1 131 738 (1968).
[45] J.A. Christie, F.S. Macbeath and J.C. Turner, Machine Dishwashing Composition Containing Oxygen Bleach and Paraffin Oil and Benzotriazole Compound Silver Tarnishing Inhibitors, US Patent No. 5 698 504 (1997).
[46] G. Waschenbach, P. Robinson, H. Maggand and W. Hoflinger, Dishwasher Product in Tablet Form, US Patent No. 6 194 368 (2001).
[47] W. Vetter, J. Lorenz, Environ. Sci. Pollut. Res. 20 (2013) 4435–4440.
[48] T. Reemtsma, U. Miehe, U. Duennbier, M. Jekel, Water Res. 44 (2) (2010) 596–604.
[49] D. Voutsa, P. Hartmann, C. Schaffner, W. Giger, Environ. Sci. Pollut. Res. 13 (5) (2006) 333–341.
[50] D.W. Kolpin, E.T. Furlong, M.T. Meyer, E.M. Thurman, S.D. Zaugg, L.B. Barber, H.T. Buxton, Environ. Sci. Technol. 36 (6) (2002) 1202–1211.
[51] R. Loos, B.M. Gawlik, G. Locoro, E. Rimaviciute, S. Contini, G. Bidoglio, Environ. Pollut. 157 (2) (2009) 561–568.
[52] D.A. Cancilla, A. Holtkamp, L. Matassa, X. Fang, Environ. Toxicol. Chem. 16 (3) (1997) 430–434.
[53] D.A. Pillard, J.S. Cornell, D.L. DuFresne, M.T. Hernandez, Water Res. 35 (2) (2001) 557–560.
[54] D.A. Cancilla, J.C. Baird, S.W. Geis, S.R. Corsi, Environ. Toxicol. Chem. 22 (1) (2003) 134–140.
[55] S.R. Corsi, S.W. Geis, J.E. Loyo-Rosales, C.P. Rice, Environ. Sci. Technol. 40 (23) (2006) 7409–7415.
[56] US Environmental Protection Agency, Technical Development Document for the Final Effluent Limitations Guidelines and New Source Performance Standards for the Airport Deicing Category, EPA-821-R-12–005, 2012.
[57] J.S. Cornell, D.A. Pillard, M.T. Hernandez, Environ. Toxicol. Chem. 19 (6) (2000) 1465–1472.
[58] L.J. Novak, K. Holtze, R.A. Kent, C. Jefferson, D. Anderson, Environ. Toxicol. Chem. 19 (7) (2000) 1846–1855.
[59] G.D. Breedveld, R. Roseth, L. Hem, M. Sparrevik, Triazoles in the Terrestrial Environment, NGI Report No. 20001103–1, 2002.
[60] T. Reemtsma, S. Weiss, J. Mueller, M. Petrovic, S. González, D. Barcelo, F. Ventura, T.P. Knepper, Environ. Sci. Technol. 40 (17) (2006) 5451–5458.
[61] S. Weiss, J. Jakobs, T. Reemtsma, Environ. Sci. Technol. 40 (23) (2006) 7193–7199.
[62] D.B. Mawhinney, B.J. Vanderford, S.A. Snyder, Environ. Sci. Technol. 46 (13) (2012) 7102–7111.
[63] A. Kiss, E. Fries, Environ. Sci. Pollut. Res. 16 (6) (2009) 702–710.
[64] W. Giger, C. Schaffner, H.-P.E. Kohler, Environ. Sci. Technol. 40 (23) (2006) 7186–7192.
[65] H. Wolschke, Z. Xie, A. Möller, R. Sturm, R. Ebinghaus, Water Res. 45 (18) (2011) 6259–6266.
[66] A.C. Chiaia-Hernandez, R. Ashauer, M. Moest, T. Hollingshaus, J. Jeon, P. Spaak, J. Hollender, Environ. Sci. Technol. 47 (18) (2013) 10667–10675.

[67] Environment Canada, Draft Screening Assessment Report, Phenol, 2-(2H-benzotriazol-2-yl)-4,6-bis(1,1-dimethylpropyl)- (BDTP), CAS Number 25973-55-1 2014. Available online at http://www.ec.gc.ca/ese-ees/default.asp?lang=En&n=78FEE504-1.

[68] European Chemicals Agency. Information on Chemicals, Registered Substances, 2-(2H-benzotriazol-2-yl)-4,6-ditertpentylphenol, http://echa.europa.eu/web/guest/information-on-chemicals/registered-substances (accessed 3/10/2014).

[69] Benzotriazoles Coalition, Benzotriazoles: Category Justification and Testing Rationale, 2001.

[70] M.-H. Li, Toxicol. Environ. Chem. 95 (6) (2013) 984–991.

[71] A. Seeland, M. Oetken, A. Kiss, E. Fries, J. Oehlmann, Environ. Sci. Pollut. Res. 19 (5) (2012) 1781–1790.

[72] M.K. Durjava, B. Kolar, L. Arnus, E. Papa, S. Kovarich, U. Sahlin, W. Peijnenburg, ATLA, Altern. Lab. Anim. 41 (1) (2013) 65–75.

[73] B.M. Phillips, B.S. Anderson, J.W. Hunt, K. Siegler, J.P. Voorhees, R.S. Tjeerdema, K. McNeill, Environ. Toxicol. Chem. 31 (2012) 1595–1603.

[74] US Environmental Protection Agency, High Production Volume Chemicals Branch, Risk Assessment Division Office of Pollution Prevention and Toxics, Washington, DC, 2007.

[75] S. Huntscha, T.B. Hofstetter, E.L. Schymanski, S. Spahr, J. Hollender, Environ. Sci. Technol. 48 (2014) 4435–4443.

Chapter 17

QSARs on the Thyroid Hormone Effects of Polybrominated Diphenyl Ether (PBDE) Derivatives

Fei Li,[1,2,3] Xianhai Yang,[1] Jingwen Chen[1,*] and Felichesmi Lyakurwa[1]
[1]Key Laboratory of Industrial Ecology and Environmental Engineering (Ministry of Education), School of Environmental Science and Technology, Dalian University of Technology, Dalian, Liaoning, China; [2]Key Laboratory of Coastal Environmental Processes and Ecological Remediation, Yantai Institute of Coastal Zone Research, Chinese Academy of Sciences, Yantai Shandong, China; [3]Shandong Provincial Key Laboratory of Coastal Environmental Processes, Yantai Institute of Coastal Zone Research, Chinese Academy of Sciences, Yantai Shandong, China
**Corresponding author: E-mail: jwchen@dlut.edu.cn*

Chapter Outline

Comprehensive Analytical Chemistry, Vol. 67. http://dx.doi.org/10.1016/B978-0-444-63299-9.00017-X

1. THYROID HORMONE EFFECTS OF POLYBROMINATED DIPHENYL ETHERS

1.1 Thyroid Hormones and Their Functions

The endocrine system consists of glands, each of which secretes different types of hormones directly into the bloodstream to maintain homeostasis. Hormones regulate various functions, including metabolism, growth and development, tissue function, healthy sleep and mood. The thyroid hormones (THs), such as triiodothyronine (T_3) and tetraiodothyronine (T_4) are produced by the thyroid glands and are tyrosine-based hormones (Figure 1).

In plasma, THs are bound to specific proteins to ensure their distribution throughout the vascular system of the target cells. Most of the known functions of THs are mediated by the interaction of T_3 with the nuclear T_3 receptor (TR), which acts as ligand-modulated transcription factors. T_3 either enters the target cell directly from plasma or is converted from T_4 by deiodinases in the target cells (Figure 2). While almost none of the genes regulated by T_3 are exclusively responsive to T_3, virtually all molecular, cellular, and metabolic events are more or less sensitive to the THs [1–3].

THs are usually involved in several important physiological processes such as regulation of the energy metabolism [4]; growth and differentiation; development and maintenance of brain function and the sympathetic nervous system [5–7]; thermoregulation [8]; osmoregulation and renal function [9]; regulation of onset and proper function of other endocrine systems including the estrogen system, sexual behavior and fertility; and cardiovascular functioning [7,10,11]. During development of organism, differentiation of cells and tissues,

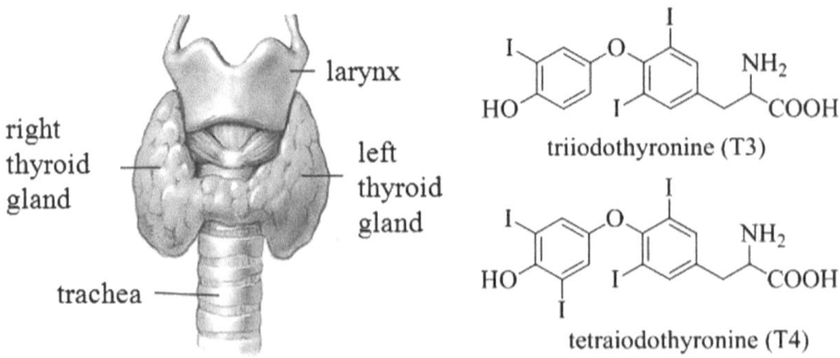

FIGURE 1 Site of thyroid gland and structures of thyroid hormones.

and maintenance or alteration of physiological functions of adult individuals, in many cases TH effects can best be characterized as "permissive hormone action." This indicates that the TH status of cells, tissues, and organisms provides the background and platform for other biological signals—hormonal, neural, immunological, nutritive, and environmental—that are critical for the maintenance of both development and homeostasis of the organism as a whole [12,13].

1.2 Thyroid-Disrupting Chemicals: Effects and Mechanism

1.2.1 Thyroid Hormones Disruption and Thyroid-Disrupting Chemicals

Under the normal physiological conditions, thyroid function is regulated by a finely-tuned negative-feedback mechanism for circulating THs at the hypothalamic and pituitary levels, which maintains stable serum levels of THs with a specific set point in each individual. Lessons learned over decades from the biomedical studies of iodide deficiency, congenital hypothyroidism, and genetic diseases related to defective TH function, and data obtained from various animal

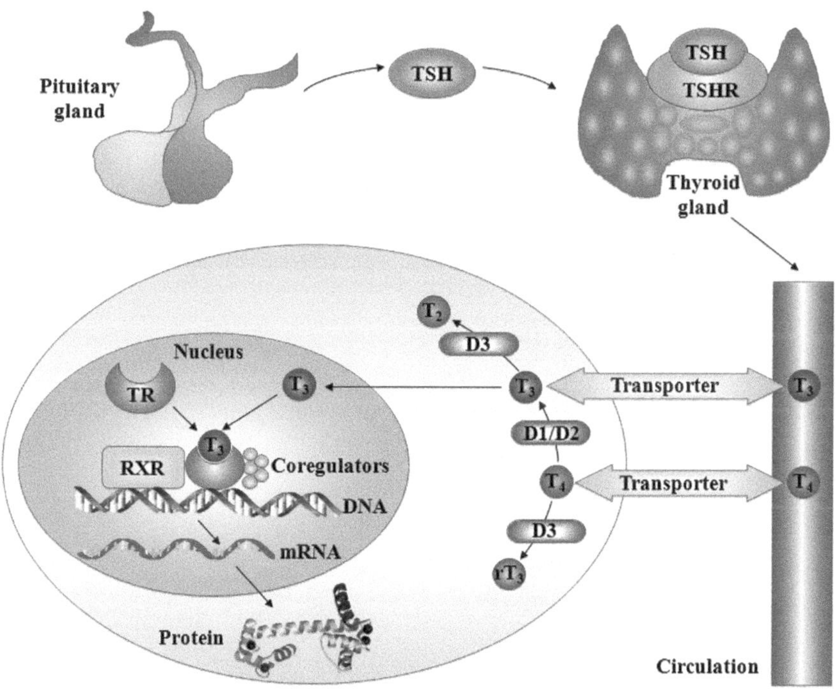

FIGURE 2 Thyroid hormone pathway (D1: type 1 iodothyronine deiodinase; D2: type 2 iodothyronine deiodinase; D3: type 3 iodothyronine deiodinase; rT_3: reverse T_3; RXR: retinoid X receptor; T_2: di-iodothyronine; TR: TH receptor; TSH: thyroid-stimulating hormone; TSHR: TSH receptor).

models corroborate the hypothesis that, transient or persistent TH disruption (THD) could alter the maintenance of homeostasis within the hypothalamic–pituitary–thyroid (HPT) axis and modulate the peripheral TH-dependent functions. Alterations of the TH system are associated with several serious human diseases (e.g., cancers [4,13], metabolic diseases [14,15], learning disabilities [16], heart rhythm problems [17], and impaired reproductive development [18]).

Many man-made chemicals that are used in various products such as pesticides, toys, food additives, and personal care products were identified to elicit THD effects—termed as the thyroid-disrupting chemicals (TDCs). Herein TDCs refer to xenobiotics that alters the structure or function of the thyroid gland, alters the regulatory enzymes associated with the TH homeostasis, or changes the circulation or tissue concentration of THs [19]. To date, hundreds of individual anthropogenic and natural chemicals with diverse routes of exposure, chemical properties, and environmental fates are known or suspected to interact with the endocrine systems in humans and wildlife. Nevertheless, some persistent organic pollutants (POPs) are one of the most important TDCs [20]. The THD effects of polybrominated diphenyl ethers (PBDEs), a class of POPs, are introduced in the next section.

1.2.2 PBDEs and Its Derivatives

PBDEs that comprise a class of substances consisting of 209 possible congeners with 1–10 bromine atoms were extensively used as flame retardants in a variety of consumer products since the 1970s. The general chemical structures of PBDE derivatives are illustrated in Figure 3.

As additive flame retardants, PBDEs are not chemically bound to consumer products, but merely dissolved, so they can leach out and evaporate into the environment over time. In 1979, DeCarlo first detected tri- to decabrominated diphenyl ether in the environment close to a plant that manufactures PBDEs [21]. In 1981, PBDEs were first detected in fish collected from a river of Sweden [22]. Since then,

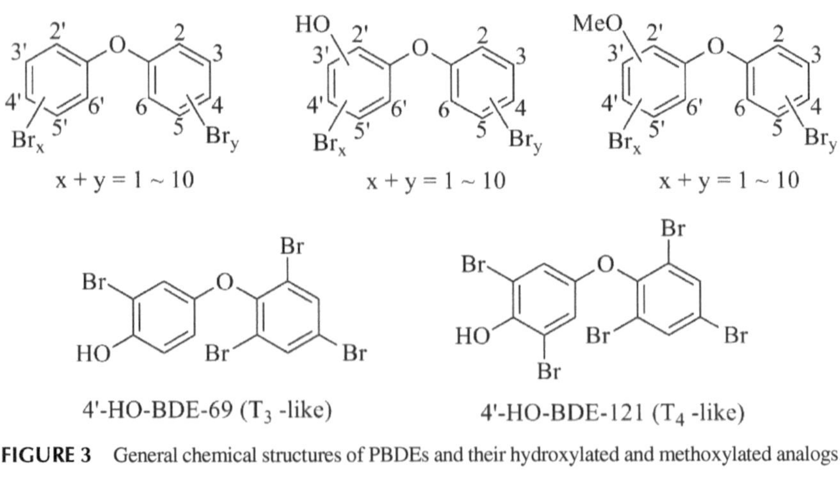

FIGURE 3 General chemical structures of PBDEs and their hydroxylated and methoxylated analogs.

PBDEs have become widespread environmental pollutants, and have been detected in water, sediments, soil, air, wildlife, and human tissues [23–28]. The observed total PBDE levels have increased exponentially. For example, the concentrations of PBDEs in humans have been doubled every 4–6 years since the 1970s in North America [24]. Figure 4 depicts the global distribution of PBDEs in human plasma, biota, and sediments, which results from a survey of various published literature.

Two commercially available PBDEs, penta- and octa-BDEs have been listed under the Stockholm Convention on Persistent Organic Pollutants [29]. For another commercially available PBDEs, deca-BDE is currently under consideration for listing under Annexes A, B, and/or C to the convention (http://chm.pops.int/).

Except for PBDEs, hydroxylated PBDEs (HO-PBDEs) and methoxylated PBDEs (MeO-PBDEs) have also been detected in the surface water, snow, rain, soils and marine sediments, plants, marine organisms (algae, mussel, and fish), and human blood [30–37]. Unlike PBDEs, neither HO-PBDEs nor MeO-PBDEs are artificially synthesized or used in industrial processes. However, both HO-PBDEs and MeO-PBDEs are known to be produced naturally in the marine environment [38–41]· It has also been reported that HO-PBDEs are biotransformation and phototransformation products of PBDEs [32,42–47]. Recent in vivo and in vitro experiments have also indicated that methoxylated polybrominated diphenyl ethers (MeO-PBDEs) can be biotransformed into hydroxylated PBDEs (HO-PBDEs) [44,48].

Serious concerns have recently arisen about the adverse effects of PBDE derivatives on humans and other species. As PBDE derivatives, for example,

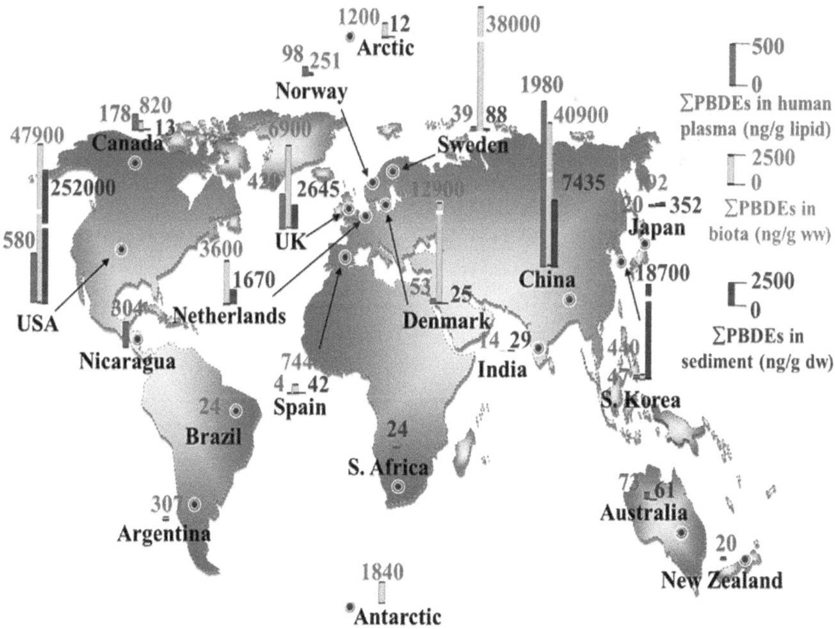

FIGURE 4 Global distributions of PBDEs in human plasma, biota, and sediments.

HO-PBDEs are structurally similar to THs (Figure 3), they may interact with a number of the molecular components of HPT axis, and impact the function of the peripheral tissues. According to in vivo and in vitro experiments, PBDE derivatives can be toxic to TH system and TH-mediated process [49–56]. Likewise, both epidemiological and field investigation results indicated that the exposure of human beings and wildlife to PBDE derivatives was associated with TH dysfunction [57–60].

1.2.3 Mechanisms of Action for TDCs

The mechanisms of action (MOA) for TDCs that are involved in TH signaling disruption include: (1) impacting HPT axis function and regulation, (2) inhibiting TH synthesis, (3) disrupting TH transport proteins, (4) activating/inhibiting TH receptors (TRα and TRβ), and (5) inhibiting TH metabolism (Figure 5) [19,61–66]:

1. *Effects of PBDE derivatives on HPT axis function and regulation.* Through a rigorous literature review, it can be found that PBDE derivatives could disrupt TH-signaling system through almost all the five identified mechanisms. PBDEs can disrupt the hypothalamic–pituitary response to the negative-feedback signal of serum T_4 [67,68]. Under normal physiological conditions, once a xenobiotic chemical causes a decrease in the circulation levels of the serum total and free T_4, the hypothalamic–pituitary will respond to this negative-feedback signal, and increase the serum thyroid-stimulating hormone (TSH). However, PBDEs can decrease the serum total and free T_4 level without any concomitant increase in serum TSH [67], and the underlying mechanism needs further investigations.
2. *Effects of PBDE derivatives on TH synthetase and metabolic enzymes.* Some chemicals can decrease the serum concentration of T_4 by inhibiting the TH synthesis process. For example, perchlorate disrupts sodium/iodide symporter (NIS) uptake of iodine, and methimazole inhibits the activation of iodine by thyroid peroxidase (TPO). As a result, these chemicals decrease the synthesis of THs [62]. There was no available information about whether PBDE derivatives can disrupt NIS and/or TPO up to now. However, 2,2′,4,4′-tetrabrominated diphenyl ether can significantly induce TPO gene in larvae of zebrafish (*Danio rerio*) and NIS gene in embryos of the same fish [69]. Except for disrupting the TH synthetase, PBDE derivatives can also inhibit the metabolic enzymes (e.g., deiodinase [70] and TH sulfotransferase isozymes [71]). For another metabolic enzyme, that is, uridine diphospho-glucuronosyl-transferase, the current results indicate that the tested PBDEs do not inhibit its activity [72].
3. *Effects of PBDE derivatives on TH receptors.* Disrupting TH receptors and TH transport proteins are the well-characterized mechanisms elicited by PBDE derivatives on the THP axis. An extensive but not exhaustive list of the available tested TH activity of PBDE derivatives on TH receptor (TRα and TRβ) are presented in Table 1 [52,56,73–81]. To date, more than 59

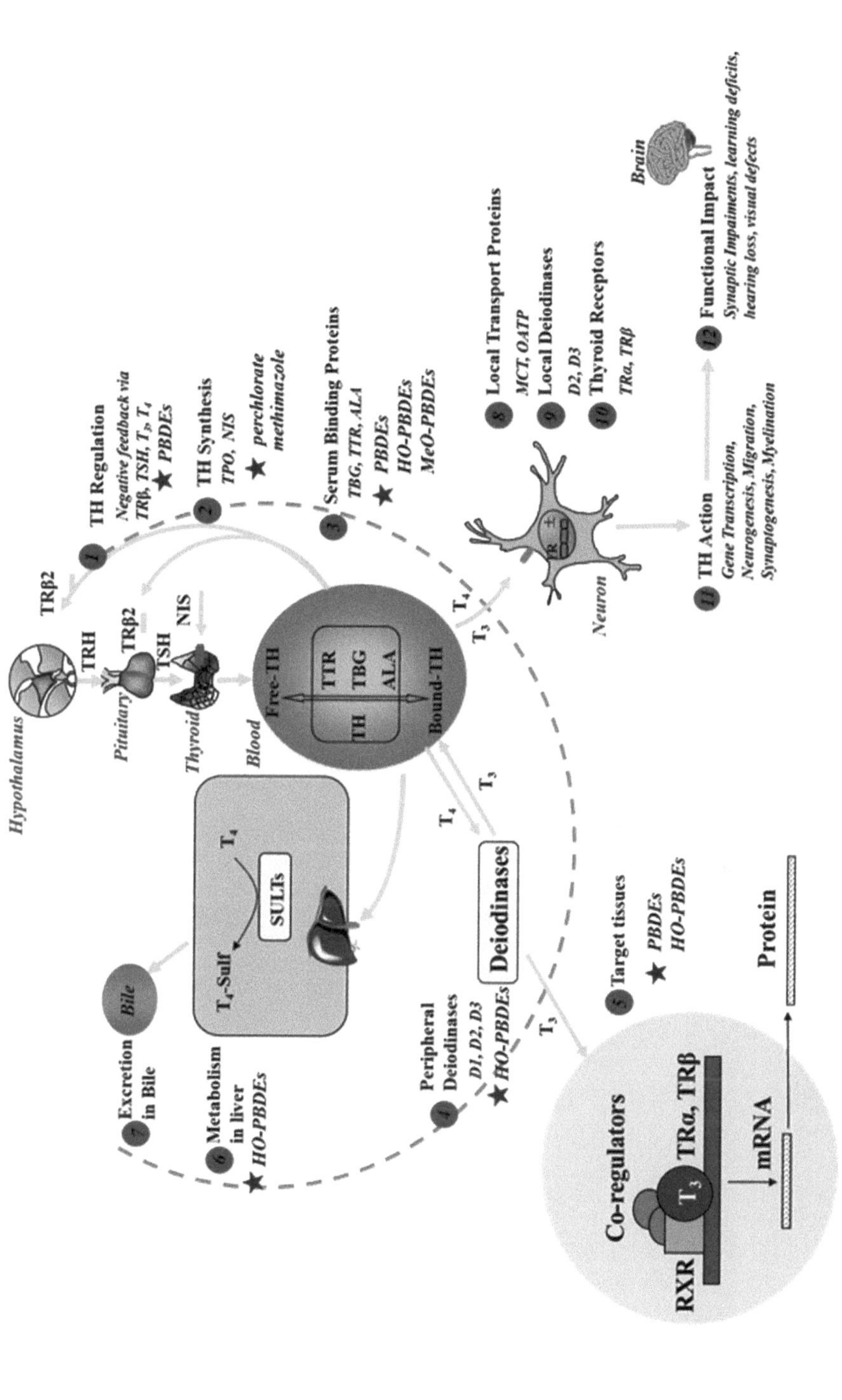

FIGURE 5 Possible sites of action of EDCs on the HPT axis (ALA: albumin; D1/2/3: deiodinases type 1, 2, or 3; NIS: sodium/iodide symporter; MCT: monocarboxylase transporter; OATP: organic anion transporter polypeptide; RXR: retinoid X receptor; T_4-Sult: T_4 Sulfate; TBG: thyroxine-binding globulin; TPO: thyroperoxidase; TRα/β: TH receptor alpha/beta; TRH: thyrotropin-releasing hormone; TTR: transthyretin; TSH: thyroid-stimulating hormone; SULTs: TH sulfotransferase).

TABLE 1 In vitro Determined Disrupting Effects of PBDE Derivatives on Thyroid Receptors (TR), Thyroid Hormone (TH) Transport Proteins, and Human TH Sulfotransferase (hTHS)

			TR			TH Transport Proteins	hTHS
NO.	Compounds	CAS	Binding Potency[a]	Agonistic Activity[b]	Antagonistic Activity[c]	Binding Potency[d]	Inhibiting Effect[e]
1	4,4′-Dibrominated diphenyl ether	002050-47-7	–	N.E. (hTRα) [79], N.E. (hTRβ) [79]	N.E. (hTRα) [79], N.E. (hTRβ) [79]	–	–
2	2,2′,6-Tribrominated diphenyl ether	147217-73-0	–	N.E. (rat TR) [74]	N.E. (rat TR) [74]	N.E. (hTTR) [74], −0.266 (sea bream TTR) [90]	–
3	2,4,4′-Tribrominated diphenyl ether	041318-75-6	–	N.E. (rat TR) [74], N.E. (hTRα) [79], N.E. (hTRβ) [79], N.E. (TRβ) [52]	N.E. (rat TR) [74], N.E.(*Xenopus laevis* TRα) [77], N.E. (*Xenopus laevis* TRβ) [77], N.E. (hTRα) [79], N.E. (hTRβ) [79], N.E. (TRβ) [52]	N.E. (hTTR) [74], 0.203 (sea bream TTR) [90]	–
4	2,4,6-Tribrominated diphenyl ether	155999-95-4	–	<6.70 (hTRβ) [56]		N.E. (hTTR) [74]	–
5	3,4,5-Tribrominated diphenyl ether	337513-63-9	–	N.E. (rat TR) [74]	N.E. (rat TR) [74]	<−2.66 (hTTR) [74], −2.89 (hTTR) [89], −0.437 (sea bream TTR) [90]	–
6	3,4′, 5-Tribrominated diphenyl ether	407606-57-9	–	N.E. (rat TR) [74]	N.E. (rat TR) [74]	N.E. (hTTR) [74], 0.399 (sea bream TTR) [90]	–

7	2,2′,4, 4′-Tetrabrominated diphenyl ether	006336-43-1	N.E. (hTRβ) [80]	N.E. (rat TR) [74], N.E. (hTRα) [79], N.E. (hTRβ) [79], N.E. (hTRβ) [80], N.E. (TRβ) [52]	N.E. (rat TR) [74], N.E. (hTRα) [79], N.E. (hTRβ) [79], N.E. (hTRβ) [80], N.E. (TRβ) [52]	<−2.66 (hTTR) [74], −2.88 (hTTR) [89], −2.60 (hTTR) [49], N.E.(hTTR) [86], 0.656 (Sea bream TTR) [90], −2.18 (Gull TTR) [91], −1.95 (Gull TTR) [55], −2.03 (Gull ALB) [55]	N.E. [71]
8	2,2′,4, 5′-Tetrabrominated diphenyl ether	243982-82-3	–	N.E. (rat TR) [74]	N.E. (rat TR) [74]	<−2.66 (hTTR) [74], −2.77 (hTTR) [89], N.E. (hTTR) [86], 1.68 (Sea bream TTR) [90]	–
9	2,3′,4, 4′-Tetrabrominated diphenyl ether	189084-61-5	–	N.E. (TRβ) [52]	N.E. (TRβ) [52]	–	–
10	2,3′,4, 5′-Tetrabrominated diphenyl ether	446263-38-2	–	–	–	N.E. (hTTR) [86]	–
11	2,3′,4, 6-Tetrabrominated diphenyl ether	327185-09-1	–	<6 (rat TR) [76]	N.E. (rat TR) [76]	–	
12	2,4,4′, 6-Tetrabrominated diphenyl ether	189084-63-7	–	N.E. (rat TR) [76]	N.E. (rat TR) [76]	–	
13	3,3′,4, 5′-Tetrabrominated diphenyl ether	446263-48-4	–	N.E. (rat TR) [74]	N.E. (rat TR) [74]	N.E. (hTTR) [74]	–

Continued

TABLE 1 In vitro Determined Disrupting Effects of PBDE Derivatives on Thyroid Receptors (TR), Thyroid Hormone (TH) Transport Proteins, and Human TH Sulfotransferase (hTHS)—cont'd

			TR			TH Transport Proteins	hTHS
NO.	Compounds	CAS	Binding Potency[a]	Agonistic Activity[b]	Antagonistic Activity[c]	Binding Potency[d]	Inhibiting Effect[e]
14	2,2′,3,4,4′-Pentabrominated diphenyl ether	182346-21-0	–	N.E. (hTRα) [79], N.E. (hTRβ) [79], N.E. (TRβ) [52]	N.E. (hTRα) [79], N.E. (hTRβ) [79], N.E. (TRβ) [52]	–	–
15	2,2′,4,4′,5-Pentabrominated diphenyl ether	060348-60-9	–	N.E. (rat TR) [74], N.E. (hTRα) [79], N.E. (hTRβ) [79], N.E. (TRβ) [52]	N.E. (rat TR) [74], N.E. (hTRα) [79], N.E. (hTRβ) [79], N.E. (TRβ) [52]	N.E.(hTTR) [74], N.E. (hTTR) [86], 0.551 (sea bream TTR) [90]	–
16	2,2′,4,4′,6-Pentabrominated diphenyl ether	189084-64-8	–	N.E. (rat TR) [74], N.E. (hTRα) [79], N.E. (hTRβ) [79], N.E. (TRβ) [52]	N.E. (rat TR) [74], N.E. (hTRα) [79], N.E. (hTRβ) [79], 45% at 1 nm (TRβ) [52]	N.E. (hTTR) [74], −0.368 (sea bream TTR) [90]	–
17	2,3,4,5,6-Pentabrominated diphenyl ether	189084-65-9	–	<6.70 (hTRβ) [56]	–	N.E. (hTTR) [74]	–
18	3,3′,4,5,5′-Pentabrominated diphenyl ether	446263-86-0	–	14% (rat TR) [74], 5.04 (rat TR) [76]	N.E. (rat TR) [74], N.E. (rat TR) [76]	−2.60 (hTTR) [74], N.E. (Sea bream TTR) [90]	–
19	2,2′,3,4,4′,5′-Pentabrominated diphenyl ether	182677-30-1	N.E.(rat TR) [78]	–	–	–	–

20	2,2′,4,4′,5,5′-Hexabrominated diphenyl ether	068631-59-2	–	N.E. (rat TR) [74], N.E. (hTRα) [79], N.E. (hTRβ) [79], N.E. (TRβ) [52]	N.E. (rat TR) [74], N.E. (hTRα) [79], N.E. (hTRβ) [79], 40% 10 nM (TRβ) [52]	N.E. (hTTR) [74], N.E. (sea bream TTR) [90]	–
21	2,2′,4,4′,5,6′-Hexabrominated diphenyl ether	207122-15-4	–	N.E. (TRβ) [52]	40% 1 nM (TRβ) [52]	–	–
22	2,2′,4,4′,6,6′-Hexabrominated diphenyl ether	035863-94-5	–	N.E. (rat TR) [74]	N.E. (rat TR) [74]	N.E. (hTTR) [74], N.E. (sea bream TTR) [90]	–
23	3,3′,4,4′,5,5′-Hexabrominated diphenyl ether	446255-18-1	–	N.E. (rat TR) [74]	N.E. (rat TR) [74]	<−2.66 (hTTR) [74], N.E. (sea bream TTR) [90]	–
24	2,2′,3,3′,4,5′,6-Heptabrominated diphenyl ether	446255-22-7	–	N.E. (TRβ) [52]	N.E. (TRβ) [52]	–	–
25	2,2′,3,4,4′,5,6-Heptabrominated diphenyl ether	189084-67-1	–	N.E. (rat TR) [74]	N.E. (rat TR) [74]	−2.11 (hTTR) [74], N.E. (sea bream TTR) [90]	–
26	2,2′,3,4,4′,5′,6-Heptabrominated diphenyl ether	207122-16-5	–	N.E. (rat TR) [74], N.E. (TRβ) [52]	N.E. (rat TR) [74], N.E. (TRβ) [52]	N.E. (hTTR) [74], N.E. (sea bream TTR) [90]	–
27	2,2′,3,4,5,5′,6-Heptabrominated diphenyl ether	405237-86-7	–	12% (rat TR) [74], 5.96 (rat TR) [76]	N.E. (rat TR) [74], N.E. (rat TR) [76]	−2.13 (hTTR) [74], N.E. (sea bream TTR) [90]	–

Continued

TABLE 1 In vitro Determined Disrupting Effects of PBDE Derivatives on Thyroid Receptors (TR), Thyroid Hormone (TH) Transport Proteins, and Human TH Sulfotransferase (hTHS)—cont'd

			TR			TH Transport Proteins	hTHS
NO.	Compounds	CAS	Binding Potency[a]	Agonistic Activity[b]	Antagonistic Activity[c]	Binding Potency[d]	Inhibiting Effect[e]
28	2,3,3′,4,4′,5,6-Heptabrominated diphenyl ether	189084-68-2	–	N.E. (rat TR) [74]	N.E. (rat TR) [74]	–2.21 (hTTR) [74], N.E. (sea bream TTR) [90]	–
29	2,2′,3,3′,4,4′,5,5′,6-Nonabrominated diphenyl ether	063387-28-0	–	N.E. (rat TR) [74], N.E. (rat TR) [76]	–28% (rat TR) [74], 6.35 (rat TR) [76], –30%(*Xenopus laevis* TRα) [77], –63%(*Xenopus laevis* TRβ) [77]	N.E. (hTTR) [74], N.E. (sea bream TTR) [90]	–
30	2,2′,3,3′,4,4′,5,5′,6,6′-Decabrominated diphenyl ether	001163-19-5	–	N.E. (rat TR) [74], N.E. (hTRα) [79], N.E. (hTRβ) [79], N.E. (TRβ) [52]	N.E. (rat TR) [74], N.E. (hTRα) [79], N.E. (hTRβ) [79], 45% at 1 nM (TRβ) [52]	N.E. (hTTR) [74], N.E. (sea bream TTR) [90]	–
31	DE-71 (PBDEs mixture)	117148-05-7	–	N.E. (TRβ) [52]	35% at 10 nM (TRβ) [52]	–	–
32	2′-HO-4-monobrominated diphenyl ether	300560-92-3	N.E. (hTRα) [75]; N.E. (hTRβ) [75]	N.E. (rat TR) [75], N.E. (hTRα) [75], N.E. (hTRβ) [75]	N.E. (rat TR) [75]	–0.824 (hTTR) [83], –1.20 (hTTR) [84], –2.40 (hTBG) [83], N.E.(hTBG) [84]	500 [71]

33	2′-HO-2,4-dibrominated diphenyl ether	602326-09-0	–	–	–	–0.523 (hTTR) [83], –0.522 (hTTR) [84], –1.63 (hTBG) [83], –2.17 (hTBG) [84]	–
34	3′-HO-2,4-dibrominated diphenyl ether	602326-18-1	N.E. (rat TR) [78], N.E. (hTRα) [75], N.E. (hTRβ) [75]	4.90 (rat TR) [75], N.E. (hTRα) [75], 7.64 (hTRβ) [56], N.E. (hTRβ) [75]	N.E. (rat TR) [75]	–0.222 (hTTR) [83], –0.159 (hTTR) [84], –1.70 (hTBG) [83], –1.97 (hTBG) [84]	410 [71]
35	4′-HO-2,2′,4-tribrominated diphenyl ether	602326-21-6	N.E.(rat TR) [78]	N.E. (hTRα) [79], 8.66 (hTRβ) [56], N.E. (hTRβ) [79]	N.E. (hTRα) [79], N.E. (hTRβ) [79]	0.176 (hTTR) [83], –1.10 (hTBG) [83]	–
36	2′-HO-2,4,4′-tribrominated diphenyl ether	004656-58-0	N.E. (hTRα) [75], N.E. (hTRβ) [75]	5.51 (rat TR) [75], $p<0.05$(hTRα) [75], 8.07 (hTRβ) [56], $p<0.05$ (hTRβ) [75]	N.E. (rat TR) [75]	0.477 (hTTR) [83], –0.539 (hTTR) [84], –1.27 (hTBG) [83], –1.95 (hTBG) [84]	–
37	3′-HO-2,4,4′-tribrominated diphenyl ether	025959-30-2	4.60 (hTRα) [75], 4.70 (hTRβ) [75]	5.20 (rat TR) [75], $p<0.05$(hTRα) [75], 7.28 (hTRβ) [56], $p<0.05$ (hTRβ) [75]	N.E. (rat TR) [75]	0 (hTTR) [83], –0.141 (hTTR) [84], –1.18 (hTBG) [83], –1.15 (hTBG) [84]	190 [71]
38	4′-HO-2,4,6-tribrominated diphenyl ether	218303-99-2	<6 (hTRα) [73], <6 (hTRβ) [73]	<5 (rat TR) [76]	N.E. (rat TR) [76]	–0.392 (hTTR) [87]	–
39	4-HO-2,2′,3,4′-tetrabrominated diphenyl ether	602326-22-7	N.E.(rat TR) [78]	N.E. (hTRα) [79], 9.72 (hTRβ) [56], N.E. (hTRβ) [79]	N.E. (hTRα) [79], N.E. (hTRβ) [79]	0.523 (hTTR) [83], 0.634 (hTTR) [49], –1.14 (hTBG) [83]	–

Continued

TABLE 1 In vitro Determined Disrupting Effects of PBDE Derivatives on Thyroid Receptors (TR), Thyroid Hormone (TH) Transport Proteins, and Human TH Sulfotransferase (hTHS)—cont'd

			TR			TH Transport Proteins	hTHS
NO.	Compounds	CAS	Binding Potency[a]	Agonistic Activity[b]	Antagonistic Activity[c]	Binding Potency[d]	Inhibiting Effect[e]
40	3-HO-2,2′,4,4′-tetrabrominated diphenyl ether	025959-31-3	10^{-6} to 10^{-4} M (rat TR) [78], 5.48 (hTRα) [75], 5.51 (hTRβ) [75]	N.E. (rat TR) [75], N.E. (hTRα) [75], 8.77 (hTRβ) [56], N.E. (hTRβ) [75]	N.E. (rat TR) [75]	0.632 (hTTR) [83], 0.374 (hTTR) [84], 0.602 (hTTR) [49], −0.0835 (hTTR) [86], −1.10 (hTBG) [83], −0.469 (hTBG) [84]	60 [71]
41	6-HO-2,2′,4,4′-tetrabrominated diphenyl ether	079755-43-4	5.25 (hTRα) [75], 5.57 (hTRβ) [75]	N.E. (rat TR) [74], 6.10 (rat TR) [75], $p<0.05$ (hTRα) [75], 10.4 (hTRβ) [56], $p<0.05$ (hTRβ) [75], N.E. (TRβ) [52]	N.E. (rat TR) [74], N.E. (rat TR) [75], N.E. (TRβ) [52]	0.301 (hTTR) [83], −0.0942 (hTTR) [84], −0.409 (hTTR) [49], −0.528 (hTTR) [88], −0.515 (hTTR) [74], −0.675 (hTTR) [86], −1.47 (sea bream TTR) [90], −1.24 (gull TTR) [91], −1.06 (gull TTR) [55], −1.18 (hTBG) [83], −0.615 (hTBG) [84], 0.0406 (gull ALB) [55]	130 [71]

42	5-HO-2,2′,4,4′-tetrabrominated diphenyl ether	602326-30-7	5.66 (hTRα) [75], 5.51 (hTRβ) [75]	5.51 (rat TR) [75], $p<0.05$ (hTRα) [75], 8.44 (hTRβ) [56], $p<0.05$ (hTRβ) [75]	N.E. (rat TR) [75]	0.875 (hTTR) [83], 0.0765 (hTTR) [84], 0.477 (hTTR) [49], −0.382 (hTTR) [86], −1 (hTBG) [83], −0.451 (hTBG) [84]	400 [71]
43	4′-HO-2,2′,4,5′-tetrabrominated diphenyl ether	602326-23-8	N.E.(rat TR) [78]	N.E. (hTRα) [79], 7.87 (hTRβ) [56], N.E.(hTRβ) [79], N.E. (TRβ) [52]	N.E. (hTRα) [79], N.E. (hTRβ) [79], N.E. (TRβ) [52]	0.477 (hTTR) [83], 0.634 (hTTR) [49], −0.768 (hTTR) [86], −0.0915 (gull TTR) [91], 0.0881 (gull TTR) [55], −1.40 (hTBG) [83], 0.152 (gull ALB) [55]	650 [71]
44	6′-HO-2,2′,4,5′-tetrabrominated diphenyl ether	597069-16-6	–	–	–	−0.555 (hTTR) [86]	–
45	2′-HO-2,3′,4,4′-tetrabrominated diphenyl ether	184174-87-6	–	–	–	−0.187 (hTTR) [49]	–
46	2′-HO-2,3′,4,5′-tetrabrominated diphenyl ether	080246-25-9	–	N.E. (TRβ) [52]	N.E. (TRβ) [52]	0.574 (hTTR) [83], −1.27 (hTBG) [83]	–
47	4′-HO-2,3′,4,6-tetrabrominated diphenyl ether	218303-98-1	6.70 (hTRα) [73], 7.39 (hTRβ) [73]	6.64 (rat TR) [76], 7.21 (TR) [81]	N.E. (rat TR) [76], N.E.(*Xenopus laevis* TRα) [77], N.E. (*Xenopus laevis* TRβ) [77]	0.0873 (hTTR) [87]	–
48	6-HO-2,2′,3,3′,4-pentabrominated diphenyl ether	1093168-88-7	-	10.4 (hTRβ) [56]	–	–	–

Continued

TABLE 1 In vitro Determined Disrupting Effects of PBDE Derivatives on Thyroid Receptors (TR), Thyroid Hormone (TH) Transport Proteins, and Human TH Sulfotransferase (hTHS)—cont'd

			TR			TH Transport Proteins	hTHS
NO.	Compounds	CAS	Binding Potency[a]	Agonistic Activity[b]	Antagonistic Activity[c]	Binding Potency[d]	Inhibiting Effect[e]
49	6-HO-2,2′,3,4,4′-pentabrominated diphenyl ether	035162-01-7	-	9.77 (hTRβ) [56]	–	0.632 (hTTR) [83], –1(hTBG) [83]	–
50	6-HO-2,2′,3,4,5′-pentabrominated diphenyl ether	1266324-64-4	-	9.29 (hTRβ) [56]	–	–	–
51	4-HO-2,2′,3,4′,5-pentabrominated diphenyl ether	602326-24-9	10^{-6} to 10^{-4} (rat TR) [78]	N.E. (hTRα) [79], 7.63 (hTRβ) [56], N.E.(hTRβ) [79]	5.092 (hTRα) [79], 5.137 (hTRβ) [79]	–	24 [71]
52	5′-HO-2,2′,4,4′,5-pentabrominated diphenyl ether	923925-51-3	–	10.3 (hTRβ) [56]	–	–	520 [71]
53	6′-HO-2,2′,4,4′,5-pentabrominated diphenyl ether	923925-52-4	–	9.62 (hTRβ) [56]	–	–	310 [71]

54	6-HO-2,2′,4,4′,5-pentabrominated diphenyl ether	297742-10-0	–	–	–	–0.657 (hTTR) [86]	–
55	3-HO-2,2′,4,4′,6-pentabrominated diphenyl ether	1093168-89-8	6.08 (hTRα) [75], 5.63 (hTRβ) [75]	N.E. (rat TR) [75], N.E. (hTRα) [75], N.E. (hTRβ) [75]	22% (rat TR) [75]	0.0745 (hTTR) [84], 0.189 (hTBG) [84]	–
56	4′-HO-2,2′,4,5,5′-pentabrominated diphenyl ether	1309567-66-5	–	–	–	–	640 [71]
57	4′-HO-2,3′,4,5′,6-pentabrominated diphenyl ether	091370-78-4	6.26 (hTRα) [73], 6.74 (hTRβ) [73]	5.85 (rat TR) [76], 6.34 (TR) [81]	N.E. (rat TR) [76]	0.151 (hTTR) [87]	–
58	6-HO-2,2′,3,4,4′,6′-hexabrominated diphenyl ether	1266324-65-5	-	11.3 (hTRβ) [56]	–	–	–
59	3-HO-2,2′,4,4′,5,6′-hexabrominated diphenyl ether	1300682-86-3	6.27 (hTRα) [75], 6.41 (hTRβ) [75]	N.E. (rat TR) [75], N.E. (hTRα) [75], 10.8 (hTRβ) [56], N.E. (hTRβ) [75]	12% (rat TR) [75]	–	–
60	3′-HO-2,2′,4,4′,5,6′-hexabrominated diphenyl ether	1171030-72-0	–	–	–	0.477 (hTTR) [83], 0.114 (hTTR) [84], –0.875 (hTBG) [83], –0.232 (hTBG) [84]	80 [71]
61	6-HO-2,3,3′,4,4′,5′-hexabrominated diphenyl ether	1244033-58-6	–	12.2 (hTRβ) [56]	–	–	–

Continued

TABLE 1 In vitro Determined Disrupting Effects of PBDE Derivatives on Thyroid Receptors (TR), Thyroid Hormone (TH) Transport Proteins, and Human TH Sulfotransferase (hTHS)—cont'd

			TR			TH Transport Proteins	hTHS
NO.	Compounds	CAS	Binding Potency[a]	Agonistic Activity[b]	Antagonistic Activity[c]	Binding Potency[d]	Inhibiting Effect[e]
62	6-HO-2,2′,3,4,4′,5,5′-heptabrominated diphenyl ether	1350848-48-4	–	–	–	–	13500 [71]
63	4-HO-2,2′,3,4′,5,6,6′-heptabrominated diphenyl ether	784619-38-4	6.50 (hTRα) [75], 6.59 (hTRβ) [75]	N.E. (rat TR) [75], N.E. (hTRα) [75], N.E. (hTRβ) [75]	33% (rat TR) [75]	0.78 (hTTR) [84], −0.277 (hTBG) [84]	–
64	4′-HO-2,2′,3,3′,4,5′,6,6′-octabrominated diphenyl ether	1350848-50-8	–	–	–	–	510 [71]
65	4′-MeO-2,2′,4-tribrominated diphenyl ether	602326-27-2	–	N.E. (hTRα) [79], N.E. (hTRβ) [79]	N.E. (hTRα) [79], N.E. (hTRβ) [79]	–	–
66	4-MeO-2,2′,3,4′-tetrabrominated diphenyl ether	602326-25-0	–	N.E. (hTRα) [79], N.E. (hTRβ) [79]	N.E. (hTRα) [79], N.E. (hTRβ) [79]	–	–
67	6-MeO-2,2′,4,4′-tetrabrominated diphenyl ether	102739-99-1	–	–	–	N.E. (hTTR) [86], −1.72 (gull TTR) [91], −1.59 (gull TTR) [55], −1.99 (gull ALB) [55]	–

68	6′-MeO-2,2′,4,5′-tetrabrominated diphenyl ether	597106-79-3	–	N.E. (hTRα) [79], N.E. (hTRβ) [79]	N.E. (hTRα) [79], N.E. (hTRβ) [79]	–	–
69	2′-MeO-2,3′,4,5′-tetrabrominated diphenyl ether	096920-28-4	–	–	–	N.E. (hTTR) [86]	–
70	4-MeO-2,2′,3,4′,5-pentabrominated diphenyl ether	602326-29-4	N.E. (rat TR) [78]	N.E. (hTRα) [79], N.E. (hTRβ) [79]	N.E. (hTRα) [79], N.E. (hTRβ) [79]	–	–

N.E., No effect; –, Respresents No Data.
[a]*The End Point is the Logarithm of Competing Concentration of a Chemical with Triiodothyronine (T_3) Binding to the TR (log IC).*
[b]*The End Point is the Logarithm of Effective Concentration of a Chemical Activating TR (log EC).*
[c]*The End Point is the Logarithm of Inhibitory Concentration of a Chemical Inhibiting TR (log IC).*
[d]*The End Point is the Logarithm of Relative Competing Potency of a Chemical with Tetraiodothyronine (T_4) Binding to hTTR (log RP); RP is Defined by Eqn (1) in the Text.*
[e]*The End Point is the Inhibitory Concentration of a Chemical when it Inhibits 50% of 3,3′-Diiodothyronine Sulfated by THS (IC_{50}).*

PBDE derivatives have been tested for their hormone activities to TR. The experimental protocols used to test the TR-disrupting effects in most cases include the receptor competitive binding assay, cell proliferation assay, reporter gene assay, and coactivator recruitment assay [66].

Both agonistic and antagonistic activity to TR was tested for 28 PBDEs. Among them, three PBDEs (2,3′,4,6-tetrabrominated diphenyl ether, 3,3′,4,5,5′-pentabrominated diphenyl ether, and 2,2′,3,4,5,5′,6-heptabrominated diphenyl ether) were found to exhibit the agonistic activity for rat TR by cell proliferation assay [74,76], and five PBDEs (2,2′,4,4′,6-pentabrominated diphenyl ether, 2,2′,4,4′,5,5′-hexabrominated diphenyl ether, 2,2′,4,4′,5,6′-hexabrominated diphenyl ether, 2,2′,3,3′,4,4′,5,5′,6-nonabrominated diphenyl ether, and 2,2′,3,3′,4,4′,5,5′,6,6′-decabrominated diphenyl ether) were found to exhibit the antagonistic activity to TR [52,74,76,77]. Except for the PBDE congeners, a PBDEs mixture named DE 71 was also confirmed as an antagonist for TRβ by the reporter gene assay in CV-1 monkey fibroblast-derived cells [52]. The results summarized above indicate that most of the tested PBDE congeners themselves do not disrupt the TH signal system through TR pathway.

The hormone activities of four MeO-PBDEs were also tested for the disrupting activity of TR by the receptor competitive binding assay and the reporter gene assay [78,79]. None of them elicited the hormone activities. In contrast, 23 of 25 tested HO-PBDEs could bind directly with TR or/and exhibited agonistic or/and antagonistic activity to TR [56,73,75,76,78,79,81]. Kitamura et al. [78] tested the binding potency of six HO-PBDEs to TR by employing the nuclear extracts of rat pituitary cells. Their results indicated that 3-HO-2,2′,4,4′-tetrabrominated diphenyl ether and 4-HO-2,2′,3,4′,5-pentabrominated diphenyl ether markedly inhibited the binding of T_3 to rat TR. For human TRα and TRβ, Marsh et al. [73] and Ren et al. [75] tested the binding potency of 13 HO-PBDEs by employing a competitive binding assay. Ten HO-PBDEs showed detectable binding potency to TRα or TRβ.

As shown in Table 1, the agonistic and antagonistic activity for HO-PBDEs seems to be inconsistent. It is noteworthy that in the previous studies, different species or host cell lines were employed. Thirteen HO-PBDEs have been tested for their agonistic and antagonistic activity with rat TR (cell proliferation assay) [73,74,76,78]. Among the 13 HO-PBDEs, eight were identified as agonist [73,76], and three (3-HO-2,2′,4,4′,6-pentabrominated diphenyl ether, 3-HO-2,2′,4,4′,5,6′-hexabrominated diphenyl ether, and 4-HO-2,2′,3,4′,5,6,6′-heptabrominated diphenyl ether) exhibited antagonistic activity [73].

Six HO-PBDEs were tested for the antagonistic activity of TR by using the reporter gene assay [52,77,79], and none of them exhibited the antagonistic activity except for 4-HO-2,2′,3,4′,5-pentabrominated diphenyl ether that exhibited the antagonistic activity for human TRα and TRβ in the Chinese hamster ovary cells [79]. For human TRα, no HO-PBDEs were observed to display agonistic activity in the reporter gene assay [73,79]. However, the coactivator recruitment assay showed that 4 of 10 HO-PBDEs activated the human TRα-mediated transcription [73]. For human TRβ, 18 of 21 HO-PBDEs exhibited agonistic activity [56,73,79]. Thus, these results indicate that activating the TR-mediated signal is a critical mechanism for the tested HO-PBDEs.

4. *Effects of PBDE derivatives on TH transport proteins.* The human TH transport proteins include transthyretin (hTTR), thyroxine-binding globulin (hTBG), and albumin. The experimental protocols for testing the TH transport protein-disrupting activities of TDCs include the radiolabeled ligand displacement method [82], the competitive fluorescence displacement assay [83–85], and the surface plasmon resonance biosensor-based method [86]. The current available data [49,55,74,83,84,86–91] on disrupting activity of PBDE derivatives to TH transport proteins are listed in Table 1. The disrupting activities of PBDE derivatives to TH transport proteins are species specific. Most of the studies tested the disrupting potency of PBDE derivatives on hTTR, as the binding potency of T_4 with hTBG was higher than that with hTTR, and it is slightly easier for TDCs to compete for T_4 binding to hTTR than hTBG. Moreover, hTTR is the major TH carrier in the brain and fetal tissues.
   PBDEs and their derivatives have distinct binding potency with the different human TH transport proteins. PBDE derivatives have different binding potency with hTTR than PBDEs. HO-PBDEs have higher binding potency to hTTR. However, introduction of a methoxy group to the parent molecule of PBDEs does not significantly increase its disrupting activities to hTTR [55,86,91]. The logarithm of the relative competing potency of a chemical with T_4 binding to a TH transport protein (log *RP*) is usually employed to characterize the binding potency of TDCs, which is defined as:

$$RP = \frac{IC_{50,\mathrm{T_4}}}{IC_{50,\mathrm{TDCs}}} \tag{1}$$

   where $IC_{50,\mathrm{T4}}$ and $IC_{50,\mathrm{TDCs}}$ are the half-maximal inhibitory concentration (IC_{50}) for T_4 and TDCs, respectively. log *RP* values for many HO-PBDEs on hTTR are >0, implying the binding potency of these HO-PBDEs to hTTR is stronger than that of T_4. While for hTBG, only 3-HO-2,2′,4,4′,6-pentabrominated diphenyl ether was reported with its log *RP* value >0 [84].

1.3 THD Assays

Concern about THD effects has resulted in strategies being developed to identify TDCs [66]. New integrated strategies for screening and testing endocrine-disrupting chemicals (EDCs) have been put forward by the Organization for Economic Co-operation and Development (OECD), US Environmental Protection Agency (US EPA), and the European Union [92–97]. A five-level conceptual framework was proposed by the OECD for the assessment and testing of EDCs [98], which includes Level 1: Existing data and non-test information; Level 2: In vitro assays providing data about selected endocrine mechanisms/pathways; Level 3: In vivo assays providing data about selected endocrine mechanisms/pathways; Level 4: In vivo assays providing data on adverse effects on endocrine relevant end points; and Level 5: In vivo assays providing more comprehensive data on adverse effects on endocrine relevant end points over more extensive parts of the life cycle of the organism. The US EPA has proposed a two-tiered screening and testing process for EDCs (http://www.epa.gov/endo/). The core of newly developed integrated strategies was a shift from the dependence on the animal-based characterization of chemicals hazard to a mixture of in vivo, high-throughput in vitro tests, and in silico (computational toxicology) models [93,95,96].

Information on potential of chemicals causing THD is typically derived from the animal studies. However, more than 140,000 commercially used chemicals have been preregistered by the European REACH (Registration, Evaluation, Authorisation and Restriction of Chemicals) regulation [99]. For the majority of these chemicals, their THD information is either limited or unavailable. It is unlikely that animal experiments will be performed for all THD relevant chemicals in the near future as whole-animal tests (in vivo) have several limitations including ethical unacceptability [100,101], high-cost [92], time-consuming [102], and pleiotropic effects [103–105]. In addition, in vivo tests often do not provide information on the MOA of TDCs, making it harder to extrapolate results across species. Relevant effects may not be identified in animal studies when the effects are delayed, life stage specific, not assessed by the experimental paradigm, or only occur when an organism has to adapt to environmental factors by modulating TH levels. Thus, developing in vitro and in silico alternatives to identify potential TDCs and quantify their potency has become necessary [66].

Murk et al. recently reviewed the mechanism-based testing strategy using in vitro approaches for identification of TDCs [66]. According to the review, in vitro assays are available for many end points corresponding with the potential MOA of TDCs, and for the application of modern "-omics" technologies. New in vitro tests for screening TDCs are under development [66]. Murk et al. has recommended a battery of test methods to be able to classify chemicals as of less or high concern for further hazard and risk assessment for TDCs. There are also other reviews or governmental agency reports describing the status of in vitro and in vivo assays for TDCs [98,106–108].

Computational toxicology (in silico) applies mathematical and computer models and molecular biological and chemical approaches to explore both qualitative and quantitative relationships between chemical exposure and adverse health outcomes [109]. Typical example of in silico models is quantitative structure–activity relationships (QSARs) of organic chemicals, which correlate and predict chemical toxicities from their molecular structures. Herein QSARs on THD effects of PBDE derivatives are introduced.

2. BRIEF INTRODUCTION ON QSARs

In recent years, a vision of toxicology testing for the twenty-first century was broadly discussed [92–95,110–114], in which the in silico models was considered to be an important part of the new testing strategies. The European Union, US

EPA, and OECD supported several programs/projects on in silico technology for priority setting and screening of chemicals [114–116]. To date, many in silico models, tools, and databases have been developed [117–122] and most of the models and tools were developed based on QSARs.

QSARs are based on the fundamental concept that molecules with common structural features are thought to exhibit and elicit similar physicochemical, environmental behavioral, and toxicological effects. Thus, untested physicochemical, environmental behavioral, and toxicological effects can be predicted qualitatively or quantitatively based on their molecular structures by using the developed QSAR models [102,123]· Several papers have reviewed the progress and challenges of using QSARs by chemicals regulatory agencies and authorities [124–126]. To guide the development, validation, and application of QSAR models, the OECD has issued QSARs development and validation guideline [127]. According to the guideline, QSARs should be associated with the following information: (1) a defined end point; (2) an unambiguous algorithm; (3) a defined applicability domain (AD); (4) appropriate measures of goodness-of-fit, robustness, and predictivity; and (5) mechanistic interpretation, if possible.

The end points modeled by QSARs refer to any physicochemical property, environmental behavioral parameter, or toxicological effects that can be measured. The algorithms commonly used to construct QSARs are those used for pattern recognition or machine learning, usually include multiple linear regression (MLR), partial least squares (PLS) regression, artificial neural network (ANN), k-nearest neighbor (k-NN) clustering, genetic algorithms (GA), and so on. The AD of QSAR models describes the scope and limitations of a model [127–129]. The robustness of QSARs is usually measured by Q^2 (explained variance in prediction), while the predictive ability of a model is usually evaluated by Q^2_{EXT} (external explained variance) [127,130–132].

Understanding the mechanisms that molecular structures determine the environmental behavior and toxicology of organic chemicals is of great importance for QSAR modeling. Both experiment and molecular modeling methods can be employed to unveil the mechanisms. Molecular modeling methods (e.g., molecular docking, molecular dynamics, and quantum mechanics) can be employed to understand the molecular mechanisms of EDCs to exert their effects. For example, by molecular docking, intermolecular interactions between a small molecule (ligand) and a biomacromolecule (receptor) can be analyzed [116].

Many QSAR models have been established for xenoestrogens or xenoandrogens [119–121,133]. However, QSAR models on TDCs have been less documented. Table 2 lists some QSAR models that characterize the disrupting potency of TDCs on the end points from TH pathways. Among the QSAR models, two were developed to characterize the relationship between HO-PBDEs and human TRβ [56,134]; and seven characterize the relationship between hTTR and PBDEs/HO-PBDEs [83,135–139]. One model for TH sulfotransferase was also constructed [71]. In the following section, QSAR models for HO-PBDEs and substituted phenolic compounds exerting disrupting effects to TRβ and hTTR are detailed.

TABLE 2 Summary of QSARs Models for Human Thyroid Receptors (TR), Human Thyroid Hormone (TH) Transport Proteins and Human TH Sulfotransferase (hTHS)

No.	Mechanism of Action	TDCs[a]	End Point[b]	Model[c]	$n_{training}$[d]	n_{EXT}	N	Model Statistics	AD	References
1	hTRβ	HO-PBDEs	–log REC_{20}	QSAR	14	4	6	$R^2_{tarining}=0.91$, $Q^2_{CUM}=0.87$, $RMSE_{training}=0.42$, $Q^2_{EXT}=0.50$, $RMSE_{EXT}=0.73$	Y	[56]
		$-\log REC_{20}=5.73\times10+8.01\times10^{-1}\ n_{Br}+9.62\times10^{-1}\log K_{OW}-4.95\times10\ I_A+2.84\ E_{LUMO}-1.66\ \omega+3.26\times10^{-2}\ \mu^2$								
		n_{Br}, the number of bromine atoms; log K_{OW}, the logarithm of *n*-octanol/water partition coefficient; E_{LUMO}, energy of the lowest unoccupied molecular orbital; I_A, harmonic oscillator model of aromaticity index; ω, electrophilicity index; μ^2, square of dipole moment.								
2		HO-PBDEs,	–log REC_{20}	CoMSIA[d]	14	4	4	$R^2_{tarining}=0.95$, $Q^2_{LOO}=0.57$, $R^2_{EXT}=0.85$	N	[134]
		Steric, electrostatic, hydrogen bond donor, and hydrogen bond acceptor								
3	hTTR	2,4,6-TBP, HBCDs, HO-PBDEs, PBDEs, TBBPA, TBBPA-DBPE	log *RP*	QSAR	9	8	2	$R^2_{training}=0.96$, $Q^2_{LOO}=0.91$, $RMSE_{training}=0.29$, $R^2_{EXT}=0.93$, $Q^2_{EXT}=0.90$, $RMSE_{EXT}=0.47$	Y	[137]
		$\log RP=-9.07+4.08\times10\ qpmax+3.93\ MATS6v$								
		qpmax, maximum positive charge; *MATS6v*, Moran autocorrelation of lag 6 weighted by van der Waals volume								

4		HO-PBDEs	log *K*	–	14		2	$R^2_{training}=0.88$	N	[83]
		$\log K=-1.6\times10^{-1}\log K_{OW}^2+2.34\log K_{OW}$								
5		PBDEs, HO-PBDEs,	log *RP*	CoMSIA	22	6	4	$Q^2_{LOO}=0.75$, $R^2_{EXT}=0.93$	N	[135]
		Steric, electrostatic, hydrophobic, and hydrogen bond donor								
6		2,4,6-TBP, HBCDs, HO-PBDEs, PBDEs, TBBPA, TBBPA-DBPE	log *RP*	Classification models	20	9	2	$NER_{training}=0.95$, $S_{n\text{-}training}=1$, $S_{p\text{-}training}=0.87$, $NER_{EXT}=0.89$, $S_{n\text{-}EXT}=0.8$, $S_{p\text{-}EXT}=1$	Y	[138]
		DISPe, displacement value/weighted by Sanderson electronegativity; *nArOH*, number of aromatic hydroxyls								
7		2,4,6-TBP, HBCDs, HO-PBDEs, PBDEs, PFCs, TBBPA, TBBPA-DBPE	log *RP*	Classification models	37	16	3	$NER_{training}=0.84$, $S_{n\text{-}training}=0.91$, $S_{p\text{-}training}=0.79$, $NER_{EXT}=0.81$, $S_{n\text{-}EXT}=0.89$, $S_{p\text{-}EXT}=0.86$	Y	[139]
		nArOH; *F03[Br–Br]*, Frequency of Br–Br at topological distance 3); *HATS6m*, leverage-weighted autocorrelation of lag 6/weighted by mass								

Continued

TABLE 2 Summary of QSARs Models for Human Thyroid Receptors (TR), Human Thyroid Hormone (TH) Transport Proteins and Human TH Sulfotransferase (hTHS)—cont'd

No.	Mechanism of Action	TDCs[a]	End Point[b]	Model[c]	$n_{training}$[d]	n_{EXT}	N	Model Statistics	AD	References
8		2,4,6-TBP, HBCDs, HO-PBDEs, PBDEs, PFCs, TBBPA, TBBPA-DBPE	log *RP*	QSAR	23	9	3	$R^2_{training}=0.89$, $Q^2_{LOO}=0.81$, $RMSE_{training}=0.42$, CCC index=0.95, $Q^2_{EXT}=0.93$, $RMSE_{EXT}=0.34$	Y	[139]
		$\log RP=-4.69\times10^{-1}-3.19\ R5u+7.51\times10^{-1}\ F07[C\text{–}O]+1.58\ nArOH$								
		R5u, R autocorrelation of lag 5/unweighted; *F07*[*C–O*], Frequency of C–O at topological distance 7								
9		2,4-DBP, 2,4,6-TBP, HBCDs, HO-PBDEs, HO-PCBs, HO-PCDD/Fs, PBDEs, PBP, PCBs, PFCs,TBBPA, TBBPA-DBPE, TCBPA, TriBBPA	log *RP*	QSAR	38	9	3	$R^2_{training}=0.86$, $Q^2_{CUM}=0.84$, $RMSE_{training}=0.51$, $R^2_{EXT}=0.95$, $Q^2_{EXT}=0.93$, $RMSE_{EXT}=0.32$	Y	[136]
		$\log RP=-4.37\times10^{-1}-1.26\times10^{-1}\ pK_a-3.16\ qO^-_{adj}+1.84\times10^{-2}\log D$								
		pK_a, acid dissociation constant; qO^-_{adj}, the adjusted most negative net atomic charge on oxygen atom; log *D*, logarithm of the *n*-octanol/water distribution coefficient								

10	hTBG	HO-PBDEs	log *K*	–	14		2	$R^2_{training}=0.74$	N	[83]
		$\log K=-1.6\times10^{-1} \log K_{OW}^2+2.30 \log K_{OW}$								
11	hTHS	HO-PBDEs	IC_{50}	QSAR	13	–	3	$R^2_{training}=0.64$	N	[71]
		IC_{50} (nanomolar)$=2.30\times10^2\ pK_a+9.78\times10\ n_{Br}+3.20\times10^2\ (para\text{-}OH)-1.77\times10^3$								
		para-OH, the OH substitution pattern								

[a] *2,4-DBP, 2,4-dibromophenol; 2,4,6-TBP, 2,4,6-tribromophenol; HBCDs, hexabromocyclododecane; HO-PBDEs, hydroxylated polybrominated diphenyl ethers; HO-PCBs, hydroxylated polychlorinated biphenyls; HO-PCDD/Fs, hydroxylated polychlorinated dibenzo-p-dioxins and dibenzofurans; PBDEs, polybrominated diphenyl ethers; PBP, pentabromophenol; PCBs, polychlorinated biphenyls; PFCs, poly- and perfluorinated organic compounds; TBBPA, 3,3′,5,5′-tetrabromobisphenol A; TBBPA-DBPE, tetrabromobisphenol A bis (2,3-dibromopropyl ether); TCBPA, 3,3′,5,5′-tetrachlorobisphenol A; TriBBPA, 3,3′,5-tribromobisphenol A.*

[b] REC_{20} *is the concentration inducing 20% of the maximum effect;* RP *is the relative competing potency of a chemical with tetraiodothyronine* (T_4) *binding to hTTR,* RP *is defined by Eqn (1) in the text;* K *is the binding constants of a chemical with hTTR/hTBG;* IC_{50} *is the half-maximal inhibitory concentration.*

[c] *CoMSIA is the comparative molecular similarity indices analysis.*

[d] N *is the number of predictor variables;* Q^2_{CUM} *is the total variation of the dependent variables that can be predicted by all the extracted PLS components;* Q^2_{LOO} *is the leave-one-out cross-validated* Q^2; Q^2_{EXT} *is the external explained variance;* $n_{training}$ *and* n_{EXT} *are the number of compounds,* $R^2_{training}$ *and* R^2_{EXT} *are the squared correlation coefficient between observed and fitted values,* $RMSE_{training}$ *and* $RMSE_{EXT}$ *are the root mean square error,* $NER_{training}$ *and* NER_{EXT} *are non error rate,* $S_{n\text{-}training}$ *and* $S_{n\text{-}EXT}$ *are sensitivity, and* $S_{p\text{-}training}$ *and* $S_{p\text{-}EXT}$ *are specificity in training and validation sets, respectively.*

3. QSAR MODELS FOR hTRβ AND hTTR

3.1 QSARs on Hormone Activity of HO-PBDEs on hTRβ

3.1.1 Binding of HO-PBDEs with hTRβ and Molecular Structural Parameters

An insight on the interactions between HO-PBDEs and hTRβ is a prerequisite for developing mechanism-based QSAR models. Here, molecular docking was employed to identify the key ligand–receptor interactions between HO-PBDEs and hTRβ [56]. The docking analysis showed there are hydrogen bonding, π–π interactions, and hydrophobic interactions between HO-PBDEs and hTRβ. Two types of hydrogen bonds were observed: those formed between the hydroxyl oxygen of HO-PBDEs and the hydrogen of Arg 282 and Ile 276, and those between the hydroxyl hydrogen of HO-PBDEs and the carbonyl oxygen of Leu 341. The π–π interactions were formed between the phenyl of HO-PBDEs and Phe 272/Phe 455. All the compounds have hydrophobic interactions with hTRβ. The docking analysis provided insights into the interactions between the ligands and the receptor, which facilitated the selection of appropriate molecular parameters to characterize the interactions in the QSAR modeling. Figure 6 illustrates the interactions of HO-PBDEs with hTRβ.

The TH activities of HO-PBDEs were hypothesized to be dependent on the partition of the compounds between water and the biophase, and the interactions between the ligands and hTRβ. Thus, 12 theoretical parameters were selected and computed to characterize the processes: the logarithm of *n*-octanol/water partition coefficient (log K_{OW}), average molecular polarizability (α), molecular volume (V), dipole moment (μ), energy of the highest

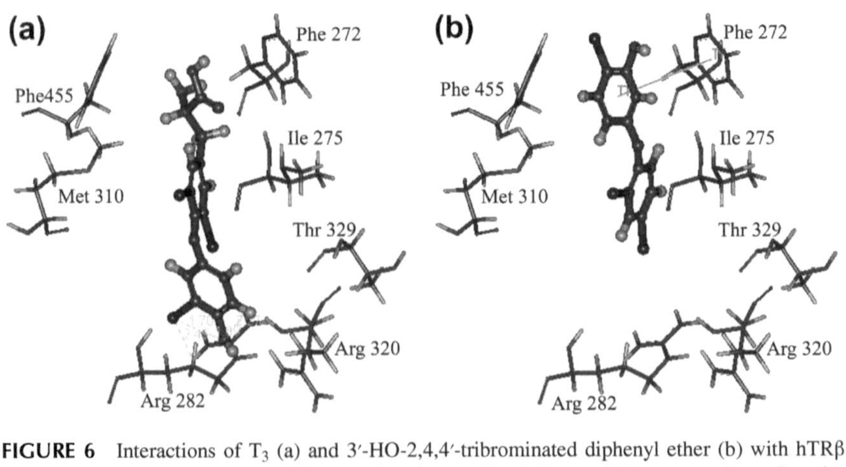

FIGURE 6 Interactions of T_3 (a) and 3′-HO-2,4,4′-tribrominated diphenyl ether (b) with hTRβ (PDB ID: 1NAX). The green (light gray in print versions) line and the orange (gray in print versions) line represent hydrogen bonds and π–π interactions, respectively.

occupied molecular orbital (E_{HOMO}) and the lowest unoccupied molecular orbital (E_{LUMO}), formal charge on hydroxyl hydrogen atoms (qO$\underline{H}$), formal charge on hydroxyl oxygen atoms ($q\underline{O}$H), formal charge on ether oxygen atoms (qO), electrophilicity index (ω), harmonic oscillator model of aromaticity index (I_A), and the number of bromine atoms (n_{Br}). log K_{OW} was purposely selected to describe the partition process. The parameters V, n_{Br}, α, and μ also partly describe the partition because many of these parameters correlate with log K_{OW}. The parameters E_{HOMO}, E_{LUMO}, qO$\underline{H}$, $q\underline{O}$H, qO, ω, and I_A were purposely selected to describe the intermolecular electrostatic interactions (including hydrogen bonding) between the ligands and hTRβ. The quantum chemical parameters E_{HOMO}, E_{LUMO}, qO$\underline{H}$, $q\underline{O}$H, and qO were proved successful in many QSAR studies for characterizing intermolecular electrostatic interactions [140]. ω measures the ability of a compound to soak up electrons. The relative binding affinity of some estrogen derivatives correlated strongly with ω [141]. The aromaticity of compounds (I_A) may influence their non-covalent interactions with the receptor, and I_A has been used to characterize halogenated biphenyls [142]. log K_{OW} was calculated by EPI Suit 4.0^{TM} (http://www.epa.gov/oppt/exposure/pubs/episuitedl.htm). I_A was calculated by DRAGON software [143]. V (defined as the volume inside a contour of 0.001 electrons/bohr^3 density) and the quantum chemical descriptors were calculated by Gaussian 03 program package [144].

3.1.2 QSARs Model for hTRβ

Here the end point modeled by QSARs is the concentration inducing 20% of the maximum effect (REC_{20}) value, which was calculated from the fitted dose–response curve. The dose–response curve was determined by the recombinant two-hybrid yeast assay that was detailed by Li et al. [145]. Partial least squares (PLS) regression was performed in developing the QSAR model because PLS can analyze data with strongly collinear, noisy, and numerous predictor variables. The optimal QSAR model was

$$-\log REC_{20} = 5.73\times10 + 8.01\times10^{-1}\, n_{Br} + 9.62\times10^{-1}\, \log K_{OW} - 4.95\times10\, I_A$$
$$+ 2.84\, E_{LUMO} - 1.66\, \omega + 3.26\times10^{-2}\, \mu^2 \qquad (2)$$

$$n_{training} = 14,\ A = 3,\ Q^2_{CUM} = 0.87,\ R^2_{training} = 0.91,\ RMSE_{training} = 0.42,\ p < 0.0001,$$

$$n_{EXT} = 4,\ Q^2_{EXT} = 0.50,\ RMSE_{EXT} = 0.73$$

where $n_{training}$ and n_{EXT} are the number of compounds in training and validation sets, respectively, A is the number of PLS components, Q^2_{CUM} is the total variation of the dependent variables that can be predicted by all the extracted PLS components, $R^2_{training}$ is the squared correlation coefficient between observed and fitted values in training set, Q^2_{EXT} is the external explained variance, and $RMSE_{training}$ and $RMSE_{EXT}$ are the root mean square errors in training and validation sets, respectively, p is the significance level. These statistical parameters indicated that the model has high goodness-of-fit, robustness, and predictive ability.

The fitting and validation result of the QSAR model is showed in Figure 7. The AD of the model was defined by Williams plot, as shown in Figure 8. The model selected six predictor variables. As can be concluded from the six predictor variables, hydrogen bonding and electrostatic interactions between HO-PBDEs and hTRβ are important factors governing TH activities. The HO-PBDEs with higher ability to accept electrons tend to have weak hydrogen bonding with TRβ and lower TH activities.

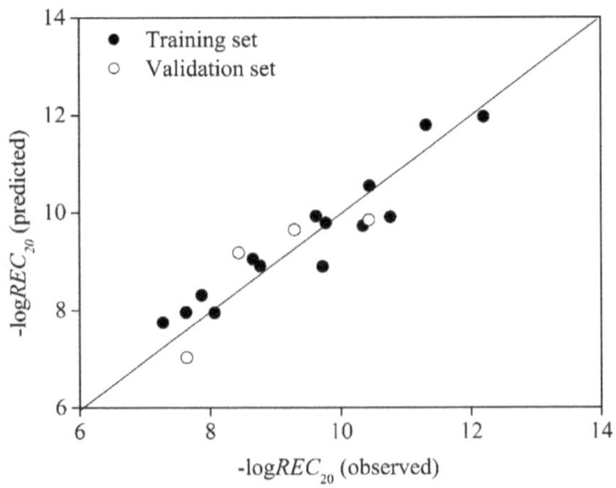

FIGURE 7 Plot of predicted versus observed –log REC_{20} values for both training and validation sets.

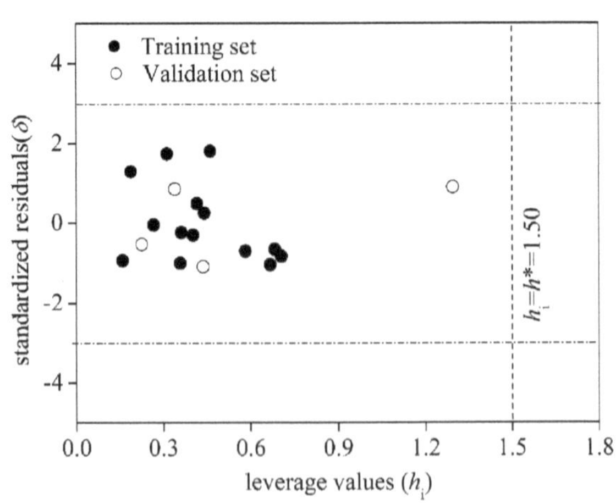

FIGURE 8 Williams plot of the developed QSAR model (the warning leverage $h^* = 1.50$).

3.2 QSARs for Substituted Phenolic Compounds Binding with hTTR

3.2.1 Binding of Phenolic Compounds with hTTR and Molecular Descriptors

It has been known that hydroxyl groups, aromatic rings, and halogens in TDCs are critical structural alerts that can affect the binding potency between TDCs and TH transport proteins. For almost all of the high potency hTTR disruptors, their molecular structures contain ionizable functional groups, for example, –OH, –COOH, and –SO_3H. It is known that compounds with the ionizable groups can ionize depending on their pK_a values and pH of the environment. For ionizable compounds, their toxicity and MOA differ greatly with respect to the ionic or neutral forms. Similarly, for disrupting hTTR transport process, neutral and anionic forms of ionizable TDCs may have different binding potency with hTTR.

Recently, molecular docking with quantum mechanics/molecular mechanics (QM/MM) optimizations has been adopted to probe the binding mechanism of phenolic compounds with hTTR [136]. The binding patterns of ionizable ligands in hTTR crystal structures have also been analyzed. The results indicated that the anionic phenolic compounds bind stronger than the corresponding neutral forms (Figure 9). Electrostatic interactions between the –O^- groups and the –NH_3^+ group of Lys 15 in hTTR are the dominant forces for most of

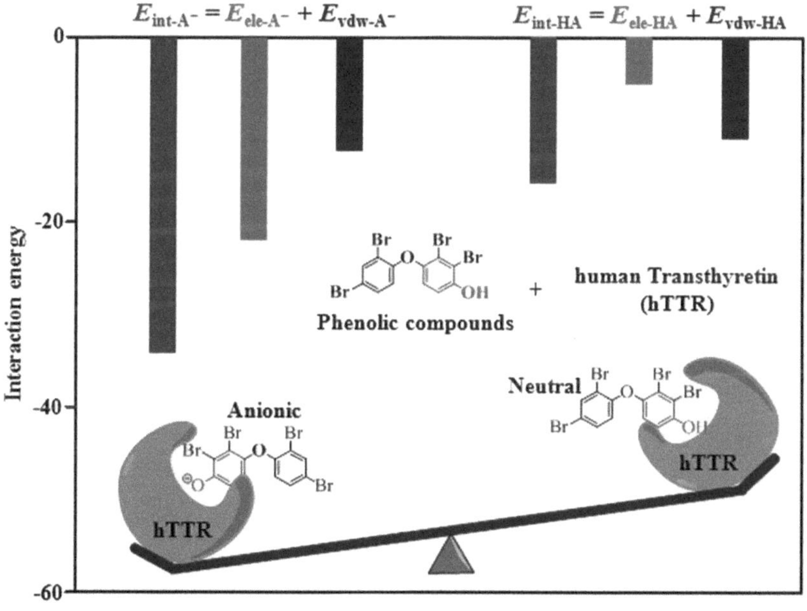

FIGURE 9 Anionic phenolic compounds bind stronger with hTTR than their neutral forms.

the anionic phenolic compounds. Due to the dominant and orientational electrostatic interactions, the $-O^-$ groups point toward the entry port of the binding site of hTTR. As the $-COO^-$ group in T_4 also point toward the entry port of hTTR binding site, it can be inferred that the ionized $-O^-$ group in TDCs plays a similar role to the $-COO^-$ group in T_4. The results suggest that the effects of ionization should not be neglected when developing in silico models for screening of potential hTTR disruptors.

The docking also indicated that the aromatic rings of TDCs can form cation–π interactions with the $-NH_3^+$ group of Lys 15 in hTTR. The strength of the cation–π interactions can be qualitatively characterized by the electrostatic potential surfaces of the relevant aromatics; that is, the more negative the electrostatic potential surface on an aromatic ring, the stronger the cation–π interaction. By analyzing the surface distribution of electrostatic potential, it was found that the electrostatic potential for the anionic forms is more negative than that of the corresponding neutral forms. Thus, the anionic forms of the phenolic compounds also have enhanced cation–π interactions with hTTR. In addition to the electrostatic and cation–π interactions, some compounds also have hydrogen bonds with hTTR and all the compounds have hydrophobic interactions with hTTR. Figure 10 illustrates the interactions of PBDE derivatives with hTTR.

The binding pattern analysis indicates that the interactions between phenolic TDCs and hTTR are relevant with the following factors: (1) ionization, (2) the electrostatic interactions, and (3) hydrophobic interactions. Thus, 16 molecular

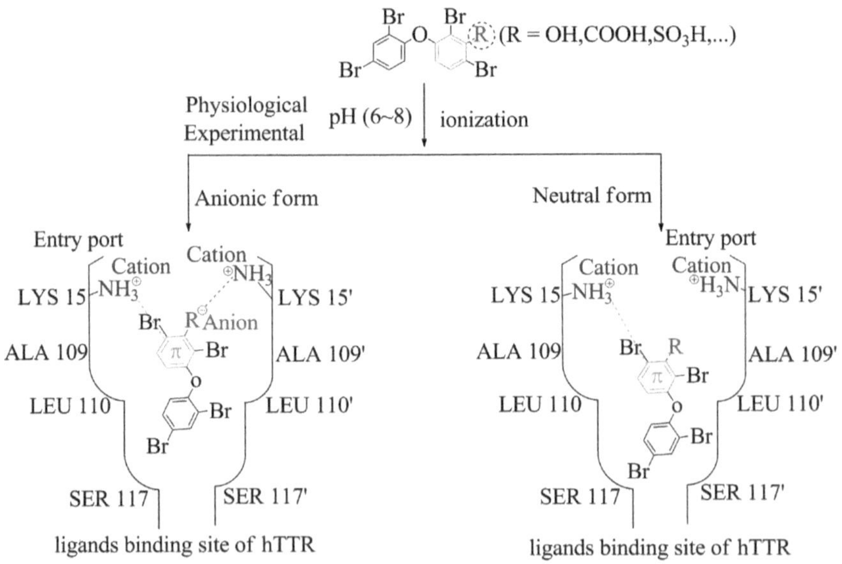

FIGURE 10 Interactions of PBDE derivatives with hTTR (the red (gray in print versions) line and the orange (light gray in print versions) line represent hydrogen bonds and cation–π interactions, respectively).

structural descriptors were selected to develop QSAR models. The logarithm of the *n*-octanol/water distribution coefficient (log *D*), molecular weight (*Mw*), and the number of halogen atoms (n_X) were adopted to characterize hydrophobic interactions. Five quantum chemical descriptors E_{HOMO}, E_{LUMO}, the most positive net atomic charge on a hydrogen atom (qH^+), the most negative net atomic charge on a carbon atom (qC^-), and on an oxygen atom (qO^-) were selected to characterize hydrogen bonding, electron donor–acceptor interactions, and electrostatic interactions. I_A was used to describe π interactions. The nonbonded interaction energy (E_{int}) between phenolic TDCs and hTTR was also included as a descriptor variable. log *D* was calculated from log K_{OW} and pK_a by the following expression:

$$\log D = \log K_{OW} - \log(1 + 10^{pH - pK_a}) \quad (3)$$

E_{int} is the sum of electrostatic (E_{ele}) and van der Waals interaction energy (E_{vdw}):

$$E_{int} = E_{ele} + E_{vdw} \quad (4)$$

To consider the contribution of both the anionic and the neutral forms, adjusted values for five descriptors (E_{int}, E_{HOMO}, E_{LUMO}, qC^-, and qO^-) were calculated by the following equation:

$$X_{adj} = \delta_{HA} \cdot X_{HA} + \delta_{A^-} \cdot X_{A^-} \quad (5)$$

$$\delta_{HA} = \frac{[HA]}{[HA]+[A^-]} = \frac{1}{1+10^{pH-pK_a}}, \quad \delta_{A^-} = \frac{[A^-]}{[HA]+[A^-]} = \frac{10^{pH-pK_a}}{1+10^{pH-pK_a}} \quad (6)$$

where X_{HA} and X_{A^-} are the descriptors for the neutral and ionized forms, respectively; δ_{HA} and δ_{A^-} are the fractions of the neutral and ionized species at a given pH condition, respectively.

The experimental and SPARC [146] calculated pK_a values were employed in the QSAR model development. log K_{OW} was calculated by EPI Suit 4.1™ (http://www.epa.gov/oppt/exposure/pubs/episuitedl.htm). I_A was calculated by DRAGON 6.0 [147]. E_{int} was calculated by Discovery Studio 2.5.5 (Accelrys Software Inc). All the quantum chemical descriptors were calculated by Gaussian 09 program package [148].

3.2.2 QSARs Model for Phenolic Compounds Binding with hTTR

The end point modeled by QSARs is log *RP* defined by Eqn (1). PLS regression was performed in developing the QSAR model. The optimal QSAR model was:

$$\log RP = -4.37 \times 10^{-1} - 1.26 \times 10^{-1} pK_a - 3.16\, qO^-_{adj} + 1.84 \times 10^{-2} \log D \quad (7)$$

$n_{training} = 38$, $A = 2$, $R^2_{training} = 0.86$, $Q^2_{CUM} = 0.84$, $RMSE_{training} = 0.51$, $p < 1.0 \times 10^{-5}$

$n_{EXT} = 9$, $Q^2_{EXT} = 0.93$, $R^2_{EXT} = 0.95$ and $RMSE_{EXT} = 0.32$

where R^2_{EXT} is the squared correlation coefficient between observed and fitted values in validation set.

The statistical parameters indicate that the QSAR model has high goodness-of-fit, robustness, and predictive ability. The model selected three predictor variables, pK_a, qO^-_{adj}, and log D. pK_a is the most important parameter in explaining log RP. pK_a has a negative coefficient in the model, which encodes the information that the anionic forms of the phenolic compounds bind stronger than the neutral forms. qO^-_{adj} is the second important predictor variable for explaining log RP. According to the QSAR model, a more negative value of qO^-_{adj} leads to a higher log RP value, as a compound with a more negative value of qO^-_{adj} has stronger electrostatic interactions between the –O⁻/–OH group and hTTR. log D is also a descriptor for proteinophilic property, and it may characterize hydrophobic interactions between the phenolic TDCs and hTTR in the model [136].

4. CONCLUSIONS

PBDE derivatives may interact with many molecular components of HPT axis and influence the function of the peripheral tissues. The disrupting effects of PBDE derivatives on those targets can be characterized by experiments and in silico models. However, only dozens of PBDE derivatives have been tested for their potential THD effects. Further high-throughput in vitro screening assays and in silico prediction models to identify TDCs and quantify their potency are needed.

Computational toxicology is the application of cheminformatics, bioinformatics, and computer models to help assess chemical hazards and risks to human health and the environment. QSARs are typical in silico models, which correlate and predict the physicochemical, environmental behavioral, and toxicological properties from molecular structural descriptors of environmental chemicals. As cases, two mechanism-based QSAR models for HO-PBDEs and substituted phenolic compounds exerting disrupting effects to TRβ and hTTR have been described.

ACKNOWLEDGMENTS

The study was supported by the High-tech Research and Development Program of China (2012AA06A301) and the National Natural Science Foundation of China (21107136, 21325729).

REFERENCES

[1] R.T. Zoeller, S.W. Tan, R.W. Tyl, Crit. Rev. Toxicol. 37 (2007) 11–53.
[2] S. Konig, V. Moura Neto, Cell. Mol. Neurobiol. 22 (2002) 517–544.
[3] C.M. Dayan, V. Panicker, Nat. Rev. Endocrinol. 5 (2009) 211–218.
[4] S.Y. Cheng, J.L. Leonard, P.J. Davis, Endocr. Rev. 31 (2010) 139–170.
[5] J. Bernal, Nat. Clin. Pract. Endocrinol. Metab. 3 (2007) 249–259.
[6] S. Horn, H. Heuer, Mol. Cell. Endocrinol. 315 (2010) 19–26.

[7] A. Warner, J. Mittag, J. Mol. Endocrinol. 49 (2012) R29–R35.
[8] M.O. Ribeiro, Thyroid 18 (2008) 197–203.
[9] F. Vargas, J.M. Moreno, I. Rodríguez-Gómez, R. Wangensteen, A. Osuna, M. Alvarez-Guerra, J. García-Estañ, Eur. J. Endocrinol. 154 (2006) 197–212.
[10] S. Danzi, I. Klein, Med. Clin. North Am. 96 (2012) 257–268.
[11] G.E. Krassas, K. Poppe, D. Glinoer, Endocr. Rev. 31 (2010) 702–755.
[12] M.J. Lopez-Espinosa, D. Mondal, B. Armstrong, M.S. Bloom, T. Fletcher, Environ. Health Perspect. 120 (2012) 1036–1041.
[13] A. Pascual, A. Aranda, Biochim. Biophys. Acta 1830 (2013) 3908–3916.
[14] M. Song, Y.J. Kim, M.K. Song, H.S. Choi, Y.K. Park, J.C. Ryu, Environ. Sci. Technol. 45 (2011) 7906–7914.
[15] G.A. Brent, J. Clin. Invest. 122 (2012) 3035–3043.
[16] J. Patel, K. Landers, H. Li, R.H. Mortimer, K. Richard, J. Endocrinol. 209 (2011) 1–8.
[17] I. Donangelo, G.D. Braunstein, Am. Fam. Physician 83 (2011) 933–938.
[18] D.R. Mann, T.M. Plant, Brain Res. 1364 (2010) 175–185.
[19] K.M. Crofton, Int. J. Androl. 31 (2008) 209–222.
[20] World Health Organization (WHO)/United Nations Environment Programme (UNEP), in: Å. Bergman, J.J. Heindel, S. Jobling, K.A. Kidd, R.T. Zoeller (Eds.), State of the Science of Endocrine Disrupting Chemicals 2012, 2013. Available: http://www.who.int/ceh/publications/endocrine/en/.
[21] V.J. DeCarlo, Ann. N.Y. Acad. Sci. 320 (1979) 678–681.
[22] O. Andersson, G. Blomkvist, Chemosphere 10 (1981) 1051–1060.
[23] A. Sjödin, D.G. Patterson Jr., A. Bergman, Environ. Int. 29 (2003) 829–839.
[24] R.A. Hites, Environ. Sci. Technol. 38 (2004) 945–956.
[25] C.A. de Wit, Chemosphere 46 (2002) 583–624.
[26] J. Doucet, B. Tague, D.L. Arnold, G.M. Cooke, S. Hayward, C.G. Goodyer, Environ. Health Perspect. 117 (2009) 605–610.
[27] A. Besis, C. Samara, Environ. Pollut. 169 (2012) 217–229.
[28] C.E. Chen, J.W. Chen, H.X. Zhao, Q. Xie, Z.Q. Yin, L.K. Ge, Environ. Int. 36 (2010) 163–167.
[29] M.A. Abdallah, S. Harrad, Environ. Int. 63 (2014) 130–136.
[30] A. Malmvärn, G. Marsh, L. Kautsky, M. Athanasiadou, A. Bergman, L. Asplund, Environ. Sci. Technol. 39 (2005) 2990–2997.
[31] D. Ueno, C. Darling, M. Alaee, G. Pacepavicius, C. Teixeira, L. Campbell, R.J. Letcher, A. Bergman, G. Marsh, D. Muir, Environ. Sci. Technol. 42 (2008) 1657–1664.
[32] X. Qiu, R.M. Bigsby, R.A. Hites, Environ. Health Perspect. 117 (2009) 93–98.
[33] P.W. Bradley, Y. Wan, P.D. Jones, S. Wiseman, H. Chang, M.H. Lam, D.T. Long, J.P. Giesy, Environ. Toxicol. Chem. 30 (2011) 1236–1242.
[34] H.S. Wang, Z.J. Chen, K.L. Ho, L.C. Ge, J. Du, M.H. Lam, J.P. Giesy, M.H. Wong, C.K. Wong, Environ. Int. 47 (2012) 66–72.
[35] M. Montaño, A.C. Gutleb, A.J. Murk, Environ. Sci. Technol. 47 (2013) 6071–6081.
[36] J. Sun, J. Liu, Y. Liu, G. Jiang, Chemosphere 92 (2013) 322–328.
[37] S. Wang, S. Zhang, H. Huang, Z. Niu, W. Han, Environ. Pollut. 184 (2014) 405–413.
[38] E.L. Teuten, L. Xu, C.M. Reddy, Science 307 (2005) 917–920.
[39] A. Malmvärn, Y. Zebühr, L. Kautsky, K. Bergman, L. Asplund, Chemosphere 72 (2008) 910–916.
[40] Y. Wan, S. Wiseman, H. Chang, X. Zhang, P.D. Jones, M. Hecker, K. Kannan, S. Tanabe, J. Hu, M.H. Lam, J.P. Giesy, Environ. Sci. Technol. 43 (2009) 7536–7542.

[41] S.B. Wiseman, Y. Wan, H. Chang, X. Zhang, M. Hecker, P.D. Jones, J.P. Giesy, Mar. Pollut. Bull. 63 (2011) 179–188.
[42] H.M. Stapleton, S.M. Kelly, R. Pei, R.J. Letcher, C. Gunsch, Environ. Health Perspect. 117 (2009) 197–202.
[43] X.B. Wang, Y. Wang, J.W. Chen, Y.Q. Ma, J. Zhou, Z.Q. Fu, Environ. Sci. Technol. 46 (2012) 5126–5133.
[44] J.D. Raff, R.A. Hites, J. Phys. Chem. A. 110 (2006) 10783–10792.
[45] Q. Xie, J.W. Chen, J.P. Shao, C.E. Chen, H.X. Zhao, C. Hao, Chemosphere 76 (2009) 1486–1490.
[46] J. Zhou, J.W. Chen, C.H. Liang, Q. Xie, Y.N. Wang, S.Y. Zhang, X.L. Qiao, X.H. Li, Environ. Sci. Technol. 45 (2011) 4839–4845.
[47] C.A. Erratico, S.C. Moffatt, S.M. Bandiera, Toxicol. Sci. 123 (2011) 37–47.
[48] M.A. McKinney, R. Dietz, C. Sonne, S. De Guise, K. Skirnisson, K. Karlsson, E. Steingrímsson, R.J. Letcher, Environ. Toxicol. Chem. 30 (7) (2011) 1506–1514.
[49] T. Hamers, J.H. Kamstra, E. Sonneveld, A.J. Murk, T.J. Visser, M.J. Van Velzen, A. Brouwer, A. Bergman, Mol. Nutr. Food Res. 52 (2008) 284–298.
[50] L.H. Tseng, M.H. Li, S.S. Tsai, C.W. Lee, M.H. Pan, W.J. Yao, P.C. Hsu, Chemosphere 70 (2008) 640–647.
[51] L. Yu, J. Deng, X. Shi, C. Liu, K. Yu, B. Zhou, Aquat. Toxicol. 97 (2010) 226–233.
[52] K. Ibhazehiebo, T. Iwasaki, J. Kimura-Kuroda, W. Miyazaki, N. Shimokawa, N. Koibuchi, Environ. Health Perspect. 119 (2011) 168–175.
[53] Y. Xiong, K. Ibhazehiebo, T. Iwasaki, N. Koibuchi, Neurotoxicology 33 (2012) 753–757.
[54] D.A. Mensching, M. Slater, J.W. Scott, D.C. Ferguson, V.R. Beasley, J. Toxicol. Environ. Health A. 75 (2012) 201–212.
[55] F. Ucán-Marin, A. Arukwe, A.S. Mortensen, G.W. Gabrielsen, R.J. Letcher, Environ. Sci. Technol. 44 (2010) 497–504.
[56] F. Li, Q. Xie, X.H. Li, N. Li, P. Chi, J.W. Chen, Z.J. Wang, C. Hao, Environ. Health Perspect. 118 (2010) 602–606.
[57] Y. Song, N. Wu, H. Tao, Y. Tan, M. Gao, J. Han, H. Shen, K. Liu, J. Lou, Environ. Toxicol. Chem. 31 (2012) 2047–2051.
[58] J. Chevrier, K.G. Harley, A. Bradman, M. Gharbi, A. Sjödin, B. Eskenazi, Environ. Health Perspect. 118 (2010) 1444–1449.
[59] M. Kiciński, M.K. Viaene, E. Den Hond, G. Schoeters, A. Covaci, A.C. Dirtu, V. Nelen, L. Bruckers, K. Croes, I. Sioen, W. Baeyens, N. Van Larebeke, T.S. Nawrot, Environ. Health 11 (2012) 86.
[60] A.R. Zota, J.S. Park, Y. Wang, M. Petreas, R.T. Zoeller, T.J. Woodruff, Environ. Sci. Technol. 45 (2011) 7896–7905.
[61] M. Boas, U. Feldt-Rasmussen, N.E. Skakkebaek, K.M. Main, Eur. J. Endocrinol. 154 (2006) 599–611.
[62] M.D. Miller, K.M. Crofton, D.C. Rice, R.T. Zoeller, Environ. Health Perspect. 117 (2009) 1033–1041.
[63] M.L. Jugan, Y. Levi, J.P. Blondeau, Biochem. Pharmacol. 79 (2010) 939–947.
[64] T.J. Woodruff, J. Steroid. Biochem. Mol. Biol. 127 (2010) 108–117.
[65] M.E. Gilbert, J. Rovet, Z. Chen, N. Koibuchi, Neurotoxicology 33 (4) (2012) 842–852.
[66] A.J. Murk, E. Rijntjes, B.J. Blaauboer, R. Clewell, K.M. Crofton, M.M. Dingemans, J.D. Furlow, R. Kavlock, J. Köhrle, R. Opitz, T. Traas, T.J. Visser, M. Xia, A.C. Gutleb, Toxicol. In Vitro 27 (2013) 1320–1346.
[67] T.E. Stoker, S.C. Laws, K.M. Crofton, J.M. Hedge, J.M. Ferrell, R.L. Cooper, Toxicol. Sci. 78 (2004) 144–155.

[68] J. Yuan, L. Chen, D. Chen, H. Guo, X. Bi, Y. Ju, P. Jiang, J. Shi, Z. Yu, J. Yang, L. Li, Q. Jiang, G. Sheng, J. Fu, T. Wu, X. Chen, Environ. Sci. Technol. 42 (2008) 2195–2200.
[69] W.K. Chan, K.M. Chan, Aquat. Toxicol. 108 (2012) 106–111.
[70] C.M. Butt, D. Wang, H.M. Stapleton, Toxicol. Sci. 124 (2011) 339–347.
[71] C.M. Butt, H.M. Stapleton, Chem. Res. Toxicol. 26 (2013) 1692–1702.
[72] S.N. Kuriyama, A. Wanner, A.A. Fidalgo-Neto, C.E. Talsness, W. Koerner, I. Chahoud, Toxicology 242 (2007) 80–90.
[73] G. Marsh, A. Bergman, L.G. Bladh, M. Gillner, E. Jakobsson, Organohalogen. Comp. 37 (1998) 305–308.
[74] T. Hamers, J.H. Kamstra, E. Sonneveld, A.J. Murk, M.H. Kester, P.L. Andersson, J. Legler, A. Brouwer, Toxicol. Sci. 92 (2006) 157–173.
[75] X.M. Ren, L.H. Guo, Y. Gao, B.T. Zhang, B. Wan, Toxicol. Appl. Pharmacol. 268 (2013) 256–263.
[76] M. Schriks, C.M. Vrabie, A.C. Gutleb, E.J. Faassen, I.M. Rietjens, A.J. Murk, Toxicol. In Vitro. 20 (2006) 490–498.
[77] M. Schriks, J.M. Roessig, A.J. Murk, J.D. Furlow, Environ. Toxicol. Pharmacol. 23 (2007) 302–307.
[78] S. Kitamura, S. Shinohara, E. Iwase, K. Sugihara, N. Uramaru, H. Shigematsu, N. Fujimoto, S. Ohta, J. Health Sci. 54 (2008) 607–614.
[79] H. Kojima, S. Takeuchi, N. Uramaru, K. Sugihara, T. Yoshida, S. Kitamura, Environ. Health Perspect. 117 (2009) 1210–1218.
[80] A. Suvorov, C. Bissonnette, L. Takser, M.F. Langlois, J. Appl. Toxicol. 31 (2011) 179–184.
[81] J. Freitas, P. Cano, C. Craig-Veit, M.L. Goodson, J.D. Furlow, A.J. Murk, Toxicol. In Vitro 25 (2011) 257–266.
[82] K.J. Van Den Berg, Chem. Biol. Interact. 76 (1990) 63–75.
[83] J. Cao, Y. Lin, L.H. Guo, A.Q. Zhang, Y. Wei, Y. Yang, Toxicology 277 (2010) 20–28.
[84] X.M. Ren, L.H. Guo, Environ. Sci. Technol. 46 (2012) 4633–4640.
[85] M. Montaño, E. Cocco, C. Guignard, G. Marsh, L. Hoffmann, A. Bergman, A.C. Gutleb, A.J. Murk, Toxicol. Sci. 130 (2012) 94–105.
[86] G.R. Marchesini, A. Meimaridou, W. Haasnoot, E. Meulenberg, F. Albertus, M. Mizuguchi, M. Takeuchi, H. Irth, A.J. Murk, Toxicol. Appl. Pharmacol. 232 (2008) 150–160.
[87] I.A. Meerts, J.J. van Zanden, E.A. Luijks, I. van Leeuwen-Bol, G. Marsh, E. Jakobsson, A. Bergman, A. Brouwer, Toxicol. Sci. 56 (2000) 95–104.
[88] J. Legler, P.H. Cenijn, T. Malmberg, Å. Bergman, A. Brouwer, Organohalog. Compd. 56 (2002) 53–56.
[89] M. Harju, T. Hamers, J.H. Kamstra, E. Sonneveld, J.P. Boon, M. Tysklind, P.L. Andersson, Environ. Toxicol. Chem. 26 (2007) 816–826.
[90] I. Morgado, T. Hamers, L. Van der Ven, D.M. Power, Chemosphere 69 (2007) 155–163.
[91] F. Ucán-Marín, A. Arukwe, A. Mortensen, G.W. Gabrielsen, G.A. Fox, R.J. Letcher, Toxicol. Sci. 107 (2009) 440–450.
[92] C.E. Willett, P.L. Bishop, K.M. Sullivan, Toxicol. Sci. 123 (2011) 15–25.
[93] T. Hartung, Nature 460 (2009) 208–212.
[94] T. Hartung, C. Rovida, Nature 460 (2009) 1080–1081.
[95] F.S. Collins, G.M. Gray, J.R. Bucher, Science 319 (2008) 906–907.
[96] Committee on toxicity testing and assessment of environmental agents, in: Toxicity Testing in the 21st Century: A Vision and a Strategy, The National Academies Press, Washington, 2007.
[97] M. Hecker, H. Hollert, Environ. Sci. Eur. 23 (2011) 15.

[98] Organisation for Economic Co-operation and Development, Guidance document on standardised test guidelines for evaluating chemicals for endocrine disruption, in: Technical Report for OECD Environment, Health and Safety Publications Series on Testing and Assessmen No. 150, Paris, 2012.

[99] C. Rudén, S.O. Hansson, Environ. Health Perspect. 118 (2010) 6–10.

[100] I.F. Purchase, P.A. Botham, L.H. Bruner, O.P. Flint, J.M. Frazier, W.S. Stokes, Toxicol. Sci. 43 (1998) 86–101.

[101] S. Pfuhler, D. Kirkland, P. Kasper, M. Hayashi, P. Vanparys, P. Carmichael, S. Dertinger, D. Eastmond, A. Elhajouji, C. Krul, A. Rothfuss, G. Schoening, A. Smith, G. Speit, C. Thomas, J. van Benthem, R. Corvi, Mutat. Res. 680 (2009) 31–42.

[102] J.W. Chen, X.H. Li, H.Y. Yu, Y.N. Wang, X.L. Qiao, Sci. China Ser. B-Chem. 51 (2008) 593–606.

[103] M.L. Scippo, C. Argiris, C. Van De Weerdt, M. Muller, P. Willemsen, J. Martial, G. Maghuin-Rogister, Anal. Bioanal. Chem. 378 (2004) 664–669.

[104] R.H. Schreurs, E. Sonneveld, P.T. van der Saag, B. van der Burg, W. Seinen, Toxicol. Lett. 156 (2005) 261–275.

[105] S. Takeuchi, M. Iida, S. Kobayashi, K. Jin, T. Matsuda, H. Kojima, Toxicology 210 (2005) 223–233.

[106] R.T. Zoeller, S.W. Tan, Crit. Rev. Toxicol. 37 (2007) 195–210.

[107] Organisation for Economic Co-operation and Development, Detailed review paper on the state of the science on Novel in vitro and in vivo screening and testing methods and endpoints for evaluating endocrine disruptors, in: Technical Report for OECD Environment, Health and Safety Publications Series on Testing and Assessmen No. 178, Paris, 2012.

[108] A. Kortenkamp, O. Martin, M. Faust, R. Evans, R. McKinlay, F. Orton, E. Rosivatz, State of the Art Assessment of Endocrine Disruptors, Final Report, 2011. Available: http://ec.europa.eu/environment/chemicals/endocrine/pdf/sota_edc_final_report.pdf.

[109] R. Kavlock, D. Dix, J. Toxicol. Environ. Health B. Crit. Rev. 13 (2010) 197–217.

[110] P.J. Bushnell, R.J. Kavlock, K.M. Crofton, B. Weiss, D.C. Rice, Neurotoxicol. Teratol. 32 (2010) 313–328.

[111] D. Krewski, M.E. Andersen, E. Mantus, L. Zeise, Risk Anal. 29 (2009) 474–479.

[112] R.J. Kavlock, C.P. Austin, R.R. Tice, Risk Anal. 29 (2009) 485–487.

[113] I. Rusyn, G.P. Daston, Environ. Health Perspect. 118 (2010) 1047–1050.

[114] E. van Vliet, ALTEX 28 (2011) 17–44.

[115] P. Matthiessen, I. Johnson, Environ. Pollut. 146 (2007) 9–18.

[116] J.R. Rabinowitz, M.R. Goldsmith, S.B. Little, M.A. Pasquinelli, Environ. Health Perspect. 116 (2008) 573–577.

[117] M.T. Cronin, J.C. Dearden, J.D. Walker, A.P. Worth, Environ. Toxicol. Chem. 22 (2003) 1829–1843.

[118] A. Worth, A. Bassan, E. Fabjan, A. Gallegos-Saliner, T. Netzeva, G. Patlewicz, M. Pavan, I. Tsakovska, The Use of Computational Methods in the Grouping and Assessment of Chemicals Preliminary Investigations, Technical Report for the Office for Official Publications of the European Communities, Luxembourg, 2007.

[119] M.T. Cronin, A.P. Worth, QSAR Comb. Sci. 27 (2008) 91–100.

[120] E. Lo Piparo, A. Worth, Review of QSAR Models and Software Tools for Predicting Developmental and Reproductive Toxicity, Technical Report for the Publications Office of the European Union, Luxembourg, 2010.

[121] P. Castello, A. Worth, Information Sources and Databases on Endocrine Active Substances, Technical report for the Office for Official Publications of the European Union, Luxembourg, 2011.

[122] OECD QSAR Toolbox, http://www.qsartoolbox.org/ (accessed Sep. 28, 2013).
[123] A.E. Soffers, M.G. Boersma, W.H. Vaes, J. Vervoort, B. Tyrakowska, J.L. Hermens, I.M. Rietjens, Toxicol. In Vitro 15 (2001) 539–551.
[124] J.S. Jaworska, M. Comber, C. Auer, C.J. Van Leeuwen, Environ. Health Perspect. 111 (2003) 1358–1360.
[125] M.T. Cronin, J.D. Walker, J.S. Jaworska, M.H. Comber, C.D. Watts, A.P. Worth, Environ. Health Perspect. 111 (2003) 1376–1390.
[126] M.T. Cronin, J.S. Jaworska, J.D. Walker, M.H. Comber, C.D. Watts, A.P. Worth, Environ. Health Perspect. 111 (2003) 1391–1401.
[127] Organisation for Economic Co-operation and Development (OECD), Guidance document on the validation of (quantitative) structure activity relationships [(Q)SAR] models, in: Technical Report for OECD Environment, Health and Safety Publications Series on Testing and Assessmen No. 69, Paris, 2007.
[128] T.I. Netzeva, A. Worth, T. Aldenberg, R. Benigni, M.T. Cronin, P. Gramatica, J.S. Jaworska, S. Kahn, G. Klopman, C.A. Marchant, G. Myatt, N. Nikolova-Jeliazkova, G.Y. Patlewicz, R. Perkins, D. Roberts, T. Schultz, D.W. Stanton, J.J. van de Sandt, W. Tong, G. Veith, C. Yang, Altern. Lab. Anim. 33 (2005) 155–173.
[129] J. Jaworska, N. Nikolova-Jeliazkova, T. Aldenberg, Altern. Lab. Anim. 33 (2005) 445–459.
[130] G. Schüürmann, R.U. Ebert, J.W. Chen, B. Wang, R. Kühne, J. Chem. Inf. Model. 48 (2008) 2140–2145.
[131] N. Chirico, P. Gramatica, J. Chem. Inf. Model. 51 (2011) 2320–2335.
[132] N. Chirico, P. Gramatica, J. Chem. Inf. Model. 52 (2012) 2044–2058.
[133] J. Devillers, N. Marchand-Geneste, A. Carpy, J.M. Porcher, SAR QSAR Environ. Res. 17 (2006) 393–412.
[134] X. Li, L. Ye, X. Wang, X. Wang, H. Liu, Y. Zhu, H. Yu, Toxicol. Appl. Pharmacol. 265 (2012) 300–307.
[135] W.H. Yang, S. Shen, L.L. Mu, H.X. Yu, Environ. Toxicol. Chem. 30 (2011) 2431–2439.
[136] X.H. Yang, H.B. Xie, J.W. Chen, X.H. Li, Chem. Res. Toxicol. 26 (2013) 1340–1347.
[137] E. Papa, S. Kovarich, P. Gramatica, Chem. Res. Toxicol. 23 (2010) 946–954.
[138] S. Kovarich, E. Papa, P. Gramatica, J. Hazard Mater. 190 (2011) 106–112.
[139] E. Papa, S. Kovarich, P. Gramatica, SAR QSAR Environ. Res. 24 (2013) 333–349.
[140] L.M. Colosi, Q. Huang, W.J. Weber Jr., J. Am. Chem. Soc. 128 (2006) 4041–4047.
[141] P.K. Chattaraj, U. Sarkar, D.R. Roy, Chem. Rev. 106 (2006) 2065–2091.
[142] M. Alonso, S. Casado, C. Miranda, J.V. Tarazona, J.M. Navas, B. Herradón, Chem. Res. Toxicol. 21 (2008) 643–658.
[143] R. Todeschini, V. Consonni, Handbook of Molecular Descriptors, Wiley-VCH, Weinheim, 2000.
[144] M.J. Frisch, G.W. Trucks, H.B. Schlegel, G.E. Scuseria, M.A. Robb, J.R. Cheeseman, J.A. Montgomery Jr., T. Vreven, K.N. Kudin, J.C. Burant, J.M. Millam, S.S. Iyengar, J. Tomasi, V. Barone, B. Mennucci, M. Cossi, G. Scalmani, N. Rega, G.A. Petersson, H. Nakatsuji, M. Hada, M. Ehara, K. Toyota, R. Fukuda, J. Hasegawa, M. Ishida, T. Nakajima, Y. Honda, O. Kitao, H. Nakai, M. Klene, X. Li, J.E. Knox, H.P. Hratchian, J.B. Cross, V. Bakken, C. Adamo, J. Jaramillo, R. Gomperts, R.E. Stratmann, O. Yazyev, A.J. Austin, R. Cammi, C. Pomelli, J.W. Ochterski, P.Y. Ayala, K. Morokuma, G.A. Voth, P. Salvador, J.J. Dannenberg, V.G. Zakrzewski, S. Dapprich, A.D. Daniels, M.C. Strain, O. Farkas, D.K. Malick, A.D. Rabuck, K. Raghavachari, J.B. Foresman, J.V. Ortiz, Q. Cui, A.G. Baboul, S. Clifford, J. Cioslowski, B.B. Stefanov, G. Liu, A. Liashenko, P. Piskorz, I. Komaromi, R.L. Martin, D.J. Fox, T. Keith, M.A. Al-Laham, C.Y. Peng, A. Nanayakkara, M. Challacombe, P.M.W. Gill, B. Johnson, W. Chen, M.W. Wong, C. Gonzalez, J.A. Pople, Gaussian 03, Revision C.02, Gaussian, Inc., Wallingford, CT, 2004.

[145] J. Li, M. Ma, Z.J. Wang, Environ. Toxicol. Chem. 27 (2008) 159–167.

[146] SPARC. v4.6, A.R.Chem Danielsville, GA. http://ibmlc2.chem.uga.edu/sparc, accessed Mar 19, 2013.

[147] DRAGON, Software for Molecular Descriptor Calculation, Version 6.0, Talete srl, Milano, 2012.

[148] M.J. Frisch, G.W. Trucks, H.B. Schlegel, G.E. Scuseria, M.A. Robb, J.R. Cheeseman, G. Scalmani, V. Barone, B. Mennucci, G.A. Petersson, H. Nakatsuji, M. Caricato, X. Li, H.P. Hratchian, A.F. Izmaylov, J. Bloino, G. Zheng, J.L. Sonnenberg, M. Hada, M. Ehara, K. Toyota, R. Fukuda, J. Hasegawa, M. Ishida, T. Nakajima, Y. Honda, O. Kitao, H. Nakai, T. Vreven, J.A. Montgomery Jr., J.E. Peralta, F. Ogliaro, M. Bearpark, J.J. Heyd, E. Brothers, K.N. Kudin, V.N. Staroverov, R. Kobayashi, J. Normand, K. Raghavachari, A. Rendell, J.C. Burant, S.S. Iyengar, J. Tomasi, M. Cossi, N. Rega, J.M. Millam, M. Klene, J.E. Knox, J.B. Cross, V. Bakken, C. Adamo, J. Jaramillo, R. Gomperts, R.E. Stratmann, O. Yazyev, A.J. Austin, R. Cammi, C. Pomelli, J.W. Ochterski, R.L. Martin, K. Morokuma, V.G. Zakrzewski, G.A. Voth, P. Salvador, J.J. Dannenberg, S. Dapprich, A.D. Daniels, O. Farkas, J.B. Foresman, J.V. Ortiz, J. Cioslowski, D.J. Fox, Gaussian 09, Revision a.02, Gaussian, Inc., Wallingford, CT, 2009.

Chapter 18

The Toxicity of Persistent Organic Pollutants to Aquatic Organisms

S. Michele Harmon
Department of Biology and Geology, University of South Carolina Aiken, Aiken, SC, USA
E-mail: micheleh@usca.edu

Chapter Outline

Comprehensive Analytical Chemistry, Vol. 67. http://dx.doi.org/10.1016/B978-0-444-63299-9.00018-1

1. INTRODUCTION

Persistent organic pollutants (POPs) are of great concern because of several common characteristics. They tend to disperse widely across ecosystems; they accumulate to high concentrations in the tissues of wildlife; and they possess certain sublethal, insidious mechanisms for toxicity that include the potential for endocrine disruption and carcinogenesis. Those POPs that have garnered the most attention fall into several contaminant classes that include polychlorinated biphenyls (PCBs), dioxins, furans, perfluorinated chemicals (PFCs), and organobromide compounds. This chapter will attempt to cover the known information on the aquatic toxicity of these classes of POPs.

To fully understand the data presented in this chapter, it is important to also understand some of the terminology and methodology used by toxicologists. Aquatic toxicity is usually measured in laboratory experiments through either acute (short-term) or chronic (long-term) exposure scenarios. The data collected from acute testing are often reported in the literature as the lethal concentration for 50% of the test organisms (LC_{50}), in other words, the concentration required to kill or significantly impair half of the exposed test organisms. Traditionally, chronic exposure experiments often measure nonlethal effects, such as impaired growth or reproduction. These types of experiments result in data that are often expressed in terms of a no observed effects concentration (NOEC) or a lowest observed effects concentration (LOEC). Theoretically, NOEC and LOEC values indicate "safe" environmental thresholds for the chemical of concern. Much of the data presented in this chapter will be presented as LC_{50}, NOEC, and/or LOEC values.

In more recent years, the development and improvement of cellular and molecular experimental methods have also provided a means for expansion on the traditional measures of sublethal effects. Researchers now have to ability to use biomarkers that allow them to expand beyond the traditional endpoints of growth and fecundity into measures of effects at the organism's cellular level. Single-strand DNA breaks are evaluated through Single Cell Gel Electrophoesis (SCGE; also called comet assays); endocrine disruption is often evaluated through vitellogenin (VTG) assays; neurological effects are evaluated through acetylcholinesterase or carboxylase assays; and overall toxic stress may be evaluated through a number of assays including upregulation of cytochrome P450, heat shock proteins or metallothioneins, as well as the activity levels of 7-ethoxyresorufin-O-deethylase (EROD), catalase (CAT), superoxide dismutase, glutathione transferase, and glutathione peroxidase. A number of genetic techniques also allow researchers to evaluate a chemical's tendency to cause genetic mutation.

The biomarker techniques discussed above are typically applied to fish species, and the most common laboratory-bred research fish species is the zebrafish, *Danio rerio*. In zebrafish, the methodology for laboratory rearing of several universal research lineages, as well as the cellular/molecular research techniques for studying toxic effects to these fish are well developed. Data from zebrafish testing are abundant and are often extrapolated to other species, including mammals, making zebrafish a common model for the effects of chemicals on the developing vertebrate embryo. In other words, this species is often used to evaluate the potential teratogenic effects of pollutants of concern. Other fish, with perhaps more environmental relevance than zebrafish, that are commonly used in toxicology research include fathead minnows (*Pimephales promelas*), bluegill sunfish (*Lepomis macrochirus*), Rainbow trout (*Oncorhynchus mykiss*), and Japanese medaka (*Oryzias latipes*).

Other groups of aquatic organisms discussed in this chapter include amphibians and invertebrates. The most common methods for modeling the developmental effects of toxicants on amphibians use embryos of the African clawed frog, *Xenopus laevis*, in a bioassay commonly called the frog embryo teratogenesis assay (FETAX). As with zebrafish assays, there are universal *X. laevis* research lineages and well-developed methodologies. Various cladoceran species, particularly those of the genus *Daphnia*, serve as the most common model for the effects of contaminants on aquatic invertebrates.

The ability of POPs to accumulate in the tissues of aquatic organisms has also been the subject of vast amounts of research. These contaminants tend to accumulate in lipid-rich cells of aquatic organisms (hence the term lipophilic) and biomagnify with each trophic level. This means that upper-tropic-level organisms in an ecosystem, like fish-eating mammals or birds, will often receive detrimentally toxic doses from their prey, while the prey items themselves seem unaffected. The ability of these compounds to accumulate in aquatic organisms is directly proportional to partition coefficients of the compounds and the overall lipid content of the organism. There have been many toxicokinetic models published which attempt to predict the accumulation of POPs by aquatic organisms based upon the octanol–water partitioning coefficient (K_{ow} value) of the compound. A high K_{ow} value indicates a high degree of lipophilicity, and thus a higher potential for bioaccumulation. Models based on measured concentrations of POP accumulation in organisms go further and take into account the complexities and metabolic capabilities of the organisms themselves, as well as some of the environmental factors that also affect uptake and bioavailability. This chapter will report some of these measured values as they relate to trophic status of the organisms in question. Values reported here will be in the form of either a bioconcentration factor (BCF) or a bioaccumulation factor (BAF). BCF values refer to the ratio of the contaminant concentration in the organism/concentration in the water. BAF vales refer to the ratio of the contaminant concentration in the organism/concentration in its food.

2. PCBs AND POLYCHLORINATED TERPHENYLS

PCBs and polychlorinated terphenyls (PCTs) are synthetically produced industrial chemicals used for a variety of purposes. They have been used worldwide in electrical transformers and capacitors, as heat exchange fluids, as paint additives, in carbonless copy paper, and in plastics. The production of PCBs stopped in 1977 due to environmental and human health concerns, but these highly persistent, toxic, and bioaccumulative compounds still remain in the environment as global contaminants. They form strong bonds with organic particles in the water column, sediments, and soil.

There are 209 individual PCB congeners depending upon the number of chlorine atoms in the molecule. Biphenyl congeners can contain up to eight chlorine atoms on a double aromatic ringed structure. Trademarked commercial mixtures of PCBs use the name "Aroclor" along with a numeric indicator. These commercial mixtures represent combinations of PCBs with varying chlorine concentrations. PCTs include a third aromatic ring to which chlorine atoms can be added, and they were produced as the commercial mixtures Arochlor 5432 and 5460.

The toxicity and persistence of PCB/PCT congeners in the environment depends upon the degree of chlorination as well as the position of the chlorine atoms. Those that are *ortho*-substituted are readily degradable and not as persistent or prevalent in the environment. Those congeners that are coplanar (non-*ortho* substituted) and mono-*ortho*-substituted are quite stable and often found in aged environmental samples [1]. The coplanar congeners are said to be "dioxin-like" in their toxicity, meaning that they bind to aryl hydrocarbon receptor (AhR) receptors on cells and activate xenobiotic metabolism. Furthermore, PCB exposure in mammals has been linked to cardiovascular disorders, endocrine disruption, developmental effects, neurologic effects, and cancers [2]. PCBs and their hydroxylated metabolites have a high degree of structural similarity to the thyroid hormone thyroxine (T_4), and thus interfere with the normal binding of T_4 to receptors or transport proteins involved in the endocrine processes related to the thyroid system [3]. These processes include overall energy metabolism in an organism, as well as growth and development of the neurological and cardiovascular systems in the embryonic and/or juvenile stages.

The presence of PCBs in the aquatic environment is of concern, not only because of their direct toxicity, but also because of their ability to accumulate in the tissues of aquatic organisms and then biomagnify with each trophic level. BCFs and BAFs for PCBs vary depending upon the organism, the degree of conger chlorination, and the water zone where the particular organism lives and feeds. BCF values reported for lower-chlorinated PCB congeners typically range from 5×10^2 to 4×10^4 [4,5]. BCF values for tetra- to hexaPCBs range from 1×10^3 to 3×10^5, while the more highly chlorinated congeners and coplanar PCBs can have BCF values as high as 2×10^6 [4,6]. BCF values for Aroclor mixtures in fish range from 6×10^2 for the bioconcentration of Aroclor 1242 in Atlantic salmon (*Salmo salar*) to 2.74×10^5 for Aroclor 1242 in male fathead minnows (*P. promelas*) [4,7]. Fish species with high lipid contents, such as lake trout (*Salvelinius namaycush*) and coho salmon (*Oncorhynchus kisutch*), demonstrate high BAFs, especially for the higher-chlorinated congeners. BAF values reported for Salmonids ranged from 2.93×10^5 for tri-chlorinated congeners up to 1.3×10^7 for octa-chlorinated PCB congeners [4,8].

2.1 Toxicity of PCBs to Invertebrates

Various species of *Daphnia* typically serve as models for the effects of contaminants on freshwater invertebrates, and there has been much research on the effects of PCBs to these commonly-used indicator organisms. Acute bioassays reported in the literature for *Daphnia magna* include a 14-day LC_{50} of 24 µg/L for Aroclor 1254, as well as the following 3-week LC_{50} values: 180 µg/L for Aroclor 1221, 72 µg/L for Aroclor 1232, 67 µg/L for Aroclor 1242, 25 µg/L for Aroclor 1248, 21 µg/L for Aroclor 1254, 36 µg/L for Aroclor 1260, 43 µg/L for Aroclor 1262, and 253 µg/L for Aroclor 1268 [9]. Other effects include altered behavioral response of *D. magna* to biotic threats (predators and cyanobacteria) after exposure to PCBs [10], delayed molting [11], decreases in mean swimming activity and changes in swimming depth [12].

Data reported from acute toxicity tests with other crustacean species exposed to Aroclor mixtures include testing with the amphipod *Gammarus pseudolimnaeus* that has resulted in 96-h LC_{50} values of 73 µg/L and 10 µg/L for Aroclor 1242 [9,13] and 29 µg/L for Aroclor 1248 [9]. Aroclor tests with the freshwater shrimp *Gammarus fasciatus* resulted in 96-h LC_{50} values of 52 µg/L and 2400 µg/L for Aroclors 1248 and 1254, respectively [13]. Bioassays with another freshwater shrimp species, *Palaemonetes kadiakensis*, resulted in a 168-h LC_{50} of 0.3 µg/L. Data from 96-h testing with *Palaemonetes pugio* exposed to Aroclor 1016 range from 12 µg/L of for adults to 150 µg/L for 1-day-old juveniles. When *P. pugio* was exposed to Aroclor 1242, 96-h LC_{50} values ranged from 15 µg/L to 57 µg/L [13]. Testing with early instars of two species of crayfish, *Orconectes nais* and *Procambarus* sp., resulted in respective 168-h LC_{50} values of 30 and 100 µg/L for Aroclors 1242 and 1254 [13]. Acute values for other crustacean species have been reported in the literature and thoroughly reviewed elsewhere [14]. These species include amphipods (*Corophium insidiosum*, *Ampelisca abdita*, *Elasmopus bampo*), common shrimp (*Crangon crangon*), bay shrimp (*Crangon septemspinosa*), mysid shrimp (*Mysidopsis bahia*), fiddler crab (*Uca pugilator*), and white shrimp (*Litopenaeus vannamei*). The 96-h LC_{50} values for these species ranged from 9 µg/L for *C. insidiosum* to 1640 µg/L for *L. vannamei* after exposure to Aroclor 1254 [14]. More recently, testing conducted with the harpacticoid copepod *Tigriopus japonicus*, an invertebrate indicator species for pollutants in Asian waters [15], resulted in a 96-h LC_{50} value of 2.83 mg/L for PCB 126. In sublethal testing, there were no observed effects on growth, development, or reproduction for the first generation (F_0) exposed. During the second generational (F_1) exposure, growth and fecundity were significantly reduced at concentrations greater than 1 µg/L [16].

In oysters and mussels, acute testing with Aroclor 1016 resulted in a 96-h LC_{50} value of 10.2 μg/L for *Crassostrea virginica* [17]. In sublethal invertebrate testing with PCBs, Aroclors 1254 and 1242 have been shown to affect the ability of common mussels (*Mytilus edulis*) to attach [18]. Aroclor 1254 also affects the health and growth of the eastern oyster (*C. virginica*) [19].

2.2 Toxicity of PCBs to Fish

Acutely toxic effects of PCBs on fish have been studied in a number of species including channel catfish (*Ictalurus punctatus*), Atlantic salmon (*S. salar*), brook trout (*Salvelinus fontinalis*), brown trout (*Salmo trutta*), cutthroat trout (*Salmo clarki*), lake trout (*S. namaycush*), rainbow trout (*Salmo gairdneri*), bluegill sunfish (*L. macrochirus*), longnose sucker (*Catostomus catostomus*), yellow perch (*Perca flavescens*) [13], harlequin fish (*Rasbora heteromorpha*) [20], fathead minnow (*P. promelas*) [21], and Cisco (chub) (*Coregonus* sp.) [22]. A complete listing of the data from all of these acute fish bioassays have been summarized and published by the World Health Organization [23]. For these studies, the 96-h LC_{50} values vary from 8 μg/L for *P. promelas* exposed to Aroclor 1254 up to values greater than 100 mg/L for *I. punctatus* exposed to Aroclor 1016. The variation observed in acute bioassay data depends upon fish species, PCB mixture, test conditions, and the chemical makeup of the water used in testing [23]. A number of additional studies have shown that LC_{50} values decrease (i.e., toxicity increases) substantially over time, indicating that the traditional 96-h acute bioassay tends to grossly underestimate the lethality of persistent pollutants such as PCBs [23].

Because PCB exposure has been linked to cardiovascular disorders, endocrine disruption and developmental effects in mammals, many studies have investigated the potential for these effects in fish. Zebrafish, *D. rerio*, have been used extensively for the investigation of PCB-induced developmental effects. A number of researchers have reported that PCB exposure caused increased mortality and decreased hatching success in *D. rerio* during embryo and early life-stage bioassays [24,25]. Other PCB-induced developmental effects in this species include pericardial sac edema, yolk sac edema, growth retardation, inhibition of swim bladder inflation, and overall skeletal malformations [25–28], as well as structural and biochemical changes in the developing central nervous system [29]. Moderate and high levels of PCB exposure (0.25–2 mg/L) have induced abnormalities in developing zebrafish retinas, resulting in visual impairment [30]. PCB 126 exposure has also caused heart malformations in *D. rerio* embryos that appear similar to hypoplastic left heart syndrome [31], a cardiovascular malformation observed in certain human cohorts that also received significant PCB exposure [32]. PCB-induced endocrine disruption has been linked to changes in regulation of zebrafish genes involved in endocrine signaling and growth including regulator genes for steroid and thyroid hormone functions (*cga*, *ncoa3*), insulin signaling and metabolic homeostasis (*pik3r1*, *pfkfb3*, *ptb1*), and p53 activation (*mdm4*) [33,34]. Life-cycle testing with *D. rerio* adults exposed through the dietary route has demonstrated considerable bioaccumulation of PCBs in males and females, as well as maternal transfer to the eggs followed by a significant reduction in fecundity [35].

Further evidence of PCB-induced thyroid system disruption in fish has been reported in studies with sculpin [36], coho salmon (*O. kisutch*) [37,38], rainbow trout (*O. mykiss*) [37,39], European flounder (*Platichthys flesus*) [40], and white suckers (*Catostomus commersoni*) [41].

Early-life-stage effects in fish species other than *D. rerio* have been reported in testing with eggs taken from lake trout (*S. namaycush*), where significant relationships were observed between embryonic mortality (eggs dying between fertilization and hatch) and the concentrations of total PCBs in both the eggs and adults [42]. Significant mortality was also reported after eggs of the common sole (*Solea solea*) were exposed to the dioxin-like polychlorinated biphenyl PCB 126 and Arochlor 1254 [43]. Mortality and developmental effects were reported during early life stage bioassays with shortnose sturgeon (*Acipenser brevirostum*) and Atlantic sturgeon (*Acipenser oxyrinchus*) exposed to PCB 126 [44]. Maternal transfer and early-life-stage toxicity was observed in Japanese medaka (*O. latipes*) [45]. Elevated EROD activity and behavioral abnormalities have been observed in *Fundulus heteroclitus* larvae after embryonic exposure to PCB 126 [46].

2.3 Toxicity of PCBs to Amphibians

Some of the earliest research on amphibian exposure to PCBs involved embryo-larval bioassays with three species of amphibians, the American toad (*Bufo americanus*), Fowler's toad (*Bufo fowleri*), and the Leopard frog (*Rana pipiens*). This early study demonstrated that toxicity of Aroclor mixtures increased with increasing chlorination. *R. pipiens* was the most sensitive of the three species, with an LC_{50} of 1.03 μg/L after exposure to Aroclor 1254. Other data from this study included the following LC_{50} values for Aroclor 1254: 2.02 and 3.74 μg/L for *B. americanus* and *B. fowleri*, respectively. Toxicity decreased for Aroclor 1242 with LC_{50} values of 2.71, 12.09, and 2.13 μg/L for *B. americanus*, *B. fowleri*, and *R. pipiens*. The lowest toxicity, and thus the highest LC_{50} values, was reported for Aroclor 1016 with LC_{50} values of 7.16, 27.72, and 6.19 μg/L for *B. americanus*, *B. fowleri*, and *R. pipiens* [47]. There have been many other studies on the acute toxicity of individual PCBs and Aroclor mixtures, all with similar results. These have been thoroughly reviewed elsewhere [48].

Bioaccumulation and maternal transfer of PCBs in amphibians has been studied in several species including the cave salamander (*Proteus anguinus*) [49], green frogs (*Rana clamitans*) [50], brown mountain frogs (*Rana ornativentris*) [51,52], Japanese brown frogs (*Rana japonica*) [51,52], common frogs (*Rana temporaria*) [53], and *X. laevis* [54]. Several other amphibian studies have been devoted to the bioaccumulation and subsequent metabolism of PCBs during key life-cycle stages of frogs. As with other aquatic organisms, frogs bioaccumulate PCBs in lipid-rich tissues. These stored xenobiotics become mobilized for metabolism during metamorphosis when lipids from the tail are being reabsorbed [55–57] and when hibernation forces the animal to utilize stored lipids as the primary energy source [58]. These processes produce measurable biomarkers of chemical activity as the animal metabolizes these PCBs, and there is concern when this increase in metabolic activity comes at critical reproductive times.

Developmental effects in amphibians are often studied using embryos of the African clawed frog, *X. laevis*, in a bioassay commonly called the FETAX. FETAX testing demonstrated that PCB exposure resulted in edema, ocular malformation, forelimb malformations [59], gut malformation [59], feminization of male gonads [60], and decreased larynges in males [61]. Other researchers have used *X. laevis* to show induction of oxidative stress after PCB exposure [62], as well as alterations in protein expression [63] and specific gene expression signatures [64,65]. Thyroid disruption, in particular, was observed when *X. laevis* exposed to 50 μg/L of Aroclor 1254 exhibited significantly delayed metamorphosis and altered gene expression of three thyroid system genes: transthyretin and types II and III deiodinase [66]. Developmental effects noted in other amphibian species included decreased tadpole survivorship, an increase in edema, decrease in growth, and decrease in swimming performance when embryos and larvae of the green frog (*R. clamitans*) and the northern leopard frog (*R. pipiens*) were exposed to a coplanar PCB (PCB 126) at concentrations of 5–50 μg/L [67]. Embryos and tadpoles of the *R. pipiens* exposed to 0.5 and 50 μg/L of mono-*ortho* and di-*ortho* PCB congeners (PCBs 101 and 70) exhibited skewed sex ratios and an increase in the proportion of intersex gonads that corresponded to increasing PCB tissue concentrations [68].

Adult northern leopard frogs (*R. pipiens*) showed signs of hepatic oxidative stress after exposure to PCB 126 [69]. When using amphibian cell lines for research, VTG-inducing activities of hydroxylated PCBs were observed in primary-cultured hepatocytes of the adult male *X. laevis* [70].

2.4 Toxicity of PCBs to Phytoplankton

Diatoms and algae are affected by PCB exposure in both freshwater [71] and marine environments [72], but this varies by PBC congener (or mixture) and by species, with diatoms and larger algal species being the most sensitive. PCB pollution, therefore, has the ability to shift phytoplankton community dynamics in favor of less sensitive phytoplankton species, thus changing lower trophic structure in polluted aquatic ecosystems [72–76]. Those phytoplankton species which remain still bioaccumulate PCBs and effectively transfer these contaminant to primary consumers such as rotifers [77], zebra mussels (*Dreissena polymorpha*) [78], and Pacific oysters (*Crassostrea gigas*) [79]. Laboratory testing has shown that PCBs can effectively inhibit the growth of phytoplankton *Selenastrum capricornutum* at concentrations of 182 nmol/L [80].

3. DIOXINS AND FURANS

Polychlorinated dibenzo-*p*-dioxins (PCDDs or dioxins) and polychlorinated dibenzofurans (PCDFs or furans) are unintentional byproducts that occur when carbon compounds are combusted in the presence of chlorinated compounds. This occurs during the incineration of municipal and medical wastes and certain industrial processes. Dioxins and furans may also be present as trace contaminants in certain herbicides, wood preservatives, and in PCB mixtures.

Both PCDDs and PCDFs contain many congeners because of variation with the numbers and positions of chlorine atoms, and their configuration directly impacts toxicity. Of the 75 PCDD congeners, seven are considered toxic. Of the 135 PCDF congeners, 10 are considered toxic [81]. Dioxin congeners that have two to eight chlorine atoms are designated as follows: di-chlorinated dioxin, tri-chlorinated dioxin, tetra-chlorinated dioxin (TCDD), penta-chlorinated dioxin, hexa-chlorinated dioxin, hepta-chlorinated dioxin, and octa-chlorinated dioxin (OCDD). Dioxins are then further named according to the number and the positions of the chlorine atoms. Of the 75 dioxin congeners, 2,3,7,8-TCDD is typically considered the most toxic to mammals; therefore, it commonly serves as a prototype for all toxic dioxin compounds [82]. Furans follow similar nomenclature patters and consist of a total of 135 PCDF congeners. As with dioxins, the congeners with chlorine atoms at the 2,3,7,8-positions of the parent dibenzofuran molecule are considered especially toxic [83].

Dioxins and furans are developmental and reproductive toxicants in mammals, birds, and fish [84]. They are also known for their status as an activator of the AhR, a ligand-activated transcription factor that regulates xenobiotic-metabolizing enzymes such as cytochrome P450 [85,86]. There are a variety of environmental contaminants that also bind to AhR, and these are now known collectively in the scientific literature as "dioxin-like" compounds.

Dioxins and furans are hydrophobic and lipophilic. When in the aquatic environment, therefore, these contaminants tend to partition from the water column to sediment and suspended particulate organic matter (detritus, humic substances). They remain sorbed and eventually associate with sediments. Because of their high degree of lipophilicity, dioxins tend to bioaccumulate in exposed aquatic biota, particularly those with a high fat content. Lower trophic level organisms such as phytoplankton and zooplankton are exposed through uptake from the water column or from porewater and then pass these contaminants to predators through trophic transfer. BCFs for dioxins tend to increase with the degree of chlorination up to TCDDs, but BCF values decrease with additional chlorination [82,87]. Highly chlorinated congers, such as OCDDs, generally have lower BCF values because they are less bioavailable due to their large size and rapid sorption to sediment.

3.1 Toxicity of Dioxins and Furans to Invertebrates

Data from invertebrate toxicity testing include a 48-h LC_{50} value of 1700 μg/L for *D. magna* [88] exposed to 2,3,7,8-tetrachlorodibenzofuran (TCDF). Testing with freshwater crayfish, *Pacifastacus leniusculus*, exposed to 2,3,7,8-TCDD has shown an LC_{50} of 30–100 μg/kg of body weight, with the induction of cytochrome P450 in treatments as low as 3 μg/kg of body weight [89]. Environmentally relevant exposure concentrations of 2–10 pg/g 2,3,7,8-TCDD had an effect on gametogenesis in the eastern oyster (*C. virginica*) [90].

3.2 Toxicity of Dioxins and Furans to Fish

Developmental effects, including impaired growth, developmental malformations, and mortality, have been recorded as a result of dioxin and furan exposures during early-life-stage bioassays for several fish species. These include Japanese medaka (*O. latipes*) [91–93], northern pike (*Esox lucius*) [93], rainbow trout (*O. mykiss*) [94–97], killifish (*F. heteroclitus*), lake trout (*S. namaycush*) [84], fathead minnow (*P. promelas*) [93], white sucker (*C. commersoni*) [93], lake herring (*Coregonus artedii*) [93], channel catfish (*I. punctatus*) [93], red seabream (*Pagrus major*) [98], shortnose sturgeon (*A. brevirostum*) [44], and Atlantic sturgeon (*A. oxyrinchus*) [44]. Developmental effects have also been recorded in numerous studies with embryos of zebrafish (*D. rerio*) [93,99–102], which serve as a general model for vertebrate developmental biology. Of the numerous observed developmental effects in the fish, the edema, hemorrhage, and craniofacial abnormalities noted in TCDD-exposed trout embryos [103,104] as well as heart deformities observed in exposed zebrafish [102] are indicative of the same types of cardiovascular dysfunction noted in mammals and birds.

Toxicity testing where fish eggs have been exposed to 2,3,7,8-TCDD has resulted in the following LC_{50} values for egg viability/mortality: 539 pg/g of egg for fathead minnow (*P. promelas*), 644 pg/g for channel catfish (*I. punctatus*), 902 pg/g for lake herring (*C. artedii*), 1110 pg/g for Japanese medaka (*O. latipes*), 1890 pg/g for white sucker (*C. commersoni*), 2460 pg/g for northern pike (*E. lucius*), and 2610 pg/g for zebrafish (*D. rerio*) [93]. Additional specific toxicity values for early life stage bioassays with 2,3,7,8-TCDD include an EC_{50} value of 14 ng/L for the prevention of hatching in *O. latipes*, as well as an LC_{50} of 9 ng/L for 3-day posthatch *O. latipes* [91]. Data from these early life stage tests indicate large differences among fish species in their sensitivities to dioxin and dioxin-like compounds. It has been hypothesized that these differences are due to differences in AhR dynamics among different classes of fishes [86].

Embryonic killifish (*F. heteroclitus*) and medaka (*O. latipes*) have been used to investigate oxidative stress and apoptosis induced by 2,3,7,8-TCDD exposure [105–107]. In older fish, elevated hepatic enzymes indicative of oxidative stress have been reported in mirror carp (*Cyprinus carpio*) after exposure to dioxin and furan congeners via intraperitoneal injection [108] and in flounder (*P. flesus*) after oral doses of 2,3,7,8-TCDD [109]. Intraperitoneal injection of 2,3,7,8-TCDD also resulted in 80-day post-treatment LD_{50} values of 2–4 μg/kg for Common carp (*C. carpio*); 4–8 μg/kg for black bullhead (*Ameiurus melas*); 11 μg/kg for largemouth bass (*Micropterus solmoides*); and 3 μg/kg for yellow perch (*P. flavescens*) [110]. Dietary exposure to 2,3,7,8-TCDD has resulted in altered gene expression in juvenile rainbow trout (*O. mykiss*) [111], and delayed spawning in adult brook trout (*S. fontinalis*) [112]. Dietary exposure to a low-level PCDD/PCDF mixture (23 ng/kg of feed) caused and significant increases in the EROD activity and AhR gene expression in gilthead seabream (*Sparus aurata*) [113]. In studies with fish cell lines, dioxin exposure resulted in AhR activation in cell lines from red seabream (*P. major*) [114].

Finally, it has also been shown that dioxins and furans induce antiestrogenic effects that reduce fecundity of exposed aquatic species due to inhibition of VTGs, hepatically-derived yolk-protein precursors required for oogenesis [115,116].

3.3 Toxicity of Dioxins and Furans to Amphibians

Amphibians are generally less sensitive to dioxin toxicity than fish; however, they do respond to these toxicants, particularly in early-life-stage testing. For example, leopard frogs (*R. pipiens*) demonstrated a significant increase (10%) in mortality after their eggs were exposed to 3 μm/L of 2,3,7,8-TCDD for 24 h [117].

Amphibian developmental tests using *X. laevis* embryos demonstrated that dioxin exposure resulted in developmental delays [118], reduced growth [119], generalized edema [120], malformations of the gut [121], and liver malformations due to degenerating hepatocytes [120]. Evidence of AhR activation after dioxin exposure has been observed in *X. laevis* larvae and cell lines [122,123], although it is generally accepted that amphibians are less sensitive to AhR activation than other vertebrate species, meaning that greater dioxin concentrations are required to produce these effects.

4. BROMINATED FLAME RETARDANTS

Brominated flame retardants (BFRs) are commercially prepared organobromide compounds that have been widely used in textiles, furniture, and electronic products as a means of reducing their flammability. BFRs are typically hydrocarbon compounds of varying size with different numbers of attached bromine atoms. These compounds are then mixed in varying commercial formulations with widely varying chemical properties. Because of their broad use and persistent nature, BFRs are now considered global contaminants throughout the aquatic environment. The BFRs that have received the most attention as environmental contaminants fall under the following categories:

1. Polybrominated diphenyl ethers (PBDEs)
2. Polybrominated biphenyl (PBB)
3. Hexabromocyclododecane (HBCD or HBCDD)
4. Tetrabromobisphenol A (TBBPA or TBBP-A).

In the aquatic environment, contamination by BFRs does not typically result in acute toxic responses of aquatic organisms. This is due to the low solubility properties of these compounds under normal environmental conditions [124,125]. In other words, concentrations of BFRs in the aquatic compartment are rarely high enough to elicit a lethal response. Because of their hydrophobic and lipophilic nature, these compounds tend to associate with the organic components of the system (particulates and sediment) and then bioconcentrate in biota via food consumption or sediment contact. Environmental problems occur when BFRs subsequently biomagnify in food chains. Additional concerns related to these compounds center around potential sublethal responses related to their endocrine-disrupting capabilities, particularly the disruption of thyroid hormonal systems in both mammalian and aquatic organisms [81,126,127]. These compounds have also been implicated in disruption of neurological function, effects on development and reproduction, as well as overall oxidative stress at the cellular level [128].

4.1 Polybrominated Diphenyl Ethers

This family of compounds includes 209 possible congeners that include numerous isomers for mono-, di-, tri-, tetra-, penta-, hexa-, hepta-, octa-, nona-, and decabromodiphenyl ethers. Broadly used as flame retardants in the past, most of these compounds are no longer manufactured. Commercial PBDE products typically consist of a mixture of congeners that were marketed under different trade names. Tetrabromodiphenyl ether was sold as PBDE 47; pentaBDE mixtures were sold as DE-60F, DE-61, DE-62, and DE-71; octaBDE mixtures were sold as DE-79; and decaBDE were marketed as DE 83R and Saytex 102E. The lower brominated PBDEs, those with one to five bromine atoms, are considered the most toxic because they tend to bioaccumulate more efficiently. Even though production of penta- and octaPBDEs technical mixtures ceased in 2004, there are many products that still contain high levels of PBDEs. These compounds are highly persistent in the environment.

4.1.1 Toxicity of PBDEs to Invertebrates

Acute laboratory toxicity tests with *D. magna* have resulted in 48-h LC_{50} values of 110.7, 7.9, 2.6, and 11.1 μg/L for PBDEs-28, -47, -99, and -100, respectively. Sublethal studies indicated that PBDEs-28 and -47 can delay molting at concentrations in the μg/L range [129]. Other researchers have shown that PBDE-153 impaired reproduction in *D. magna* at 12.5 and 25 μg/L [130], and BDE-47 impaired *D. magna* reproduction at concentrations greater than 14 μg/L [131]. Acute toxicity testing with the meiobenthic harpacticoid copepod *Nitocra spinipes* resulted in a 96-h LC_{50} value of 4.4 mg/L for BDE-47 [132], and juvenile development in *N. spinipes* was sensitive to nominal concentrations of BDE-47 greater than 0.013 mg/L and concentrations of BDE-99 greater than 0.03 mg/L. The authors also observed that particle-adsorbed PBDEs in ingested food were the predominant route of exposure. In a multigenerational test with *N. spinipes* fed a diet of algae laced with PBDE (BDE-47), population structure was altered via a diminished proportion of nauplii and an increased proportion of copepodites. RNA content was significantly reduced in exposed individuals, indicating declined growth and a reduction in overall genetic variability due to toxicant exposure [133]. The marine copepod *Calanus finmarchicus* will accumulate PBDE-99 from food with a reported BAF of 7.6 [134].

4.1.2 Toxicity of PBDEs to Fish

Acute aquatic toxicity to fish from PBDEs is not an environmentally relevant concern, because these compounds are not particularly soluble. Some researchers have even reported an inability to measure LC_{50} concentrations for fish exposed to commercial pentaBDPE mixtures in laboratory experiments because the LC_{50} and subsequent NOEC values were greater than the water solubility of the substance. The greatest, and perhaps most relevant, effects in lab studies occurred when fish were exposed to PBDEs via contaminated food. When lake trout (*S. namaycush*) were exposed to a mixture of 13 environmentally relevant PBDEs through food, a decrease in T_4 plasma levels were observed after 56 days of exposure to 2.5 and 25 ng/g food [135]. Similar thyroid hormone decreases were observed in fathead minnows (*P. promelas*) after dietary exposure to PBDE-47 [136] and BDE-209 [137] [138]. Exposure to PBDE-47 also significantly reduced fecundity in *P. promelas* due to a 50% reduction in mature sperm in sexually mature males [139]. Exposure to a commercial PeBDE mixture (Bromkal 70-5DE) through the ingestion of contaminated food reduced egg production and caused hepatic lipidosis in the three-spined stickleback, *Gasterosteus aculeatus* [140]. Rainbow trout (*O. mykiss*) exposed to BDE-47 via food have also shown impaired physiological mechanisms, such as inhibited EROD activity in the liver [141]. Significant bioaccumulation and maternal transfer of PBDE-47 was observed in marine medaka (*Oryzias melastigma*) following dietary exposure [142]. When larval zebrafish (*D. rerio*) were fed PBDE-47 contaminated food they have shown altered locomotion behavior [143] and decreases in body mass, particularly in males [144].

Embryonic zebrafish (*D. rerio*) have been used in many studies to investigate the potential developmental effects of PBDEs. In a study of the lethality and developmental effects on embryonic zebrafish by six PBDE congeners (BDE 28, BDE 47, BDE 99, BDE 100, BDE 153, BDE 183), a relationship was found between log K_{OW} and LC_{50}. The LC_{50} values for BDE 28, 47, 99, and 100 were reported as 3.6, 4.2, 5.2, and 8.4 mg/L, respectively. The higher brominated congeners (BDEs 153 and 183) did not induce mortality at any concentration tested, meaning that the LC_{50} values were at least greater than 11.7 and 8.4 mg/L, respectively. This trend supported the long-held hypothesis that lower-brominated congeners elicit a greater toxic effect [145]. Exposure of zebrafish embryos to lower-brominated PBDE congeners, such as BDE 28, 47, 49, 99, and 100, has also resulted in developmental toxicity that includes morphological abnormalities [145–147], impaired cardiovascular function and cerebrospinal fluid flow [146,147], and alterations in behavior [145,147,148].

Exposure to DE-71 has resulted in disruption of neuron development [149], as well as biochemical and structural changes in the embryonic eye that may explain some of the abnormal behaviors that have been observed as these embryos develop [150]. Parental exposure of *D. rerio* to DE-71 results in maternal transfer of the contaminant along with its endocrine-disrupting developmental effects [151,152].

Potential reproductive and developmental effects have been explored in several other fish species. Killifish embryos (*F. heteroclitus*) exposed to DE-71 (0.001–100 μg/L) resulted in a hatching delay of up to 4.5 days [153]. Microinjection of pentaBDPE into the eggs of rainbow trout (*O. mykiss*) resulted in changes in liver morphology and slight increases in EROD activity [154]. PBDE exposure produced gill injury, intestinal alterations, and hepatic lesions in juvenile marine flatfish (*Psetta maxima*) exposed to BDE-47 at concentrations of 0.03 and 0.3 μg/L [155]. Researchers working with both embryonic and larval *P. maxima* exposed to BDE-47, have reported LC_{50} values of 27.3 and 14.1 μg/L for embryos and larvae, respectively [156].

4.1.3 Toxicity of PBDEs to Amphibians

The bioaccumulation of PBDEs in amphibians has been studied by a number of researchers both in the field [53,157,158] and in the laboratory [159,160], with some researchers suggesting a possible role of these chemicals in the decline of amphibian populations [53]. Exposure to PBDEs tends to slow or disrupt amphibian metamorphosis [159,161–163], a process controlled by thyroid hormones [164]. Research such as this has prompted development of an amphibian metamorphosis assay by The Organization for Economic Co-operation and Development and the United States Environmental Protection Agency (US EPA) specifically for evaluation of amphibian thyroid effects [165]. There have also been indications of altered sex ratios when tadpoles of the northern leopard frog (*R. pipiens*) were exposed to environmentally relevant concentrations of DE-71 [166].

4.1.4 Toxicity of PBDEs to Phytoplankton

Finally, PBDE congeners inhibit the growth of marine algae *Isochrysis galbana* at concentrations greater than 15 μg/L in the laboratory [167], and growth of the marine diatom *Skeletonema costatum* was inhibited by concentrations of BDE-47 greater than 6.6 μg/L [131].

4.2 Polybrominated Biphenyls

PBBs are a group of manufactured chemicals that were once widely produced as commercial flame retardants. PBBs were produced as three primary homologs: hexabromobiphenyl (hexaBB), octabromobiphenyl (octaBB), and decabromobiphenyl (decaBB) [168]. The two primary commercial formulations were FireMaster BP-6 and FireMaster FF-1, the principal component of which was 2,2′,4,4′,5,5′-hexabromobiphenyl or BB-153 [169]. These compounds are widespread and persistent in the global environment. They are toxic, and they easily bioaccumulate in the tissues of living organisms. These chemicals are no longer produced in most countries, as alternatives are available for all uses. The bioconcentration of PBBs in fish shows variation depending upon the level of bromination. Those PBBs with six or fewer bromine substitutions tend to bioconcentrate in fish [168], but the octabromo- and decabromobiphenyls do not significantly bioconcentrate [168]. This is likely due to the fact that higher brominated compounds are not soluble in water, and therefore, not bioavailable for uptake.

4.3 Hexabromocyclododecane

HBCD ($C_{12}H_{18}Br_6$) is a BFR commonly used in extruded and expanded polystyrene foams for the building and construction industry. It is also present in some other consumer goods such as upholstered furniture, automobile interior textiles, and car cushions. Alpha (α-HBCD), beta (β-HBCD), and gamma (γ-HBCD) are the three main diastereoisomers used in commercial mixtures. These, therefore, are of the greatest environmental relevance. As with other BFRs, HBCD has low water solubility and thus tends to partition to the organic phases of the aquatic environment. It will bioaccumulate in aquatic organisms and biomagnify into upper trophic levels [170]. The toxicological concerns surrounding HBCD in mammals include its effects on thryoid function, neurological function, development and reproduction, as well as overall oxidative stress at the cellular level [128].

Acute aquatic toxicity is not an environmentally relevant concern for this compound because of its low solubility (2–3 μg/L at 25 °C) in natural environments [128]. Chronic exposure to HBCD, however, has resulted in altered thryoid function in Atlantic salmon smolts (*S. salar* L.) [171], rainbow trout (*O. mykiss*) [172,173], and in the amphibian model, *X. laevis* [174]. Developmental toxicity has been observed in *D. rerio* embryos exposed to all three diastereoisomers, with toxicity rankings of γ-HBCD > β-HBCD > α-HBCD [175]. Developmental delays and abnormalities have also been reported for the marine sea urchin (*Psammechinus miliaris*) [176] after exposure to HBCD in an early-life-stage-exposure test.

Accumulation and deposition of HBCD diastereoisomers has been explored in juvenile rainbow trout (*O. mykiss*) [177,178]. All three diastereoisomers have been shown to bioaccumulate in aquatic species in a heavily contaminated environment [179]. In other studies, dietary exposure resulted in elevated markers for oxidative stress in *O. mykiss* and in eelpout (*Zoarces viviparous*) [180]. Waterborne HBCD exposure resulted in elevated oxidative stress biomarkers in zebrafish embryos [181,182], the Chinese rare minnow (*Gobiocypris rarus*) [183], and the benthic marine clam *Macomoa balthica* (L.) [184]. Oxidative stress was also noted in carp (*Carassius auratus*) after intraperitoneal injection [185].

4.4 Tetrabromobisphenol A

TBBPA became another widely-used flame retardant after PBDEs met with prohibition in Europe and the United States. Its primary use is its application to printed circuit boards; therefore, the manufacture and waste disposal of electronic components has contributed much TBBPA contamination to aquatic environments, particularly in Asia. This compound accumulates in aquatic biota and has shown endocrine disrupting effects in other animals [125].

Acute effects of TBBPA are not of particular concern in the aquatic environment because of its low solubility. There has, however, been a series of acute laboratory toxicity tests that were conducted in support of a 1995 risk assessment published by the World Health Organization [186]. The acute toxicity values included in this report are as follows: 48-h LC_{50} for *D. magna* of 0.96 mg/L, 96-h LC_{50} of 1.2 mg/L for 10-day-old mysid shrimp (*M. bahia*), 96-h LC_{50} of 0.51 mg/L for bluegill sunfish (*L. macrochirus*), 96-h LC_{50} of 0.40 mg/L for rainbow trout (*S. gairdneri*), LC_{50} of 0.54 mg/L for fathead minnows (*P. promelas*).

More recently, a number of acute toxicity tests were conducted in support of water quality criteria determination for TBBPA in China. These tests used native organisms with the following results: 96-h LC_{50} values of 3.13, 1.25, 1.22, 24.6, 7.85, 0.86, and 1.20 mg/L for *C. auratus*, *Rana limnocharis*, *Macrobrachium nipponense*, *Chironomus plumosus*, *Limnodrilus hoffmeisteri*, *Pseudorasbora parva*, and *Scenedesmus quadricanda,* respectively. There was also a reported 48-h LC_{50} for native Chinese *D. magna* of 1.34 mg/L [187].

Sublethal effects have also been investigated in several other amphibians and aquatic organisms. Thyroid disruption has been noted in the Pacific tree frog (*Pseudacris regilla*) [188] and European flounder (*P. flesus*) [189]. Elevated biomarkers of oxidative stress have been noted in *C. auratus* after chronic TBBPA expsosure [185,190]. Dietary exposure to TBBPA has also resulted in elevated markers for oxidative stress in rainbow trout (*O. mykiss*) and in eelpout (*Z. viviparous*) [180]. Developmental delays and abnormalities have been reported for the marine sea urchin (*P. miliaris*) [176] after exposure to TBBPA in an early-life-stage-exposure test.

TBBPA was also shown to inhibit the growth and cell viability of the freshwater microalgae *Pseudokirchneriella subcapitata* and *Nitzschia palea* after 72 h of exposure to 9.2, 12.9, and 16.5 μM concentrations [191]. It also affected the growth of *Chlorella pyrenoidosa* with a reported 96-h EC_{50} value of 45.5 μg/L [26].

4.5 Polybrominated Dibenzo-*p*-Dioxins and Polybrominated Dibenzofurans

Polybrominated dibenzo-*p*-dioxins (PBDDs) and polybrominated dibenzofurans (PBDFs) are similar in structure to chlorinated PCDDs and PCDFs, and they come in a variety of congeners depending upon the number and positions of bromine atoms. In general, environmental concentrations of PBDDs and PBDFs are lower than those of PCDD and PCDFs, but they are still of concern because of the heavy use of BFRs in commercial products [192]. PBDDS and PBDFs can be produced as a by-product of the combustion and disposal of these BFRs, particularly those found in electronic waste. As the use of BFRs increases, so does the potential for widespread PBDD and PBDF contamination. Despite this, data on the toxicity of these brominated compounds in aquatic systems is limited.

PBDDs and PBDFs are toxic to the early life stages of fish, with 2,3,7,8-tetrabromodibenzo-p-dioxin (TBDD) being the most toxic congener. The indications of toxicity in fish embryos were identical to those produced by chlorinated dioxins like TCDD, including yolk sac edema, pericardial edema, multifocal hemorrhages, reduced growth, and craniofacial malformations [97]. Bioaccumulation and maternal transfer of PBDDs and PBDFs were observed zebrafish (*D. rerio*) after dietary exposure [193,194]. These exposures also induced AhR-remediated response systems in the fish, leading to reduced spawning success, altered ovarian morphology, and reduced hepatic VTG gene expression, thus indicating that sublethal exposures to brominated dioxins may impair reproductive physiology in fish [193].

5. PERFLUORINATED CHEMICALS

Another category of POPs of environmental concern includes the PFCs. These anthropogenic compounds are anionic, fluorine-containing compounds that are widely used as surfactants and as processing additives during fluoropolymer production. In some instances, PFCs are produced intentionally, while at other times, they are unintended degradation products of related anthropogenic chemicals. The current intentional use of PFCs is widespread and includes: electric and electronic parts, fire fighting foam, photo imaging, hydraulic fluids and textiles. They are most widely known for their use in stain- and water-resistant chemical treatments for textiles. Numerous studies have shown that PFCs are toxic, persistent, bioaccumulative, biomagnifying, and present in both freshwater and marine environments around the world [195–199]. The PFCs that have received the most attention are perfluorooctane sulfonic acid (PFOS) and perfluorooctanoate (PFOA), with PFOS generally showing greater aquatic toxicity than PFOA [200,201]. PFOS also exhibits a higher tendency to bind to organic matter and bioaccumulate than PFOA due to its longer perfluoroalkyl chain length [202], and it has been shown to accumulate to levels of human health concern in fish tissue.

5.1 Perfluorooctane Sulfonic Acid

PFOS was added to Annex B of the Stockholm Convention on POPs in May 2009. While there are US drinking water limits for the protection of human health (0.2 μg/L for PFOS and 0.4 μg/L for PFOA) [203,204], there are no regulatory limits at this time for the protection of aquatic life. At least one conservative model-based estimate of ecosystem toxicity has placed aquatic toxicity thresholds for PFOS in the range of 0.61–6.66 μg/L [205] using data from a number of toxicity studies. Researchers in China used US EPA methods to recommend a criterion maximum concentration of 3.78 mg/L for PFOS and 45.54 mg/L for PFOA [206]. As with most POPs, these compounds have a great affinity for aquatic sediment and biota.

5.1.1 Toxicity of PFOS to Invertebrates

Acute effects of PFOS to cladoceran species have been demonstrated in numerous studies, with 48-h LC_{50} values of 17.95 mg/L for *Moina macrocopa* [200]; 134 mg/L for *Daphnia pulicaria* [207]; and values for *D. magna* that include 37.4 mg/L [200], 67.2 mg/L [207], 61 mg/L [208], 63 mg/L [201]. Sublethal effects for *M. macrocopa* included impaired reproduction at exposures of 0.31 mg/L for 7 days [200]. Concentrations of PFOS ranging from 1 to 10 mg/L decreased activity levels in aquatic larvae of the damsel fly (*Enallagma cyathigerum*) [209] and also resulted in reduced hatching success, slowed larval development, greater larval mortality, and a decrease in successful metamorphosis of this species [210]. Mortality data for the marine species, *M. bahia*, indicated a 48-h LC_{50} value of 57 mg/L and an NOEC value of 0.25 mg/L based on a 35-day reproduction test [208].

5.1.2 Toxicity of PFOS to Fish

Acute toxicity values for PFOS exposure to freshwater fish for 96 h has been reported at 9.1 mg/L for fathead minnow (*P. promelas*), 7.8 mg/L for bluegill sunfish (*L. macrochirus*) [208], 22 mg/L for rainbow trout (*O. mykiss*) [208], and 2.5 mg/L for trout parr [211].

In a lethality test with adult zebrafish (*D. rerio*) exposed to PFOS, Sharpe et al. [211] reported a 96-h LC_{50} value of 22.2 mg/L. Sublethal studies with zebrafish embryos have shown that PFOS will affect growth, alter sex ratios, and inhibit male gonad development at concentrations of 250 μg/L [212]. PFOS exposure has also resulted in delayed hatching and gross developmental malformations in embryos exposed to concentrations as low as 1 mg/L [213,214]. Maternal exposure results in offspring deformity and mortality [212,215], and up to 10% of the adult body burden was transferred to developing embryos when fish underwent their reproductive cycles in the presence of PFOS [211]. Exposure to 0.5 mg/L resulted in impaired antioxidant defense mechanisms that subsequently increased the toxicity of cadmium to zebrafish embryos [216]. PFOS exposure also enhanced production of reactive oxygen species in zebrafish embryos, followed by subsequent induction of antioxidative enzymes [217].

Tests with the swordtail (*Xiphophorus helleri*) indicated endocrine disruption at PFOS exposures of 2.5 mg/L, as well as reduced growth and histopathological changes in liver [218]. PFOS has also been reported to increase DNA single-strand breaks in cells of the common carp, *C. carpio* [219]. Exposure of *C. carpio* for 14 days at concentrations of 0.1, 0.5 and 1 mg/L PFOS resulted in altered expression of genes involved in energy metabolism, reproduction, and stress response [220]. Parental exposure of Japanese medaka (*O. latipes*) to PFOS at concentrations of 1 mg/L resulted in elevated mortality and histopathological changes in offspring [200]. A 21-day EC_{50} based on reproduction in *P. promelas* was reported at 0.23 mg PFOS/L [221].

5.1.3 Toxicity of PFOS to Amphibians

PFOS exposure resulted in a 96-h LC_{50} value of 13.8 mg/L for embryos of the African clawed frog (*X. laevis*) [208], and low doses (0.01 mg/L) of PFOS induced significant disrupting effects in thyroid hormone and sex hormone production for this same species [222]. Tadpoles of the northern leopard frog (*R. pipiens*) exhibited reduced growth and delayed metamorphosis after exposure to 3 mg/L of PFOS, and significant mortality was observed when tadpoles were exposed to 10 mg/L [223].

Multigenerational toxicity toward the growth of the algae *P. subcapitata* was also determined, with EC_{50} values ranging from 43 mg/L [224] to 82 mg/L [207]. NOEC values based on 96-h growth tests for *P. subcapitata* ranged from 5.3 mg/L [207] to 44 mg/L [208]. PFOS was also shown to inhibit algal growth in *Scenedesmus obliquus* in a concentration-dependent manner [225] and in *Chlorella vulgaris* with a reported NOEC value of 8.2 mg/L [207].

5.2 Perfluorooctanoate

Acute tests with PFOA have resulted in 48-h LC_{50} values of 199.5 mg/L for *M. macrocopa* [200] and LC_{50} values ranging from 476.5 mg/L [200] to 181 mg/L [201] for *D. magna*. Sublethal tests with fish have shown that exposure to PFOA caused endocrine disruption in rare minnows (*G. rarus*) at exposure concentrations as low as 3 mg/L [226]. Exposure to PFOA also resulted in significant increases in VTG concentration and CAT activity in *C. carpio* [219]. The parental exposure of Japanese medaka (*O. latipes*) to PFOA at concentrations of 10 mg/L resulted in elevated mortality and histopathological changes in offspring [200].

6. CONCLUSIONS

There has been much research on the fate, transport, and toxicity of POPs in aquatic environments. Most of the data presented here indicate that acute ecological impacts to aquatic biota from exposure to these types of contaminants occur at relatively high concentrations that are not environmentally relevant. The most significant toxicological aspects of these compounds, however, lies in their environmental persistence and their ability to bioaccumulate and biomagnify through aquatic food chains. As a result of several decades of research, many of these pollutants have been banned or heavily controlled in the United States and other Western countries. Much of the knowledge regarding POPs was integrated during the 1990s by the United Nations Environment Programme resulting in an international United Nations treaty, *The Stockholm Convention on POPs*, which was adopted in 2001. Contamination from these compounds, however, is global and persistent due to their very nature. While extensive research has been conducted over the last 30 years, there is still much to learn, especially as new compounds are generated and put into use.

REFERENCES

[1] A. Beyer, M. Biziuk, in: D.M. Whitacre (Ed.), Environmental Fate and Global Distribution of Polychlorinated Biphenyls, vol. 201, Springer, New York, 2009, pp. 137–158.
[2] Agency for Toxic Substances and Disease Registry (ATSDR), Addendum to the Toxicological Profile for Polychlorinated Biphenyls (PCBs), U.S. Department of Health and Human Services, November 2000.
[3] M. Boas, U. Feldt-Rasmussen, N.E. Skakkebaek, K.M. Main, Eur. J. Endocrinol. 154 (2006) 599–611.
[4] Agency for Toxic Substances and Disease Registry (ATSDR), Toxicological Profile for Polychlorinated Biphenyls (PCBs), U.S. Department of Health and Human Services, November 2000.
[5] H.J. Geyer, G.G. Rimkus, I. Scheunert, A. Kaune, K. Schramm, A. Kettrup, M. Zeeman, D.C.G. Muir, L.G. Hansen, D. Mackay, in: O. Hutzinger, B. Beek (Eds.), Bioaccumulation and Occurrence of Endocrine-Disrupting Chemicals (EDCs), Persistent Organic Pollutants (POPs), and other Organic Compounds in Fish and other Organisms Including Humans, Springer Berlin Heidelberg, Berlin, 1999.
[6] L.G. Hansen, The Ortho Side of PCBs: Occurrence and Disposition, Kluwer Academic Publishers, Boston, MA, 1999.
[7] ASTER (Assessment Tools for the Evaluation of Risk), Ecotoxicity Profile, U.S. Environmental Protection Agency, Environmental Research Laboratory, Duluth, MN, 1998.
[8] B.G. Oliver, A.J. Niimi, Environ. Sci. Technol. 22 (1988) 388–397.
[9] A.V. Nebeker, F.A. Puglisi, Trans. Am. Fish. Soc. 103 (1974) 722–728.
[10] P. Bernatowicz, J. Pijanowska, Ecotoxicol. Environ. Saf. 74 (2011) 711–718.
[11] E.M. Zou, M. Fingerman, Ecotoxicol. Environ. Saf. 38 (1997) 281–285.
[12] K. Schmidt, S. Pflugmacher, G.B.O. Staaks, C.E.W. Steinberg, J. Freshwater Ecol. 21 (2006) 109–120.
[13] F.L. Mayer, M.R. Ellersick, Manual of Acute Toxicity: Interpretation and Database for 410 Chemicals and 66 Species of Freshwater Animals, U.S. Department of the Interior, Fish and Wildlife Service, Washington, DC, 1986.
[14] P.C. Fuchsman, T.R. Barber, J.C. Lawton, K.B. Leigh, Environ. Toxicol. Chem. 25 (2006) 2601–2612.
[15] S. Raisuddin, K.W.H. Kwok, K.M.Y. Leung, D. Schlenk, J.S. Lee, Aquat. Toxicol. 83 (2007) 161–173.
[16] F. Guo, L. Wang, W. Wang, Environ. Toxicol. Chem. 31 (2012) 639–645.
[17] P.R. Parrish, D.J. Hansen, J.N. Couch, J.M. Patrick Jr., G.H. Cook, Assoc. Southeastern Biol. Bull. 21 (1974) 74.
[18] D. Roberts, Environ. Pollut. 8 (1975) 241–254.
[19] T.W. Duke, J.I. Lowe, A.J. Wilson, Florida Bull. Environ. Contam. Toxicol. 5 (1970) 171–180.

[20] T.E. Tooby, P.A. Hursey, J.S. Alabaster, Chem. Ind. (1975) 523–526.
[21] A.V. Nebeker, F.A. Puglisi, D.L. Defoe, Trans. Am. Fish. Soc. 103 (1974) 562–568.
[22] D.R.M. Passino, J.M. Kramer, Bull. Environ. Contam. Toxicol. 24 (1980) 527–534.
[23] WHO, Environmental Health Criteria 140, Polychlorinated Biphenyls and Terphenyls (Second Edition), International Programme on Chemical Safety (IPCS), World Health Organization, Geneva, Switzerland, 1993.
[24] P.E. Olsson, L. Westerlund, S.J. Teh, K. Billsson, A.H. Berg, M. Tysklind, J. Nilsson, L.O. Eriksson, D.E. Hinton, AMBIO 28 (1999) 100–106.
[25] T. Sisman, F. Geyikoglu, M. Atamanalp, Toxicol. Ind. Health 23 (2007) 529–536.
[26] G. Na, T. Chen, Y. Zhang, L. Zhang, C-y. Liu, J. Gu, Z. Yao, Asian J. Ecotoxicol. 5 (2010) 375–381.
[27] M.E. Jonsson, A. Kubota, A.R. Timme-Laragy, B. Woodin, J.J. Stegeman, Toxicol. Appl. Pharm. 265 (2012) 166–174.
[28] L. Ju, K. Tang, X.R. Guo, Y. Yang, G.Z. Zhu, Y. Lou, Mol. Med. Rep. 5 (2012) 1227–1231.
[29] J.A. Kreiling, R. Creton, C. Reinisch, J. Toxicol. Environ. Health, Part A 70 (2007) 1005–1013.
[30] Y.P. Wang, Q. Hong, D.N. Qin, C.Z. Kou, C.M. Zhang, M. Guo, X.R. Guo, X. Chi, M.L. Tong, J. Appl. Toxicol. 32 (2012) 186–193.
[31] A.C. Grimes, K.N. Erwin, H.A. Stadt, G.L. Hunter, H.A. Gefroh, H.J. Tsai, M.L. Kirby, Toxicol. Sci. 106 (2008) 193–205.
[32] K.S. Kuehl, C.A. Loffredo, Pediatr. Cardiol. 27 (2006) 25–31.
[33] R. Nourizadeh-Lillabadi, J.L. Lyche, C. Almaas, B. Stavik, S.J. Moe, M. Aleksandersen, V. Berg, K.S. Jakobsen, N.C. Stenseth, J.U. Skare, P. Alestrom, E. Ropstad, J. Toxicol. Environ. Health, Part A 72 (2009) 112–130.
[34] J.L. Lyche, R. Nourizadeh-Lillabadi, C. Almaas, B. Stavik, V. Berg, J.U. Skare, P. Alestrom, E. Ropstad, J. Toxicol. Environ. Health, Part A 73 (2010) 1032–1057.
[35] T. Daouk, T. Larcher, F. Roupsard, L. Lyphout, C. Rigaud, M. Ledevin, V. Loizeau, X. Cousin, Aquat. Toxicol. 105 (2011) 270–278.
[36] N.K. Brar, C. Waggoner, J.A. Reyes, R. Fairey, K.M. Kelley, Aquat. Toxicol. 96 (2010) 203–215.
[37] J.F. Leatherland, R.A. Sonstegard, J. Fish. Res. Board Can. 35 (1978) 1285–1289.
[38] L.C. Folmar, W.W. Dickhoff, W.S. Zaugg, H.O. Hodgins, Aquat. Toxicol. 2 (1982) 291–299.
[39] J.F. Leatherland, R.A. Sonstegard, J. Fish. Dis. 3 (1980) 115–124.
[40] H.T. Besselink, S. vanBeusekom, E. Roex, A.D. Vethaak, J.H. Koeman, A. Brouwer, Environ. Pollut. 92 (1996) 267–274.
[41] K.R. Munkittrick, C.B. Portt, G.J. Vanderkraak, I.R. Smith, D.A. Rokosh, Can. J. Fish. Aquat. Sci. 48 (1991) 1371–1380.
[42] M.J. Mac, T.R. Schwartz, C.C. Edsall, A.M. Frank, J. Great Lakes Res. 19 (1993) 752–765.
[43] E.M. Foekema, M.L. Parron, M.T. Mergia, E.R.M. Carolus, J.H.J. vd Berg, C. Kwadijk, Q. Dao, A.J. Murk, Ecotoxicol. Environ. Saf. 101 (2014) 14–22.
[44] R.C. Chambers, D.D. Davis, E.A. Habeck, N.K. Roy, I. Wirgin, Environ. Toxicol. Chem. 31 (2012) 2324–2337.
[45] K. Nakayama, Y. Oshima, K. Nagafuchi, T. Hano, Y. Shimasaki, K. Nakayamaj, Environ. Toxicol. Chem. 24 (2005) 591–596.
[46] C.M. Couillard, B. Legare, A. Bernier, Z. Dionne, Mar. Environ. Res. 71 (2011) 257–265.
[47] W.J. Birge, J.A. Black, A.G. Westerman, Effects of Polychlorinated Biphenyl Compounds and Proposed PCB-Replacement Products on Embryo-Larval Stages of Fish and Amphibians, U.S. Department of the Interior, Fish and Wildlife Service, Washington, DC, 1978.
[48] D.W. Sparling, Ecotoxicology of Organic Contaminants to Amphibians, CRC Press, Boca Raton, Florida, 2000.

[49] M. Pezdirc, E. Heath, L.B. Mali, B. Bulog, Chemosphere 84 (2011) 987–993.
[50] A.R. Loveridge, C.A. Bishop, J.E. Elliott, C.J. Kennedy, Bull. Environ. Contam. Toxicol. 79 (2007) 315–318.
[51] K. Kadokami, M. Takeishi, M. Kuramoto, Y. Ono, Environ. Toxicol. Chem. 21 (2002) 129–137.
[52] K. Kadokami, M. Takeishi, M. Kuramoto, Y. Ono, Chemosphere 57 (2004) 383–389.
[53] A.F.H. Ter Schure, P. Larsson, J. Merila, K.I. Jonsson, Environ. Sci. Technol. 36 (2002) 5057–5061.
[54] R.B. Zhao, D.Y. Sun, S. Fu, X.F. Wang, R.S. Zhao, J. Environ. Sci.-China 19 (2007) 374–384.
[55] J.L. Leney, K.C. Balkwill, K.G. Drouillard, G.D. Haffner, Environ. Toxicol. Chem. 25 (2006) 1627–1634.
[56] J.L. Leney, K.G. Drouillard, G.D. Haffner, Environ. Sci. Technol. 40 (2006) 1491–1496.
[57] J.L. Leney, K.G. Drouillard, G.D. Haffner, Environ. Toxicol. Chem. 25 (2006) 2971–2980.
[58] R.A. Angell, G.D. Haffner, Environ. Toxicol. Chem. 29 (2010) 700–707.
[59] Z.F. Qin, J.M. Zhou, L. Cong, X.B. Xu, Environ. Toxicol. Chem. 24 (2005) 2573–2578.
[60] Z.F. Qin, J.M. Zhou, S.G. Chu, X.B. Xu, Environ. Health Persp. 111 (2003) 553–556.
[61] Z.F. Qin, X.F. Qin, L. Yang, L. Han-Ting, X.R. Zhao, X.B. Xu, Aquat. Toxicol. 84 (2007) 321–327.
[62] V. Gillardin, F. Silvestre, C. Divoy, J.P. Thome, P. Kestemont, Ecotoxicol. Environ. Saf. 72 (2009) 546–551.
[63] V. Gillardin, F. Silvestre, M. Dieu, E. Delaive, M. Raes, J.P. Thome, P. Kestemont, Mol. Cell. Proteomics 8 (2009) 596–611.
[64] A.M. Jelaso, E. Lehigh-Shirey, A. Predenkiewicz, J. Means, C.F. Ide, Environ. Mol. Mutagen. 40 (2002) 24–35.
[65] A.M. Jelaso, E. Lehigh-Shirey, J. Means, C.F. Ide, Environ. Mol. Mutagen. 42 (2003) 1–10.
[66] E.A.L. Shirey, A.J. Langerveld, D. Mihalko, C.F. Ide, Environ. Res. 102 (2006) 205–214.
[67] M.L. Rosenshield, M.B. Jofre, W.H. Karasov, Environ. Toxicol. Chem. 18 (1999) 2478–2486.
[68] M.B. Jofre, W.H. Karasov, Chemosphere 70 (2008) 1609–1619.
[69] Y.W. Huang, D.J. Hoffman, W.H. Karasov, J. Toxicol. Environ. Health, Part A 70 (2007) 676–681.
[70] K. Nomiyama, Y. Nomura, T. Takahashi, Y. Uchiyama, K. Arizono, R. Shinohara, Chemosphere 78 (2010) 800–806.
[71] A. Sodergren, C. Gelin, Bull. Environ. Contam. Toxicol. 30 (1983) 191–198.
[72] J.L. Mosser, T.C. Teng, C.F. Wurster, N.S. Fisher, Science 175 (1972) 191.
[73] S.A. Moore, R.C. Harriss, Nature 240 (1972) 356.
[74] P. Lundy, C.F. Wurster, R.G. Rowland, Water Res. 18 (1984) 187–194.
[75] D.C. Biggs, R.G. Rowland, H.B. Oconnors, C.D. Powers, C.F. Wurster, Environ. Pollut. 15 (1978) 253–263.
[76] D.C. Biggs, R.G. Rowland, C.F. Wurster, Bull. Environ. Contam. Toxicol. 21 (1979) 196–201.
[77] C. JoaquimJusto, V. Gosselain, J.P. Descy, J.P. Thome, Hydrobiologia 313 (1995) 249–257.
[78] E.J. Willman, J.B. Manchester-Neesvig, C. Agrell, D.E. Armstrong, Environ. Toxicol. Chem. 18 (1999) 1380–1389.
[79] Q. Jo, S.K. Kim, C.S. Lee, P.Y. Lee, J. Environ. Biol. 32 (2011) 787–791.
[80] P. Mayer, B. Halling-Sorensen, D. Sijm, N. Nyholm, Environ. Toxicol. Chem. 17 (1998) 1848–1851.
[81] D. Wikoff, L. Fitzgerald, L. Birnbaum in: A. Schecter (Ed.), Dioxins and Health: Including Other Persistent Organic Pollutants and Endocrine Disruptors, 3rd Edition, John Wiley & Sons, Inc., Hoboken, New Jersey, 2012, pp. 1–36.

[82] Agency for Toxic Substances and Disease Registry (ATSDR), Toxicological Profile for Chlorinated Dibenzo-p-dioxins (CDDs), U.S. Department of Health and Human Services, 1998.
[83] Agency for Toxic Substances and Disease Registry (ATSDR), Toxicological Profile for Chlorinated Dibenzofurans (CDFs), U.S. Department of Health and Human Services, 1994.
[84] R.E. Peterson, H.M. Theobald, G.L. Kimmel, Crit. Rev. Toxicol. 23 (1993) 283–335.
[85] K. Sogawa, Y. FujiiKuriyama, J. Biochem. 122 (1997) 1075–1079.
[86] J.A. Doering, J.P. Giesy, S. Wiseman, M. Hecker, Environ. Sci. Pollut. R. 20 (2013) 1219–1224.
[87] H. Loonen, M. Tonkes, J.R. Parsons, Sci. Total Environ. (S) (1993) 491–498.
[88] G.A. Leblanc, Bull. Environ. Contam. Toxicol. 24 (1980) 684–691.
[89] C.M. Ashley, M.G. Simpson, D.M. Holdich, D.R. Bell, Aquat. Toxicol. 35 (1996) 157–169.
[90] M.L. Wintermyer, K.R. Cooper, Aquat. Toxicol. 81 (2007) 10–26.
[91] J.D. Wisk, K.R. Cooper, Environ. Toxicol. Chem. 9 (1990) 1159–1169.
[92] C.M. Chen, K.R. Cooper, Bull. Environ. Contam. Toxicol. 63 (1999) 423–429.
[93] G.E. Elonen, R.L. Spehar, G.W. Holcombe, R.D. Johnson, J.D. Fernandez, R.J. Erickson, J.E. Tietge, P.M. Cook, Environ. Toxicol. Chem. 17 (1998) 472–483.
[94] M.K. Walker, R.E. Peterson, Aquat. Toxicol. 21 (1991) 219–238.
[95] M.K. Walker, L.C. Hufnagle, M.K. Clayton, R.E. Peterson, Aquat. Toxicol. 22 (1992) 15–37.
[96] E.W. Zabel, M.K. Walker, M.W. Hornung, M.K. Clayton, R.E. Peterson, Toxicol. Appl. Pharm. 134 (1995) 204–213.
[97] M.W. Hornung, E.W. Zabel, R.E. Peterson, Toxicol. Appl. Pharm. 140 (1996) 227–234.
[98] M. Yamauchi, E.-Y. Kim, H. Iwata, Y. Shima, S. Tanabe, Aquat. Toxicol. 80 (2006) 166–179.
[99] T.R. Henry, J.M. Spitsbergen, M.W. Hornung, C.C. Abnet, R.E. Peterson, Toxicol. Appl. Pharm. 142 (1997) 56–68.
[100] A.G. Berends, E.J. Boelhouwers, J.L.G. Thus, J. deGerlache, C.G. deRooij, Chemosphere 35 (1997) 853–865.
[101] D.S. Antkiewicz, C.G. Burns, S.A. Carney, R.E. Peterson, W. Heideman, Toxicol. Sci. 84 (2005) 368–377.
[102] J.S. Marit, L.P. Weber, Aquat. Toxicol. 106 (2012) 131–139.
[103] M.K. Walker, J.M. Spitsbergen, J.R. Olson, R.E. Peterson, Can. J. Fish. Aquat. Sci. 48 (1991) 875–883.
[104] P.D. Guiney, R.M. Smolowitz, R.E. Peterson, J.J. Stegeman, Toxicol. Appl. Pharm. 143 (1997) 256–273.
[105] S.M. Cantrell, L.H. Lutz, D.E. Tillitt, M. Hannink, Toxicol. Appl. Pharm. 141 (1996) 23–34.
[106] S.M. Cantrell, J. Joy-Schlezinger, J.J. Stegeman, D.E. Tillitt, M. Hannink, Toxicol. Appl. Pharm. 148 (1998) 24–34.
[107] B.H. Toomey, S. Bello, M.E. Hahn, S. Cantrell, P. Wright, D.E. Tillitt, R.T. Di Giulio, Aquat. Toxicol. 53 (2001) 127–138.
[108] M.E.J. Van Der Weiden, L.P. De Vries, K. Fase, M. Celander, W. Seinen, M. Van Den Berg, Mar. Environ. Res. 39 (1995) 375.
[109] H.T. Besselink, E. vanSanten, W. Vorstman, A.D. Vethaak, J.H. Koeman, A. Brouwer, Environ. Toxicol. Chem. 16 (1997) 816–823.
[110] J.M. Kleeman, J.R. Olson, R.E. Peterson, Fund. Appl. Toxicol. 10 (1988) 206–213.
[111] Q. Liu, M.L. Rise, J.M. Spitsbergen, T.S. Hori, M. Mieritz, S. Geis, J.E. McGraw, G. Goetz, J. Larson, R.J. Hutz, M.J. Carvan III, Aquat. Toxicol. 140 (2013) 356–368.
[112] J.E. Tietge, R.D. Johnson, K.M. Jensen, P.M. Cook, G.E. Elonen, J.D. Fernandez, G.W. Holcombe, D.B. Lothenbach, J.W. Nichols, Environ. Toxicol. Chem. 17 (1998) 2395–2407.

[113] M. Abalos, E. Abad, A. Estevez, M. Sole, A. Buet, L. Quiros, B. Pina, J. Rivera, Chemosphere 73 (2008) S303–S310.
[114] S.-M. Bak, M. Iida, M. Hirano, H. Iwata, E.-Y. Kim, Environ. Sci. Technol. 47 (2013) 2877–2885.
[115] T.C. King-Heiden, V. Mehta, K.M. Xiong, K.A. Lanham, D.S. Antkiewicz, A. Ganser, W. Heideman, R.E. Peterson, Mol. Cell. Endocrinol. 354 (2012) 121–138.
[116] S.M. Bugel, L.A. White, K.R. Cooper, Aquat. Toxicol. 126 (2013) 1–8.
[117] R.E. Jung, M.K. Walker, Environ. Toxicol. Chem. 16 (1997) 230–240.
[118] N. Dell'Orto, D. Cantelli, C. Urani, Chemosphere 37 (1998) 2809–2821.
[119] A. Collier, L. Orr, J. Morris, J. Blank, Int. J. Environ. Res. Public Health 5 (2008) 368–377.
[120] M.K. Sakamoto, S. Mima, T. Tanimura, J. Environ. Pathol. Toxcol. 14 (1995) 69–82.
[121] M.K. Sakamoto, S. Mima, T. Tanimura, J. Environ. Pathol. Toxcol. 18 (1999) 289–295.
[122] A.L. Zimmermann, E.A. King, S.R. Scogin, W.H. Powell, Mar. Environ. Res. 66 (2008) 30.
[123] L.B. Laub, B.D. Jones, W.H. Powell, Chem. Biol. Interact. 183 (2010) 202–211.
[124] C.A. de Wit, Chemosphere 46 (2002) 583–624.
[125] I. Watanabe, S. Sakai, Environ. Int. 29 (2003) 665–682.
[126] J. Legler, A. Brouwer, Environ. Int. 29 (2003) 879–885.
[127] J. Legler, Chemosphere 73 (2008) 216–222.
[128] C.H. Marvin, G.T. Tomy, J.M. Armitage, J.A. Arnot, L. McCarty, A. Covaci, V. Palace, Environ. Sci. Technol. 45 (2011) 8613–8623.
[129] R. Davies, E.M. Zou, Ecotoxicology 21 (2012) 1371–1380.
[130] T. Nakari, S. Huhtala, Ecotoxicol. Environ. Safe. 71 (2008) 514–518.
[131] T. Kallqvist, M. Grung, K.-E. Tollefsen, Environ. Toxicol. Chem. 25 (2006) 1657–1662.
[132] M. Breitholtz, L. Wollenberger, Aquat. Toxicol. 64 (2003) 85–96.
[133] J. Gardestrom, E. Gorokhova, M. Gilek, M. Grahn, B.-E. Bengtsson, M. Breitholtz, Aquat. Toxicol. 79 (2006) 41–48.
[134] K. Magnusson, M. Magnusson, P. Ostberg, M. Granberg, P. Tiselius, Mar. Environ. Res. 63 (2007) 67–81.
[135] G.T. Tomy, V.P. Palace, T. Halldorson, E. Braekevelt, R. Danell, K. Wautier, B. Evans, L. Brinkworth, A.T. Fisk, Environ. Sci. Technol. 38 (2004) 1496–1504.
[136] S.C. Lema, J.T. Dickey, I.R. Schultz, P. Swanson, Environ. Health Persp. 116 (2008) 1694–1699.
[137] P.D. Noyes, D.E. Hinton, H.M. Stapleton, Toxicol. Sci. 122 (2011) 265–274.
[138] P.D. Noyes, S.C. Lema, L.J. Macaulay, N.K. Douglas, H.M. Stapleton, Environ. Sci. Technol. 47 (2013) 10012–10021.
[139] E.K. Muirhead, D. Skillman, S.E. Hook, I.R. Schultz, Environ. Sci. Technol. 40 (2006) 523–528.
[140] G. Holm, L. Norrgren, T. Andersson, A. Thuren, Aquat. Toxicol. 27 (1993) 33–50.
[141] U. Tjarnlund, G. Ericson, U. Orn, C. de Wit, L. Balk, Mar. Environ. Res. 46 (1998) 107–112.
[142] J.P. van de Merwe, A.K.Y. Chan, E.N.Y. Lei, M.S. Yau, M.H.W. Lam, R.S.S. Wu, Aquat. Toxicol. 103 (2011) 199–204.
[143] C.-T. Chou, Y.-C. Hsiao, F.-C. Ko, J.-O. Cheng, Y.-M. Cheng, T.-H. Chen, Aquat. Toxicol. 98 (2010) 388–395.
[144] L. Torres, C.E. Orazio, P.H. Peterman, R. Patino, Fish Physiol. Biochem. 39 (2013) 1115–1128.
[145] C.Y. Usenko, E.M. Robinson, S. Usenko, B.W. Brooks, E.D. Bruce, Environ. Toxicol. Chem. 30 (2011) 1865–1872.
[146] S.C. Lema, I.R. Schultz, N.L. Scholz, J.P. Incardona, P. Swanson, Aquat. Toxicol. 82 (2007) 296–307.

[147] V. McClain, H.M. Stapleton, F. Tilton, E.P. Gallagher, Comp. Biochem. Phys. C 155 (2012) 253–258.
[148] L.G. Chen, C.J. Huang, C.Y. Hu, K. Yu, L.H. Yang, B.S. Zhou, Environ. Toxicol. Chem. 31 (2012) 2338–2344.
[149] X.J. Chen, C.J. Huang, X.C. Wang, J.F. Chen, C.L. Bai, Y.H. Chen, X.P. Chen, Q.X. Dong, D.R. Yang, Aquat. Toxicol. 120 (2012) 35–44.
[150] L.G. Chen, Y.B. Huang, C.J. Huang, B. Hu, C.Y. Hu, B.S. Zhou, Environ. Toxicol. Chem. 32 (2013) 1370–1375.
[151] L. Chen, K. Yu, C. Huang, L. Yu, B. Zhu, P.K.S. Lam, J.C.W. Lam, B. Zhou, Environ. Sci. Technol. 46 (2012) 9727–9734.
[152] L. Yu, J.C.W. Lam, Y. Guo, R.S.S. Wu, P.K.S. Lam, B. Zhou, Environ. Sci. Technol. 45 (2011) 10652–10659.
[153] A.R. Timme-Laragy, E.D. Levin, R.T. Di Giulio, Chemosphere 62 (2006) 1097–1104.
[154] L. Norrgren, T. Andersson, M. Bjork, Aquat. Toxicol. 26 (1993) 307–316.
[155] S. Barja-Fernandez, J.M. Miguez, R. Alvarez-Otero, Ecotoxicol. Environ. Safe. 95 (2013) 60–68.
[156] L. Mhadhbi, J. Fumega, M. Boumaiza, R. Beiras, Environ. Sci. Pollut. Res. 19 (2012) 708–717.
[157] J.-P. Wu, X.-J. Luo, Y. Zhang, S.-J. Chen, B.-X. Mai, Y.-T. Guan, Z.-Y. Yang, Environ. Sci. Technol. 43 (2009) 5212–5217.
[158] P.-Y. Liu, G.-D. Du, Y.-X. Zhao, Y.-S. Mu, A.-Q. Zhang, Z.-F. Qin, X.-Y. Zhang, S.-S. Yan, Y. Li, R.-G. Wei, X.-F. Qin, Y.-J. Yang, Chemosphere 84 (2011) 972–978.
[159] G. Carlsson, P. Kulkarni, P. Larsson, L. Norrgren, Aquat. Toxicol. 84 (2007) 71–79.
[160] T.L. Cary, W.H. Karasov, Environ. Toxicol. Chem. 32 (2013) 1631–1640.
[161] G.C. Balch, L.A. Velez-Espino, C. Sweet, M. Alaee, C.D. Metcalfe, Chemosphere 64 (2006) 328–338.
[162] X. Qin, X. Xia, Z. Yang, S. Yan, Y. Zhao, R. Wei, Y. Li, M. Tian, X. Zhao, Z. Qin, X. Xu, J. Environ. Sci.-China 22 (2010) 744–751.
[163] T.L.C. Coyle, W.H. Karasov, Environ. Toxicol. Chem. 29 (2010) 133–141.
[164] J.M. Wong, Y.B. Shi, J. Biol. Chem. 270 (1995) 18479–18483.
[165] K. Miyata, K. Ose, J. Toxicol. Path. 25 (2012) 1–9.
[166] N.D. Van Schmidt, T.L. Cary, M.E. Ortiz-Santaliestra, W.H. Karasov, Environ. Toxicol. Chem. 31 (2012) 347–354.
[167] L. Mhadhbi, J. Fumega, R. Beiras, Water Air Soil Pollut. 223 (2012) 4007–4016.
[168] Agency for Toxic Substances and Disease Registry (ATSDR), Toxicological profile for polybrominated diphenyl ethers and polybrominated biphenyls, U.S. Department of Health and Human Services, 2004.
[169] L.W. Robertson, B. Chittim, S.H. Safe, M.D. Mullin, C.M. Pochini, J. Agric. Food Chem. 31 (1983) 454–457.
[170] A. Covaci, S. Harrad, M.A.E. Abdallah, N. Ali, R.J. Law, D. Herzke, C.A. de Wit, Environ. Int. 37 (2011) 532–556.
[171] N. Lower, A. Moore, Mar. Freshwat. Behav. Physiol. 40 (2007) 267–284.
[172] V.P. Palace, K. Pleskach, T. Halldorson, R. Danell, K. Wautier, B. Evans, M. Alaee, C. Marvin, G.T. Tomy, Environ. Sci. Technol. 42 (2008) 1967–1972.
[173] V. Palace, B. Park, K. Pleskach, B. Gemmill, G. Tomy, Chemosphere 80 (2010) 165–169.
[174] M. Schriks, E. Zvinavashe, J.D. Furlow, A.J. Murk, Chemosphere 65 (2006) 1904–1908.
[175] M. Du, D. Zhang, C. Yan, X. Zhang, Aquat. Toxicol. 112 (2012) 1–10.
[176] H.M.R. Anselmo, L. Koerting, S. Devito, J.H.J. van den Berg, M. Dubbeldam, C. Kwadijk, A.J. Murk, Ecotoxicol. Environ. Saf. 74 (2011) 2182–2192.

[177] K. Law, V.P. Palace, T. Halldorson, R. Danell, K. Wautier, B. Evans, M. Alaee, C. Marvin, G.T. Tomy, Environ. Toxicol. Chem. 25 (2006) 1757–1761.
[178] M. Haukas, E. Mariussen, A. Ruus, K.E. Tollefsen, Aquat. Toxicol. 95 (2009) 144–151.
[179] J.P. Wu, Y.T. Guan, Y. Zhang, X.J. Luo, H. Zhi, S.J. Chen, B.X. Mai, Environ. Int. 37 (2011) 210–215.
[180] D. Ronisz, E.F. Finne, H. Karlsson, L. Forlin, Aquat. Toxicol. 69 (2004) 229–245.
[181] J. Hu, Y. Liang, M. Chen, X. Wang, Environ. Toxicol. 24 (2009) 334–342.
[182] J. Deng, L. Yu, C. Liu, K. Yu, X. Shi, L.W.Y. Yeung, P.K.S. Lam, R.S.S. Wu, B. Zhou, Aquat. Toxicol. 93 (2009) 29–36.
[183] X. Zhang, F. Yang, X. Zhang, Y. Xu, T. Liao, S. Song, H. Wang, Aquat. Toxicol. 86 (2008) 4–11.
[184] K. Smolarz, A. Berger, Aquat. Toxicol. 95 (2009) 239–247.
[185] M. Feng, R. Qu, C. Wang, L. Wang, Z. Wang, Aquat. Toxicol. 140 (2013) 314–323.
[186] WHO, Environmental Health Criteria 172, Tetrabromobisphenol a and Derivatives. The International Programme on Chemical Safety (IPCS). World Health Organization, Geneva, Switzerland, 1995.
[187] S. Yang, Z. Yan, F. Xu, S. Wang, F. Wu, Environ. Pollut. 169 (2012) 59–63.
[188] N. Veldhoen, A. Boggs, K. Walzak, C.C. Helbing, Aquat. Toxicol. 78 (2006) 292–302.
[189] R.V. Kuiper, R.F. Canton, P.E.G. Leonards, B.M. Jenssen, M. Dubbeldam, P.W. Wester, M. van den Berg, J.G. Vos, A.D. Vethaak, Ecotoxicol. Environ. Saf. 67 (2007) 349–360.
[190] S. Yang, F. Xu, B. Zheng, F. Wu, S. Wang, Aquat. Toxicol. 142 (2013) 248–256.
[191] T. Debenest, A.-N. Petit, F. Gagne, M. Kohli, N. Nien, C. Blaise, Chemosphere 85 (2011) 50–55.
[192] K. Kannan, C. Liao, H. Moon, in: A. Schecter (Ed.), Polybrominated Dibenzo-p-dioxins and Dibenzofurans, John Wiley & Sons, Inc., Hoboken, NJ, 2012.
[193] A.N. Halden, K. Arnoldsson, P. Haglund, A. Mattsson, E. Ulleras, J. Sturve, L. Norrgren, Aquat. Toxicol. 102 (2011) 150–161.
[194] K. Arnoldsson, A.N. Halden, L. Norrgren, P. Haglund, Environ. Toxicol. Chem. 31 (2012) 804–812.
[195] S. Fujii, C. Polprasert, S. Tanaka, N.P.H. Lien, Y. Qiu, J. Water Supply Res. Tech. 56 (2007) 313–326.
[196] J. Cao, B. Xi, Q. Xu, Z. Wang, T. Jiang, L. Jiang, J.- Mao, Huanjing Kexue 32 (2011) 2817–2826.
[197] Y.Y. Pan, Y.L. Shi, J.M. Wang, X.L. Jin, Y.Q. Cai, Bull. Environ. Contam. Toxicol. 87 (2011) 152–157.
[198] T.Y. Wang, Y.L. Lu, C.L. Chen, J.E. Naile, J.S. Khim, J.P. Giesy, Environ. Geochem. Health 34 (2012) 301–311.
[199] S.K. Boontanon, C. Kunacheva, N. Boontanon, N. Musirat, S. Fujii, S. Tanaka, J. Environ. Eng. 139 (2013) 588–593.
[200] K. Ji, Y. Kim, S. Oh, B. Ahn, H. Jo, K. Choi, Environ. Toxicol. Chem. 27 (2008) 2159–2168.
[201] M.-H. Li, Environ. Toxicol. 24 (2009) 95–101.
[202] J.M. Conder, R.A. Hoke, W. De Wolf, M.H. Russell, R.C. Buck, Environ. Sci. Technol. 42 (2008) 995–1003.
[203] US EPA, Provisional Health Advisories for Perfluorooctanaoic Acid (PFOA) and Perfluorooctyl Sulfonate (PFOS), 2009.
[204] US EPA, Perfluorooctanoic Acid (PFOA) and Fluorinated Telomers, 2013. www.epa.gov/oppt/pfoa/.
[205] P. Qi, Y. Wang, J. Mu, J. Wang, Environ. Toxicol. Chem. 30 (2011) 836–842.

[206] S.W. Yang, F.F. Xu, F.C. Wu, S.R. Wang, B.H. Zheng, Sci. Total Environ. 470 (2014) 677–683.
[207] T.M. Boudreau, C.J. Wilson, W.J. Cheong, P.K. Sibley, S.A. Mabury, D.C.G. Muir, K.R. Solomon, Environ. Toxicol. Chem. 22 (2003) 2739–2745.
[208] Organization for Economic Co-operation and Development (OECD), Hazard Assessment of Perfluorooctane Sulfonate (PFOS) and Its Salts, ENV/JM/RD(2002)17/FINAL, France, Paris, 2002.
[209] H. Van Gossum, J. Bots, T. Snijkers, J. Meyer, S. Van Wassenbergh, W. De Coen, L. De Bruyn, Environ. Pollut. 157 (2009) 1332–1336.
[210] J. Bots, L. De Bruyn, T. Snijkers, B. Van den Branden, H. Van Gossum, Environ. Pollut. 158 (2010) 901–905.
[211] R.L. Sharpe, J.P. Benskin, A.H. Laarman, S.L. MacLeod, J.W. Martin, C.S. Wong, G.G. Goss, Environ. Toxicol. Chem. 29 (2010) 1957–1966.
[212] M. Wang, J. Chen, K. Lin, Y. Chen, W. Hu, R.L. Tanguay, C. Huang, Q. Dong, Environ. Toxicol. Chem. 30 (2011) 2073–2080.
[213] X.J. Shi, Y.B. Du, P.K.S. Lam, R.S.S. Wu, B.S. Zhou, Toxicol. Appl. Pharm. 230 (2008) 23–32.
[214] H.H. Huang, C.J. Huang, L.J. Wang, X.W. Ye, C.L. Bai, M.T. Simonich, R.L. Tanguay, Q.X. Dong, Aquat. Toxicol. 98 (2010) 139–147.
[215] Y. Du, X. Shi, C. Liu, K. Yu, B. Zhou, Chemosphere 74 (2009) 723–729.
[216] S. Kim, K. Ji, S. Lee, J. Lee, J. Kim, S. Kim, Y. Kho, K. Choi, Environ. Toxicol. Chem. 30 (2011) 870–877.
[217] X.J. Shi, B.S. Zhou, Toxicol. Sci. 115 (2010) 391–400.
[218] J. Han, Z. Fang, Aquat. Toxicol. 99 (2010) 281–290.
[219] W.-K. Kim, S.-K. Lee, J. Jung, J. Hazard. Mater. 180 (2010) 395–400.
[220] A. Hagenaars, D. Knapen, I.J. Meyer, K. van der Ven, P. Hoff, W. De Coen, Aquat. Toxicol. 88 (2008) 155–163.
[221] G.T. Ankley, D.W. Kuehl, M.D. Kahl, K.M. Jensen, A. Linnum, R.L. Leino, D.A. Villeneuve, Environ. Toxicol. Chem. 24 (2005) 2316–2324.
[222] Q.-P. Liu, J. Qian Li, S.-Z. Guo, H.-H. Shi, Asian J. Ecotoxicol. 3 (2008) 464–472.
[223] G.T. Ankley, D.W. Kuehl, M.D. Kahl, K.M. Jensen, B.C. Butterworth, J.W. Nichols, Environ. Toxicol. Chem. 23 (2004) 2745–2755.
[224] R. Rosal, I. Rodea-Palomares, K. Boltes, F. Fernandez-Pinas, F. Leganes, A. Petre, Chemosphere 81 (2010) 288–293.
[225] W. Liu, S. Chen, X. Quan, Y.-H. Jin, Environ. Toxicol. Chem. 27 (2008) 1597–1604.
[226] Y. Wei, J. Dai, M. Liu, J. Wang, M. Xu, J. Zha, Z. Wang, Environ. Toxicol. Chem. 26 (2007) 2440–2447.

Chapter 19

Health-Based Risk Assessment of PBDEs

Li Chen, Lucy Li and Yunhui Zhang*
Department of Environmental Health, School of Public Health, Fudan University, Shanghai, China
Corresponding author: E-mail: yhzhang@shmu.edu.cn

Chapter Outline

1. INTRODUCTION

Polybrominated diphenyl ethers (PBDEs) are a group of brominated flame retardants (BFRs) commonly used in consumer products such as electronics, textiles, and polyurethane foams [1]. PBDEs have been marketed as three commercial mixtures: pentabromodiphenyl ether (penta-BDE), octabromodiphenyl ether (octa-BDE), and decabromodiphenyl ether (deca-BDE). The global demand for

Comprehensive Analytical Chemistry, Vol. 67. http://dx.doi.org/10.1016/B978-0-444-63299-9.00019-3

PBDEs was estimated to be 67,490 tons in 2001 (11, 6, and 83% were produced as penta-, octa-, and deca-BDE, respectively), of which 49% was used in North America, 37% in Asia, and 12% in Europe [2]. Among the PBDEs used in China, BDE-47, -99, -209 are the most common congeners. The multiple exposure sources and wide use of PBDE have aroused great concern on PBDE-exposure on people's health, especially the exposures to BDE-47, -99, -209, in China.

PBDEs have been shown to have toxicological effects, including developmental neurotoxicity and interference with thyroid hormone homeostasis, in laboratory animals [3–5]. Due to these effects, penta-and octa-BDE have been banned in the European Union since 2004, and were classified as persistent organic pollutants (POPs) due to their bioaccumulation, toxicity, and persistence, and have been banned under a Stockholm convention [6]. Since these compounds are continued to be found in the environment, reservoir sources may continue to contaminate air, water, soil, sediment and biotic organisms [7,8], all of which are believed to contribute substantially to the human intake of these compounds. Hites [9] found that the PBDE concentrations in people had increased exponentially by a factor of approximately 100 during the last 30 years.

2. ASSESSMENT OF EXPOSURE TO POLYBROMINATED DIPHENYL ETHERS VIA INHALATION AND DIET IN CHINA

Although there is a wealth of data on PBDE distributions in environmental media, including surface water, air, soil and sediment, and in the food chain, until now no nationwide systematic monitoring and tracking system for PBDE distribution has been established and no nationwide assessment of PBDE-exposure has been conducted in China. Given that dozens of papers on regional PBDE concentrations have been published, the assessment of nonoccupational exposure to PBDEs in humans has become possible in this study.

Geographic information system (GIS) is a powerful tool in the exposure assessment of environmental contaminants, which is not only feasible but also can provide a visualization of contamination in the environment [10]. The combination of GIS use and statistical analysis allows for better understanding of the distribution of the contaminants in the environment of study population.

GIS was used to visualize the published data reviewed in this study in order to evaluate PBDE concentrations and distributions in environmental media. Ni's equations [11] were used to determine the major routes of PBDE exposure, and to identify the geographical areas with high PBDE exposure levels. The results might facilitate further health-based risk assessments of PBDEs and might be helpful for the decision making on the management and control of PBDEs in China.

2.1 Data Sources and Research Methods

A literature retrieval was carried out by using MEDLINE. The MeSH terms "PBDEs," "environmental exposure," "air," "water," "soil," "sediment," "food," "geographic information system" and "China" were used. We also retrieved the references listed in the related published papers and reviews included in PubMed and China National Knowledge Infrastructures. For this analysis, we selected papers which: (1) were original ecological studies, (2) were published in English or Chinese, (3) defined all or subgroups of PBDEs as the target chemicals, and (4) studied the environmental or human exposure status by using measured concentrations of PBDEs. Moreover, we retrieved conference proceedings in the ISI Web of Science for abstracts of other unpublished studies, using the same MeSH terms mentioned above, but we did not find any other related study.

A total of 184 papers were selected for this analysis, but only 80 papers qualified were used in this assessment. These papers were published from January 2001 to October 2013, and reported PBDE concentrations in environmental media and food in China.

2.2 GIS Mapping of the Data

For all the 80 studies, the sample sites could be indicated on the map precisely by using GOOGLE EARTH® via their longitudes and latitudes. By using the ArcView GIS 3.2 software, the sampling sites were geocoded. Boundaries of municipalities and provinces were obtained from the State Bureau of Surveying and Mapping. All the data were entered into ArcView GIS 3.2 operated on a PC workstation, and the data were incorporated into several maps to indicate the PBDE distributions in different sampling sites. PBDE concentration data in different environmental media were shown as separate map layers, with levels of PBDE concentrations being color coded.

2.3 Regional Exposure Assessment of PBDEs in China

Given the availability of the environmental and food concentrations of PBDEs in the Yangtze River Delta and Pearl River Delta, the exposure levels in these two regions can be estimated. According to Ni et al. [11], the daily intake levels of PBDEs in humans were estimated according to the collective environmental and food data by using the following formula: $I = \sum(C_i \times IR_i)$, where I is the total intake of PBDEs (ng/day); C_i is the concentration of PBDEs in the environmental medium and in various foods (ng/m^3 for air; ng/L for drinking water; ng/g for foods); and IR_i is the intake rate from the environmental medium and various foods (m^3/day for air; L/day for drinking water; g/day for foods).

2.4 Data Analysis

The normality of the distribution was tested by using a nonparametric test (Kolmogorov-Smironov Z). Geometric mean (GM) was used to reflect the average

PBDE-concentration if log-transformed data was normal, otherwise a median value was used. Statistical analysis was conducted with the software SAS 9.1.

2.5 Geographic Distributions and Environmental Concentrations of PBDEs Contamination in China

PBDEs exist in both gaseous and solid phases in air. Due to frequent contact of brominated fire retardants in daily life, the important pathway of human exposure to PBDEs is through inhalation of air. Figure 1 shows BDE-47, -99, -209 (particle plus gas phase) concentrations in the 20 sampling sites. The concentration of BDE-47 ranged from 0.04 to 4446.5 pg/m^3 with the GM of 49.37 pg/m^3. Higher BDE-47 contamination in air was found in Guiyu in Guangdong Province. The concentration of BDE-99 in Guiyu was also in a high level (3283.85 pg/m^3). The GM of BDE-99 concentrations in air was 45.36 pg/m^3. Air concentrations of BDE-209 were higher than other PBDEs in China, ranging from 37.1 to 7161.5 pg/m^3 and with the GM of 544.25 pg/m^3.

Surface water has been sampled for PBDE-detection in China, and in total there were 42 sampling sites. PBDEs exist in both aqueous phase and suspended particle phases in water. During water treatment processes, most of the particles can be effectively removed. Therefore, PBDEs in aqueous phases was more focused in this assessment.

Figure 2 shows BDE-47, -99, -209 concentrations in different rivers, mainly in the Yangtze River and the Pearl River. The BDE-47 concentrations in surface water varied from 0.68 to 10,700 pg/L, with the GM of 55.11 pg/L. The concentrations of BDE-99 were lower than that of BDE-47, ranging from 0.01 to 3220 pg/L) and with the GM of 28.66 pg/L. BDE-209 was found to have high concentrations in water, ranging from 0.36 to 18,000 pg/L and with the GM of 1946.79 pg/L.

PBDEs are substances that hardly decompose and easily accumulate in soil, resulting in the high contamination of PBDEs in China (Figure 3). In the 44 sampling sites, the Yellow River Delta was found to have the lowest BDE-47 concentration of 0.003 ng/g. The concentration of BDE-47 in Hong Kong was high (2287 ng/g). The GM of BDE-47 concentrations in soil was 1.30 ng/g. The lowest level of BDE-99 (0.002 ng/g) was detected in one sample site at the Yellow River Delta. The GM of BDE-99 concentrations was 1.14 ng/g, and an e-waste site in Hong Kong was reported to have the highest BDE-99 concentration (1410 ng/g) in soil. BDE-209 concentrations ranged from 0.18 to 18519.6 ng/g, with GM of 47.66 ng/g.

Studies on sediment PBDEs were productive, and a total of 195 sampling sites were found with sediment PBDE concentration data (Figure 4). The BDE-47 concentrations ranged from 0.004 to 3054 ng/g (GM: 0.23 ng/g). And the BDE-99 concentrations ranged from 0.002 to 2827 ng/g (GM: 0.25 ng/g). The concentrations of BDE-209 in sediment were also high, ranging from 0.02 to 7341 ng/g and with the GM of 8.22 ng/g.

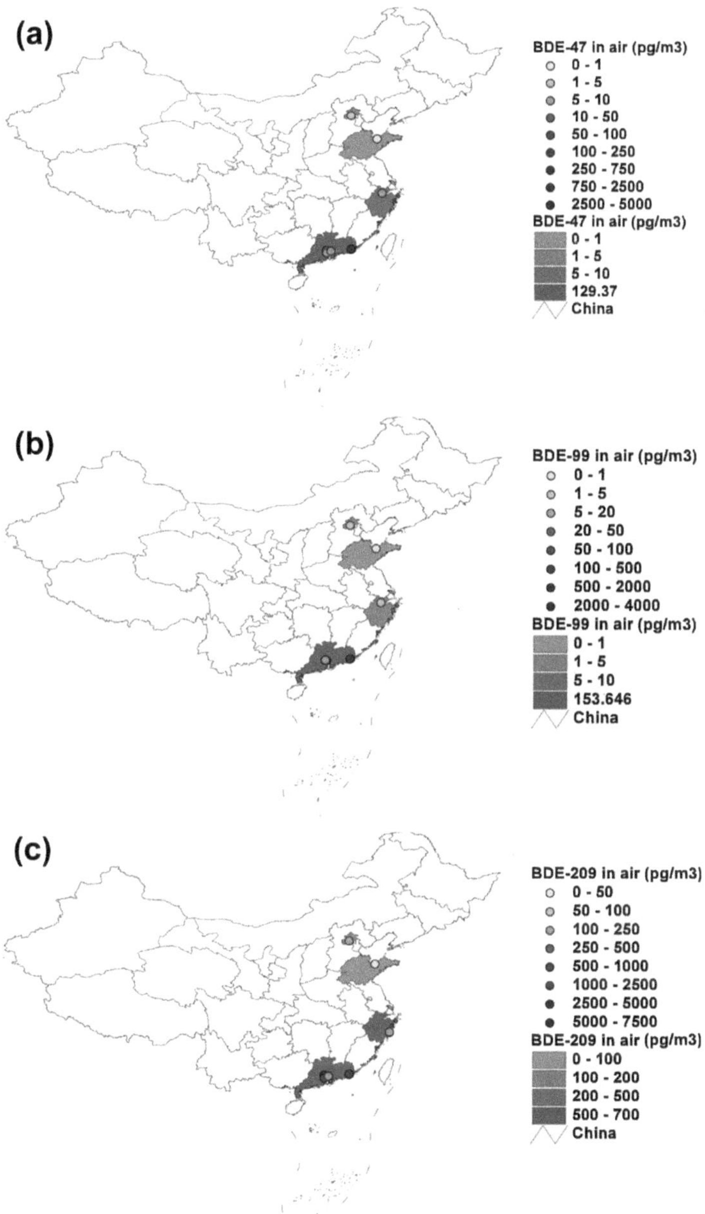

FIGURE 1 Geographic distributions and concentrations of PBDEs contamination in air. (a) BDE-47; (b) BDE-99; (c) BDE-209.

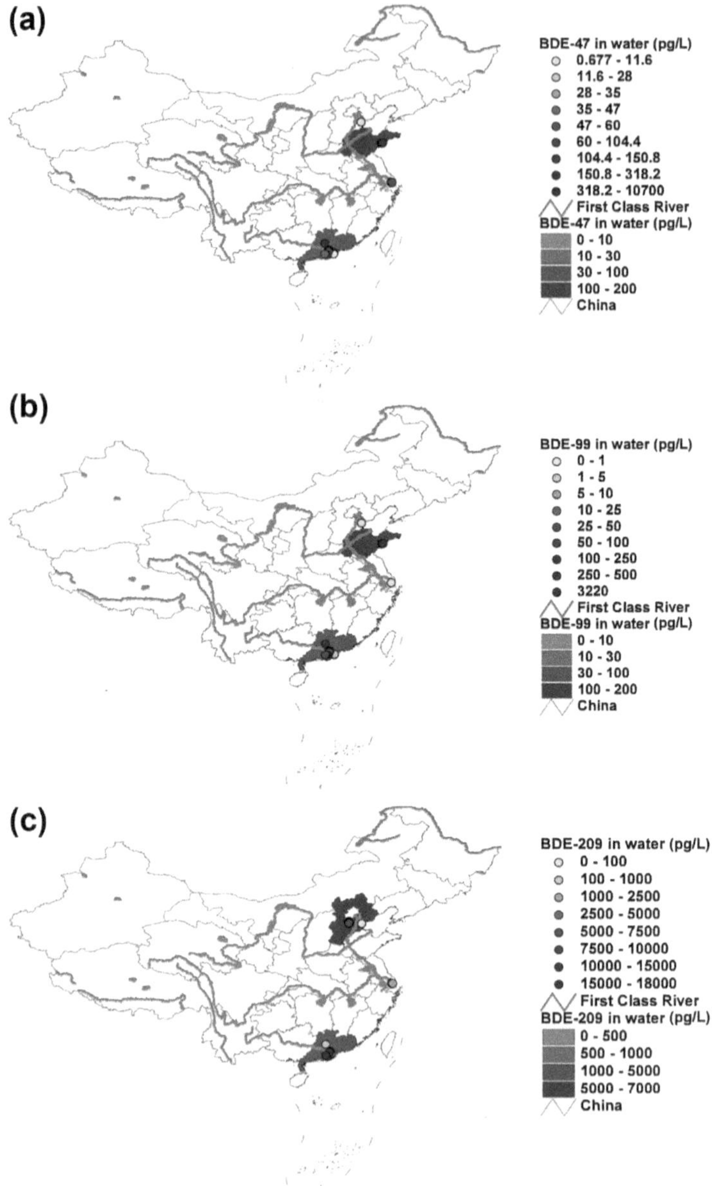

FIGURE 2 Geographic distributions and concentrations of PBDEs contamination in surface water. (a) BDE-47; (b) BDE-99; (c) BDE-209.

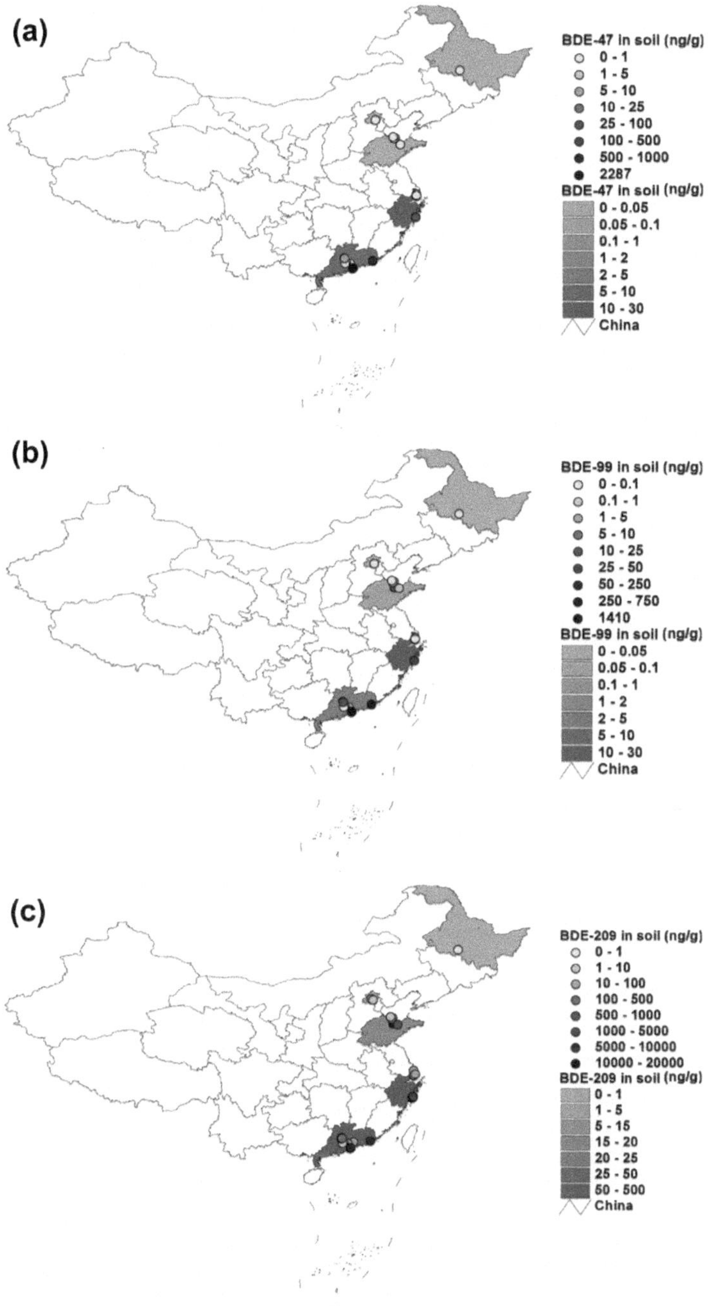

FIGURE 3 Geographic distributions and concentrations of PBDEs contamination in soil. (a) BDE-47; (b) BDE-99; (c) BDE-209.

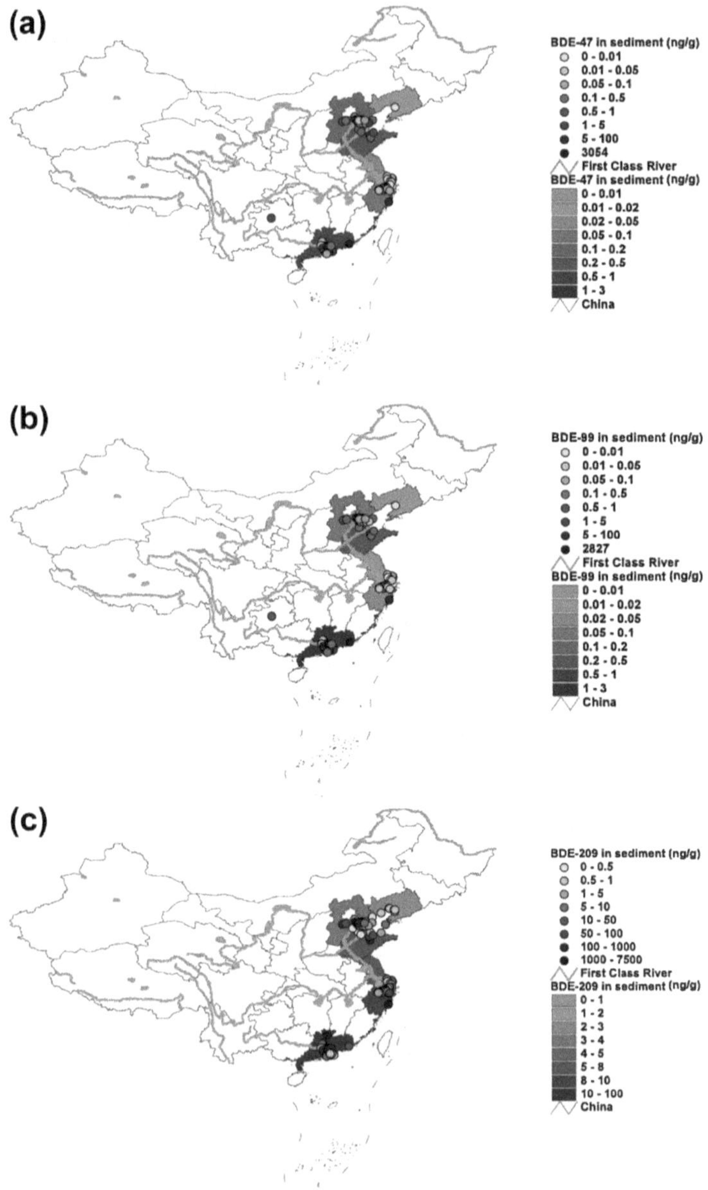

FIGURE 4 Geographic distributions and concentrations of PBDEs contamination in sediment. (a) BDE-47; (b) BDE-99; (c) BDE-209.

2.6 Estimated PBDEs Daily Dietary Intake

Due to the characteristics of lipophilicity, bioaccumulation and biomagnifications, PBDEs can accumulate in the food chain and in foodstuffs (i.e., fish from high trophic levels or with lipid-rich tissues) [11]. In this study, most of the data on PBDE concentrations in food came from eastern or southern China, such as Shanghai, Shenzhen, Hong Kong, etc. In this study we used these data to assess the daily dietary exposure to PBDEs of residents living in areas of the Yangtze River Delta and the Pearl River Delta. Together with the Chinese Diet Guidelines from the Chinese Nutrition Society [12], PBDE-exposure via food in the areas mentioned above was calculated.

Results showed that PBDE concentrations varied with the type of food (Table 1). Both in the Pearl River Delta and the Yangtze River Delta, fish and seafood were shown to be the main dietary source for human exposure to PBDEs. In addition, lipid-rich foods such as meat, eggs and dairy products contributed significantly to human exposure to PBDEs.

As shown in Table 1, in the Pearl River Delta, daily average PBDE-intakes from food were estimated to be 95.88 and 7.86 ng/day for BDE-47 and BDE-99 and estimated to be 41.26 ng/day for BDE-47 and 3.13 ng/day for BDE-99 in the Yangtze River Delta, respectively.

In China, the concentration data on BDE-209 in food are limited, resulting in the failure to calculate dietary BDE-209 intake level in our study. To date, BDE-209 has only been measured in certain European food surveys. The UK Food Standards Agency estimated a daily BDE-209 intake of 318.6 ng/day (4.5 ng/kg BW/day, assuming a body weight of 70.8 kg [13]) for adults based on average consumption across their whole diet [14]. Moreover, Knutsen et al. [15] reported the range of BDE-209 intakes in a group of 184 Norwegians with a wide range of seafood consumption (4–455 g/day). Since we failed to figure out dietary BDE-209 intake for our Chinese samples, we used dietary BDE-209 intakes of 318.6 ng/day (UK) and 97.7 ng/g (1.38 ng/kg BW/day, assuming a body weight of 70.8 kg [13]) (Norway) for our calculations.

3. ASSESSMENT OF HUMAN EXPOSURE TO PBDEs

The exposure of PBDEs levels in the Pearl River Delta and the Yangtze River Delta were evaluated by using Ni's models. Table 2 shows the estimated levels of PBDE exposure via different exposure routes in these two areas. Human exposure levels to the total PBDEs ($\sum$PBDEs: the sum of congeners of BDE-47, BDE-99 and BDE-209) were 225.1–446.0 ng/day in the Pearl River Delta and 148.9–369.8 ng/day in the Yangtze River Delta. Considering the average body weight of Chinese adults (62.7 kg) [16], the intake levels of $\sum$PBDEs were estimated to be 3.6–7.1 ng/kg BW/day for adults living in the Pearl River Delta, and 2.4–5.9 ng/kg BW/day for adults living in the Yangtze River Delta, respectively. Dietary exposure was found to be the predominant exposure route in the Chinese adult population, accounting for >90% of total PBDE exposures. In the Pearl River

TABLE 1 Estimated PBDEs Daily Dietary Intakes in the Pearl River Delta and Yangtze River Delta (ng/day)

		Concentrations (ng/g)			Intake of PBDEs (ng/day)	
PBDEs	**Food**	**Pearl River Delta**	**Yangtze River Delta**	**Intake Rate (g/day)**	**Pearl River Delta**	**Yangtze River Delta**
BDE-47	Oil	0.017	/	27.5	0.470	/
	Dairy	0.016	/	300	4.751	/
	Bean/nut	0.030	0.042	40	1.180	1.685
	Meat	0.035	0.055	62.5	2.202	3.446
	Aquatic	1.121	0.160	75	84.072	11.982
	Egg	0.038	0.088	37.5	1.442	3.314
	Vegetable	0.001	0.008	700	0.798	5.662
	Grain	0.003	0.047	325	0.961	15.172
	Total	/	/	/	95.88	41.26
BDE-99	Oil	0.011	/	27.5	0.307	/
	Dairy	/	/	300	/	/
	Bean/nut	0.0104	0.013	40	0.416	0.523
	Meat	0.011	0.008	62.5	0.711	0.476
	Aquatic	0.059	0.010	75	4.456	0.777
	Egg	0.041	/	37.5	1.519	/
	Vegetable	0.0004	0.002	700	0.277	1.354
	Grain	0.001	/	325	0.168	/
	Total	/	/	/	7.86	3.13
BDE-209	Total	/	/	/	97.7–318.6 [14,15]	97.7–318.6 [14,15]

TABLE 2 PBDE Exposure Levels in the Yangtze River Delta and Pearl River Delta

			Pearl River Delta		Yangtze River Delta	
PBDEs	**Environmental Media and Food**	**Intake Rate**[a]	C_i[b]	I[c]	C_i[b]	I[c]
BDE-47	Air	19 [16]	0.129	2.45	0.001	0.02
	Water	1.7 [16]	0.029	0.05	0.055	0.09
	Food	/	/	95.88[d]	/	41.26[d]
	Total	/	/	98.38	/	41.37
BDE-99	Air	19 [16]	0.153	2.91	0.002	0.04
	Water	1.7 [16]	0.024	0.04	0.031	0.05
	Food	/	/	7.86[d]	/	3.13[d]
	Total	/	/	10.81	/	3.22
BDE-209	Air	19 [16]	0.685	13.02	0.327	6.21
	Water	1.7 [16]	3.033	5.16	0.258	0.44
	Food	/	/	97.7–318.6[d]	/	97.7–318.6[d]
	Total	/	/	115.9–336.8	/	104.4–325.3
∑PBDEs	Total	/	/	225.1–446.0		148.9–369.8

[a]*Units of intake rate were m^3/day in air and L/day in water, respectively.*
[b]*Units of PBDE concentrations were ng/m^3 in air and ng/L in water, respectively.*
[c]*Unit of PBDEs intake is ng/day.*
[d]*Intake of PBDEs via food consumption was showed in Table 1.*

Delta, the estimated exposure levels of BDE-47, BDE-99 and BDE-209 were 2.4, 3.4 and 1.2-fold higher than those in the Yangtze River Delta, respectively.

4. DEVELOPMENT OF A REFERENCE DOSE FOR BDE-47, -99, AND -209 USING BENCHMARK DOSE METHODS

PBDEs have increasingly captured the attention of scientists and policymakers. However, there are few recommendations for PBDEs intake and health risks may not be easily quantified. US Environmental Protection Agency (USEPA) conducted health assessments on several PBDEs congeners after a comprehensive review of toxicity data and proposed oral Reference Dose (RfD) latest updated in 2008. Since all the USEPA oral RfD were focused on neurobehavioral effects, this paper critically evaluated toxicological database on other endpoints along with recent neurotoxicity study results. USEPA's benchmark dose software (BMDS) Version 1.4.1 were used to identify BMD and BMDL based on data from selected literature. According to methods and feature of each study, different uncertainty factor (UF) value was decided and subsequently RfD were proposed.

The selection of appropriate studies is based on animal studies whose route of exposure were similar to human, the quality of the studies, and the relevance and reporting adequacy of the endpoints. More specifically, the following requirements should be met: (1) Only studies in which animals are administered of oral exposure are included as this study addresses human oral exposure of PBDEs. (2) Studies should show a statistically or biologically significant graded monotonic dose-related trend in the selected endpoints.

As almost all toxicology studies of PBDEs are consisted of continuous data, when individual data are unavailable (which is usually the case in published reports), the number of subjects, means of response variable, and measure of response variability (e.g., standard deviation (SD), standard error (SE), or variance) are needed.

For the nine selected studies, we used USEPA's BMDS Version 1.4.1 to calculate BMD which following the six-step process for BMD analysis developed by USEPA, including: (1) choice of a benchmark response (BMR), (2) selecting a set of models, (3) assessing model fit, (4) model selection when BMDLs are divergent, (5) model selection when BMDLs are not divergent, and (6) data reporting.

A total of 91 endpoints across nine studies for three major PBDE congeners (BDE-47, -99, -209) were evaluated for applicability to BMD modeling (Table 3). Based on the nature of the significant endpoints, the following categories of endpoints were identified across the studies: (1) thyroid, (2) neurobehavior, and (3) reproductive and developmental. After conducting modeling for each of these 91 endpoints, 68 endpoints were eliminated from further consideration because of they could not be successfully modeled. For each of the remaining 23 endpoints, the BMD and BMDL values were chosen to represent these endpoints.

The studies considered for quantitative evaluation included all multidose studies discussed below: one reproductive/developmental study for BDE-47

TABLE 3 Endpoints Evaluated Using the Benchmark Dose Modeling Software

Critical Effect	Study Reference
BDE-47	
Reproductive and developmental endpoints	
Testicle/body weight	[17]
Epididymis/body weight	[17]
Ovaries/body weight	[17]
Uterus/body weight	[17]
Thyroid/body weight	[17]
Cerebrum/body weight	[17]
BDE-99	
Thyroid endpoints	
Parent (F0) generation endpoints	
TSH	[18]
T4	[18,19]
FT4	[18,19]
T3	[18,19]
FT3	[19]
First (F1) generation endpoints (male/female)	
T3	[19]
FT3	[19]
T4	[19]
FT4	[19]
Neurobehavior endpoints	
Brain region volumes (HVC, RA, mass)	[23]
Reproductive and developmental endpoints	
Body weight	[20]
Liver weight	[20]
Thymus weight	[20]
Thymus/body weight	[20]

Continued

TABLE 3 Endpoints Evaluated Using the Benchmark Dose Modeling Software—cont'd

Critical Effect	Study Reference
Spleen weight	[20]
Spleen/Body weight	[20]
Testis weight	[20]
Testis/body weight	[20]
Epididymis weight	[20]
Epididymis/body weight	[20]
Seminal vesicle weight empty	[20]
Seminal vesicle/body weight	[20]
Prostate weight	[20]
Prostate/body weight	[20]
Spermatid	[20]
Daily sperm production	[20]
Sperm number	[20]
Abnormal sperm	[20]
LH	[20]
Testosterone	[20]
Uterine weight	[20]
Implantations per litter	[20]
Viable fetuses per litter	[20]
Fetal weight per litter	[20]
Mounting latency	[20]
Intromission latency	[20]
Ejaculatory latency	[20]
Intromission frequency	[20]
Number of penetrations before the first ejaculation	[20]
BDE-209	
Neurobehavior endpoints	
Fetal brain weight (g)	[21]
Fetal brain/body weight (g)	[21]

TABLE 3 Endpoints Evaluated Using the Benchmark Dose Modeling Software—cont'd

Critical Effect	Study Reference
Hemisphere height (PND 72)	[24]
Cortex vertical thickness (PND 72)	[24]
Cortex radial thickness (PND 72)	[24]
Vertical height between hippocampal pyramidal neuron layers (PND 72)	[24]
Dentate hilus vertical height (PND 72)	[24]
Length of ventral limb of dentate hilus (PND 72)	[24]
Pons vertical thickness (PND 72)	[24]
Habituation ratio locomotion (2-month-old and 4-month-old)	[22]
Reproductive and developmental endpoints	
Gestation length	[24]
Number of implantation sites	[24]
Number of unaccounted for sites	[24]
Number born/litter	[24]
Number live/litter	[24]
Body weight	[25,26]
Liver weight	[25]
Kidney weight	[25]
Testes (or ovary) weight	[25]
Epididymis (or uterus) weight	[25]
Prostate weight	[25]
Thyroid weight	[25]
Adrenal weight	[25]
Liver/body weight	[25]
Kidney/body weight	[25]
Testes/body weight	[25]
Epididymis/body weight	[25]
Prostate/body weight	[25]

Continued

TABLE 3 Endpoints Evaluated Using the Benchmark Dose Modeling Software—cont'd

Critical Effect	Study Reference
Ovaries/body weight	[25]
Uterus/body weight	[25]
Thyroid/body weight	[25]
Adrenal/body weight	[25]
Gestational length	[26]
Live pups per litter (PND 1, PND 4)	[26]
Live pup weight (g) (at birth, at weaning)	[26]

TSH, Thyroid stimulating hormone; T4, Thyroxine; FT4, Free thyroxine; T3, Tri-iodothyronine; FT3, Free tri-iodothyronine; HVC, Hyperstriatum ventrale, pars caudalis; RA, Robust nucleus of the arcopallium; LH, Luteinizing hormone.

[17], three studies for BDE-99 (two thyroid studies [18,19] and one reproductive study [20], and two neurobehavior studies for BDE-209 [21,22]. For each of these studies, endpoints were selected for BMD modeling on the basis of toxicological significance, relevance to humans, whether effects were test-related, evidence of a dose–response, statistical significance, and severity of effect.

Because all the data are continuous measurements, we use 1 standard deviation as a BMR in BMD modeling for each type of endpoint. BMDs and BMDLs for each of the 23 successfully modeled endpoints were identified and are presented in Table 4. A total of four reproductive/developmental endpoints of BDE-47 were compiled, along with nine thyroid and seven reproductive/developmental endpoints of BDE-99 across two generations and three neurobehavior endpoints of BDE-209. There is a widespread range of values across endpoints. Results of significant BMDLs are separately discussed by different PBDE congeners below and shown in Table 2, and the lowest BMDLs of each congener were selected to develop an RfD.

4.1 BDE-47 Reproductive and Developmental Endpoints

In this study, SD rats were exposed to a single oral dose of BDE-47 (1, 5, and 10 mg/g, three rats each group) on postnatal day 10. Organ-to-body weight ratios were measured in 2-month-old rats. This study found that some doses of BDE-47 decreased the organ (testicle, epididymis, ovaries and uterus)-to-body weight ratios of the thyroid and uterus, and increased the ratio of ovaries ($P<0.05$) [21].

Animal researches studying on toxicity of BDE-47 were rare. Reproductive endpoints are often the most sensitive endpoints of BDE-47, and provide important information regarding multigenerational effects of this chemical. He et al. [17]

TABLE 4 Summary of BMD and BMDL Estimates for PBDEs

Endpoint	Study Reference	BMD (µg/kg)	BMDL (µg/kg)
BDE-47			
Reproductive/ developmental			
Testicle/body weight	[17]	13.94	5.13
Epididymis/body weight	[17]	435.36	69.53
Ovaries/body weight	[17]	3.47	2.46
Uterus/body weight	[17]	3.93	**2.10**
BDE-99			
Thyroid			
I. Parent (F0) Generation			
TSH	[18]	1127.76	654.21
T4	[18]	8949.97	1378.94
FT4	[18]	3273.13	1084.64
FT3	[19]	127.12	90.72
II. First (F1) Generation			
Male			
FT3	[19]	112.08	**81.77**
T4	[19]	574.41	230.36
T3	[19]	226.51	134.05
Female			
T3	[19]	319.50	172.46
FT3	[19]	202.66	129.10
Reproductive/ developmental			
Body weight	[20]	104.96	80.90
Thymus/body weight	[20]	164.24	117.37
Testis weight	[20]	283.73	173.94
Seminal vesicle weight empty	[20]	234.52	152.94

Continued

TABLE 4 Summary of BMD and BMDL Estimates for PBDEs—cont'd

Endpoint	Study Reference	BMD (μg/kg)	BMDL (μg/kg)
Prostate weight	[20]	158.33	114.03
Intromission frequency	[20]	108.61	83.34
No. of penetrations before the first ejaculation	[20]	103.61	79.99
BDE-209			
Neurobehavior			
Fetal brain weight	[21]	609,749	496,873
Habituation ratio locomotion (2-month-old)	[22]	2352	1990
Habituation ratio locomotion (4-month-old)	[22]	2000	**1698**

Note: Bold values represent recommended BMDL for each type of endpoint.
BMD = Benchmark dose; BMDL = Benchmark dose low.

analyzed six reproductive and developmental endpoints for BDE-47, and only four of which seemed to be well modeled. Three of the four BMDLs for BDE-47 ranged from 2.1 to 5.13 μg/kg, and the lowest BMDL obtained for uterus/body weight is recommended for consideration in developing an RfD for BDE-47.

4.2 BDE-99 Thyroid Endpoints

In Alonso et al. [18] study, 10 adult male rats received BDE-99 by gavage at single doses of 0, 0.6 and 1.2 mg/kg, respectively. Forty-five days after exposure, serum samples were collected for thyroid-stimulating hormone (TSH), thyroxine (T4), and triiodothyronine (T3) analysis. A tendency to enhance TSH was noted in the BDE-99 exposed groups, while no significant differences were observed in serum T3 and T4 (total and free) in comparison to the control group.

In the other study, thyroid hormone concentrations were evaluated in Wistar rats (dams and offspring) after treatment by gavage on gestation day (GD) 6 with a single low dose of either 60 or 300 μg BDE-99/kg body weight (bw), respectively. The dose of 300 μg BDE-99/kg bw reduced T4 concentration in dams at the beginning of lactation (postgestational day (PGD) 1), and caused a slight reduction in T4 on PGD 22, although not statistically significant. In offspring, reduced T4 was observed only at PND 22, probably due to cumulative effects of BDE-99 during lactation. PBDEs have been shown to reduce T4 concentrations in several studies, but this is the first study demonstrating endocrine disruption at low doses [19].

4.3 Reproductive and Developmental Endpoints

This study assessed the effect of developmental exposure to BDE-99 on adult male reproductive health. Wistar rat dams were treated by gavage on GD 6 with a single low dose of 60 or 300 μg BDE-99/kg bw (12 rats each group). In offspring, reproductive performance was assessed in male pups at adulthood. The exposure to low-dose BDE-99 during development caused permanently impaired spermatogenesis by the means of reduced sperm and spermatid counts. The doses used in this study are relevant to human exposure levels, being approximately six and 29 times, respectively [20].

Reviewing modeling results presented in Table 4, the lowest BMDL obtained for thyroid endpoints of BDE-99 was 81.77 μg/kg, and 79.99 μg/kg for reproductive endpoints. Kuriyama et al. [19] conducted a two-generation study, analyzing total 12 thyroid endpoints for F0 and F1 generation, and obtained the lowest BMDL of 81.77 μg/kg for FT3 level of male offspring. Due to the reproducibility of this endpoint, the rigorous study protocol and thorough documentation [19], the BMDL of 81.77 μg/kg is recommended to represent the toxicity of BDE-99 for purposes of developing an RfD instead of the lower value obtained for reproductive endpoint. In view of the lack of mechanistic data on potential *trans*-generational effects of PBDEs, that use of data from the F1 generation was appropriate and protective.

4.4 BDE-209 Neurobehavior Endpoints

In Chi et al. [21] study, four groups of pregnant C57 mice were administrated with BDE-209 in 20% fat emulsion at a dose of 150, 750, 1500, or 2500 g/kg body weight via gastric intubation on GD 7–9, while a control group was given 20% fat emulsion (10 mice in each group). Maternal mice were euthanized on GD 16 and examined for external malformations of the fetus. Results indicated the weight of brain was decreased in the four dosed groups, of which the 2500 mg/kg group exhibited a statistically significant decline.

The other study was undertaken to explore the dose–response effects of BDE-209 on spontaneous behavior and habituation. Neonatal male NMRI mice were given 1.3, 2.2, 13.4, or 20.1 mg BDE-209/kg bw (10 mice in each group) when 3 days old. The agent was administered as a single oral dose via a metal gastric tube. Spontaneous behavior was observed in adult mice at 2 and 4 months of age, showed reduced or lack of habituation, effects that worsen with age. This indicated that BDE-209 can be as potent as the lower brominated PBDEs in causing developmental neurotoxic defects [22].

The lowest BMDL based on the Johansson et al. [22] study was 1698 μg/kg for habituation ratio locomotion of adult mice at four months old. However, compared with the 4-month study, BMDL of Chi et al. study [21] was too high to be taken into consideration in developing an RfD, as shown in Table 4, which may indicate the developmental effect is less sensitive than the neurobehavior endpoints, so that 1698 μg/kg for habituation ratio locomotion at 4-month-old was selected to represent toxicity of BDE-209.

The BMDLs recommended for each of the three major PBDE congeners (BDE-47, -99, -209) are shown in bold font in Table 4. We tested all the models (linear, polynomial, power, and hill) for continuous data, only to find that linear model was the optimal one. And UF was applied to extrapolate animal experimental results to human population, which means to derive RfD from BMDLs.

Uncertainties generally result from two sources. The first one is extrapolation, both from nonhuman species to humans and from general human population to sensitive individuals. For this study, species differences arise from differences in toxicokinetics and toxicodynamics of PBDEs and a 10-fold default is assigned. Another 10-fold is intended to cover the difference in sensitivity between an average person in the population to a sensitive human in the population (WHO, 1994).

The other is uncertainties relating to deficiencies in the database, consisting of absence of an NOAEL, exposure route difference, exposure duration difference and insufficient database. In some studies NOAEL are not available and LOAEL may be used as an alternative and uncertainty needs to be accounted. While in this study, BMD is calculated and now recognized as more scientifically credible approach thus negates the need for an LOAEL to NOAEL UF. Meanwhile, selected studies and human exposure are via the same route, orally through the diet, indicating that no extra UF should be used here. A threefold UF was used to adjust for exposure duration for BDE-47 and BDE-209 as the selected studies used short-term administration. While the principal study for BDE-99 was two-generation study which can be recognized as chronic exposure, no extra UF was counted. Another factor called modifying factor (MF), generally no higher than 3, is typically also used to account for other considerations such as perceived adequacy of the scientific data. We assigned 1, 3, and 1 to BDE-47, -99, and -209, respectively. The UFs and the MFs are then multiplied together to provide a total UF, which is used to derive a chronic RfD. UFs figured out for BDE-47, -99, and -209 were 1000, 300, 1000, respectively.

Consistent with USEPA, the lowest BMDLs of 2.10, 81.77, and 1698 μg/kg were used to develop RfDs for BDE-47, -99, and -209, respectively. These BMDLs were derived from data on decreased uterus/body weight, FT3 (F1 generation), and habituation ratio locomotion (4-month-old) provided in animal experiments [17,19,21]. These endpoints are relevant to humans, represent sensitive endpoints.

TABLE 5 Summary of RfDs for Three Types of Endpoints

PBDE Congener	Type	RfD (mg/kg day)	Reference
BDE-47	Reproductive/developmental	2.1×10^{-6}	This study
	Neurobehavior	0.0001	USEPA
BDE-99	Thyroid	2.7×10^{-4}	This study
	Neurobehavior	0.0001	USEPA
BDE-209	Neurobehavior	1.7×10^{-3}	This study
	Neurobehavior	0.007	USEPA

Due to the large toxicology database, which encompasses varying stages of the life cycle, varying exposure periods, and studies on several rat strains, an uncertainty factor is not necessary to protect healthy population. Uncertainty factors are used to address lack of knowledge regarding the relative sensitivity of a chemical in humans versus animals. As shown in Table 5, acceptable RfD based on this study evaluating the entire PBDEs data for BDE-47, -99, and -209 were 2.1×10^{-6} mg/kg day, 2.7×10^{-4} mg/kg day and 1.7×10^{-3} mg/kg day, respectively. RfDs for BDE-99 and BDE-209 were comparable to EPA results. RfD of BDE-47 was much lower than that of EPA and there may be two reasons accounting for this: one is that reproductive/developmental proves to be more sensitive than neurobehavior for BDE-47; the other reason is that the principal study uses very low-dose exposure and thus BMDL obtained is relatively low.

5. SUMMARY

Human exposure to PBDEs has been of great concern, as it has become evident that PBDEs are ubiquitously found in human tissues [9,27–29]. There were several studies on human exposure to PBDEs worldwide. However, in China, most studies focused on specific exposure sources [30], and little is known about the PBDE-exposure through multiple pathways. This study provided comprehensive information on the current contamination status of PBDEs in China and discussed the potential sources of human exposure to PBDEs. Based on these sources, the first comprehensive assessment of PBDE-exposure in the nonoccupational population of China was conducted.

In the environmental media contaminated with PBDEs, BDE-209 congener was found to have the highest concentration. The remarkable levels and dominance of BDE-209 may be related with the significant production, usage, or disposal of deca-containing products in China. BDE-209 is the only commercial PBDE mixture still permitted currently [31], which is contained in more than 95% of the commercial deca-BDE products [32]. Moreover, in China, the estimated annual production of the predominant commercial deca-BDE mixture increased from 10,000 to about 30,000 metric tons between 2000 and 2005 [33]. Our results were also consistent with those of the European Food Safety Authority Panel on Contaminants in the Food Chain, which revealed BDE-209 and BDE-47 to be the most common dietary contaminants [34].

Since dioxins' and similar POPs routes of entry into the general population is almost exclusively from food, it was also hypothesized that the predominant route of human exposure to PBDEs would be through dietary intake [35]. Consistent with existing literature, dietary exposure was found in our study to be a significant route of human PBDE-intake and was largely attributed to consumption of aquatic foods. This result was supported by studies conducted in the UK and Canada, where the dietary exposure accounted for 93% and 96% of the total exposures, respectively [36,37]. Studies in Finland, Sweden, and Belgium also identified fish consumption to be the major route for human exposure to PBDEs [38–40].

Until now, the data concerning PBDE concentrations in food, as well as human exposure to PBDEs through the diet has been available. However, studies on PBDE exposure via multiple routes are rare. Thus, we compared PBDE exposure data via ingestion in other countries to understand the PBDE exposure in China. ∑PBDEs exposure level via ingestion in Spain was reported to be 97.3 ng/day [37], while in Japan and North American the ∑PBDEs exposure levels were 72 and 52–88 ng/day among the populations, respectively [41,42]. Schuhmacher et al. [43] also reported the ∑PBDEs exposure levels in Spain, which were 72 ng/day for adults living in industrial area and 63 ng/day for adults in urban district. Taking into consideration of exposure via ingestion and inhalation, the ∑PBDEs exposure level in British people was 127 ng/day [38]. All of these data were lower than the estimated intake of ∑PBDEs in Chinese, especially for people living in the Pearl River Delta, which might be due to the high concentrations of PBDEs in food and soil in China.

After assessing 91 endpoints across the nine studies, a total of 23 of these endpoints were identified for BMD modeling, and BMDL estimates corresponding to various dose–response models were compiled for these separate endpoints. Thyroid, neurobehavior and reproductive endpoints for BDE-47, -99, -209 were quantitatively evaluated. According to methods and feature of each study, different uncertainty factor value was decided and subsequently reference doses (RfDs) were proposed. Consistent with USEPA, the lowest BMDLs of 2.10, 81.77, and 1698 μg/kg were used to develop RfDs for BDE-47, -99, and -209, respectively. RfDs for BDE-99 and BDE-209 were comparable to EPA results, and however, RfD of BDE-47 was much lower than that of EPA, which may result from that reproductive/developmental proves to be more sensitive than neurobehavior for BDE-47 and the principal study uses very-low-dose exposure.

The oral RfD for the deca-BDE homolog (e.g., BDE-209), tetrabromodiphenyl ether (tetra-BDE) homolog (e.g., BDE-47), and penta-BDE homolog (e.g., BDE-99) released by EPA are 7×10^{-3}, 1×10^{-4}, and 2×10^{-3} mg/kg BW/day, respectively [44–46]. In this study, the estimated intake levels of $\sum$PBDEs were 2.4–5.9 ng/kg BW/day in the Yangtze River Delta and 3.6–7.1 ng/kg BW/day in the Pearl River Delta, respectively, which were lower than the EPA RfD value, indicating that the environmental PBDE-exposure might not induce significant health hazards in Chinese adults.

In summary, PBDEs are ubiquitous in the environmental media in China. The PBDE concentrations varied in geographical patterns, and Guangdong Province was the most serious contaminated area. BDE-209 was the major PBDE congener in the environmental media and food, and BDE-47 also had a high level in food. Daily intake levels of $\sum$PBDEs were 148.9–369.8 and 225.1–446.0 ng/day for adults in the Yangtze River Delta and the Pearl River Delta, respectively. Moreover, dietary intake is the major route of PBDE exposure. Decision makers should take the potential health effects of PBDEs on people into consideration, especially on the people in areas with a high PBDE concentration, such as the Pearl River Delta. Furthermore, a comprehensive risk assessment for Chinese (including the sensitive population) exposed to PBDEs is needed in future research.

ACKNOWLEDGMENTS

We thank Dr Bingheng Chen at Fudan University for her critical comments. We also wish to express our gratitude to Dr Q. Jay Zhao of the US Environmental Protection Agency for his suggestions on uncertainty factors' selection.

REFERENCES

[1] M. Alaee, P. Arias, A. Sjodin, et al., An overview of commercially-used brominated flame retardants, their applications, their use patterns in different countries/regions and possible modes of release, Environ. Int. 29 (2003) 683–689.

[2] R.J. Law, C.R. Allchin, J. de Boer, et al., Levels and trends of brominated flame retardants in the European environment, Chemosphere 64 (2006) 187–208.

[3] L.S. Birnbaum, D.F. Staskal, Brominated flame retardants: cause for concern? Environ. Health Perspect. 112 (2004) 9–17.

[4] L.G. Costa, G. Giordano, Developmental neurotoxicity of polybrominated diphenyl ether (PBDE) flame retardants, Neurotoxicology 28 (2007) 1047–1067.

[5] P.R.S. Kodavanti, C.G. Coburn, V.C. Moser, et al., Development exposure to a commercial PBDE mixture, DE-71: neurobehavioral, hormonal, and reproductive effects, Toxicol. Sci. 116 (2010) 297–312.

[6] International Institute for Sustainable Development, in: Summary of the Fourth Conference of the Parties to the Stockholm Convention on Persistent Organic Pollutants, 2009. http://www.unido.org/fileadmin/user_media/Services/Environmental_Management/Stockholm_Convention/POPs/SummaryReportofCOP4_01.pdf. (26.08.14).

[7] C.A. de Wit, An overview of brominated flame retardants in the environment, Chemosphere 46 (2002) 583–624.

[8] A. Sjodin, R.S. Jones, J.F. Focant, et al., Retrospective time-trend study of polybrominated diphenyl ether and polybrominated and polychlorinated biphenyl levels in human serum from the United States, Environ. Health Perspect. 112 (2004) 654–658.

[9] R.A. Hites, Polybrominated diphenyl ethers in the environment and in people: a meta-analysis of concentrations, Environ. Sci. Technol. 38 (2004) 945–956.

[10] J.R. Nuckols, M.H. Ward, L. Jarup, Using geographic information systems for exposure assessment in environmental epidemiology studies, Environ. Health Perspect. 112 (2004) 1007–1115.

[11] K. Ni, Y.L. Lu, T.Y. Wang, et al., A review of human exposure to polybrominated diphenyl ethers (PBDEs) in China, Hum. Ecol. Risk Assess. 17 (2011) 923–965.

[12] Chinese Nutrition Society, in: Dietary Pagoda, 2007. http://www.cnsoc.org/cn/. (26.08.14).

[13] Wikipedia, the Free Encyclopedia, in: Body Weight, 2014. http://en.wikipedia.org/wiki/Body_weight. (26.08.14).

[14] Food Standard Agency (FSA), in: Brominated Chemicals: UK Dietary Intakes (Food Survey Information Sheet 10/06), 2006. http://www.food.gov.uk/science/research/surveillance/fsisbranch2006/fsis1006. (26.08.14).

[15] H.K. Knutsen, H.E. Kvalem, C. Thomsen, et al., Dietary exposure to brominated flame retardants correlates with male blood levels in a selected group of Norwegians with a wide range of seafood consumption, Mol. Nutr. Food Res. 52 (2008) 217–227.

[16] Z.S. Wang, X.L. Duan, P. Liu, et al., Human exposure factors of Chinese people in environmental health risk assessment, Res. Environ. Sci. 22 (2009) 1164–1170 (in Chinese).

[17] P. He, A. Wang, Q. Niu, et al., Toxic effect of PBDE-47 on thyroid development, learning, and memory, and the interaction between PBDE-47 and PCB 153 that enhances toxicity in rats, Toxicol. Ind. Health 27 (2011) 279–288.

[18] V. Alonso, V. Linares, M. Belles, et al., Effects of BDE-99 on hormone homeostasis and biochemical parameters in adult male rats, Food Chem. Toxicol. 48 (2010) 206–211.

[19] S.N. Kuriyama, A. Wanner, A.A. Fidalgo-Neto, et al., Developmental exposure to low-dose PBDE-99: tissue distribution and thyroid hormone levels, Toxicology 242 (2007) 80–90.

[20] S.N. Kuriyama, C.E. Talsness, K. Grote, et al., Developmental exposure to low dose PBDE 99: effects on male fertility and neurobehavior in rat offspring, Environ. Health Perspect. 113 (2005) 149–154.

[21] Y. Chi, H.F. Xia, M.M. Su, et al., Metabonomic phenotyping reveals an embryotoxicity of deca-brominated diphenyl ether in mice, Chem. Res. Toxicol. 24 (2011) 1976–1983.

[22] N. Johansson, H. Viberg, A. Fredriksson, et al., Neonatal exposure to deca-brominated dihenyl ether (PBDE 209) causes dose-response changes in spontaneous behavior and cholinergic susceptibility in adult mice, Neuro Toxicol. 29 (2008) 911–919.

[23] M.L. Eng, J.E. Elliott, S.A. MacDougall-Shackleton, et al., Early exposure to 2,2′,4,4′,5-pentabromodiphenyl ether (BDE-99) affects mating behavior of zebra finches, Toxicol. Sci. 127 (2012) 269–276.

[24] J.A. Biesemeier, M.J. Beck, H. Silberberg, et al., An oral developmental neurotoxicity study of decabromodiphenyl ether (DecaBDE) in rats, Birth Defects Res. (Part B) 92 (2011) 17–35.

[25] T.H. Kim, Y.J. Lee, E. Lee, et al., Effects of gestational exposure to decabromodiphenyl ether on reproductive parameters, thyroid hormone levels, and neuronal development in Sprague-Dawley rats offspring, J. Toxicol. Environ. Health A 72 (2009) 1296–1303.

[26] L.H. Tseng, M.H. Li, S.S. Tsai, et al., Developmental exposure to decabromodiphenyl ether (PBDE 209): effects on thyroid hormone and hepatic enzyme activity in male mouse offspring, Chemosphere 70 (2008) 640–647.

[27] D.M. Guvenius, A. Aronsson, G. Ekman-Ordeberg, et al., Human prenatal and postnatal exposure to polybrominated diphenyl ethers, polychlorinated biphenyls, polychlorobiphenyls and pentachlorophenyl, Environ. Health Perspect. 111 (2003) 1235–1241.
[28] C.S. Kang, J.H. Lee, S.K. Kim, et al., Polybrominated diphenyl ethers and synthetic musks in umbilical cord Serum, maternal serum, and breast milk from Seoul, South Korea, Chemosphere 80 (2010) 116–122.
[29] H.-B. Moon, D.-H. Lee, Y.S. Lee, et al., Polybrominated diphenyl ethers, polychlorinated biphenyls and organochlorine pesticides in adipose tissues of Korean women, Arch. Environ. Contam. Toxicol. 62 (2012) 176–184.
[30] S. Lee, K. Kannan, H.-B. Moon, Assessment of exposure to polybrominated diphenyl ethers (PBDEs) via seafood consumption and dust ingestion in Korea, Sci. Total Environ. 443 (2013) 24–30.
[31] J. He, D. Yang, C. Wang, et al., Chronic zebrafish low dose decabrominated diphenyl ether (BDE-209) exposure affected parental gonad development and locomotion in F1 offspring, Ecotoxicology 20 (2011) 1813–1822.
[32] W.Y. Qu, X.H. Bi, G.Y. Sheng, et al., Exposure to polybrominated diphenyl ethers among workers at an electronic waste dismantling region in Guangdong, China, Environ. Int. 33 (2007) 1029–1034.
[33] Z.M. Zhou, Implement of administrative measure on the control of pollution caused by electronic information products and the exemption of deca-BDE mixture, Flame Retard. Meter Technol. 4 (2006) 15–16 (in Chinese).
[34] European Food Safety Authority (EFSA), in: Panel on Contaminants in the Food Chain; Scientific Opinion on Polybrominated Diphenyl Ethers (PBDEs) in Food, 2011. http://www.efsa.europa.eu/efsajournal. (27.08.14).
[35] A. Bocio, J.M. Llobet, J.L. Domingo, et al., Polybrominated diphenyl ethers (PBDEs) in foodstuffs: human exposure through the diet, J. Agric. Food Chem. 51 (2003) 3191–3195.
[36] S. Harrad, R. Wijesekera, S. Hunter, et al., Preliminary assessment of UK human dietary and inhalation exposure to polybrominated diphenyl ethers, Environ. Sci. Technol. 38 (2004) 2345–2350.
[37] B.H. Wilford, T. Harner, J. Zhu, et al., Passive sampling survey of polybrominated diphenyl ether flame retardants in indoor and outdoor air in Ottawa, Canada: implications for sources and exposure, Environ. Sci. Technol. 38 (2004) 5312–5318.
[38] H. Kiviranta, M.L. Ovaskainen, T. Vartiainen, Market basket study on dietary intake of PCDD/Fs, PCBs, and PBDEs in Finland, Environ. Int. 30 (2004) 923–932.
[39] P.O. Darnerud, S. Atuma, M. Aune, et al., Dietary intake estimations of organohalogen contaminants (dioxins, PCB, PBDE and chlorinated pesticides, e.g., DDT) based on Swedish market basket data, Food Chem. Toxicol. 44 (2006) 1597–1606.
[40] S. Voorspoels, A. Covaci, H. Neels, et al., Dietary PBDE intake: a market-basket study in Belgium, Environ. Int. 33 (2007) 93–97.
[41] K. Akutsuab, S. Takatori, H. Nakazawa, et al., Dietary intake estimations of polybrominated diphenyl ethers (PBDEs) based on a total diet study in Osaka, Japan, Food Addit. Contam. Part B: Surveill. 1 (2008) 58–68.
[42] A. Schecter, O. Päpke, T.R. Harris, et al., Polybrominated diphenyl ether (PBDE) levels in an expanded market basket survey of US food and estimated PBDE dietary intake by age and sex, Environ. Health Perspect. 114 (2006) 1515–1520.
[43] M. Schuhmacher, H. Kiviranta, T. Vartiainen, et al., Concentrations of polychlorinated biphenyls (PCBs) and polybrominated diphenyl ethers (PBDEs) in milk of women from Catalonia, Spain, Chemosphere 67 (2007) 295–300.

[44] Environmental Protection Agency (EPA) Integrated Risk Information System (IRIS), in: Decabromodiphenyl Ether (BDE-209) (CASRN 1163-19-5, 2008. www.epa.gov/IRIS/subst/0035.htm. (27.08.14).

[45] EPA IRIS, in: 2,2′,4,4′-Tetrabromodiphenyl Ether (CASRN 5436-43-1), 2008. www.epa.gov/iris/subst/1010.htm. (27.08.14).

[46] EPA IRIS, in: Pentabromodiphenyl Ether (CASRN 32534-81-9), 1990. www.epa.gov/ncea/iris/subst/0184.htm. (27.08.14).

Index

Note: Page numbers followed by, "f" and "t" indicate, figures and tables respectively.

C

H

I

J

K

L

Q

R

S

T

U

V

W

X

Z

FSC
www.fsc.org
MIX
Papier | Fördert gute Waldnutzung
FSC® C083411

Kolibri 360 GmbH
Ferdinand-Jühlke-Straße 7
99095 Erfurt, Deutschland
produktsicherheit@kolibri360.de